Structure Determination by
X-ray Crystallography

Structure Determination by X-ray Crystallography

Fourth Edition

Mark Ladd

Department of Chemistry
University of Surrey
Guildford, England

and

Rex Palmer

Department of Crystallography
Birkbeck College
University of London
London, England

Kluwer Academic/Plenum Publishers
NEW YORK, BOSTON, DORDRECHT, LONDON, MOSCOW

Library of Congress Cataloging-in-Publication Data

Ladd, M. F. C. (Marcus Frederick Charles)
  Structure and determination by X-ray crystallography/by Mark Ladd and Rex Palmer.—4th
    p. cm.
  Includes bibliographical references and index.
  ISBN HB: 0-306-47453-0; PB: 0-306-47454-9
    1. X-ray crystallography. I. Palmer, R. A. (Rex Alfred), 1936–. II. Title.

QD945 .L32 2003
548—dc21

                                                                                  2002040

ISBN HB: 0-306-47453-0
     PB: 0-306-47454-9

©2003 Kluwer Academic/Plenum Publishers, New York
233 Spring Street, New York, N.Y. 10013

http://www.wkap.nl/

10 9 8 7 6 5 4 3 2 1

A C.I.P. record for this book is available from the Library of Congress

Foreword

I was highly flattered when I was asked by Mark Ladd and Rex Palmer if I would write the Foreword to this Fourth Edition of their book. "Ladd & Palmer" is such a well-known and classic book on the subject of crystal structure determination, one of the standards in the field: I did feel daunted by the prospect, and wondered if I could do justice to it.

The determination of crystal structures by X-ray crystallography has come a long way since the 1912 discoveries of von Laue and the Braggs. In the intervening years great advances have been made, so that today it is almost taken for granted that crystal structures can be determined in which hundreds, if not thousands, of separate atomic positions can be found with apparent ease. In the early years the structures of relatively simple materials, such as the alkali halides, were often argued over and even disputed, whereas today we routinely see published structures of most complex molecular crystals, including the structures of viruses and proteins.

How has this phenomenal development come about? Partly it is the advance in theories of diffraction, particularly made in order to solve the well-known phase problem. If one knows all the phases and amplitudes of the scattered x-rays it is then a straightforward matter to find the positions of the atoms in the crystal: indeed this information tells us everything we need to know. However, because of the absence of suitable lenses for X-rays, one normally only measures the scattered amplitudes and the phase information is lost. The advent of techniques such as the use of direct methods and anomalous scattering measurements has meant that usually reliable estimates of the phases can be made, thus completing the information needed to 'see' the atoms in the structure.

Another advance has been in the area of instrumentation. For many years after the discovery of X-ray diffraction the usual way in which X-ray diffraction data were collected was by using a photographic film around the crystal and then the intensities of the resulting spots were estimated by eye (As I write this I am startled by the realisation that such crude techniques were widely used even when I was a research student in crystallography back in the 1960's!). Fortunately, this time-consuming and very imprecise method has now been superseded by the development of the commercial diffractometer. Originally such machines had only two circles of movement and then later were replaced by more sophisticated machines with four circles of movement. These latter diffractometers freed us from having to orient the crystal before collecting data, as they were capable of searching for x-ray spots and working out the orientation, unit cell and symmetry for themselves. More recently, area detectors in the form of image plates and CCD systems have become available so that huge amounts of data can be collected in as little as one hour. There have been advances also in the x-ray sources, with x-ray synchrotrons in several locations around the world providing highly intense and wavelength-tuneable x-rays. And then one must not forget the incredible development of modern computers. When I started in research we were still using Facit calculators (these mechanical devices were capable only of adding, subtracting, multiplying and dividing). Then computers as large as a room came along with storage capacities of just 8K. The first computer that I used is now on exhibit in the Science Museum in London! Today we can do almost everything on a desktop PC. It is perhaps not commonly realised that many of the routines and algorithms that have now became standard on modern computers owe their very existence to the huge computational demands made by crystallographers, since for many years they were just about the only scientists with a need to process very large amounts of data.

Today, we have computer power, automatic diffractometers to collect the data and computer programs that in principle analyse the data and produce for us the crystal structures. Nowadays, there are many scientists around the world using these systems and publishing crystal structures completely automatically. So why do we need Ladd & Palmer? After all, when we drive a car these days, do we generally need to know how the engine works? Is it not case that crystal structure determination has become so automatic that we need only press buttons?

Unfortunately (or fortunately for we crystallographers) it is not so simple. The literature is full of incorrectly determined crystal structures made by people without experience, despite the sophistication of the techniques available to them. For example Richard Marsh in the United States has made a career out of exposing false space groups in lots of published papers. I have seen instances where protein structures have been determined and accepted, but a look through a polarising microscope immediately showed that the crystals were twinned and therefore of lower symmetry than had originally been assumed. And then, there are many, many instances where molecules do not sit in well-defined positions in a crystal, but are disordered over two or more sites. In the last ten or so years there has been a veritable explosion in protein crystal structure determination in which people, mainly with training in biology, but with little or no crystallographic training, have come into the subject. For them it is the molecule that is important and not the crystal per se. For them the crystal is simply a necessary and sometimes annoying way of getting molecules aligned in a regular arrangement, so that diffraction can be used as a means to find out about them. This lack of basic knowledge can be dangerous as it raises questions about the validity of their results.

Crystal structure determination is full of pitfalls to trap the unwary. The International Union of Crystallography several years ago even coined the phrase "Practicing Crystallographic Idiot" or PCI for short, to describe those who treat structure determination as automatic. This is why we need Ladd & Palmer. We need to convert the PCI to a PCE ("Practicing Crystallographic Expert")! Now there are many books on crystallography and structure determination out there, but Ladd & Palmer is probably unique in being the most thorough treatment you are going to find. The book takes you right through from simple beginnings up to the most recent ideas in macromolecular crystallography. There are lots and lots of worked examples and tutorials, and I see that in this Fourth Edition even computer software has been included. I note with pleasure that now Ladd & Palmer has a chapter on structure determination using powders. The use of powder diffraction took off back in 1969 with the publication of the seminal paper by Hugo Rietveld, in which he showed that it was possible to refine the atomic positions in a structure by fitting the total profile of the powder pattern. As a result, after having languished many years in the crystallographic doldrums, powder diffraction suddenly took off. In recent years the idea of *ab initio* determination of structures using powder methods has become a 'hot' topic, and recently success has even been gained in determining structural information about proteins. As there are many instances where single crystals cannot be obtained or where twinning is a severe phenomenon, this opens possibilities of still getting atomic positions. This Fourth Edition is a substantial and scholarly work that deserves to be on the shelves of anyone wishing to determine crystal structures. I am very pleased to have had the opportunity to recommend it to you.

A. M. Glazer
Clarendon Laboratory
Oxford University

Preface to the Fourth Edition

There have been many advances in x-ray crystallography since the production of the third edition of this book, and we have endeavoured to introduce a number of them into this new edition. The overall plan of the book and the important additions to the previous two editions have been maintained because we believe that they have been well received in the academic community, but substantial revisions have now been carried out and new material and chapters added.

In particular, we have extended the discussion of the theory of x-ray diffraction and added new chapters on structure determination from powder data, on macromolecular crystallography and on computational procedures in x-ray crystallography. We consider that x-ray crystallography is a universal tool for studying molecular structure, a view upheld by the pioneers in the subject, notably W. H. & W. L. Bragg, J. D. Bernal, Dorothy Hodgkin (neé Crowfoot), Kathleen Lonsdale (neé Yardley) and Linus Pauling, so that the broadening of the scope of the text in this way is fully justified.

We have maintained the practice of devising problems to illustrate the work of each chapter, and have provided detailed, tutorial solutions. The appendices contain mostly mathematical procedures related to the material of the main text.

This edition is accompanied by a suite of computer programs on a compact disc. The programs are available also at the web address <www.wkap.nl/subjects/crystallography> from which they can be downloaded, and which, except for ESPOIR, ITO12 and LEPAGE, may be amended from time to time. The first set will be labelled Version 1, dated 1 April 2003.

These programs enable the reader to participate fully in many of the aspects of x-ray crystallography discussed in the book. In particular, the program system XRAY* is interactive, and enables the reader to follow through, at the monitor screen, computational techniques involved in single-crystal structure determination, albeit in two dimensions. Several sets of x-ray data are provided for practice with this system.

The text, particularly the later chapters, refers to a number of programs that are essential in current structure analysis, by both single-crystal and powder methods. Abbreviated references to these programs are listed in Appendix 10, where authors' names and Journal and/or Internet adDr.esses are provided, so that through the goodwill of the authors they may be obtained from the appropriate sources. In order to distinguish between these programs and those in the CD suite with the book, the latter are marked with an asterisk, as in XRAY*.

We have listed appropriate bibliographic references at the end of each chapter. Although it leads to duplication in some cases, we feel that it is most useful to have these references close to the material to which they are relevant. In addition, we call attention to the International Union of Crystallography, which publishes journals, books and other information of importance to crystallographers. A convenient reference to this material is <http://www.iucr.ac.uk>

Our first acknowledgement is to Professor A. M. Glazer of the Clarendon Laboratory, Oxford who kindly wrote the Foreword to this book. We are grateful also to the various copyright holders for permission to reproduce those figures that carry appropriate acknowledgements; a number of figures, particularly in Chapters 7 and 10, has been prepared by Effective Graphics U.K. The original program which formed the basis of the XRAY* package included with the CD suite was prepared by Dr. Neil Bailey of the University of Sheffield and his colleagues, and we are grateful to him for permission to use it in the present context. It has been modified (M.F.C.L.) for PC operation and several enhancements made, including the contouring of Fourier maps on the monitor screen. We thank Dr. Jan Visser of the Technisch Physische Dienst, Delft for permission to include the powder indexing program IT012*; Dr. Armel Le Bail of Laboratoire des Fluorures, Université du Maine, Le Mans for permission to include the program ESPOIR*; Dr. A. L. Spek of the University of Utrecht for making available the program LEPAOE*; Dr. Lynne McCusker of the Department of Physics, ETH Zurich for reading the chapter on powder methods and for helpful comments; Mr. Robin Shirley of the Department of Psychology, University of Surrey for helpful discussions on aspects of powder indexing; and also Dr. Lachlan Cranswick and Dr. Jeremy Cockcroft of the School of Crystallography, Birkbeck College, London for further useful comments. Finally, we thank Kluwer Academic/Plenum Publishers for inviting this edition and for bringing it to a state of completion.

University of Surrey   Mark Ladd
Birkbeck College, London   Rex Palmer

Preface to the First Edition

Crystallography may be described as the science of the structure of materials, using this word in its widest sense, and its ramifications are apparent over a broad front of current scientific endeavor. It is not surprising, therefore, to find that most universities offer some aspects of crystallography in their undergraduate courses in the physical sciences. It is the principal aim of this book to present an introduction to structure determination by x-ray crystallography that is appropriate mainly to both final-year undergraduate studies in crystallography, chemistry, and chemical physics, and introductory postgraduate work in this area of crystallography. We believe that the book will be of interest in other disciplines, such as physics, metallurgy, biochemistry, and geology, where crystallography has an important part to play.

In the space of one book, it is not possible either to cover all aspects of crystallography or to treat all the subject matter completely rigorously. In particular, certain mathematical results are assumed in order that their applications may be discussed. At the end of each chapter, a short bibliography is given, which may be used to extend the scope of the treatment given here. In addition, reference is made in the text to specific sources of information.

We have chosen not to discuss experimental methods extensively, as we consider that this aspect of crystallography is best learned through practical experience, but an attempt has been made to simulate the interpretive side of experimental crystallography in both examples and exercises.

During the preparation of this book, we have tried to keep in mind that students meeting crystallography for the first time are encountering a new discipline, and not merely extending a subject studied previously. In consequence, we have treated the geometry of crystals a little more fully than is usual at this level, for it is our experience that some of the difficulties which students meet in introductory crystallography lie in the unfamiliarity of its three-dimensional character.

We have limited the structure-determining techniques to the three that are used most extensively in present-day research, and we have described them in depth, particularly from a practical point of view. We hope that this treatment will indicate our belief that crystallographic methods can reasonably form part of the structural chemist's repertoire, like quantum mechanics and nmr spectroscopy.

Each chapter is provided with a set of problems, for which answers and notes are given. We recommend the reader to tackle these problems; they will provide a practical involvement which should be helpful to the understanding of the subject matter of the book. From experience in teaching this subject, the authors are aware of many of the difficulties encountered by students of crystallography, and have attempted to anticipate them in both these problems and the text. For any reader who has access to crystallographic computing facilities, the authors can supply copies of the data used to solve the structures described in Chapters 6 and 8. Certain problems have been marked with an asterisk. They are a little more difficult than the others and may be omitted at a first reading.

The Hermann–Mauguin system of symmetry notation is used in crystallography, but, unfortunately, this notation is not common to other disciplines. Consequently, we have written the Schoenflies symbols for point groups on some of the figures that depict point-group and molecular symmetry in three dimensions, in addition to the Hermann–Mauguin symbols. The Schoenflies notation is described in Appendix A3. General symbols and constants are listed in the Notation section.

We wish to acknowledge our colleague, Dr P. F. Lindley, of Birkbeck College, London, who undertook a careful and critical reading of the manuscript and made many valuable suggestions. We acknowledge an unknown number of past students who have worked through many of the problems given in this book, to our advantage and, we hope, also to theirs. We are grateful to the various copyright holders for permission to reproduce those figures that carry appropriate acknowledgments. Finally, we thank the Plenum Publishing Company for both their interest in this book and their ready cooperation in bringing it to completion.

University of Surrey   Mark Ladd
Birkbeck College, London   Rex Palmer

Disclaimer

Every effort has been made to ensure the correct functioning of the software associated with this book. However, the reader planning to use the software should note that, from the legal point of view, there is no warranty, expressed or implied, that the programs are free from error or will prove suitable for a particular application; by using the software the reader accepts full responsibility for all the results produced, and the authors and publisher disclaim all liability from any consequences arising from the use of the software. The software should not be relied upon for solving a problem, the incorrect solution of which could result in injury to a person or loss of property. If you do use the programs in such a manner, it is at your own risk. The authors and publisher disclaim all liability for direct or consequential damages resulting from your use of the programs.

Contents

Chapter 3
I X-rays, X-ray Diffraction, and Structure Factors
II Intensities and Intensity Statistics 117

Chapter 4
I Optical and X-ray Examination of Crystals
II Measurement of Intensity Data from Single
Crystals 213

I Optical and X-ray Examination of Crystals 213

II Measurement of Intensity Data from Single Crystals 260

Chapter 6
Fourier Techniques in X-ray Structure
Determination 335

Chapter 7
Direct Methods and Refinement **421**

Chapter 8
Examples of Crystal Structure Determination 519

Chapter 9
X-ray Structure Determination with Powders 567

Chapter 11
Computer-Aided Crystallography 681

Physical Constants and Other Numerical Data

Atomic mass unit	u	1.6605×10^{-27} kg
Avogadro constant	L	6.0221×10^{23} $C\,mol^{-1}$
Bohr radius for hydrogen	a_0	5.2918×10^{-11} m
Elementary charge	e	1.6021×10^{-19} C
Planck constant	h	6.6261×10^{-34} $J\,Hz^{-1}$
Rest mass of the electron	m_e	9.1094×10^{-31} kg
Rest mass of the neutron	m_n	1.6750×10^{-27} kg
Speed of light in a vacuum	c	2.9979×10^8 ms^{-1}

Conversions

1 eV (electron-volt) $= 1.6021 \times 10^{-19}$ J

1 Å (Ångström unit) $= 10^{-10}$ m $= 0.1$ nm

Prefixes to Units

femto	pico	nano	micro	milli	centi	deci	kilo	mega	giga
f	p	n	μ	m	c	d	k	M	G
10^{-15}	10^{-12}	10^{-9}	10^{-6}	10^{-3}	10^{-2}	10^{-1}	10^3	10^6	10^9

Notation

These notes provide a key to the main symbols and constants used throughout the book. Inevitably, some symbols have more than one use. This feature arises partly from general usage in crystallography, and partly from a desire to preserve a mnemonic character in the notation wherever possible. It is our belief that, in context, no confusion will arise. Where several symbols are closely linked, they are listed together under the first member of the set.

$A'(hkl)$, $B'(hkl)$	Components of the structure factor, measured along the real and imaginary axes, respectively, in the complex plane (Argand diagram)
$A(hkl)$, $B(hkl)$	Components of the geometric structure factor, measured along the real and imaginary axes, respectively, in the complex plane
A	A-face-centered unit cell; absorption correction factor
Å	Ångström unit
a, b, c	Unit-cell edges parallel to the x, y, and z axes, respectively, of a crystal; intercepts made by the parametral plane on the x, y, and z axes respectively; glide planes with translational components of $a/2$, $b/2$, and $c/2$, respectively
a, b, c	Unit-cell edge vectors parallel to the x, y, and z axes, respectively
a^*, b^*, c^*	Edges in the reciprocal unit cell associated with the x^*, y^*, and z^* axes, respectively
a*, b*, c*	Reciprocal unit-cell vectors associated with the x^*, y^*, and z^* axes, respectively
B	B-face-centered unit cell; overall isotropic temperature factor
B_j	Isotropic temperature factor for the jth atom
C	C-face-centered unit cell
¢	"Not constrained by symmetry to equal"

c	Speed of light; as a subscript: calculated, as in $\lvert F_c \rvert$
D_m	Experimentally measured crystal density
D_c, D_x	Calculated crystal density
d	Interplanar spacing
$d(hkl)$	Interplanar spacing of the (hkl) family of planes
d^*	Distance in reciprocal space
$d^*(hkl)$	Distance from the origin to the hklth reciprocal lattice point
Da	Dalton; equivalent to u
E, $E(hkl)$	Normalized structure factor (centrosymmetric crystals)
$\mathcal{E}(hkl)$	Total energy of the hklth diffracted beam from one unit cell
e	Electron charge
e, exp	Exponential function
$F(hkl)$	Structure factor for the hklth spectrum referred to one unit cell
$F^*(hkl)$	Conjugate of $F(hkl)$
$\lvert F \rvert$	Modulus, or amplitude, of the structure factor $F(hkl)$
f	Atomic scattering factor
$f_{j,\theta}$, f_j	Atomic scattering factor for the jth atom
g	Glide line in two-dimensional space groups
g_j	Atomic scattering factor for the jth atom, in a crystal, corrected for thermal vibrations
H	Hexagonal (triply primitive) unit cell
(hkl), $(hkil)$	Miller, Miller–Bravais indices associated with the x, y, and z axes or the x, y, u, and z axes, respectively; any single index containing two digits has a comma placed *after* such an index
$\{hkl\}$	Form of (hkl) planes
hkl	Reciprocal lattice point corresponding to the (hkl) family of planes
$\mathbf{h}$	Vector with components h, k, l in reciprocal space
h	Planck's constant
I	Body-centered unit cell; intensity of reflection
$I(hkl)$	Intensity of reflection from the (hkl) planes referred to one unit cell
$\mathcal{I}$	Imaginary axis in the complex plane
i	$\sqrt{-1}$; an operator that rotates a vector in the complex plane through $90°$ in a right-handed (counterclockwise) sense
$J(hkl)$	Integrated reflection
K	Scale factor for $\lvert F_o(hkl) \rvert$ data
L	Lorentz correction factor
M_r	Relative molecular mass ('weight')
m	Mirror plane
N	Number of atoms per unit cell

n	Glide plane, with translational component of $(a+b)/2, (b+c)/2$, or $(c+a)/2$
n_1, n_2, n_3	Principal refractive indices in a biaxial crystal
o	subscript: observed, as in $|F_o(hkl)|$
P	Probability; Patterson function
$P(u, v, w)$	Patterson function at the fractional coordinates u, v, w in the unit cell
p	Polarization correction factor
R	Rhombohedral unit cell; rotation axis (of degree R); reliability factor (several R parameters exist)
$\overline{R}$	Inversion axis
$\mathcal{R}$	Real axis in the complex plane
RU	Reciprocal lattice unit
$s, s(hkl), s(\mathbf{h})$	Sign of a centric reflection
$T_{j,\theta}$	Thermal vibration parameter for the jth atom
$[UVW]$	Zone or direction symbol
$\langle UVW \rangle$	Form of zone axes or directions
u	Atomic mass unit
(u, v, w)	Components of a vector in Patterson space
$\overline{U^2}$	Mean-square amplitude of vibration
v_n	Spacing between the zeroth- and nth-layer lines
V	Volume
V_c	Volume of a unit cell
w	Weight factor
x, y, u, z	Crystallographic reference axes descriptors
X, Y, Z	Spatial coordinates, in absolute measure, of a point, parallel to the x, y, and z axes, respectively
x, y, z	Spatial fractional coordinates in a unit cell
x_j, y_j, z_j	Spatial fractional coordinates of the jth atom in a unit cell
$[x, \beta, \gamma]$	Line parallel to the x axis and intersecting the y and z axes at β and γ, respectively
(x, y, γ)	Plane normal to the z axis and intersecting it at γ
$\pm\{x, y, z; \ldots\}$	$x, y, z; \bar{x}, \bar{y}, \bar{z}; \ldots$
Z	Number of formula entities of mass M_r per unit cell
Z_j	Atomic number of the jth atom in a unit cell
α, β, γ	Angles between the pairs of unit-cell edges bc, ca, and ab, respectively
$\alpha^*, \beta^*, \gamma^*$	Angles between the pairs of reciprocal unit-cell edges b^*c^*, c^*a^*, and a^*b^*, respectively
δ	Path difference
$\varepsilon, \varepsilon(hkl)$	Statistical weight of a reflection (epsilon factor)
ε, ω	Principal refractive indices for a uniaxial crystal
θ	Bragg angle

κ	Reciprocal space constant
λ	Wavelength
μ	Linear absorption coefficient
ν	Frequency
ρ	Radius of stereographic projection
$\rho(x, y, z)$	Electron density at the point x, y, z
Φ	Interfacial (internormal) angle
$\phi(hkl), \phi(h), \varphi$	Phase angle associated with a structure factor
χ, ψ, ω	$(\cos \chi, \cos \psi, \cos \omega)$ direction cosines of a line with respect to the $x, y,$ and z axes
ω	Angular frequency
Ω	Azimuthal angle in experimental methods
$\bar{X}, \langle X \rangle$	Average value of X

Structure Determination by
X-ray Crystallography

1

Crystal Morphology and Crystal Symmetry

1.1 Introduction

Crystallography grew up empirically as a branch of mineralogy. It was supported by laws deduced from observations, such as the law of constancy of interfacial angles and the law of rational intercepts, and involved mainly the recognition, description, and classification of naturally occurring crystal species, that is, it was a study of the *morphology*, or external form, of crystals. By the end of the 19th century, it was believed that crystals were composed of orderly arrays of atoms and molecules and, on this basis, Federov, Schönflies, and Barlow, independently, concluded that there were only 230 ordered spatial patterns, or space groups, based on the 14 crystal lattices deduced earlier by Bravais.

X-ray crystallography is a relatively new discipline, dating from the discovery in 1912 of the diffraction of x-rays by crystals. The classic experiment, suggested by von Laue, and performed by Friedrich and Knipping, demonstrated the diffraction of x-rays from a crystal of copper sulphate. This material was, perhaps, not the best choice because of its low (triclinic) symmetry. Nevertheless, the diffraction effects showed conclusively both that crystals were periodic, in three dimensions, and that x-rays possessed wave-like properties.

The results from this experiment may be said to be a landmark in the development of modern science. The diffraction technique that was initiated by Laue was improved quickly by Bragg, in his work on the crystals of the alkali-metal halides and zinc sulphide. Barlow, one of those who, earlier, had derived the space groups, had also developed structure models for some metallic elements, and simple binary compounds such as sodium chloride, cesium chloride, and zinc blende. None of his results was proved at that time: all was speculation, but remarkably accurate, as it turned out.

Bragg investigated the diffraction patterns expected from Barlow's structures for zinc blende and sodium chloride and found that the models were correct. Other structures proposed by Barlow were confirmed, and the number of structure

1

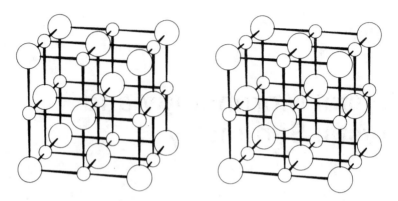

FIGURE 1.1.   Stereoview of the face-centered cubic unit cell and environs of the crystal structure of sodium chloride: ⦿= Na^+,○= Cl^-.

analyses grew very rapidly. Figure 1.1 is a stereoview of the structure of sodium chloride, NaCl.

Several stereoviews are used in this book in order to demonstrate clearly the three-dimensional nature of crystal structures; Appendix A1 describes the process of stereoviewing, and the construction of a simple stereoviewer.

These early structure analyses were carried out with the aid of an x-ray ionization spectrometer, the forerunner of the modern single-crystal x-ray diffractometer, designed largely by W H Bragg. Generally, the name Bragg refers to W L Bragg, but his father (W H) played an important role at the very beginning of the technique of crystal structure analysis by x-ray diffraction.

X-ray diffraction provides the most powerful techniques for probing the internal structures of crystals, and determining with high precision the actual atomic arrangement in space. Figure 1.2 shows a three-dimensional contour map of the electron density in a medium sized molecule, euphenyl iodoacetate,[a] $C_{32}H_{53}O_2I$. The contour lines join points of equal electron density in the structure; hydrogen atoms are not revealed in this map because of their relatively small scattering power for x-rays. If we assume that the centers of atoms are located at the maxima in the electron density map, we can deduce the molecular model in Figure 1.3; the chemical structural formula is shown for comparison. The iodine atom is represented by the large number of contours at the extreme left of Figure 1.2. The carbon and oxygen atoms are depicted by approximately equal numbers of contours, except for the atoms in the side chain, shown on the extreme right of the figure. Thermal vibrations of the atoms are most severe in this portion of the molecule, and they have the effect of smearing out the electron density, so that its

[a] C. H. Carlisle and M. F. C. Ladd, *Acta Crystallographica* **21**, 689 (1966).

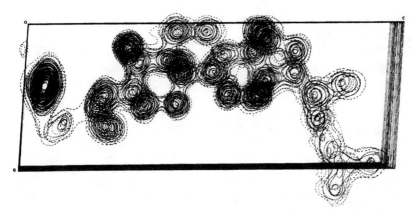

FIGURE 1.2.   Three-dimensional electron density contour map for euphenyl iodoacetate, as seen along the *b* direction of the unit cell; the contours connect points of equal electron density.

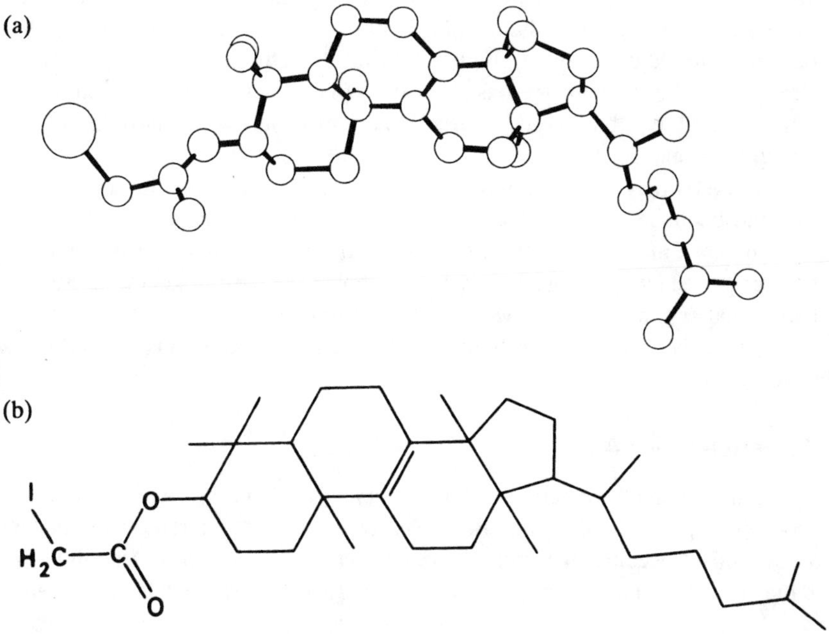

(a)

(b)

H_2C-C

FIGURE 1.3.   Euphenyl iodoacetate, $C_{32}H_{53}O_2I$: (a) molecular model, excluding hydrogen atoms; (b) chemical structural formula. The orientations (standard numbering) at C_{14} and C_{17} are both α, and that at C_{10} is β.

gradient, represented by the closeness of the contours, is less steep than in other parts of the molecule.

Molecules of much greater complexity than this example are now being investigated; the structures of proteins, enzymes, and nucleic acids—the "elements" of life itself—are being revealed by powerful x-ray diffraction techniques.

1.2   The Crystalline State

A crystalline substance may be defined as a homogeneous solid having an ordered internal atomic arrangement and a definite, though not necessarily stoichiometric, overall chemical composition. In addition to the more obvious manifestations of crystalline materials, like sugar and common salt, other substances, such as cellophane sheet and fibrous asbestos, which reveal different degrees of long-range order (extending over many atomic dimensions), may be described as crystalline.

With the unaided eye, fragments of glass and of quartz look similar to each other, yet quartz is crystalline and glass is noncrystalline, or amorphous. Glass has an atomic arrangement that displays only very short-range order (extending over a few atomic dimensions). Figure 1.4 illustrates the structures of quartz and silica glass; both of them are based on the same atomic group, the tetrahedral SiO_4 structural unit, but in quartz these groups are arranged regularly throughout three-dimensional space.

A crystal may be defined as a substance that is crystalline in three dimensions and bounded by plane faces. The word crystal is derived from the Greek $\kappa\rho\upsilon\sigma\tau\alpha\lambda\lambda o\varsigma$, meaning *ice*, used to describe quartz, which once was thought to be water permanently congealed by intense cold. We have made the useful distinction that crystalline substances exhibit long-range order in three dimensions or less, whereas crystals have both this three-dimensional regularity and plane bounding faces.

1.2.1   Reference Axes

In describing the external features of crystals, we make use of relationships in coordinate geometry. It is important to set up a system of reference axes, and we consider this procedure first in the more familiar two dimensions. A straight line AB may be referred to rectangular axes (Figure 1.5) and described by the equation

$$Y = mX + b \tag{1.1}$$

where m $(= \tan \Phi)$ is the *slope* of the line and b is the *intercept* made by AB on the y axis. Any point $P(X, Y)$ on the line satisfies (1.1). If the line had been referred

(a)

(b)

FIGURE 1.4.   Arrangements of SiO_4 structural units (the darker spheres represent Si) in (a) α-quartz; (b) silica glass. [Crown copyright. Reproduced from *NPL Mathematics Report Ma62* by R. J. Bell and P. Dean, with the permission of the Director, National Physical Laboratory, Teddington, Middlesex, UK.]

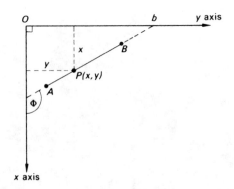

FIGURE 1.5.   Line *AB* referred to rectangular axes.

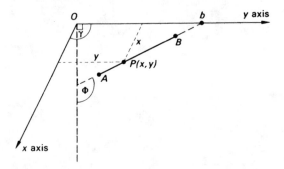

FIGURE 1.6.   Line *AB* referred to oblique axes.

to oblique axes (Figure 1.6), its equation would have been

$$Y = MX + b \tag{1.2}$$

where b has the same value as before, and M is given by

$$M = \tan \Phi \sin \gamma - \cos \gamma \tag{1.3}$$

Evidently, oblique axes are less convenient in this case.

We may describe the line in another way. Let *AB* intersect the x axis at a and the y axis at b (Figure 1.7) and have slope m. At $X = a$, we have $Y = 0$, and, using (1.1),

$$ma + b = 0 \tag{1.4}$$

whence

$$Y = (-b/a)X + b \tag{1.5}$$

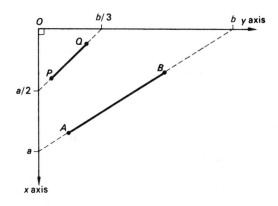

FIGURE 1.7.   Lines *AB* and *PQ* referred to rectangular axes.

or

$$(X/a) + (Y/b) = 1 \tag{1.6}$$

Equation (1.6) is the *intercept form* of the equation of the straight line *AB*. This line will be used as a reference, or *parametral*, line. Consider next any other line, such as *PQ*; let its intercepts on the *x* and *y* axes be, for example, $a/2$ and $b/3$, respectively. The line may be identified by two numbers *h* and *k* defined such that *h* is the ratio of the intercept made on the *x* axis by the parametral line to that made by the line *PQ*, and *k* is the corresponding ratio for the *y* axis. Thus

$$h = a/(a/2) = 2 \tag{1.7}$$

$$k = b/(b/3) = 3 \tag{1.8}$$

PQ is described as the line (23)—two-three. It follows that *AB* is (11). Although the values of *a* and *b* are not specified, once the parametral line is chosen, any other line can be defined uniquely by its indices *h* and *k*.

In Figure 1.8, common sense (and convention, as we shall see) dictates the choice (c) of reference axes *x* and *y* for the rectangle; these lines are parallel to the perimeter lines, which are important features of the rectangle. If *AB* is (11), then *PQ*, *QR*, *RS*, and *SP* are (10), (01), ($\bar{1}$0),[a] and (0$\bar{1}$), respectively. A zero value for *h* or *k* indicates a parallelism of the line with the corresponding axis (its intercept is at infinity); a negative value for *h* or *k*, indicated by a bar over the symbol, implies an intercept on the negative side of the corresponding reference axis.

This simple description of the perimeter lines in *PQRS* is not obtained with either the orientation (a) or the oblique axes (b) in Figure 1.8. In considering a

---

[a] Read as "bar-one zero," or "one-bar zero" in the United States.

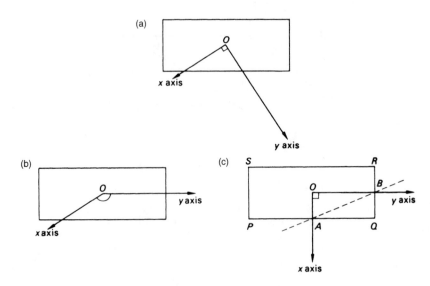

FIGURE 1.8. Rectangle referred to rectangular and oblique axes: (a) arbitrary orientation of rectangular axes; (b) oblique axes; (c) standard orientation of rectangular axes.

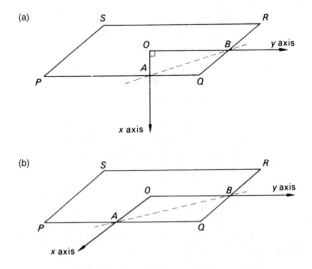

FIGURE 1.9. Parallelogram: (a) referred to rectangular axes; (b) referred to oblique axes.

parallelogram, however, oblique axes are more convenient for our purposes. It is left as an exercise for the reader to confirm that, if AB in Figure 1.9 is (11), then PQ, QR, RS, and SP are again (10), (01), ($\bar{1}$0), and (0$\bar{1}$), respectively, provided that the reference axes are chosen parallel to the sides of the figure, as shown.

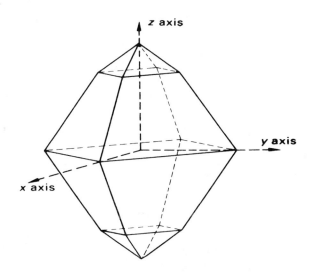

FIGURE 1.10.   Idealized   tetragonal   crystal   with   orthogonal   (mutually   perpendicular) crystallographic axes drawn in.

Crystallographic Axes

Three reference axes are needed in the description of a crystal (Figure 1.10). An extension of the above arguments leads to the adoption of x, y, and z axes parallel to important directions in the crystal. We shall see later that these directions (crystal edges, or possible crystal edges) are related closely to the symmetry of the crystal; in some cases, a choice of non-orthogonal axes then will arise naturally.

It is usual to work with right-handed axes. In Figure 1.11, $+y$ and $+z$ are in the plane of the paper, as shown, and $+x$ is directed forward; the succession $+x \rightarrow +y \rightarrow +z$ simulates an anticlockwise screw motion, which is one way of describing right-handed axes. Notice the selection of the interaxial angles α, β, and γ, and the mnemonic connection between their positions and the directions of the x, y, and z axes.

1.2.2   Equation of a Plane

In Figure 1.12, the plane ABC intercepts the x, y, and z axes (which need not be orthogonal) at A, B, and C, respectively. ON is the perpendicular from the origin O to the plane; it has the length d, and its direction cosines (Appendix A2) are $\cos \chi$, $\cos \psi$, and $\cos \omega$ with respect to OA, OB, and OC, respectively. OA, OB, and OC have the lengths a, b, and c, respectively, and P is any point X, Y, Z in the plane ABC. Let PK be parallel to OC and meet the plane AOB at K, and let KM be parallel to OB and meet OA at M. Then the lengths of OM, MK, and KP

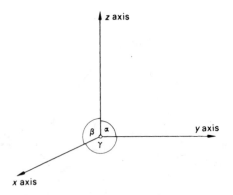

FIGURE 1.11.   Right-handed, general (non-orthogonal) crystallographic axes, and the interaxial angles.

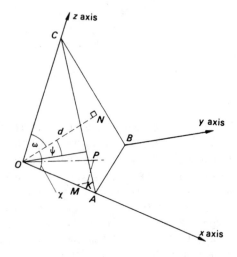

FIGURE 1.12.   Plane ABC in three-dimensional space; ON, of length d, is the normal to the plane from the origin O.

are X, Y, and Z, respectively. Since ON is the projection of OP on to ON, it is equal to the sum of the projections OM, MK, and KP all on to ON. Hence,

$$d = X \cos \chi + Y \cos \psi + Z \cos \omega \qquad (1.9)$$

In $\triangle OAN$, $d = OA \cos \chi = a \cos \chi$. Similarly, $d = b \cos \psi = c \cos \omega$, and, hence, dividing by d,

$$(X/a) + (Y/b) + (Z/c) = 1 \qquad (1.10)$$

Equation (1.10) is the intercept form of the equation of the plane ABC, and may be compared to (1.6).

1.2.3 Miller Indices

The faces of a crystal are planes in three-dimensional space. Once the crystallographic axes are chosen, a parametral plane may be defined and any other plane described in terms of three numbers h, k, and l. It is an experimental fact that, in crystals, if the parametral plane is designated by integral values of h, k, and l, normally (111), then the indices of all other crystal faces are integers, generally *small* integers. This result is known as the *law of rational intercepts (indices)*, and has a basis in lattice theory. The notation for describing the faces of a crystal was introduced first by Miller in 1839, and h, k, and l are called Miller indices.

In Figure 1.13, let the parametral plane (111) be ABC, making intercepts a, b, and c on the crystallographic axes x, y, and z, respectively. Another plane LMN makes corresponding intercepts of lengths a/h, b/k, and c/l. The Miller indices of plane LMN are expressed by the ratios of the intercepts of the parametral plane to those of the plane LMN. If in the figure, $a/h = a/4$, $b/k = b/3$, and $c/l = c/2$, then LMN is (432) (see also Section 2.3). If fractions occur in formulating h, k, or l,

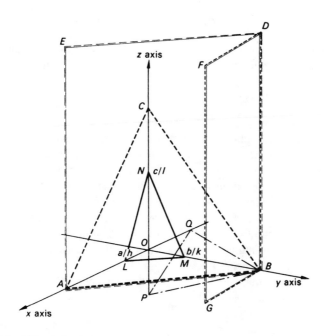

FIGURE 1.13.   Miller indices of planes: $OA = a$, $OB = b$, $OC = c$; ABC is the parametral plane (111), and LMN is the plane (hkl).

they are cleared by multiplication throughout by the lowest common denominator. Conditions of parallelism to axes and intercepts on the negative sides of the axes lead respectively to zero or negative values for h, k, and l. Thus, $ABDE$ is (110), $BDFG$ is (010), and PBQ is ($\bar{2}1\bar{3}$). It may be noted that it has not been necessary to assign numerical values to either a, b, and c or α, β, and γ in order to describe the crystal faces by their Miller indices. In the next chapter we shall identify a, b, and c with the edges of the crystal unit cell in a lattice, but this relationship is not needed at present.

The preferred choice of the parametral plane leads to small numerical values for the Miller indices of crystal faces; rarely are h, k, and l greater than 4. If LMN had been chosen as (111), then ABC would have been (346). Summarizing, we may say that the plane (hkl) makes intercepts a/h, b/k, and c/l along the crystallographic x, y, and z axes, respectively, where a, b, and c are the corresponding intercepts made by the parametral plane.

From (1.9) and (1.10), the intercept equation of the general plane (hkl) may be written as

$$(hX/a) + (kY/b) + (lZ/c) = 1 \qquad (1.11)$$

The equation of the parallel plane passing through the origin is

$$(hX/a) + (kY/b) + (lZ/c) = 0 \qquad (1.12)$$

since it must satisfy the condition $X = Y = Z = 0$. Note, however, that the Miller indices of a crystal plane cannot be determined if the origin is chosen on that plane.

Miller–Bravais Indices

In crystals that exhibit 6-fold symmetry, or 3-fold symmetry referred to hexagonal axes (see Table 1.3), four axes of reference may be used. The axes are designated x, y, u, and z; the x, y, and u axes lie in one plane, at 120° to one another, and the z axis is perpendicular to the xyu plane (Figure 1.14); the sequence x, y, u, z is right-handed. As a consequence, planes in these crystals are described by four numbers, the Miller–Bravais indices h, k, i, and l. The index i is not independent of h and k: thus, if the plane ABC in Figure 1.14 intercepts the x and y axes at $a/2$ and $b/3$, for example, then the u axis is intercepted at $-u/5$. If the z axis is intercepted at $c/4$, the plane is designated ($23\bar{5}4$). In general, $i = -(h + k)$, and the parametral plane is ($11\bar{2}1$).

We can show that $i = -(h + k)$ with reference to Figure 1.15. From the definition of Miller indices,

$$OA = a/h$$
$$OB = b/k \qquad (1.13)$$

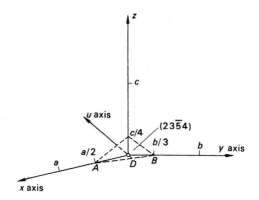

FIGURE 1.14. Miller–Bravais indices $(hkil)$. The crystallographic axes are labeled x, y, u, z, and the plane $(23\bar{5}4)$ is shown; the parametral plane is $(11\bar{2}1)$.

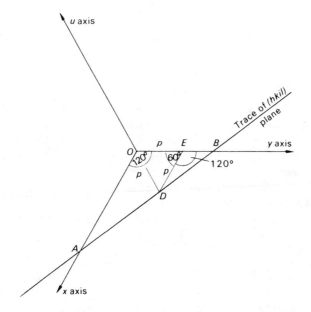

FIGURE 1.15. Equivalence of the Miller–Bravais indices i and $-(h + k)$.

Let the plane $(hkil)$ intercept the u axis at p; draw DE parallel to AO. Since OD bisects $\angle AOB$, $AOD = 60°$, so that $\triangle ODE$ is equilateral; hence

$$OD = DE = OE = p \qquad (1.14)$$

Triangles EBD and OBA are similar; hence

$$EB/DE = OB/OA = (b/k)/(a/h) \qquad (1.15)$$

Since $EB = b/k - p$, combining (1.14) and (1.15) leads to

$$p = ab/(ak + bh) \qquad (1.16)$$

Since $a = b = u$, from the symmetry, $u/p = h + k$. We write u/p as $-i$, since p lies on the negative side of the u axis $(OD = -u/p)$, so that

$$i = -(h + k) \qquad (1.17)$$

1.2.4   Axial Ratios

If both sides of (1.12) are multiplied by b, we obtain

$$\frac{hX}{a/b} + kY + \frac{lZ}{c/b} = 0 \qquad (1.18)$$

The quantities a/b and c/b are termed *axial ratios*; they can be deduced from an analysis of the crystal morphology.

1.2.5   Zones

Most well formed crystals have their faces arranged in groups of two or more with respect to certain directions in the crystal. In other words, crystals exhibit symmetry; this feature is an external manifestation of the ordered arrangement of atoms in the crystal. Figure 1.16 illustrates zircon, $ZrSiO_4$, an example of a highly symmetric crystal. It is evident that several faces have a given direction in common. Such faces are said to lie in a *zone*, and the common direction is called a *zone axis*. Any two faces, $(h_1k_1l_1)$ and $(h_2k_2l_2)$, define a zone. The zone axis is the line of intersection of the two planes, and is given by the solution of the equations

$$(h_1X/a) + (k_1Y/b) + (l_1Z/c) = 0$$
$$(h_2X/a) + (k_2Y/b) + (l_2Z/c) = 0 \qquad (1.19)$$

for the two planes passing through the origin (since we are concerned here only with the directionality). The solution is given by the line

$$\frac{X}{a(k_1l_2 - k_2l_1)} = \frac{Y}{b(l_1h_2 - l_2h_1)} = \frac{Z}{c(h_1k_2 - h_2k_1)} \qquad (1.20)$$

which must also pass through the origin. It may be written as

$$X/(aU) = Y/(bV) = Z/(cW) \qquad (1.21)$$

where $[UVW]$ is called the *zone symbol*.

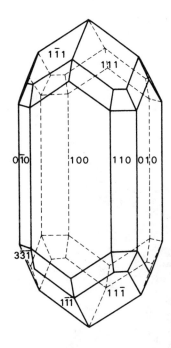

FIGURE 1.16.   A highly symmetric crystal (zircon, $ZrSiO_4$), showing the Miller indices of some of its faces. What are the Miller indices of the other faces on this crystal?

If any other face (hkl) lies in the same zone as $(h_1k_1l_1)$ and $(h_2k_2l_2)$, then it follows from (1.12) and (1.21), that

$$hU + kV + lW = 0 \tag{1.22}$$

which is an expression of the *Weiss zone law*.

In the zircon crystal, the vertical (prism) faces lie in one zone. If the prism faces are indexed in the usual manner (see Figure 1.16), then, from (1.20) and (1.21), the corresponding zone symbol is [001]. The symbols $[UVW]$ and $[\bar{U}\,\bar{V}\,\bar{W}]$ refer to lines that are collinear, but of opposite sense. From (1.22), we see that (110) and $(\bar{1}\bar{1}0)$ are faces in the [001] zone, but (111) is not. Other relationships follow from (1.22) in a similar way. In the manipulation of these equations, it may be noted that a zone axis is described by $[UVW]$, the simplest symbol; the axes that may be described as $[nU, nV, nW]$ $(n = 0, \pm1, \pm2, \ldots)$ are coincident with $[UVW]$ in crystal morphology.

Interfacial Angles

The *law of constant interfacial angles* states that in all crystals of the same substance, angles between corresponding faces have a constant value. Interfacial angles are measured by a *goniometer*. Its simplest form is the contact goniometer,

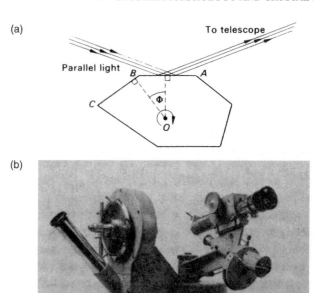

FIGURE 1.17.   Optical goniometry: (a) principle of the reflecting goniometer; AB and BC represent two adjacent faces on the crystal; (b) two-circle optical goniometer; the crystal rotates about the vertical circle, and the telescope and collimator about the horizontal circle.

which is a semicircular protractor with an arm pivoted at its center, but its use requires large crystals, a condition not always easily obtainable in practice.

An improvement in technique was brought about by the reflecting goniometer. The principle of this instrument is shown in Figure 1.17a, and forms the basis of modern optical goniometers. A crystal is arranged to rotate about a zone axis O, which is set perpendicular to a plane containing the incident and crystal-reflected light beams. Parallel light reflected from the face AB is received by a telescope. If the crystal is rotated in a clockwise direction, a reflection from the face BC is received next when the crystal has been turned through the angle Φ and the interfacial angle is 180 − Φ degrees. Accurate goniometry brought a quantitative significance to observable angular relationship in crystals. Figure 1.17b illustrates a two-circle optical goniometer.

1.3 Stereographic Projection

The study of crystal morphology in terms of the analytical description of planes and zones is inadequate for an overall appreciation of the many faces exhibited by a crystal. It is necessary to be able to represent a crystal by means of a two-dimensional drawing, while preserving certain essential properties. In crystal morphology, the interfacial angles, which are a fundamental feature of crystals, must be maintained in plane projection, and the *stereographic projection* is useful for this purpose. Furthermore, with imperfectly formed crystals, the true symmetry may not be apparent by inspection. In favorable cases, the symmetry may be revealed completely by a stereographic projection of the crystal. We shall develop this projection with reference to the idealized crystal shown in Figure 1.18.

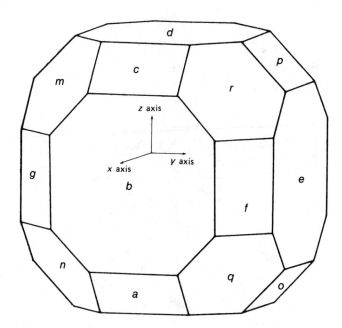

FIGURE 1.18. Cubic crystal showing three forms of planes: cube—b, e, d, and parallel faces; octahedron—r, m, n, q, and parallel faces; rhombic dodecahedron—f, g, p, o, c, a, and parallel faces. The x, y, and z axes are parallel to important (symmetry) directions in the crystal.

This crystal belongs to the cubic system (see Table 1.3): the crystallographic reference axes x, y, and z are orthogonal, and the parametral plane (111) makes equal intercepts ($a = b = c$) on these axes. The crystal shows three forms of planes. In crystallography, a *form* of planes, represented by $\{hkl\}$, refers to the set of planes that are equivalent under the point-group symmetry (see Section 1.4) of the crystal. The crystal under discussion shows the cube form $\{100\}$—six faces (100), ($\bar{1}$00), (010), (0$\bar{1}$0), (001), and (00$\bar{1}$); the octahedron $\{111\}$—eight faces; and the rhombic dodecahedron $\{110\}$—12 faces. Each face on the crystal drawing has a related parallel face on the actual crystal, for example, b (shown) and b'. The reader may care to list the sets of planes in the cubic forms $\{111\}$ and $\{110\}$; the answer will evolve from the discussion of the stereographic projection of the crystal.

From a point within the crystal, lines are drawn normal to the faces of the crystal. A sphere of arbitrary radius is described about the crystal, its center O being the point of intersection of the normals which are then produced to cut the surface of the sphere (Figure 1.19).

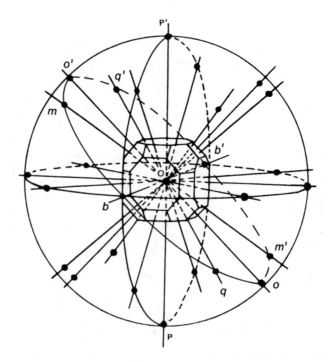

FIGURE 1.19. Spherical projection of the crystal in Figure 1.18; the radius (ρ) of the sphere is arbitrary. The inclined circle projects as $G_3 G'_3$ in Figure 1.21.

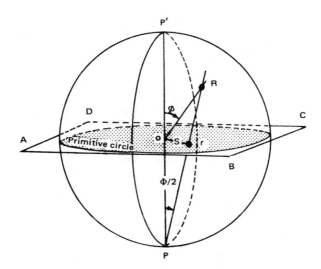

FIGURE 1.20. Development of the stereographic projection (stereogram) from the spherical projection. Points on the upper hemisphere, such as R, are joined to the lowermost point, P.

In Figure 1.20, the plane of projection is $ABCD$, and it intersects the sphere in the *primitive circle*. The portion of the plane of projection enclosed by the primitive circle is the *primitive plane*, or *primitive*. The point of intersection of each normal with the upper hemisphere is joined to the lowest point P on the sphere. The intersection of each such line with the primitive is the stereographic projection, or *pole*, of the corresponding face on the crystal, and is indicated by a dot on the stereographic projection, or *stereogram*. In particular, R is the intersection of the normal to the face r with the sphere, and r (in this figure) is the corresponding pole.

If the crystal is oriented such that the normal to face d (and d') coincides with PP' in the sphere, then the normals to the zone $e, f, b, \ldots, g'$ lie in the plane of projection and intersect the sphere on the primitive circle. In order to avoid increasing the size of the stereogram unduly, the intersections of the face normals with the lower hemisphere are joined to the uppermost point P' on the sphere and their poles are indicated on the stereogram by open circles.

The completed stereogram is illustrated by Figure 1.21. The poles now should be compared with the corresponding faces on the crystal drawing. A fundamental property of the stereogram is that all circles drawn on the sphere project as circles.[a] Thus, the curve $G_1 G_1'$ is an arc of a circle; specifically, it is the projection of a great circle that is inclined to the plane of projection. A great circle is the

[a] See Bibliography (Phillips).

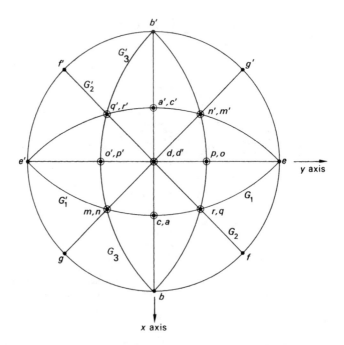

FIGURE 1.21.   Stereogram of the crystal in Figure 1.18. The zone circle (great circle) G_1G_1', symbol [101], passes through e,q,a,n,e',q',a',n'; the zone circle G_2G_2' symbol [1$\bar{1}$0], passes through f,r,d,q', f',r',d',q; the zone circle G_3G_3', symbol [011], passes through b,m,o',q',b',m',o,q (see Figure 1.19).

trace, on the sphere, of a plane that passes through the center of the sphere; it may be likened to a meridian on the globe of the world. Limiting cases of inclined great circles are the primitive circle, which lies in the plane of projection, and straight lines, such as G_2G_2', which are projections of great circles lying normal to the plane of projection. All poles on a great circle represent faces lying in one and the same zone.

Circles formed on the surface of the sphere by planes that do not pass through the center of the sphere are called small circles; they may be likened to parallels of latitude on the globe.

In order to construct Figure 1.21, the following practical principles must be followed. The interfacial angles are measured in zones. If an optical goniometer is used, the angle Φ (see Figure 1.17) is plotted directly on the stereogram. *Although Φ is the angle between the normals to planes, it is often called the interfacial angle in this context.* Next, the crystal orientation with respect to the sphere is chosen: for example, let zone b, f, e,... be on the primitive circle, and zone $b,c,d,...$

run from bottom to top on the projection. Since $\widehat{bf}$, the angle between face b and f, is 45°, zone $f, r, d, \ldots$ can be located on the stereogram.

The distance S of the pole r from the center of the stereogram (Figure 1.20) is given by

$$S = \rho \tan(\Phi/2) \tag{1.23}$$

where ρ is the radius of the stereogram and Φ is the interfacial angle $\widehat{dr}$ (cf. Figure 1.18). A simple graphical method, employing a Wulff net,[a] is often sufficiently accurate to locate poles on a stereogram.

Triangles such as dcr can be solved by spherical trigonometry. If the angles $\widehat{dc}$, $\widehat{cr}$, and $\widehat{qd}$ are represented here by the letters a, b, and c respectively, and the angles within the triangle and opposite them by A, B, and C respectively, then from the sine and cosine rules in spherical trigonometry,[a] we have

$$\frac{\sin A}{\sin a} = \frac{\sin B}{\sin b} = \frac{\sin C}{\sin c} \tag{1.24}$$

and

$$\cos c = \cos a \cos b - \sin a \sin b \cos C \tag{1.25}$$

Thus, in the triangle dcr, $\widehat{cr} = b = 35.26°$, $\widehat{dc\,dr} = B = 45°$, $\widehat{dr} = c$, $\widehat{dc\,cr} = C = 90°$. Hence, from (1.24) or (1.25),

$$c = \widehat{dr} = 54.74°$$

The completed stereogram (Figure 1.21) may now be indexed. The parametral plane is chosen as face r (the parametral plane must intersect all three crystallographic axes), and the remaining faces are then allocated h, k, and l values (Figure 1.22). We may note here that if two zone axes are $[U_1 V_1 W_1]$ and $[U_2 V_2 W_2]$, then the face (hkl) that contains these two zone axes is given by solving two equations of the form of (1.22), following (1.19) to (1.20). Thus, $h = V_1 W_2 - V_2 W_1$, and similarly for k and l. It is not necessary to write the indices for both poles at the same point on the stereogram. If the dot is hkl, then we know that the open circle is $hk\bar{l}$. Figure 1.23 shows the crystal of Figure 1.18 again, but with the Miller indices inserted for direct comparison with its stereogram.

We shall not be concerned here with any further development of the stereogram.[a] The angular truth of the stereographic projection makes it very suitable for representing not only interfacial angles, but also symmetry directions, point groups, and bond directions in molecules and ions.

---

[a] See Bibliography (Phillips).

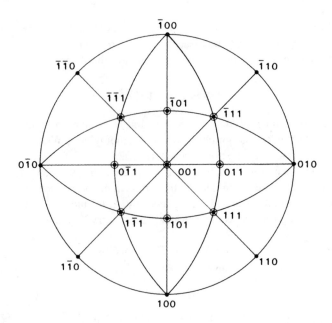

FIGURE 1.22.   Stereogram in Figure 1.21 indexed, taking r as 111. The zone containing (100) and (111) is $[0\bar{1}1]$, and that containing (010) and (001) is [100]; the face p common to these two zones is (011)—see (1.20) to (1.22).

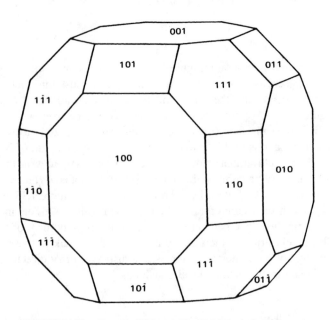

FIGURE 1.23.   Crystal of Figure 1.18 with Miller indices inserted.

1.4 External Symmetry of Crystals

The existence of faces on a crystal in groups of two or more, in a similar orientation with respect to some line or plane in the crystal, is a manifestation of symmetry. The crystal drawing of zircon in Figure 1.16 shows several sets of symmetrically arranged faces.

Few of us have difficulty in recognizing symmetry in two-dimensional objects such as the outline of a shield, a Maltese cross, the three-legged emblem of the Isle of Man, a five-petaled Tudor rose, or the six-pointed Star of David. But it is a rather different matter when we are dealing with three-dimensional objects.

The difficulty arises first from the fact that we can see all parts of a two-dimensional object simultaneously, and thus we take in the relation of the parts to the whole; but we cannot do that so easily with three-dimensional objects. Second, while some three-dimensional objects, such as flowers, pencils, and architectural columns, are simple enough for us to visualize and to rotate in our mind's eye, few of us have a natural gift for mentally perceiving and manipulating more complex three-dimensional objects. Nevertheless, the art of doing so can be developed with suitable aids and patience. If, initially, you have problems, take heart. You are not alone and, like many before you, you will be suprised at how swiftly the required facility can be acquired. Engineers, architects, and sculptors may be blessed with a natural three-dimensional visualization aptitude, but they have learned to develop it—particularly by making and handling models.

Standard practice in the past was to reduce three-dimensional objects to one or more two-dimensional drawings (projections and elevations): it was cheap, well suited to reproduction in books, and less cumbersome than handling three-dimensional models. In this book, we shall continue to use such two-dimensional representations, but to rely on them exclusively only delays the acquisition of a three-dimensional visualization aptitude. Fortunately, we can now also use stereoscopic image pairs, such as that shown in Figure 1.1. These illustrations are a great help, but, because they provide a view from only one standpoint, they are not always quite the equal of models that can be examined by hand.

Symmetry may be defined as *that spatial property of a body (or pattern) by which the body (or pattern) can be brought from an initial state to another indistinguishable state by means of a certain operation—a symmetry operation.* For our purposes, the operation will be considered to take place in n-dimensional space ($n = 1, 2,$ or 3) and to represent an action with respect to a symmetry element.

A *symmetry element* is a geometrical entity (point, line, or plane) in a body or assemblage, with which is associated an appropriate symmetry operation. The symmetry element is strictly conceptual, but it is convenient to accord it a sense of reality. The symmetry element connects all parts of the body or assemblage as a number of symmetrically related parts. The term *assemblage* is often useful because it describes more obviously a bundle of radiating face normals (Figure 1.19), or a number of bonds emanating from a central atom in the case of a molecule or ion (Figure 1.38), to which these symmetry concepts equally apply.

The *symmetry operation* corresponding to a symmetry element, when applied to a body, converts it to a state that is indistinguishable from the initial state of that body, and thus the operation *reveals* the symmetry inherent in the body. In numerous cases, different symmetry operations can reveal one and the same symmetry element. Thus 3^1 ($=3$), 3^2, and 3^3($=3$) may be regarded as either multiple steps of 3 (a 3-fold rotation operation, q.v.) or single-step operations in their own right, but all are contained within the same single symmetry element, 3. The latter idea is of particular importance in the study of group theory.

Symmetry elements may occur singly in a body (for example, Figures 1.24, 1.25) or in certain combinations (for example, Figure 1.26). A set of interacting symmetry elements in a finite body, or just one such element, is referred to as a *point group*. A point group may be defined as *a set of symmetry elements all of which pass through a single fixed point*: this point is taken as the origin of the reference axes for the body. It follows that the symmetry operations of a point group must leave *at least* one point unmoved: in some cases it will be a line (symmetry axis) or a plane (reflection plane) that is invariant under the action of the point group.

It can be contended that, in real objects, since they are imperfect, even if only on a microscopic scale, a second indistinguishable state can be obtained only by a rotation of $360°$ (or $0°$); this operation is *identity*, or "doing nothing." For practical purposes, however, the effects of most imperfections are small, and although our discussion of symmetry will be set up in terms of ideal geometrical objects, the extension of the results to real situations is scientifically rewarding.

The observable symmetry may depend upon the nature of the examining technique. Thus, different results for a given material may arise from a study of its optical, magnetic, and photoelastic properties (see also Section 6.6). Here, we shall be concerned with the symmetry shown by *directions in space*, such as the normals to the faces on crystals, or the bond directions in chemical species. Such angular relationships can be presented conveniently on stereograms (Section 1.3), and we shall draw fully on this method of representation in the ensuing discussion.

Several concepts in symmetry can be introduced with two-dimensional objects; subsequently, the third dimension can be introduced mainly as a geometrical extension to the two-dimensional ideas. There is a single one-dimensional point group; it is more difficult conceptually, and we are not particularly concerned with it in this book.

1.4.1 Two-Dimensional Point Groups

If we examine the two-dimensional objects in Figure 1.24, we can discover two types of symmetry elements that can bring an object from one state to another

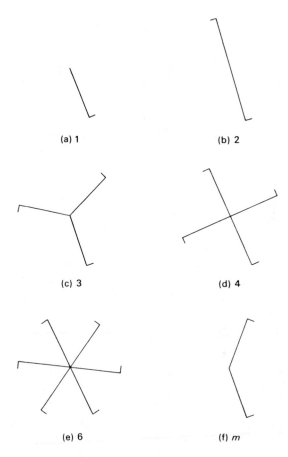

FIGURE 1.24.   Two-dimensional objects and their point groups. The motifs are built up from the asymmetric unit (a), by operating on it according to the point-group symbol. Note that the symmetry element must intersect the asymmetric unit, and has been set at the uppermost part of (a).

indistinguishable state: parts (a) to (e) of Figure 1.24 depict rotational symmetry, whereas (f) shows reflection symmetry.

Rotation Symmetry

An object possesses two-dimensional rotational symmetry of degree R (or R-fold symmetry) about a *point* if it can be brought from one state to another indistinguishable state by each and every rotation of $(360/R)$ degrees about that symmetry point. Figures 1.24a–e illustrate the rotational symmetry elements R of

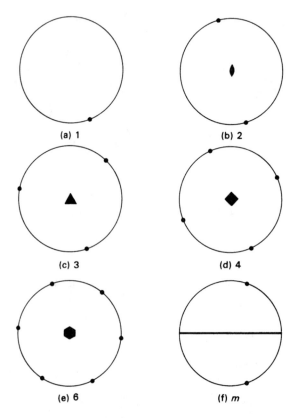

FIGURE 1.25.   Stereograms of the point groups of the objects in Figure 1.24; the conventional graphic symbols for R ($R = 1, 2, 3, 4, 6$) and m are shown.

1, 2, 3, 4, and 6, respectively. The onefold element is the identity element, and is crystallographically trivial; every object has onefold symmetry.

Reflection Symmetry

An object possesses reflection symmetry, symbol m, in two dimensions if it can be brought from one state to another indistinguishable state by reflection across the symmetry *line*. The operation is not one that we can perform physically with an object, unlike rotation, but we can appreciate from the object (and its stereogram) that m symmetry is present. The m line divides the figure into its *asymmetric unit*, ⌄, and a mirror image or enantiomorph of this part, ╱, which situation is characteristic of reflection symmetry (Figure 1.24f).

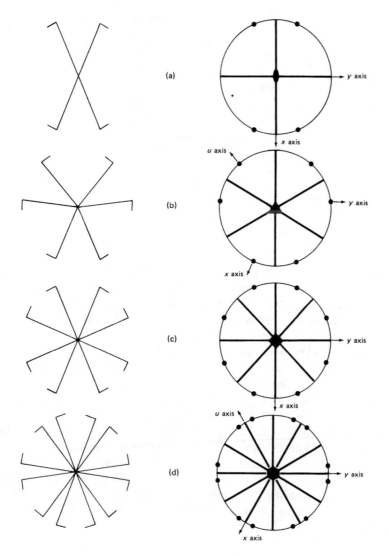

FIGURE 1.26.   Further two-dimensional objects with their stereograms and point groups: (a) 2mm, (b) 3m, (c) 4mm, (d) 6mm.

Each of the objects in Figure 1.24 has a symmetry pattern that can be described by a two-dimensional point group, and it is convenient to illustrate these point groups by stereograms. Figure 1.25 shows stereograms for the two-dimensional point groups 1, 2, 3, 4, 6, and m. It should be noted that in using stereogram-like drawings to illustrate two-dimensional symmetry, the

TABLE 1.1.   Two-Dimensional Point Groups and Notation

System	Point groups	Symbol meaning, appropriate to position occupied		
		First position	Second position	Third position
Oblique	1, 2	Rotation about a point	—	—
Rectangular	$1m^a$	⎧ Rotation about a point ⎫	$m \perp x$	—
	$2mm$	⎩ ⎭	$m \perp x$	$m \perp y$
Square	4	⎧ Rotation about a point ⎫	—	—
	$4mm$	⎩ ⎭	$m \perp x, y$	m at $45°$ to x, y
Hexagonal	3		—	—
	$3m$	⎧ ⎫	$m \perp x, y, u$	—
	6	⎨ Rotation about a point ⎬	—	—
	$6mm$	⎩ ⎭	$m \perp x, y, u$	m at $30°$ to x, y, u

^{a} This point group is usually called m, but the full symbol is given here in order to clarify the location of the symmetry elements in the symbol.

representative points (poles) are placed on the perimeter; such situations may represent *special* forms (q.v.) on the stereograms of three-dimensional objects.

Combinations of R and m lead to four more point groups; they are illustrated in Figure 1.26. We have deliberately omitted point groups in which $R = 5$ and $R \geq 7$, for a reason that will be discussed later.

It is convenient to allocate the 10 two-dimensional point groups to two-dimensional *systems*, and to choose reference axes in close relation to the directions of the symmetry elements. Table 1.1 lists these systems, together with the meanings of the positions in the point-group symbols. It should be noted that combinations of m with R ($R \geq 2$) introduce additional reflection lines of a different crystallographic form. In the case of $3m$, however, these additional m lines are coincident with the first set; the symbol $3mm$ is not meaningful.

It is important to *remember* the relative orientations of the symmetry elements in the point groups, and the variations in the meanings of the positions in the different systems. In the two-dimensional hexagonal system, three axes are chosen in the plane; this selection corresponds with the use of Miller–Bravais indices in three dimensions.

1.4.2   Three-Dimensional Point Groups

The symmetry elements encountered in three dimensions are rotation axes (R), inversion axes ($\bar{R}$), and a reflection (mirror) plane (m). A *center of symmetry* can be invoked also, although neither this symmetry element nor the m plane is independent of $\bar{R}$ ($R = 1, 2, 3, 4, 6$).

The operations of rotation and reflection are similar to those in two dimensions, except that the geometric extensions of the operations are now increased to rotation about a *line* and reflection across a *plane*.

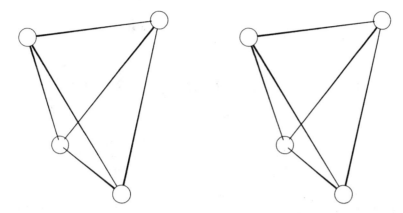

FIGURE 1.27. Stereoview of a hypothetical C_4 molecule; the $\bar{4}$ axis is in the vertical direction. (See also Appendix A1.)

Inversion Axes

An object is said to possess an inversion[a] axis $\bar{R}$ (read as bar-R; R-bar in U.S. convention), if it can be brought from one state to another indistinguishable state by the combined actions of rotation by $(360/R)$ degrees and inversion through a point on the axis that also serves as the origin; the two actions form a *single* symmetry operation. Like the mirror plane, the inversion axis depicts a nonperformable symmetry operation, and it may be represented conveniently on a stereogram. It is a little more difficult to envisage this operation than those of rotation and reflection. Figure 1.27 illustrates a hypothetical molecule having a vertical $\bar{4}$ axis: the stereoscopic effect can be created by using a stereoviewer (see Appendix A1, which also contains instructions for making a model with $\bar{4}$ symmetry.).

In pictorial representations of the three-dimensional point groups, it is helpful to indicate the third dimension on the stereogram and, in addition, to illustrate the change-of-hand relationship that occurs with the $\bar{R}$ (including m) symmetry operations. For example, referring to Figure 1.28, the element 2 lying in the plane of projection, and the element $\bar{4}$ normal to the plane of projection, when acting on a point derived from the upper hemisphere (symbol ●) both move the point into the lower hemisphere region (symbol ○). Both operations involve a reversal of the sign of the vertical coordinate, but only $\bar{4}$ involves also a change of hand, and this distinction is not clear from the conventional notation. Consequently, we shall adopt a symbolism, common to three-dimensional space groups, which will effect the necessary distinction.

---

[a] Strictly, *roto*-inversion.

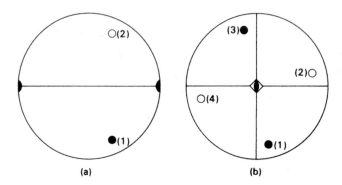

FIGURE 1.28.   Stereograms of general forms: (a) point group 2 (axis horizontal and in the plane of the stereogram); (b) point group $\bar{4}$ (axis normal to the plane of the stereogram). In (a), the point ● is rotated through 180° to O: (1) → (2). In (b), the point ● is rotated through 90° and then inverted through the origin to O; this combined operation generates, in all, four symmetry-equivalent points: (1) → (4) → (3) → (2).

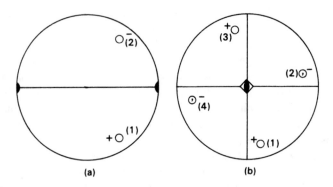

FIGURE 1.29.   (a) Stereograms from Figure 1.28 in the revised notation; the different natures of points (2) in (a) and (2) and (4) in (b), all with respect to point (1), are clear.

A representative point in the l-positive hemisphere will be shown by o^+, signifying, for example, the face (hkl), or its pole. A change of hemisphere to $(hk\bar{l})$ will be indicated by o^-, and a change of hand on reflection or inversion by $\odot^+$ or $\odot^-$ (see Figure 1.29). This notation may appear to nullify partially the purpose of a stereogram. However, although the stereogram is a two-dimensional diagram, it should convey a three-dimensional impression, and this notation is used as an aid to this end.

TABLE 1.2.   Three-Dimensional Symmetry Symbols

Symbol	Name	Action for indistinguishability	Graphic symbol
1	Monad	$360°(0°)$ rotation; identity	None
2	Diad	$180°$ rotation	⬮ ⊥ projection ❙ ∥ projection
3	Triad	$120°$ rotation	▲ ⊥ or inclined to projection
4	Tetrad	$90°$ rotation	◆ ⊥ projection ■ ∥ projection
6	Hexad	$60°$ rotation	⬤ ⊥ projection
$\bar{1}$	Inverse monad	Inversion[a]	o ⊥
$\bar{3}$	Inverse triad	$120°$ rotation + inversion	△ ⊥ or inclined to projection
$\bar{4}$	Inverse tetrad	$90°$ rotation + inversion	◈ ⊥ projection ▯ ∥ projection
$\bar{6}$	Inverse hexad	$60°$ rotation + inversion	⬠ ⊥ projection
m	Mirror plane[b]	Reflection across plane	▬ ⊥ projection ❙ ∥ projection

[a] $\bar{R}$ is equivalent to R plus $\bar{1}$ only where R is an odd number: $\bar{1}$ represents the center of symmetry, but $\bar{2}, \bar{4}$, and $\bar{6}$ are not centrosymmetric point groups. For R even, $R + \bar{1} \equiv R/m$.

[b] The symmetry elements m and $\bar{2}$ produce an equivalent operation, with $\bar{2}$ oriented perpendicularly to the mirror plane.

Crystal Classes

There are 32 crystal symmetry classes, each characterized by a point group. They comprise the symmetry elements R and $\bar{R}$, taken either singly or in combinations, with R restricted to the values 1, 2, 3, 4, and 6. A simple explanation for this restriction is that figures based only on these rotational symmetries can be stacked together to fill space completely, as Figure 1.30 shows. A further discussion of these values for R is given in Section 2.6.

We shall not be concerned here to derive the crystallographic point groups— and there are several ways in which it can be done[a]—but to give, instead, a scheme which allows them to be worked through simply and adequately for present purposes. In addition, use of the program EULR* (see Chapter 11) shows how the combinations of symmetry operations based on R and $\bar{R}$ ($R = 1, 2, 3, 4, 6$) can lead to the 32 crystallographic point groups.

The symbols for rotation and reflection symmetry are similar to those used in two dimensions. Certain additional symbols are required in three dimensions, and Table 1.2 lists them all.

Figure 1.31a shows a stereogram for point group m. The inverse diad is lying normal to the m plane. A consideration of the two operations in the given relative orientations shows that they produce equivalent actions. It is conventional to use the symbol m for this operation, although sometimes it is helpful to employ $\bar{2}$ instead. Potassium tetrathionate (Figure 1.31b) crystallizes in point group m.

[a] See Bibliography (Ladd).

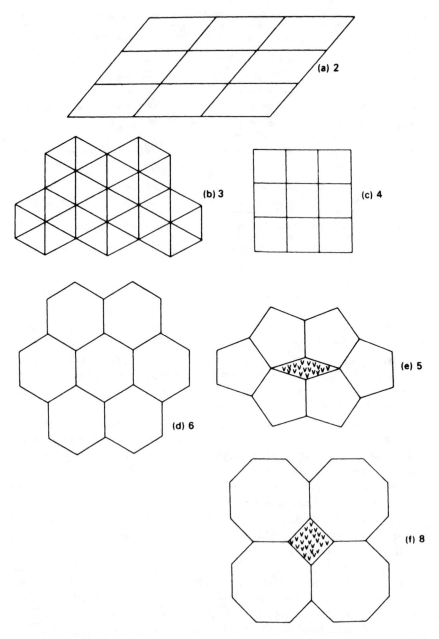

FIGURE 1.30. Sections of three-dimensional figures and the rotational symmetries of their smallest structural units; (a)–(d) are space-filling patterns. In (e) and (f) the v-marks represent voids in the pattern (see also Problem 1.10a).

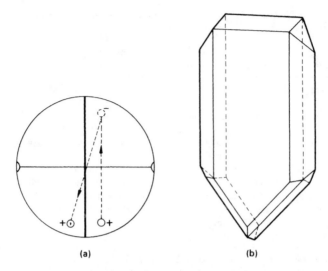

(a)   **(b)**

FIGURE 1.31.   Point group m: (a) stereogram showing equivalence of m and $\bar{2}$ (the symbol $\mathfrak{l}$ for $\bar{2}$ is not conventional); (b) crystal of potassium tetrathionate ($K_2S_4O_6$), point group m.

Crystal Systems and Point-Group Scheme

Crystals are grouped into seven *systems* according to the *characteristic* symmetry listed in Table 1.3. The characteristic symmetry refers to the minimum necessary for classification of a crystal in a system; a given crystal may contain more than the characteristic symmetry of its system. The conventional choice of crystallographic reference axes leads to special relationships between the intercepts a, b, and c of the parametral plane (111) and between the interaxial angles α, β, and γ, in all systems other than triclinic.

A crystallographic point-group scheme is given in Table 1.4, under the seven crystal systems as headings. The main difficulty in understanding point groups lies not in knowing the action of the individual symmetry elements, but rather in appreciating both the relative orientation of the different elements in a point-group symbol and the fact that this orientation changes among the crystal systems according to the principal symmetry axis, that is, the rotation axis R of highest degree. These orientations must be learned: *they are the key* to point-group and space-group studies.

Table 1.5 lists the meanings of the three positions in the three-dimensional point-group symbols. Tables 1.4 and 1.5 should be studied carefully in conjunction with Figure 1.36. For example, consider carefully point groups 222 and 422, and note how and why the orientations represented by the three positions in the symbol change their meanings. In 222, the three symmetry axes are along x, y, and z,

TABLE 1.3. Crystal Systems and their Characteristics

System	Characteristic symmetry axes, with their orientation	Parametral plane intercepts and interaxial angles, assuming the simplest indexing of faces[a,b]
Triclinic	None	a ⊄ b ⊄ c; α ⊄ β ⊄ γ ⊄ $90°, 120°$
Monoclinic	One 2 or $\bar{2}$ axis[c] along y	a ⊄ b ⊄ c; $\alpha = \gamma = 90°$; β ⊄ $90°, 120°$
Orthorhombic	Three mutually perpendicular 2 or $\bar{2}$ axes along x, y, and z	a ⊄ b ⊄ c; $\alpha = \beta = \gamma = 90°$
Tetragonal	One 4 or $\bar{4}$ axis along z	$a = b$ ⊄ c; $\alpha = \beta = \gamma = 90°$
Trigonal[d]	One 3 axis along z	$a = b$ ⊄ c; $\alpha = \beta = 90°$;
Hexagonal	One 6 or $\bar{6}$ axis along z	$\gamma = 120°$
Cubic	Four 3 axes inclined at $54.74°(\cos^{-1} 1/\sqrt{3})$ to x, y, and z	$a = b = c$; $\alpha = \beta = \gamma = 90°$

[a] We shall see in Chapter 2 that the same relationships apply to conventional unit cells in lattices.
[b] The special symbol ⊄ should be read as "not constrained by symmetry to equal".
[c] It must be remembered that $\bar{2}$ is equivalent to an m plane normal to the $\bar{2}$ axis.
[d] For convenience, the trigonal system is referred to hexagonal axes (but see also Table 1.5).

TABLE 1.4. Crystallographic Point-Group Scheme[a]

Type	Triclinic	Monoclinic	Trigonal	Tetragonal	Hexagonal	Cubic[b]
R	1	2	3	4	6	23
$\bar{R}$	$\bar{1}$	m	$\bar{3}$	$\bar{4}$	$\bar{6}$	$m3$
$R + $ center	—	$2/m$	—	$4/m$	$6/m$	—
		Orthorhombic				
$R2$		222	32	422	622	432
Rm		$mm2$	$3m$	$4mm$	$6mm$	$\bar{4}3m$
$\bar{R}m$		—	$\bar{3}m$	$\bar{4}2m$	$\bar{6}m2$	$m3m$
$R2 + $ center		mmm	—	$\frac{4}{m}mm$	$\frac{6}{m}mm$	—

[a] The reader should consider the implications of the spaces (marked —) in this table.
[b] The cubic system is characterized by its four threefold axes; R refers here to the element 2, 4, or $\bar{4}$, but 3 is always present along $\langle 111 \rangle$.

respectively. In 422, 4 is taken along z, by convention; the first symbol 2 (second position) represents both the x and y directions, because they are equivalent under 4-fold symmetry. This combination of 4 and 2 introduces symmetry along [110] and [1$\bar{1}$0], so that the second symbol 2 represents this symmetry. Similar situations exist among other point groups where the principal symmetry axis is of a degree greater than 2.

The reader should not be discouraged by the wealth of convention which surrounds this part of the subject. It arises for two main reasons. There are many different, equally correct ways of describing crystal geometry. For example, the

TABLE 1.5. Three-Dimensional Point Groups and Hermann–Mauguin Notation

System	Point groups[a]	Symbol meaning for each position		
		1st position	2nd position	3rd position
Triclinic	$1, \bar{1}$	All directions in crystal	—	—
Monoclinic[b]	$2, m, \dfrac{2}{m}$	2 and/or $\bar{2}$ along y	—	—
Orthorhombic	$222, mm2, mmm$	2 and/or $\bar{2}$ along x	2 and/or $\bar{2}$ along y	2 and/or $\bar{2}$ along z
Tetragonal	$4, \bar{4}, \dfrac{4}{m}$,	4 and/or $\bar{4}$ along z	—	—
	$422, 4mm, \bar{4}2m, \dfrac{4}{m}mm$	4 and/or $\bar{4}$ along z	2 and/or $\bar{2}$ along x, y	2 and/or $\bar{2}$ at 45° to x,y and in xy plane, i.e., along $\langle 110\rangle$
Cubic[d]	$23, m3$	2 and/or $\bar{2}$ along x, y, z	3 and/or $\bar{3}$ at 54°44′[c] to x, y, z, i.e., along $\langle 111\rangle$	—
	$432, \bar{4}3m, m3m$	4 and/or $\bar{4}$ along x, y, z	3 and/or $\bar{3}$ at 54°44′[c] to x, y, z, i.e., along $\langle 111\rangle$	2 and/or $\bar{2}$ at 45° to x, y, z, i.e., along $\langle 110\rangle$
Hexagonal	$6, \bar{6}, \dfrac{6}{m}$	6 and/or $\bar{6}$ along z	—	—
	$622, 6mm, \bar{6}m2, \dfrac{6}{m}mm$	6 and/or $\bar{6}$ along z	2 and/or $\bar{2}$ along x, y, u	2 and/or $\bar{2}$ perpendicular to x, y, u and in xy plane
Trigonal[e]	$3, \bar{3}$	3 and/or $\bar{3}$ along z	—	—
	$32, 3m, \bar{3}m$	3 and/or $\bar{3}$ along z	2 and/or $\bar{2}$ along x, y, u	—

[a] R/m occupies a single position in a point-group symbol because only *one* direction is involved.

[b] In the monoclinic system, the y axis is taken as the unique 2 or $\bar{2}$ axis. Since $\bar{2} \equiv m$, then if $\bar{2}$ is along y, the m plane represented by the same position in the point-group symbol is perpendicular to y. The latter comment applies *mutatis mutandis* in other crystal systems. (It is best to specify the orientation of a plane by that of its normal.)

[c] Actually $\cos^{-1}(1/\sqrt{3})$.

[d] Other notation uses $m\bar{3}$ and $m\bar{3}m$ for $m3$ and $m3m$, respectively.

[e] For convenience; the trigonal system is referred to hexagonal axes; on the axes of a rhombohedral unit cell (q,v.), the orientations of the first and second positions of the symbol are [111] and $(1\bar{1}0)$, respectively.

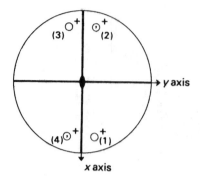

x axis

FIGURE 1.32. Stereogram, symmetry elements, and general form for point group $mm2$.

unique axis in the monoclinic system could be chosen as x or z instead of y, or along some arbitrary direction. Second, a strict system of notation is desirable for the purposes of concise, unambiguous communication of crystallographic material. With familiarity, the conventions cease to be a problem.

We shall now consider two point groups in a little more detail in order to elaborate the topics discussed so far.

Point Group mm2.   We shall see that once we fix the orientations of two of the symmetry elements in this point group, the third is introduced in a unique orientation. Referring to Figure 1.32, we start with mm as shown. Point (1), in a general position, is reflected across the m plane perpendicular to the x axis (m_x) to give point (2). This point is now reflected across the second m plane (m_y) to (3). Then either (3) across m_x or (1) across m_y produces (4). It is evident now that the points in each of the pairs (1), (3) and (2), (4) are related by the 2-fold rotation axis along z.

Point Group 4mm.   If we start with the 4 axis along z and m perpendicular to x, we see straightaway that another m plane (perpendicular to y) is required (Figure 1.33a and b); the 4-fold axis acts on other symmetry elements in the crystal as well as on the faces. A general point operated on by the symmetry $4m$ produces eight points in all (Figure 1.33c). The stereogram shows that a second form of (vertical) m planes, lying at $45°$ to the first set,[a] is introduced (Figure 1.33d). No further points are introduced by the second set of m planes: a 4-fold rotation (1) $\rightarrow$ (2), followed by reflection across the mirror plane normal to the x axis, (2) $\rightarrow$ (3), is equivalent to reflection of the original point across the mirror at $45°$ to x, (1) $\rightarrow$ (3). The reader should now refer again to Table 1.5 for the relationship between the positions of the symmetry elements and the point-group symbols, particularly for the tetragonal and orthorhombic systems, from which these detailed examples have been drawn.

---

[a] More concisely, we may say that the normals to the two forms of m planes are at $45°$ to one another.

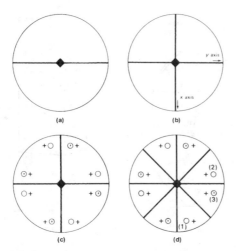

FIGURE 1.33. Intersecting symmetry elements: (a) one m plane intersecting 4 is inconsistent; (b) consistent; (c) general form of points generated by $4m$; (d) complete stereogram, point group $4mm$.

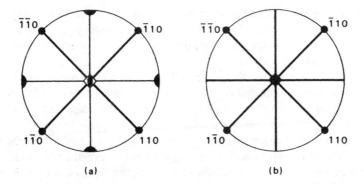

FIGURE 1.34. {110} form in tetragonal point groups: (a) point group $\bar{4}2m$; (b) point group $4mm$.

In this discussion, we have used a *general* form, which we may think of as {*hkl*}, to illustrate the point group. Each symmetry-equivalent point lies in a general position (point-group symmetry 1) on the stereogram of the point group. Certain crystal planes may coincide with symmetry planes or lie normal to (or contain) symmetry axes. These planes constitute *special* forms, and their poles lie in special positions on the stereogram; the forms {110} and {010} in $4mm$ are examples of special forms. The need for the general form in a correct description of a point group is illustrated in Figure 1.34. The poles of the faces on each of the two stereograms shown are identical, although they may be derived from crystals

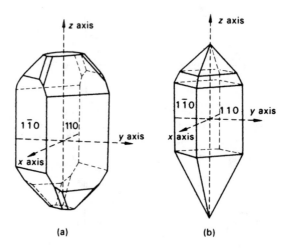

(a)   (b)

FIGURE 1.35.   Tetragonal crystals showing, among others, the {110} form: (a) copper pyrites ($\bar{4}2m$); (b) iodosuccinimide, apparent point group $4mm$; x-ray photographs revealed that the true point group is 4.

in different classes, $\bar{4}2m$ and $4mm$ in this example. Figure 1.35 shows crystals of these two classes with the {110} form, among others, developed. In Figure 1.35b, the presence of *only* special forms led originally to an incorrect deduction of the point group of this crystal.

The stereograms for the 32 crystallographic point groups are shown in Figure 1.36. The conventional crystallographic axes are drawn once for each system. Two comments on the notation are necessary at this stage. The symbol -⊕+ indicates two points, O^+ and $\odot^-$, related by a mirror plane in the plane of projection. In the cubic system, the four points related by a 4-fold axis in the plane of the stereogram lie on a small circle (Figure 1.37). In general, two of the points are projected from the upper hemisphere and the other two points from the lower hemisphere. We can distinguish them readily by remembering that 2 is a subgroup (q.v.) of both 4 and $\bar{4}$.

The use of the program SYMM* for assisting with point-group recognition is described in Chapter 11, and the reader may wish to refer forward at this stage. Appendix A3 discusses the Schönflies symmetry notation for point groups. Because this notation is also used, we have written the Schönflies symbols in Figure 1.36, in parentheses, after the Hermann–Mauguin symbols.

Subgroups and Laue Groups

A subgroup of a given point group is a point group of lower symmetry than the given group, contained within it and capable of separate existence as a point

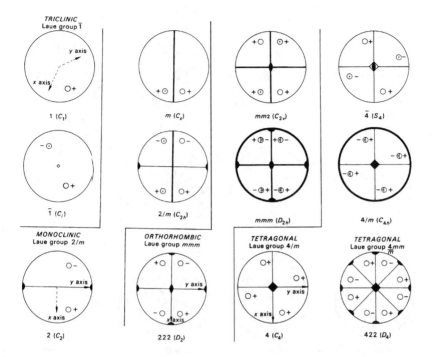

FIGURE 1.36.   Stereograms showing both the symmetry elements and the general form $\{hkl\}$ in the 32 crystallographic point groups. The arrangement is by system and common Laue group. The crystallographic axes are named once for each system and the z axis is chosen normal to the stereogram. The Schönflies symbols are given in parentheses.

group. For example, 32 is a subgroup of $\bar{3}m$, 622, $\bar{6}m2$, $\frac{6}{m}mm$, 432, and $m3m$, whereas $\bar{4}$ is a subgroup of $\frac{4}{m}$, $\bar{4}2m$, $\frac{4}{m}mm$, $\bar{4}3m$, and $m3m$. The subgroup principle provides a rationale for some of the graphic symbols for symmetry elements. Thus, $\bar{4}$ is shown by a square (4-fold rotation), unshaded (to distinguish it from 4), and with a 2-fold rotation symbol inscribed (2 is a subgroup of $\bar{4}$).

Point group $\bar{1}$ and point groups that have $\bar{1}$ as a subgroup are centrosymmetric. Since x-ray diffraction patterns are, in the absence of significant anomalous dispersion (q.v.), effectively centrosymmetric, the arrangement of spots on an x-ray diffraction photograph obtained from any crystal can exhibit only the symmetry that would be found from a crystal having the corresponding centrosymmetric point group. In the case of a crystal belonging to a non-centrosymmetric point group, the corresponding centrosymmetric point group is simply the given group combined with a centre of symmetry.

There are 11 such point groups; they are called *Laue groups*[a], since symmetry is often investigated by the Laue x-ray method (see Section 4.4.1). In Table 1.6, the

---

[a] Strictly, the term Laue group should be *Laue class*, but the former is in general use.

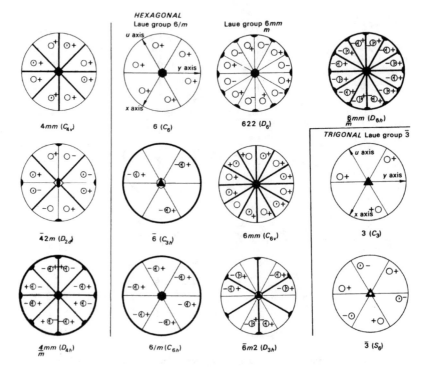

FIGURE 1.36.   (Continued).

point groups are classified according to their Laue group, and the symmetry of the
Laue flat-plate film photographs is given for directions of the x-ray beam normal
to the crystallographic forms listed. The *Laue-projection symmetry* corresponds
to one of the 10 two-dimensional points groups.

What is the Laue-projection symmetry on {110} for a crystal of point group
4*mm*? This question can be answered with the stereogram of the corresponding
Laue group, $\frac{4}{m}mm$. Reference to the appropriate diagram in Figure 1.36 shows that
an x-ray beam traveling normal to {110} encounters 2*mm* symmetry. The entries
in Table 1.6 can be deduced in this way. The reader, should refer again to Table 1.5
and compare corresponding entries between Tables 1.5 and 1.6.

Point-group projection symmetry is the symmetry of the projection of the
general form of a point group on to a plane. Thus, the point-group projection
symmetry of 4*mm* on to {110} is *m*.

Noncrystallographic Point Groups

We have stated that in crystals the elements R and $\bar{R}$ are limited to the
numerical values 1, 2, 3, 4, and 6. However, there are *molecules* that exhibit

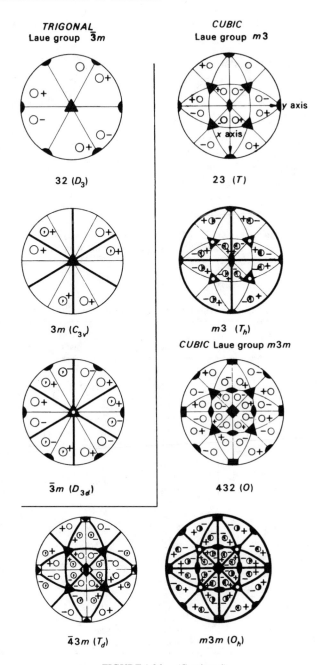

FIGURE 1.36. (Continued).

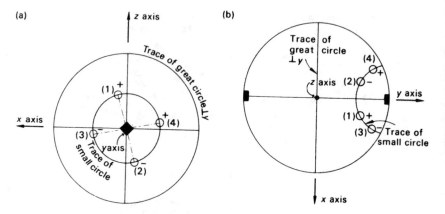

FIGURE 1.37.   Stereogram notation for points related by a 4-fold axis (y) lying in the plane of a stereogram. The $\pm$ signs refer to the z direction: (a) vertical section normal to the y axis, (b) corresponding stereogram; the pairs of points (1)–(2) and (3)–(4) are related by 2-fold symmetry (subgroup of 4).

TABLE 1.6.   Laue Groups and Laue-Projection Symmetry

System	Point groups	Laue group	Laue-projection symmetry normal to the given form		
			{100}	{010}	{010}
Triclinic	$1, \bar{1}$	$\bar{1}$	1	1	1
Monoclinic	$2, m, 2/m$	$2/m$	m	2	m
Orthorhombic	$222, mm2, mmm$	mmm	$2mm$	$2mm$	$2mm$
			{001}	{100}	{110}
Tetragonal	$4, \bar{4}, 4/m$	$4/m$	4	m	m
	$422, 4mm,$ $\bar{4}2m, \dfrac{4}{m}mm$	$\dfrac{4}{m}mm$	$4mm$	$2mm$	$2mm$
			{0001}	{10$\bar{1}$0}	{11$\bar{2}$0}
Trigonal[a]	$3, \bar{3}$	$\bar{3}$	3	1	1
	$32, 3m, \bar{3}m$	$\bar{3}m$	$3m$	m	2
Hexagonal	$6, \bar{6}, 6/m$	$6/m$	6	m	m
	$622, 6mm,$ $\bar{6}m2, \dfrac{6}{m}mm$	$\dfrac{6}{m}mm$	$6mm$	$2mm$	$2mm$
			{100}	{111}	{110}
Cubic	$23, m3$	$m3$	$2mm$	3	m
	$432, \bar{4}3m, m3m$	$m3m$	$4mm$	$3m$	$2mm$

[a] Referred to hexagonal axes.

symmetries other than those of the crystallographic point groups. Indeed, R could, in principle, take any integer value between one and infinity. The statement $R = \infty$ implies cylindrical symmetry; the molecule of carbon monoxide has an ∞ axis along the C–O bond, if we assume spherical atoms.

In uranium heptafluoride (Figure 1.38) a 5-fold symmetry axis is present, and the point-group symbol may be written as $\overline{10}m2$, or $\frac{5}{m}m$. The stereogram of this point group is shown in Figure 1.39; the graphic symbol for $\frac{5}{m}(\overline{10})$ is not standard.

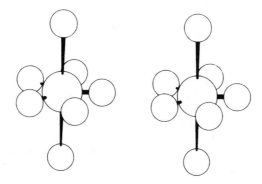

FIGURE 1.38.   Stereoview of the molecule of uranium heptafluoride, UF$_7$.

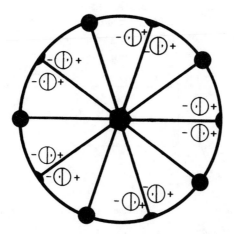

FIGURE 1.39.   Stereogram of the noncrystallographic point group $\overline{10}m2$ (D_{5h}) showing the general form (20 poles), and a special form of 5 poles lying on the m planes that can be used to represent the five F atoms in one plane in UF$_7$. The poles for the remaining two F atoms lie at these center of the stereogram, on the $\overline{10}$ axis.

Other examples of noncrystallographic point groups will be encountered among chemical molecules, and a stereogram can always be used to represent the point-group symmetry. In every such example, however, the substance will crystallize in one of the seven crystal systems and the crystals will belong to one of the 32 crystal classes.

Bibliography

General and Historical Study of Crystallography

BRAGG, W. L., *A General Survey* (*The Crystalline State*, Vol. I), London, Bell (1949).
EWALD, P. P. (Editor), *Fifty Years of X-ray Diffraction*, Utrecht, Oosthoek (1962).

Crystal Morphology and Stereographic Projection

PHILLIPS, F. C., *An Introduction to Crystallography*, London, Longmans (1971).

Crystal Symmetry and Point Groups

HAHN, T. (Editor), *International Tables for Crystallography*, Vol. A, 5th ed., Kluwer Academic (2002).
HENRY, N. F. M. and LONSDALE, K. (Editors), *International Tables for X-ray Crystallography*, Vol. I, Birmingham, Kynoch Press (1965).
LADD, M. F. C., *Symmetry in Molecules and Crystals*, Chichester, Ellis Horwood (1989).
LADD, M., *Symmetry and Group Theory in Chemistry*, Chichester, Horwood Publishing (1998).

Problems

1.1. The line AC (Figure P1.1) may be indexed as (12) with respect to the rectangular axes x and y. What are the indices of the same line with respect to the axes x' and y, where the angle $x'Oy = 120°$? PQ is the parametral line for both sets of axes, and $OB/OA = 2$.

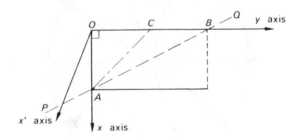

FIGURE P1.1.   Line referred to rectangular and oblique axes.

1.2. Write the Miller indices for planes that make the intercepts given below:

(a) $a, -b/2, ||c$.   (b) $2a, b/3, c/2$.

(c) $||a, ||b, -c$.   (d) $a, -b, 3c/4$.

(e) $||a, -b/4, c/3$.   (f) $-a/4, b/2, -c/3$.

(a)

(b)

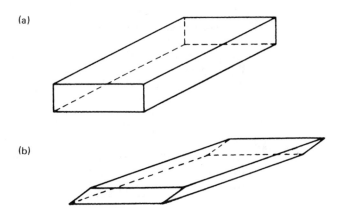

FIGURE P1.2.   Matchbox: (a) Normal; (b) Squashed.

1.3. Evaluate zone symbols for the pairs of planes given below:

    (a) (123), $(0\bar{1}1)$.   (b) $(20\bar{3})$, (111).
    (c) $(41\bar{5})$, $(1\bar{1}0)$.   (d) $(\bar{1}1\bar{2})$, (001).

1.4. What are the Miller indices of the plane that lies in both of the zones $[123]$ and $[\bar{1}1\bar{1}]$? Why are there, are apparently, two answers to this problem and to each part of Problem 1.3?

1.5. How many different, unique point groups can be obtained from the symbol 422 by replacing one or more of the rotation axes by roto-inversion axes of the same degree. Write the standard symbols for the unique point groups so derived.

1.6. Take the cover of a matchbox (Figure P1.2a).

    (a) Ignore the label, and write down its point group.
        Squash it diagonally (Figure P1.2b).
    (b) What is the point group now?
    (c) In each case, what is the point group if the label is not ignored?

1.7. Draw stereograms to show the general form in each of the point groups deduced in Problems 1.6a and 1.6b. Satisfy yourself that in 1.6a three, and in 1.6b two, symmetry operations carried out in sequence produce a resultant action that is equivalent to another operation in the group.

1.8. How many planes are there in the forms $\{010\}$, $\{\bar{1}10\}$, and $\{11\bar{3}\}$ in each of the point groups $2/m$, $\bar{4}2m$, and $m3$?

1.9. What symmetry would be revealed by the Laue flat-film photographs where the x-ray beam is normal to a plane in the form given in each of the examples below?

	Point group	Orientation
(a)	$\bar{1}$	{100}
(b)	$mm2$	{011}
(c)	m	{010}
(d)	422	{120}
(e)	3	{10$\bar{1}$0}
(f)	$3m$	{11$\bar{2}$0}
(g)	$\bar{6}$	{0001}
(h)	$\bar{6}m2$	{0001}
(i)	23	{111}
(j)	432	{110}

In some examples, it may help to draw stereograms.

1.10. (a) What is the nontrivial symmetry of the figure obtained by packing a number of equivalent but irregular quadrilaterals in one plane?

(b) What is the symmetry of the Dobermann in Figure P1.3? This example illustrates how one can study symmetry by means of everyday objects.

FIGURE P1.3. "Vijentor Seal of Approval at Valmara."

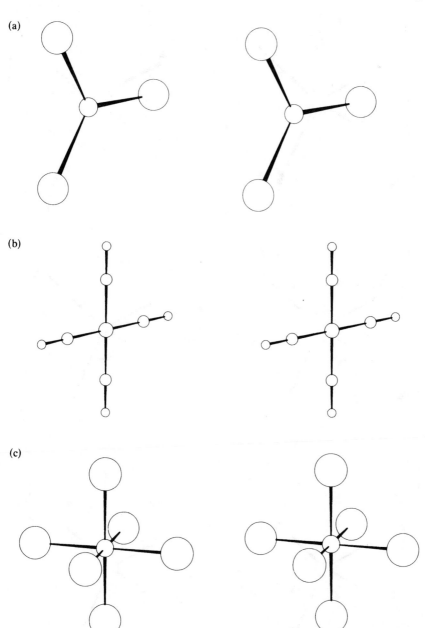

FIGURE P1.4. (a) $[CO_3]^{2-}$. Planar. (b) $[Ni(CN)_4]^{2-}$. Planar. (c) $[PtCl_6]^{2-}$. All Cl–Pt–Cl angles are 90°. (d) CH_4. All H–C–H angles are 109.47°. (e) $CHCl_3$. Pyramidal. (f) CHBrClI. (g) C_6H_6. Planar. (h) C_6H_5Cl. Planar. (i) $C_6H_4Cl_2$. Planar. (j) C_6H_4BrCl. Planar.

(d)

(e)

(f)

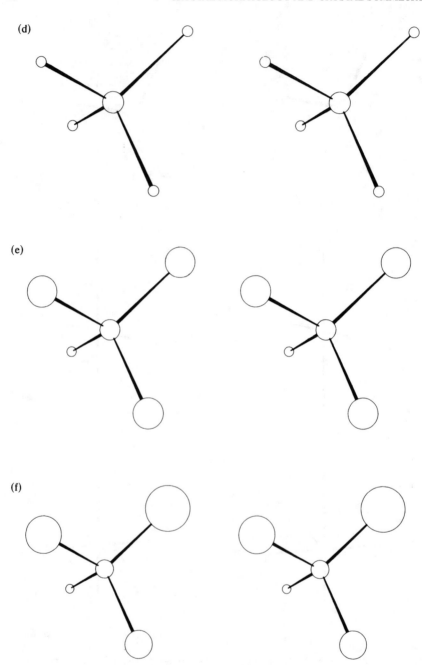

FIGURE P1.4.   (Continued).

(g)

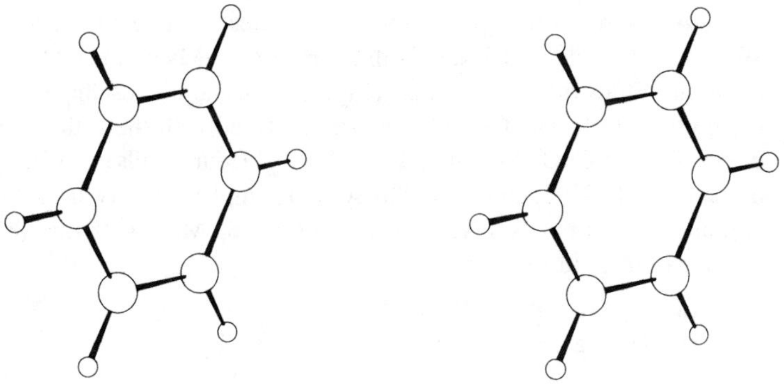

(h)

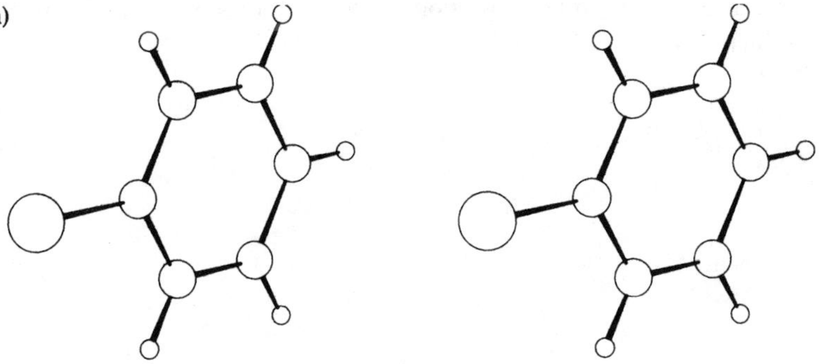

(i)

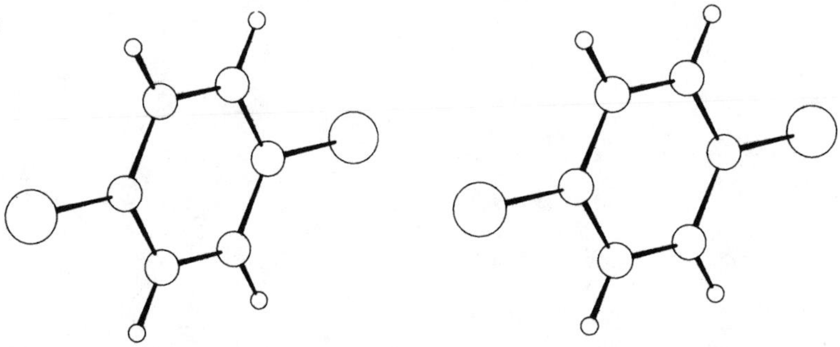

(j)

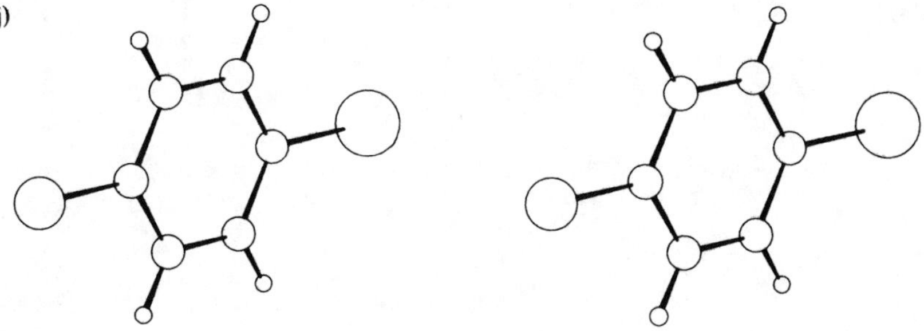

FIGURE P1.4.   (Continued).

1.11. Write the point-group symbol for each species of molecule or ion in the 10 drawings of Figure P1.4, in both the Hermann–Mauguin and Schönflies notations. Study the stereograms in Figure 1.36, and suggest the form $\{hkl\}$, the normals to faces of which may be identified with the following bond directions: (a) C–O (three bonds), (b) Ni–C (four bonds), (c) Pt–Cl (six bonds), (d) C–H (four bonds). In some examples, it may help to make ball-and-spoke models. If you use the program SYMM* with this question, allocate model numbers as follow:

   (a) 90   (b) 49   (c)   3   (d) 18   (e) 42
   (f) 91   (g) 30   (h) 16   (i) 65   (j) 71

1.12. What is the point-group projection symmetry for each of the examples in Problem 1.9?

2

Lattices and Space-Group Theory

2.1 Introduction

In this chapter, we continue our study of crystals by investigating the internal arrangements of crystalline materials. Crystals are characterized by periodicities in three dimensions. An atomic grouping, or pattern motif, which, itself, may or may not be symmetrical, is repeated over and over again by a certain symmetry mechanism that corresponds to the *space group* of the crystal. Altogether, there are 230 space groups, and each crystalline substance will belong to one or other of them. In its simplest form, a space group may be derived from repeating the pattern motif by the translations of a lattice, as discussed below. It can be developed further by incorporating additional symmetry elements, as demonstrated through Problem 2.1. We now enlarge on these ideas, starting with an examination of lattices.

2.2 Lattices

Every crystal has a lattice as its geometrical basis. A lattice may be described as a regular, infinite arrangement of points in which every point has the same environment as any other point. This description is applicable, equally, in one-, two-, and three-dimensional space.

Lattice geometry in three-dimensional space is described in relation to three noncoplanar basic repeat (translation) vectors **a**, **b**, and **c**. Any lattice point may be chosen as an origin, whence a vector **r** to any other lattice point is given by

$$\mathbf{r} = U\mathbf{a} + V\mathbf{b} + W\mathbf{c} \tag{2.1}$$

where U, V, and W are positive or negative integers or zero, and represent the coordinates of the given lattice point. The *direction* (directed line) joining the origin to the points U, V, W; $2U, 2V, 2W$; $\ldots$; nU, nV, nW defines the row $[UVW]$.

A set of such rows, or directions, related by the lattice symmetry constitutes a form of directions $\langle UVW \rangle$ (compare zone symbols, Section 1.2.5). The magnitude r can be evaluated *mutatis mutandis* by (2.16).

2.2.1   Two-Dimensional Lattices

We begin our detailed study of lattices in two dimensions rather than three. A two-dimensional lattice is called a *net*; it may be imagined as being formed by aligning, in a regular manner, *rows* of equally spaced points (Figure 2.1a). The net (lattice) is the array of points; the connecting lines are a convenience, drawn to aid our appreciation of the lattice geometry.

Since nets exhibit symmetry, they can be allocated to the two-dimensional systems (Section 1.4.1). The most general net is shown in Figure 2.1b. A sufficient and representative portion of the lattice is the *unit cell*, outlined by the vectors **a** and **b**; an infinite number of such unit cells stacked side by side builds up the net.

The net under consideration exhibits 2-fold rotational symmetry about each point; consequently, it is placed in the oblique system. The chosen unit cell is primitive (symbol p), which means that one lattice point is associated with the area of the unit cell: each point is shared equally by four adjacent unit cells. In the oblique unit cell, $a \not\subset b$, and $\gamma \not\subset 90°$ or 120°; angles of 90° and 120° *in a lattice* imply symmetry higher than 2.

Consider next the stacking of unit cells in which $a \not\subset b$ but $\gamma = 90°$ (Figure 2.2). The symmetry at every point is $2mm$, and this net belongs to the rectangular system. The net in Figure 2.3 may be described by a unit cell in which $a' = b'$ and $\gamma' \not\subset 90°$ or 120°. It may seem at first that such a net is oblique, but careful inspection shows that each point has $2mm$ symmetry, and so this net, too, is allocated to the rectangular system.

In order to display this fact clearly, a centered (symbol c) unit cell is chosen, shown in Figure 2.3 by the vectors **a** and **b**. This cell has two lattice points per

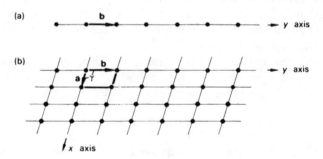

FIGURE 2.1.   Formation of a net: (a) row of equally spaced points (a one-dimensional lattice); (b) regular stack of rows to form a net.

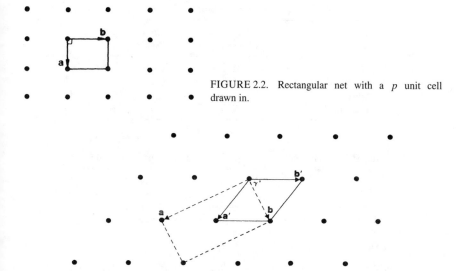

FIGURE 2.2.   Rectangular net with a *p* unit cell drawn in.

FIGURE 2.3.   Rectangular net with *p* and *c* unit cells drawn in; the *c* unit cell is the standard choice.

unit-cell area. It is left as an exercise to the reader to show that a centered, oblique unit cell does not represent a net with a fundamentally different arrangement of points from that in Figure 2.1b.

2.2.2   Choice of Unit Cell

From the foregoing discussion, it will be evident that the choice of unit cell is somewhat arbitrary. We shall follow a universal crystallographic convention in choosing a unit cell: the unit cell is the smallest repeat unit for which its delineating vectors are parallel to, or coincide with, important symmetry directions in the lattice. Returning to Figure 2.3, the centered cell is preferred because **a** and **b** coincide with the symmetry (*m*) lines in the net. The primitive unit cell (**a**′, **b**′) is, of course, a possible unit cell, but it does not, in isolation, reveal the lattice symmetry clearly. The symmetry is still there; it is invariant under choice of unit cell. The following equations show the necessary equivalence of *a*′ and *b*′:

$$a'^2 = a^2/4 + b^2/4 \tag{2.2}$$

$$b'^2 = a^2/4 + b^2/4 \tag{2.3}$$

the value of γ' depends only on the ratio a/b.

Two other nets are governed by the unit-cell relationships $a = b$, $\gamma = 90°$ and $a = b$, $\gamma = 120°$; their study constitutes the second problem at the end of this chapter. The five two-dimensional lattices are summarized in Table 2.1. A lattice

TABLE 2.1.   The Five Two-dimensional Lattices

System	Unit-cell symbol(s)	Symmetry at lattice points	Unit-cell edges and angles
Oblique	p	2	$a \neq b, \gamma \neq 90°, 120°$
Rectangular	p, c	$2mm$	$a \neq b, \gamma = 90°$
Square	p	$4mm$	$a = b, \gamma = 90°$
Hexagonal	p	$6mm$	$a = b, \gamma = 120°$

TABLE 2.2.   Notation for Conventional Crystallographic Unit Cells

Centering site(s)	Symbol	Miller indices of centered faces of the unit cell	Fractional coordinates[a] of centered sites in the unit cell
None	P	—	—
bc faces	A	100	$0, \frac{1}{2}, \frac{1}{2}$
ca faces	B	010	$\frac{1}{2}, 0, \frac{1}{2}$
ab faces	C	001	$\frac{1}{2}, \frac{1}{2}, 0$
Body center	I	—	$\frac{1}{2}, \frac{1}{2}, \frac{1}{2}$
All faces	F	$\begin{cases} 100 \\ 010 \\ 001 \end{cases}$	$\begin{cases} 0, \frac{1}{2}, \frac{1}{2} \\ \frac{1}{2}, 0, \frac{1}{2} \\ \frac{1}{2}, \frac{1}{2}, 0 \end{cases}$

[a] A fractional coordinate x is given by X/a, where X is the coordinate in absolute measure and a is the unit-cell repeat distance in the same direction and in the same units.

has the highest point-group symmetry of its system at each lattice point (compare Table 1.1 with 2.1 and 1.5 with 2.3).

2.2.3 Three-Dimensional Lattices

The three-dimensional lattices, or Bravais lattices, may be imagined as being developed by the regular stacking of nets. There are 14 ways in which this can be done, and the Bravais lattices are distributed, unequally, among the seven crystal systems, as shown in Figure 2.4. Each lattice is represented by a unit cell, outlined by three vectors **a**, **b**, and **c**. In accordance with convention, these vectors are chosen so that they both form a parallelepipedon of smallest volume in the lattice and are parallel to, or coincide with, important symmetry directions in the lattice, so that not all conventional unit cells are primitive. In three dimensions, we encounter unit cells centered on a pair of opposite faces, body-centered, or centered on all faces. Table 2.2 lists the unit cell types and their notation.

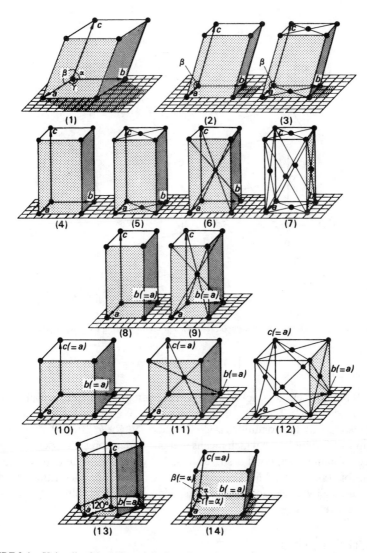

FIGURE 2.4. Unit cells of the 14 Bravais lattices; interaxial angles are 90° unless indicated otherwise by a numerical value or symbol: (1) triclinic P, (2) monoclinic P, (3) monoclinic C, (4) orthorhombic P, (5) orthorhombic C, (6) orthorhombic I, (7) orthorhombic F, (8) tetragonal P, (9) tetragonal I, (10) cubic P, (11) cubic I, (12) cubic F, (13) hexagonal P, (14) trigonal R. Note that (13) shows three P hexagonal unit cells. A hexagon of lattice points (without the central point in the basal planes shown) does not lead to a lattice. Why?

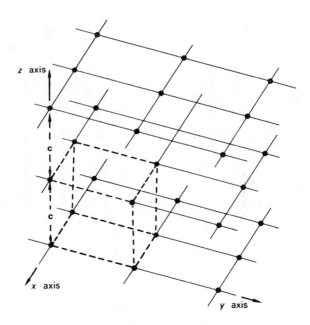

FIGURE 2.5.   Oblique nets stacked regularly at a vector spacing **c** to form a triclinic lattice.

Triclinic Lattice

If oblique nets are stacked in a general, but regular, manner, a triclinic lattice is obtained (Figure 2.5). The unit cell is characterized by $\bar{1}$ symmetry at each lattice point, with the conditions $a \not\!\!\,c\, b \not\!\!\,c\, c$ and $\alpha \not\!\!\,c\, \beta \not\!\!\,c\, \gamma \not\!\!\,c\, 90°, 120°$. This unit cell is primitive (symbol P),[a] which means that one lattice point is associated with the unit cell volume; each point is shared equally by eight adjacent unit cells in three dimensions (see Figure 2.6). There is no symmetry direction to constrain the choice of the unit cell vectors, and a parallelepipedon of smallest volume can always be chosen conventionally.

Monoclinic Lattices

The monoclinic system is characterized by one diad (rotation or inversion), with the y axis (and b) chosen parallel to it. The conventional unit cell is specified by the conditions $a \not\!\!\,c\, b \not\!\!\,c\, c$, $\alpha = \gamma = 90°$, and $\beta \not\!\!\,c\, 90°, 120°$. Figure 2.6 illustrates a stereoscopic pair of drawings of a monoclinic lattice, showing eight P unit cells; according to convention, the β angle is chosen to be oblique.

---

[a] Capital letters are used for lattices in three dimensions.

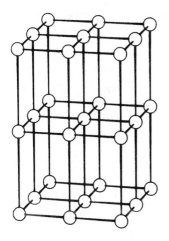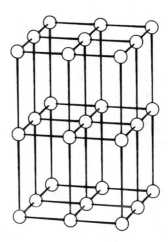

FIGURE 2.6. Stereoview showing eight adjacent P unit cells in a monoclinic lattice. The sharing of corner points can be seen readily by focusing attention on the central lattice point in the drawings. A similar sharing occurs with P unit cells of lattices in all systems.

Reference to Figure 2.4 shows that there are two conventional monoclinic lattices, symbolized by the unit-cell types P and C.

A monoclinic unit cell centered on the A faces is equivalent to that described as C; the choice of the b axis[a] is governed by symmetry, but a and c are interchangeable labels; the direction of $\mathbf{b}$ must be reversed in order to preserve right-handed axes.

The centering of the B faces is illustrated in Figure 2.7. In this situation a new unit cell, $\mathbf{a}', \mathbf{b}', \mathbf{c}'$, can be defined by the following equations:

$$\mathbf{a}' = \mathbf{a} \tag{2.4}$$

$$\mathbf{b}' = \mathbf{b} \tag{2.5}$$

$$\mathbf{c}' = \mathbf{a}/2 + \mathbf{c}/2 \tag{2.6}$$

If β is not very obtuse, the equivalent transformation $\mathbf{c}' = -\mathbf{a}/2 + \mathbf{c}/2$ can ensure that β' is obtuse. Since $\mathbf{c}'$ lies in the ac plane, $\alpha' = \gamma' = 90°$, but $\beta' \not\subset 90°$ or $120°$. The new monoclinic cell is primitive; symbolically we may write $B \equiv P$. Similarly, it may be shown that $I \equiv F \equiv C \equiv (A)$ (Figures 2.8 and 2.9).

If the C cell (Figure 2.10) is reduced to primitive, it no longer displays the characteristic monoclinic symmetry *clearly* (see Table 2.3); neither α' nor γ'

[a] We often speak of the b axis (to mean the y axis) because our attention is usually confined to the unit cell.

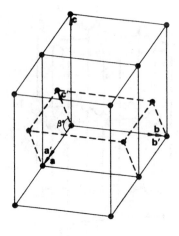

FIGURE 2.7.   Monoclinic lattice showing that $B \equiv P$; β is the angle between $\mathbf{c}$ and $\mathbf{a}$, and β' the angle between $\mathbf{c}'$ and $\mathbf{a}'$.

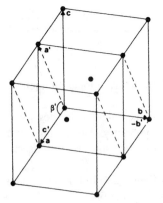

FIGURE 2.8.   Monoclinic lattice showing that $I \equiv C$.

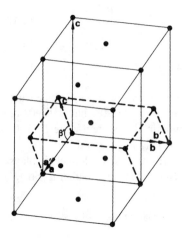

FIGURE 2.9.   Monoclinic lattice showing that $F \equiv C$.

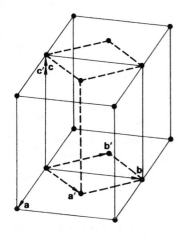

FIGURE 2.10.   Monoclinic lattice showing that $C \not\equiv P$.

TABLE 2.3.   The 14 Bravais Lattices and their Notation

System	Unit cell(s)	Symmetry at lattice points	Axial relationships
Triclinic	P	$\bar{1}$	$a \not\subset b \not\subset c; \alpha \not\subset, \beta \not\subset \gamma \not\subset 90°, 120°$
Monoclinic	P, C	$2/m$	$a \not\subset b \not\subset c; \alpha = \gamma = 90°; \beta \not\subset 90°, 120°$
Orthorhombic	P, C, I, F	mmm	$a \not\subset b \not\subset c; \alpha = \beta = \gamma = 90°$
Tetragonal	P, I	$\dfrac{4}{m}mm$	$a = b \not\subset c; \alpha = \beta = \gamma = 90°$
Cubic	P, I, F	$m3m$	$a = b = c; \alpha = \beta = \gamma = 90°$
Hexagonal	P	$\dfrac{6}{m}mm$	$a = b \not\subset c; \alpha = \beta = 90°; \gamma = 120°$
Trigonal[a]	R or P	$3m$	$a = b = c; \alpha = \beta = \gamma \not\subset 90°, <120°$

[a] On hexagonal axes, column 4 is the same as for the hexagonal system, but the symmetry at each lattice points remains $\bar{3}m$. This table may be compared with Table 1.3.

is 90°. We may conclude that there are two monoclinic lattices, described by the unit cell types P and C.

It may be necessary to calculate the new dimensions of a transformed unit cell. Consider the example $B \to P$, (2.4)–(2.6). Clearly, $a' = a$ and $b' = b$. Taking the scalar product[a] of (2.6) with itself, we obtain

$$\mathbf{c}' \cdot \mathbf{c}' = (\mathbf{a}/2 + \mathbf{c}/2) \cdot (\mathbf{a}/2 + \mathbf{c}/2) \tag{2.7}$$

Hence

$$c'^2 = a^2/4 + c^2/4 + ac(\cos \beta)/2 \tag{2.8}$$

[a] The scalar (dot) product of two vectors $\mathbf{p}$ and $\mathbf{q}$ is denoted by $\mathbf{p} \cdot \mathbf{q}$, and is equal to $pq \cos \widehat{\mathbf{pq}}$, where $\widehat{\mathbf{pq}}$ represents the angle between the (positive) directions of $\mathbf{p}$ and $\mathbf{q}$.

The new angle β' is given by

$$\cos \beta' = \mathbf{a}' \cdot \mathbf{c}'/a'c' \tag{2.9}$$

To make β' obtuse, it may be necessary to begin with $-\mathbf{a}/2$ in (2.6).

Using (2.6) again, we obtain

$$\cos \beta' = [-a/2 + c(\cos \beta)/2]/c' = (-a + c \cos \beta)/(2c') \tag{2.10}$$

where c' is given by (2.8). This type of calculation can be carried out in any crystal system, giving due consideration to any nontrivial relationships between a, b, and c and between α, β, and γ.

Orthorhombic Lattices

The monoclinic system was treated in some detail. It will not be necessary here to give such an extensive discussion for either this system or the remaining crystal systems. Remember always to think of the unit cell as a representative portion of its lattice and not as a finite body.

The orthorhombic system is characterized by three mutually perpendicular diads (rotation and/or inversion); the unit-cell vectors are chosen to be parallel to, or to coincide with, these symmetry axes. The orthorhombic unit cell is specified by the relationships $a \not\subset b \not\subset c$ and $\alpha = \beta = \gamma = 90°$. It will not be difficult for the reader to verify that the descriptions P, C, I, and F are necessary and sufficient in this system. One way in which this exercise may be carried out is as follows. After centering the P unit cell, three questions must be asked, in the following order:

1. Does the centered cell represent a lattice?
2. If so, is its symmetry different from that of the P unit cell?
3. If the symmetry is unchanged, is the lattice different in type (arrangement of points) from the lattice or lattices already determined for the given system, and has the unit cell been chosen correctly?

Note that we answered these questions implicitly in discussing the monoclinic lattices.

The descriptions A, B, and C do not necessarily remain equivalent for orthorhombic space groups in the class $mm2$; it is necessary to distinguish C from A (or B). The reader may like to consider now, or later, why this distinction is necessary.

Tetragonal Lattices

The tetragonal system is characterized by one tetrad (rotation or inversion) along z (and c); the unit-cell conditions are $a = b \not\subset c$ and $\alpha = \beta = \gamma = 90°$.

There are two tetragonal lattices, specified by the unit-cell symbols P and I (Figure 2.4); C and F tetragonal unit cells may be transformed to P and I, respectively (see also Problem 2.4).

Cubic Lattices

The symmetry of the cubic system is characterized by four triad axes at angles of $\cos^{-1}(\frac{1}{3})$ to one another; they are the body diagonals $\langle 111 \rangle$ of a cube. The four 3-fold axes, in this orientation, introduce 2-fold axes along $\langle 100 \rangle$. There are three cubic Bravais lattices (Figure 2.4) with conventional unit cells P, I, and F.

Hexagonal Lattice

The basic feature of a hexagonal lattice is that it should be able to accommodate a 6-fold symmetry axis. This requirement is achieved by a lattice based on a P unit cell, with $a = b \neq c$, $\alpha = \beta = 90°$, and $\gamma = 120°$, the c direction being taken along a 6-fold axis in the lattice (Figure 2.4).

Lattices in the Trigonal System

A two-dimensional unit cell in which $a = b$ and $\gamma = 120°$ is compatible with either 6-fold or 3-fold symmetry (see Figure 2.22, plane groups $p6$ and $p3$). For this reason, the hexagonal lattice (P unit cell) may be used for certain crystals which belong to the trigonal system. However, as shown in Figure 2.11, the presence of two 3-fold axes within a unit cell, with x, y coordinates of $\frac{2}{3}, \frac{1}{3}$ and $\frac{1}{3}, \frac{2}{3}$, respectively, and parallel to the z axis, introduces the possibility of a lattice which, although belonging to the trigonal system, has a triply primitive unit cell R_{hex} with points at $\frac{2}{3}, \frac{1}{3}, \frac{1}{3}$ and $\frac{1}{3}, \frac{2}{3}, \frac{2}{3}$ (in addition to 0, 0, 0) in the unit cell. Thus, for some trigonal crystals the unit cell will be P, and for others it will be R_{hex}, the latter being distinguished by systematically absent x-ray reflections (Table 3.2). The R_{hex} cell can be transformed to a primitive rhombohedral unit cell R, with $a = b = c$ and $\alpha = \beta = \gamma \neq 90°$, $< 120°$; the 3-fold axis is then along [111]. The R cell may be thought of as a cube extended or squashed along one of its 3-fold axes.

The lattice based on an R unit cell is the only truly exclusive trigonal lattice, the trigonal lattice based on a P unit cell being borrowed from the hexagonal system (see Table 2.3).

We note in passing that the symbols P, R, A, B, C, I, and F cannot apply, strictly, to lattices;[a] they are unit-cell symbols, and refer to the *types* of unit cells already chosen to represent their lattices. However, terminology such as "P lattice" is common and, as long as it is used with understanding, is perfectly acceptable.

[a] M. F. C. Ladd, *Journal of Chemical Education* **74**, 461 (1997).

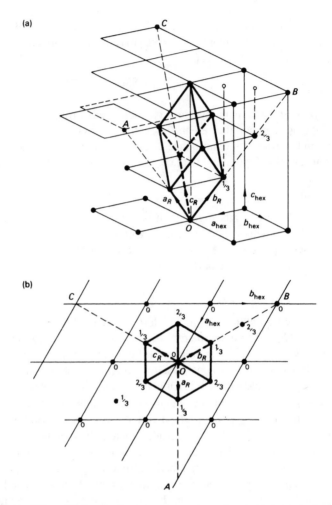

FIGURE 2.11.   Trigonal lattice (the fractions refer to values of c_{hex}): (a) rhombohedral (R) unit cell (obverse setting) developed from a triply primitive hexagonal (R_{hex}) unit cell. (In the *reverse* setting, the rhombohedral lattice and unit cell are rotated about [111] 60° clockwise with respect to the R_{hex} axes.) The ratio of the volumes of any two unit cells in one and the same lattice is equal to the ratio of the numbers of lattice points in the two unit cell volumes. (b) Plan view of (a) as seen along c_{hex}.

2.3   Families of Planes and Interplanar Spacings

Figure 2.12 shows one unit cell of an orthorhombic lattice projected on to the *ab* plane. The trace of the (110) plane nearest the origin O is indicated by a dashed line, and the perpendicular distance of this plane from O is $d(110)$. By

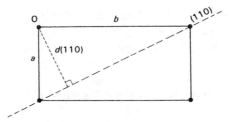

FIGURE 2.12. One P unit cell in an orthorhombic lattice in projection on (001), showing the trace of the (110) plane.

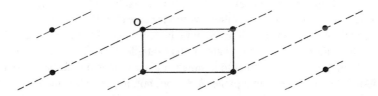

FIGURE 2.13. Family of (110) planes in an orthorhombic lattice, as seen in projection along c.

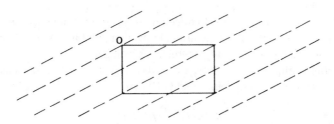

FIGURE 2.14. Family of (220) planes in an orthorhombic lattice, as seen in projection along c.

repeating the operation of the translation $\pm\mathbf{d}(110)$ on the plane (110), a series, or *family*, of parallel, equidistant planes is generated, as shown in Figure 2.13. Our discussion of the external symmetry of crystals led to a description of the external faces of crystals by Miller indices (Section 1.2.3), which are by definition prime to one another. In discussing x-ray diffraction effects, however, it is necessary to consider planes for which the indices h, k, and l may contain a common factor while still making intercepts $a/h, b/k$, and c/l on the x, y, and z axes, respectively, as required by the definition of Miller indices. It follows that the plane with indices (nh, nk, nl) makes intercepts $a/nh, b/nk$, and c/nl along x, y, and z, respectively, and that this plane is nearer to the origin by a factor of $1/n$ than is the plane (hkl). In other words, $d(nh, nk, nl) = d(hkl)/n$. In general, we denote a family of planes as (hkl) where h, k, and l may contain a common factor. For example, the (220) family of planes is shown in Figure 2.14 with interplanar spacing $d(220) = d(110)/2$;

alternate (220) planes therefore coincide with (110) planes. Note, that an *external* crystal face normal to $d(hh0)$ would always be designated (110), since external observations reveal the shape but not the size of the unit cell.

2.4 Reciprocal Lattice–Geometrical Treatment

Although we shall discuss the reciprocal lattice in detail in the next chapter, it is useful to introduce it here, because there exists a reciprocal lattice for each of the Bravais lattices. The reciprocal lattice, a lattice in reciprocal (diffraction) space, is derived here graphically from the Bravais lattice, a lattice in real (direct) space, and we choose the monoclinic system for an example.

Figure 2.15a represents a monoclinic lattice as seen in projection along the y axis, the normal to the (010) plane in this example. From the origin O of a P unit cell, lines are drawn normal to families of planes (hkl) in real space. We note in passing that the normal to a plane (hkl) does not, in general, coincide with the *direction* (see Section 2.2) of the same indices, $[hkl]$. However, there are special cases, such as [010] and the normal to (010) in the present example, in which the two directions do coincide.

Along each line, reciprocal lattice points hkl (no parentheses) are marked off such that the distance from the origin to the first point in any line is inversely proportional to the corresponding interplanar spacing $d(hkl)$.

In three dimensions, we refer to $d^*(100), d^*(010)$, and $d^*(001)$ as a^*, b^*, and c^*, respectively, and so define a unit cell in the reciprocal lattice. In general,

$$d^*(hkl) = \kappa/d(hkl) \tag{2.11}$$

where κ is a constant. Hence, for the monoclinic system,

$$a^* = \kappa/d(100) = \kappa/(a \sin \beta) \tag{2.12}$$

From Figure 2.15a, the scalar product $\mathbf{a} \cdot \mathbf{a}^*$ is given by

$$\mathbf{a} \cdot \mathbf{a}^* = aa^* \cos(\beta - 90°) = a\kappa \frac{\cos(\beta - 90°)}{a \sin \beta} = \kappa \tag{2.13}$$

The mixed scalar products, such as $\mathbf{a} \cdot \mathbf{c}^*$ are identically zero, because the angle between a and c^* is 90°.

The reciprocal lattice points form a true lattice with a representative unit cell outlined by $\mathbf{a}^*, \mathbf{b}^*$, and $\mathbf{c}^*$ which, therefore, involves six reciprocal cell parameters in the most general case—three sides a^*, b^*, and c^*, and three angles α^*, β^*, and γ^*. The size of the reciprocal cell is governed by the choice of the constant κ. In practice, κ may be taken as the wavelength λ of the x-radiation used, in which

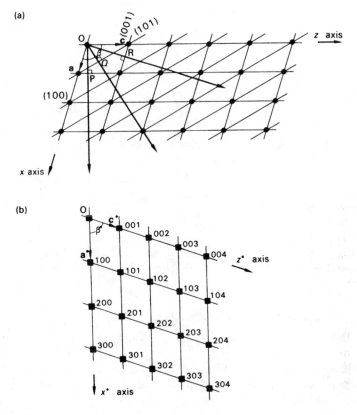

FIGURE 2.15.   Direct and reciprocal lattices: (a) monoclinic P, as seen in projection along b, showing three families of planes; (b) corresponding reciprocal lattice showing the points representing these three and other families of planes.

case reciprocal lattice units are dimensionless. Alternatively, κ may be taken as unity, in which case reciprocal lattice units have the dimensions of length^{-1}. The different situations where one or other convention is used will become clear shortly.

A reciprocal lattice row $hkl; 2h, 2k, 2l; \ldots$ may be considered to be derived from the families of planes (nh, nk, nl) with $n = 1, 2, \ldots$, since $d(nh, nk, nl) = d(hkl)/n$. Hence,

$$d^*(nh, nk, nl) = nd^*(hkl) \tag{2.14}$$

where $d^*(hkl)$ is the distance of the reciprocal lattice point hkl from the origin, expressed in reciprocal lattice units (RU). Since h, k, and l are the coordinates of

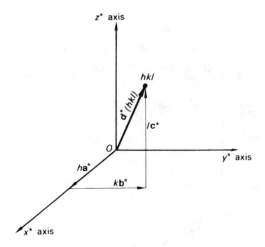

FIGURE 2.16.   Vector components of $\mathbf{d}^*(hkl)$ in a reciprocal lattice. Following the appropriate vector paths parallel to the three reciprocal lattice axes x^*, y^*, and z^* gives the result $\mathbf{d}^*(hkl) = h\mathbf{a}^* + k\mathbf{b}^* + l\mathbf{c}^*$.

reciprocal lattice points, the vector $\mathbf{d}^*(hkl)$ is given by

$$\mathbf{d}^*(hkl) = h\mathbf{a}^* + k\mathbf{b}^* + l\mathbf{c}^* \tag{2.15}$$

This equation is illustrated in Figure 2.16. Equation (2.15) provides a straightforward method for deriving expressions for d^* and d. Thus, taking the dot product of $\mathbf{d}^*(hkl)$ with itself, we have

$$\mathbf{d}^*(hkl) \cdot \mathbf{d}^*(hkl) = d^{*2}(hkl)$$

$$= h^2 a^{*2} + k^2 b^{*2} + l^2 c^{*2} + 2kl b^* c^* \cos \alpha^*$$

$$+ 2lh c^* a^* \cos \beta^* + 2hk a^* b^* \cos \gamma^* \tag{2.16}$$

Now $d(hkl)$ may be obtained from (2.11) and (2.16). Simplifications of (2.16) arise through symmetry constraints on the unit cell vectors in different crystal systems. The reader should check the entries in Table 2.4, starting with Table 2.3 and Equation (2.16).

2.5   Unit-Cell Transformations

In this section, we consider the transformations of unit-cell vectors, zone symbols and directions, Miller indices, reciprocal unit-cell vectors, and fractional coordinates of sites in the unit cell, all without involving a change in the origin of the unit cell.

TABLE 2.4.   Expressions for $d^*(hkl)$, and $d(hkl)$ with $\kappa = 1$

System	$d^{*2}(hkl)$	$d^2(hkl)$
Triclinic	$h^2 a^{*2} + k^2 b^{*2} + l^2 c^{*2} +$ $2klb^* c^* \cos \alpha^* +$ $2lhc^* a^* \cos \beta^* +$ $2hka^* b^* \cos \gamma^*$	$1/d^{*2}(hkl)$
Monoclinic	$h^2 a^{*2} + k^2 b^{*2} + l^2 c^{*2} +$ $2hla^* c^* \cos \beta^*$	$\left\{ \dfrac{1}{\sin^2 \beta} \left[\dfrac{h^2}{a^2} + \dfrac{l^2}{c^2} - \dfrac{2hl \cos \beta}{ac} \right] + \dfrac{k^2}{b^2} \right\}^{-1}$
Orthorhombic	$h^2 a^{*2} + k^2 b^{*2} + l^2 c^{*2}$	$\left\{ \dfrac{h^2}{a^2} + \dfrac{k^2}{b^2} + \dfrac{l^2}{c^2} \right\}^{-1}$
Tetragonal	$(h^2 + k^2) a^{*2} + l^2 c^{*2}$	$\left\{ \dfrac{h^2 + k^2}{a^2} + \dfrac{l^2}{c^2} \right\}^{-1}$
Hexagonal and trigonal (P)	$(h^2 + k^2 + hk) a^{*2} + l^2 c^{*2}$	$\left\{ \dfrac{4(h^2 + k^2 + hk)}{3a^2} + \dfrac{l^2}{c^2} \right\}^{-1}$
Trigonal (R) (rhombohedral)	$[h^2 + k^2 + l^2 +$ $2(hk+kl+hl)(\cos \alpha^*)]a^{*2}$	$a^2 (TR)^{-1}$, where $T = h^2 + k^2 + l^2 +$ $2(hk + kl + hl)[(\cos^2 \alpha - \cos \alpha)/\sin^2 \alpha]$ and $R = (\sin^2 \alpha)/(1 - 3\cos^2 \alpha + 2\cos^3 \alpha)$
Cubic	$(h^2 + k^2 + l^2) a^{*2}$	$\left\{ \dfrac{h^2 + k^2 + l^2}{a^2} \right\}^{-1} = \dfrac{a^2}{h^2 + k^2 + l^2}$

2.5.1   Bravais Unit-Cell Vectors

Let $\mathbf{a}$, $\mathbf{b}$, and $\mathbf{c}$ be transformed to $\mathbf{a}'$, $\mathbf{b}'$, and $\mathbf{c}'$, such that

$$\mathbf{a}' = s_{11}\mathbf{a} + s_{12}\mathbf{b} + s_{13}\mathbf{c}$$
$$\mathbf{b}' = s_{21}\mathbf{a} + s_{22}\mathbf{b} + s_{23}\mathbf{c} \qquad (2.17)$$
$$\mathbf{c}' = s_{31}\mathbf{a} + s_{32}\mathbf{b} + s_{33}\mathbf{c}$$

which may be written in matrix notation as

$$\begin{bmatrix} \mathbf{a}' \\ \mathbf{b}' \\ \mathbf{c}' \end{bmatrix} = \begin{bmatrix} s_{11} & s_{12} & s_{13} \\ s_{21} & s_{22} & s_{23} \\ s_{31} & s_{32} & s_{33} \end{bmatrix} \cdot \begin{bmatrix} \mathbf{a} \\ \mathbf{b} \\ \mathbf{c} \end{bmatrix} \qquad (2.18)$$

or, more concisely, as

$$\mathbf{a}' = \mathbf{S} \cdot \mathbf{a} \qquad (2.19)$$

where the dot $\cdot$ symbolizes matrix multiplication; $\mathbf{a}$ and $\mathbf{a}'$ represent the two sets of column vectors $\mathbf{a}$, $\mathbf{b}$, $\mathbf{c}$, and $\mathbf{a}'$, $\mathbf{b}'$, $\mathbf{c}'$, and $\mathbf{S}$ is the 3×3 matrix of elements s_{ij}.

The inverse transformation may be written as

$$\mathbf{a} = \mathbf{S}^{-1} \cdot \mathbf{a}' \qquad (2.20)$$

where $\mathbf{S}^{-1}$ is the matrix

$$\mathbf{S}^{-1} = \begin{bmatrix} t_{11} & t_{12} & t_{13} \\ t_{21} & t_{22} & t_{23} \\ t_{31} & t_{32} & t_{33} \end{bmatrix} \qquad (2.21)$$

The elements t_{ij} may be obtained by rearranging (2.17), or by the following equation:

$$t_{ij} = (-1)^{i+j} |\mathbf{M}_{ji}| / |\mathbf{S}| \qquad (2.22)$$

where $|\mathbf{M}_{ji}|$ is the minor determinant of $\mathbf{S}$ obtained by striking out its jth row and ith column, and $|\mathbf{S}|$ is the determinant value of the matrix $\mathbf{S}$.

2.5.2   Zone Symbols and Directions

From Section 2.2, we have

$$\mathbf{r} = U\mathbf{a} + V\mathbf{b} + W\mathbf{c} \qquad (2.23)$$

and for the transformed cell

$$\mathbf{r} = U'\mathbf{a}' + V'\mathbf{b}' + W'\mathbf{c}' \qquad (2.24)$$

Thus, using the form of (2.20) with (2.23),

$$\mathbf{r} = [U\ V\ W] \cdot \begin{bmatrix} \mathbf{a} \\ \mathbf{b} \\ \mathbf{c} \end{bmatrix} = [U\ V\ W] \cdot \mathbf{S}^{-1} \cdot \begin{bmatrix} \mathbf{a}' \\ \mathbf{b}' \\ \mathbf{c}' \end{bmatrix} \qquad (2.25)$$

so that, on equating (2.25) to $\mathbf{r}$ from (2.24), we obtain

$$[U'\ V'\ W'] = [U\ V\ W] \cdot \mathbf{S}^{-1} \qquad (2.26)$$

By transposition (interchanging columns and rows),

$$\begin{bmatrix} U' \\ V' \\ W' \end{bmatrix} = (\mathbf{S}^{-1})^{\mathrm{T}} \cdot \begin{bmatrix} U \\ V \\ W \end{bmatrix} \qquad (2.27)$$

or, concisely,

$$\mathbf{U}' = (\mathbf{S}^{-1})^{\mathrm{T}} \cdot \mathbf{U} \qquad (2.28)$$

where $(S^{-1})^T$ is the transpose of matrix (2.21); U and U' refer to the triplet zone symbols, or directions, $[U\,V\,W]$ and $[U'\,V'\,W']$, respectively. In a similar manner, or by pre-multiplication of (2.28) by S^T we can show that

$$U = S^T \cdot U' \tag{2.29}$$

Since, for any matrix $S, S \cdot S^{-1} = 1$.

2.5.3 Coordinates of Sites in the Unit Cell

For any point x, y, z in a unit cell, the vector $\mathbf{r}$ from the origin to that point is given by

$$\mathbf{r} = x\mathbf{a} + y\mathbf{b} + z\mathbf{c} \tag{2.30}$$

Comparison of this equation with (2.23) shows that coordinates transform as do zone symbols. Thus, in the usual notation, we write

$$\mathbf{x}' = (S^{-1})^T \cdot \mathbf{x} \tag{2.31}$$

2.5.4 Miller Indices

From (2.15) and (2.23), it follows that

$$\mathbf{d}^*(hkl) \cdot \mathbf{r} = hU + kV + lW \tag{2.32}$$

Thus, with (2.29),

$$\mathbf{d}^*(hkl) \cdot \mathbf{r} = [h\ k\ l] \cdot \begin{bmatrix} U \\ V \\ W \end{bmatrix} = [h\ k\ l] \cdot S^T \cdot \begin{bmatrix} U' \\ V' \\ W' \end{bmatrix} \tag{2.33}$$

But also

$$\mathbf{d}^*(h'k'l') \cdot \mathbf{r} = \begin{bmatrix} h'\ k'\ l' \end{bmatrix} \cdot \begin{bmatrix} U' \\ V' \\ W' \end{bmatrix} \tag{2.34}$$

because $\mathbf{d}^*(hkl)$ and $\mathbf{d}^*(h'k'l')$ are one and the same vector. Hence

$$\begin{bmatrix} h'\ k'\ l' \end{bmatrix} = [h\ k\ l] \cdot S^T \tag{2.35}$$

Transposing

$$\begin{bmatrix} h' \\ k' \\ l' \end{bmatrix} = S \cdot \begin{bmatrix} h \\ k \\ l \end{bmatrix} \tag{2.36}$$

or

$$\mathbf{h}' = S \cdot \mathbf{h} \tag{2.37}$$

where $\mathbf{h}$ and $\mathbf{h}'$ are column vectors with components h, k, l and h', k', l', respectively. Thus, Miller indices transform in the same way as do unit-cell vectors in real space. If we operate on both sides of (2.37) by $\mathbf{S}^{-1}$, then

$$\mathbf{S}^{-1} \cdot \mathbf{h}' = \mathbf{S}^{-1} \cdot \mathbf{S} \cdot \mathbf{h}$$

or

$$\mathbf{h} = \mathbf{S}^{-1} \cdot \mathbf{h}' \tag{2.38}$$

2.5.5   Reciprocal Unit-Cell Vectors

From 2.15, we develop

$$\mathbf{d}^*(hkl) = \begin{bmatrix} \mathbf{a}^* & \mathbf{b}^* & \mathbf{c}^* \end{bmatrix} \cdot \begin{bmatrix} h \\ k \\ l \end{bmatrix}$$

$$= \begin{bmatrix} \mathbf{a}^* & \mathbf{b}^* & \mathbf{c}^* \end{bmatrix} \cdot \mathbf{S}^{-1} \cdot \begin{bmatrix} h' \\ k' \\ l' \end{bmatrix} \tag{2.39}$$

In the transformed reciprocal unit cell

$$\mathbf{d}^*(hkl) = \begin{bmatrix} \mathbf{a}'^* & \mathbf{b}'^* & \mathbf{c}'^* \end{bmatrix} \cdot \begin{bmatrix} h' \\ k' \\ l' \end{bmatrix} \tag{2.40}$$

so that

$$\begin{bmatrix} \mathbf{a}'^* & \mathbf{b}'^* & \mathbf{c}'^* \end{bmatrix} = \begin{bmatrix} \mathbf{a}^* & \mathbf{b}^* & \mathbf{c}^* \end{bmatrix} \cdot \mathbf{S}^{-1} \tag{2.41}$$

Transposing

$$\begin{bmatrix} \mathbf{a}'^* \\ \mathbf{b}'^* \\ \mathbf{c}'^* \end{bmatrix} = (\mathbf{S}^{-1})^{\mathrm{T}} \cdot \begin{bmatrix} \mathbf{a}^* \\ \mathbf{b}^* \\ \mathbf{c}^* \end{bmatrix} \tag{2.42}$$

or

$$\mathbf{a}'^* = (\mathbf{S}^{-1})^{\mathrm{T}} \cdot \mathbf{a}^* \tag{2.43}$$

so that reciprocal unit-cell vectors transform in the same way as do zone symbols.

As an example of the transformations that we have just derived, a transformation matrix from unit cell 1 to unit cell 2 may be written as

$$\mathbf{S} = \begin{bmatrix} 1 & 0 & 1 \\ 0 & 1 & \bar{2} \\ 1 & 2 & 1 \end{bmatrix}$$

Given the plane $(1\bar{3}5)$ and the site $-0.10, 0.15, 0.25$ in unit cell 1, determine the corresponding values for unit cell 2.

Miller indices: $h_2 = h_1 + l_1 = 6$

$$k_2 = k_1 - 2l_1 = \overline{13}$$

$$l_2 = h_1 + 2k_1 + l_1 = 0$$

that is, the plane is $(6\ \overline{13}\ 0)$ in unit cell 2.

For the coordinates we need the matrix $(\mathbf{S}^{-1})^{\mathrm{T}}$. The determinant $|\mathbf{S}|$ is 4. Then, applying (2.22),

$$\mathbf{S}^{-1} = \begin{bmatrix} 5/4 & 1/2 & -1/4 \\ -1/2 & 0 & 1/2 \\ -1/4 & -1/2 & 1/4 \end{bmatrix}$$

whereupon the transpose becomes

$$(\mathbf{S}^{-1})^{\mathrm{T}} = \begin{bmatrix} 5/4 & -1/2 & -1/4 \\ 1/2 & 0 & -1/2 \\ -1/4 & 1/2 & 1/4 \end{bmatrix}$$

Coordinates: $x_2 = 5x_1/4 + y_1/2 - z_1/4 = -0.2625$

$$y_2 = x_1/2 - z_1/2 = -0.1750$$

$$z_2 = -x_14 + y_1/2 + z_1/4 = 0.1625$$

that is, the site $-0.2625, -0.1750, 0.1625$.

The reciprocal lattice has the same symmetry as the Bravais lattice from which it was deduced. This fact may be appreciated from a comparison of the constructions of the reciprocal lattice and the stereogram. Both of these constructions are built up from normals to planes, so that the symmetry expressed through the poles of a stereogram is the same as that at the reciprocal lattice points.

2.6   Rotational Symmetries of Lattices

We can now discuss analytically the permissible rotational symmetries in lattices, already stated to be of degrees 1, 2, 3, 4, and 6. In Figure 2.17, let A and B represent two adjacent lattice points, of repeat distance t, in any row. An R-fold rotation axis is imagined to act at each point and to lie normal to the plane of the diagram. An anticlockwise rotation of Φ about A maps B on to B', and a clockwise rotation of Φ about B maps A on to A'. It follows from the geometry of the figure that AB is parallel to $A'B'$ and, from the property of lattices, $A'B' = Jt$, where J is an integer.

Lines $A'S$ and $B'T$ are drawn perpendicular to AB, as shown.
Hence,

$$A'B' = TS = AB - (AT + BS) \tag{2.44}$$

or

$$Jt = t - 2t \cos \Phi \tag{2.45}$$

whence

$$\cos \Phi = (1 - J)/2 = M/2 \tag{2.46}$$

where M is another integer. Since $-1 \leqslant \cos \Phi \leqslant 1$, it follows from (2.46) that the only admissible values for M are $0, \pm 1, \pm 2$; these values give rise to the rotational symmetries already discussed. This treatment gives a quantitative aspect to the packing considerations mentioned previously (see Section 1.4.2).

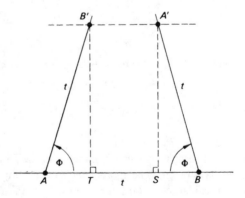

FIGURE 2.17. Rotational symmetry in crystal lattices. Permissible values of Φ are $360°(0°), 180°, 120°, 90°,$ and $60°$ (1-, 2-, 3-, 4-, and 6-fold rotations).

2.7 Space Groups

In order to extend our study of crystals into the realm of atomic arrangements, we must consider now the symmetry of extended, ideally infinite, patterns in space. We recall that a point group describes the symmetry of a finite body, and that a lattice constitutes a mechanism for repetition, to an infinite extent, by translations parallel to three noncoplanar directions. We may ask, therefore, what is the result of repeating a point-group pattern by the translations of a Bravais lattice. We shall see that it is like an arrangement of atoms in a crystal.

A space group can be described as an infinite set of symmetry elements, the operation with respect to any of which brings the infinite array of points to which they refer into a state that is indistinguishable from that before the operation. We may apply space-group rules to crystals because the dimensions of crystals used in experimental investigations are large with respect to the repeat distances of the pattern. For example, the dimension a of the face-centered cubic unit cell of sodium chloride is 0.564 nm. Thus, in a crystal of experimental size (ca 0.2 × 0.2 × 0.2 mm^3), there are approximately 5 × 10^{16} unit cells.

A space group may be considered to be made up of two parts, a pattern motif and a repeat mechanism. An analogy can be drawn with a wallpaper pattern (Victorian style), a simple example of which is shown in Figure 2.18a. We shall analyze this pattern.

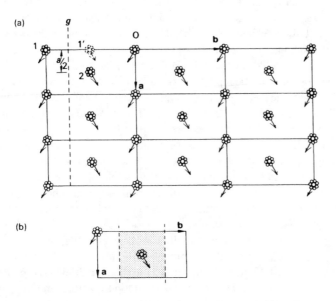

FIGURE 2.18. Wallpaper pattern: (a) extended pattern, (b) asymmetric unit; the space-group symmetry applied to the asymmetric unit generates the whole infinite pattern.

The conventional unit cell for this pattern is indicated by the vectors **a** and **b**. If we choose a pattern motif consisting of two flowers (Figure 2.18b) and continue it indefinitely by the repeat vectors **a** and **b**, the plane pattern is generated. However, we have ignored the symmetry between the two flowers in the chosen pattern motif. If one flower (1) is reflected across the dashed line (g) to ($1'$) and then translated by **a**/2, it then occupies the position of the second flower (2), or the pattern represented by Figure 2.18a is brought from one state to another indistinguishable state by the symmetry operation. This operation takes place across a *glide line*, a symmetry element that occurs in some extended two-dimensional patterns.

We say that the necessary and sufficient pattern motif is a single flower, occupying the *asymmetric unit*—the unshaded (or shaded) portion of Figure 2.18b. If the single flower is repeated by both the glide-line symmetry and the unit-cell translations, then the extended pattern is again generated. Thus, to use our analogy, if we know the asymmetric unit of a crystal structure, which need not be the whole unit-cell contents, and the space-group symbol for the crystal, we can generate the whole structure.

2.7.1   Two-Dimensional Space Groups

Our discussion leads naturally into two-dimensional space groups, or *plane groups*. Consider a pattern motif showing 2-fold symmetry (Figure 2.19a)—the point-group symbolism is continued into the realm of space groups. Next, consider a primitive oblique net (Figure 2.19b); it is of infinite extent in the plane, and the framework of lines divides the field, conceptually, into a number of identical primitive (p) unit cells. An origin is chosen at any lattice point.

Now, let the motif be repeated around each point in the net, and in the same orientation, with the 2-fold rotation points of the motif and the net in coincidence (Figure 2.19c). It will be seen that additional 2-fold rotation points are introduced, at the unique fractional coordinates (see footnote to Table 2.2) $0, \frac{1}{2}; \frac{1}{2}, 0;$ and $\frac{1}{2}, \frac{1}{2}$ in each unit cell. We must always look for such induced symmetry elements after the point-group motif has been operated on by the unit-cell repeats. Ultimately, this will be found to be straightforward. Meanwhile, a simple check consists in ensuring that any point on the unit-cell diagram can be reached from any other point by means of a *single* symmetry operation, including translational repeats. This plane group is given the symbol $p2$.

In general, we shall not need to draw several unit cells; one cell will suffice provided that the pattern motif is completed around all lattice points intercepted by the given unit cell. Figure 2.20 illustrates the standard drawing of $p2$: the origin is taken on a 2-fold point, the x axis runs from top to bottom, and the y axis runs from left to right. Thus, the origin is considered to be in the top left-hand corner of the cell, but each corner could be an equivalent origin; we must remember that

(a)

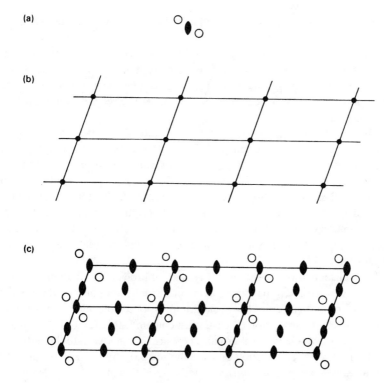

(b)

(c)

FIGURE 2.19. Plane group $p2$: (a) 2-fold symmetry motif, (b) oblique net with p unit cells outlined, (c) extended pattern of plane group $p2$ obtained by a combination of (a) with (b).

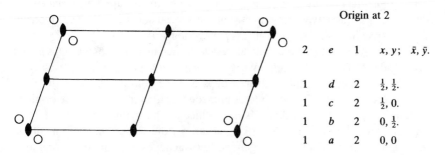

				Origin at 2
2	e	1	$x, y;$	$\bar{x}, \bar{y}.$
1	d	2	$\frac{1}{2}, \frac{1}{2}.$	
1	c	2	$\frac{1}{2}, 0.$	
1	b	2	$0, \frac{1}{2}.$	
1	a	2	$0, 0$	

FIGURE 2.20. Standard drawing and description of plane group $p2$. The lines which divide the unit cell into four quadrants are, as usual, drawn for convenience only.

the drawing is a representative portion of an infinite array, whether in two or three dimensions.

The asymmetric unit, represented by O, may be placed anywhere in the unit cell (for convenience, near the origin), and then repeated by the symmetry $p2$ to build up the picture, taking care to complete the arrangements around each corner. The additional 2-fold points can then be identified. The reader should now carry out this construction.

The list of fractional coordinates in Figure 2.20 refers to the unique symmetry-related sites in the unit cell. The first row of these sites, related by the space-group symmetry, lists the *general equivalent positions*. In $p2$ they are given the coordinates x, y and $\bar{x}, \bar{y}$. We could use $1 - x, 1 - y$ instead of $\bar{x}, \bar{y}$, but it is more usual to work with a set of coordinates near one and the same origin.

Each coordinate line in the space-group description lists, in order from left to right, the number of positions in each set, the Wyckoff[a] notation (for reference purposes only), the symmetry at each site in the set, and the fractional coordinates of all sites in the set.

In a conceptual two-dimensional crystal, or projected real atomic arrangement, the asymmetric unit may contain either a single atom or a group of atoms. If it consists of part (half, in this plane group) of one molecule, then the whole molecule, as seen in projection at least, must contain 2-fold rotational symmetry, or a symmetry of which 2 is a subgroup. There are four unique 2-fold points in the unit cell; in the Wyckoff notation they are the sets (a), (b), (c), and (d), and they constitute the sets of *special equivalent positions* in this plane group. Notice that general positions have symmetry 1, whereas special positions have a higher crystallographic point-group symmetry. Where the unit cell contains fewer (an integral submultiple) of a species than the number of general equivalent positions in its space group, then it may be assumed that the species are occupying special equivalent positions and have the symmetry of the special site, at least. Exceptions to this rule arise in disordered structures (see Section 7.9).

We move now to the rectangular system, which includes point groups m and $2mm$, and both p and c unit cells. We shall consider first plane groups pm and cm.

The formation of these plane groups may be considered along the lines already described for $p2$, and we refer immediately to Figure 2.21a. The origin is chosen on m, but its y coordinate is not defined by this symmetry element. (In a structure, the origin is defined by fixing arbitrarily the y coordinate of one of the atoms in the unit cell.) In pm, the general equivalent positions are two in number, and there are two sets of special equivalent positions on m lines.

Plane group cm (Figure 2.21b) introduces several new features. The coordinate list is headed by the expression $(0, 0; \frac{1}{2}, \frac{1}{2})+$; this means that two

---

[a] See Bibliography to Chapter 8.

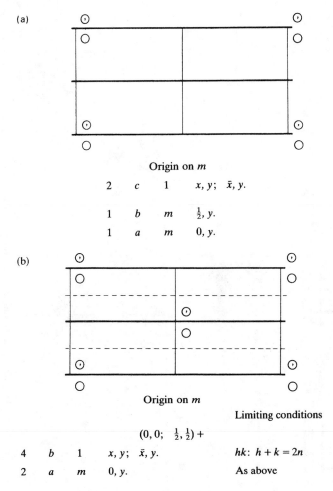

(a)

Origin on m

2	c	1	x, y; $\bar{x}, y$.
1	b	m	$\frac{1}{2}, y$.
1	a	m	$0, y$.

(b)

Origin on m

Limiting conditions

$(0, 0;\ \frac{1}{2}, \frac{1}{2}) +$

4	b	1	x, y; $\bar{x}, y$.	hk: $h + k = 2n$
2	a	m	$0, y$.	As above

FIGURE 2.21. Plane groups in the rectangular system: (a) pm, (b) cm; glide lines (g) are indicated by the dashed lines.

translations—0, 0 and $\frac{1}{2}, \frac{1}{2}$—are added to all the listed coordinates. Hence, the full list of general positions reads

$$x, y; \quad \bar{x}, y; \quad \tfrac{1}{2} + x, \tfrac{1}{2} + y; \quad \tfrac{1}{2} - x, \tfrac{1}{2} + y$$

Given x, the distance $\frac{1}{2} - x$, for example, is found by first moving $\frac{1}{2}$ along the a axis from the origin and then moving back along the same line by the amount x.

The centering of the unit cell in conjunction with the m lines introduces the glide-line symmetry element, symbol g and graphic symbol - - -. The glide

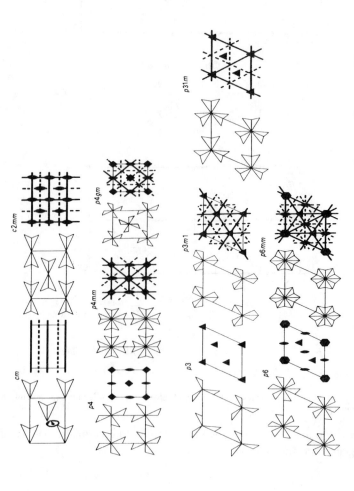

FIGURE 2.22.   Unit cells of the 17 plane groups. In each pair of drawings, that showing the general equivalent positions uses a scalene triangle motif instead of the usual circle. It is noteworthy that the conventional unit-cell drawing shows the unit cell and its immediate environment.

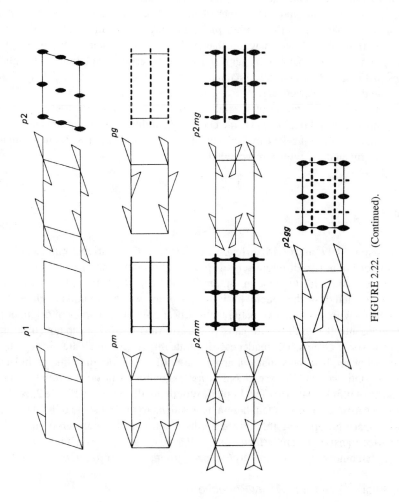

FIGURE 2.22.   (Continued).

lines interleave the mirror lines, and their action is a combination of reflection and translation, the two movements comprising a *single* symmetry operation. The translational component is one half of the repeat distance in the direction of the glide line. Thus, a pair of general positions x, y and $\frac{1}{2} - x, \frac{1}{2} + y$ are related by the g line. We shall encounter glide lines in any centered unit cell where m lines are present, and in certain other plane groups. For example, we may ask if there is any meaning to the symbol pg, a glide-symmetry motif repeated by the unit-cell translations. The answer is that pg is a possible plane group; in fact, it is the symmetry of the pattern in Figure 2.18. The differing orientations of the glide lines in Figures 2.18 and 2.22 (standard) may be expressed by the *full* symbols $p11g$ and $p1g1$, respectively.

There is only one set of special positions in cm, in contrast to two sets in pm. This situation arises because the centering condition in cm requires that both mirror lines in the unit cell be included in one and the same set. If we try to postulate two sets, by analogy with pm, we obtain

$$0, y; \quad \tfrac{1}{2}, \tfrac{1}{2} + y \tag{2.47}$$

and

$$\tfrac{1}{2}, y; \quad 0(\text{or} 1), \tfrac{1}{2} + y \tag{2.48}$$

Expressions (2.47) and (2.48) differ only in the value of the variable y and therefore do not constitute two different sets of special equivalent positions.

We could refer to plane group cm by the symbol cg. If we begin with the origin on g and draw the general positions as before, we should find the glide lines interleaved with m lines. Two patterns that differ only in the choice of origin or in the values attached to the coordinates of the equivalent positions do not constitute different space groups. The reader can illustrate this statement by drawing cg, and, by drawing pg, can show that pm and pg are different. The glide line or, indeed, any translational symmetry element is not encountered in point groups; it is a property of infinite patterns. The 17 plane groups are illustrated in Figure 2.22. The asymmetric unit is represented therein by a scalene triangle instead of by the usual circle. Space groups that are derived by the repetition of a point-group motif by the lattice translations are termed *symmorphic* space groups, as with $p2$, pm, and $c2mm$, but otherwise *non-symmorphic* space groups, as with pg, $p2mg$, and $p2gg$.

Conditions Governing X-Ray Reflection

Our main reason for studying space-group symmetry is that it provides information about the repeat patterns of atoms in crystal structures. X-ray diffraction spectra are characterized partly by the indices of the families of planes from which, in the Bragg treatment of diffraction (see Chapter 3), the x-rays are considered to be reflected. The pattern of indices reveals information about the space group of the crystal. Where a space group contains translational symmetry, certain sets of

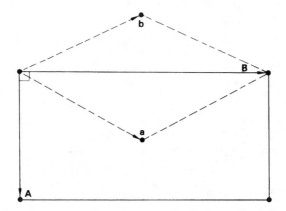

FIGURE 2.23. Centered rectangular unit cell (**A**, **B**) and primitive unit cell (**a**, **b**) within the same lattice.

reflections will be systematically absent from the experimental diffraction data record. We meet this situation for the first time in *cm* (Figure 2.21b); reflections *hk* are limited to those for which the sum $h + k$ is an even number.

Figure 2.23 illustrates a rectangular lattice. Two unit cells are depicted on this lattice, a centered cell with vectors **A** and **B**, and a primitive cell with vectors **a** and **b**. The relationship between them is summarized by the equations

$$\mathbf{A} = \mathbf{a} - \mathbf{b}$$
$$\mathbf{B} = \mathbf{a} + \mathbf{b} \tag{2.49}$$

Since Miller indices of planes transform in the same way as unit-cell vectors, it follows that

$$H = h - k$$
$$K = h + k \tag{2.50}$$

where H and K apply to the unit cell **A**, **B**. Adding equations (2.50), we obtain

$$H + K = 2h \tag{2.51}$$

which is even for all values of h. Thus, in the centerd unit cell, reflections can occur only when the sum of the indices, $H + K$, is an even integer. This topic is discussed more fully in Chapter 3, whereupon the significance of the right-hand column of data in figures such as Figures 2.21 or 2.24 will become clear.

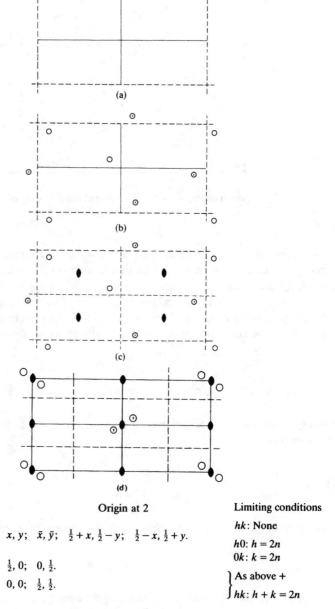

Origin at 2

4 c 1 x, y; $\bar{x}, \bar{y}$; $\tfrac{1}{2}+x, \tfrac{1}{2}-y$; $\tfrac{1}{2}-x, \tfrac{1}{2}+y$.

2 b 2 $\tfrac{1}{2}, 0$; $0, \tfrac{1}{2}$.

2 a 2 $0, 0$; $\tfrac{1}{2}, \tfrac{1}{2}$.

Limiting conditions

hk: None

$h0$: $h = 2n$

$0k$: $k = 2n$

$\left.\begin{array}{l} \\ \end{array}\right\}$ As above +

hk: $h + k = 2n$

FIGURE 2.24. Formation and description of plane group $p2gg$.

2.7.2   Plane Groups Related to 2*mm*

Point group 2*mm* belongs to the rectangular system and, as a final example in two dimensions, we shall study plane group $p2gg$. It is often helpful to recall the parent point group of any space group. All that we need to do is to ignore the unit-cell symbol, and replace any translational symmetry elements by the corresponding nontranslational symmetry elements. Thus, pg is derived from point group m, and $p2gg$ from 2*mm*.

In point group 2*mm*, we know that the two m lines intersect in the 2-fold rotation point, and this remains true for plane group $p2mm$. In $p2gg$, however, we may not assume that 2 lies at the intersection of the two g lines. In our study of point groups, we saw that the symmetry elements in a given symbol have a definite relative orientation with respect to the crystallographic axes; this is preserved in the corresponding space groups. Thus, we know that the g lines are normal to the x and y axes, and we can take an origin, initially, at their intersection (Figure 2.24a). In Figure 2.24b, the general equivalent positions have been inserted; this diagram reveals the positions of the 2-fold points, inserted in Figure 2.24c, together with the additional g lines in the unit cell. The standard orientation of $p2gg$ places the 2-fold point at the origin; Figure 2.24d shows this setting and the description of this plane group. We see again that two interacting symmetry elements lead to a combined action which is equivalent to that of a third symmetry element, but their positions must be chosen correctly. This question did not arise in point groups because, by definition, all symmetry elements pass through a point—the origin.

There are two sets of special equivalent positions in $p2gg$; the pairs of 2-fold rotation points that constitute each set must be selected correctly. One way of ensuring a proper selection is by inserting the coordinate values of the point-group symmetry element constituting a special position into the coordinates of the general positions. Thus, by taking $x = y = 0$, for one of the 2-fold points, we obtain a set of special positions with coordinates $0, 0$ and $\frac{1}{2}, \frac{1}{2}$. If we had chosen $0, 0$ and $0, \frac{1}{2}$ as a set, the resulting pattern would not have conformed to $p2gg$ symmetry, but to pm, as Figure 2.25 shows. Special positions form a subset of the general positions, under the same space-group symmetry.

The general equivalent positions give rise to two conditions limiting reflections, because the structure is "halved" with respect to b for the reflections $0k$, and with respect to a for the reflections $h0$. The special positions take both of these conditions, and the extra conditions shown because occupancy of the special positions[a] in this plane group gives rise to centered arrangements. After the development of the structure factor (see Chapter 4), some of the different limiting conditions will be derived analytically.

---

[a] The entities occupying special positions must, themselves, conform to the space-group symmetry.

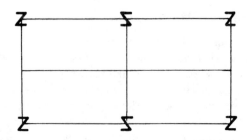

FIGURE 2.25.   Occupation of the special positions 0, 0 and 0, $\frac{1}{2}$ in $p2gg$ leads to pm symmetry.

What plane group would be formed if we do set the 2-fold rotation point at the intersection of the glide lines?

2.7.3   Three-Dimensional Space Groups

The principles that have emerged from the discussion on plane groups can be extended to three dimensions. Whereas the plane groups are limited to 17 in number, there are 230 space groups. We shall limit our discussion to a few space groups mainly in the monoclinic and orthorhombic systems. We believe this will prove adequate because many of the important principles will evolve and, from a practical point of view, a large percentage of crystals belong to these two systems.

Monoclinic Space Groups

In the monoclinic system, the lattices are characterized by P and C unit-cell descriptions, and the point groups are 2, m, and $2/m$. We consider first space groups $P2$ and $C2$.

As with the plane groups, we may begin with a motif, which has 2-fold symmetry, but now about a line (axis) in three-dimensional space. This motif is arranged in a fixed orientation with respect to all the points of a monoclinic lattice. Figure 2.26 shows a stereoscopic pair of illustrations for a unit cell in $C2$, drawn with respect to the conventional right-handed axes.

In Figure 2.27, $P2$ and $C2$ are shown in projection. The standard drawing of space-group diagrams is on the ab plane of the unit cell, with $+x$ running from top to bottom, $+y$ from left to right, both in the plane of the paper, and $+z$ coming up from the paper. The positive or negative signs attached to the representative points indicate the z coordinates, that is, in $\bigcirc^{+}$ and $\bigcirc^{-}$, the signs stand for z and $\bar{z}$, respectively. The relationship with the preferred stereogram notation (see Section 1.4.2) will be evident here.

In both $P2$ and $C2$, the origin is chosen on 2, and is, thus, defined with respect to the x and z axes, but not with respect to y (compare pm and cm). The graphic symbol for a diad axis in the plane of the diagram is $\rightarrow$.

In space group $P2$, the general and special equivalent positions may be derived quite readily. The special sets (b) and (d) should be noted carefully; they are sometimes forgotten by the beginner because symmetry elements distant $c/2$ from those drawn in the ab plane are not indicated on the conventional diagrams. The diad along y and at $x = 0, z = \frac{1}{2}$, for example, relates x, y, z to a point at $\bar{x}, y, 1 - z$; its presence, and that of the diad at $x = z = \frac{1}{2}$, may be illustrated by drawing the space group in projection on the ac plane of the unit cell. The reader should make this drawing and compare it with Figure 2.27a.

It is often useful to consider a structure in projection on to one of the principal planes (100), (010), or (001). The symmetry of a projected space group corresponds with a plane group, and the symmetries of the principal projections are included with the space-group description (Figure 2.27). The full plane-group symbols, given in parentheses, indicate the orientations of the symmetry elements (see Table 1.5). In $C2$, certain projections produce more than one repeat in some directions; the projected cell dimensions, represented by $a', b',$ and c', may then be halved with respect to their original values. The Miller indices transform like unit cell vectors, according to Section 2.5.4; thus, for example, with b halved, $220 \rightarrow 210$ and $210 \rightarrow 410$ (which is equivalent to halving the k index in each case).

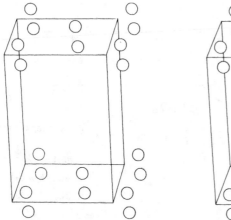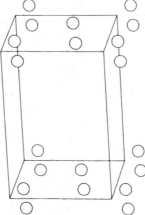

FIGURE 2.26.   Stereoscopic pair of illustrations of the environs of one unit cell of space group $C2$; general equivalent positions are shown. The *diagram* contains nine axes of symmetry 2, and six axes of symmetry 2_1. Can you identify their positions?

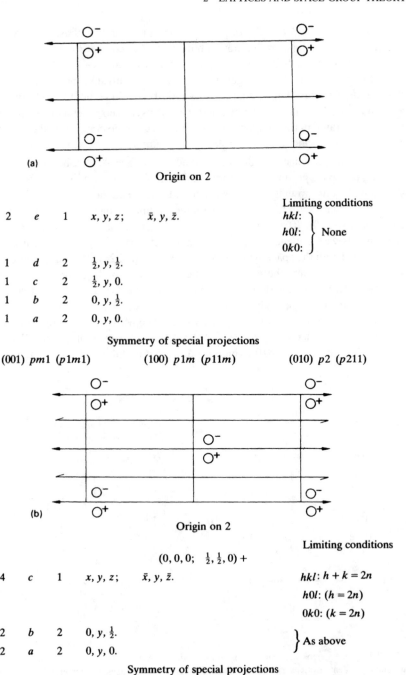

Origin on 2

					Limiting conditions
2	e	1	$x, y, z;$	$\bar{x}, y, \bar{z}.$	$hkl:$
					$h0l:$ } None
					$0k0:$
1	d	2	$\frac{1}{2}, y, \frac{1}{2}.$		
1	c	2	$\frac{1}{2}, y, 0.$		
1	b	2	$0, y, \frac{1}{2}.$		
1	a	2	$0, y, 0.$		

Symmetry of special projections

(001) $pm1$ $(p1m1)$ (100) $p1m$ $(p11m)$ (010) $p2$ $(p211)$

Origin on 2

$(0, 0, 0; \quad \frac{1}{2}, \frac{1}{2}, 0) +$

					Limiting conditions
4	c	1	$x, y, z;$	$\bar{x}, y, \bar{z}.$	$hkl: h + k = 2n$
					$h0l: (h = 2n)$
					$0k0: (k = 2n)$
2	b	2	$0, y, \frac{1}{2}.$		
2	a	2	$0, y, 0.$		} As above

Symmetry of special projections

(001) $cm1$ $(c1m1)$ (100) $p1m$ $(p11m)$ $b' = b/2$ (010) $p2$ $(p211)$ $a' = a/2$

FIGURE 2.27.   Monoclinic space groups in the standard setting: (a) $P2$, (b) $C2$.

The projection of $C2$ on to (100) is shown by Figure 2.28 in three stages, starting from the y and z coordinates of the set of general equivalent positions. The question is sometimes asked, how do two points of the same hand, such as x, y, z and $\bar{x}, y, \bar{z}$ in $C2$, become of opposite hand, such as y, z and $y, \bar{z}$ in $p1m$, after projection? The difficulty may be associated with the use of the highly symmetric circle as a representative point. It is suggested that the reader make a drawing of $C2$, and of the stages of the projection on to (100), using the symbol ♥ instead of $\bigcirc^+$, and ♀ instead of $\bigcirc^-$. It is important to remember that, in the plane groups, as in the two-dimensional point groups, all symmetry operations take place wholly *within* the plane of the figure.

Space group $C2$ may be obtained by adding the translation $\frac{1}{2}, \frac{1}{2}, 0$, that associated with a C cell (Table 2.2), to the equivalent positions of $P2$. This operation

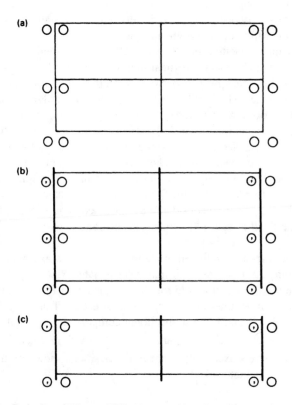

FIGURE 2.28. Projection of $C2$ on to (100): (a) y, z positions from $C2$ (z axis left to right), (b) two-dimensional symmetry elements added, (c) one unit cell—$p1m$ ($p11m$), $b' = b/2, c' = c$. (Plane groups $p11m$ and $p1m1$ are equivalent because they correspond only to an interchange of the x and y axes; 1 is a trivial symmetry element.)

is equivalent to repeating the original 2-fold motif at the lattice points of the C monoclinic unit cell. This simple relationship between P and C cells is indicated by the heading $(0, 0, 0; \frac{1}{2}, \frac{1}{2}, 0)+$ of the coordinate list in $C2$; it may be compared with that for cm (Figure 2.21b).

There are four sets of special positions in $P2$, but only two sets in $C2$; the reason for this has been discussed in relation to plane groups pm and cm (see Section 2.7.1).

2.7.4   Screw Axes

Screw axes are symmetry elements that can relate points in an infinite, three-dimensional, regular array: thus, they are not a feature of point groups. A screw axis operation may be thought of as a combination of rotation and translation, although it is a *single* symmetry operation: an infinitely long spiral staircase would give an indication of the nature of the symmetry operation.

Imagine that the bottom step (Figure 2.29) is rotated (anticlockwise, looking down the stairs) by 60° about the vertical support (axis) and then translated upward by one sixth of the repeat distance between steps in similar orientations, so that it takes the place of the second step, which itself moves upward in a similar manner. Clearly, if this procedure were repeated six times, the bottom step would reach the position and orientation of the sixth step up; we symbolize this screw axis as 6_1. Infinite length is, theoretically, a requirement because as the bottom step is rotated and translated upward, so another step, below the figure, comes up into its position in order that indistinguishability is maintained. The spiral staircases of the Monument in London and of the Statue of Liberty in New York seem to be of infinite length, and might be considered as macroscopic near-examples of screw axes. Examine them carefully on your next visit and determine their symmetry nature.

The centering of the unit cell in $C2$ introduces screw axes which interleave the diads (Figure 2.27). A screw axis may be designated R_p ($p < R$) and a screw-axis operation consists of an R-fold rotation coupled with a translation parallel to the screw axis of p/R times the repeat in that direction. Thus, in $C2$, the screw axis is of the type 2_1 and has a translational component of $\frac{1}{2}$ parallel to b. The general equivalent positions x, y, z and $\frac{1}{2} - x, \frac{1}{2} + y, \bar{z}$ are related by a 2_1 axis along $[\frac{1}{4}, y, 0]$.[a] Screw axes are present in the positions shown by their graphic symbol ⟶ (see also Table 2.5).

---

[a] We use this notation to describe lines, in this example the line parallel to the y axis through $x = \frac{1}{4}, z = 0$.

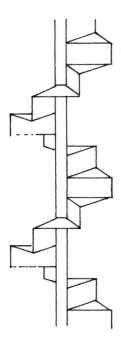

FIGURE 2.29.   Spiral staircase: An illustration of 6_1 screw axis symmetry.

Limiting Conditions in C2

The limiting conditions for this space group are given in Figure 2.27. Two of them are placed in parentheses; this notation is used to indicate that they are dependent upon a more general condition. Thus, since the hkl reflections are limited by the condition $h + k = 2n$ (even), because the cell is C-centered, it follows that $h0l$ are limited by $h = 2n$ (0 is an even number). There are several other nonindependent conditions that could have been listed. For example, $0kl$: $k = 2n$ and $h00$: $h = 2n$. However, in the monoclinic system, in addition to the hkl reflections, we are concerned particularly only with $h0l$ and $0k0$, because the symmetry plane is parallel to (010) and the symmetry axis is parallel to [010]. This feature is discussed more fully in Chapter 3.

Space Group P2₁

Space groups $C2$ and $C2_1$ are equivalent (compare cm and cg). On the other hand, $P2$ contains no translational symmetry, so $P2_1$ is a new space group (Figure 2.30). There are no special positions in $P2_1$. Special positions cannot exist on a translational symmetry element, since it would mean that the entity placed on such an element consisted of an infinite repeating pattern.

TABLE 2.5. Notation for Symmetry Axes in Space Groups, and
Limiting Conditions for Screw Axes

Symbol	Graphic symbol	Screw-axis orientation and translation	Limiting condition
1	None		
$\bar{1}$	o		
2	(normal to paper) (parallel to paper)		
2_1	(normal to paper)	$[100]$ $a/2$	$h00: h = 2n$
	(parallel to paper)	$[010]$ $b/2$	$0k0: k = 2n$
		$[001]$ $c/2$	$00l: l = 2n$
3	▲		
$\bar{3}$	△		
3_1	▲	$[0001]$ $c/3$	$000l: l = 3n$
3_2	▲	$[0001]$ $2c/3$	$000l: l = 3n$
4	◆		
$\bar{4}$	◈		
$4_1, 4_3$		$[100]$ $a/4, 3a/4$	$h00: h = 4n$
		$[010]$ $b/4, 3b/4$	$0k0: k = 4n$
		$[001]$ $c/4, 3c/4$	$00l: l = 4n$
4_2	◆		
6	●		
$\bar{6}$	⬡		
$6_1, 6_5$		$[0001\}$ $c/6, 5c/6$	$000l: l = 6n$
$6_2, 6_4$		$[0001]$ $2c/6, 4c/6$	$000l: l = 3n$
6_3		$[0001]$ $3c/6$	$000l: l = 2n$

Notes:
1. The 3_1 and 3_2 axes are referred to the hexagonal setting of the trigonal system.
2. Compare the $2_1, 4_2$, and 6_3 axes, the 4_1 and 4_3 axes, and $3_1, 3_2, 6_2, 6_4$ axes.

2.7.5 Glide Planes

Consider again Figure 2.18, but let each dashed line be the trace of a glide plane normal to b. Whereas in two dimensions, the direction of translation, after the reflection part of the operation, is unequivocal, in three dimensions there are four possibilities, although each of them will not necessarily give rise to a different space group.

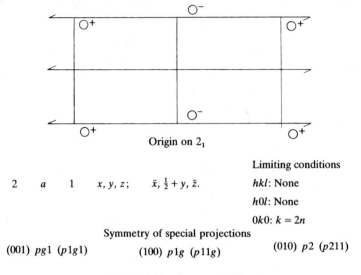

Origin on 2_1

<table>
<tr><td></td><td></td><td></td><td></td><td></td><td>Limiting conditions</td></tr>
<tr><td>2</td><td>a</td><td>1</td><td>x, y, z;</td><td>$\bar{x}, \tfrac{1}{2}+y, \bar{z}$.</td><td>$hkl$: None</td></tr>
<tr><td></td><td></td><td></td><td></td><td></td><td>$h0l$: None</td></tr>
<tr><td></td><td></td><td></td><td></td><td></td><td>$0k0$: $k = 2n$</td></tr>
</table>

Symmetry of special projections

(001) $pg1$ $(p1g1)$ (100) $p1g$ $(p11g)$ (010) $p2$ $(p211)$

FIGURE 2.30. Space group $P2_1$.

In the case of the glide plane normal to b, the direction of translation could be along a (amount $a/2$), along c (amount $c/2$), along a diagonal direction n [amount $(a + c)/2$], or along a diagonal direction d [amount $(a \pm c)/4$]. The d-glide plane is not often encountered in practice, and will not be discussed in detail here.[a]

The translations form mnemonics for the glide-plane names. Thus, in Figure 2.36, an a-glide plane at $c/4$ is shown. The symbol n may refer to more than one orientation (Table 2.6), but the space-group symbol here, relating to the corresponding point-group symbol in Table 1.5, provides the necessary information. Thus, if the n-glide plane is normal to a, the translation component of the n-glide symmetry operation must be $(b + c)/2$. It is imperative to understand fully the Hermann–Mauguin point-group notation (Table 1.5), because that for space groups follows in a logical manner.

The translational components for screw axes and for glide planes are integer fractions of the repeat distances. A 2_1 axis parallel to b has a component of translation of $b/2$. A 2_2, or in general an R_R, axis is a combination of 2 (or R) and the corresponding unit-cell translation, in this case b, and corresponds to simple rotation. The term *translation* is preferable in the context of screw axis or glide plane, and the term *repetition* for the unit-cell (Bravais) translation.

If a space group is formed from the combination of a point group with m planes and a lattice of centered unit cells, glide planes are introduced into the

[a] See Bibliography (Ladd).

TABLE 2.6. Notation for Symmetry Planes in Space Groups, and Limiting
Conditions for Glide Planes

Symbol	Graphic symbol	Glide plane orientation and translation		Limiting condition
m	—————— or ╱	⊥ paper ‖ paper	— —	— —
a	- - - - - - ↓	⊥ paper ‖ paper	$(h0l)$ $a/2$ $(hk0)$ $a/2$	$h0l: h = 2n$ $hk0: h = 2n$
b	- - - - - - - ←	⊥ paper ‖ paper	$(0kl)$ $b/2$ $(hk0)$ $b/2$	$0kl: k = 2n$ $hk0: k = 2n$
c	· · · · · · · · · · ·	{ ⊥ paper ⊥ paper	$(0kl)$ $c/2$ $(h0l)$ $c/2$	$0kl: l = 2n$ $h0l: l = 2n$
n	- · - · - · - · - ↑	{ ⊥ paper ⊥ paper ‖ paper	$(0kl)$ $(b+c)/2$ $(h0l)$ $(c+a)/2$ $(hk0)$ $(a+b)/2$	$0kl: k + l = 2n$ $h0l: l + h = 2n$ $hk0: h + k = 2n$
d	· - · ◄ · - · - ◄ · - · ↗	{ ⊥ paper ⊥ paper ‖ paper	$(0kl)$ $(b \pm c)/4$ $(h0l)$ $(c \pm a)/4$ $(hk0)$ $(a \pm b)/4$	$0kl: k + l = 4n$ $h0l: l + h = 4n$ $hk0: h + k = 4n$

Notes:
1. The trigonal system is here referred to hexagonal axes.
2. An arrow shows the direction of the glide translation. A fraction indicates the z height of the plane.
3. The condition $(a + b + c)/4$ exists for d-glide planes parallel to $\{1\bar{1}0\}$ in the tetragonal and cubic systems.

space group. They are the three-dimensional analog of glide lines. The glide-plane operation consists of reflection across the plane plus a translation parallel to the plane. The direction of translation is indicated by the glide-plane symbol (Table 2.6).

Space Group $P2_1/c$

As an example of a space group with a glide plane, we shall study $P2_1/c$, a space group encountered frequently in practice. This space group is derived from point group $2/m$, and must, therefore, be centrosymmetric. However, the center of symmetry does not lie at the intersection of 2_1 and c. It is convenient to take

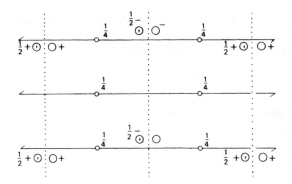

FIGURE 2.31. Space group $P2_1/c$ with the origin at an intersection of 2_1 and c.

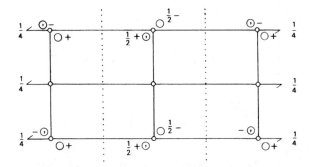

FIGURE 2.32. Space group $P2_1/c$ with the origin on $\bar{1}$ (standard setting).

the origin on a center of symmetry[a] in centrosymmetric space groups and, in this example, we must determine the correct positions of the symmetry elements in the unit cell. We shall approach the solution of this problem in two ways, the first of which is similar to our treatment of plane group $p2gg$.

Since the screw axis must intersect the glide plane normally, according to the space-group symbol, the point of intersection will be taken as an origin and the space group drawn (Figure 2.31). We see now that the centers of symmetry lie at points such as $0, \frac{1}{4}, \frac{1}{4}$. This point may be taken as a new origin, and the space group redrawn (Figure 2.32); a fraction ($\frac{1}{4}$, for example) placed next to a symmetry element indicates the position of that symmetry element with respect to the ab plane.

It is desirable, however, to be able to draw the standard space-group illustration at the outset. From a choice of origin, and using the full meaning of the

[a] Sometimes the origin will have a point symmetry higher than $\bar{1}$, for example, $2/m$ or mmm, but $\bar{1}$ is a subgroup of such point symmetries.

space-group symbol, we can obtain the positions of the symmetry elements by means of a simple scheme.

Let the symmetry elements be placed as follows:

$\bar{1}$ at $0, 0, 0$ (choice of origin)

2_1 along $[p, y, r]$ (parallel to the y axis)

c the plane (x, q, z) (normal to the y axis)

It is important to note that we have employed only the standard choice of origin and the information contained in the space-group symbol. Next, we carry out the symmetry operations:

$$(1) \quad x, y, z \xrightarrow{\;2_1\;} 2p - x, \tfrac{1}{2} + y, 2r - z \quad (2)$$

$$\Big\downarrow {-c}$$

$$2p - x, 2q - \tfrac{1}{2} - y, -\tfrac{1}{2} + 2r - z \quad (3)$$

$$\xrightarrow{\;\bar{1}\;} -x, -y, -z \quad (4)$$

The symbol $-c$ is used to indicate that the c-glide translation of $\tfrac{1}{2}$ is subtracted, which is crystallographically equivalent to being added.[a]

We now use the fact that the combined effect of two operations is equivalent to a third operation, starting from the original point (1). Symbolically, $c \cdot 2_1 \equiv \bar{1}$, or 2_1 followed by c is equivalent to $\bar{1}$. Thus, points (3) and (4) are one and the same, whence, by comparing coordinates, $p = 0$ and $q = r = \tfrac{1}{4}$. Comparison with Figure 2.32 shows that these conditions lead to the desired positions of the symmetry elements in $P2_1/c$.

The change in the x coordinate in the operation (1) → (2) is illustrated in Figure 2.33; the argument can be applied to any similar situation in monoclinic and orthorhombic space groups, and we can always consider one coordinate at a time. The completion of the details of this space group forms the basis of a problem at the end of this chapter.

We shall not discuss centered monoclinic space groups, but they do not present difficulty once the primitive space groups have been mastered. Figure 2.34 shows a stereoscopic pair of illustrations of the zinc and iodine atoms in the structure of diiodo-$(N, N, N', N'$-tetramethylethylenediamine)zinc(II). It crystallizes in space group $C2/c$ with four molecules per unit cell; the zinc atoms lie on

[a] ± 1 may be added to any coordinate to give a crystallographically equivalent position.

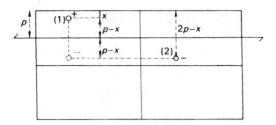

FIGURE 2.33. Operation about a 2_1 axis along the line $[p, y, 0]$: The x coordinate of point 2 relative to that of point 1 (x) is $2p - x$. A similar construction may be used for the y coordinate in the c-glide operation.

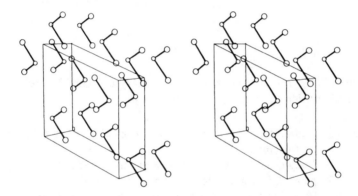

FIGURE 2.34. Stereoview of the unit cell for the structure of diiodo-$(N, N, N', N'$-tetramethylethylenediamine)zinc(II), showing the zinc and iodine (larger circles) atoms.

twofold axes. The reader should make a drawing of $C2/c$, putting in all the symmetry elements and a set of general equivalent positions, for comparison with Figure 2.34.[a]

2.7.6 Analysis of the Space-Group Symbol

In this section we consider the general interrelationship between space-group symbols and point-group symbols. On encountering a space-group symbol, the first problem is to determine the parent point group. This process has been discussed (see Section 2.7.2); here are a few more examples. It is not necessary to have

[a] S. Htoon and M. F. C. Ladd, *Journal of Crystal and Molecular Structure* **4**, 357 (1974).

explored all space groups in order to carry out this exercise:

$$P2_1/c \rightarrow (2_1/c) \rightarrow (2/c) \rightarrow 2/m$$

$$Ibca \rightarrow mmm$$

$$P4_12_12 \rightarrow 422$$

$$F\bar{4}3c \rightarrow \bar{4}3m$$

Next we must identify a crystal system for each point group:

$$2/m \rightarrow \text{monoclinic}$$

$$mmm \rightarrow \text{orthorhombic}$$

$$422 \rightarrow \text{tetragonal}$$

$$\bar{4}3m \rightarrow \text{cubic}$$

Now, from Table 1.5, we can associate certain crystallographic directions with each symmetry element in the space group symbol:

$P2_{1/c}$: Primitive, monoclinic unit cell; c-glide plane $\perp b$; 2_1 axis $\parallel b$; centrosymmetric.

$Ibca$: Body-centered, orthorhombic unit cell; b-glide plane $\perp a$; c-glide plane $\perp b$; a-glide plane $\perp c$; centrosymmetric.

$P4_12_12$: Primitive, tetragonal unit cell; 4_1 axis $\parallel c$; 2_1 axes $\parallel a$ and b; twofold axes at $45°$ to a and b, in the ab plane; noncentrosymmetric.

$F\bar{4}3c$: Face-centered, cubic unit cell; $\bar{4}$ axes $\parallel a, b$, and c; threefold axes $\parallel \langle 111 \rangle$; c-glide planes $\perp \langle 110 \rangle$; noncentrosymmetric.

It should be noted carefully that the unique symmetry elements (where there are more than two present) given in a space-group symbol may not intersect in the third, equivalent symmetry element, and the origin must always be selected with care. Appropriate procedures for the monoclinic and orthorhombic systems have been discussed; in working with higher symmetry space groups, similar rules can be drawn up.

Because of the similarities between space groups and their parent point groups, a reflection symmetry, for example, in the same orientation with respect to the crystallographic axes always produces the same changes in the *signs* of the coordinates. Thus, the m plane perpendicular to z in point group mmm changes x, y, z to $x, y, \bar{z}$. The a-glide plane in $Pnma$ changes x, y, z to $\frac{1}{2} + x, y, \frac{1}{2} - z$; the translational components of $\frac{1}{2}$ are a feature of this space group, but the signs of x, y, and z are still $+, +$, and $-$ after the operation.

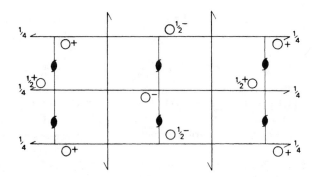

Origin halfway between three pairs of nonintersecting screw axes

Limiting conditions

4   a 1 $x, y, z;$ $\frac{1}{2} - x, \bar{y}, \frac{1}{2} + z;$ $\frac{1}{2} + x, \frac{1}{2} - y, \bar{z};$ $\bar{x}, \frac{1}{2} + y, \frac{1}{2} - z.$ $hkl:$
$0kl:$
$h0l:$ None
$hk0:$

$h00: h = 2n$
$0k0: k = 2n$
$00l: l = 2n$

Symmetry of special projections

(001) $p2gg$ (100) $p2gg$ (010) $p2gg$

FIGURE 2.35.   Space group $P2_12_12_1$: In space-group diagrams, ♪ represents a 2_1 axis normal to the plane of projection.

2.7.7   Orthorhombic Space Groups

We shall consider two orthorhombic space groups, $P2_12_12_1$ and *Pnma*. The first is illustrated in Figure 2.35; it should be noted that the three mutually perpendicular 2_1 axes do *not* intersect one another in this space group. Although $P2_12_12_1$ is a noncentrosymmetric space group, the three principal projections are centric; each has the two-dimensional space group $p2gg$.

Change of Origin. Considering the projection of $P2_12_12_1$ on to (001), we obtain from the general equivalent positions the two-dimensional set

$$x, y; \quad \frac{1}{2} - x, \bar{y}; \quad \frac{1}{2} + x, \frac{1}{2} - y; \quad \bar{x}, \frac{1}{2} + y$$

It is convenient to change the origin to a 2-fold rotation point, say at $\frac{1}{4}, 0$. To carry out this transformation, the coordinates of the new origin are subtracted from the original coordinates:

$$x - \frac{1}{4}, y; \quad \frac{1}{4} - x, \bar{y}; \quad \frac{1}{4} + x, \frac{1}{2} - y; \quad -x - \frac{1}{4}, \frac{1}{2} + y$$

Next, new variables x_0 and y_0 are chosen such that, for example, $x_0 = x - \frac{1}{4}$ and $y_0 = y$. Then, by substituting, we obtain

$$x_0, y_0; \quad \bar{x}_0, \bar{y}_0; \quad \tfrac{1}{2} + x_0, \tfrac{1}{2} - y_0; \quad \tfrac{1}{2} - x_0, \tfrac{1}{2} + y_0$$

If the subscript is dropped, these coordinates are exactly those given already for $p2gg$ (Figure 2.24d), which is the plane group of the projection of $P2_12_12_1$ on (001). This type of change of origin is useful when studying projections.

Space group $Pnma$ is shown with the origin on $\bar{1}$ (Figure 2.36). The symbol tells us that the unit cell is primitive, with an n-glide plane normal to the x axis (see Table 2.6), an m plane normal to y, and an a-glide plane normal to z. Although this space group is derived from point group mmm, we cannot assume that the three planes in $Pnma$ intersect in a center of symmetry. We are, therefore, faced with a problem similar to that discussed with $P2_1/c$. The solution of this type of problem depends upon the fact that $m \cdot m \cdot m \equiv \bar{1}$, and is illustrated fully in Problem 2.10 at the end of this chapter.

The coordinates of the general and the special equivalent positions can be derived easily from the diagram. The translational symmetry elements n and a give rise to the limiting conditions shown. Nonindependent conditions are shown in parentheses; in the orthorhombic system, all of the classes of reflection listed should be considered, as will be discussed in Chapter 3.

It is useful to remember that among the triclinic, monoclinic, and orthorhombic space groups, at least, pairs of coordinates which have one *sign* change of x, y, or z indicate a symmetry plane normal to the axis of the coordinate with the changed sign. If two sign changes exist, a symmetry axis lies parallel to the axis of the coordinate that has *not* changed sign. Three sign changes indicate a center of symmetry. In these three systems, where any coordinate, say x, is related by symmetry to another at $t - x$, the symmetry element intersects the x axis at $t/2$ (by virtue of Figure 2.33, mutatis mutandis).

2.7.8 Relative Orientations of Symmetry Elements in Space Groups

Earlier in this chapter, we looked briefly at the problem of choosing the relative positions of the symmetry elements in space groups while keeping a particular symmetry element at a given site, such as a center of symmetry at the origin in space groups of class $2/m$. We now discuss some simple rules whereby this task can be accomplished readily, with due regard to the relative orientations of the symmetry elements given by the space-group symbol itself (see Tables 1.5 and 2.5). We shall consider here the symmetry planes and symmetry axes in classes mmm and $2/m$, although the rules can be applied more widely, as we shall see.

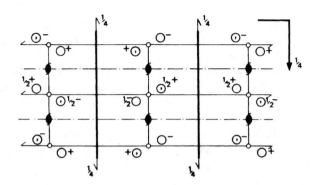

Origin at $\bar{1}$

8 d 1 $x, y, z;$ $\frac{1}{2}+x, \frac{1}{2}-y, \frac{1}{2}-z;$ $\bar{x}, \frac{1}{2}+y, \bar{z};$ $\frac{1}{2}-x, \bar{y}, \frac{1}{2}+z;$ Limiting conditions

       $\bar{x}, \bar{y}, \bar{z};$ $\frac{1}{2}-x, \frac{1}{2}+y, \frac{1}{2}+z;$ $x, \frac{1}{2}-y, z;$ $\frac{1}{2}+x, y, \frac{1}{2}-z.$ hkl: None

                                                 $0kl$: $k + l = 2n$

                                                 $h0l$: None

                                                 $hk0$: $h = 2n$

                                                 $h00$: $(h = 2n)$

                                                 $0k0$: $(k = 2n)$

                                                 $00l$: $(l = 2n)$

4 c m $x, \frac{1}{4}, z;$ $\bar{x}, \frac{3}{4}, \bar{z};$ $\frac{1}{2}-x, \frac{3}{4}, \frac{1}{2}+z;$ $\frac{1}{2}+x, \frac{1}{4}, \frac{1}{2}-z.$ As above

4 b $\bar{1}$ $0,0,\frac{1}{2};$ $0,\frac{1}{2},\frac{1}{2};$ $\frac{1}{2},0,0;$ $\frac{1}{2},\frac{1}{2},0.$ $\left.\rule{0pt}{18pt}\right\}$ As above +

4 a $\bar{1}$ $0,0,0;$ $0,\frac{1}{2},0;$ $\frac{1}{2},0,\frac{1}{2};$ $\frac{1}{2},\frac{1}{2},\frac{1}{2}.$ hkl: $h + l = 2n; k = 2n$

Symmetry of special projections

(011) $p2gm$ (100) $c2mm$ (010) $p2gg$

FIGURE 2.36.   Space group $Pnma$; the full space-group symbol is $P\frac{2_1}{n}\frac{2_1}{m}\frac{2_1}{a}$.

Half-Translation Rule

Location of Symmetry Planes.   Consider space group $Pnna$; the translations associated with the three symmetry planes are $(b + c)/2, (c + a)/2,$ and $a/2,$ respectively. If they are summed, the result T is $(a + b/2 + c)$. We disregard the *whole* translations a and c, because they refer us to neighboring unit cells; thus, T becomes $b/2$, and the center of symmetry is therefore displaced by $T/2$, or $b/4$, from the point of intersection of the three symmetry planes $n, n,$ and a. As a second example, consider $Pmma$. The only translation is $a/2$; thus, $T = a/2$, and the center of symmetry is displaced by $a/4$ from mma.

Space group $Imma$ may be formed from $Pmma$ by introducing the body-centering translation $\frac{1}{2}, \frac{1}{2}, \frac{1}{2}$ (see Figure 6.18b). Alternatively, the half-translation rule may be applied to the complete space-group symbol. In all, *Imma* contains the translations $(a+b+c)/2$ and $a/2$, and $T = a+(b+c)/2$, or $(b+c)/2$; hence, the center of symmetry is displaced by $(b+c)/4$ from *mma*. This center of symmetry is one of a second set of eight introduced, by the body-centering translation, at $\frac{1}{4}, \frac{1}{4}, \frac{1}{4}$ (half the I translation) from a *Pmma* center of symmetry. This alternative setting is given in the *International Tables for X-Ray Crystallography*;[a] it corresponds to that in Figure 6.18b with the origin shifted to the center of symmetry at $\frac{1}{4}, \frac{1}{4}, \frac{1}{4}$. Space groups in class *mmm* based on A, B, C, and F unit cells similarly introduce additional sets of centers of symmetry. The reader may care to apply these rules to space group *Pnma* and then check the result with Figure 2.36.

Type and Location of Symmetry Axes. The quantity T, reduced as above to contain half-translations only, readily gives the types of 2-fold axes parallel to a, b, and c. Thus, if T contains an $a/2$ component, then 2_x (parallel to a) $\equiv 2_1$, otherwise $2_x \equiv 2$. Similarly for 2_y and 2_z, with reference to the $b/2$ and $c/2$ components. Thus, in *Pnna*, $T = b/2$, and so $2_x \equiv 2, 2_y \equiv 2_1$, and $2_z \equiv 2$. In *Pmma*, $T = a/2$; hence, $2_x \equiv 2_1, 2_y \equiv 2$, and $2_z \equiv 2$.

The location of each 2-fold axis may be obtained from the symbol of the symmetry plane perpendicular to it, being displaced by half the corresponding glide translation (if any). Thus, in *Pnna*, we find 2 along $[x, \frac{1}{4}, \frac{1}{4}], 2_1$ along $[\frac{1}{4}, y, \frac{1}{4}]$, and another 2 along $[\frac{1}{4}, 0, z]$. In *Pmma*, 2_1 is along $[x, 0, 0]$, 2 is along $[0, y, 0]$, and another 2 is along $[\frac{1}{4}, 0, z]$. The reader may care to continue the study of *Pnma*, and then check the result, again against Figure 2.36.

In the monoclinic space groups of class $2/m$, a 2_1 axis, with a translational component of $b/2$, shifts the center of symmetry by $b/4$ with respect to the point of intersection of 2_1 with m (see Figure S6.4b). In $P2/c$, the center of symmetry is shifted by $c/4$ with respect to $2/c$, and in $P2_1/c$ the corresponding shift is $(b+c)/4$ (see Figure 2.32).

General Equivalent Positions

Once we know the positions of the symmetry elements in a space-group pattern, the coordinates of the general equivalent positions in the unit cell follow readily.

Consider again *Pmma*. From the above analysis, we may write

$\bar{1}$ at $0, 0, 0$ (choice of origin)

m_x the plane $(\frac{1}{4}, y, z)$, m_y the plane $(x, 0, z)$, a the plane $(x, y, 0)$

---

[a] See Bibliography, Chapter 1.

Taking a point x, y, z across the three symmetry planes in turn, we have (from Figure 2.33)

$$x, y, z \xrightarrow{m_x} \tfrac{1}{2} - x, y, z$$
$$\xrightarrow{m_y} x, \bar{y}, z$$
$$\xrightarrow{a} \tfrac{1}{2} + x, y, \bar{z}$$

These four points are now operated on by $\bar{1}$ to give the total of eight equivalent positions for $Pmma$:

$$\pm\{x, y, z; \quad \tfrac{1}{2} - x, y, z; \quad x, \bar{y}, z; \quad \tfrac{1}{2} + x, y, \bar{z}\}$$

The reader may now like to complete the example of $Pnma$.

A similar analysis may be carried out for the space groups in the $mm2$ class, with respect to origins on 2 or 2_1 (consider, e.g., Figure 3.32), although we have not discussed many of these space groups in this book.

2.7.9 Tetragonal and Hexagonal Space Groups

We shall examine one space group from each of the systems tetragonal and hexagonal because new features arise on account of the higher rotational symmetry in these two systems.

Tetragonal Space Group $P4nc$

It is evident that this space group is based in the point group $4mm$. Reference to Table 1.5 shows that the symbol has the following interpretation: a 4-fold axis along z; a n-glide planes normal to x (and y, because of the fourfold symmetry); c-glide planes normal to $[110]$ and its 4-fold symmetry-related direction $[1\bar{1}0]$. The orientation of the n glides can be handled in the manner already discussed. In the case of the c glide, it is straightforward to show that, if the glide plane intercepts the x and y axes at the value q, then a point x, y, z is reflected and translated to the position $q - y, q - x, \tfrac{1}{2} + z$. Thus, as in Section 2.7.5, we can set up the interpretation of the symbol $P4nc$, again using Euler's theorem (from the determination of point groups—program EULR*), that the combination of any two operations is equivalent to a third operation. Thus, $n \cdot 4 \equiv c$, but, in contradistinction to the point group $4mm$, the three operators need not pass through the origin point.

Let the symmetry elements be placed as follows:

4 along the z axis, that is, the line $[0, 0, z]$

n normal to x, being the plane $(\alpha y z)$

c normal to $[010]$, the plane $(q q z)$

$P4nc$
C^6_{4v}

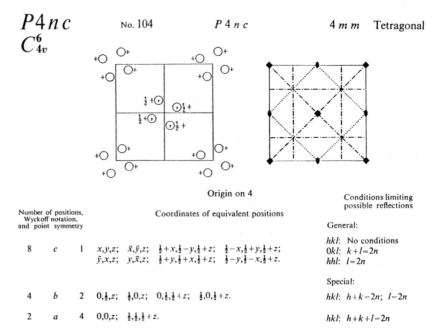

No. 104 $P\,4\,n\,c$ $4\,m\,m$ Tetragonal

Origin on 4

Number of positions, Wyckoff notation, and point symmetry			Coordinates of equivalent positions	Conditions limiting possible reflections
				General:
8	c	1	$x,y,z;\;\; \bar{x},\bar{y},z;\;\; \tfrac{1}{2}+x,\tfrac{1}{2}-y,\tfrac{1}{2}+z;\;\; \tfrac{1}{2}-x,\tfrac{1}{2}+y,\tfrac{1}{2}+z;$ $\bar{y},x,z;\;\; y,\bar{x},z;\;\; \tfrac{1}{2}+y,\tfrac{1}{2}+x,\tfrac{1}{2}+z;\;\; \tfrac{1}{2}-y,\tfrac{1}{2}-x,\tfrac{1}{2}+z.$	hkl: No conditions $0kl$: $k+l=2n$ hhl: $l=2n$
				Special:
4	b	2	$0,\tfrac{1}{2},z;\;\; \tfrac{1}{2},0,z;\;\; 0,\tfrac{1}{2},\tfrac{1}{2}+z;\;\; \tfrac{1}{2},0,\tfrac{1}{2}+z.$	hkl: $h+k=2n;\;\; l=2n$
2	a	4	$0,0,z;\;\; \tfrac{1}{2},\tfrac{1}{2},\tfrac{1}{2}+z.$	hkl: $h+k+l=2n$

FIGURE 2.37. Diagrams to show the general equivalent positions and symmetry elements for the tetragonal space group $P4nc$. [Reproduced from *International Tables for X- Ray Crystallography*, Volume I (Kynoch Press, Birmingham, England, 1965) by permission of the International Union of Crystallography.]

A point x, y, z (1) rotated clockwise about 4 becomes $y, \bar{x}, z$ (2); this point is taken across the n glide to $2\alpha - y, \tfrac{1}{2} - x, \tfrac{1}{2} + z$ (3). If we now operate on the original point (1) by the c glide, then x, y, z is reflected to $q - y, q - x, \tfrac{1}{2} + z$ (4). Now, points (3) and (4) are one and the same, so that $q = \tfrac{1}{2}$ and $\alpha = \tfrac{1}{4}$. This setting of the symmetry elements gives rise to the standard diagram for $P4nc$, shown in Figure 2.37. The positions of the additional symmetry elements, not apparent from the symbol, should again be noted. The diagram of the unit cell and its environs is complete, because any point shown can be reached from any other point on the diagram by a single symmetry operation. The reader may like to consider how the above argument is modified if a symmetry-equivalent n-glide at (x, β, z) is used.

Hexagonal Space Group $P6_3/m$

In this space group we encounter 6-fold (and 3-fold) rotation operations. In Appendix 4, we show that a point x, y, z rotated anticlockwise about a 6-fold axis along z is moved first to the position $x - y, x, \frac{1}{2} + z$. The translation of $\frac{1}{2}$ accompanying the z coordinate arises from that associated with the 6_3 axis, namely, a translation of 3/6, or 1/2. The sequence of points obtained by the successive operations of 6_3 along $[00*1]$ are:

$$x, y, z; \quad x - y, x, \tfrac{1}{2} + z; \quad \bar{y}, x - y, z; \quad \bar{x}, \bar{y}, \tfrac{1}{2} + z; \quad y - x, \bar{x}, z; \quad y, y - x, \tfrac{1}{2} + z$$

(1)   (2)   (3)   (4)   (5)   (6)

Points (1) and (3) are related by a 3-fold rotation ($3 \equiv 6_3^2$, that is, two successive operations of 6_3), whereas points (1) and (4) are related by 2_1 symmetry. The space group is completed by introducing the m plane at $z = \frac{1}{4}$: this position ensures that the center of symmetry is at the origin; actually the symmetry at the origin is $\bar{3}$ ($\bar{1}$ is a subgroup of $\bar{3}$). Other important symmetry elements now in evidence include $\bar{6}, 3,$ and $\bar{1}$.

Figure 2.38 illustrates space group $P6_3/m$. The twelve general equivalent positions, comprise the six listed above and another six obtained by inverting across the center of symmetry at the origin; all coordinates change sign. Consider point (2) reflected across the m plane to $x - y, x, \frac{1}{2} - z$. How may this point be reached from x, y, z in a single operation? Either a *clockwise* $\bar{3}$ operation or an anticlockwise $\bar{3}^2$ (equivalent to two successive anticlockwise $\bar{3}$ operations) operation relates these two points; we note in passing that both $\bar{3}$ and $\bar{3}^2$ are symmetry *operators* in this group, related to the single symmetry *element* $\bar{3}$.

2.8 Matrix Representation of Symmetry Operations

The representation of symmetry operations by matrices has a certain inherent elegance, and is useful for displaying the close relationship between point groups and space groups. In this discussion, we shall use the triplet x, y, z to represent a point in three-dimensional space. It could lie on the normal to the face of a crystal or be an atom in a crystal structure, and we can indicate it concisely by the vector $\mathbf{x}$.

A symmetry operation may be written as

$$\mathbf{R} \cdot \mathbf{x} + \mathbf{t} = \mathbf{x}' \tag{2.52}$$

where $\mathbf{x}$ and $\mathbf{x}'$ are the triplets before and after the operation, $\mathbf{R}$ is a matrix representing the symmetry operation, and $\mathbf{t}$ is a translation vector with components parallel to $x, y,$ and z.

2.8.1 Matrices in Point-Group Symmetry

From the definition of point group (see Section 1.4), it follows that $\mathbf{t}$ is identically zero. This condition is achieved as long as all symmetry elements pass

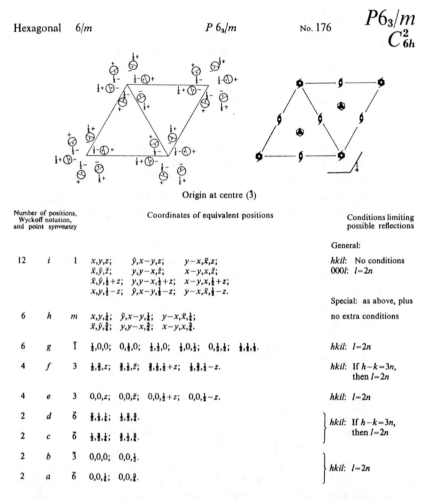

Hexagonal   6/m   P 6₃/m   No. 176 $P6_3/m$
C_{6h}^2

Origin at centre (3̄)

Number of positions, Wyckoff notation, and point symmetry			Coordinates of equivalent positions	Conditions limiting possible reflections
				General:
12	i	1	x,y,z; $\bar{y},x-y,z$; $y-x,\bar{x},z$; $\bar{x},\bar{y},\bar{z}$; $y,y-x,\bar{z}$; $x-y,x,\bar{z}$; $\bar{x},\bar{y},\tfrac{1}{2}+z$; $y,y-x,\tfrac{1}{2}+z$; $x-y,x,\tfrac{1}{2}+z$; $x,y,\tfrac{1}{2}-z$; $\bar{y},x-y,\tfrac{1}{2}-z$; $y-x,\bar{x},\tfrac{1}{2}-z$.	hkil: No conditions 000l: l=2n
				Special: as above, plus
6	h	m	$x,y,\tfrac{1}{4}$; $\bar{y},x-y,\tfrac{1}{4}$; $y-x,\bar{x},\tfrac{1}{4}$; $\bar{x},\bar{y},\tfrac{3}{4}$; $y,y-x,\tfrac{3}{4}$; $x-y,x,\tfrac{3}{4}$.	no extra conditions
6	g	1̄	$\tfrac{1}{2},0,0$; $0,\tfrac{1}{2},0$; $\tfrac{1}{2},\tfrac{1}{2},0$; $\tfrac{1}{2},0,\tfrac{1}{2}$; $0,\tfrac{1}{2},\tfrac{1}{2}$; $\tfrac{1}{2},\tfrac{1}{2},\tfrac{1}{2}$.	hkil: l=2n
4	f	3	$\tfrac{1}{3},\tfrac{2}{3},z$; $\tfrac{2}{3},\tfrac{1}{3},\bar{z}$; $\tfrac{2}{3},\tfrac{1}{3},\tfrac{1}{2}+z$; $\tfrac{1}{3},\tfrac{2}{3},\tfrac{1}{2}-z$.	hkil: If h−k=3n, then l=2n
4	e	3	$0,0,z$; $0,0,\bar{z}$; $0,0,\tfrac{1}{2}+z$; $0,0,\tfrac{1}{2}-z$.	hkil: l=2n
2	d	6̄	$\tfrac{1}{3},\tfrac{2}{3},\tfrac{1}{4}$; $\tfrac{2}{3},\tfrac{1}{3},\tfrac{3}{4}$.	hkil: If h−k=3n, then l=2n
2	c	6̄	$\tfrac{1}{3},\tfrac{2}{3},\tfrac{3}{4}$; $\tfrac{2}{3},\tfrac{1}{3},\tfrac{1}{4}$.	
2	b	3̄	$0,0,0$; $0,0,\tfrac{1}{2}$.	hkil: l=2n
2	a	6̄	$0,0,\tfrac{1}{4}$; $0,0,\tfrac{3}{4}$.	

FIGURE 2.38.   Diagrams to show the general equivalent positions and symmetry elements for the hexagonal space group $P6_3/m$. [Reproduced from *International Tables for X- Ray Crystallography*, Volume I (Kynoch Press, Birmingham, England, 1965) by permission of the International Union of Crystallography.]

through a single point, the origin. If it were not the case, then two 2-fold axes, for example, could be parallel. The consequence of this arrangement is shown in Figure 2.39. Thus, for point groups, (2.52) reduces to

$$\mathbf{R} \cdot \mathbf{x} = \mathbf{x}'$$ (2.53)

Let $\mathbf{R}_1$ represent an m plane perpendicular to the x axis, as in the orthorhombic system, for example. Then, we have

$$
\begin{bmatrix} \bar{1} & 0 & 0 \\ 0 & 1 & 0 \\ 0 & 0 & 1 \end{bmatrix} \cdot \begin{bmatrix} x \\ y \\ z \end{bmatrix} = \begin{bmatrix} \bar{x} \\ y \\ z \end{bmatrix} \qquad (2.54)
$$

$$
\underset{\mathbf{R}_1}{} \qquad \underset{\mathbf{x}}{} \qquad \underset{\mathbf{x}'}{}
$$

The multiplication is carried out, as usual, *along the row* and *down the column*, with the result at the intersection $*$; that is,

$$
\begin{bmatrix} \cdots\cdots\rightarrow \\ \\ \end{bmatrix} \cdot \begin{bmatrix} \\ \downarrow \\ \end{bmatrix} = \begin{bmatrix} * \\ \\ \end{bmatrix}
$$

$$
\underset{\mathbf{R}_1}{} \qquad \underset{\mathbf{x}}{} \qquad \underset{\mathbf{x}'}{} \qquad (2.55)
$$

$$
\bar{x} = -1 \times x + 0 \times y + 0 \times z \qquad (2.56)
$$

and similarly for y and z.

Let the triplet $\mathbf{x}'$ now suffer reflection across a mirror plane normal to y, using matrix $\mathbf{R}_2$:

$$
\begin{bmatrix} 1 & 0 & 0 \\ 0 & \bar{1} & 0 \\ 0 & 0 & 1 \end{bmatrix} \cdot \begin{bmatrix} \bar{x} \\ y \\ z \end{bmatrix} = \begin{bmatrix} \bar{x} \\ \bar{y} \\ z \end{bmatrix} \qquad (2.57)
$$

$$
\underset{\mathbf{R}_2}{} \qquad \underset{\mathbf{x}'}{} \qquad \underset{\mathbf{x}''}{}
$$

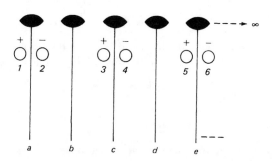

FIGURE 2.39.   Points 1 and 2 rotated about axis b produce points 4 and 3. But 3 and 4 are now related by another diad, c. The effect of diad c on points 1 and 2 is to produce points 6 and 5. But these points are related to 3 and 4 by diad d and to each other by diad e. Now 3 and 4, for example, can be rotated about e, and so on. Clearly, this process would lead to an infinite number of parallel, equidistant diad axes, together with the symmetry-related points, which is totally incompatible with a point group.

It should be clear that the relationship between $\mathbf{x}$ and $\mathbf{x}''$ is that of a 2-fold rotation about the z axis. Thus, for the two m planes,

$$m \cdot m = 2 \tag{2.58}$$

as we have seen already (Section 1.4.2).

Another way of reaching the same final result is first to combine the two matrices $\mathbf{R}_1$ and $\mathbf{R}_2$,

$$\begin{bmatrix} 1 & 0 & 0 \\ 0 & \bar{1} & 0 \\ 0 & 0 & 1 \end{bmatrix} \cdot \begin{bmatrix} \bar{1} & 0 & 0 \\ 0 & 1 & 0 \\ 0 & 0 & 1 \end{bmatrix} = \begin{bmatrix} \bar{1} & 0 & 0 \\ 0 & \bar{1} & 0 \\ 0 & 0 & 1 \end{bmatrix} \tag{2.59}$$
$$\qquad \mathbf{R}_2 \qquad\qquad \mathbf{R}_1 \qquad\qquad \mathbf{R}_3$$

and then to use the right-hand side of (2.59) in (2.53):

$$\begin{bmatrix} \bar{1} & 0 & 0 \\ 0 & \bar{1} & 0 \\ 0 & 0 & 1 \end{bmatrix} \cdot \begin{bmatrix} x \\ y \\ z \end{bmatrix} = \begin{bmatrix} \bar{x} \\ \bar{y} \\ z \end{bmatrix} \tag{2.60}$$
$$\qquad \mathbf{R}_3 \qquad\quad \mathbf{x} \qquad \mathbf{x}''$$

Equation (2.59) corresponds to the order of operation $\mathbf{R}_1$ first, followed by $\mathbf{R}_2$. For $\mathbf{R} \leqslant 2$ this order of multiplication need not be followed, but it is good practice to multiply the matrices in the standard manner; we can highlight this feature by considering point group $4mm$. The matrices for a 4-fold rotation along the z axis and an m plane perpendicular to x are, in order,

$$\overset{m \perp x}{\begin{bmatrix} \bar{1} & 0 & 0 \\ 0 & 1 & 0 \\ 0 & 0 & 1 \end{bmatrix}} \cdot \overset{4 \text{ along } z}{\begin{bmatrix} 0 & \bar{1} & 0 \\ 1 & 0 & 0 \\ 0 & 0 & 1 \end{bmatrix}} = \begin{bmatrix} 0 & 1 & 0 \\ 1 & 0 & 0 \\ 0 & 0 & 1 \end{bmatrix} \tag{2.61}$$
$$\qquad \mathbf{R}_2 \qquad\qquad\quad \mathbf{R}_1 \qquad\qquad\quad \mathbf{R}_3$$

Hence, $\mathbf{x}(x, y, z)$ operated on first by $\mathbf{R}_1$ and then by $\mathbf{R}_2$ becomes $\mathbf{x}'(y, x, z)$, and $\mathbf{R}_3$ represents an m plane normal to $[1\bar{1}0]$. Multiplying in the reverse order, that is,

$$\mathbf{R}_1 \cdot \mathbf{R}_2 \cdot \mathbf{x} = \mathbf{x}'' \tag{2.62}$$

whence

$$\mathbf{R}_1 \cdot \mathbf{R}_2 = \mathbf{R}_4 \tag{2.63}$$

so that $\mathbf{x}''$ is, $\bar{y}$, $\bar{x}$, z, and $\mathbf{R}_4$ is a matrix representing an m plane normal to $[110]$. (What is this matrix?) The m planes represented by $\mathbf{R}_3$ and $\mathbf{R}_4$ are equivalent under the symmetries $\mathbf{R}_1$, or $\mathbf{R}_2$, but lead to physically different sites. Thus, if we are expecting $\mathbf{x}'$ from $\mathbf{x}$ and obtain instead $\mathbf{x}''$, it may be confusing and, in considering some physical properties, could be significantly different. All other point groups may be considered in the manner just described.

2.8.2   Matrices in Space-Group Symmetry

In space-group symmetry, $\mathbf{t}$ (2.52) is generally not equal to zero. This situation will exist whenever the space group contains translational symmetry. We will consider first space group $P2_1/c$ (see Section 2.7.5). As before, we take the origin on $\bar{1}$ ($\mathbf{R}_3$), 2_1($\mathbf{R}_1$) along $[p, y, r]$, and c ($\mathbf{R}_3$) the plane (x, q, z). The operation $\mathbf{R}_1$ followed by $\mathbf{R}_2$ is, from our previous discussion,

$$
\underbrace{\begin{bmatrix} 1 & 0 & 0 \\ 0 & \bar{1} & 0 \\ 0 & 0 & 1 \end{bmatrix}}_{\mathbf{R}_2} + \underbrace{\begin{bmatrix} 0 \\ 2q \\ \frac{1}{2} \end{bmatrix}}_{\mathbf{t}_2} \cdot \underbrace{\begin{bmatrix} \bar{1} & 0 & 0 \\ 0 & 1 & 0 \\ 0 & 0 & \bar{1} \end{bmatrix}}_{\mathbf{R}_1} + \underbrace{\begin{bmatrix} 2p \\ \frac{1}{2} \\ 2r \end{bmatrix}}_{\mathbf{t}_1} = \underbrace{\begin{bmatrix} \bar{1} & 0 & 0 \\ 0 & \bar{1} & 0 \\ 0 & 0 & \bar{1} \end{bmatrix}}_{\mathbf{R}_3} + \underbrace{\begin{bmatrix} 0 \\ 0 \\ 0 \end{bmatrix}}_{\mathbf{t}_3} \tag{2.64}
$$

Matrix $\mathbf{R}_1$ is just that for 2-fold rotation about the y axis,[a] and $\mathbf{R}_2$ is the matrix for an m plane normal to the y axis[a] (compare $\mathbf{R}_2$ in (2.59)). The translation vectors $\mathbf{t}_1$ and $\mathbf{t}_2$ are obtained by following the arguments relating to Figure 2.33. Matrix $\mathbf{R}_3$ is the multiplication $\mathbf{R}_2 \cdot \mathbf{R}_1$ and, clearly, is equivalent to a center of symmetry ($\bar{1}$) at the origin. Since, by definition of the standard origin, $\mathbf{t}_3$ must be zero, we have

$$
\mathbf{t}_2 + \mathbf{t}_1 = \mathbf{t}_3 = 0 \tag{2.65}
$$

It follows that $p = 0$, $q = \frac{1}{4}$ and $r = \frac{1}{4}$, as before. These results may be regarded as a matrix justification of the scheme used in Section 2.7.5, and expressed in the half-translation rule (see Section 2.7.8).

As a final example, we shall consider space group $Pnma$ (see Section 2.7.7). From the symbol, we can write

$$
\begin{aligned}
&\mathbf{R}_1 \quad n \text{ is } (p, y, z) \quad n\text{-translation } 0, \tfrac{1}{2}, \tfrac{1}{2} \\
&\mathbf{R}_2 \quad m \text{ is } (x, q, z) \quad m \text{ (no translation)} \\
&\mathbf{R}_3 \quad a \text{ is } (x, y, r) \quad a\text{-translation } \tfrac{1}{2}, 0, 0 \\
&\mathbf{R}_4 \quad \bar{1} \text{ is } 0, 0, 0 \quad\ \text{no translation}
\end{aligned}
$$

We know that, in space groups of the mmm class,

$$
\mathbf{R}_3 \cdot \mathbf{R}_2 \cdot \mathbf{R}_1 = \mathbf{R}_4 \tag{2.66}
$$

---

[a] y normal to the x, z plane.

Hence,

$$
\begin{bmatrix} \bar{1} & 0 & 0 \\ 0 & 1 & 0 \\ 0 & 0 & 1 \end{bmatrix} + \begin{bmatrix} \frac{1}{2} \\ 0 \\ 2r \end{bmatrix} \cdot \begin{bmatrix} 1 & 0 & 0 \\ 0 & \bar{1} & 0 \\ 0 & 0 & 1 \end{bmatrix} + \begin{bmatrix} 0 \\ 2q \\ 0 \end{bmatrix} \cdot \begin{bmatrix} 1 & 0 & 0 \\ 0 & 1 & 0 \\ 0 & 0 & \bar{1} \end{bmatrix} + \begin{bmatrix} 2p \\ \frac{1}{2} \\ \frac{1}{2} \end{bmatrix}
$$
$$
\underset{\mathbf{R}_3}{} \quad \underset{\mathbf{t}_3}{} \quad \underset{\mathbf{R}_2}{} \quad \underset{\mathbf{t}_2}{} \quad \underset{\mathbf{R}_1}{} \quad \underset{\mathbf{t}_1}{}
$$

(2.67)

$$
= \begin{bmatrix} \bar{1} & 0 & 0 \\ 0 & \bar{1} & 0 \\ 0 & 0 & \bar{1} \end{bmatrix} + \begin{bmatrix} 0 \\ 0 \\ 0 \end{bmatrix}
$$
$$
\underset{\mathbf{R}_4}{} \qquad \underset{\mathbf{t}_4}{}
$$

And we have

$$\mathbf{t}_3 + \mathbf{t}_2 + \mathbf{t}_1 = \mathbf{t}_4 = 0 \tag{2.68}$$

(multiplying the matrices and adding the translation vectors), so that $p = \frac{1}{4}, q = \frac{1}{4}$, and $r = \frac{1}{4}$ as given in Figure 2.36. The full symbol of point group mmm is $\frac{2}{m}\frac{2}{m}\frac{2}{m}$, so that in $Pnma$ there are 2 or 2_1 axes normal to the symmetry planes. We can obtain the results readily from (2.67), inserting the values of p, q, and r into the translation vectors; if the fraction $\frac{1}{2}$ appears in line with the x coordinate in a plane normal to x, then the axis is 2_1, and similarly for the y and z positions. Hence, the full symbol for this space group is $\frac{P2_1 \, 2_1 \, 2_1}{n \quad m \quad a}$. The same result could be achieved with the scheme used for solving Problem 2.10, but with less elegance.

The essential difference between point groups and space groups has been shown to rest in the translation vectors, and the infinite space to which they refer. Symmorphic space groups such as $Pm, C2/m$, and $Imm2$, some of which contain translational symmetry elements, do not need any special treatment to determine the orientation of the symmetry elements with respect to the origin. Since they contain the point-group symbol, the origin is given immediately, for example, on m in Pm, at $2/m$ ($\bar{1}$) in $C2/m$, and along $mm2$ in $Imm2$. The half-translation rule, once understood, is the simplest method of locating the origin, certainly for the space groups in the monoclinic and orthorhombic systems, which represent the majority of known crystals.

2.9 Diffraction Symbols

We look ahead briefly to the results in later chapters, and note that after a crystal has been examined to the extent that indices can be assigned to the x-ray diffraction spectra, the totality of the diffraction information can be assembled into a *diffraction symbol*. This parameter includes the Laue group and the symmetry determined through the systematic absences.

TABLE 2.7.   Orthorhombic Space Group Diffraction Symbols

Diffraction symbol				222	$mm2$	mmm
$mmm\,P$	.	.	.	$P222$	$Pmm2$	$Pmmm$
$mmm\,P$	.	.	2_1	$P222_1$		
$mmm\,P$	2_1	2_1	.	$P2_12_12$		
$mmm\,P$	2_1	2_1	2_1	$P2_12_12_1$		
$mmm\,P$	c	.	.		$Pc2m = \begin{pmatrix} Pma2 \\ Pmc2_1 \end{pmatrix}$ $Pcm2_1 =$	$Pcmm = Pmma$
$mmm\,P$	n	.	.		$Pnm2_1 = Pmn2_1$	$Pnmm = Pmmn$
$mmm\,P$	c	c	.		$Pcc2$	$Pccm$
$mmm\,P$	c	a	.		$Pca2_1$	$Pcam = Pbcm$
$mmm\,P$	b	a	.		$Pba2$	$Pbam$
$mmm\,P$	n	c	.		$Pnc2$	$Pncm = Pmna$
$mmm\,P$	n	a	.		$Pna2_1$	$Pnam = Pnma$
$mmm\,P$	n	n	.		$Pnn2$	$Pnnm$
$mmm\,P$	c	c	a			$\textbf{Pcca}$
$mmm\,P$	b	c	a			$\textbf{Pbca}$
$mmm\,P$	c	c	n			$\textbf{Pccn}$
$mmm\,P$	b	a	n			$\textbf{Pban}$
$mmm\,P$	b	c	n			$\textbf{Pbcn}$
$mmm\,P$	n	n	a			$\textbf{Pnna}$
$mmm\,P$	n	n	n			$\textbf{Pnnn}$
$mmm\,C$	.	.	.	$C222$	$Cmm2 = \begin{pmatrix} Cmm2 \\ Amm2 \end{pmatrix}$ $Cm2m =$	$Cmmm$
$mmm\,C$	.	.	2_1	$C222_1$		
$mmm\,C$	.	c	.		$Cmc2_1 = \begin{pmatrix} Cmc2_1 \\ Ama2 \end{pmatrix}$ $C2cm =$	$Cmcm$
$mmm\,C$	.	.	a		$C2ma = Abm2$	$Cmma$
$mmm\,C$	.	c	a		$C2ca = Aba2$	$Cmca$
$mmm\,C$	c	c	.		$Ccc2$	$Cccm$
$mmm\,C$	c	c	a			$\textbf{Ccca}$
$mmm\,I$	.	.	.	$\begin{bmatrix} I222 \\ I2_12_12_1 \end{bmatrix}$	$Imm2$	$Immm$
$mmm\,I$	.	a	.		$Ima2$	$Imam = Imma$
$mmm\,I$	b	a	.		$Iba2$	$Ibam$
$mmm\,I$	b	c	a			$\textbf{Ibca}$
$mmm\,F$	.	.	.	$F222$	$Fmm2$	$Fmmm$
$mmm\,F$	d	d	.		$\textbf{Fdd2}$	
$mmm\,F$	d	d	d			$\textbf{Fddd}$

In Table 2.7, we list the diffraction symbols for the orthorhombic space groups, and the following particular features of this table should be noted.

1. Space groups listed in bold type are determined uniquely when the Laue group is known. Examples are $\textbf{P222}_1$ or $\textbf{Pbcn}$; otherwise the space groups are shown in normal italic type, such as $Pmm2$ or $Pbam$.

TABLE 2.8.   The 230 Three-dimensional Space Groups Arranged by
Crystal Class and System

System/class	Space group/s
Triclinic	
1	$P1$
$\bar{1}$	$P\bar{1}$
Monoclinic	
2	$P2, P2_1, C2$
m	Pm, Pc, Cm, Cc
$2/m$	$P2/m, P2_1/m, C2/m, P2/c, \mathbf{P2_1/c}, C2/c$
Orthorhombic	
222	$P222, \mathbf{P222_1}, \mathbf{P2_12_12}, \mathbf{P2_12_12_1}, \mathbf{C222_1}, C222, F222, [I222, I2_12_12_1]$
$mm2$	$Pmm2, (Pmc2_1, Pma2), Pcc2, Pca2_1, Pnc2, Pmn2_1, Pba2, Pna2_1, Pnn2,$ $(Cmm2, Amm2), (Cmc2_1, Ama2), Ccc2, Abm2, Aba2, Fmm2, \mathbf{Fdd2}, Imm2,$ $Iba2, Ima2$
mmm	$Pmmm, \mathbf{Pnnn}, Pccm, \mathbf{Pban}, Pmma, \mathbf{Pnna}, Pmna, \mathbf{Pcca}, Pbam, \mathbf{Pccn}, Pbcm, Pnnm,$ $Pmmn, \mathbf{Pbcn}, \mathbf{Pbca}, Pnma, Cmcm, Cmca, Cmmm, Cccm, Cmma, \mathbf{Ccca}, Fmmm,$ $\mathbf{Fddd}, Immm, Ibam, \mathbf{Ibca}, Imma$
Tetragonal	
4	$P4, \{\mathbf{P4_1, P4_3}\}, P4_2, I4, I4_1,$
$\bar{4}$	$P\bar{4}, I\bar{4}$
$4/m$	$P4/m, P4_2/m, \mathbf{P4/n}, \mathbf{P4_2/n}, I4/m, \mathbf{I4_1/a}$
422	$P422, P42_12, \{\mathbf{P4_122, P4_322}\}, \{\mathbf{P4_12_12, P4_32_12}\}, \mathbf{P4_222}, \mathbf{P4_22_12}, I422, \mathbf{I4_122},$
$4mm$	$P4mm, P4bm, P4_2cm, P4_2nm, P4cc, P4nc, P4_2mc, P4_2bc, I4mm, I4cm, I4_1md,$ $\mathbf{I4_1cd}$
$\bar{4}2m$	$(P\bar{4}2m, P\bar{4}m2), P\bar{4}2c, P\bar{4}2_1m, \mathbf{P\bar{4}2_1c}, P\bar{4}c2, P\bar{4}b2, P\bar{4}n2, (I\bar{4}m2, I\bar{4}2m), I\bar{4}c2, I\bar{4}2d$
$4/mmm$	$P4/mmm, P4/mcc, \mathbf{P4/nbm}, \mathbf{P4/nnc}, P4/mbm, P4/mnc, \mathbf{P4/nmm}, \mathbf{P4/ncc},$ $P4_2/mmc, P4_2/mcm, \mathbf{P4_2/nbc}, \mathbf{P4_2/nnm}, P4_2/mbc, P4_2/mnm, \mathbf{P4_2/nmc},$ $\mathbf{P4_2/ncm}, I4/mmm, I4/mcm, \mathbf{I4_1/amd}, \mathbf{I4_1/acd}$
Trigonal	
3	$P3, \{\mathbf{P3_1, P3_2}\}, R3$
$\bar{3}$	$P\bar{3}, R\bar{3}$
32	$P312, P321, (\{\mathbf{P3_112, P3_212}\}, \{\mathbf{P3_121, P3_221}\}), R32$
$3m$	$P3m1, P31m, P3c1, P31c, R3m, R3c$
$\bar{3}m$	$P\bar{3}1m, P\bar{3}1c, P\bar{3}m1, P\bar{3}c1, R\bar{3}m, R\bar{3}c$
Hexagonal	
6	$P6, \{\mathbf{P6_1, P6_5}\}, \{\mathbf{P6_2, P6_4}\}, P6_3$
$\bar{6}$	$P\bar{6}$
$6/m$	$P6/m, P6_3/m$
622	$P622, \{\mathbf{P6_122, P6_522}\}, \{\mathbf{P6_222, P6_422}\}, \mathbf{P6_322}$
$6mm$	$P6mm, P6cc, P6_3cm, P6_3mc$
$\bar{6}m2$	$P\bar{6}m2, P\bar{6}c2, P\bar{6}2m, P\bar{6}2c$
$6/mmm$	$P6/mmm, P6/mcc, P6_3/mcm, P6_3/mmc$
Cubic	
23	$P23, F23, \mathbf{P2_13}, [I23, I2_13]$
$m3$	$Pm3, \mathbf{Pn3}, Fm3, Fd3, Im3, \mathbf{Pa3}, Ia3$
432	$P432, \mathbf{P4_232}, F432, \mathbf{F4_132}, \{\mathbf{P4_132, P4_332}\}, I432, \mathbf{I4_132},$
$\bar{4}3m$	$P\bar{4}3m, F\bar{4}3m, I\bar{4}3m, P\bar{4}3n, F\bar{4}3c, \mathbf{I\bar{4}3d}$
$m3m$	$Pm3m, \mathbf{Pn3n}, Pm3n, \mathbf{Pn3m}, Fm3m, Fm3c, \mathbf{Fd3m}, \mathbf{Fd3c}, Im3m, \mathbf{Ia3d}$

Notes:
1. Space groups that are uniquely determinable from the diffraction symbol are shown in boldface italic: **Pbca**.
2. Enantiomorphous pairs that are uniquely determinable, as a pair, from the diffraction symbol are shown in boldface italic within braces: $\{\mathbf{P4_1}, \mathbf{P4_3}\}$. The space groups in any such pair can be distinguished by x-ray structure analysis; see, for example, Section 10.4.7ff.
3. Space groups that are not uniquely determinable from the diffraction symbol are shown in italic type: $Iba2$.
4. Special pairs are enclosed in brackets: $[I222, I2_12_12_1]$. The symbol 2 is used in the groups in which the three 2-fold axes intersect in a point.
5. Groups that can be determined only when the orientation of the point group and its symbol are known are enclosed in parentheses: $(Pma2, Pmc2_1)$.

2. Space groups contained within parentheses may be distinguished once the orientation of the point group is known. For example, the diffraction symbol *mmm P c . .* could relate to $Pc2m$, which is the **bca** setting of $Pma2$. Alternatively, it could relate to $Pcm2_1$, which is the **bac̄** setting of $Pmc2_1$. Note that nonstandard settings of space groups are listed in small italic type, as with $Pc2m$ and $Pcm2_1$.

3. Space groups enclosed between brackets, such as $I222$ and $I2_12_12_1$, cannot be distinguished by the diffraction data, even if the point group is known. In the case shown in Table 2.7, both space groups contain sets of mutually perpendicular 2- and 2_1-axes. The group with mutually intersecting 2-axes and mutually intersecting 2_1-axes, is designated $I222$.

A fuller discussion of diffraction symbols may be found in the *International Tables for X-Ray Crystallography*, Volume A (or Volume I). A list of the 230 space groups, highlighting those that may be determined from their diffractions symbols, is presented in Table 2.8.

Bibliography

Lattices and Space Groups

BURNS, G., and GLAZER, A. M., *Space Groups for Solid State Scientists*, Academic Press (1978).

HAHN, T. (Editor), *International Tables for Crystallography*, Vol. A, 5th ed., Kluwer Academic (2002).

HENRY, N. F. M., and LONSDALE, K. (Editors), *International Tables for X-Ray Crystallography*, Vol. I, Birmingham, Kynoch Press (1965).

LADD, M. F. C., *Symmetry in Molecules and Crystals*, Chichester, Ellis Horwood (1989).

LADD, M. F. C., *Symmetry and Group Theory in Chemistry*, Chichester, Horwood Publishing (1998).

LADD, MARK, *Crystal Structures: Lattices and Solids in Stereoview*, Chichester, Horwood Publishing (1999).

SHMUELI, U. (Editor), *International Tables for X-Ray Crystallography*, Vol. B, 2nd ed., Kluwer Academic (2001).

Problems

2.1. Figure P2.1a shows the molecule of cyclosporin H repeated by translations in two dimensions. In Figure P2.1b, the molecules are related also by 2-fold rotation operations, while still subjected to the same translations as in Figure P2.1a. Four parallelogram-shaped, adjacent repeat units of pattern from an ideally infinite array are shown in each diagram. Convince yourself that Figure P2.1a is formed by repeating a single molecule by the (lattice) translations shown, and that Figure P2.1b follows from it by the addition of a single 2-fold operation acting at any parallelogram corner. Furthermore, for Figure P2.1b state in words

(i) the locations of all 2-fold operators belonging to a single parallelogram unit, and
(ii) how many of these 2-fold operators are unique to a single parallelogram unit.

2.2. Two nets are described by the unit cells (i) $a = b$, $\gamma = 90°$ and (ii) $a = b, \gamma = 120°$. In each case (a) what is the symmetry at each net point, (b) to which two-dimensional system does the net belong, and (c) what are the results of centering the unit cell?

2.3. A monoclinic F unit cell has the dimensions $a = 6.000$, $b = 7.000$, $c = 8.000$ Å, and $\beta = 110.0°$. Show that an equivalent monoclinic C unit cell, with an *obtuse* β angle, can represent the same lattice, and calculate its dimensions. What is the ratio of the volume of the C cell to that of the F cell?

2.4. Carry out the following exercises with drawings of a tetragonal P unit cell.

(a) Center the B faces. Comment on the result.
(b) Center the A and B faces. Comment on the result.
(c) Center all faces. What conclusions can you draw now?

2.5. Calculate the length of $[31\bar{2}]$ (see Section 2.2ff) for both unit cells in Problem 2.2.

2.6. The relationships $a \mathrel{\mathcal{C}} b \mathrel{\mathcal{C}} c$, $\alpha \mathrel{\mathcal{C}} \beta \mathrel{\mathcal{C}} 90°, 120°$, and $\gamma = 90°$ may be said to define a diclinic system. Is this an eighth system? Give reasons for your answer.

2.7. (a) Draw a diagram to show the symmetry elements and general equivalent positions in $c2mm$ (origin on $2mm$). Write the coordinates and point symmetry of the general and special positions, in their correct sets, and give the conditions limiting x-ray reflection in this plane group. (b) Draw a diagram of the symmetry elements in plane group $p2mg$ (origin on 2); take care not to put the 2-fold point at the intersection of m and g (why?). On the diagram, insert each of the motifs P, V, and Z in turn, each letter drawn in its most symmetrical manner, using the *minimum* number of motifs consistent with the space-group symmetry.

2.8. (a) Continue the study of space group $P2_1/c$ (Section 2.7.5). Write the coordinates of the general and special positions, in their correct sets. Give the limiting conditions for all sets of positions, and write the plane-group symbols for the three principal projections. Draw a diagram of the space group

(a)

(b)

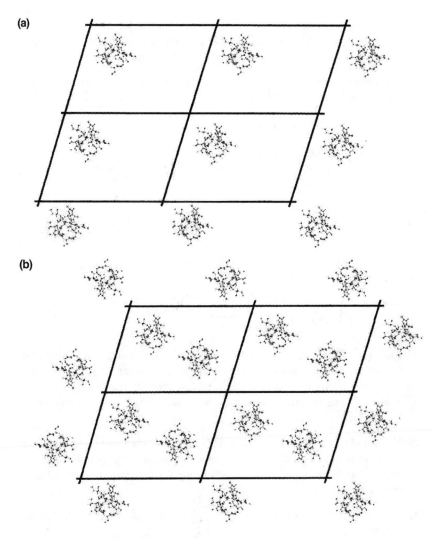

FIGURE P2.1.   (a) The molecule of cyclosporin H repeated by translations in two dimensions. (b) The molecules are related also by 2-fold rotation operations.

as seen along the b axis. (b) Biphenyl, 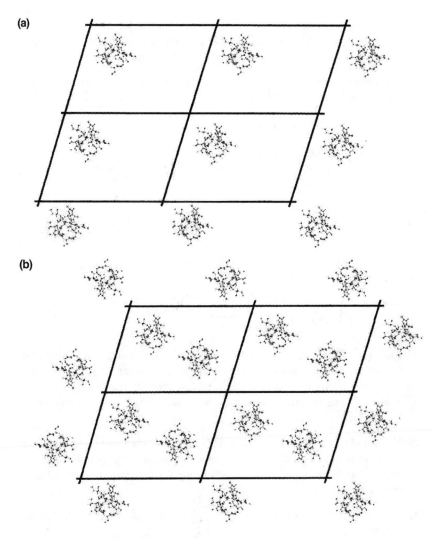 crystallizes in space group $P2_1/c$, with two molecules per unit cell. What can be deduced about both the positions of the molecules in the unit cell and the molecular conformation? (The planarity of each benzene ring in the molecule may be assumed.)

2.9. Write the coordinates of the vectors between all pairs of general equivalent positions in $P2_1/c$ with respect to the origin, and note that they are of

two types. Remember that $-\frac{1}{2}$ and $+\frac{1}{2}$ in a coordinate are crystallographically equivalent, because we can always add or subtract 1 from a fractional
coordinate without altering its crystallographic implication.

2.10. The orientation of the symmetry elements in the orthorhombic space group
Pban may be written as follows:

$$\bar{1} \text{ at } 0,0,0 \text{ (choice of origin)}$$

$$\left.\begin{array}{l} b\text{-glide} \parallel (p, y, z) \\ a\text{-glide} \parallel (x, q, z) \\ n\text{-glide} \parallel (x, y, r) \end{array}\right\} \text{ (from the space-group symbol}^{a})$$

Determine p, q, and r from the following scheme, using the fact that
$n \cdot a \cdot b \equiv \bar{1}$:

$$x, y, z \xrightarrow{\;-b\;} 2p - x, \; -\tfrac{1}{2} + y, z$$

$$\left\downarrow \bar{1} \qquad\qquad\qquad\qquad\qquad \right\downarrow a$$

$$\left\{\begin{array}{l} \cdots, \cdots, \cdots \\ \cdots, \cdots, \cdots \end{array}\right. \xleftarrow{\;\;-n\;\;} \cdots, \cdots, \cdots$$

2.11. Construct a space-group diagram for *Pbam*, with the origin at the intersection of the three symmetry planes. List the coordinates of both the general
equivalent positions and the centers of symmetry. Derive the standard coordinates for the general positions by transforming the origin to a center of
symmetry.

2.12. Show that space groups *Pa, Pc,* and *Pn* represent the same pattern, but that
Ca is different from *Cc* (*Cn*). What is the more usual symbol for space
group *Ca*? What would be the space group for *Cc* after an interchange of
the x and z axes?

2.13. For each of the space groups $P2/c, Pca2_1, Cmcm, P\bar{4}2_1c, P6_322,$ and
$Pa3$:

(a) Write down the parent point group and crystal system.
(b) List the full meaning conveyed by the symbol.
(c) State the independent conditions limiting x-ray reflection.

2.14. Consider Figure 2.25. What would be the result of constructing this diagram
with **Z** alone, and not using its mirror image?

---

^{a} In general, the symbol $\parallel$ in this context will mean the plane (or line) specified; that is, the b-glide
plane, for example, will be the plane (p, y, z).

2.15. (a) Draw a P unit cell of a cubic lattice in the standard orientation.

(b) Center the A faces. What system and standard unit-cell type now exist?

(c) From the position at the end of (b), let c and all other lines parallel to it be angled backward a few degrees in the ac plane. What system and standard unit-cell type now exist?

(d) From the position at the end of (c), let c and all other lines parallel to it be angled sideways a few degrees in the bc plane. What system and standard unit-cell type now exist?

For (b) to (d), write the transformation equations that take the unit cell as drawn into its standard orientation.

2.16. Set up matrices for the following symmetry operations: $\bar{4}$ along the z axis, m normal to the y axis. Hence, determine the Miller indices of a plane obtained by operating on (hkl) by $\bar{4}$, and on the result of this operation by m. What are the nature and orientation of the symmetry element represented by the given combination of $\bar{4}$ and m?

2.17. The matrices for an n-glide plane normal to a and an a-glide plane normal to b in an orthorhombic space group are as follows:

$$\begin{bmatrix} 1 & 0 & 0 \\ 0 & \bar{1} & 0 \\ 0 & 0 & 1 \end{bmatrix} + \begin{bmatrix} \frac{1}{2} \\ 0 \\ 0 \end{bmatrix} \qquad \begin{bmatrix} \bar{1} & 0 & 0 \\ 0 & 1 & 0 \\ 0 & 0 & 1 \end{bmatrix} + \begin{bmatrix} 0 \\ \frac{1}{2} \\ \frac{1}{2} \end{bmatrix}$$
$$\qquad\qquad a \qquad\qquad\qquad\qquad\qquad n$$

What are the nature and orientation of the symmetry element arising from the combination of n and a? What is the space-group symbol and its class?

2.18. Deduce the matrices for the symmetry operations 6_3 and m in space group $P6_3/m$, origin on center of symmetry. From the results, deduce the coordinates of the general equivalent positions for this space group. (A 6-fold right-handed rotation converts the point x, y, z to $x - y$, x, z.)

2.19. A unit cell is determined as $a = b = 3$ Å, $c = 9$ Å, $\alpha = \beta = 90°$, $\gamma = 120°$. Later, it proves to be a triply-primitive hexagonal unit cell. With reference to Figure 2.11, determine the equations for the unit cell transformation $R_{\text{hex}} \rightarrow R_{\text{obv}}$, and calculate the parameters of the rhombohedral unit cell.

2.20. In relation to Problem 2.19, given the the plane $(13*4)$ and zone symbol $[1\bar{2}*3]$ in the hexagonal unit cell, determine these parameters in the obverse rhombohedral unit cell. (The $*$ indicates that the three numbers relate to the x, y and z axes, respectively.)

2.21. By means of a diagram, or otherwise, show that a site x, y, z reflected across the plane (qqz) in the tetragonal system has the coordinates $q - y$, $q - x$, z after reflection.

2.22. Deduce a diffraction symbol table for the monoclinic space groups.

2.23. Draw the projection of an orthorhombic unit cell on (001), and insert the trace of the (210) plane and the parallel plane through the origin.

    (a) Consider the transformation $\mathbf{a}' = \mathbf{a}/2$, $\mathbf{b}' = \mathbf{b}$, $\mathbf{c}' = \mathbf{c}$. Using the appropriate transformation matrix, write the indices of the plane with respect to the new unit cell. Draw the new unit cell and insert the planes at the same perpendicular spacing, starting with the plane through the origin. Does the geometry of the diagram confirm the indices obtained from the matrix?.

    (b) Make a new drawing, like the first, but now consider the transformation $\mathbf{a}' = \mathbf{a}, \mathbf{b}' = \mathbf{b}/2, \mathbf{c}' = \mathbf{c}$. What does (210) become under this transformation? Draw the new unit cell and insert the planes as before. Does the geometry confirm the result from the matrix?

2.24. Why are space groups $Cmm2$ and $Amm2$ distinct, yet $Cmmm$ and $Ammm$ are equivalent?

3

I X-rays, X-ray Diffraction, and Structure Factors
II Intensities and Intensity Statistics

I X-rays, X-ray Diffraction, and Structure Factors

3.1 Generation and Properties of X-rays

X-rays are an electromagnetic radiation of short wavelength, and can be produced by the sudden deceleration of rapidly moving electrons at a target material. If an electron falls through a potential difference of V volt, it acquires an energy of eV electron-volt (eV), where e is the charge on an electron. This energy may be expressed as quanta of x-rays of wavelength λ, where each quantum is given by

$$\lambda = hc/(eV) \tag{3.1}$$

h being Planck's constant and c the speed of light in vacuum. Substitution of numerical values into (3.1) leads to

$$\lambda = 12.4/V \tag{3.2}$$

where V is measured in kilovolt and λ is given in Ångström units (Å). The wavelength range of x-rays is approximately 0.1–100 Å, but for the purposes of practical x-ray crystallography, the range used is approximately 0.6–3.0 Å.

3.1.1 X-rays and White Radiation

X-rays are produced through the impact of electrons on a metal target (anode). With the exception of synchrotron radiation, discussed in Section 3.1.5,

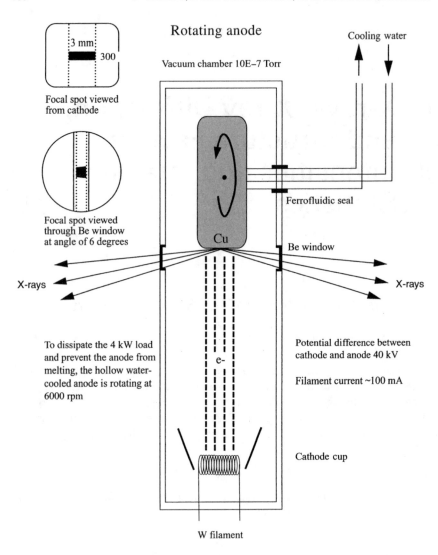

FIGURE 3.1.   Schematic diagram of a sealed crystallographic x-ray tube. The target anode is provided with a means of rotation, so as to aid the dissipation of heat generated by the electron impact on the target and to prolong the life of the target.

the most widely used source of x-rays in conventional crystallography laboratories is the sealed hot-cathode tube, illustrated diagrammatically in Figure 3.1. In this device, electrons emitted from a heated tungsten filament (cathode) are accelerated by a high voltage (40 kV or more) toward a water-cooled, target anode,

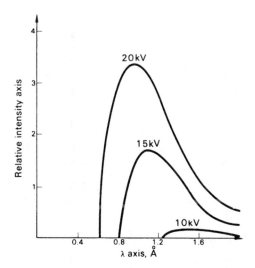

FIGURE 3.2.   Variation of intensity with wavelength for a sealed x-ray tube, for three different operating voltages; as V increases, the maximum wavelength in the continuous spectrum moves to shorter wavelengths.

usually made of copper or molybdenum. A large proportion of the energy reaching the target is dissipated as heat on account of multiple collisions within the target material, but about 10% of it is converted usefully for crystallographic purposes. In order to dissipate the heat rapidly and efficiently, the water-cooled anode is rotated, as indicated in Figure 3.1. As a consequence, a higher accelerating voltage can be applied to the tube, resulting in a more powerful x-ray source. If the energy eV is not too high, there will be a continuous distribution of x-ray wavelengths, or "white" radiation, as shown by Figure 3.2. As the accelerating voltage V is increased, the intensity of the radiation increases, and the maximum of the curve moves to shorter wavelengths.

3.1.2   Characteristic Radiation

At a certain, higher value of V, the impinging electrons excite inner electrons in the target atoms. Then, other electrons from higher energy levels fall back to the inner levels, and their transitions are accompanied by the emission of x-radiation of high intensity, *characteristic* of the material of the target. In this case, the x-ray wavelength depends on the energies of the two levels involved, E_1 and E_2, such that

$$\lambda = hc/|E_2 - E_1| \tag{3.3}$$

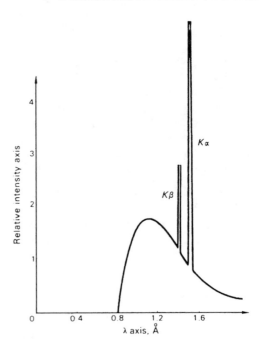

FIGURE 3.3. Characteristic K spectrum from an x-ray tube, superimposed upon the "white" radiation (continuous spectrum).

Figure 3.3 illustrates the curve of radiation intensity against x-ray wavelength, when the accelerating voltage is sufficient to excite the K spectrum of the target metal. The K spectrum consists of the $K\alpha$ and $K\beta$ wavelengths, which are always produced together, and correspond to electrons falling back from the L and M levels, respectively to the K shell. Two slightly different L energy levels exist, so that the important $K\alpha$ spectrum consists of two components of closely similar wavelength, $K\alpha_1$ and $K\alpha_2$. The wavelengths of the K radiations for a target material of copper are: $K\beta = 1.3926$ Å, $K\alpha_1 = 1.54056$ Å, and $K\alpha_2 = 1.54437$ Å respectively. The mean value for $K\alpha$ is obtained by averaging the $K\alpha_1$ and $K\alpha_2$ wavelengths in their intensity ratio of 2:1, thus giving the average value of 1.54183 Å for $K\alpha$; the α_1–α_2 doublet is resolved when the angle of scatter is large, that is, at high values of the Bragg angle θ (q.v.).

3.1.3 Absorption of X-rays

All materials absorb x-rays, and the transmitted intensity is attenuated according to an exponential law

$$I = I_0 \exp(-\mu t) \tag{3.4}$$

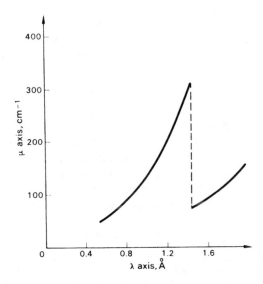

FIGURE 3.4. Variation with wavelength of the linear absorption coefficient μ for nickel; the discontinuity at approximately 1.49 Å corresponds with the L absorption edge of the element.

where I_0 and I are, respectively, the incident and transmitted intensities, μ is the *linear absorption coefficient* of the material, and t is the path length for x-rays through the material. The absorption of x-rays increases with the atomic numbers of the elements in the absorbing material.

The variation of μ with wavelength is illustrated by Figure 3.4, which refers to elemental nickel. The absorption coefficient μ of any material decreases approximately as $\lambda^{5/2}$; as λ falls, the energy of the radiation (hc/λ) becomes greater and so is more penetrating. With continuing decrease in wavelength, a position is reached where the energy of the radiation is sufficient to eject an electron from the L energy level of an atom of the material (nickel). At this point, known as the *absorption edge*, or *resonance level*, the value of μ is greatly enhanced. As the wavelength decreases further, the absorption coefficient continues to fall off as before. In the case of nickel, this particular L absorption edge occurs at a wavelength of 1.4886 Å.

If the material consists of a single elemental species, μ in (3.4) may be termed the atomic absorption coefficient[a] μ_a, given by

$$\mu_a = M_r\mu/(DL) \tag{3.5}$$

[a] See Bibliography, *International Tables for X-Ray Crystallography*, Vol. 4.

where M_r is the relative atomic mass and D the density of the material; L is the Avogadro constant. Of more general applicability is the *mass absorption coefficient* μ_m, given by $\mu_m = \mu/D$, so that $\mu_a = \mu_m(M_r/L)$. Hence, for a compound, we have

$$\mu = \sum_i \mu_{m,i} D_i \qquad (3.6)$$

where $\mu_{m,i}$ is the mass absorption coefficient for the ith species of partial density D_i in the compound; D_i is calculated for the ith species as $DM_{r,i}/M_r$. For example, sodium chloride, NaCl, has a density of 2165 kg m^{-3}, and the relative atomic masses and mass absorption coefficients for Na and Cl are 22.98 and 3.01 m^2 kg^{-1}, and 35.45 and 10.6 m^2 kg^{-1} for Na and Cl, respectively. Hence, the linear absorption coefficient for NaCl is given by

$$\mu = 2165[(3.01 \times 22.98/58.43) + (10.6 \times 35.45/58.43)] = 1.65 \times 10^4 \text{ m}^{-1}$$

and this parameter is needed in the correction of x-ray intensities (see Section 3.9.3). The attenuation factor I/I_0, for a crystal of NaCl of thickness 0.1 mm in the path of the x-ray beam is then $\exp(-1.65 \times 10^4 \times 0.1 \times 10^{-3})$, or 0.192.

3.1.4 Monochromatic (Filtered) Radiation

The majority of x-ray structure analyses require the use of monochromatic radiation, but we have seen from Figures 3.2 and 3.3 that x-ray sources contain a range of wavelengths. However, and in particular for radiation from a copper target, we note from the wavelength values given in Sections 3.1.2 and 3.1.3 that the absorption edge for nickel (1.4886 Å) lies between the wavelengths for Cu $K\alpha$ ($\lambda = 1.5418$ Å) and Cu $K\beta$ (1.3926 Å) radiation. The effect of passing the x-rays from a copper target through a thin (ca 0.018 mm) nickel foil is shown by Figure 3.5, a superposition of Figures 3.3 and 3.4. The $K\beta$ radiation is almost totally absorbed by the nickel, and the "white" radiation is decreased significantly in intensity. There is also a loss in intensity of the $K\alpha$ radiation, but the intense part of the beam behaves as a closely monochromatic, or *filtered*, radiation. Evidence for the remaining presence of some $K\beta$ and white radiation may be seen on Figure 8.5, for example.

Another important source of monochromatic radiation makes use of a crystal to select a single wavelength from the output of an x-ray tube, by Bragg reflection from a plane (hkl) that produces a high intensity. While this technique produces a monochromatic beam, reference to Section 3.3.2 shows that the planes (nh, nk, nl) will simultaneously be in the correct orientation to reflect the overtone wavelengths λ/n. Generally, the overtones are weaker in intensity with a carefully chosen crystal, and may be reduced further with a suitable filter, so that the resulting

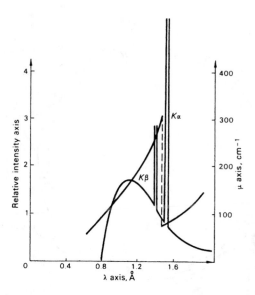

FIGURE 3.5.   Diagrammatic superposition of the curves of Figures 3.3 and 3.4, for Cu $K\alpha$ x-radiation; the strongly preferential absorption of the $K\beta$ radiation result in an almost monochromatic $K\alpha$ radiation.

beam is very closely monochromatic. The use of the crystal monochromator is discussed further in Chapter 9.

Absorption and absorption edges are also important in choosing the correct radiation for a particular application. For example, copper x-radiation would be unsuitable for materials containing a high percentage of iron. The K absorption edge for iron is 1.7433 Å, so that radiation of this wavelength would be strongly absorbed by the iron moiety and subsequently re-emitted as the characteristic K spectrum of iron. In such a case, molybdenum radiation, $\lambda(K\alpha) = 0.71073$ Å, would be a satisfactory alternative, whereupon zirconium is employed as a satisfactory β-filter.

3.1.5   Synchrotron Sources

A synchrotron is a large-scale particle accelerator designed primarily as a tool for fundamental studies in particle physics. It has many additional applications, and in x-ray crystallography, it functions as a very powerful source of x-rays, with intensity several orders of magnitude greater than that of the sealed tube. The orbit of particles in the synchrotron is produced by means of magnetic fields that increase, with time, in proportion to the increased momentum of the particles, while the radius of the orbit remains constant. This design requires that

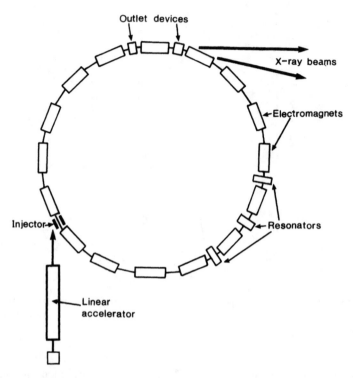

FIGURE 3.6.   Diagrammatic representation of a synchrotron device; the diameter of the storage ring is typically 20 m.

the magnetic force operate only over narrow ranges around the orbit, alternating between radially increasing and decreasing flux fields. Particles are accelerated in a linear accelerator prior to injection into the synchrotron, Figure 3.6. Acceleration is achieved by means of resonators placed in the gaps between the magnets. In an electron synchrotron, the injection of the electrons takes place at relativistic energies, in the region of 10 MeV; at this energy their speed approaches that of light. The higher the given energy of the accelerated particles, the greater the radius needed for the ring electromagnet of the synchrotron. For an energy of 100 MeV, the radius of the ring is 200 m. The high-energy acceleration of negatively charged electrons leads to the emission of a strong electromagnetic radiation, synchrotron radiation (SR), with a wide spectrum extending from radio waves to x-rays.

An important difference between x-radiation from sealed tube or rotating anode sources and x-rays produced by a synchrotron is their physical state of polarization. X-rays generated from conventional laboratory sources are totally non-polarized unless a crystal monochromator (Section 4.9) is being used. In contrast, synchrotron radiation is 100% linearly polarized in the plane of the electron beam orbit and elliptically polarized above and below the plane. In addition, the output radiation is pulsed, because the electrons do not form a uniform stream.

Figure 3.7a considers a three-dimensional x-ray wave at the origin, with components of oscillation vibrating in the y and z directions. The beam from a synchrotron is plane-polarized, with the component in the z direction being an order of magnitude less in intensity. X-ray beams from crystal monochromators, because they have undergone Bragg diffraction, are also polarized, but the extent is considerably less. The effect of this primary or beam polarization on the intensity of the diffraction pattern must be taken into account during data processing. This is achieved through application of a *polarization factor P* that includes a source-dependent property, with components normal and parallel to the plane of polarization:

$$P = (I_\parallel - I_\perp)/(I_\parallel + I_\perp)$$

where $I_\parallel$ and $I_\perp$ are the intensities of the electrical fields of the x-ray beam along and normal to the axis of a rotation camera, respectively. The value of P is

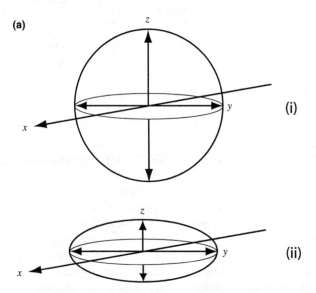

FIGURE 3.7. (a) A three-dimensional wave traversing the origin, with oscillatory components vibrating along the x and y directions, (i) conventional x-ray tube radiation, (ii) synchrotron radiation (not to scale). The radiation from a synchrotron is plane-polarized, with the z component an order of magnitude less than that in the y direction; the spread of radiation in the vertical (z) direction is given as $\Delta = m_e c^2/E$, where E is the electron-beam energy. (b) Spectral curves in the x-ray region from a normal bending magnet and a wiggler for a 2 GeV 1 A beam in the synchrotron radiation source, and the types of experiment used in the wavelength regions specified. The peak of the curve is approximately $1.4\lambda_c$, corresponding to the maximum output of energy per unit wavelength.

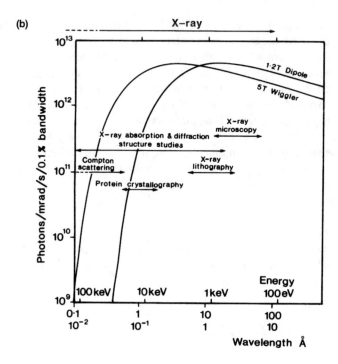

FIGURE 3.7.   (Continued).

calibrated for each workstation at a synchrotron installation. For station 9.6 at the CLRC, Daresbury UK, P is 0.84, whereas for a graphite monochromator, P is equal to 0.11.

A typical SR spectrum is shown in Figure 3.7b, and may be compared with that from the sealed tube. The photon intensity is given in units of photon per second for a horizontal angular aperture of 1 mrad (1 mrad = 3.4 min of arc), with a 1 A beam current and a 0.1% spectral bandwidth, after performing vertical integration over the full angular divergence of the radiation above and below the orbital plane. The peak intensity in these units is approximately 5×10^{13}, and lies at a wavelength that is approximately $1.4\lambda_c$, where λ_c is a characteristic, maximum photon wavelength given by $\lambda_c / \text{Å} = 18.64/(BE^2)$, the electron-beam energy being in giga electron-volt and the dipole magnetic field in tesla, T.

The flux attainable in practice depends on the multiplying factors set by the values of the dependent parameters listed above. Thus, the horizontal aperture of an experimental workstation may be less than 1 mrad for topography, typically 5–10 mrad for the majority of spectroscopy experiments, and up to 40 mrad for the "high-aperture" port used for time-resolved measurements. A bandwidth of

0.1% represents a good resolution (0.1 Å at 100 Å) for many procedures. The flux available will change proportionally if this resolution is varied. The stored current and, hence, the photo flux, gradually decline as electrons are lost by scattering from closed electron orbits. The beam lifetime, that is, the time of fall to approximately 1/e of the initial intensity, is approximately 8 hr.

The output characteristics of the synchrotron can be modified by devices inserted into the straight sections between the magnets. One insertion device is termed a "wiggler": it comprises a series of dipolar magnets of alternating polarity that cause the beam to undergo a series of oscillations perpendicularly to its general direction. A single wiggle produces only a wavelength shift, but multiple wiggles, each producing its own radiation, combine to increase the intensity of the output beam by 10-fold. The wiggler is important in experiments with wavelengths of 1 Å or less, and so finds application in x-ray crystallography.

The synchrotron source may be said to have revolutionized x-ray crystallography in certain applications. It has enabled a rapid collection of data to be achieved, and so is of great value in dealing with unstable samples, such as proteins, with poorly diffracting crystals, or polymers, in time-resolved studies, or solid-state reactions and other transformations, including enzyme-catalyzed processes, or in x-ray topographical studies of crystal defects.

3.2   X-ray Scattering

Scattering occurs generally when electromagnetic radiation interacts with matter. Some everyday examples of scattering are the blue of the sky, the haloes around distant car lights at night, the pattern seen when looking at a sodium street-lamp through a stretched handkerchief and the interference patterns in light reflected from the surface of a compact disc. The type of scattering of interest here is that of x-rays interacting with the electron density distribution in crystals. The x-ray scattering from a crystal may be described in terms of the intensity of the scattering function and the angle of scatter, 2θ.

3.2.1   Scattering by a Single Electron

Consider a scattering electron at an origin O in three-dimensional space. A plane wave of monochromatic x-rays of frequency ν and amplitude $\mathcal{A}$ may be represented in its transverse displacement y by the equation

$$y = \mathcal{A}\exp[\mathrm{i}2\pi(\nu t - R/\lambda)] \qquad (3.7)$$

where $2\pi R/\lambda$ is the phase shift at a distance R from the origin, attained after a time t. There is also a phase difference α at the origin O, arising from the scattering process itself, and a reduction in intensity, or amplitude squared, because of the inverse-square proportionality relationship. Hence, the displacement y at a point R may be recast as

$$y = \psi_{2\theta}(A/R)\exp\{i[2\pi(\nu t - R/\lambda) - \alpha]\} \tag{3.8}$$

where $\psi_{2\theta}$ is a function of the material at the scattering angle 2θ. This equation shows that the displacement amplitude Ψ_1 for a single electron at the point R is

$$\Psi_1 = \psi_{2\theta}A/R \tag{3.9}$$

and the phase difference, or lag, of the displacement at P behind the incident wave at O is $-[(2\pi R/\lambda) + \alpha]$. The intensity $I_{2\theta}$ of the scattered beam, given by $|yy^*|$, for unit solid angle around the direction of scatter 2θ, is then

$$I_{2\theta} = \Psi_1^2 R^2 = \psi_{2\theta}^2(A/R)^2 \times R^2 = K\psi_{2\theta}^2 A^2 = \psi_{2\theta}^2 I_0 \tag{3.10}$$

where y^* is conjugate (q.v.) to y, K is the constant relating intensity to the square of the amplitude and I_0 is the incident intensity.

3.2.2   Scattering by Two or More Electrons

Figure 3.8 represents a plane train of waves incident upon two electrons (scattering centers) O and A in three-dimensional space. The electrons O and A scatter the incident x-ray wave, and we need to determine the phase difference at a point P along the forward direction, the distance OP being very much greater than the distance OA. The path difference between the two wavelets scattered by O and A is $OY - AX$; thus the phase difference ϕ is

$$\phi = (2\pi/\lambda)(OY - AX) \tag{3.11}$$

The incident and diffracted waves may be defined by the unit vectors $\mathbf{s}_0$ and $\mathbf{s}$, respectively, such that $AX = \mathbf{r}\cdot\mathbf{s}_0$ and $OY = \mathbf{r}\cdot\mathbf{s}$. Thus,

$$\phi = 2\pi(\mathbf{r}\cdot\mathbf{s}_0 - \mathbf{r}\cdot\mathbf{s})/\lambda = 2\pi\mathbf{r}\cdot(\mathbf{s}/\lambda - \mathbf{s}_0/\lambda) = 2\pi\mathbf{r}\cdot\mathbf{S} \tag{3.12}$$

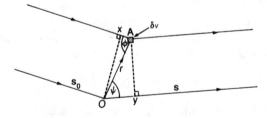

FIGURE 3.8.   Combined scattering at two centers O and A; $\mathbf{s}_0$ and $\mathbf{s}$ are unit vectors in the incident and scattered beams, respectively.

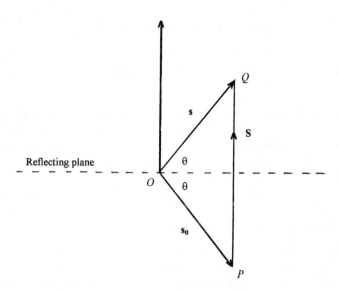

FIGURE 3.9.   Relationship of the scattering vector $\mathbf{S}$, or $(\mathbf{s} - \mathbf{s}_0)/\lambda$, to the reflecting plane (hkl); $\mathbf{S}$ is normal to the plane (hkl), and its magnitude $|\mathbf{S}|$, or S, is $2(\sin\theta)/\lambda$.

From Figure 3.9, it is clear that $\mathbf{S}$ is a vector normal to a plane through O that may be regarded conveniently as a *reflecting plane*; furthermore,

$$S = |\mathbf{s} - \mathbf{s}_0|/\lambda \qquad (3.13)$$

so that the magnitude $|\mathbf{S}|$, or S, is equal to $2(\sin\theta)/\lambda$. The displacement amplitude of the wave scattered by an electron at the origin, say A, and reaching the external point P after a time t is $(\mathcal{A}/R)\exp(i2\pi\nu t)$, where $\mathcal{A}$ is the amplitude of the scattered wave at unit distance ($R = 0$) from the origin in the direction of $\mathbf{s}$. Following (3.8), the displacement y at P after a time t occasioned by including the second electron, distant $\mathbf{r}$ from the origin, is given by

$$y = \psi_{2\theta}(\mathcal{A}/R)\exp\{i[2\pi(\nu t - R/\lambda) - \alpha]\}$$
$$+ \psi_{2\theta}(\mathcal{A}/R)\exp\{i[2\pi(\nu t - R/\lambda + \mathbf{r} \cdot \mathbf{S}) - \alpha]\}$$
$$= \psi_{2\theta}(\mathcal{A}/R)\exp\{i[2\pi(\nu t - R/\lambda) - \alpha]\}[1 + \exp(i2\pi\mathbf{r} \cdot \mathbf{S})] \qquad (3.14)$$

so that the resultant amplitude for two electrons is, from (3.9),

$$\Psi_2 = \psi_{2\theta}(\mathcal{A}/R)[1 + \exp(i2\pi\mathbf{r} \cdot \mathbf{S})] \qquad (3.15)$$

In the case that neither of the electrons of the previous example occupies the origin, then the number 1 on the right-hand side of (3.15) would be replaced by another exponential term. In general, the result for n electrons is given by

$$\Psi_n = \psi_{2\theta}(A/R) \sum_{j=1}^{n} \exp(i2\pi \mathbf{r}_j \cdot \mathbf{S}) \qquad (3.16)$$

This equation assumes that the n scatterers have equal power, which would be true if they were all electrons. In the event that the scatterers are unequal, then the function $\psi_{2\theta}$ would be unique to each species and included within the summation as $\psi_{2\theta,j}$. We shall consider this situation more fully later on in this chapter. We note the assumption of the identity of the α terms for all j scatterers, emerging from the use of (3.8), which has been shown to be valid in the case of x-ray scattering.

3.2.3   Representation of Waves and Wave Sums

In the preceding two sections, we have discussed traveling waves and the resultant effect of the combination of such waves, for which the frequency, and wavelength, is unaltered in the scattering process. The general exponential wave (3.7) can be expressed in its components through de Moivre's theorem: thus, the wave $\exp[\pm i2\pi(\nu t - x/\lambda)]$, traveling along the x axis, may be written as $\cos 2\pi(\nu t - x/\lambda) \pm i \sin 2\pi(\nu t - x/\lambda)$. We consider the real part of this expression, the cosine term, for the purposes of illustration. Thus, we shall write

$$y' = f \cos(\omega t - \phi) \qquad (3.17)$$

where ϕ is the phase $2\pi x/\lambda$ of the wave at time t, and $\omega = 2\pi\nu$. This wave is illustrated in Figure 3.10, where the phase at the first positive maximum is $2\pi \times 0.7$ rad, or $252°$.

When two waves of the same frequency combine, the resultant amplitude depends on their relative phase, as we have shown. Figure 3.11 shows two waves of the same frequency, exactly in phase ($x = n\lambda; n = 0, 1, 2, \ldots$), but of different amplitude: the resultant amplitude is the sum of the two amplitudes. If the waves are exactly out of phase $x = (n + 1)\lambda/2$, the resultant is zero if the amplitudes

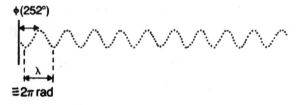

FIGURE 3.10.   The wave $y' = f \cos(\omega t - \phi)$, showing the phase ϕ relative to an arbitrary origin. In the illustration, $\phi = 0.7 \times 2\pi$ rad, or $252°$.

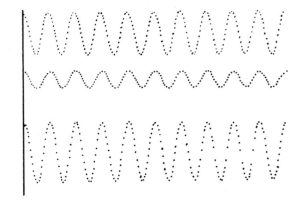

FIGURE 3.11. Combination of two waves that are in exact register (path difference $= n\lambda$). The resultant wave has the same wavelength, but with an amplitude that is equal to the sum of the amplitudes of the contributing waves.

are equal, Figure 3.12a, or the difference of the amplitudes if they are unequal, Figure 3.12b.

All cases between these extremes are possible, depending on the value of the phase angle ϕ. In Figure 3.13, we illustrate the combination of two waves, one of amplitude 100 (f_1) and phase zero, and the other of amplitude 50 (f_2) and phase 240°, with respect to the origin. From (3.17), we write for the transverse displacements of two waves

$$y_1 = f_1 \cos(\omega t - \phi_1) \quad y_2 = f_2 \cos(\omega t - \phi_2) \tag{3.18}$$

Their sum Y is given by

$$Y = \cos \omega t\, (f_1 \cos \phi_1 + f_2 \cos \phi_2) + \sin \omega t\, (f_1 \sin \phi_1 + f_2 \sin \phi_2) \tag{3.19}$$

Let the resultant Y take the form

$$Y = F(\cos \omega t - \phi) = F(\cos \omega t \cos \phi + \sin \omega t \sin \phi) \tag{3.20}$$

By comparing coefficients between (3.19) and (3.20), it follows that

$$F \cos \phi = f_1 \cos \phi_1 + f_2 \cos \phi_2$$
$$F \sin \phi = f_1 \sin \phi_1 + f_2 \sin \phi_2 \tag{3.21}$$

Hence, the amplitude $|F|$ of the resultant wave is given by

$$|F| = [(f_1 \cos \phi_1 + f_2 \cos \phi_2)^2 + (f_1 \sin \phi_1 + f_2 \sin \phi_2)^2]^{1/2} \tag{3.22}$$

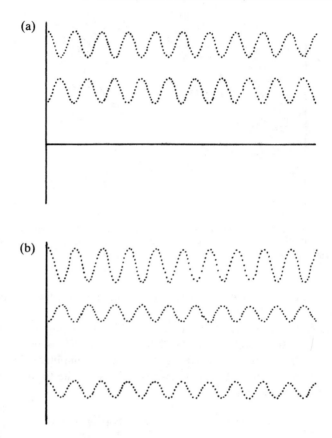

FIGURE 3.12. Combination of two waves that are exactly out of register; path difference $=(n + 1)\lambda/2$. (a) Both waves have the same amplitude, so that the resultant wave has an
amplitude of zero; (b) the two waves have different amplitudes, so that the resultant wave has an
amplitude that is equal to the difference of the amplitudes of the contributing waves.

and its phase angle ϕ by

$$\phi = \tan^{-1}[(f_1 \sin \phi_1 + f_2 \sin \phi_2)/(f_1 \cos \phi_1 + f_2 \cos \phi_2)] \qquad (3.23)$$

Thus, the combination of the two waves of equal frequency, but with different
amplitudes and phases, results in a third wave of the same frequency, but with
its own amplitude and phase, as shown by Figure 3.13. This process could be
continued for as many waves as desired. However, we shall describe a neater
approach to such summations later, in Section 3.5.2.

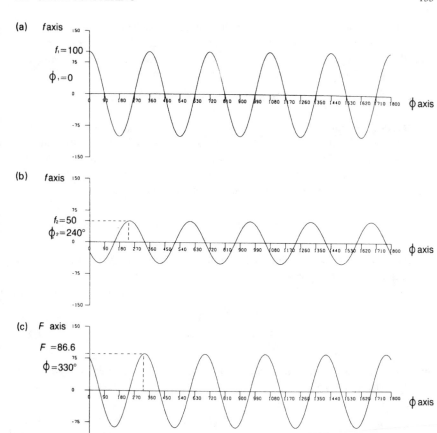

FIGURE 3.13.   Combination of two waves of amplitudes 100 (f_1) and 50 (f_2) with phases 0° (ϕ_1) and 240° (ϕ_2), respectively. The resultant wave has an amplitude F and a phase ϕ; no change of wavelength (or frequency) takes place. (a) First wave, (b) second wave, (c) resultant wave. $F = [(100+50\cos 240°)^2 + (50\sin 240°)^2]^{1/2} = 86.6; \phi = \tan^{-1}[(50\sin 240°)/(100+50\cos 240°)] = -30° (330°)$. The reader may wish to confirm these results from equations (3.22) and (3.23).

3.2.4   Coherent and Incoherent Scattering

In coherent scattering, or Thomson scattering, the incident and scattered waves have the same wavelength. When x-rays fall upon an electron, the alternating electric-field vector imparts an alternating acceleration to the electron. Classical electromagnetic wave theory shows that an accelerated charged particle emits radiation, through a process of absorption and re-emission, the emitted radiation traveling in all directions for a given angle of scatter.

If an electron of charge e at an origin O interacts with monochromatic x-rays it executes a simple harmonic motion. If the amplitude of the acceleration is $\mathbf{a_e}$, then theory shows that the amplitude E of the electric vector of the scattered radiation is given by

$$E = \mathbf{a_e}\, e \sin\varphi / (4\pi \varepsilon_0 c^2 r) \tag{3.24}$$

where r is the distance of the electron normal to the plane containing the electric vector and φ is the angle between the acceleration vector $\mathbf{a_e}$ at the electron and the direction of scatter, Figure 3.14. The incident electric vector $\mathbf{E}$ is shown resolved at O, the electron, into components perpendicular and normal to the plane XOP. The corresponding components of acceleration amplitudes are

$$\mathbf{a}_{e,\perp} = E_{e,\perp} e/m_e \quad \mathbf{a}_{e,\parallel} = E_{e,\parallel} e/m_e \tag{3.25}$$

where m_e is the mass of the electron. Eliminating $\mathbf{a_e}$ through (3.24) for the scattered wave distant r from O gives

$$E_{s,\perp} = E_{e,\perp} e^2 / (4\pi \varepsilon_0 c^2 m_e r)$$
$$E_{s,\parallel} = E_{s,\parallel} e^2 \cos 2\theta / (4\pi \varepsilon_0 c^2 m_e r) \tag{3.26}$$

Since the perpendicular and parallel components of the incident radiation are equally distributed, we have

$$E_{e,\perp}^2 = E_{e,\parallel}^2 = \tfrac{1}{2} I_0 \tag{3.27}$$

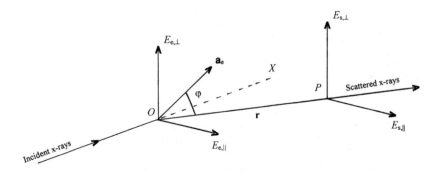

FIGURE 3.14. Relationship between the components $E_{e,\perp}$ and $E_{e,\parallel}$ of the electric vector (of acceleration $\mathbf{a_e}$) of the incident radiation at a point O to the components $E_{s,\perp}$ and $E_{s,\parallel}$ of the scattered radiation at a point of observation P. The angle $\angle XOP$ is the scattering angle 2θ, and $\mathbf{r}$ is the vector distance from the point O to P.

The intensity $I_{2\theta}$ of the scattered radiation, defined as power per unit solid angle, is then

$$I_{2\theta} = \tfrac{1}{2}r^2(E_{s,\perp}^2 + E_{s,\parallel}^2) = \tfrac{1}{2}[e^2/(4\pi\varepsilon_0 c^2 m_e)]^2(1 + \cos^2 2\theta)I_0$$
$$= R_e^2 \tfrac{1}{2}(1 + \cos^2 2\theta)I_0 \tag{3.28}$$

where R_e is the classical radius of the electron, and $\tfrac{1}{2}(1 + \cos^2 2\theta)$ is a geometrical factor that we shall encounter further.

In Compton, or incoherent, scattering, the wavelength of the scattered radiation is longer than that of the incident radiation, which implies a loss of energy in the scattering process. When x-rays of energy hc/λ impinge on an electron, an elastic collision takes place: the electron recoils with an energy of $\tfrac{1}{2}m_e v^2$, and the radiation scattered at an angle 2θ has an energy $hc/(\lambda + d\lambda)$, such that, by conservation of energy,

$$hc/\lambda = hc/(\lambda + d\lambda) + \tfrac{1}{2}m_e v^2 \tag{3.29}$$

By simple manipulation, ignoring $\lambda\,d\lambda$ in comparison to λ^2, we have

$$hc\,d\lambda/\lambda^2 = \tfrac{1}{2}m_e v^2 \tag{3.30}$$

Conservation of momentum requires the condition

$$2(h/\lambda)\sin\theta = m_e v \tag{3.31}$$

Eliminating v through Equations (3.30) and (3.31) gives

$$d\lambda = h/(m_e c)(1 - \cos 2\theta)$$

or, by inserting the fundamental constants,

$$d\lambda/\text{Å} = 0.0243(1 - \cos 2\theta) \tag{3.32}$$

Thomson scattering illustrates the particle property of the electron, whereas Compton scattering shows its wave nature.

3.2.5 Scattering by an Atom

In an atom, electrons are bound in levels of distinct energies, and in the scattering of x-rays by an atom both coherent and incoherent scattering are involved. A full analysis of the scattering process requires a wave-mechanical treatment, involving both modes of scattering, from which we obtain $\rho = \Psi\Psi^*$, where Ψ^* is conjugate to Ψ, or $\rho = |\Psi|^2$ if we assume a real nature for the density function ρ. These expressions may be interpreted such that $|\Psi|^2 d\tau$, or $\rho\,d\tau$, represents the probability of finding the electron in a volume element $d\tau$. We shall use the Thomson formula alone because incoherent scattering contributes to the background radiation, and is but a small fraction of the total intensity in the case of crystalline materials.

Consider a plane of atoms in a crystal. We demonstrate in the ensuing sections that all atoms on this plane scatter in phase with one another and with the atoms in parallel planes, for a given scattering vector $\mathbf{S}$. Thus, we need to consider how the electrons in any one atom combine to give a total scattering amplitude for the atom.

Let $\rho(\mathbf{r})d\tau$ be the probability that an electron in the chosen atom lies in a small volume element $d\tau$ distant r from the origin, the center of the atom, where r is the magnitude of the vector $\mathbf{r}$. Let $f(\mathbf{S})$ represent the scattering power of the atom in the direction $\mathbf{S}$; then we have, from the foregoing,

$$f(\mathbf{S}) = \int \rho(\mathbf{r}) \exp(i2\pi \mathbf{r} \cdot \mathbf{S}) \, d\tau \qquad (3.33)$$

Let $\mathbf{S}$ make an angle ψ with the direction of r (Figure 3.15). Then

$$2\pi \mathbf{r} \cdot \mathbf{S} = (4\pi/\lambda) \sin \theta \, r \cos \psi = mr \cos \psi$$

where $m = 4\pi(\sin \theta)/\lambda$. Since spherical symmetry has been assumed, the volume element $d\tau$ is a spherical annulus of radius r and thickness dr on $\mathbf{S}$ as axis, so that $d\tau = 2\pi r^2 \sin \psi \, d\psi \, dr$. Let $mr \cos \psi = x$, so that $dx = -mr \sin \psi \, d\psi$. Now (3.33) may be expressed as

$$f(\mathbf{S}) = 2\pi \int_0^\infty r^2/(mr)\rho(r) \, dr \int_{mr}^{-mr} -\exp(ix) \, dx$$

$$= 4\pi \int_0^\infty r^2 \rho(r)(\sin mr)/mr \, dr \qquad (3.34)$$

Since $m = (4\pi \sin \theta)/\lambda$, (3.34) may be recast as the function of $\mathbf{S}$:

$$f(\mathbf{S}) = 4\pi \int_0^\infty r^2 \rho(r)(\sin 2\pi r S)/(2\pi r S) \, dr \qquad (3.35)$$

where S is $(2 \sin \theta)/\lambda$. The atomic scattering factor may be defined as the ratio of the amplitude of coherent scattering from an atom to that scattered by a single electron at the center of the atom. It follows from (3.35) that, for scattering in the forward direction, when $(\sin 2\pi r S)/(2\pi r S) = 1$, the expression $4\pi \int_0^\infty r^2 \rho(r) \, dr$ becomes the total electron density for the atom. Hence, we may write

$$f(\mathbf{S})_{S=0} = Z \qquad (3.36)$$

where Z is the total number of electrons in the atom, or atomic number.

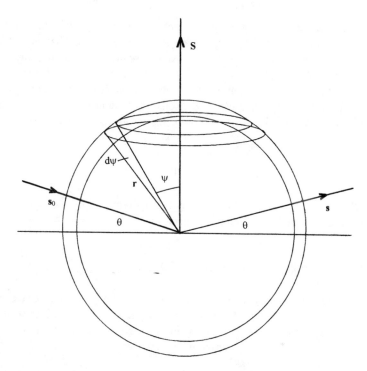

FIGURE 3.15. Scattering by a single atom. Section of a sphere of radius r, showing an annular ring of thickness dr at an angle ψ to the scattering vector $\mathbf{S}$. The volume of the annular ring is $[\pi(r+dr)^2 - \pi r^2]r \sin\psi\,d\psi = 2\pi r^2 \sin\psi\,d\psi\,dr$ (neglecting second order terms in dr).

As an example calculation, we consider the contribution of a 1s electron to $f(\mathbf{S})$. We equate $\rho(r)$ to $|\psi_{1s}|^2$ and use Slater's analytical wavefunctions. The standard one-electron 1s wavefunction ψ_{1s} may be written as $(1/\sqrt{\pi})c_1^{3/2}\exp(-c_1 r)$; c_1 is $(Z-\sigma)/a_0$, where σ is Slater's quantum mechanical screening constant[a] and a_0 is the Bohr radius for hydrogen. With lithium, for example, $\sigma_{1s} = 0.30$, and from (3.35), we have

$$f_{1s}(\mathbf{S}) = 2c_1^3/(\pi S) \int_0^\infty r \exp(-c_1 r)\sin(2\pi Sr)\,dr$$

From a table of standard integrals, $\int_0^\infty x\exp(-ax)\sin(bx)\,dx = 2ab/(a^2+b^2)^2$, so that

$$f_{1s}(\mathbf{S}) = c_1^4/(c_1^2 + \pi^2 S^2)^2 \tag{3.37}$$

[a]See, for example, Mark Ladd, *Introduction to Physical Chemistry*, Cambridge University Press (1998).

In lithium, there are two contributions from (3.37) and one contribution from a similar expression for the 2s electron (see also Problem 3.3). They are added to obtain the value of f at a given value of $\mathbf{S}$. Atomic scattering factor data are readily available, quoted normally as functions of $(\sin\theta)/\lambda$. Such data refer to systems of electrons at rest: at a finite temperature, the effective scattering from an atom is less than the value at rest, and we shall discuss this situation later in this chapter. A satisfactory calculation of (rest) atomic scattering factors is afforded by the equation

$$f(S) = \sum_{j=1}^{4} a_j \exp(-b_j S^2) + c \qquad (3.38)$$

where S here is $(\sin\theta)/\lambda$, and the nine constants required by the equation have been recorded for all atomic and ionic species.[a]

3.3 Scattering by Regular Arrays of Atoms

The interaction of x-rays with a crystal is a complex process, often described as a diffraction phenomenon, although, strictly speaking, it is a combined scattering and interference effect. Two treatments, those of von Laue and Bragg, describe the process, and we shall consider them in that order.

3.3.1 Laue Equations

We shall consider the development of the Laue equations first through a geometrical argument. Figure 3.16 represents a regular, one-dimensional array of atoms of spacing b, imagined in three-dimensional space. Parallel x-rays are incident at an angle ϕ_2 and scattered at an angle ψ_2 to the direction of $\mathbf{b}$. The path difference for rays scattered by neighboring centers is represented by $AQ - BP$, or $b(\cos\psi_2 - \cos\phi_2)$; this difference must be equal to an integral number of wavelengths for reinforcement to occur, so that

$$b(\cos\psi_2 - \cos\phi_2) = k\lambda \qquad (3.39)$$

This equation may be written alternatively as

$$\mathbf{b} \cdot (\mathbf{s} - \mathbf{s}_0)/\lambda = \mathbf{b} \cdot \mathbf{S} = k \qquad (3.40)$$

It is satisfied by the generators of a cone of semivertical angle ψ_2 coaxial with the row (Figure 3.16). For a given value of ψ_2 there will be a series of cones corresponding to the orders of k $(k = 0, 1, 2, \ldots)$, as shown in Figure 3.17. The

---

[a] *International Tables for X-ray Crystallography*, Vol. IV, Kynoch Press (1962).

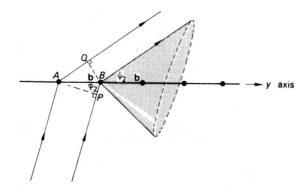

FIGURE 3.16.   Diffraction from a row of scattering centers, of spacing b along the y axis. The Laue equation $b(\cos \psi_2 - \cos \phi_2) = k\lambda$, or $\mathbf{b} \cdot \mathbf{S} = k$, is satisfied by any generator of the cone.

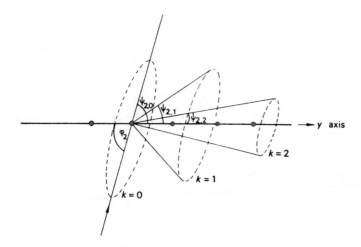

FIGURE 3.17.   Several orders of diffraction can arise from a row of scattering centers, for a given value of ϕ_2, corresponding to integral values for k.

discussion is extended readily to a net a, b as shown in Figure 3.18, so giving rise to a second condition

$$\mathbf{a} \cdot (\mathbf{s} - \mathbf{s}_0)/\lambda = \mathbf{a} \cdot \mathbf{S} = h \qquad (3.41)$$

A second cone intersects the first cone generally in two lines, BR and BS, but for the special case that both (3.40) and (3.41) hold simultaneously, the two lines coincide and the atoms of the net scatter in phase, with the incident and diffracted beams lying in the plane of the net.

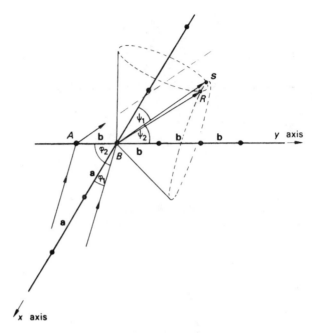

FIGURE 3.18.   Diffraction from a net of scattering centers; for clarity, only one row parallel to each of the axes has been drawn. When the directions BR and BS coincide, the two Laue equations $\mathbf{a} \cdot \mathbf{S} = h$ and $\mathbf{b} \cdot \mathbf{S} = k$ hold simultaneously, and scattering takes place in the plane of the net.

Generalizing to three dimensions, we now have the complete Laue equations

$$\mathbf{a} \cdot \mathbf{S} = h$$
$$\mathbf{b} \cdot \mathbf{S} = k \qquad\qquad (3.42)$$
$$\mathbf{c} \cdot \mathbf{S} = l$$

Any of the total of three possible pairs of equations define scattering from the corresponding net, but for the particular case that all three equations apply simultaneously, the three-dimensional array scatters in phase and produces the hklth spectrum.

Next, we discuss this scattering process in a more general manner. Consider a small parallelopipedal crystal completely enveloped by a train of plane, monochromatic x-ray waves, and, for convenience, let the scattering matter be concentrated at the points of the Bravais lattice of a crystal. Upon irradiation, each lattice point becomes the center of a scattered wavelet, and we need to find the combined effect of these wavelets at a point P distant R from the crystal, where R is very large compared to the dimensions of the crystal. We shall consider the interaction between the wavelets and the incident train as negligible, because the refractive index of crystals for x-rays is very small, unity to within 10^{-5}. We assume also that the wavelets travel through the crystal without themselves being scattered and suffer no absorption, which conditions are valid for a very small crystal bathed in the x-ray beam.

Let the lattice array of atoms be represented by a unit cell of basic vectors **a**, **b**, and **c**, parallel to the x, y, and z reference axes, respectively. Let there be N_a, N_b, and N_c lattice points parallel to a, b, and c, respectively, so that the total number of lattice points in the crystal is $N_a N_b N_c$. The total displacement amplitude Y, following (3.7) and (3.16), is a sum over $N_a N_b N_c$ lattice points. For any lattice point, the vector **r** to it from the origin is given by $\mathbf{r} = U\mathbf{a} + V\mathbf{b} + W\mathbf{c}$, where U, V, and W are integers or zero; hence

$$Y = (A/R)\exp(i\omega t)\sum_{U=0}^{U=N_a-1}\exp(i2\pi U\mathbf{a}\cdot\mathbf{S})$$

$$\times \sum_{V=0}^{V=N_b-1}\exp(i2\pi V\mathbf{b}\cdot\mathbf{S})\sum_{W=0}^{W=N_c-1}\exp(i2\pi W\mathbf{c}\cdot\mathbf{S}) \qquad (3.43)$$

The intensity of the scattered radiation is again given by $|YY^*|$. The terms on the right-hand side of (3.43) are uncorrelated, so that they can be considered separately. Each of them is a geometric progression: that in W leads to the sum

$$S_W = \{1 - \exp(i2\pi N_W\mathbf{c}\cdot\mathbf{S})\} \div \{1 - \exp(i2\pi\mathbf{c}\cdot\mathbf{S})\} \qquad (3.44)$$

and

$$S_W S_W^* = [1 - \cos(2\pi N_W\mathbf{c}\cdot\mathbf{S})] \div [1 - \cos(2\pi\mathbf{c}\cdot\mathbf{S})]$$

$$= \sin^2(N_W\psi_W)/\sin^2(\psi_W) \qquad (3.45)$$

where

$$\psi_W = \pi\mathbf{c}\cdot\mathbf{S} = (2\pi/\lambda)c\sin\theta\cos\gamma \qquad (3.46)$$

and γ is the angle between **c** and **S**. Similar equations exist for the directions of U and V, so that the total intensity $|Y|^2$ is given by

$$|Y|^2 = I_0 = (A/R)^2[(\sin^2 N_U\psi_U/\sin^2\psi_U)$$

$$\times (\sin^2 N_V\psi_V/\sin^2\psi_V)(\sin^2 N_W\psi_W/\sin^2\psi_W)] \qquad (3.47)$$

The type of term on the right-hand side of (3.47) arises in the treatment of a plane diffraction grating. The final term, for example, has a maximum value of N_W^2 when $\psi_W = l\pi$, where l is an integer or zero. It is straightforward to plot the function $\sin^2 N_W \psi_W / \sin^2 \psi_W$ for $\psi_W = 0$ to π and show that there are $(N_W - 2)$ subsidiary maxima between the main maxima, with successive subsidiary maxima falling off in intensity more rapidly as N_W is increased. In any practical case N_W is very large, so that the term has significant values only for ψ_W at or very close to $l\pi$. Thus, the maximum value of the intensity becomes

$$I_{\max} = (A/R)^2 N_U^2 N_V^2 N_W^2 \qquad (3.48)$$

when the following (Laue) conditions for diffraction, as in (3.42), hold:

$$\mathbf{a} \cdot \mathbf{S} = 2a \, \sin\theta/\lambda \cos\alpha = h$$
$$\mathbf{b} \cdot \mathbf{S} = 2b \, \sin\theta/\lambda \cos\beta = k \qquad (3.49)$$
$$\mathbf{c} \cdot \mathbf{S} = 2c \sin\theta/\lambda \cos\gamma = l$$

where h, k, and l are integers. The direction cosines $\cos\alpha$, $\cos\beta$, and $\cos\gamma$ of the vector $\mathbf{S}$, normal to the reflecting plane, with respect to the directions of $\mathbf{a}$, $\mathbf{b}$, and $\mathbf{c}$ are, therefore, proportional to a/h, b/k, and c/l, respectively. Successive planes (hkl) in the crystal intersect the x, y, and z axes at a/h, b/k, and c/l, respectively, so that they are parallel to the reflecting plane (hkl). Thus, equations (3.49) show that a scattered beam is derived from the incident beam by "reflection" from the (hkl) family of planes. Furthermore, if $d(hkl)$ is the interplanar spacing, then from Section 1.2.2.

$$d(hkl) = a/h \cos\alpha = b/k \cos\beta = l/c \cos\gamma \qquad (3.50)$$

and from (3.49) it follows that

$$2d(hkl) \sin\theta = \lambda \qquad (3.51)$$

a relation deduced by Bragg, originally in the form $2d(hkl) \sin\theta = n\lambda$.

3.3.2 Bragg Equation

The deduction of the Bragg equation, often thought to be ad hoc, was occasioned by the observation that if a crystal in a position that produced a scattered x-ray beam was rotated through an angle φ to another scattering position, then the scattered beam had been rotated through 2φ, as in the reflection of light from a plane mirror.

In Figure 3.19, two planes from a family of planes (hkl) are shown, together with the incident and reflected rays. The part of the incident beam that is not reflected at a given level passes on to be reflected from a deeper level in the crystal. Furthermore, all rays reflected from a given level remain in phase after reflection, because there is no path difference between them.

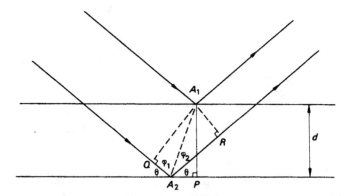

FIGURE 3.19. Geometry of x-ray reflection. The path difference between the two typical rays reflected from successive planes is $(QA_2 + A_2R)$. When this difference is equal to an integral number of wavelengths λ, a reflection is obtained, according to the Bragg equation.

The path difference δ between the two reflected rays shown is given by

$$\delta = QA_2 + A_2R = A_1A_2 \cos\phi_1 + A_1A_2 \cos\phi_2$$
$$= A_1A_2(\cos\phi_1 + \cos\phi_2) = 2A_1A_2 \cos[(\phi_1 - \phi_2)/2]\cos[(\phi_1 + \phi_2)/2] \tag{3.52}$$

which, by simple manipulation, becomes

$$\delta = 2d \sin\theta \tag{3.53}$$

Since δ is independent of ϕ_1 and ϕ_2, (3.53) applies to all rays in the bundle reflected from the adjacent planes. By the usual rules that apply to the combination of waves, the reflected rays will interfere with one another, the interference being at least partially destructive unless the path difference δ is equal to an integral number of wavelengths. Hence, the Bragg equation:

$$2d \sin\theta = n\lambda \tag{3.54}$$

where n is an integer. The mirror-reflection analogy breaks down because this equation must be satisfied for a reflection to occur, but the treatment is, nevertheless, a very useful, geometrical way of looking at the x-ray diffraction process.

In (3.54), n is the order of the Bragg reflection. We can write this equation in another form if we recall (see Section 2.4) that $d(hkl)/n = d(nh, nk, nl)$, with h, k, and l taking common factors as necessary. Thus, n is included in the crystallographic definition of $d(hkl)$, with the Bragg equation now written as

$$2d(hkl) \sin\theta(hkl) = \lambda \tag{3.55}$$

TABLE 3.1.   Nomenclature for Interplanar
Spacings

Original Bragg formulation			Current usage	
hkl	Order	d/Å	hkl	d/Å
120	1	2.236	120	2.236
	2	1.118	240	1.118
	3	0.7453	360	0.7453
	4	0.5590	480	0.5590

Each Bragg reflection from a crystal is considered, effectively, as first-order from the (hkl) family of planes, specified uniquely by these Miller indices. To illustrate this argument further, Table 3.1 lists data for planes parallel to (120) in a cube of side 5 Å.

The Bragg and von Laue treatments are equivalent, and we shall use them as the occasion demands. It has been convenient for this discussion, although not necessary, to consider that the scattering material is concentrated at lattice points, or on lattice planes. In general, electron density is a continuous function, but each part of it has a lattice-like distribution in the crystal.

3.4   Reciprocal Lattice–Analytical Treatment

We considered a geometrical derivation of the reciprocal lattice in Chapter 2, as we believe that that treatment forms a straightforward introduction to it. Here, we shall discuss the reciprocal lattice in greater detail.

In considering the stereographic projection, we showed that the morphology of a crystal could be represented by a bundle of lines, drawn from a point, normal to the faces of the crystal. This description, although angle-true, lacks linear definition. The representation may be extended by giving each normal a length that is inversely proportional to the corresponding interplanar spacing in real space, and applying it to all possible lattice planes, so forming a reciprocal lattice.

Let a Bravais (real-space) lattice be represented by the unit cell vectors $\mathbf{a}$, $\mathbf{b}$, $\mathbf{c}$. The reciprocal lattice unit cell is defined by the vectors $\mathbf{a}^*$, $\mathbf{b}^*$, $\mathbf{c}^*$, such that $\mathbf{a}^*$ is perpendicular to $\mathbf{b}$ and $\mathbf{c}$, and so on. Then,

$$\mathbf{a}^* \cdot \mathbf{b} = \mathbf{a}^* \cdot \mathbf{c} = \mathbf{b}^* \cdot \mathbf{a} = \mathbf{b}^* \cdot \mathbf{c} = \mathbf{c}^* \cdot \mathbf{a} = \mathbf{c}^* \cdot \mathbf{b} = 0 \qquad (3.56)$$

The magnitudes of the reciprocal unit cell vectors are defined by

$$\mathbf{a}^* \cdot \mathbf{a} = \mathbf{b}^* \cdot \mathbf{b} = \mathbf{c}^* \cdot \mathbf{c} = \kappa \qquad (3.57)$$

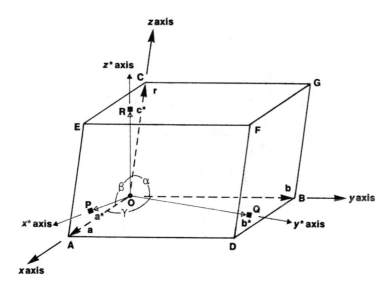

FIGURE 3.20.   Triclinic unit cell, showing its vectors **a**, **b**, and **c**, and the corresponding reciprocal unit-cell vectors, **a***, **b***, and **c***.

where κ is a constant, normally equal to unity in theoretical discussions, and to an x-ray wavelength in practical applications, where the size of the reciprocal lattice is important.

In Figure 3.20, the z^* (c^*) axis is normal to the plane a, b. Since $\mathbf{c} \cdot \mathbf{c}^* = cc^* \cos \angle COR$, that is, taking $\kappa = 1$ in this discussion,

$$c^* = |\mathbf{c}^*| = 1/(c \cos \angle COR) \tag{3.58}$$

so that the magnitude of $\mathbf{c}^*$, in reciprocal space, is inversely proportional to the c-spacing in real, or Bravais, space; similar deductions can be made for both $\mathbf{a}^*$ and $\mathbf{b}^*$. Since $\mathbf{c}^*$ is normal to both $\mathbf{b}$ and $\mathbf{c}$, it lies in the direction of their vector product:

$$\mathbf{c}^* = \eta(\mathbf{a} \times \mathbf{b}) \tag{3.59}$$

where η is a constant. Let V be the unit cell volume in real space. Then,

$$V = \mathbf{c} \cdot (\mathbf{a} \times \mathbf{b}) \tag{3.60}$$

Now $\mathbf{a} \times \mathbf{b}$ is a *vector* of magnitude $ab \sin \gamma$, the area of $OADB$, directed normal to the plane of a, b and forming a right-handed set of directions with **a** and **b**. Then,

$$\mathbf{c} \cdot \mathbf{c}^* = \eta \mathbf{c} \cdot (\mathbf{a} \times \mathbf{b}) = \eta V = 1 \tag{3.61}$$

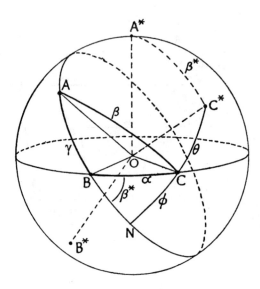

FIGURE 3.21.   Spherical triangle ABC and its polar triangle $A^*B^*C^*$ (see Appendix A5).

Hence,

$$c^* = |\mathbf{c}^*| = (ab \sin \gamma)/V \qquad (3.62)$$

with values for a^* and b^* obtained by cyclic permutation.

The angle β^* between $\mathbf{c}^*$ and $\mathbf{a}^*$ can be obtained by the equations of spherical trigonometry (see Appendix A5). In Figure 3.21, the spherical triangle ABC, where OA, OB, and OC are the directions of the x, y, and z axes respectively, has sides equal to the interaxial angles in real space α, β, and γ. The directions OA^*, OB^*, and OC^* are along $\mathbf{a}^*$, $\mathbf{b}^*$, and $\mathbf{c}^*$ respectively; OC^* is normal to the great circle through A and B, and similarly with OA^* and OB^*. Hence, $A^*B^*C^*$ is a spherical triangle polar to ABC (see Appendix A5), so that β^* is the angle between the great circles through AB and BC, or $\pi - B$, where B is the angle of the spherical triangle ABC at B. Then, from the formulae of spherical trigonometry,

$$\cos \beta^* = -\cos B = (\cos \gamma \cos \alpha - \cos \beta)/ \sin \gamma \sin \alpha \qquad (3.63)$$

with corresponding expressions for α^* and γ^* obtained by cyclic permutation. Simplified expressions obtain for (3.62) and (3.63) when the crystal symmetry is higher than triclinic.

3.4.1   Unit Cell Volumes in Real and Reciprocal Space

We may use (3.60) to evaluate an expression for the volume of a unit cell in terms of scalar quantities. Let $\mathbf{a}$, $\mathbf{b}$, and $\mathbf{c}$ be expressed in terms of a set of orthogonal unit vectors $\mathbf{i}$, $\mathbf{j}$, and $\mathbf{k}$ (see also Appendix A7):

$$\mathbf{a} = a_1\mathbf{i} + a_2\mathbf{j} + a_3\mathbf{k}$$
$$\mathbf{b} = b_1\mathbf{i} + b_2\mathbf{j} + b_3\mathbf{k} \qquad (3.64)$$
$$\mathbf{c} = c_1\mathbf{i} + c_2\mathbf{j} + c_3\mathbf{k}$$

Then, using (3.60) with expansion of the vector product $\mathbf{a} \times \mathbf{b}$, the volume may be written

$$V = (c_1\mathbf{i}+c_2\mathbf{j}+c_3\mathbf{k})\cdot(a_1b_2\mathbf{k}-a_1b_3\mathbf{j}-a_2b_1\mathbf{k}+a_2b_3\mathbf{i}+a_3b_1\mathbf{j}-a_3b_2\mathbf{i}) \qquad (3.65)$$

which, after simplification, may be expressed as the determinant

$$\begin{vmatrix} a_1 & a_2 & a_3 \\ b_1 & b_2 & b_3 \\ c_1 & c_2 & c_3 \end{vmatrix} \qquad (3.66)$$

Since rows and columns of a determinant can be interchanged without altering its value, we can write

$$V^2 = \begin{vmatrix} a_1 & a_2 & a_3 \\ b_1 & b_2 & b_3 \\ c_1 & c_2 & c_3 \end{vmatrix} \cdot \begin{vmatrix} a_1 & b_1 & c_1 \\ a_2 & b_2 & c_2 \\ a_3 & b_3 & c_3 \end{vmatrix} \qquad (3.67)$$

Multiplying the determinants, according to the rules for matrices, leads to

$$V^2 = \begin{vmatrix} a_1a_1 + a_2a_2 + a_3a_3 & a_1b_1 + a_2b_2 + a_3b_3 & a_1c_1 + a_2c_2 + a_3c_3 \\ b_1a_1 + b_2a_2 + b_3a_3 & b_1b_1 + b_2b_2 + b_3b_3 & b_1c_1 + b_2c_2 + b_3c_3 \\ c_1a_1 + c_2a_2 + c_3a_3 & c_1b_1 + c_2b_2 + c_3b_3 & c_1c_1 + c_2c_2 + c_3c_3 \end{vmatrix}$$
$$(3.68)$$

which may be expressed in vector notation as

$$V^2 = \begin{vmatrix} \mathbf{a}\cdot\mathbf{a} & \mathbf{a}\cdot\mathbf{b} & \mathbf{a}\cdot\mathbf{c} \\ \mathbf{b}\cdot\mathbf{a} & \mathbf{b}\cdot\mathbf{b} & \mathbf{b}\cdot\mathbf{c} \\ \mathbf{c}\cdot\mathbf{a} & \mathbf{c}\cdot\mathbf{b} & \mathbf{c}\cdot\mathbf{c} \end{vmatrix} \qquad (3.69)$$

Evaluating (3.69), we obtain

$$V^2 = a^2b^2c^2 + ab\cos\gamma\, bc\cos\alpha\, ca\cos\beta + ac\cos\beta\, ba\cos\gamma\, bc\cos\alpha$$
$$- ca\cos\beta\, b^2ca\cos\beta - bc\cos\alpha\, bc\cos\alpha\, a^2 - c^2ab\cos\gamma\, ab\cos\gamma \qquad (3.70)$$

which simplifies to

$$V = abc(1 - \cos^2\alpha - \cos^2\beta - \cos^2\gamma + 2\cos\alpha\cos\beta\cos\gamma)^{1/2} \quad (3.71)$$

Now VV^* follows from (3.67) as

$$VV^* = \begin{vmatrix} a_1 & a_2 & a_3 \\ b_1 & b_2 & b_3 \\ c_1 & c_2 & c_3 \end{vmatrix} \begin{vmatrix} a_1^* & b_1^* & c_1^* \\ a_2^* & b_2^* & c_2^* \\ a_3^* & b_3^* & c_3^* \end{vmatrix}$$

$$= \begin{vmatrix} a_1^*a_1 + a_2^*a_2 + a_3^*a_3 & a_1^*b_1 + a_2^*b_2 + a_3^*b_3 & a_1^*c_1 + a_2^*c_2 + a_3^*c_3 \\ b_1^*a_1 + b_2^*a_2 + b_3^*a_3 & b_1^*b_1 + b_2^*b_2 + b_3^*b_3 & b_1^*c_1 + b_2^*c_2 + b_3^*c_3 \\ c_1^*a_1 + c_2^*a_2 + c_3^*a_3 & c_1^*b_1 + c_2^*b_2 + c_3^*b_3 & c_1^*c_1 + c_2^*c_2 + c_3^*c_3 \end{vmatrix}$$

$$= \begin{vmatrix} \mathbf{a}\cdot\mathbf{a} & \mathbf{a}\cdot\mathbf{b} & \mathbf{a}\cdot\mathbf{c} \\ \mathbf{b}\cdot\mathbf{a} & \mathbf{b}\cdot\mathbf{b} & \mathbf{b}\cdot\mathbf{c} \\ \mathbf{c}\cdot\mathbf{a} & \mathbf{c}\cdot\mathbf{b} & \mathbf{c}\cdot\mathbf{c} \end{vmatrix}$$

$$= \begin{vmatrix} 1 & 0 & 0 \\ 0 & 1 & 0 \\ 0 & 0 & 1 \end{vmatrix} \quad (3.72)$$

Hence,

$$VV^* = 1 \quad (3.73)$$

3.4.2 Some Properties of the Reciprocal Lattice

The reciprocal lattice is particularly useful in that it enables the geometry of planes to be represented by the simpler geometry of points. We shall consider here some properties of the reciprocal lattice that are useful in calculations on crystal structures.

Interplanar Spacings

Let the plane (hkl) in Figure 3.22 be that in the family of such planes that is nearest to the origin. Then it intercepts the x, y, and z axes in A, B, and C as shown. The vector $\mathbf{d}^*(hkl)$ in reciprocal space is defined by

$$\mathbf{d}^*(hkl) = h\mathbf{a}^* + k\mathbf{b}^* + l\mathbf{c}^* \quad (3.74)$$

The vector $\mathbf{AB}$ (we use $\mathbf{AB}$ to represent the vector from A to B) is equal to $(\mathbf{a}/h - \mathbf{b}/k)$, so that, with (3.56) and (3.57), the scalar product $\mathbf{d}^*(hkl)\cdot(\mathbf{b}/k - \mathbf{a}/h)$ becomes

$$(h\mathbf{a}^* + k\mathbf{b}^* + l\mathbf{c}^*) \cdot (\mathbf{a}/h - \mathbf{b}/k) = \mathbf{a}\cdot\mathbf{a}^* - \mathbf{b}\cdot\mathbf{b}^* = 0$$

so that $\mathbf{d}^*(hkl)$ is normal to $\mathbf{AB}$. Similarly $\mathbf{d}^*(hkl)$ is normal to $\mathbf{BC}$ and $\mathbf{CA}$.

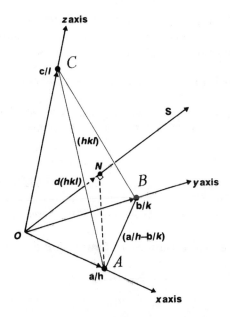

FIGURE 3.22.   A plane (*hkl*) in a Bravais lattice, showing the unit-cell vectors **a**, **b**, and **c**; N is the foot of the perpendicular from the origin O to the plane, so that ON is the interplanar spacing $d(hkl)$.

The interplanar spacing in the Bravais lattice $d(hkl)$ is the distance ON. If the unit vector along $d(hkl)$ is **n**, then $d(hkl) = \mathbf{OA} \cdot \mathbf{n}$. But $\mathbf{n} = \mathbf{d}^*(hkl)/|\mathbf{d}^*(hkl)|$; so that

$$d(hkl) = \mathbf{n} \cdot \mathbf{a}/h \ (= \mathbf{n} \cdot \mathbf{b}/k = \mathbf{n} \cdot \mathbf{c}/l) \qquad (3.75)$$

For $\kappa = 1$,

$$d(hkl) = \mathbf{n} \cdot \mathbf{a}/h = \frac{\mathbf{d}^*(hkl)}{d^*(hkl)} \cdot \mathbf{a}/h$$

$$= \frac{1}{d^*(hkl)}(h\mathbf{a}^* + k\mathbf{b}^* + l\mathbf{c}^*) \cdot \mathbf{a}/h = 1/d^*(hkl)$$

that is, the magnitude $d(hkl)$ is equal to $1/d^*(hkl)$. In this discussion, the indices h, k, and l are unique, and any or all of them may contain a common factor (see Section 3.1). From (3.74), we have

$$|\mathbf{d}^*(hkl)|^2 = (h\mathbf{a}^* + k\mathbf{b}^* + l\mathbf{c}^*) \cdot (h\mathbf{a}^* + k\mathbf{b}^* + l\mathbf{c}^*)$$

$$= h^2 a^{*2} + k^2 b^{*2} + l^2 c^{*2} + 2kl b^* c^* \cos \alpha^*$$

$$+ 2lh c^* a^* \cos \beta^* + 2hk a^* b^* \cos \gamma^* \qquad (3.76)$$

whence $d(hkl)$ and $d^*(hkl)$ can be calculated. As before, simplifications of (3.76) arise in the presence of symmetry higher than triclinic (see also Table 2.4).

Introducing the Bragg equation (3.55)

$$\sin^2 \theta(hkl) = \lambda^2/[4d^2(hkl)] = \lambda^2 d^{*2}(hkl)/4 \qquad (3.77)$$

where $d^{*2}(hkl)$ is given by (3.76), so that (3.77) can be used to express $\sin \theta$ for a reflection hkl in terms of the reciprocal unit cell parameters.

Angle Between Planes

Given any two planes $h_1 k_1 l_1$ and $h_2 k_2 l_2$, the angle between them can be found as the supplement of the angle between the two normals, $d^*(h_1 k_1 l_1)$ and $d^*(h_2 k_2 l_2)$; this angle is the *interfacial angle* of the stereographic projection (see Section 1.3).

In general, the angle ϕ between the forward directions of two vectors $\mathbf{p}$ and $\mathbf{q}$ is given by

$$\cos \phi = (\mathbf{p} \cdot \mathbf{q})/pq \qquad (3.78)$$

Following (3.76)

$$\cos \phi = [h_1 h_2 a^{*2} + k_1 k_2 b^{*2} + l_1 l_2 c^{*2}$$
$$+ (k_1 l_2 + k_2 l_1)b^* c^* \cos \alpha^* + (l_1 h_2 + l_2 h_1)c^* a^* \cos \beta^*$$
$$+ (h_1 k_2 + h_2 k_1)a^* b^* \cos \gamma^*]/[d^*(h_1 k_1 l_1)d^*(h_2 k_2 l_2)] \qquad (3.79)$$

where $d^*(h_1 k_1 l_1)$ and $d^*(h_2 k_2 l_2)$ are given by (3.76).

Weiss Zone Law

We discussed the concept of a zone briefly in Section 1.2.5. In a Bravais lattice, any row of lattice points is common to an indefinite number of lattice planes. These planes contain a common direction, so that they belong to one and the same zone. A vector $\mathbf{r}$ from the origin to the lattice point UVW is defined by

$$\mathbf{r} = U\mathbf{a} + V\mathbf{b} + W\mathbf{c}$$

and the symbol for the zone in which $\mathbf{r}$ is the common direction is $[U V W]$, where U, V, and W contain no common factor. If a plane (hkl) lies in this zone, then the vector $\mathbf{r}$ lies in or is parallel to the plane (hkl). Thus, a vector $\mathbf{d}^*$ normal to the plane is also normal to $\mathbf{r}$, that is, $\mathbf{r} \cdot \mathbf{d}^* = 0$. From the definition of $\mathbf{r}$ and using (3.74), it follows that

$$hU + kV + lW = 0 \qquad (3.80)$$

which is the *Weiss zone law*.

Equation of a Plane

Consider any plane (hkl) and a point $\mathbf{r}(x, y, z)$ in it; the normal to the plane is $\mathbf{d}^*(hkl)$. If the unit vector along $\mathbf{d}^*(hkl)$ is $\boldsymbol{\delta}$, the equation of the plane is $p = \mathbf{r} \cdot \boldsymbol{\delta}$, where p is the perpendicular distance from the origin to the plane. We may rewrite this equation as

$$\mathbf{r} \cdot \mathbf{d}^*(hkl)/|\mathbf{d}^*(hkl)| = p = nd(hkl) = n/|\mathbf{d}^*(hkl)| \qquad (3.81)$$

where (hkl) is the nth plane, in the family, from the origin. It follows from (3.81) that

$$hx + ky + lz = n \qquad (3.82)$$

When $n = 0$, the plane passes through the origin, and when $n = 1$ it is the first plane in the family from the origin. These results may be compared with those derived in Section 1.2.3.

3.4.3 Reciprocity of F and I Unit Cells

In Figure 3.23, we select a primitive unit cell from the face-centered unit cell by the transformation

$$\mathbf{a}_P = \tfrac{1}{2}\mathbf{b}_F + \tfrac{1}{2}\mathbf{c}_F$$
$$\mathbf{b}_P = \tfrac{1}{2}\mathbf{c}_F + \tfrac{1}{2}\mathbf{a}_F \qquad (3.83)$$
$$\mathbf{c}_P = \tfrac{1}{2}\mathbf{a}_F + \tfrac{1}{2}\mathbf{b}_F$$

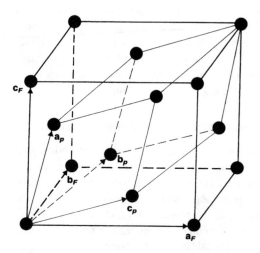

FIGURE 3.23. An F unit cell, with a P unit cell outlined within it. The number of lattice point per unit cell shows that $V_F/V_P = 4$.

Using (3.59) and (3.60), with appropriate cyclic permutation

$$
\begin{aligned}
\mathbf{a}_P^* &= (\mathbf{b}_P \times \mathbf{c}_P)/V_P = \left[\left(\tfrac{1}{2}\mathbf{c}_F + \tfrac{1}{2}\mathbf{a}_F \right) \times \left(\tfrac{1}{2}\mathbf{a}_F + \tfrac{1}{2}\mathbf{b}_F \right) \right] / V_P \\
&= [(\mathbf{c}_F \times \mathbf{b}_F) + (\mathbf{c}_F \times \mathbf{a}_F) + (\mathbf{a}_F \times \mathbf{b}_F)]/V_F \\
&= \frac{cb}{V} \sin \alpha + \frac{ca}{V} \sin \beta + \frac{ab}{V} \sin \gamma
\end{aligned}
$$

since $V_F = 4V_P$. Hence, we obtain

$$
\mathbf{a}_P^* = -\mathbf{a}_F^* + \mathbf{b}_F^* + \mathbf{c}_F^* \tag{3.84}
$$

with similar expressions for $\mathbf{b}_P^*$ and $\mathbf{c}_P^*$. The negative sign in front of $\mathbf{a}_F^*$ in (3.84) is introduced in order to preserve right-handed axes from the product $(\mathbf{c}_F \times \mathbf{b}_F)$.

In the case of the body-centered unit cell, the equations similar to (3.83) are

$$
\mathbf{a}_P = -\tfrac{1}{2}\mathbf{a}_I + \tfrac{1}{2}\mathbf{b}_I + \tfrac{1}{2}\mathbf{c}_I \tag{3.85}
$$

with similar expressions for $\mathbf{b}_P$ and $\mathbf{c}_P$. Writing (3.84) as

$$
\mathbf{a}_P^* = -2\mathbf{a}_F^*/2 + 2\mathbf{b}_F^*/2 + 2\mathbf{c}_F^*/2 \tag{3.86}
$$

we see that an F unit cell in a Bravais lattice reciprocates into an I unit cell in the corresponding reciprocal lattice, where the I unit cell is defined by the vectors $2\mathbf{a}_F^*$, $2\mathbf{b}_F^*$, and $2\mathbf{c}_F^*$. If, as is customary in practice, we define the reciprocal of an F unit cell by vectors $\mathbf{a}_F^*$, $\mathbf{b}_F^*$, and $\mathbf{c}_F^*$, then only those reciprocal lattice points for which each of $h + k$, $k + l$ (and $l + h$) are even integers belong to the reciprocal of the F unit cell. In other words, Bragg reflections from an F unit cell have indices of the same parity (see also Section 3.7.1 and Table 3.2).

3.4.4   Reciprocal Lattice and the Reflection Condition

Figure 3.24 shows a section of a reciprocal lattice, with the vectors $\mathbf{s}_0/\lambda$ and $\mathbf{s}/\lambda$ lying in the incident and scattered x-ray beams respectively; a crystal is situated at the point P. The vector $\mathbf{OQ}$, which is also the direction of $\mathbf{S}$, must be normal to an (hkl) plane and have the magnitude $(2/\lambda) \sin \theta$. When the conditions for diffraction are satisfied, according to (3.55), $S = |\mathbf{d}^*(hkl)|$. A sphere of radius $1/\lambda$ is described on P as center, and passing through O, the origin of the reciprocal lattice. If another reciprocal lattice point, such as Q, lies on the sphere, then vector $\mathbf{PQ}$ is the direction of the diffracted beam; the incident beam is along $\mathbf{PO}$. The sphere is known as the *sphere of reflection*, and will be used freely in later chapters.

TABLE 3.2.   Limiting Conditions for Centered Unit Cells

Unit-cell-type	Limiting conditions	Associated translations	Structure factor multiplier G
P	None	None	1
A	$hkl : k + l = 2n$	$b/2 + c/2$	2
B	$hkl : l + h = 2n$	$c/2 + a/2$	2
C	$hkl : h + k = 2n$	$a/2 + b/2$	2
I	$hkl : h + k + l = 2n$	$a/2 + b/2 + c/2$	2
F	$hkl : h + k = 2n$	$a/2 + b/2$	
	$hkl : k + l = 2n$	$b/2 + c/2$	
	$hkl : (l + h = 2n)^a$	$c/2 + a/2$	4
R_{hex}^b	$hkl : -h + k + l = 3n_{obv}$	$a/3 + 2b/3 + 2c/3$	
		$2a/3 + b/3 + c/3$	3
	or		
	$hkl : h - k + l = 3n_{rev}$	$a/3 + 2b/3 + c/3$	
		$2a/3 + b/3 + 2c/3$	3

[a] This condition is not independent of the other two.
[b] See Section 2.2.3 and Table 2.3.

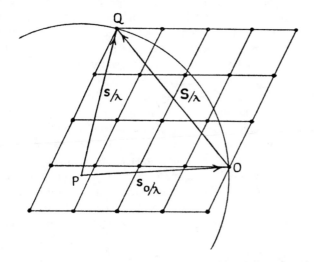

FIGURE 3.24.   Sphere of reflection, radius $1/\lambda$, with the crystal at P, the centre. The origin of the reciprocal lattice is at O. When a reciprocal lattice point, such as Q, lies on the sphere, a reflection arises along the direction PQ.

3.5   Scattering by a Crystal Structure

In Chapter 2, we discussed the geometrical properties of the lattices on which all crystal structures are based, and showed how the arrays of atoms or molecules may be arranged according to space-group symmetry. We need next to consider the diffraction of x-rays by a crystal. We refer the process of scattering to a conventional unit cell, and determine the resultant effect of all atoms in the unit cell, which leads directly to the equation for the *structure factor*. The scattering of x-rays by a crystal comprises two parts: a geometrical part that depends on the symmetry of the arrangement of its components, and a structural part that depends upon both the nature of the atoms or molecules comprising those entities and their relative positions in the unit cell.

Earlier in this chapter, we introduced the Laue equations for x-ray scattering from a crystal and showed their equivalence to the Bragg equation. In Figure 3.25a, we consider again Bragg reflection of x-rays from the (hkl) family of planes. The plane through O passes through the origin of the unit cell, and may be called the zero (hkl) plane. The plane through O' is the first (hkl) plane from the origin, and is, therefore, at a perpendicular distance $d(hkl)$ from the plane through O. The path difference between x-rays reflected from the planes through O and the adjacent plane through O' is, from (3.55), $2d(hkl) \sin \theta(hkl)$.

Consider an atom lying between the planes through O and O'; we can imagine a plane, parallel to (hkl), passing through this atom. For an incident x-ray beam in the correct reflecting position, the planes through O and O' will reflect in phase with each other but, in general, not in phase with the reflection from the plane through A. We need to determine a general expression for the path difference between waves reflected by planes such as those through O and A, in terms of the coordinates of the atoms in the structure.

A further illustration of Bragg scattering (reflection) is provided by Figure 3.25b. It indicates how individual incident wavelets may be considered to be scattered by the constituent atoms of a molecule, in this example a mono-substitued derivative of benzene. The scattered wavelet from each atom has an amplitude f characteristic of that atom, and a phase that depends on the position of the atom in the unit cell, in relation to the plane (hkl). Atoms related by lattice translations scatter in phase with one another for the given reflection. The combined scattering from atoms related by other symmetry operations can be evaluated through a knowledge of the space group, and this matter will be discussed shortly.

3.5.1   Path Difference

In Figure 3.22, the plane ABC, at a perpendicular distance $d(hkl)$ from the origin O, makes intercepts a/h, b/k, and c/l with the general crystallographic axes x, y, and z, respectively. The perpendicular ON to the plane makes angles χ, ψ, and ω with the same axes (see also Appendix A2). The intercept form of the equation of the plane is, from (1.11),

$$hX/a + kY/b + lZ/c = 1 \qquad (3.87)$$

(a)

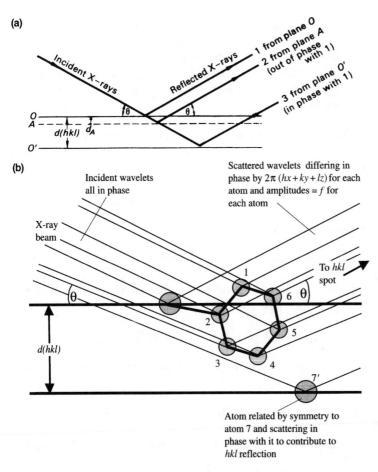

(b)

FIGURE 3.25.   Constructions for Bragg reflection from crystal planes: (a) Bragg reflection from (hkl) planes, taking into account the variation of atomic distribution throughout the unit cell. Planes O and O' are similar, but contain a different sample of atoms from that on the conceptual, parallel plane A. The ray 2 reflected from plane A is out of phase with those (1 and 3) reflected from planes O and O'. (b) Model of the scattering of incident x-rays by individual atoms of the species in a crystal structure. Atoms related by lattice translations scatter in phase; the phase relationships of atoms related by other symmetry depend upon the space group of the crystal.

where $a/h = d(hkl)/\cos\chi$, $b/k = d(hkl)/\cos\psi$, and $c/l = d(hkl)/\cos\omega$. Hence, equation (3.87) may be written as

$$X\cos\chi + Y\cos\psi + Z\cos\omega = d(hkl) \tag{3.88}$$

where X, Y, and Z may be taken as the coordinates of the point N, and $\cos\chi$, $\cos\psi$, and $\cos\omega$ are the direction cosines of ON with respect to x, y, and z, respectively. This equation represents the plane ABC, and others parallel to it, in terms of the perpendicular distance $d(hkl)$. In Figure 3.25a, the plane through O' may be identified with plane ABC; the plane through O has a zero value for $d(hkl)$, because it is a plane through the origin. The plane through A has the perpendicular distance d_A, and since plane ABC has the Miller indices (hkl), it follows from the foregoing that

$$\cos \chi = d(hkl)/(a/h) \quad \cos \psi = d(hkl)/(b/k) \quad \cos \omega = d(hkl)/(c/l)$$
$$(3.89)$$

Thus, the equation of the plane through A is

$$X \cos \chi + Y \cos \psi + Z \cos \omega = d_A \qquad (3.90)$$

Eliminating the direction cosines through (3.89), we have

$$(hX/a + kY/b + lZ/c)d(hkl) = d_A \qquad (3.91)$$

Let an atom on the plane through A have fractional coordinates $x_A = X_A/a$, $y_A = Y_A/b$, and $z_A = Z_A/c$. Then,

$$d_A = (hx_A + ky_A + lz_A)\, d(hkl) \qquad (3.92)$$

Using (3.53), the path difference δ_A for waves reflected from the planes through O and A is

$$\delta_A = 2d_A \sin \theta(hkl) = 2(hx_A + ky_A + lz_A)\, d(hkl) \sin \theta(hkl) \qquad (3.93)$$

Eliminating $2d(hkl) \sin \theta(hkl)$ through the Bragg equation (3.55), it follows that

$$\delta_A = \lambda(hx_A + ky_A + lz_A) \qquad (3.94)$$

so that the corresponding phase angle ϕ_A is

$$\phi_A = (2\pi/\lambda)\delta_A = 2\pi(hx_A + ky_A + lz_A) \qquad (3.95)$$

This relationship expresses a quantitative measure of the phase difference between rays reflected from different (conceptual) planes of atoms, all parallel to any given (hkl) plane, with respect to the parallel plane through the origin. It must be realized that, in a real situation, there need be no atom lying on any given (hkl) plane; the electron density associated with each atom pervades the whole of the crystal space.

We can reach the result in (3.95) in another way. In (3.16) we expressed the resultant wave from n scattering centeres, there thought of as electrons. Now, we can consider them as atoms each specified in scattering power by $f(\mathbf{S})$, or $f(\theta)$.

The contribution to a wave scattered by the Ath atom in a unit cell is $f_A(\mathbf{S})\exp(i2\pi\mathbf{r}_A\cdot\mathbf{S})$, where $\exp(i2\pi\mathbf{r}_A\cdot\mathbf{S})$ is the phase of the contribution from the Ath atom. Now $\mathbf{r}_A$ is given by

$$\mathbf{r}_A = x_A\mathbf{a} + y_A\mathbf{b} + z_A\mathbf{c} \qquad (3.96)$$

where x_A, y_A, and z_A are the fractional coordinates of atom A, as before. Since $|\mathbf{S}| = 2\sin\theta(hkl)/\lambda$, which from the Bragg equation is $1/d(hkl)$, or $d^*(hkl)$, $\mathbf{S}$ is the reciprocal lattice vector $\mathbf{d}^*(hkl)$. Applying (3.74), we have

$$\mathbf{r}_A \cdot \mathbf{S} = (x_A\mathbf{a} + y_A\mathbf{b} + z_A\mathbf{c}) \cdot (h\mathbf{a}^* + k\mathbf{b}^* + l\mathbf{c}^*) = hx_A + ky_A + lz_A \quad (3.97)$$

Thus, we have again, following (3.16), the phase angle for an atomic species A as $2\pi(\mathbf{r}\cdot\mathbf{S})$, or

$$\phi_A = 2\pi(hx_A + ky_A + lz_A) \qquad (3.98)$$

and the phase contribution $\exp[i2\pi(hx_A + ky_A + lz_A)]$.

3.5.2 Argand Diagram

In Sections 3.2.2 and 3.2.3, we considered the summation of waves of equal frequency, each wave being characterized by an amplitude and a phase. In formulating an equation to represent the combined scattering from all atoms in a unit cell, we allocate to the jth atom an amplitude f_j, its atomic scattering factor, and a phase ϕ_j, measured with respect to the origin of the unit cell. The scattering from the unit cell is then the sum of the scattering by the individual atoms, at any value of $\mathbf{S}$. Each wave, then, has the form $f_j\exp(i\phi_j)$, where $\phi_j = (\mathbf{r}_j\cdot\mathbf{S})$. Then each atomic contribution takes the form $f_j\exp(i\phi_j)$. In Section 3.2.3, we showed how this term may be expanded, by de Moivre's theorem, into cosine (real) and sine (imaginary) components. Therefore, a simple way of representing this wave is as a vector in the complex plane, on an Argand diagram.

In Figure 3.26, we show the combination of two waves $\mathbf{f}_1$ and $\mathbf{f}_2$ to give the resultant $\mathbf{F}$:

$$\mathbf{F} = \mathbf{f}_1 + \mathbf{f}_2 = f_1\exp(i\phi_1) + f_2\exp(i\phi_2) \qquad (3.99)$$

$\mathbf{F}$, $\mathbf{f}_1$, and $\mathbf{f}_2$ are vectors in the complex plane, having both a magnitude and direction; $\exp(i\phi)$ may be regarded as an operator that rotates a vector $\mathbf{f}$ counterclockwise on an Argand diagram by the angle ϕ from the positive, real axis.

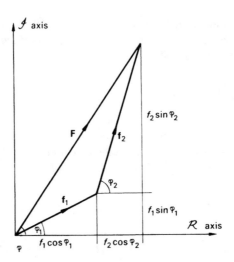

FIGURE 3.26.   Combination of the two waves $f_1 \exp(i\phi_1)$ and $f_2 \exp(i\phi_2)$, shown as vectors on an Argand diagram. The resultant is $\mathbf{F}$, and its phase is expressed by the angle ϕ between $\mathbf{F}$ and the $\mathcal{R}$ (real) axis.

3.5.3   Combination of n Waves

The foregoing analysis may be extended to n waves. The resultant sum $\mathbf{F}$ is, from (3.99),

$$\mathbf{F} = f_1 \exp(i\phi_1) + f_2 \exp(i\phi_2) + \cdots + f_j \exp(i\phi_j) + \cdots = \sum_{j=1}^{n} f_j \exp(i\phi_j)$$

(3.100)

On an Argand diagram, (3.100) expresses a polygon of vectors, shown in Figure 3.27 for the combination of six waves, where the resultant $\mathbf{F}$ may be represented as

$$\mathbf{F} = |\mathbf{F}| \exp(i\phi) \tag{3.101}$$

where the amplitude $|\mathbf{F}|$ is obtained from

$$|\mathbf{F}| = (\mathbf{F}\,\mathbf{F}^*)^{1/2} \tag{3.102}$$

$\mathbf{F}^*$ is the complex conjugate of $\mathbf{F}$, that is, $|\mathbf{F}| \exp(-i\phi)$ (Figure 3.28). Following Section 3.2.3,

$$|\mathbf{F}| = (A'^2 + B'^2)^{1/2} \tag{3.103}$$

where

$$A' = \sum_{j=1}^{n} f_j \cos \phi_j \tag{3.104}$$

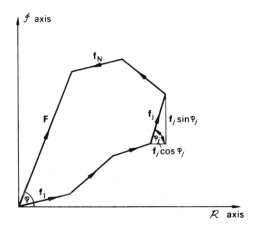

FIGURE 3.27.   Combination of N waves ($N = 6$) on an Argand diagram: $\mathbf{F} = \sum\limits_{j=1}^{6} f_j \exp(\mathrm{i}\phi_j)$.

and

$$B' = \sum_{j=1}^{n} f_j \sin \phi_j \qquad (3.105)$$

The phase ϕ of the resultant $\mathbf{F}$ is given by

$$\phi = \tan^{-1}(B'/A') \qquad (3.106)$$

3.5.4 Structure Factor Equation

We need to express the equation that has been deduced for $\mathbf{F}$ in a manner that includes the coordinates of the atoms in the unit cell. It follows from (3.96) to (3.98) that the phase for the jth atom can be represented by $2\pi(hx_j + ky_j + lz_j)$. Hence, we use this expression in (3.100) to give the structure factor equation for the hklth reflection:

$$F(hkl) = \sum_{j=1}^{n} f_{j,\theta} \exp[\mathrm{i}2\pi(hx_j + ky_j + lz_j)] \qquad (3.107)$$

Although $F(hkl)$ has the quality of a vector, it is conventional not to represent it normally in bold (vector) type; $F(hkl)$ refers to the combined scattering from the n atoms in the unit cell to give the hklth spectrum, or the wave from the (hkl) planes, relative to the scattering by a single electron at the origin. The atomic scattering factor $f_{j,\theta}$ for the jth atom expresses its dependence on θ, an alternative parameter to $\mathbf{S}$ in this context. Again, frequently the θ dependence is not expressed in the formula, although it is implicitly always present. The atomic coordinates in

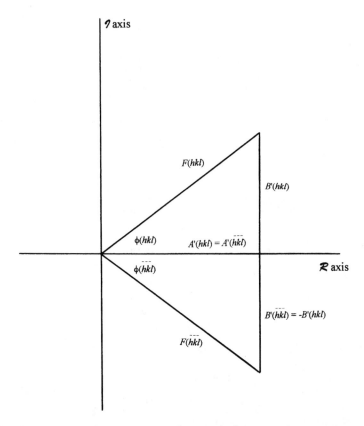

FIGURE 3.28. $F(hkl)$ and its components $A'(hkl)$ and $B'(hkl)$; $\phi(hkl)$ is the phase angle for $F(hkl)$. Its conjugate $F^*(hkl)$, or $F(\bar{h}\bar{k}\bar{l})$, has the same value for A', and the same magnitude for B' but with its sign reversed; hence $\phi(hkl) = -\phi(\bar{h}\bar{k}\bar{l})$. In a centrosymmetric structure, $F(hkl)$ lies along the real axis, so that $F(hkl) = A'(hkl)$ and its phase is either 0 or π. Since, $A'(hkl) = |F(hkl)| \cos \phi$, it is common to speak of the *sign* of $F(hkl)$ in a centrosymmetric structure.

(3.100) are fractional, as applied in Section 3.5.1, and so are independent of the size of the unit cell.

3.6   Using the Structure Factor Equation

In the next two sections, we explore some of the properties and appli-cations of the structure factor equation that are encountered in practical x-ray crystallography. The trigonometrical relations in Appendix A6 may be helpful in some of the ensuing arguments.

3.6.1   Friedel's Law

Except where anomalous scattering (see Section 6.5) is significant, x-ray diffraction spectra form a centrosymmetric array. The diffraction spectra from a crystal may be thought of as its reciprocal lattice, with each point hkl weighted by the corresponding value of the amplitude $|F(hkl)|$, or the intensity $I(hkl)$. Friedel's law expresses the centrosymmetric property as

$$|I(hkl)| = |I(\bar{h}\bar{k}\bar{l})| \qquad (3.108)$$

within the limits of experimental error, and may be derived as follows.

Since the atomic scattering factor is a function of $(\sin\theta/\lambda)$, it will have the same value for both the hkl and $\bar{h}\bar{k}\bar{l}$ reflections. Thus, $f_\theta = f_{-\theta}$, because reflections from opposite sides of any plane occur at the same value of the Bragg angle θ.

From (3.107)

$$F(\bar{h}\bar{k}\bar{l}) = \sum_{j=1}^{n} f_j \exp[-\mathrm{i}2\pi(hx_j + ky_j + lz_j)] \qquad (3.109)$$

and from Figure 3.28

$$\begin{aligned}
F(hkl) &= A'(hkl) + \mathrm{i}B'(hkl) \\
F(\bar{h}\bar{k}\bar{l}) &= A'(\bar{h}\bar{k}\bar{l}) + \mathrm{i}B'(\bar{h}\bar{k}\bar{l}) = A'(hkl) - \mathrm{i}B'(hkl)
\end{aligned} \qquad (3.110)$$

Hence, the following relations hold:

$$\phi(hkl) = -\phi(\bar{h}\bar{k}\bar{l})$$

$$|F(hkl)| = |F(\bar{h}\bar{k}\bar{l})| = [A'^2(hkl) + B'^2(hkl)]^{1/2} \qquad (3.111)$$

and

$$I(hkl) = I(\bar{h}\bar{k}\bar{l})$$

which is Friedel's law.

3.6.2   Structure Factor for a Centrosymmetric Crystal

One of the questions that frequently arises at the outset of a crystal structure determination is whether or not the space group is centrosymmetric. In a centrosymmetric structure, with the origin on a center of symmetry, the n atoms in the unit cell lie in related pairs, with coordinates $\pm(x, y, z)$. From (3.104) and (3.105), expressing the phase angle as shown in (3.98), we write for the two parts of the structure factor equation

$$A'(hkl) = \sum_{j=1}^{n/2} f_j[\cos 2\pi(hx_j + ky_j + lz_j) + \cos 2\pi(-hx_j - ky_j - lz_j)]$$

$$= 2\sum_{j=1}^{n/2} f_j \cos 2\pi(hx_j + ky_j + lz_j) \tag{3.112}$$

$$B'(hkl) = \sum_{j=1}^{n/2} f_j[\sin 2\pi(hx_s + ky_s + lz_s) + \sin 2\pi(-hx_s - ky_s - lz_s)]$$

$$= 0 \tag{3.113}$$

because $\sin(-\phi) = -\sin(\phi)$, for all ϕ. In this case, $A'(hkl) = F(hkl)$, and $\phi(hkl)$ can take only the values 0 or π, so that the phase angle attaches itself to $|F(hkl)|$ as a positive or negative sign. We often speak of the *signs* of reflections in centrosymmetric crystals, so that $F(hkl) = s(hkl)|F(hkl)|$. These results apply only when the origin of the unit cell is taken on $\bar{1}$; in any other setting of the origin in a centrosymmetric space group, there will be a nonzero component in $B'(hkl)$. Centrosymmetric crystals usually present fewer difficulties to the structure analyst than do non-centrosymmetric crystals, in which the phase angles range from 0 to 2π.

3.7   Limiting Conditions and Systematic Absences

We considered limiting conditions briefly in Chapter 2; here we investigate them more fully through the structure factor equation, and show how they are determined for several different example symmetries.

An x-ray diffraction pattern can be used to determine the type of unit cell that corresponds to the chosen system of reference axes. From (3.107), it would be an unexpected coincidence for many intensities to be zero. With unit cells having no translational symmetry, the intensity of a reflection is not usually zero for any particular combinations of h, k, and l: no *limiting conditions* apply in such a case. Vanishingly weak intensities may arise for certain reflections because of the particular structure under investigation; we call these reflections *accidental absences*, and we shall discuss them further shortly. In centered unit cells or in the presence of translational symmetry, glide planes and screw axes, reflections of certain combinations of h, k, and l are totally absent; we call such unobservable reflections *systematic absences*.

3.7.1 Body-Centered Unit Cell

As a first example, we know that in a body-centered (I) unit cell, the atoms are related in pairs as x, y, z and $\frac{1}{2} + x, \frac{1}{2} + y, \frac{1}{2} + z$. Using (3.107), we have

$$
F(hkl) = \sum_{j=1}^{n/2} f_j \{\exp[\mathrm{i}2\pi(hx_j + ky_j + lz_j)]
$$

$$
+ \exp[\mathrm{i}2\pi(hx_j + ky_j + lz_j + h/2 + k/2 + l/2)]\} \tag{3.114}
$$

The term within the braces $\{\ldots\}$ may be expressed as $\exp[\mathrm{i}2\pi(hx_j+ky_j+lz_j)]\{1+\exp[\mathrm{i}2\pi(h+k+l)/2]\}$. Since $h+k+l$ is integral, $\{1+\exp[\mathrm{i}2\pi(h+k+l)/2]\} = 1+\cos[2\pi(h+k+l)/2] = 2\cos^2[2\pi(h+k+l)/4] = G$, where G is a multiplying factor for the *reduced* structure factor equation in centered unit cells; in the body centered unit cell $G = 2$, so that

$$
F(hkl) = 2\cos^2[2\pi(h+k+l)/4] \sum_{j=1}^{n/2} f_j \exp[\mathrm{i}2\pi(hx_j + ky_j + lz_j)] \tag{3.115}
$$

This equation may be broken down into its two components, $A'(hkl)$ and $B'(hkl)$, in the usual way. Further simplification is possible: in this example, G takes the value 2 if $h+k+l$ is even, and 0 if $h+k+l$ is odd. Hence, we write the *limiting condition* that shows which reflections are permitted by the geometry of an I unit cell as

$$
hkl: h+k+l = 2n \quad n = 0, \pm 1, \pm 2, \ldots
$$

The same situation expressed as *systematic absences*, the condition under which reflections are forbidden by the space-group geometry, is

$$
hkl: h+k+l = 2n+1 \quad n = 0, \pm 1, \pm 2, \ldots
$$

Both terms are in common use, and the reader should distinguish between them carefully.

Analogous expressions can be derived for any centered unit cell. The G factors for all types of centering have been summarized in Table 3.2. It is evident that, where a reflection arises in a centered unit cell, the structure factor equation has the same form as that for the corresponding primitive unit cell, but multiplied by the G factor appropriate to the unit-cell type. The summation in the reduced structure factor equation is then taken over that fraction of atoms *not* related by the centering symmetry. In practice, the diffraction pattern is recorded, indices allocated to the spectra and then scrutinized for systematic absences, so as to determine the unit-cell type. The reader may care to work through the derivations of $F(hkl)$ for, say, a C and an F unit cell, and determine the limiting conditions for each unit cell type.

3.7.2   Screw Axes and Glide Planes

As we are concerned in this discussion with the geometry of the unit cell rather than the chemical nature of its contents, it is convenient to introduce the following nomenclature. Let N be the total number of atoms in the unit cell, and let n of them be the number in the asymmetric unit, with the number of asymmetric units being m, so that $N = nm$. Symbolically, we may write

$$\sum_{j=1}^{N} \equiv \sum_{r=1}^{n}\sum_{s=1}^{m}$$

where the sum over r refers to the symmetry-independent atoms, and that over s to the symmetry-related atoms. Thus, the structure factor equation contains two parts that may be considered separately. The sum over m symmetry-related atoms is expressed through the coordinates of a set of general equivalent positions. Thus,

$$A(hkl) = \sum_{s=1}^{m} \cos 2\pi(hx_s + ky_s + lz_s)$$

$$B(hkl) = \sum_{s=1}^{m} \sin 2\pi(hx_s + ky_s + lz_s)$$

(3.116)

Extending to the n atoms in the asymmetric unit, with one such term for each atom,

$$A'(hkl) = \sum_{r=1}^{n} f_r A_r(hkl)$$

$$B'(hkl) = \sum_{r=1}^{n} f_r B_r(hkl)$$

(3.117)

We shall not attach the subscript θ to the atomic scattering factor f unless the occasion demands it; we know that its value depends on θ, for a given wavelength λ.

The terms $A(hkl)$ and $B(hkl)$ are independent of the nature and arrangement of the atoms in the asymmetric unit; they are a property of the space-group symmetry, and are called *geometrical structure factors*. We shall consider some examples taken from the monoclinic and orthorhombic systems, to show how glide-plane and screw-axis symmetries give rise to limiting conditions with special classes of reflections. For this discussion, the subscript s in (3.116) need not be retained, because all m positions are related to the position x, y, z by symmetry.

Space Group $P2_1$

General equivalent positions: x, y, z; $\bar{x}, \frac{1}{2} + y, \bar{z}$ (see Figure 2.30).

Geometric structure factors:

$$A(hkl) = \cos 2\pi(hx + ky + lz) + \cos 2\pi(-hx + ky - lz + k/2)$$
$$= 2\cos 2\pi(hx + lz - k/4)\cos 2\pi(ky + k/4) \tag{3.118}$$

In a similar way,

$$B(hkl) = \sin 2\pi(hx + kx + lz) + \sin 2\pi(-hx + ky - lz + k/2)$$
$$= 2\cos 2\pi(hx + lz - k/4)\sin 2\pi(ky + k/4) \tag{3.119}$$

Limiting Conditions in $P2_1$

Geometric structure factors enable one to determine limiting conditions, that is, to predict which reflections are capable of arising in an x-ray diffraction pattern. If we can show, for given values of h, k, and l, that both $A(hkl)$ and $B(hkl)$ are systematically zero, then $I(hkl)$ will be zero, regardless of the atomic positions.

For $P2_1$, we can cast (3.118) and (3.119) in the following forms, according to the parity (evenness or oddness) of k. Expanding (3.118), we have (see Appendix A6)

$$A(hkl)/2 = [\cos 2\pi(hx + lz)\cos 2\pi(k/4) + \sin 2\pi(hx + lz)\sin 2\pi(k/4)]$$
$$\times [\cos 2\pi(ky)\cos 2\pi(k/4) - \sin 2\pi(ky)\sin 2\pi(k/4)] \tag{3.120}$$

In expanding the right-hand side of (3.120), terms such as

$$\cos 2\pi(hx + lz)\cos 2\pi(k/4)\sin 2\pi(ky)\sin 2\pi(k/4)$$

occur. This particular term is equivalent to

$$\tfrac{1}{2}\cos 2\pi(hx + lz)\sin 2\pi(ky)\sin 4\pi(k/4)$$

which is zero, because k is an integer. Hence, (3.120) becomes

$$A(hkl)/2 = [\cos 2\pi(hx + lz)\cos 2\pi(ky)\cos^2 2\pi(k/4)]$$
$$- [\sin 2\pi(hx + lz)\sin 2\pi(ky)\sin^2 2\pi(k/4)] \tag{3.121}$$

In a similar manner, we find

$$B(hkl)/2 = [\cos 2\pi(hx + lz)\sin 2\pi(ky)\cos^2 2\pi(k/4)]$$
$$+ [\sin 2\pi(hx + lz)\cos 2\pi(ky)\sin^2 2\pi(k/4)] \tag{3.122}$$

Separating for k even and k odd, we obtain

$$k = 2n \qquad A(hkl) = 2\cos 2\pi(hx + lz)\cos 2\pi(ky) \qquad (3.123)$$

$$B(hkl) = 2\cos 2\pi(hx + lz)\sin 2\pi(ky) \qquad (3.124)$$

$$k = 2n + 1 \qquad A(hkl) = -2\sin 2\pi(hx + lz)\sin 2\pi(ky) \qquad (3.125)$$

$$B(hkl) = 2\sin 2\pi(hx + lz)\cos 2\pi(ky) \qquad (3.126)$$

Only one systematic condition can be extracted from these equations: if both h and l are zero, then from (3.125) and (3.126)

$$A(hkl) = B(hkl) = 0$$

In other words, the limiting condition associated with a 2_1 axis is

$$0k0\colon k = 2n$$

The example of the 2_1 axis has been treated in detail; it shows again how a diffraction record may be used to reveal information about the translational symmetry elements of a space group. We can show how the limiting conditions for a 2_1 axis arise from a consideration of the Bragg equation. Figure 3.29 is a schematic illustration of a 2_1 symmetry pattern; the motif ❡ represents a structure at a height z, and $\circlearrowleft$ the structure at a height $\bar{z}$ after operating on it with the 2_1 axis. The planes MM' represent the family $(0k0)$ and NN' the family $(02k, 0)$.

Reflections of the type $(0k0)$ from MM' planes are canceled by the reflections from the NN' planes, because their phase change relative to MM' is $180°$. Clearly this result is not obtained with the $02k, 0$ reflections. Although the figure illustrates

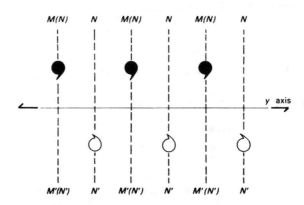

FIGURE 3.29.   Pattern of a structure containing a 2_1 screw axis: $d(NN') = d(MM')/2$, so that the MM' planes are *halved* by the NN' family.

TABLE 3.3.   Limiting Conditions for Screw Axes

Screw axis	Orientation	Limiting condition	Translational component
2_1	$\parallel x$	$h00 : h = 2n$	$a/2$
2_1	$\parallel y$	$0k0 : k = 2n$	$b/2$
2_1	$\parallel z$	$00l : l = 2n$	$c/2$
3_1 or 3_2	$\parallel z$	$000l : l = 3n$	$c/3$ or $2c/3$
4_1 or 4_3	$\parallel z$	$00l : l = 4n$	$c/4$ or $3c/4$
4_2	$\parallel z$	$00l : l = 2n$	$2c/4\ (c/2)$
6_1 or 6_5	$\parallel z$	$000l : l = 6n$	$c/6$ or $5c/6$
6_2 or 6_4	$\parallel z$	$000l : l = 3n$	$2c/6\ (c/3), c/6(2c/3)$
6_3	$\parallel z$	$000l : l = 2n$	$3c/6\ (c/2)$

the situation for $k = 1$, the same argument can be applied to any pair of values k and $2k$, where k is an odd integer.

Limiting conditions for other screw axes, and in other orientations, can be written down by analogy. Try to decide what the conditions for the following screw axes are, and then check your findings with Table 3.3:

2_1 parallel to x
3_2 parallel to z
4_3 parallel to y
6_5 parallel to z

Notice that pure rotation axes, as in space group $P2$, do not introduce any limiting conditions.

Centric Zones

Centric zones, sometimes termed, loosely, centrosymmetric zones (see also Section 3.10.3), are of particular importance in crystal structure determination. In space group $P2_1$ and other space groups of crystal class 2, the $h0l$ reflections are of special interest. Among equations (3.123)–(3.126), only (3.123) is relevant here because zero behaves as an even number, and $\sin(2\pi 0y) = 0$. Hence,

$$A(h0l) = 2\cos 2\pi (hx + lz)$$
$$B(h0l) = 0$$

(3.127)

From (3.106), $\phi(h0l)$ is either 0 or π; in other words, the [010] zone is centric, which is important for a structure analysis in this space group. Centric zones occur in the non-centrosymmetric space groups that have 2 as a subgroup of their point groups (see Sections 1.4.2 and 2.7.3ff).

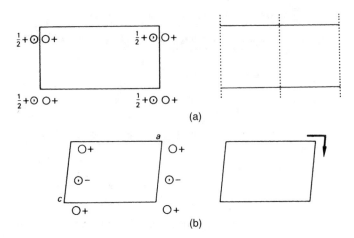

FIGURE 3.30.   Space group Pc: (a) viewed along c; (b) viewed along b.

Space Group Pc

General equivalent positions: $x, y, z; x, \bar{y}, \frac{1}{2} + z$ (see Figure 3.30). Geometric structure factors: proceeding as before, we obtain

$$A(hkl) = 2 \cos 2\pi (hx + lz + l/4) \cos 2\pi (ky - k/4)$$
$$B(hkl) = 2 \sin 2\pi (hx + lz + l/4) \cos 2\pi (ky - l/4)$$

(3.128)

If we expand these equations as with the example for $P2_1$, and then separate the terms for l even and l odd, we shall find systematic absences only for the $h0l$ reflections. Thus, the limiting condition for Pc is

$$h0l: l = 2n$$

Again, there is a relationship between the index (l) involved in the condition and the symmetry translation ($c/2$).

Space Group P2₁/c

This space group, which is often encountered in practice, contains the two translational symmetry operations already discussed, namely, 2_1 parallel to y and c-glide normal to y (see Figure 2.32 and Problem 2.8a).

$P2_1/c$ is a centrosymmetric space group, and the general equivalent positions may be summarized as

$$\pm\{x, y, z; \ x, \tfrac{1}{2} - y, \tfrac{1}{2} + z\}$$

Geometric structure factors: in the standard setting of this space group, the origin is on $\bar{1}$, so that we can immediately apply equations (3.113) and write

$$A(hkl) = 2\{\cos 2\pi[hx + ky + lz] + \cos 2\pi[hx - ky + lz + (k + l)/2]\}$$
$$B(hkl) = 0 \qquad\qquad (3.129)$$

Combining the two cosine terms

$$A(hkl) = 4\cos 2\pi[hx + lz + (k + l)/4]\cos 2\pi[ky - (k + l)/4]$$

Separating for $k + l$ even and odd, we obtain

$$k + l = 2n \qquad A(hkl) = 4\cos 2\pi(hx + lz)\cos 2\pi(ky)$$
$$k + l = 2n + 1 \quad A(hkl) = -4\sin 2\pi(hx + lz)\sin 2\pi(ky) \qquad (3.130)$$

We now deduce the limiting conditions as

$$
\begin{array}{lll}
hkl: & \text{None} & P \text{ unit cell} \\
h0l: & l = 2n & c\text{-glide normal to } y \\
0k0: & k = 2n & 2_1 \text{ axis parallel to } y
\end{array}
$$

These three classes of reflections are important in monoclinic reciprocal space, because only with them can we determine the characteristic systematic absences. Despite Friedel's law, the diffraction symmetry reveals the true space group in this example. Figure 3.31 illustrates weighted reciprocal space levels for a monoclinic crystal of space group Pc, $P2/c$, or $P2_1/c$.

Space Group *Pma2*

From the data in Figure 3.32 we can write down expressions for the geometric structure factors:

$$
\begin{aligned}
A(hkl) = {} & \cos 2\pi(hx + ky + lz) + \cos 2\pi(-hx - ky + lz) \\
& + \cos 2\pi(-hx + ky + lz + h/2) + \cos 2\pi(hx - ky + lz + h/2)
\end{aligned}
$$
$$(3.131)$$

Combining the first and third, and second and fourth terms, we have

$$
\begin{aligned}
A(hkl) = {} & 2\cos 2\pi(ky + lz + h/4)\cos 2\pi(hx - h/4) \\
& + 2\cos 2\pi(-ky + lz + h/4)\cos 2\pi(hx + h/4)
\end{aligned}
$$
$$(3.132)$$

Further simplification of this expression requires the separate parts to contain a common factor. We return to (3.131) and make a minor alteration to the term

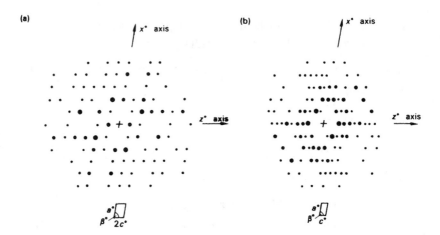

FIGURE 3.31. Reciprocal nets in the x^*, z^* plane appropriate to space groups Pc, $P2/c$, and $P2_1/c$: (a) $k = 0$ and (b) $k > 0$. The c-glide plane, which is perpendicular to b, causes a halving of the rows parallel to x^* when $k = 0$, so that only the rows with $l = 2n$ are present. Hence, the true c^* spacing is not observed on the reciprocal lattice level $k = 0$, but can be determined from higher levels. The symmetry on both levels is 2, in keeping with the diffraction symmetry $2/m$; $|F(hkl)| = |F(\bar{h}k\bar{l})|$. The reciprocal lattice points are *weighted* according to $F(hkl)$, or to $|F(hkl)|^2$, and we speak of such diffraction patterns as *weighted reciprocal lattices*.

$\cos 2\pi(hx - ky + lz + h/2)$. Since h is an integer, we may write this term as the crystallographically equivalent term $\cos 2\pi(hx - ky + lz - h/2)$. Another way of looking at this process is that the fourth general equivalent position has been changed to $-\frac{1}{2} + x, \bar{y}, z$, which is equivalent to moving through one repeat a in the negative direction to a crystallographically equivalent position, a perfectly valid and generally applicable tactic.

Returning to $Pma2$, (3.132) now becomes

$$A(hkl) = 2\cos 2\pi(ky + lz + h/4)\cos 2\pi(hx - h/4)$$

$$+ 2\cos 2\pi(-ky + lz - h/4)\cos 2\pi(hx - h/4) \qquad (3.133)$$

which simplifies to

$$A(hkl) = 2[\cos 2\pi(hx - h/4)][\cos 2\pi(ky + lz + h/4) + \cos 2\pi(-ky + lz - h/4)]$$
$$(3.134)$$

Combining again:

$$A(hkl) = 4[\cos 2\pi(hx - h/4)]\cos 2\pi(ky + h/4)\cos 2\pi lz \qquad (3.135)$$

Similarly,

$$B(hkl) = 4[\cos 2\pi(hx - h/4)]\cos 2\pi(ky + h/4)\sin 2\pi lz \qquad (3.136)$$

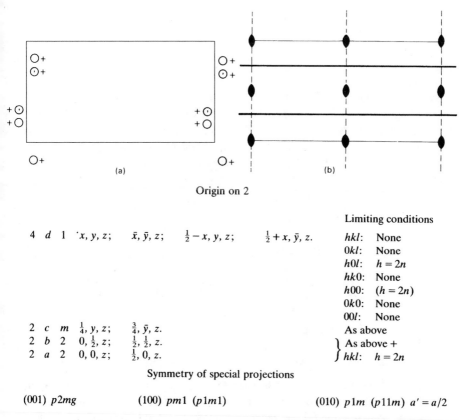

Origin on 2

				Limiting conditions	
4	d	1	x, y, z; $\bar{x}, \bar{y}, z$; $\frac{1}{2} - x, y, z$; $\frac{1}{2} + x, \bar{y}, z$.	hkl:	None
				$0kl$:	None
				$h0l$:	$h = 2n$
				$hk0$:	None
				$h00$:	$(h = 2n)$
				$0k0$:	None
				$00l$:	None
2	c	m	$\frac{1}{4}, y, z$; $\frac{3}{4}, \bar{y}, z$.	As above	
2	b	2	$0, \frac{1}{2}, z$; $\frac{1}{2}, \frac{1}{2}, z$.	} As above +	
2	a	2	$0, 0, z$; $\frac{1}{2}, 0, z$.	} hkl: $h = 2n$	

Symmetry of special projections

(001) $p2mg$ (100) $pm1$ $(p1m1)$ (010) $p1m$ $(p11m)$ $a' = a/2$

FIGURE 3.32. General equivalent positions and symmetry elements in space group $Pma2$, with the origin on 2; the origin is not fixed in the z direction by the symmetry elements. The diagram shows *inter alia* the coordinates of the special equivalent positions and the limiting conditions.

In the orthorhombic system, the (seven) regions of reciprocal space of particular importance are listed under Figure 3.32. Separating (3.135) and (3.136) for even and odd values of h, we obtain

$$h = 2n \qquad A(hkl) = 4\cos 2\pi hx \cos 2\pi ky \cos 2\pi lz$$

$$B(hkl) = 4\cos 2\pi hx \cos 2\pi ky \sin 2\pi lz \qquad (3.137)$$

$$h = 2n + 1 \qquad A(hkl) = -4\sin 2\pi hx \sin 2\pi ky \cos 2\pi lz$$

$$B(hkl) = -4\sin 2\pi hx \sin 2\pi ky \sin 2\pi lz \qquad (3.138)$$

from which we find the limiting conditions

$$hkl: \quad \text{None}$$
$$h0l: \quad h = 2n$$

The condition $h00$: $(h = 2n)$ should be considered carefully. One might be excused for thinking at first that it implies the existence of a 2_1 axis parallel to the x axis, but for the knowledge that there are no symmetry axes parallel to the x axis in class $mm2$. This particular limiting condition is dependent upon the previous one. We must emphasize here that confusion can very easily arise if the limiting conditions are interpreted in other than the following hierarchal order:

hkl:	Unit cell type	$\downarrow$ Order of inspection
$0kl$:	Glide plane $\perp x$	}
		}
$h0l$:	Glide plane $\perp y$	}
		}
$hk0$:	Glide plane $\perp z$	}
$h00$:	2_1 axis $\parallel x$	}
		}
$0k0$:	2_1 axis $\parallel y$	}
		}
$00l$:	2_1 axis $\parallel z$	}

One should proceed to a lower level in this list only after considering the full implications of the conditions at higher levels. Conditions such as that for $h00$ in $Pma2$ are called *redundant* or *dependent*, and are placed in parentheses. Reflections involved in such conditions are certainly absent from a diffraction record, but do not, necessarily, contribute to the determination of a space-group symmetry. Table 3.4 summarizes the limiting conditions for glide-plane symmetry.

Space Group *Pman*

This space group may be derived from $Pma2$ by the addition of an n-glide plane perpendicular to the z axis, with a translational component of $(a + b)/2$. We have now seen on several occasions that it is advantageous to set the origin at $\bar{1}$ wherever possible; Figure 3.33(a,b) shows *Pman* drawn in this orientation. It is left to the reader to show that the geometric structure factors are

$$A(hkl) = 8 \cos 2\pi hx \cos 2\pi [ky - (h + k)/4] \cos 2\pi [lz + (h + k)/4]$$
$$B(hkl) = 0 \tag{3.139}$$

and subsequently to derive the limiting conditions for this space group.

TABLE 3.4. Limiting Conditions for Glide Planes

Glide plane	Orientation	Limiting condition	Translational component
a	$\perp b$	$h0l : h = 2n$	$a/2$
a	$\perp c$	$hk0 : h = 2n$	$a/2$
b	$\perp a$	$0kl : k = 2n$	$b/2$
b	$\perp c$	$hk0 : k = 2n$	$b/2$
c	$\perp a$	$0kl : l = 2n$	$c/2$
c	$\perp b$	$h0l : l = 2n$	$c/2$
n	$\perp a$	$0kl : k + l = 2n$	$(b + c)/2$
n	$\perp b$	$h0l : l + h = 2n$	$(c + a)/2$
n	$\perp c$	$hk0 : h + k = 2n$	$(a + b)/2$
d	$\perp a$	$0kl : k + l = 4n \ (k, l = 2n)$	$(b \pm c)/4$
d	$\perp b$	$h0l : l + h = 4n \ (l, h = 2n)$	$(c \pm a)/4$
d	$\perp c$	$hk0 : h + k = 4n \ (h, k = 2n)$	$(a \pm b)/4$

3.8 Practical Determination of Space Groups from Diffraction Data

The determination of the space group of a crystal is an important and early feature in the x-ray analysis of its structure. We shall assume that we have available the x-ray diffraction record for the several examples of monoclinic and orthorhombic crystals to be examined. It is necessary to bear in mind that x-ray techniques can reveal the presence of that translational symmetry which can arise through one or more of four symmetry operations:

1. Translations relating to the unit cell (a and/or b and/or c);
2. Translations relating to centering of the unit cell ($\frac{1}{2}a$ and/or $\frac{1}{2}b$ and/or $\frac{1}{2}c$);
3. Translations relating to glide planes;
4. Translations relating to screw axes.

All four categories lead to systematic absences, and the totality of the translational symmetry, together with the Laue group, forms the diffraction symbol (see Section 2.9).

3.8.1 Monoclinic Space Groups

Single crystal x-ray photographs taken with a monoclinic crystal showed typically the reflections listed in Table 3.5. From the important reflection types,

(a)

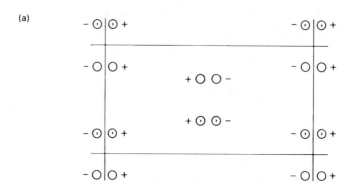

(b)

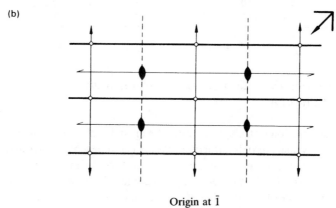

Origin at $\bar{1}$

Limiting conditions

$\pm\{x, y, z;\ \ \bar{x}, y, z;\ \ \tfrac{1}{2}+x, \tfrac{1}{2}-y, z;\ \ \tfrac{1}{2}-x, \tfrac{1}{2}-y, z\}$

hkl:	None
$0kl$:	None
$h0l$:	$h = 2n$
$hk0$:	$h + k = 2n$
$h00$:	$(h = 2n)$
$0k0$:	$(k = 2n)$
$00l$:	None

FIGURE 3.33.   General equivalent positions (a) and symmetry elements (b) in space group $Pman$, with the origin at $\bar{1}$. The diagram shows also the coordinates of the special equivalent positions and the limiting conditions.

$hkl, h0l$, and $0k0$, we deduce the limiting conditions:

$$hkl :\quad h + k = 2n$$
$$h0l :\quad (h = 2n)$$
$$0k0 :\quad (k = 2n)$$

TABLE 3.5.   Some Reflection Data for
Monoclinic Crystal I

hkl			
200	401	112	510
201	402	113	020
202	600	114	040
203	110	310	060
400	111	311	080

TABLE 3.6.   Some Reflection Data for
Monoclinic Crystal II

hkl			
100	204	111	322
200	402	122	020
300	502	113	040
400	110	311	060
202	310	123	080

from which we must conclude, using Table 3.7, that the space group is one of $C2, Cm$, or $C2/m$. The diffraction data alone do not distinguish between these three possible space groups; we show in Section 3.10.3 how the ambiguity might be resolved.

Table 3.6 provides the next list of diffraction data for inspection. There is no condition on hkl, but $h0l$ are restricted by l being even, and $0k0$ by k being even: this space group is identified uniquely as $P2_1/c$.

The limiting conditions for the 13 monoclinic space groups are listed in Table 3.7, in their standard orientations. In practice, it is possible, by an inadvertent choice of axes, to find oneself working with a nonstandard space-group symbol. Generally, a fairly straightforward transformation of axes will provide the standard setting (see Problems 2.12 and 3.17).

3.8.2   Orthorhombic Space Groups

We begin with the sample data in Table 3.8. From these data, we deduce the following conditions:

hkl: none $h00$: $h = 2n$

$0kl$:   none $0k0$: $k = 2n$

$h0l$:   none $00l$: $l = 2n$

$hk0$:   none

TABLE 3.7.   Limiting Conditions for
the Monoclinic Space Groups

Conditions limiting possible x-ray reflections	Space groups
hkl: none	
$h0l$: none	$P2, Pm, P2/m$
$0k0$: none	
hkl: none	
$h0l$: none	$P2_1, P2_1/m$
$0k0$: $k = 2n$	
hkl: none	
$h0l$: $l = 2n$	$Pc, P2/c$
$0k0$: none	
hkl: none	
$h0l$: $l = 2n$	$P2_1/c$
$0k0$: $k = 2n$	
hkl : $h + k = 2n$	
$h0l$: none	$C2, Cm, C2/m$
$0k0$: none	
hkl : $h + k = 2n$	
$h0l$: $l = 2n$ $(h = 2n)$	$Cc, C2/c$
$0k0$: none	

TABLE 3.8.   Some Reflection Data for an
Orthorhombic Crystal

hkl:	111	011	110	020
	112	021	120	040
	212	012	310	060
	312	101	200	002
	322	203	400	004
	332	303	600	006

Examining in the prescribed hierarchy, we find only 2_1 axes parallel to x, y, and z: the space group is determined uniquely as $P2_12_12_1$ (see Section 2.7.7 and Table 2.7).

In the final two examples, we consider only the conclusions drawn from an inspection of the diffraction records. In the first instance, we have:

hkl:	none	$h00$:	none
$0kl$:	$k = 2n$	$0k0$:	$(k = 2n)$
$h0l$:	$l = 2n$	$00l$:	$(l = 2n)$
$hk0$:	none		

These conditions apply to space groups $Pbc2_1$ and $Pbcm$; the distinction between them depends upon the presence, or otherwise, of a center of symmetry.

In the second example, we have:

hkl:	none	$h00$:	$(h = 2n)$
$0kl$:	$k = 2n$	$0k0$:	$(k = 2n)$
$h0l$:	$l = 2n$	$00l$:	$(l = 2n)$
$hk0$:	$h = 2n$		

and space group $Pbca$ is uniquely determined.

These results seem quite reasonable and straightforward, but nevertheless, one might be tempted to question their validity. For example, in the first orthorhombic crystal, is there a space group in class mmm that would give the same systematic absences as those in Table 3.8? Experience tells us that there is not. Since no glide planes are indicated by the systematic absences, the three symmetry planes would have to be m-planes. Three m-planes could not be involved with three 2_1 axes unless the unit cell were centered, for example, as in $Immm$. Hence, our original conclusion is correct.

The practicing x-ray crystallographer is assisted by the information on space groups in Volume A (and the earlier Volume I) of the *International Tables for Crystallography*.[a] Combined with a working knowledge of symmetry, these tables enable most situations arising in the course of a structure analysis to be treated correctly.

---

[a] See Bibliography, Chapter 1.

II   Intensities and Intensity Statistics

3.9   Intensity Expressions and Factors Affecting Intensities

The measurement of the intensity of a diffracted x-ray beam can be carried out both photographically and by quantum counting. We can measure either the peak intensity or the integrated intensity, the latter parameter being preferred for the expression of the intensity of x-ray reflection.

Real crystals are not geometrically perfect, so that a given reflection will be observed over a small, finite angular range. Hence, we need to be able to determine the area under a curve such as that shown in Figure 3.34 in order to represent a total intensity. In the photographic method, the peak intensity is recorded over a grid of points and the integrated result imposed on the photographic film. In collecting intensities with a diffractometer, a scintillation counter sweeps through the pre-set angular range $\pm \delta \theta_0$, so recording the total number of counts, or integrated intensity. We discuss some of the practical implications of such techniques in the next chapter, but much of the ensuing discussion in this chapter will have the collection of intensity data by an x-ray diffractometer in mind.

The total energy of a given diffraction spectrum $E(hkl)$ at an angle θ_0, for a crystal sufficiently small that absorption may be neglected, completely bathed in an x-ray beam of incident intensity I_0, and rotating with a uniform angular velocity ω, is given, for unpolarized incident radiation of intensity I_0, by

$$E(hkl)\omega/I_0 = Q\delta v \qquad (3.140)$$

where $E(hkl)\omega/I_0$ is known as the *integrated reflection*, δv is the volume of the crystal, and Q is given by

$$Q = (N^2\lambda^3/\sin 2\theta_0)|F(hkl)|^2[e^2/(4\pi\varepsilon_0 m_e c^2)]^2(1 + \cos^2 2\theta_0)/2 \qquad (3.141)$$

where N is the number of unit cells per unit volume of the crystal, and the other terms have their conventional meanings. The derivation of these expressions has been considered in detail elsewhere.[a]

Since the value of the integrated reflection does not actually depend upon the angular velocity, we let $R(\theta)I_0$ be the radiation reflected at the angle θ_0 by the crystal, so that $R(\theta)$ may be called the reflecting power. Then,

$$E = \int R(\theta)I_0/\omega\,d\theta$$

[a] *See* Bibliography, James (1958); *International Tables for X-ray Crystallography*, Vol. C., 2nd Ed. (1999).

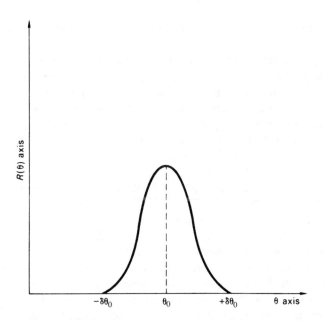

FIGURE 3.34.   Variation of reflection power $R(\theta)$ with θ; finite intensity at the Bragg angle θ_0 is recorded over the angular range $\pm\delta\theta_0$.

so that

$$E\omega/I_0 = \int R(\theta)\mathrm{d}\theta = Q\,\delta v \qquad (3.142)$$

The term $\int R(\theta)\mathrm{d}\theta$ expresses the area under the curve in Figure 3.34. From (3.141), we can write

$$E\omega/I_0 = KC(hkl)(1/\sin 2\theta)\tfrac{1}{2}(1 + \cos^2 2\theta)|F(hkl)|^2 \qquad (3.143)$$

where K is a scaling factor and $C(hkl)$ is a factor that depends upon absorption and extinction, both of which we shall discuss shortly. Essentially, the area under the curve of Figure 3.34 may be expressed as

$$\int R(\theta)\,\mathrm{d}\theta = KC(hkl)Lp|F(hkl)|^2 \qquad (3.144)$$

where the terms L and p are discussed next.

3.9.1   Polarization and Lorentz Factors

In equation (3.144), L and p represent the Lorentz and polarization factors, respectively, thus linking the quantity measured, the reflecting power, to the quantity sought, the corrected $|F(hkl)|^2$ value, which we may refer to as the *ideal intensity*.

The polarization factor p, developed in Section 3.2.4, takes into account the fact that the output of a conventional x-ray tube is unpolarized radiation, whereas the radiation after reflection from a crystal plane is polarized, thus decreasing the intensity of the diffracted beam as a function of the scattering angle 2θ.

Where the incident beam is polarized, for example, after reflection from a crystal monochromator, the polarization factor is modified to $(1 + \cos^2 2\theta \cos^2 2\theta_m)/(1 + \cos^2 2\theta_m)$, where $2\theta_m$ is the angle between the incident and scattered beams at the monochromator.

The Lorentz factor L depends on the diffraction geometry, and expresses a *time-of-reflection opportunity* for a crystal plane in the x-ray beam. For a rotating crystal with the x-ray beam normal to the reflecting plane the L factor is $1/(2\sin\theta)$; in a powder specimen, it takes the form $2/(\sin\theta \sin 2\theta)$.

In order to give expression to the Lorentz factor in a particular case, let P be a point on the zero level of the reciprocal lattice, normal to a rotation axis that passes through its origin O; the crystal is at Q and the incident x-ray beam direction is QO, as shown by Figure 3.35. The constant angular velocity of the crystal is ω, so that the reciprocal lattice point P has a linear velocity $|S|\omega$. The speed with which P moves through the surface of the Ewald sphere is the component of its velocity along the radius QP. Since $Q\hat{O}P = Q\hat{P}O = 90 - \theta$, the velocity v of P as it

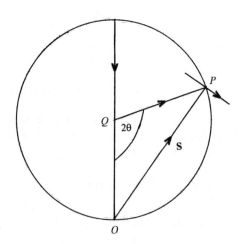

FIGURE 3.35.   Lorentz factor: sphere of reflection with the crystal at its center Q; O is the origin of the reciprocal lattice, P is a reciprocal lattice point *hkl* in the position for a reflection from the corresponding plane. The distance OP is $|S|$, and the normal to it represents the velocity vector of P; $\angle QOP = \angle QPO = (90 - \theta)°$.

passes through the sphere is given by

$$v = |\mathbf{S}|\omega \cos \theta \qquad (3.145)$$

Since $|\mathbf{S}| = (2\sin\theta)/\lambda$, the velocity v is equal to $(\omega/\lambda)\sin 2\theta$. The time t taken for P to pass through the reflecting position is proportional to $1/v$, so that this time-of-reflection is given by

$$t = \kappa/v = (\kappa/\omega)(|\mathbf{S}|\cos\theta)^{-1} \qquad (3.146)$$

where k is a constant depending on the size of the reciprocal lattice (in practice, the wavelength of x-radiation) and the limits $\pm\delta\theta$ for finite reflection.

The first term on the right-hand side of (3.146) depends on the experimental arrangement: the second term depends on the time-of-reflection opportunity for the given crystal plane; it is the Lorentz factor when the rotation axis is normal to the reflecting plane. From (3.145), $\omega/v = 1/(|\mathbf{S}|\cos\theta)$, so that

$$L = \omega/v = \lambda/(2\sin\theta\cos 2\theta) = \lambda/(\sin 2\theta) \qquad (3.147)$$

Since both ω and λ remain constant, L is equal to $(\sin 2\theta)^{-1}$ for the given experimental arrangement.

3.9.2 Extinction

We consider first a crystal bathed in the x-ray beam, under the conditions for normal Bragg reflection, with all unit cells stacked together in a regular manner. Figure 3.36 shows a family of planes, all in the same orientation θ with respect to the x-ray beam. It is clear that the first-reflected ray BC is in the correct orientation for a second reflection CD, and so on. Since there is always an inherent phase change of $\pi/2$ on reflection, the doubly reflected ray CD has a phase difference of π with respect to the incident ray AB. We note in passing that the phase change of $\pi/2$ is neglected in crystal-structure calculations since it occurs equally for all reflections.

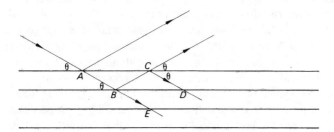

FIGURE 3.36.   Primary extinction: the phase changes on reflection at B and C are each $\pi/2$, so that between the directions BE and CD, the total phase change is π. Thus, there is an attenuation of the incident x-ray beam reaching planes deeper into the crystal.

FIGURE 3.37.   Mosaic character in a crystal: the angular misalignment between blocks may vary from 2 to 30 min of arc.

In general, rays that are reflected n and $(n - 2)$ times differ in phase by π, so that the net result is a reduction in intensity of the incident x-ray beam and, hence, the diffracted beam, in passing through the crystal. Energy is effectively conserved in this process, because each beam is depleted in energy by scattering into another beam, while being enhanced in energy by that which is scattered into the beam itself from other beams.[a] This effect is termed *primary extinction*, but it is very much reduced if the crystal is not perfect in its stacking. In fact, very few crystals are perfect: they are composed of an array of slightly misaligned blocks, constituting the so-called *mosaic character* of the crystal (Figure 3.37). The ranges of geometric perfection are generally very small, less than about 10^{-3} mm, and even crystals that show primary extinction possess some mosaic character. For the ideally perfect crystal, $I \propto |F|$, whereas for the ideally imperfect crystal $I \propto |F|^2$. Since perfection is rare and very difficult to produce in a specimen, the imperfect state with $I \propto |F|^2$ is the normal state in x-ray crystallography.

Primary extinction is most noticeable with low-order, high intensity reflections and, if it is suspected, its effect may be very substantially reduced by the thermal shock occasioned by dipping the crystal in liquid air, thereby increasing the imperfect, mosaic structure of the crystal.

In Section 3.1.2, we considered the absorption of x-rays by materials, a process that is, quite independent of the mechanism of diffraction. However, under Bragg reflection, another feature may arise with an attendant attenuation of the energy of the incident x-ray beam; this effect is known as *secondary extinction*.

Consider a situation in which the first planes encountered by the x-ray beam reflect a high proportion of the incident x-ray beam. Then, parallel planes deeper into the crystal receive less incident intensity, so that they reflect less than would

---

[a] W. H. Zachariasen, *Acta Crystallographica* **23**, 558 (1967).

be expected. The effect is most noticeable with large crystals and intense (often low-order) reflections. Crystals that have a high degree of imperfection generally show very little secondary extinction, because only a relatively small number of planes are in the reflecting position at a given time. Again, the ideally imperfect crystal shows least secondary extinction, and often only a few very strong reflections are affected, and they will not materially affect the structure determination. Nevertheless, it is possible to bring secondary extinction into a least-squares refinement (see Section 7.4) in terms of an additional variable, the *extinction parameter* ζ. The quantity then minimized in the refinement of the atomic and scale parameters is

$$\sum_{hkl} w[|F_0| - (1/K\zeta)|F_c|]^2 \qquad (3.148)$$

where F_c is the calculated structure factor.

3.9.3 Absorption Measurement and Correction

We considered absorption in Section 3.1.2, and here we consider how it may be treated in obtaining a value for the ideal intensity $|F(hkl)|^2$.

From (3.4), it follows that the transmission factor T for the x-ray beam through a crystal is given by

$$T = I/I_0 = \exp[-\mu(t_i + t_d)] \qquad (3.149)$$

where t_i and t_d are path lengths through the crystal for the incident and diffracted beams, respectively. If the shape of the crystal is known exactly, then it is possible to correct for absorption:

$$T = (1/V) \int_V \exp[-\mu(t_i + t_d)] \, dV \qquad (3.150)$$

where dV is an infinitesimal portion of the volume V of the crystal.[a] Frequently however, the crystal faces are not sufficiently well defined for this method, and an empirical procedure may be preferred.

Empirical Absorption Correction with Diffractometer Data

An empirical absorption is easily applied to data collected with a diffractometer (see also Section 4.7). Consider Figure 3.38: the incident and diffracted x-rays for a general reflection with $\phi = \phi_0$ will intersect the transmission profile at $\phi_0 - \delta$ and $\phi_0 + \delta$, where

$$\delta = \tan^{-1}(\tan\theta \cos\chi)$$

---

[a] W. R. Busing and H. A. Levy, *Acta Crystallographica* **10**, 180 (1957).

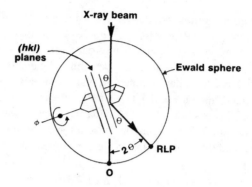

FIGURE 3.38.   Geometry of the empirical absorption correction: the crystal rotates on the ϕ-circle of a diffractometer with the χ-circle at $\pm 90°$.

Hence, $\delta = 0$ at $\chi = \pm 90°$. The transmission profile used is that with θ nearest to the equi-inclination angle ν where

$$\nu = \sin^{-1}(\sin \theta \sin \chi)$$

The transmission T is given either as the arithmetic mean or as the geometric mean of the estimated incident and reflected ray transmissions:

$$T = [T_\nu(\phi - \delta) + T_\nu(\phi + \delta)]/2$$

or

$$T = [T_\nu(\phi - \delta) \times T_\nu(\phi + \delta)]^{1/2}$$

Transmission Profiles

The transmission is measured for axial reflections ($\chi = 90°$) as a function of ϕ (Figure 3.38). The transmission is given by

$$T_\theta(\phi) = I_\theta(\phi)/I_\theta(\max) \tag{3.151}$$

The variation of T with θ is neglected as it has the same effect as a small isotropic temperature factor.

A set of profiles of T as a function of ϕ is obtained for different values of θ, and applied in data processing as detailed above.

Absorption Correction with Area Detector Data

The empirical method just described requires single Bragg reflections to be isolated and scanned so as to produce absorption correction curves. This procedure is not possible with intensity data collected with an Area Detector (see

Section 4.8.3), and other methods for applying absorption corrections have been developed. One commonly used procedure[a] uses least squares to model an empirical transmission surface as sampled by multiple symmetry-equivalent and/or azimuth rotation-equivalent intensity measurements. The fitting functions are sums of real spherical harmonics of even order:

$$Y_{lm}[-\mathbf{u}(0)] + Y_{lm}[-\mathbf{u}(1)] \qquad (2 \le l = 2n \le 8)$$

The arguments of the functions are the components of unit direction vectors, $-u(0)$ relating to the reverse incident beam and $-u(1)$ to the scattered beam, with respect to crystal-fixed Cartesian axes. The procedure had been verified against standard absorption correction data.

3.9.4 Scaling

Fluctuations in the incident x-ray beam intensity and possible radiation damage to the crystal may be monitored on a diffractometer by measuring four standard reflections of moderate intensity at regular intervals, say, hourly. Two of these reflections should have χ at approximately $0°$, and two at χ near $90°$, with each pair approximately $90°$ apart in ϕ. The average of these intensities relative to the average of their starting values is smoothed, and used to rescale the raw intensity data. If S is this scale factor, different from the scale factor K applied to $|F_0|$, then the correction applied to the measured intensity I_{meas} now becomes

$$I_{\text{corr}} = I_{\text{meas}}(Lp)^{-1}T^{-1}S^{-1} \qquad (3.152)$$

with a standard deviation given by

$$\sigma(I_{\text{corr}}) = \sigma I_{\text{meas}}(Lp)^{-1}T^{-1}S^{-1} \qquad (3.153)$$

3.9.5 Merging Equivalent Reflections

Where more than the symmetry-independent region of weighted reciprocal space is measured for any given reflection, a weighted mean intensity is calculated:

$$\bar{I} = \sum_j w_j I_j \bigg/ \sum_j w_j \qquad (3.154)$$

where the sum is over all n measured symmetry-equivalent values of the given reflection, and w_j is given by

$$w_j = \sigma_j^{-2} \qquad (3.155)$$

[a] R. H. Blessing, *Acta Crystallographica* A**51**, 33 (1995); *idem. Journal of Applied Crystallography* **30**, 421 (1997).

A chi-squared test may be used to detect equivalents that have a systematic error:

$$\chi^2 = \sum_j [(I_j - \bar{I}_j)/\sigma_j]^2 \tag{3.156}$$

where the sum is again over n symmetry equivalent reflections and the number of degrees of freedom are $(n-1)$. If χ^2 exceeds χ^2_{n-1} (at a probability level of 0.001), then the symmetry-equivalent reflection with the highest weighted deviation from the mean, $w_j(I_j - \bar{I}_j)$, is rejected and the test repeated on the remaining equivalents. If $n = 2$, the smaller intensity value is rejected. The merging R_m value, also called R_{int}, is defined by

$$R_m = \sum_{hkl} \left(\sum_j |I_j - \bar{I}_j| \right) \bigg/ \sum_{hkl} \left(\sum_j I_j \right) \tag{3.157}$$

3.9.6 Practical Intensity Expression and its Standard Deviation

We have developed the necessary theory to express the intensity of reflection and the corrections that need to be applied to it in order to obtain the ideal intensity $|F|^2$. A measurement of intensity involves values for both the intensity of the reflection, over a range $\pm\delta\theta_0$, and the background. These parameters are measured in diffractometry, by a step-scan moving-window method.[a]

The standard deviation $\sigma(I)$ in I arising from statistical fluctuations is given by

$$\sigma(I) = (I + rB + r^2 B)^{1/2} \tag{3.158}$$

where r is the ratio of the time spent in measuring the intensity I to that spent in measuring the background B; typically a value of r is 1.5.

We now express the ideal intensity in a practical form, assuming the absence of primary extinction, as

$$|F(hkl)|^2 = I(hkl)T^{-1}S^{-1}L^{-1}p^{-1} \tag{3.159}$$

where $I(hkl)$ represents the intensity of the hklth reflection that has been adjusted for fluctuations in the incident x-ray beam, corrected for the background B and merged with symmetry-equivalent reflections, then further corrected for absorption (and extinction) T, for scaling S (see Section 3.9.5), and for Lorentz and polarization factors, to give ideal intensity values I on a correct *relative* scale, with standard deviations $\sigma(I)$.

---

[a] I. J. Tickle, *Acta Crystallographica* **B31**, 329 (1975).

All the corrections to intensity values that we have considered so far have been concerned with adjustments to the experimentally measured expression of the intensity of reflection. There are other related correcting factors, one of which is the secondary extinction parameter already discussed, that are applied to *calculated* structure factors: they are the scale and temperature factors, and we consider them now.

3.9.7 Scale Factor for $|F_0|$

In the initial stage of a structure analysis, the scaling factor K for $|F_0|$ can be calculated through Wilson's method, which we describe more fully in the context of intensity statistics, in Section 3.10.1. We write the scaling factor K in terms of $|F_0|$ and $|F_c|$ as

$$|F_c| = K|F_0| \tag{3.160}$$

and a simplistic calculation of K during a structure analysis is evidently

$$K = \sum_{hkl} |F_c| / \sum_{hkl} |F_0| \tag{3.161}$$

where the sums are taken over all data for which $|F_0|$ and $|F_c|$ are available. Normally, K is adjusted in a least-squares refinement, where the scale factor is applied inversely to $|F_c|$, as indicated in (3.148).

3.9.8 Thermal Vibrations and the Temperature Factor

The picture of atoms rigidly in positions (x, y, z) needs to be modified to take into account their thermal motion arising from the vibrational, thermal energy that the atoms possess at any finite temperature. Bonding forces permit small degrees of random, relative movement of atoms, dependent upon the temperature, so that a crystal contains atoms that are vibrating about their mean positions. The effect of thermal vibration is that the electron density is smeared out over a finite volume, rather than it being concentrated at the atomic sites (x, y, z). Since the frequencies of vibration are low relative to the time taken for an x-ray beam to traverse a crystal under normal experimental conditions, the crystal may be pictured as a time average of atoms randomly displaced from their mean positions, and this condition is imposed upon the diffraction pattern of the crystal.

Thermal Vibration in One Dimension

Consider first a one-dimensional periodic arrangement of scattering centers in a row of repeat distance a, and let the jth scatterer of mean fractional position x_j be displaced by a small, absolute distance u_j. Since all unit cells in this structure are not identical, the structure factor $F(h)$, using (3.107) in the x dimension alone, is given by the time and space average

$$F(h) = \sum_j \overline{f_j \exp[\mathrm{i}2\pi h(x_j + u_j/a)]}$$

$$= \sum_j f_j \overline{\exp(\mathrm{i}2\pi h u_j/a)} \exp(\mathrm{i}2\pi h x_j) \qquad (3.162)$$

Since the displacements u_j are small, the exponential term may be expanded to three terms and, remembering that for symmetrical vibrations $\overline{u_j} = 0$, the average value of $\exp(\mathrm{i}2\pi h u_j/a)$ is approximately $(1 - 2\pi^2 h^2 \overline{u_j^2}/a^2)$ to the third term, expressed conveniently as $\exp(-2\pi^2 h^2 \overline{u_j^2}/a^2)$. In the one-dimensional analysis, $h/a = 2(\sin\theta)/\lambda$, from Table 2.4 and (3.14); hence, from (3.162), we obtain

$$F(h) = \sum_j f_{j,\theta} \exp(-8\pi^2 \overline{u_j^2} \lambda^{-2} \sin^2\theta) \exp(\mathrm{i}2\pi h x_j) \qquad (3.163)$$

The factor $\exp(-8\pi^2 \overline{u_j^2} \lambda^{-2} \sin^2\theta)$, where $\overline{u_j^2}$ is the mean square atomic displacement in the x direction, and modifies f_j, strictly $f_{j,\theta}$, to take account of thermal vibration. Normally, $8\pi^2 \overline{u_j^2}$ is written as the isotropic temperature factor B_j, known as the Debye–Waller factor, but initially in a structure determination an average value B may be applied to all atoms.

Thermal Vibration in Three Dimensions

We extend the discussion now to three dimensions, that is, to a lattice of scattering centers, or atoms, so as to obtain an expression analogous to (3.163). From (3.16), we derive an expression for the intensity I_0 for a lattice of atoms at rest by multiplying this equation by its conjugate:

$$I_0 = \psi_{2\theta}^2 (\mathcal{A}/R)^2 \sum_n \sum_m \exp\{\mathrm{i}2\pi[(\mathbf{r}_n - \mathbf{r}_m) \cdot \mathbf{S}]\} \qquad (3.164)$$

where each summation extends over the total number of atoms in the unit cell. Small, vector displacements $\mathbf{u}_j$ are now applied to each atom, so that $\mathbf{r}_j$ is replaced by $\mathbf{r}_j + \mathbf{u}_j$, and (3.164) becomes

$$I = \psi_{2\theta}^2 (\mathcal{A}/R)^2 \sum_n \sum_m \exp\{\mathrm{i}2\pi[(\mathbf{r}_n - \mathbf{r}_m) \cdot \mathbf{S}]\} \exp\{\mathrm{i}2\pi[(\mathbf{u}_n - \mathbf{u}_m) \cdot \mathbf{S}]\}$$

$$(3.165)$$

The isotropic vibration of the lattice of atoms is expressed by the mean value of the second exponential term in (3.165). Let $2\pi[(\mathbf{u}_n - \mathbf{u}_m) \cdot \mathbf{S}]$ be written as $p_{n,m}$; then, for any particular value of $p_{n,m}$, we can write its mean value as

$$\overline{\exp(\mathrm{i}p)} = 1 + \overline{\mathrm{i}p} - \overline{p^2}/2! - \overline{\mathrm{i}p^3}/3! + \overline{p^4}/4! + \cdots = 1 - \overline{p^2}/2! + \overline{p^4}/4! \qquad (3.166)$$

the mean values of the odd powers of p are zero, because positive and negative displacements are equally probable. A satisfactory approximation to (3.166) is then

$$\overline{\exp(\mathrm{i}p)} = \exp{-\tfrac{1}{2}\overline{p^2}} \qquad (3.167)$$

so that the mean value of (3.165) becomes

$$\bar{I} = \psi_{2\theta}^2 (\mathcal{A}/R)^2 \sum_n \sum_m \exp\{\mathrm{i}2\pi[(\mathbf{r}_n - \mathbf{r}_m) \cdot \mathbf{S}]\} \exp\left(-\tfrac{1}{2}\overline{p_{n,m}^2}\right) \qquad (3.168)$$

Now $p_{n,m} = 4\lambda^{-1}\pi \sin\theta (u_{n,S} - u_{m,S})$, where $u_{j,S}$ is the component of the jth displacement vector in the direction of the vector $\mathbf{S}$. Hence, we need to evaluate the mean value $\overline{(u_{n,S} - u_{m,S})^2}$, which is equivalent to $\overline{u_{n,S}^2} + \overline{u_{m,S}^2} - 2\overline{u_{n,S}u_{m,S}}$. We make the approximation that the coupling of the vibrations of atoms in a lattice is negligible, whereupon $\overline{u_{n,S}u_{m,S}} = 0$, and $\overline{u_{n,S}^2} = \overline{u_{m,S}^2} = \overline{u_S^2}$.

In (3.168), the double summation contains N^2 terms. Those with $n = m$, a total of N, have an exponential factor of unity and $p_{n,m}$ is equal to zero. Where $n \neq m$, $\tfrac{1}{2}\overline{p_{n,m}^2}$ is constant, because the vibration has been taken to be isotropic, and is equal to $2B$, where B is now given as

$$B = 8\pi^2 \overline{u_S^2} \qquad (3.169)$$

We can now write (3.165) as

$$I = \psi_{2\theta}^2 (\mathcal{A}/R)^2 \left\{ \sum_{\substack{n \quad m \\ n \neq m}} \sum \exp\{\mathrm{i}2\pi[(\mathbf{r}_n - \mathbf{r}_m) \cdot \mathbf{S}]\} \exp[-2B(\sin^2\theta)/\lambda^2] + N \right\}$$

$$(3.170)$$

In the expression for the mean isotropic temperature factor B, $\overline{u_S^2}$ is the mean square atomic displacement in the direction of vector $\mathbf{S}$, that is, normal to the reflecting plane to which $\sin\theta$ corresponds. Table 3.9 shows the effect of the exponential factor on the atomic scattering factor of carbon, for two values of B and from $(\sin\theta)/\lambda = 0$ to 0.7.

A better approximation for temperature correction assumes that the motion remains isotropic, but allows B to take a particular value B_j for each atom j in a unit cell of a structure. This procedure is used in the least-squares routine in the XRAY* program system included with this book. In general, however, each atom in a structure vibrates anisotropically, and the time-averaged electron density for an atom has the form of a triaxial ellipsoid. This ellipsoid is represented by a 3×3 tensor, where six B_{ij} components are needed in the most general case (triclinic symmetry), because the tensor is symmetric, that is, $B_{ij} = B_{ji}$. The B_{ij} values can be calculated from the isotropic B_j or B values, but normally are allowed to evolve in a least-squares refinement of atomic parameters.

TABLE 3.9.   Debye–Waller Corrections for a Carbon Atom

$\lambda^{-1}\sin\theta$	$(\lambda^{-1}\sin\theta)^2$	f	$\exp(-B\lambda^{-2}\sin^2\theta)$	
			$B = 2\,\text{Å}^2$	$B = 4\,\text{Å}^2$
0.00	0.00	6.000	6.000	6.000
0.10	0.01	5.126	5.024	4.925
0.20	0.04	3.581	3.306	3.052
0.30	0.09	2.502	2.090	1.746
0.40	0.16	1.950	1.416	1.028
0.50	0.25	1.685	1.022	0.620
0.60	0.36	1.536	0.748	0.364
0.70	0.49	1.426	0.535	0.201

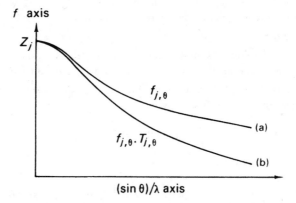

FIGURE 3.39.   Atomic scattering factors: (a) stationary atom, $f_{j,\theta}$; (b) atom corrected for thermal vibration, $f_{j,\theta}\cdot T_{j,\theta}$, also called $g_{j,\theta}$, where $T_{j,\theta} = \exp(-B\lambda^{-2}\sin^2\theta)$.

Thermal vibrations increase the effective volume of the atom, so that interference within the atom becomes more noticeable. Consequently, f falls off with increasing $(\sin\theta)/\lambda$ more rapidly than with that calculated for an atom at rest, as shown by Figure 3.39. The thermal vibrations of less rigidly retained atoms in a structure often have higher thermal vibrations than atoms that are more constrained by the stereochemistry. An example of this effect may be seen in Figure 1.2: the carbon atoms in the eight-membered side chain have a greater freedom of movement than do those in the ring system; consequently, their thermal vibrations are larger and their electron density contours more diffuse.

Statistical Expectation Value of the Debye–Waller Factor

If the unit-cell distribution of the mean square displacement parameters of the atoms is assumed to be normal, with a mean μ equal to $\overline{B}$, and a variance σ^2

equal to $(B - \overline{B})^2$, then the expectation value $\overline{W^2}$ for the Debye–Waller factor $[W^2 = \exp(-2Bs^2)^2$ where $s = (\sin\theta)/\lambda]$ is given as $\overline{W^2} = \exp[-2(\mu - \sigma^2 s^2)^2]$. This result has been incorporated into procedures for scaling and normalizing measured intensities to the Wilson expectation values. The procedures can be used to determine both isotropic μ_B and σ_B, and anisotropic $\mu_{U_{ij}}$ and $\sigma_{U_{ij}}$ distribution parameters. Tests with experimental data and refined structural models for several protein crystals have yielded reliable normalized structure factors (see Section 3.10.4), with $\Sigma_{\mathbf{h}}||E_o| - |E_c||/\Sigma_{\mathbf{h}}|E_o| \approx 5\%$.

3.10 Intensity Statistics

Statistics form an important adjunct to many aspects of x-ray crystallography. They are used in assessing the precision of unit-cell and atomic parameters, for predicting the phase angles of reflections by direct methods (see Section 7ff), for determining scale and temperature factors, to name but three. In this section, we shall be concerned with intensity statistics, and we consider first the Wilson statistics, and show how they may be used to obtain scale and temperature factors for a crystal.

3.10.1 Determining Scale and Temperature Factors

An important and familiar aspect of the statistics of the weighted reciprocal lattice is based on the equation developed by Wilson[a] for the average ideal intensity. We write (3.107), for convenience, in a compact form:

$$F(\mathbf{h}) = \sum_j g_{j,\theta} \exp[i2\pi(\mathbf{h} \cdot \mathbf{r}_j)] \tag{3.171}$$

where $\mathbf{h}$ represents the reciprocal lattice point hkl, $\mathbf{r}_j$ is the position vector of the jth atom, that is, $\mathbf{r}_j = x_j\mathbf{a} + y_j\mathbf{b} + z_j\mathbf{c}$; $g_{j,\theta}$ is the atomic scattering factor for the jth atom f_j modified by a temperature factor, such as $\exp(-B\lambda^{-2}\sin^2\theta)$, and the sum is over all atoms in the unit cell.

If we now multiply (3.171) by its conjugate, we find an expression for the ideal intensity $|F(\mathbf{h})|^2$:

$$|F(\mathbf{h})|^2 = \sum_j g_{j,\theta}^2 + \sum_j \sum_{\substack{k \\ j \neq k}} g_{j,\theta} g_{k,\theta} \exp(i2\pi\mathbf{h} \cdot \mathbf{r}_{j,k}) \tag{3.172}$$

where $\mathbf{r}_{j,k}$ is the vector distance $\mathbf{r}_j - \mathbf{r}_k$. If the distribution of atoms is uniform over the unit cell, then the second term on the right-hand side of (3.172) will tend

---

[a] A. J. C. Wilson, *Nature* **150**, 152 (1942).

to a negligible value because the many vectors $\mathbf{r}_{j,k}$ will tend to cancel one another; then, the average ideal intensity is given by

$$\overline{|F(\mathbf{h})|^2} = \sum_j g_{j,\theta}^2 \qquad (3.173)$$

and is the basis for obtaining a preliminary scale factor for $|F_0|$ and a temperature factor for f.

Methodology

Equation (3.173) has been found to hold satisfactorily over a wide range of structures, provided that the values of $|F|^2$ are averaged over small, local ranges of $(\sin\theta)/\lambda$, so that f is not varying rapidly within any range.

Applying the scale and temperature factors to (3.173), we have

$$K^2\overline{|F(\mathbf{h})|^2} = \exp[-2B\lambda^{-2}\sin^2\theta_r]\sum_j f_{j,\theta_r}^2 \qquad (3.174)$$

where θ_r is a representative value of θ for each range. Taking logarithms of both sides, we write

$$\ln q_r = 2\ln K + 2B(\sin^2\theta_r)/\lambda^2 \qquad (3.175)$$

where q_r is given by

$$q_r = \left(\sum_j f_{j,\theta_r}^2\right) \big/ \overline{|F_0(\mathbf{h})|_{\theta_r}^2} \qquad (3.176)$$

and the sum is taken over all j atoms in the unit cell. If $\ln q_r$ is plotted against $(\sin^2\theta_r)/\lambda^2$ and the best straight line drawn, the slope is equal to $2B$ and the intercept on the ordinate is equal to $2\ln K$. This graph is often called a Wilson plot, and is best obtained through the following procedure.

1. Three-dimensional space is divided into a number of spherical shells (Figure 3.40a), such that there are 80–100 reflections in each range. Although the plot (Figure 3.40b) is against $(\sin^2\theta)/\lambda^2$, it is convenient to form the range demarcations in terms of $(\sin^3\theta)/\lambda^3$, since this parameter has the dimensions of reciprocal volume $(\sin\theta \propto 1/d)$; the demarcations can be converted into the equivalent values of $(\sin^2\theta)/\lambda^2$ later.
2. The average value of $|F_0(\mathbf{h})|^2$ is calculated over the whole of the available reciprocal space: either the data set is expanded to include the symmetry-equivalent reflections, or each reflection in an asymmetric unit is accorded its correct multiplicity of planes. It is necessary to allocate values to the accidental absences (unobserved reflections within the Ewald sphere).

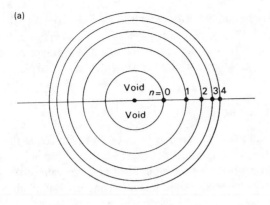

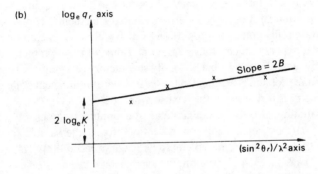

FIGURE 3.40.   Scale and temperature factors. (a) Division of reciprocal space into spherical shells; the void region contains data for which $h, k, l \leq 2$. (b) Wilson plot: the intercept is $2 \ln K$, and the slope is $2B$.

Wilson has shown that the most probable values are $0.55 F_{\min}$ for centrosymmetric crystals, $0.66 F_{\min}$ for non-centrosymmetric crystals, and $0.59 F_{\min}$ where this symmetry is undetermined; $F_{\min}$ is the minimum value of $|F_0(\mathbf{h})|$ in the range under consideration. Systematic absences are always ignored, as are those in a region up to the second order on each axis because they are atypical of the general distribution of intensities. In addition, the *average intensity multiple* (ε-factor) should be applied as a divisor of each intensity value; this parameter is discussed and tabulated in Section 3.10.3.

3. The mean values of $(\sin^2 \theta_r)/\lambda^2$ may be obtained as

$$(\sin^2 \theta_r)/\lambda^2 = \tfrac{1}{2}[(\sin^2 \theta_n)/\lambda^2 + (\sin^2 \theta_{n+1})/\lambda^2]$$

where $n+1$ is the number of the outer boundary shell defining the rth shell, starting at $n = 0$ for which value $\sin\theta = \sin\theta_{min}$ (Figure 3.40a). Alternatively, $(\sin^2\theta_r)/\lambda^2$ values may be obtained as averages of $(\sin^2\theta)/\lambda^2$ over each range. Atomic scattering factor data are tabulated and readily available.[a]

3.10.2   Pathological Cases

Assuming that all possible care has been taken in collecting the intensity data, there may still remain some situations in which the Wilson plot is nonlinear, for a variety of reasons: there may be significant nonzero values of the double summation in (3.172) or the temperature factor is non-Gaussian, both conditions leading to a partial breakdown of (3.173) on which the Wilson plot is based; the number of atoms in the unit cell may be too low to provide a uniform distribution; a proportion of atoms may lie on symmetry elements, and so contribute only to certain reflections; the proportion of heavy atoms in the unit cell may be unduly large; hypersymmetry may be present. Some of these problems have been addressed by several workers, but no totally satisfactory procedure has emerged. In most cases, the problem is resolved in the least-squares refinement of the structure; hypersymmetry is discussed in the next section.

In handling two-dimensional data, the annular regions of equal area $(\propto \lambda^{-2}\sin^2\theta)$ may contain too few reflections for a true sample of the reciprocal lattice. Then it is possible to employ a batch procedure, that is, to combine groups $(1, 2), (2, 3), (3, 4), \ldots$ before the averages are taken.

In the program XRAY* the routine for the Wilson plot uses this batch technique for the two-dimensional data sets. Not all of the example data sets provided give equally satisfactory results; this problem is considered again in Section 3.10.4.

3.10.3   Statistics of Reciprocal Space

The weighted reciprocal lattice exhibits four types of regularity and one type that may be described as irregular; we shall consider them in turn.

Accidental Absences

Accidental absences occur in the diffraction pattern of most structures, and they are disposed in an *irregular* manner within the Ewald sphere. From (3.107), it is not surprising to find that there are some instances where the sum of the vectors $f_j \exp[i2\pi(hx_j+ky_j+lz_j)]$ tend to cancel to a negligible value. The result depends upon the particular atomic arrangement in the structure rather than the symmetry of the structure. Such permitted but absent reflections can be estimated in the manner

---

[a] See *Bibliography*, Ibers and Hamilton (1974).

discussed in Section 3.10.1: it is not uncommon to omit these reflections from a structure analysis, but without real justification.

Laue Symmetry

We now consider the *regular* features of the weighted reciprocal lattice. The positions of the reciprocal lattice points and the intensities associated with them conform to one of the Laue groups, that is, one of the eleven centrosymmetric point groups discussed in Section 1.4.2. This situation arises because of Friedel's Law (see Section 3.6.1), and holds in all normal situations, that is, in the absence of resonance excitation (see Section 3.1.3).

Systematic Absences

In the presence of translational symmetry, that is, centered unit cells, glide planes, or screw axes, certain characteristic groups of reflections are absent from the diffraction records (see Section 3.7ff). The diffracted energy that is, so excluded is redistributed over other reflections. For example in a C-centered unit cell, hkl reflections are absent for $h + k = 2n + 1$. However, the structure factor equation (3.107) now takes the form

$$F(hkl) = 2[\cos^2 \pi (h + k)/4] \sum_{j=1}^{n} g_j \exp[i2\pi(hx_j + ky_j + lz_j)] \qquad (3.177)$$

from which it is evident that for the reflections present, $(h + k)$ even, $F(hkl)$ has twice the value that it would have for a corresponding primitive unit cell.

Abnormal Averages

We have shown in Section 3.5.2 that the component vectors in (3.171) can be represented in phase and amplitude on an Argand diagram. All types of symmetry link $\mathbf{r}_j$ vectors ($\mathbf{r}_j = x_j\mathbf{a} + y_j\mathbf{b} + z_j\mathbf{c}$) into groups of two or more; equation (3.112) represents the simplest example of this feature. Thus, $\Sigma_j g_j^2$, which we shall write as Σ, is enhanced, and becomes a *distribution parameter S*:

$$S = \varepsilon \Sigma = \varepsilon \Sigma_j g_j^2 \qquad (3.178)$$

Consider space group Pm, where the mirror plane is normal to y and cuts this axis at $y = 0$; then, atoms are related in pairs x, y, z and $x, \bar{y}, z$. Simple manipulation shows that (3.107) for this example becomes

$$F(hkl) = \sum_{j=1}^{n/2} g_j \exp[i2\pi(hx_j + ky_j)](2\cos 2\pi l z_j)$$

$$= A'(hkl) + i B'(hkl) \tag{3.179}$$

where

$$A'(hkl) = 2\sum_{j=1}^{n/2} g_j \cos 2\pi(hx_j + lz_j) \cos 2\pi ky_j \tag{3.180}$$

and

$$B'(hkl) = 2\sum_{j=1}^{n/2} g_j \sin 2\pi(hx_j + lz_j) \cos 2\pi ky_j \tag{3.181}$$

We need now to invoke the *central limit theorem* which states that *in a sequence of independent random variables* $x_1, x_2, \ldots, x_j, \ldots, x_n$, *where the mean values are expressed generally by* m_j *and the variances by* σ_j^2, *the sum* $x = \Sigma_j x_j$ *tends to the normal (Gaussian) distribution, with a mean* m *equal to* $\Sigma_j m_j$ *and a variance* σ^2 *equal to* $\Sigma_j \sigma_j^2$, *as the number of terms* (n) *in the sequence tends, ideally, to infinity.*

In our application, the mean values $\overline{A'(hkl)}$ and $\overline{B'(hkl)}$ both tend to zero, since the positive and negative values of these terms are equally probable and will tend to cancel one another in a normal distribution. The variance for a large sample is given generally by

$$\sigma^2 = (1/n)\sum_j (x_j - \overline{x})^2 = (1/n)\sum_j x_j^2 - \overline{x}^2 \tag{3.182}$$

since, in our example, $\overline{x} = 0$, so that $\sigma^2 = \overline{x_j^2}$. Applying this result first to $A'(hkl)$, the jth individual variance is given by $2f_j^2 \cos^2 2\pi(hx_j + lz_j)\cos^2 2\pi(ky_j)$. Thus, the variance of $A'(hkl)$ will be equal to $\overline{A'(hkl)^2}$ which, by the central limit theorem, is given by

$$\overline{A'(hkl)^2} = \sum_{j=1}^{n/2} 4g_j^2 \overline{\cos^2 2\pi(hx_j + lz_j)\cos^2 2\pi ky_j}$$

$$= \sum_{j=1}^{n/2} 4g_j^2 \overline{\cos^2 2\pi(hx_j + lz_j)}\,\overline{\cos^2 2\pi ky_j} \tag{3.183}$$

It is straightforward to show $[\overline{\cos^2\theta} = (1/\pi)\int_0^\pi \cos^2\theta\, d\theta]$ that the average value of $\cos^2\theta$ is $\frac{1}{2}$. Hence,

$$\overline{A'(hkl)^2} = \sum_{j=1}^{n/2} g_j^2 = \frac{1}{2}\Sigma \qquad (3.184)$$

In a similar manner, we can show that the average $\overline{B'(hkl)^2}$ is also equal to $\frac{1}{2}\Sigma$, so that

$$\overline{F(hkl)^2} = \Sigma \qquad (3.185)$$

However, if we consider the zone of reflections for which $k = 0$, a similar analysis shows that

$$\overline{F(h0l)^2} = 2\Sigma \qquad (3.186)$$

Hence, the ε-factor for the *intensities* in this zone in $P2/m$ is 2. The ε-factor is dependent on the *crystal class*, and Table 3.10 lists the ε-factors that arise in the 32 point groups. Another way of looking at these ε-factors is by means of stereograms. Consider Figure 1.36, point group $\overline{4}2m$, and imagine the radiating normals that give rise to the poles as vectors. When *projected* on to the z axis there is a 4-fold *superposition* of the g_j vectors, but when projected on to the plane normal to z there is no such superposition; hence, 4/1 arises for the first direction, along z, that is, $\varepsilon(00l) = 4$ and $\varepsilon(hk0) = 1$.

Acentric and Centric Distributions

The measured intensities of the whole reciprocal lattice or of certain two- or one-dimensional regions of it may conform to an *acentric*, a *centric*, or a *hypercentric* distribution, and we shall consider the properties and uses of their distribution functions.

Acentric Distribution. In the acentric distribution, typically for space group $P1$, the components $A'(\mathbf{h})$ and $B'(\mathbf{h})$ of the structure factor must be considered separately. Following the discussions above, $A'(\mathbf{h})$ will be given by

$$A'(\mathbf{h}) = \sum_{j=1}^{n} f_j \cos 2\pi(\mathbf{h}\cdot\mathbf{r}_j)$$

so that

$$\overline{A'(\mathbf{h})^2} = \sum_{j=1}^{n} f_j^2 \overline{\cos^2 2\pi(\mathbf{h}\cdot\mathbf{r}_j)} = \frac{1}{2}\Sigma \qquad (3.187)$$

with $\overline{A'(\mathbf{h})}$ again being zero. Similarly, $\overline{B'(\mathbf{h})} = 0$, and $\overline{B'(\mathbf{h})^2} = \frac{1}{2}\Sigma$. The probabilities that A' lies between A' and $A' + dA'$, and that B' lies

TABLE 3.10. Centric Reflections and ε-Factors for
Intensities in the 32 Crystal Classes

Crystal class	Centric reflections	ε-Factors
1	None	1/1
$\bar{1}$	All	1/1
(1)2(1)	$(h0l)$	2/1
(1)m(1)	$(0k0)$	1/2
(1)2/m(1)	All	2/2
222	$(0kl)$; $(h0l)$; $(hk0)$	2/1; 2/1; 2/1
$mm2$	$(hk0)$–$(h00)$–$(0k0)$	2/2; 2/2; 4/1
mmm	All	4/2; 4/2; 4/2
4	$(hk0)$	4/1
$\bar{4}$	$(hk0)$; $(00l)$	2/1
4/m	All	4/2
422	$(hk0)$; $\{h0l\}$; $\{hhl\}$	4/2; 2/1; 2/1
$4mm$	$(hk0)$–$\{h00\}$–$\{hh0\}$	8/1; 2/2; 2/2
$\bar{4}2m$	$(hk0)$–$\{hh0\}$; $\{h0l\}$–$(00l)$	4/1; 2/1; 2/2
4/m mm	All	8/2; 4/2; 4/2
23(1)	$\{hk0\}$	2/1; 3/1; 1/1
m3(1)	All	4/2; 3/1; 2/1
432	$\{hk0\}$; $\{hhl\}$	4/1; 3/1; 2/1
$\bar{4}3m$	$\{hk0\}$–$\{hh0\}$	4/1; 6/1; 2/2
m3m	All	8/1; 6/1; 4/2
6	$(hk0)$	6/1
$\bar{6}$	$(00l)$	3/2
6/m	All	6/2
622	$(hk0)$; $(h0l)$; (hhl)	6/1; 2/1; 2/1
$6mm$	$(hk0)$–$\{hh0\}$–$\{h00\}$	12/1; 2/2; 2/2
$\bar{6}m2$	$\{hhl\}$–$\{hh0\}$–$(00l)$	6/2; 2/2; 4/1
6/m mm	All	12/2; 4/2; 4/2
3	None	3/1
$\bar{3}$	All	6/1
32(1)	$\{h0\bar{h}l\}$	3/1; 2/1; 1/1
3m(1)	$\{h0\bar{h}0\}$	6/1; 1/2; 2/1
$\bar{3}m$ (1)	All	6/1; 2/2; 2/1

Notes:
1. In column 2, the horizontal bar–indicates that reflections on the right-hand side of the bar are included in one of the sets to the left of the bar.
2. In column 3, the entries p/q relate to the symbol positions in **Table 1.5**, with p referring to a row and q referring to the zone normal to that row. Thus, all space groups in class m have $\varepsilon = 2$ for the $(h0l)$ reflections and $\varepsilon = 1$ for $(0k0)$ and all others. For space groups in class 4 mm, $\varepsilon(00l) = 8$, $\varepsilon(hk0) = 1, \varepsilon < h00 >= 2, \varepsilon\{0kl\} = 2, \varepsilon < hh0 >= 2, \varepsilon < hhl >= 2$; for all other reflections, $\varepsilon = 1$.
3. Full point-group symbols have been indicated in certain cases so that columns 1 and 3 relate correctly. Thus, (1)m(1) reminds us that m is perpendicular to y, so that $\varepsilon(0k0) = 1$ and $\varepsilon(h0l) = 2$.

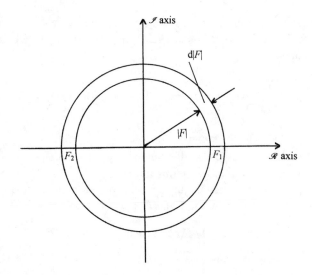

FIGURE 3.41.   Area on an Argand diagram for structure amplitudes lying between $|F|$ and $|F|+\mathrm{d}|F|$ in an acentric distribution. In a centric distribution, $|F|$ can have only two possible values, shown at F_1 and F_2.

between B' and $B' + dB'$, following a normal distribution of the type $P(x) = (2\pi\sigma^2)^{-1/2}\exp[-(x - \bar{x})^2/2\sigma^2]$, are

$$P_{\mathrm{a}}(A')\mathrm{d}A' = 1/(\pi\,\Sigma)^{1/2}\exp(-A'^2/\Sigma)\,\mathrm{d}A'$$
$$P_{\mathrm{a}}(B')\mathrm{d}B' = 1/(\pi\,\Sigma)^{1/2}\exp(-B'^2/\Sigma)\,\mathrm{d}B' \tag{3.188}$$

The region of area defined by $\mathrm{d}A'\mathrm{d}B'$ is an infinitesimal portion of an annular ring on an Argand diagram (Figure 3.41) lying between $|F|$ and $|F| + \mathrm{d}|F|$, and may be equated to $2\pi|F|\mathrm{d}|F|$. Since A' and B' are not correlated, the joint probability that the structure amplitude $|F|$ lies between $|F|$ and $|F| + \mathrm{d}|F|$ is

$$P_{\mathrm{a}}(|F|)\mathrm{d}|F| = P_{\mathrm{a}}(A')P_{\mathrm{a}}(B')\mathrm{d}A'\mathrm{d}B' = (1/\pi\,\Sigma)\exp[-(A'^2 + B'^2)/\Sigma]\mathrm{d}A'\mathrm{d}B'$$
$$= (1/\pi\,\Sigma)\exp[(-|F|^2)/\Sigma]\mathrm{d}|S| \tag{3.189}$$

where $\mathrm{d}|S|$ represents the area $\mathrm{d}A'\mathrm{d}B'$ on an Argand diagram. Thus, the joint probability refers to the area of an annular ring on the Argand diagram with radii $|F|$ and $|F| + \mathrm{d}|F|$, that is, $\mathrm{d}|S| = 2\pi|F|\mathrm{d}|F|$, so that the acentric distribution function is

$$P_{\mathrm{a}}(|F|) = (2|F|/\Sigma)\exp[(-|F|^2)/\Sigma] \tag{3.190}$$

Centric Distribution.   Space group $P\bar{1}$ provides a typical centric distribution of intensity data. The structure factors are real, and are given by the A' component of (3.107), that is,

$$F(\mathbf{h}) = A'(\mathbf{h}) = \sum_{j=1}^{n/2} 2f_j \cos 2\pi(\mathbf{h} \cdot \mathbf{r}_j)$$ (3.191)

where $\mathbf{h}$ and $\mathbf{r}_j$ have the meanings as before. From the central limit theorem, if the set of $A'(\mathbf{h})$ follows a normal distribution, the mean $\overline{A'(\mathbf{h})}$ is zero, and the variance $\overline{A'(\mathbf{h})^2}$ is the sum of $n/2$ terms of the form $4f_j^2\cos^2 2\pi(\mathbf{h} \cdot \mathbf{r}_j)$, which evaluates to Σ, the distribution parameter defined above. Hence, the probability that a structure factor lies between F and $F + dF$ is given by

$$P_c(F)dF = 1/(2\pi\Sigma)^{1/2} \exp(-F^2/2\Sigma)dF$$ (3.192)

and the centric distribution function becomes

$$P_c(F) = (2\pi\Sigma)^{-1/2} \exp(-F^2/2\Sigma)$$ (3.193)

We note that, if, in the centric distribution, we wish to consider structure *amplitudes*, then

$$P_c(|F|) = 2P_c(F) = (2/\pi\Sigma)^{-1/2} \exp(-|F|^2/2\Sigma)$$ (3.194)

since $|F|$ can derive from $\pm F$, as at F_1 and F_2 (Figure 3.41). We use this function in deriving $N_c(|E|)$ shortly, because we are concerned only with positive values in $|E|$.

Mean Values

We are now in a position to derive mean values for $|F|$ and $|F|^2$ and other parameters in the two distributions derived. The mean value for any distribution $\phi(x)$ is given generally by

$$\bar{x} = \int x\phi(x)dx \Big/ \int \phi(x)\,dx$$

but, because we are dealing with a normal distribution, $\int \phi(x)dx = 1$, so that the average value of x is given simply by

$$\bar{x} = \int x\phi(x)\,dx$$ (3.195)

The two intensity distributions are plotted in terms of $|F|$ in Figure 3.42. It is evident that the centric distribution is characterized by a significant proportion of

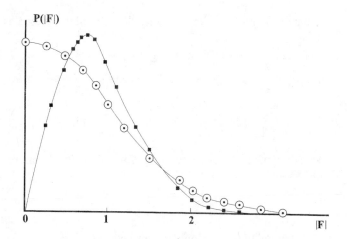

FIGURE 3.42.   Distribution function for structure amplitudes: (a) acentric, (b) centric. The acentric distribution has a low dispersion of $|F|$ values, whereas the centric distribution has significant proportions of both small and large $|F|$ values.

both strong and weak intensities, whereas the acentric distribution has a low dispersion of intensities. These features can sometimes be recognized in Weissenberg or precession x-ray photographs (see Chapter 4).

Because the centric and acentric distributions differ, it is reasonable to suppose that the distributions may be used to distinguish between centrosymmetric and non-centrosymmetric crystals. We have noted (see Section 3.6.2) that the determined space group of a crystal is often ambiguous. The cause of the ambiguity lies frequently in Friedel's law, that is, the inability to determine from the positions alone of the diffraction spectra whether or no the crystal itself is centrosymmetric.

A parameter that may be used as a discriminator is the ratio of the square of the average $|F|$ to the average of $|F|^2$:

$$M = \overline{|F|}^2 / \overline{|F|^2} \qquad (3.196)$$

For the acentric distribution, we have

$$\overline{|F|} = (2/\Sigma) \int_0^\infty |F|^2 \exp(-|F|^2/\Sigma)\, \mathrm{d}|F|$$

This integral, and many similar to it, can be solved readily by making use of the properties of the gamma (Γ) function (see Appendix A9). Let $|F|^2/\Sigma = t$, so that $2|F|\mathrm{d}|F| = \Sigma \mathrm{d}t$. Then,

$$\overline{|F|} = \Sigma^{1/2} \int_0^\infty t^{1/2} \exp(-t)\mathrm{d}t$$

The term $t^{1/2}$ may be written as $t^{(3/2-1)}$, so that the value of the integral is $\Gamma(3/2)$ or $\frac{1}{2}\pi^{1/2}$, so that

$$\overline{|F|} = \tfrac{1}{2}(\pi\,\Sigma)^{1/2} \tag{3.197}$$

In a similar manner,

$$\overline{|F|^2} = (2/\Sigma)\int_0^\infty |F|^3 \exp(-|F|^2/\Sigma)\mathrm{d}|F| \tag{3.198}$$

By making substitutions as before, it is straightforward to show that the integral in (3.198) equates to Σ. Thus,

$$M_\mathrm{a} = \tfrac{1}{4}\pi\,\Sigma/\Sigma = \pi/4 = 0.785$$

In a centric distribution, the corresponding parameter M_c is

$$M_\mathrm{c} = (2\Sigma/\pi)/\Sigma = 2/\pi = 0.637$$

A disadvantage inherent in these discriminators, even when the data are divided into ranges in which the variation of f with θ is small, is that that variation is imposed on the results. It is preferable, therefore, to use a parameter that is not dependent upon f.

3.10.4 Normalized Structure Factors

In the previous section, we stressed the importance of placing intensity data on a common statistical scale, and we discussed the ε-factor for the 32 crystal classes. For improved statistical results, either *unitary* structure factors $U(\mathbf{h})$ or *normalized* structure factors $E(\mathbf{h})$ are employed $[|U(\mathbf{h})|^2 = |E(\mathbf{h})|^2\Sigma_j f_j^2/(\Sigma_j f_j)^2]$; we shall use the parameter $|E|$.

The normalized structure factor E may be given for general reflections ($\varepsilon = 1$) by the equation

$$|E|^2 = |F|^2/\Sigma \tag{3.199}$$

For special classes of reflections, the ε-factor must be applied in accordance with Table 3.10.

From (3.189), the acentric distribution function follows as

$$P_\mathrm{a}(|E|) = 2|E|\exp(-|E|^2)| \tag{3.200}$$

and from (3.192), that for the centric distribution is

$$P_\mathrm{c}(E) = (2\pi)^{-1/2}\exp(-E^2/2) \tag{3.201}$$

Again, as with the distributions of $|F|$, the distribution of $|E|$ in the centric case is twice the value given in (3.201), and for the same reason:

$$P_c(|E|) = (2/\pi)^{1/2} \exp(-|E|^2/2) \tag{3.202}$$

It will be evident that these distribution equations do not involve any function of the atomic scattering factors, so that they are independent of the particular structure. As in the previous section, we can calculate mean values related to the new variable $|E|$. For both the acentric and the distributions, the average value of $|E|^2$ is unity. For the average value of $|E|$ in the acentric distribution, we have

$$\overline{|E|} = 2 \int_0^\infty |E|^2 \exp(-|E|^2) \, d|E|$$

Making the substitution $|E|^2 = t$, the integral becomes

$$\overline{|E|} = \int_0^\infty t^{1/2} \exp(-t) \, dt$$

Since $t^{1/2}$ may be written as $t^{3/2-1}$, the integral becomes $\Gamma(3/2)$, or $\frac{1}{2}\Gamma(1/2)$, which is $\frac{1}{2}\sqrt{\pi}$. Hence, $\overline{|E|} = 0.886$. In the case of the centric distribution, a similar calculation shows that $\overline{E} = 0.798$.

The parameter $|E^2 - 1|$ offers another useful discriminant between acentric and centric distributions. Here, we evaluate this parameter for the centric distribution.

$$\overline{|E^2 - 1|} = (2/\pi)^{1/2} \int_0^\infty |E^2 - 1| \exp(-E^2/2) dE$$

$$= (2/\pi)^{1/2} \int_0^1 (1 - E^2) \exp(-E^2/2) dE$$

$$+ (2/\pi)^{1/2} \int_1^\infty (E^2 - 1) \exp(-E^2/2) dE$$

Since, generally,

$$\int (1 - X^2) \exp(-X^2/2) dX = \int d[X[\exp(-X^2/2)] \tag{3.203}$$

$$\overline{|E^2 - 1|} = (2/\pi)^{1/2} \left\{ \int_0^1 d[E \exp(-E^2/2)] + \int_\infty^1 d[E \exp(-E^2/2)] \right\}$$

$$= (2/\pi)^{1/2} \left\{ E \exp(-E^2/2) \Big|_0^1 + E \exp(-E^2/2) \Big|_\infty^1 \right\}$$

$$= (2/\pi)^{1/2} 2e^{-1/2} = 0.968$$

A range of parameters can be determined from the probability functions for the two distributions; a few of them are listed in Table 3.11.

TABLE 3.11.   Parameters in the
Acentric and Centric Distributions of
$|E|$ values

Parameter	Acentric	Centric
$\overline{|E|}$	0.886	0.798
$|E|^2$	1.000	1.000
$\overline{|E|^2 - 1}$	0.736	0.968
$\overline{(|E|^2 - 1)^2}$	1.000	2.000

Cumulative Distributions

Rather than considering individual parameters, such as M or $\overline{|E|}$, the determination of the centricity of the distribution may be approached by means of cumulative distributions of E-values. In the acentric distribution, the fractional number of $|E|$ values less than or equal to a given value of $|E|$ is the integral of the probability function from the lower limit to that given value. Thus, we write

$$N_a(|E|) = 2 \int_0^{|E|} |E| \exp(-|E|^2) \, d|E|$$

$$= 1 - \exp(-E^2) \qquad (3.204)$$

Similarly, for the centric distribution, we have

$$N_c(|E|) = (2/\pi)^{-1/2} \int_0^{|E|} \exp(|E|^2/2) \, d|E|$$

$$= \mathrm{erf}(|E|/\sqrt{2}) \qquad (3.205)$$

where erf(...) represents the statistical error function, which is tabulated in most texts on statistics. Table 3.12 list the values of $N(|E|)$ for the two distributions up to $|E| = 3.0$, and Figure 3.43 illustrates the distributions; the region of greatest discrimination is $0 \le |E| \le 1$.

Hypersymmetry

Hypersymmetry, or hypercentrosymmetry, can arise when noncrystallographic centers of symmetry are present in the asymmetric unit of a structure. Pyrene is one example of a molecule that is, itself, centrosymmetric. The degree of hypersymmetry depends upon the number of additional centers of symmetry. Figure 3.43 includes the curve for $N_c(|E|)$ when one additional center is present in the asymmetric unit. Further discussions on hypersymmetry may be found in the literature.[a]

---

[a] D. Rogers and A. J. C. Wilson, *Acta Crystallographica* **6**, 439 (1953).

TABLE 3.12.   Acentric and
Centric Cumulative Distributions

$|E|$	$N_a(|E|)$	$N_c(|E|)$
0.0	0.000	0.000
0.2	0.039	0.159
0.4	0.148	0.311
0.6	0.302	0.451
0.8	0.473	0.576
1.0	0.632	0.683
1.2	0.763	0.770
1.4	0.859	0.838
1.6	0.923	0.890
1.8	0.961	0.928
2.0	0.982	0.954
2.2	0.992	0.972
2.4	0.997	0.984
2.6	0.999	0.991
2.8	1.000	0.995
3.0	1.000	0.997

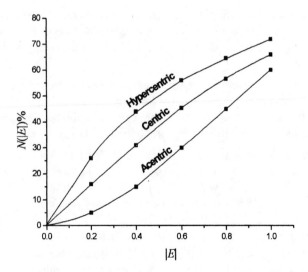

FIGURE 3.43.   Cumulative distributions $N(|E|)$: (a) acentric, (b) centric, (c) hypersymmetric. The region $0 \leq |E| < 1$ is the most discriminatory, as the curves tend to converge and actually cross over at higher values of $|E|$.

Bibliography

Synchrotron Radiation

HELLIWELL, J. R., *Macromolecular Crystallography with Synchrotron Radiation*, Cambridge University Press (1992).

Structure Factor and Intensity

JAMES, R. W., *Optical Principles of the Diffraction of X-rays: The Crystalline State*, Vol. 2, Bell, London (1958).
PRINCE, E., and WILSON, A. J. C., International Tables for X-ray Crystallography, Volume C, 2nd Edition. Kluwer Academic (1999).
WOOLFSON, M. M., *An Introduction to X-Ray Crystallography*, 2nd ed., Cambridge, Cambridge University Press (1977).

Atomic Scattering Factors

IBERS, J. A. and HAMILTON, W. C. (Editors), *International Tables for X-ray Crystallography*, Vol. IV, Birmingham, Kynoch Press (1974).

Problems

3.1. What is the change in wavelength of an x-ray photon scattered incoherently by a free electron at $45°$ to the forward direction of the incident beam? If the wavelength of the incident photon is 1 Å, what is the energy of the scattered photon?

3.2. Two identical coherent scattering centers are separated by the distance 2λ, and x-rays fall normally on to the line joining the two centers. For $2\theta = 0°$ to $180°$ in steps of $30°$, calculate the scattered amplitudes and intensities as fractions of the results with both scatterers at one point.

3.3. Calculate the atomic scattering factor f for beryllium at $(\sin\theta)/\lambda = 0.0,\ 0.2,$ and 0.5. The expression for $f(1s)$ has been given in the text. The Slater wavefunction for the 2s electron may be given as $\psi_{2s} = (c_2^5/96\pi)^{1/2} r \exp(c_2 r/2)$; you may need the general result $\int_0^\infty x^n \exp(-ax)\sin bx\, dx = n![(a+ib)^{n+1} - (a-ib)^{n+1}]/[2i(a^2 + b^2)^{n+1}]$, from which the similar expression given in the text for the 1s wavefunction applies for the case $n = 1$. The screening constants for beryllium are $\sigma_{1s} = 0.3$ and $\sigma_{2s} = 2.05$. Compare the results that are obtained for f with those from the expression $f = \sum_j^4 a_j \exp(-b_j s^2) + c$, where s is $(\sin\theta)/\lambda$ and the values of a, b, and c for beryllium are listed below:

a_1	b_1	a_2	b_2	a_3	b_3	a_4	b_4	c
1.5919	43.6427	1.1278	1.8623	0.5391	103.483	0.7029	0.5420	0.0385

3.4. An x-ray tube is operated at 30 kV. What is the energy, in J, associated with each x-ray photon produced by the tube?

3.5. Calculate the attenuation factor for a 0.2 mm plate of wollastonite ($CaSiO_3$), the density of which is 2720 kg m^{-3}. The mass absorption coefficient for Cu $K\alpha$ x-radiation and the relative atomic masses are as follow:

	Ca	Si	O
$\mu/m^2 kg^{-1}$	172	60.3	12.7
M_r	40.08	28.09	16.00

3.6. There are eight combinations of one to three negative signs with the indices *hkl* for any general reflection. With the aid of the geometrical structure factors given in the text, derive the relationships between the eight forms of the phase angle $\phi(hkl)$ for (a) space group $P2_1$, and (b) space group $Pma2$.

3.7. A triclinic unit cell has the dimensions $a = 7.36, b = 9.21, c = 13.47$ Å, and $\alpha = 101.22°, \beta = 110.62°, \gamma = 123.41°$. Calculate the six parameters of the reciprocal unit cell for Cu $K\alpha$ radiation ($\lambda = 1.5418$ Å), and (b) the volumes of the real and reciprocal unit cells.

3.8. In the direct unit cell of Problem 3.7, two atoms are situated at the fractional coordinates 0.10, 0.30, 0.20 and 0.10, 0.15, 0.35 for x, y, z, respectively. By means of vector expressions, calculate the distance between the two atoms, and the angle subtended at the origin by the two vectors from the origin to each of the two atoms.

3.9. Three atoms have the following amplitudes and phases with respect to the real axis of an Argand diagram:

(a) 13.1, 16.23°
(b) 21.4, 154.87°
(c) 37.9, −113.26°

Calculate the amplitude and phase of the resultant sum.

3.10. Express the structure factor equation in a reduced form for an A-face centered unit cell. Hence, deduce the limiting conditions associated with A centering.

3.11. A two-dimensional structure has four atoms per unit cell, two of type P and two of type Q, with the following fractional coordinates:

	x	y
P_1	0.1	0.2
P_2	0.9	0.8
Q_1	0.2	0.7
Q_2	0.8	0.3

Calculate $|F|$ for the reflections 5, 0; 0, 5; 5, 5, and 5, 10 in terms of the scattering factors g_P and g_Q for the two species. If $g_P = 2g_Q$, what are the phase angles for these reflections?

3.12. α-Uranium crystallizes in the orthorhombic system with four uranium atoms in special positions:

$$\pm\left\{0, y, \tfrac{1}{4}; \tfrac{1}{2}, \tfrac{1}{2} + y, \tfrac{1}{4}\right\}$$

Use the data below to decide whether y is better chosen as 0.10 or 0.15.

hkl	$|F(hkl)|$	$g_U(hkl)$
020	88.5	70.0
110	268.9	80.0

3.13. The unit-cell dimensions of α-uranium are $a = 2.85, b = 5.87, c = 5.00\,\text{Å}$. Use the value of y_U from Problem 3.12 to determine the shortest U–U distance in the structure. It may be helpful to plot the uranium atom positions in a few neighboring unit cells.

3.14. In the examples listed below for monoclinic crystals, the conditions limiting possible x-ray reflections are given. In each case, write the possible space groups corresponding to the information given.

    (a) hkl : none
        $h0l$: none
        $0k0 : k = 2n$
    (b) hkl : none
        $h0l : h = 2n$
        $0k0$: none

(c) $hkl : h + k = 2n$

$h0l : l = 2n \ (h = 2n)$

$0k0: (k = 2n)$

(d) hkl : none

$h0l$: none

$0k0$: none

3.15. Repeat Problem 3.14, but for the limiting below relating to orthorhombic crystals.

(a) hkl : none $h00: h = 2n$

$0kl$: none $0k0: k = 2n$

$h0l$: none $00l$: none

$hk0$: none

(b) hkl : none $h00$: none

$0kl: k = 2n$ $0k0: k = 2n$

$h0l$: none $00l$: none

$hk0$: none

(c) $hkl : h + k + l = 2n$ $h00: h = 2n$

$0kl : k = 2n, l = 2n$ $0k0: k = 2n$

$h0l : h + l = 2n$ $00l : l = 2n$

$hk0: h + k = 2n$

3.16. (a) Write the independent conditions limiting possible x-ray reflections for the following space groups: (i) $P2_1/a$; (ii) Pc; (iii) $C2$; (iv) $P2_122$; (v) $Pcc2$; (vi) $Imam$

In each case, write the symbols of the space groups, if any, in the same crystal system with the same limiting conditions.

(b) Write the conditions limiting possible x-ray reflections in the monoclinic space group $P2_1/n$ (non-standard setting).

(c) Give the conventional symbols for the space groups $A2/a$ and $B2_122_1$.

3.17. (a) Space group $Pcab$ corresponds to the nonstandard setting $a\bar{c}b$, that is, a along x, $-c$ along y, and b along z. What is the symbol in the standard $(a\ b\ c)$ setting?

(b) What is the essential difference between the space groups represented by the standard symbols $Pmna$ and $Pnma$? What are their full symbols?

3.18. The absorption correction for a crystal ground into a sphere of radius r is dependent on r, μ, and θ. Assume that extinction effects are negligible, and determine the ideal intensity for an hkl reflection, given that the measured intensity less background is 56.3, and that $r = 0.11$ mm

Transmission Factors A for a Sphere of
Radius R and Linear Absorption
Coefficient μ

μR	θ		
	25°	30°	35°
1	3.88	3.79	3.70
2	10.9	10.0	9.26
3	22.4	19.5	17.1
4	37.2	31.0	26.3

and $\mu = 18.2 \times 10^3$ m^{-1}. For this reflection, $\theta = 30°$, and some tabulated data are listed above, corresponding to the numerical integration $A = \{(1/V) \int dx \int dy \int \exp[-\mu(r_0 + r)]\,dz\}^{-1}$, where r_0 and r are, respectively, the incident and diffracted paths lengths in the crystal. Include the θ-dependent Lorentz and polarization corrections.

3.19. An organic crystal has a large overall isotropic temperature factor of 6.8 Å^2. What is the percentage reduction of the atomic scattering factor of a carbon atom at room temperature for a reflection at $\theta = 27.55°$, with Cu $K\alpha$ radiation ($\lambda = 1.5418$ Å) compared to that for a carbon atom at rest scattering under the same conditions? What is the root mean square amplitude of vibration of the atom in a direction normal to the given reflecting plane? How might the data collection process for this crystal be improved?

$(\sin\theta)/\lambda$	f_C
0.0	6.000
0.1	5.108
0.2	3.560
0.3	2.494
0.4	1.948
0.5	1.686

3.20. (a) Show that, in the centric distribution, $\overline{|F|}^2/\overline{|F|^2} = 0.637$.

(b) The value of $\overline{|E|^3}$ for the acentric distribution is 1.329. Find the value of $\overline{|E|^3}$ for the centric distribution.

(c) The value of $\overline{|E^2 - 1|}$ in the centric distribution has been shown to be 0.968. Find the corresponding value in the acentric distribution.

3.21. A synchrotron SRS source (at Daresbury, Cheshire) has an electron-beam energy of 2.0 GeV and a dipole magnetic field of 1.2 T. Calculate the maximum (characteristic) wavelength and the angular spread of the synchrotron radiation.

3.22. Calculate the ideal intensities, $|F|^2$, for the 111 and 222 reflections for NaCl and KCl. Hence, discuss these reflections in the light of Figure 9.9. The necessary data are as follows: $a(\text{NaCl}) = 5.627\,\text{Å}, a(\text{KCl}) = 6.278$. Calculate f values from the equation $\sum_1^4 a_i \exp(-b_i s^2) + c_i$, where $s = (\sin\theta)/\lambda$, and the constants for the equation are:

	a_1	b_1	a_2	b_2	a_3	b_3	a_4	b_4	c
Na$^+$	3.2565	2.6671	3.9362	6.1153	1.3998	0.2001	1.0032	14.0390	0.4040
K$^+$	7.9578	12.6331	7.4917	0.7674	6.3590	−0.0020	1.1915	31.9128	−4.9978
Cl$^-$	18.2915	0.0066	7.2084	1.1717	6.5337	19.5424	2.3386	60.4486	−16.3780

4

I Optical and X-ray Examination of Crystals
II Measurement of Intensity Data from Single Crystals

I Optical and X-ray Examination of Crystals

4.1 Introduction

The optical examination of crystals is interesting in its own right. However, in structure determinations with modern equipment, it is not uncommon to proceed immediately with x-ray studies. In many cases, the technique is straightforward, particularly with the single-crystal x-ray diffractometer (Section 4.7), and the desired results are readily obtained. There are other situations though, where complications arise because of an unusual habit (Section 4.3.3), pseudosymmetry (Section 6.4.4) or twinning (Section 4.11). In such cases, it may be possible to extract useful information from an optical examination of a crystal before using the more detailed x-ray methods.

In this chapter we shall discuss aspects of the interaction between crystals and two different electromagnetic radiations, light and x-rays. Light, with its longer wavelength (5000–6000 Å), can reveal only limited information about crystal structures, whereas x-rays with wavelengths of less than about 2 Å can be used to determine the relative positions of atoms in crystals. A preliminary examination of a crystal aims to determine its space group and unit-cell dimensions, and may be carried out by a combination of optical and x-ray techniques. The optical methods described here are simple but, nevertheless, often very effective; they should be regarded as a desirable prerequisite to an x-ray structure determination, particularly where automated intensity measurement is used.

4.2 Polarized Light

An ordinary light source emits wave trains, or pulses of light, vibrating in all directions perpendicular to the direction of propagation (Figure 4.1); the light is said to be unpolarized. The vibrations of interest to us are those of the electric vector associated with the waves. Any one of these random vibrations can be resolved into two mutually perpendicular components, and the resultant vibration may, therefore, be considered as the sum of all components in these two perpendicular directions. In order to study the optical properties of crystals, we need to restrict the resultant vibration of the light source to one direction only by eliminating the component at right angles to it.

Let us consider that a polarizer (P), consisting of a sheet of Polaroid, transmits light vibrating in the horizontal direction LM and absorbs all components vibrating in the direction perpendicular to LM. Thus, light passing through the polarizer vibrates in one plane only, and is said to be plane polarized. The plane contains the vibration direction, which is perpendicular to the direction of propagation, and the direction of propagation itself. A second Polaroid, the analyzer (A), is placed after the polarizer and rotated so that its vibration transmission direction (MN) is at 90° to that of the polarizer. It receives no component parallel to its transmission direction and, therefore, absorbs all the light transmitted by the polarizer. The two Polaroids are then said to be crossed. This effect may be demonstrated by cutting a Polaroid sheet marked with a straight line LMN into two sections, P and A (Figure 4.1). When superimposed, the two halves will not transmit light if the reference lines LM and MN are exactly perpendicular. In intermediate positions, the intensity of light transmitted varies from a maximum, where they are parallel, to zero (crossed). The production and use of plane-polarized light by this method is used in the polarizing microscrope.

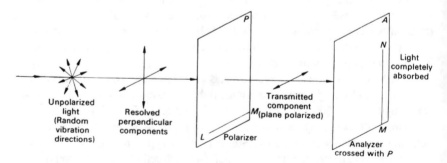

FIGURE 4.1. Production of plane-polarized light by passing unpolarized light through a sheet of Polaroid film (the polarizer, P). A second, identical sheet of Polaroid (the analyzer, A), rotated through 90° with respect to P, completely absorbs all light transmitted by P. The lines LM and MN were parallel on the sheet from which P and A were cut.

4.3   Optical Classification of Crystals

Crystals may be grouped, optically, under two main headings, isotrcpic crystals and anisotropic (birefringent) crystals. All crystals belonging to the cubic system are optically isotropic; the refractive index of a cubic crystal is independent of the direction of the light incident upon it, and its optical characteristics are similar to those of glass. Noncubic crystals exhibit a dependence on direction in their interaction with light.

Anisotropic crystals are divided into two groups, uniaxial crystals, which have one optically isotropic section and include the tetragonal, hexagonal, and trigonal crystal systems, and biaxial crystals, which have two optically isotropic sections and belong to the orthorhombic, monoclinic, and triclinic crystal systems.

A preliminary optical examination of a crystal will usually show whether it is isotropic, uniaxial, or biaxial. Distinction between the three biaxial crystal systems is often possible in practice and, depending on how well the crystals are developed, a similar differentiation may also be effected for the uniaxial crystals. Even if an unambiguous determination of the crystal system is not forthcoming, the examination should, at least, enable the principal symmetry directions to be identified; Table 4.1 summarizes this information.

4.3.1   Uniaxial Crystals

As an example of the use of the polarizing microscope, we shall consider a tetragonal crystal, such as potassium dihydrogen phosphate, lying on a microscope slide with its y axis parallel to the axis of the optical path through a microscope (Figure 4.2). The microscope is fitted with a polarizer (P), and an analyzer (A) which is crossed with respect to P and may be removed from the optical path. The crystal can be rotated on the microscope stage between P and A. With the Polaroids

TABLE 4.1.   Crystal Directions Readily Derivable from an Optical Study

Optical classification	Crystal system	Information relating to crystal axes likely to be revealed
Isotropic	Cubic	Axes may be assigned from the crystal morphology
Anisotropic, uniaxial	Tetragonal	Direction of the z axis
	Hexagonal	Direction of the z axis
	Trigonal[a]	Direction of the z axis
Anisotropic, biaxial	Orthorhombic	Direction of at least the x, y, or z axis, possibly all three axes
	Monoclinic	Direction parallel to the y axis
	Triclinic	No special relationship between the crystal axes and the vibration directions

[a] Referred to hexagonal axes.

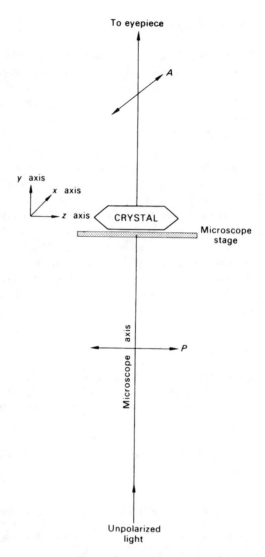

FIGURE 4.2.   Schematic experimental arrangement for examining extinction directions. A tetragonal crystal is shown on the microscope stage, and the incident light is perpendicular to the z axis of the crystal.

crossed and no crystal in between, the field of view is uniformly dark. However, with the crystal interposed, this situation will not necessarily be obtained.

The tetragonal crystal is lying with (010) on the microscope slide; both the x and z axes are, therefore, perpendicular to the microscope axis. In general, some of the light passing through the crystal will be transmitted by the analyzer, even though P and A are crossed. The intensity of the transmitted light varies as the crystal is rotated on the microscope stage between the polarizer and the analyzer. During a complete revolution of the stage, the intensity of transmitted light passes through four maxima and four minima. At the minimum positions, the crystal is usually only just visible. These positions are called extinction positions, and they occur at exactly 90° intervals of rotation. Maximum intensity is observed with the crystal at 45° to these directions.

These changes would be observed if the crystal itself were replaced by a sheet of Polaroid. Extinction would occur when the vibrations of the "crystal Polaroid" were perpendicular to those of P or A. A simple explanation of these effects is that the crystal behaves as a polarizer. Incident plane-polarized light from P is resolved by the crystal into two perpendicular components (Figure 4.3). In our tetragonal crystal, the vibration directions associated with this polarizing effect are parallel to its x and z axes. Rotating the crystal on the microscope stage will, therefore, produce extinction whenever x and z are parallel to the vibration directions of P and A. The x and z axes of a tetragonal crystal correspond to its extinction directions. It should be remembered that the x and y directions are equivalent under the 4-fold symmetry of the crystal.

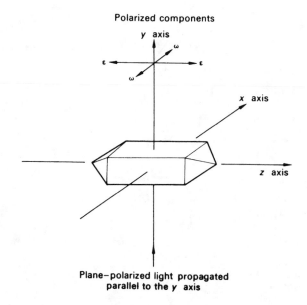

FIGURE 4.3.   Resolution of incident light into components vibrating parallel to the x and z axes of a tetragonal crystal lying with its y axis parallel to the incident beam; ω and ε are the refractive indices for light vibrating, respectively, perpendicular and parallel to z.

4.3.2   Birefringence

The vibration components produced by a crystal are associated with its different refractive indices. With reference to Figure 4.3, a tetragonal crystal with light vibrating parallel to the 4-fold symmetry axis (z) has a refractive index ε, whereas light vibrating perpendicular to z has a different refractive index, ω; the crystal is said to be birefringent, or optically anisotropic.

Figure 4.4 represents plane-polarized light incident in a general direction with respect to the crystallographic axes. It is resolved into two components, one with an associated refractive index ω and the other with an associated refractive index ε', both vibrating perpendicular to each other and to the direction of

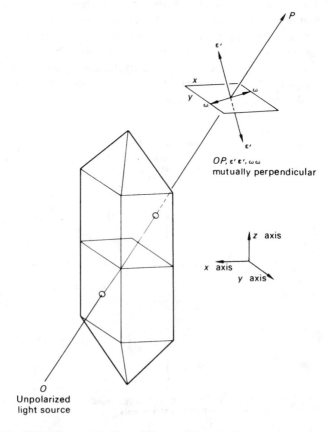

FIGURE 4.4.   Uniaxial crystal showing a light ray OP resolved into two components. One component, with refractive index ω, vibrates in the xy plane, the other, with refractive index ε', vibrates parallel to both ω and the ray direction.

incidence. In general, the value of ε' lies between those of ω and ε. Two special cases arise: one, already discussed, where the incident light is perpendicular to z, for which $\varepsilon' = \varepsilon$, and the other where the incident light is parallel to z, for which $\varepsilon' = \omega$. It follows that where the direction of incidence is parallel to the z axis, the refractive index is always ω for any vibration direction in the xy plane. Plane-polarized light incident parallel to the z axis will pass through the crystal unmodified. In this particular direction, the crystal is optically isotropic, and if rotated on the microscope stage between crossed Polaroids, it remains in extinction. The z direction of a uniaxial crystal is called the optic axis, and there is only one such direction in the crystal; it is the 4-fold symmetry axis in the example that we are using.

Identification of the z Axis of a Uniaxial Crystal

A polarizing microscope is usually fitted with eyepiece cross-wires arranged parallel and perpendicular to the vibration directions of the polarizer, and therefore we can relate the crystal vibration directions to its morphology. There are two important optical orientations for a tetragonal crystal, namely with the z axis either perpendicular or parallel to the axis of the microscope. These orientations are, in fact, important for all uniaxial crystals, and will be described in more detail.

z Axis Perpendicular to the Microcsope Axis.   In this position, a birefringent orientation is always presented to the incident light beam (Figure 4.5). Extinction will occur whenever the z axis is parallel to a cross-wire, no matter how the crystal is rotated, or flipped over, *while keeping z parallel to the microscope slide.* The success of this operation depends to a large extent on having a crystal with well-developed $(hk0)$ faces. The term *straight extinction* is used to indicate that the field of view is dark when a crystal edge is aligned with a cross-wire. A face of a uniaxial crystal for which one edge is parallel to z, an $(hk0)$ face, or to its trace on a crystal face, for example, an $(h0l)$ face, will show straight extinction.

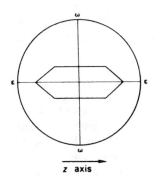

FIGURE 4.5. Extinction position for a tetragonal crystal lying with its z axis parallel to the microscope slide. Any $[UV0]$ direction could be parallel to the microscope axis; extinction will always be straight with respect to the z axis or its trace.

(a)   (b)   (c)

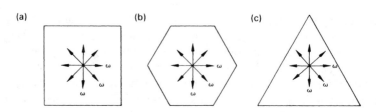

FIGURE 4.6. Idealized uniaxial crystals as seen along the z axis: (a) tetragonal, (b) hexagonal, (c) trigonal. The refractive index for light vibrating perpendicular to the z axis is given the symbol ω, and the crystals appear isotropic in this orientation.

z Axis parallel to the Microscope Axis. The crystal now presents an isotropic section to the incident light beam, and will remain extinguished for all rotations of the crystal, *while keeping z along the microscope axis.* A reasonably thin section of the crystal is required in order to observe this effect. Because of the needle-shaped habit (external development) of the crystal (KH_2PO_4), it is necessary to cut the crystal carefully so as to obtain the desired specimen.

The section of a uniaxial crystal normal to the z axis, if well developed, may provide a clue to the crystal system. Tetragonal crystals often have edges at 90° to one another, whereas hexagonal and trigonal crystals often exhibit edges at 60° or 120° to one another. These angles are external manifestations of the internal symmetry; idealized uniaxial crystal section are shown in Figure 4.6.

4.3.3 Biaxial Crystals

Biaxial crystals have two optic axes and, correspondingly, two isotropic directions. The reason for this effect lies in the low symmetry associated with the orthorhombic, monoclinic, and triclinic systems, which, in turn, results in less symmetric optical characteristics. Biaxial crystals have three principal refractive indices, n_1, n_2, and n_3 ($n_1 < n_2 < n_3$), associated with light vibrating parallel to three mutually perpendicular directions in the crystal. The optic axes that derive from this property are not necessarily directly related to the crystallographic axes. We shall not concern ourselves here with a detailed treatment of the optical properties of biaxial crystals, but will concentrate on relating the vibration (and extinction) directions to the crystal symmetry.

Orthorhombic Crystals

In the orthorhombic system, the vibration directions associated with n_1, n_2, and n_3 are parallel to the crystallographic axes, but any combination of x, y, and z with n_1, n_2, and n_3 may occur. Consequently, recognition of the extinction directions facilitates identification of the directions of the crystallographic axes. For a crystal with x, y, or z perpendicular to the microscope axis, the extinction,

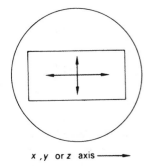

x ,y or z axis ⟶

FIGURE 4.7. Extinction directions in an orthorhombic crystal viewed along the x, y, or z axis.

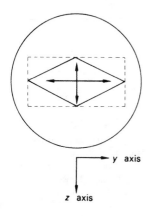

➤ y axis

z axis

FIGURE 4.8. Extinction directions as seen along the x axis of an orthorhombic crystal with {011} development—an example of symmetric extinction.

directions will be parallel (or perpendicular) to the axis in question, as shown in Figure 4.7. If the crystal is a well developed orthorhombic prism, the three crystallographic axes may be identified by this optical method. A common alternative habit of orthorhombic crystals has one axis, x, for example, as a needle axis with the {011} form prominent. The appearance of such a crystal viewed along x is illustrated in Figure 4.8, and is an example of a symmetric extinction.

Monoclinic Crystals

The lower symmetry of monoclinic crystals results in a corresponding modification of the optical properties in this system. The symmetry axis y is, conventionally, to be parallel to one of the vibration directions; x and z are related arbitrarily to the other two mutually perpendicular vibration directions. Hence, two directions are of importance in monoclinic crystals, namely, perpendicular to and parallel to the y axis.

When viewed between crossed Polaroids, a monoclinic crystal lying with its y axis perpendicular to the microscope axis will always show straight extinction,

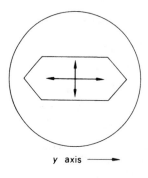

y axis ────▶

FIGURE 4.9. Extinction directions in a monoclinic crystal viewed perpendicular to the y axis—an example of straight extinction.

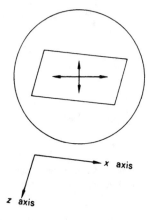

────▶ x axis

z axis

FIGURE 4.10. Extinction directions in a monoclinic crystal viewed along the y axis—an example of oblique extinction. If the forms developed are pinacoids, that is, {100}, {010}, and {001}, then the *extinction angle* (the angle between a crystal edge and a cross-wire) will be related in a simple way to the β angle.

with the cross-wires parallel (and perpendicular) to y. Often, the y axis is a well developed needle axis; rotation of the crystal about this axis while keeping it perpendicular to the microscope axis will not cause any change in the position of extinction (Figure 4.9).

If, on the other hand, the monoclinic crystal is arranged so that y is parallel to the microscope axis, the (010) plane will lie on the microscope slide. Extinction in this position will, in general, be *oblique*, as shown in Figure 4.10, thus giving further evidence for the position of the y-axis direction. The appearance of extinction in a monoclinic crystal in this orientation may be somewhat similar to that of an orthorhombic crystal showing prominent {011} development (compare Figures 4.8 and 4.10), and confusion may sometimes occur in practice.

Triclinic Crystals

The mutually perpendicular vibration directions associated with n_1, n_2, and n_3 are arbitrarily related to the crystallographic axes, which are selected initially from morphological and x-ray studies.

Reference again to Table 4.1 should now enable the reader to consolidate the ideas presented in the discussion of extinction directions in the seven crystal systems. Although it gives only limited information[a] on the optical properties of crystals, a practical study of a crystal along these lines can often provide useful information about both its system and its axial directions.

4.3.4 Interference Figures

The effects which we have discussed so far may be observed when the crystal specimen is illuminated by a more or less parallel beam of plane-polarized light. There is another technique worthy of mention, in which the crystal is examined in a convergent beam of polarized light; it produces characteristic interference figures for uniaxial and biaxial crystals. This examination may be effected, at high magnification and between crossed Polaroids, either by removing the microscope eyepiece or by inserting a Bertrand lens[b] into the microscope system, below the eyepiece. Figure 4.11a shows an idealized interference figure from a section of a uniaxial crystal cut perpendicular to the optic axis, while Figure 4.11b and 4.11c are interference figures for a biaxial crystal section cut perpendicular to a bisector of the two optic axes.

If optical figures of good quality can be obtained, the distinction between uniaxial and biaxial specimens may be achieved with the one orientation of the crystal. It may be confirmed by rotation of the crystal specimen about the microscope axis, which causes the dark brushes or isogyres in the biaxial figure to break up, as in Figure 4.11c, while those for the uniaxial interference figure remain intact.

4.4 Single-Crystal X-ray Techniques

With the development of ever more sophisticated technology there has been a gradual move away from the traditional methods of introducing the beginner to the theory and practices associated with the reciprocal lattice. However, we are of the opinion that this is still best achieved through x-ray photography, and we are also aware that laboratories in many parts of the world do still use x-ray cameras for this purpose, and, in some cases, for crystal structure determination. Consequently we retain the section on cameras in the present edition. The introduction of various types of area detectors does in fact swing methodology away from the serial ("one reflection at a time") diffractometer, back toward photographic recording principles. An understanding of x-ray camera geometry will therefore be invaluable to

---

[a] For a fuller description, see Bibliography.
[b] A Bertrand lens is a normal accessory with a good polarizing microscope.

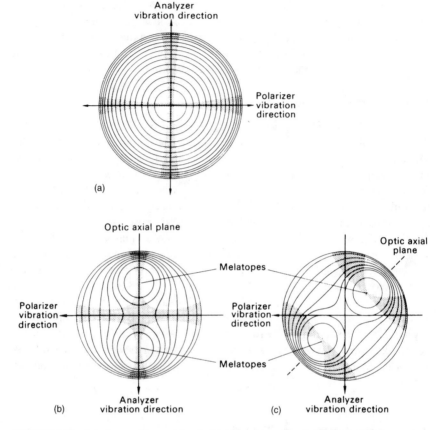

FIGURE 4.11.   Interference figures: (a) uniaxial, along the optic axis; (b) biaxial, along a bisector of the optic axes and with the polaroids crossed (a thin flake of mica gives a good biaxial interference figure); (c) as in (b), but with the polarizer rotated by 45° (position of maximum transmitted intensity), and showing the isogyres that form the cross in (b) now broken—no such effect occurs for (a). [Reproduced from *An Introduction to Crystal Optics* by P. Gay, with the permission of Longmans Group Ltd., London.]

those taking the almost imperative step of progressing from in-house facilities to the use of the synchrotron.

X-ray Photography

Over the past 10 years the role of powder crystallography has been transformed from one of very minor use in structure analysis to one in which it is rapidly becoming a major alternative choice. Chapter 9 provides details of the advances that have taken place to revolutionize this method, while this chapter attempts to cover all or most of the great variety of x-ray recording techniques for single crystals.

4.4.1 Laue Method

The three variables in the Bragg equation (3.55) provide a basis for the interpretation of x-ray crystallographic experiments. In the Laue method (Figure 4.12), the Bragg equation is satisfied by effectively varying λ, utilizing the beam of continuous (white) radiation. Since the crystal is stationary with respect to the x-ray beam, it acts as a sort of filter, selecting the correct wavelength for each reflection according to (3.55).

The spots on a Laue photograph lie on ellipses, all of which have one end of their major axis at the center of the photographic film (Figure 4.13). All spots on one ellipse arise through reflections from planes that lie in one and the same zone. In Figure 4.14, a zone axis for a given Bragg angle θ is represented by ZZ'. A reflected ray is labeled R, and we can simulate the effect of the zone by imagining the crystal to be rotated about ZZ', taking the reflected beam with it. The rays, such as R, generate a cone, coaxial with ZZ' and with a semivertical angle θ. The lower limit, in the diagram, of R is the direction (xy) of the x-ray beam, and the general intersection of a circle with a plane (the flat film) is an ellipse. Hence, we can understand the general appearance of the Laue photograph (Figure 4.13). On each ellipse, discrete spots appear instead of continuous bands because only those orientations parallel to zone axes, such as ZZ', that actually exist for crystal planes can give rise to x-ray reflections.

Symmetry in Laue Photographs. One of the useful features of Laue photographs is the symmetry observable in them. The crystal orientation with respect to the x-ray beam is selected by the experimenter from morphological and optical considerations. This orientation, together with the crystal point group, controls the symmetry on the Laue photograph.

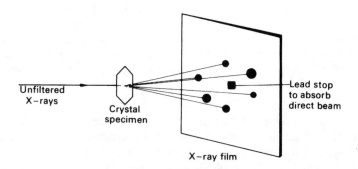

FIGURE 4.12. Schematic experimental arrangement for taking a Laue photograph on a flat-plate film.

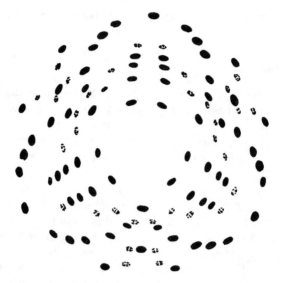

FIGURE 4.13.   Sketch of a Laue photograph of α-Al_2O_3; the 3-fold symmetry direction is normal to the photograph (along the x-ray beam).

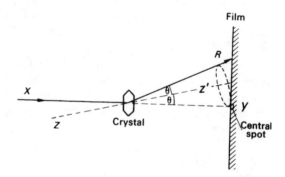

FIGURE 4.14.   Geometry of the Laue method: XY, x-ray beam direction; ZZ', a zone axis; R, generator of the cone of diffracted rays of semi-angle θ, the Bragg angle; Y, central spot on the film and extremity of the major axis of the ellipse formed by the intersection of the cone with the film.

In practice, a complication arises by the introduction of a center of symmetry into an x-ray diffraction pattern, in normal circumstances, whether or no the crystal is centrosymmetric. This situation is embodied in Friedel's law, the theoretical grounds for which have been discussed in Chapter 3. As a result of this law, the diffraction pattern may not reveal the true point-group symmetry of a crystal.

Table 1.6 shows the classification of the 32 crystallographic point groups according to Laue diffraction symmetry.

It cannot be overemphasized that the Laue group assigned to a crystal describes the symmetry of the *complete* x-ray diffraction pattern from that crystal. No single x-ray photograph can exhibit the complete diffraction symmetry, only that of a selected portion which is a projection on to the film, along the direction of the x-ray beam, of the symmetry information that would be encountered in that direction in a crystal having the Laue group of the given crystal.

It follows that in the triclinic system, no symmetry higher than 1 is ever *observable* on a Laue photograph (Table 1.6). In other crystal systems, the Laue-projection symmetry depends on the orientation of the crystal with respect to the x-ray beam. Rotation axes of any order reveal their true symmetry when the x-ray beam is parallel to the symmetry axis. Even-order rotation axes, 2, 4, or 6, give rise to mirror diffraction symmetry in the plane normal to the rotation axis when the x-ray beam is normal to that axis. A mirror plane itself shows m symmetry parallel to the mirror plane when the x-ray beam is contained by the plane. Various combinations of these effects may be observable, depending upon the Laue group in question.

The supplementary nature of the x-ray results to those obtained in the optical examination should now be evident. Uniaxial crystals can be allocated to their correct systems by a Laue photograph taken with the x-ray beam along the z axis. The Laue photograph in Figure 4.13 exhibits the Laue symmetry $3m$. Distinction between the monoclinic and orthorhombic systems, which is not always possible in an optical examination, is fairly straightforward with Laue photographs, as Table 1.6 shows. Cubic crystals can exhibit a variety of symmetries, but with the x-ray beam along $\langle 100 \rangle$ differentiation between Laue groups $m3$ and $m3m$ is obvious.

In practice, the symmetry pattern on a Laue photograph is very sensitive to the precise orientation of the crystal with respect to the x-ray beam.[a] Slight deviation from the ideal position will result in a distortion of the relative positions and intensities of the spots on the photographs.

Laue Method and Synchrotron Radiation

The synchrotron is an extremely powerful source of x-rays and produces a very wide range of wavelengths (Section 3.1.5); it is ideally suited to the Laue method of recording diffraction patterns. Since the crystal is in a fixed orientation, the angle of incidence of the x-ray beam is thus set for each (hkl) plane. For a reflection to take place at a preset θ angle, the plane must effectively *select* the wavelength required to satisfy Bragg's equation. A reflection on a Laue photograph

[a] See Bibliography (Jeffrey).

thus comprises four parameters, the usual hkl indices and the wavelength selected by the crystal.

Consider a fixed crystal in a Laue diffraction experiment. For planes (hkl) the preset angle of incidence is θ, and the wavelength required to be selected for Bragg's equation to be satisfied for the reflection hkl is $\lambda(hkl, \theta)$, given by

$$\lambda(hkl, \theta) = 2d(hkl) \sin \theta \qquad (4.1)$$

For planes $(2h, 2k, 2l)$, the wavelength to be selected is

$$\lambda(2h, 2k, 2l; \theta) = 2d(2h, 2k, 2l) \sin \theta \qquad (4.2)$$

From Section 2.3,

$$\lambda(2h, 2k, 2l; \theta) = 2d(hkl) \sin \theta / 2 = \lambda(hkl, \theta)/2 \qquad (4.3)$$

which is easily generalized to

$$\lambda(nh, nk, nl, \theta) = \lambda(hkl, \theta)/n \qquad (4.4)$$

The Bragg angle θ being common to the sets of reflections $hkl; 2h, 2k, 2l; \ldots; nh, nk, nl$, these sets will be superimposed (or multiply) on the Laue photograph. The extent of $n = 1, 2, \ldots$ will depend on the range of values of λ available from the x-ray source, which may be selected to minimize this overlap of reflections.

The interpretation of a Laue photograph may thus be complicated both by the possible existence of multiple reflection orders and by the need to assign the correct value of λ to each $I(hkl)$ prior to use in structure analysis. The latter requirement is, of course, necessary in view of the dependence of the atomic scattering factor f on both θ and λ (Section 3.2.5). Other factors requiring special attention include sensitivity characteristics and absorption of photographic film, both of which are wavelength dependent.

As an example, consider the following situation: an orthorhombic crystal with $a = 10.0$, $b = 15.0$, $c = 20.0$ Å is mounted with c vertical and perpendicular to the x-ray beam, such that b makes an angle ϕ of $30°$ with the beam direction in the horizontal plane. A diffraction spot P occurs on a flat-plate film such that its coordinates are X mm (horizontal) and Y mm (vertical), the plate being placed at a distance R mm from the crystal (Figure 4.15). Since the wavelength is variable, we define $d^*(hkl)$ by (2.11) with $\kappa = 1$, which results in the Ewald sphere (Section 4.4.4) having a wavelength-dependent radius of $1/\lambda$. Thus, if the extremes of λ used in the experiment are λ_{min} and λ_{max}, the corresponding Ewald spheres will have radii $1/\lambda_{min}$ and $1/\lambda_{max}$, as shown in Figure 4.16. The reciprocal lattice in this treatment has fixed dimensions, $a^* = 1/a$, $b^* = 1/b$ and $c^* = 1/c$. We can predict possible Laue reflections at $Y = 0$ with the aid of Figure 4.16. The a^*b^* reciprocal

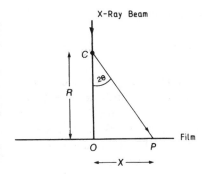

FIGURE 4.15. Geometry of Laue diffraction at the level $Y = 0$ on a film placed perpendicularly to the incident x-ray beam: C, crystal; P, position of a Bragg reflection; 2θ, scattering angle.

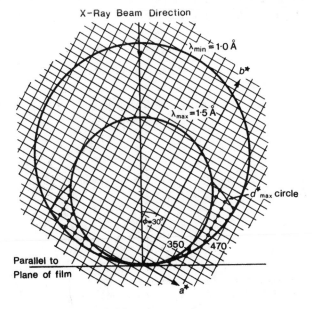

FIGURE 4.16. Reciprocal net a^*b^* for an orthorhombic crystal with $a = 10$ and $b = 15$ Å, tilted at $\phi = 30°$ to the x-ray beam; c is vertical. The wavelength range chosen is 1.0 Å (λ_{min}) to 1.5 Å (λ_{max}). Reciprocal lattice points that can give rise to diffraction spots on the right-hand side of the film (including the 350 and 470 used in the example calculations) and on the left-hand side have been emphasized. They lie between the limiting arcs of the Ewald spheres as shown. The range of $hk0$ reflections is limited also by the resolution limit d^*_{max}, which is governed by the crystal. For the reciprocal net chosen, $a^* = 1/a = 0.1$ RU and $b^* = 1/b = 0.0667$ RU; the scale is 1 RU = 3 cm.

lattice net is shown rotated as described above at 30° to the x-ray beam. Traces of the outer and inner Ewald spheres contain a reciprocal lattice area which includes all reciprocal lattice points (intensified) that are able to give rise to Laue reflections with the appropriate wavelength. The recording geometry relevant to the above experimental setup is shown in Figure 4.15, from which it can be seen that for $Y = 0$ (corresponding to $l = 0$ in the example)

$$\tan 2\theta (hk\theta) = X/R$$

where X is the horizontal distance of spot P from the origin O, and R is the crystal-film constant. Reciprocal lattice points lying within the allowed region (Figure 4.16) include 350, 470, 480, 490, ... , 312, for the limiting wavelengths $\lambda_{min} = 1$ Å, $\lambda_{max} = 1.5$ Å used in the example. We now determine the coordinates of Laue diffraction spots and the wavelengths used in producing them.

The wavelength selected for a particular $d^*(hk0)$ can be calculated with reference to Figure 4.17. In this diagram

$$\tan \varepsilon = ha^*/kb^* = hb/ka$$

and

$$\theta(hk0) = 90° - (\phi + \varepsilon)$$

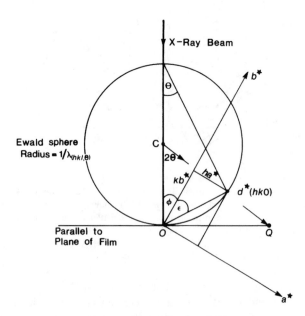

FIGURE 4.17.   Geometry of wavelength selection through a reciprocal lattice vector $d^*(hk0)$. The angle ϕ defines the crystal orientation: C, crystal; O, origin of reciprocal lattice; CQ, direction of diffracted beam.

Thus,

$$\lambda(hk0, \theta) = 2d(hk0) \sin \theta (hk0)$$
$$= 2d(hk0) \sin[90 - (\phi + \varepsilon)]$$
$$= 2d(hk0) \cos(\phi + \varepsilon)$$

Taking $\phi = 30°$:

1. Reflection 350

$$\tan \varepsilon = 3 \times 15/(5 \times 10) = 0.900$$
$$\varepsilon = 41.99°$$
$$d(350) = \left(\frac{9}{10^2} + \frac{25}{15^2} \right)^{-1/2} = 2.2299 \text{ Å}$$
$$\therefore \lambda(350, \theta) = 2 \times 2.299 \cos(30 + 41.99)$$
$$= 1.3789 \text{ Å}$$
$$\theta(350) = 90 - (30 + 41.99) = 18.01°$$

Hence for $R = 60$ mm

$$X(350) = 60 \tan(2 \times 18.01)$$
$$= 43.62 \text{ mm}$$

2. Reflection 470

$$\tan \varepsilon = 4 \times 15/(7 \times 10) = 0.8571$$
$$\varepsilon = 40.60°$$
$$d(470) = \left(\frac{16}{10^2} + \frac{49}{15^2} \right)^{-1/2} = 1.6270 \text{ Å}$$
$$\therefore \lambda(470, \theta) = 2 \times 1.6270 \cos(30 + 40.60)$$
$$= 1.0809 \text{ Å}$$
$$\theta(470) = 90 - (30 + 40.60) = 19.40°$$
$$X(470) = 60 \tan(2 \times 19.40)$$
$$= 48.24 \text{ mm}$$

Note the dependence on these results of the crystal orientation parameter ϕ. It is easy to show that for $\phi = 0°$

1. $\lambda(350, \theta) = 3.4175$ Å
2. $\lambda(470, \theta) = 2.4707$ Å

Thus, neither of these reflections would be recorded for this orientation using the given wavelength range.

Laue photographs can be optically scanned by using a densitometer, which records both intensity and position of each spot on the film. These data are then processed by computer using powerful software, which will index and also refine the unit-cell parameters if required. Interpretation of such photographs, which can contain thousands of spots, is a skilled operation. As we have seen, a knowledge of both the unit cell and crystal orientation is an advantage. With high crystal symmetry it is possible to record most of the three-dimensional diffraction pattern on a single photograph. Being independent of mechanical constraints, Laue data can be recorded very rapidly (in seconds) at a high-intensity synchrotron radiation (SR) facility using a CCD plate (q.v.) instead of a photographic film. Even the problem previously mentioned of multiplicity of orders has proved to be less of a difficulty to an application of the method than at first thought, for example, by using a wavelength range of 0.6–1.6 Å in protein crystallography. The method thus facilitates novel studies using SR, such as time-dependent solid-state reactions and enzyme-driven transformations. The latter can be synchronized by employing tailor-made photosensitive substrates to delay the biochemical reactions until the x-ray experiment is ready. Such studies provide exciting new dimensions to x-ray crystallography.

There are several important differences between SR (Section 3.1.5) and x-rays generated from a conventional laboratory source (Section 3.1 ff). The latter are usually emitted as characteristic radiation from a metallic (copper or molybdenum) target and comprise the predominant characteristic wavelengths (α and β) and a more general, less intense, polychromatic background. Use of an appropriately selected metal filter produces an effectively monochromatic beam (α) (Section 3.1.4). SR is of extremely high intensity, a property which can be exploited for the examination of weakly diffracting or very small crystals. As indicated in Section 3.1.5, SR has a continuous range of wavelengths. A particular wavelength can be selected as required by use of an appropriate filter, for either rotation photography or powder diffraction. Alternatively, the continuous polychromatic beam can be used for Laue photography with a stationary crystal, to record diffraction data efficiently and rapidly. Finally, SR has a very low beam divergence which results in very sharp diffraction spots (Figure 4.18). This can be particularly useful in providing good intensity data from poorly diffracting crystals, such as proteins, resulting in greatly improved resolution of their x-ray Fourier images.

4.4.2 Oscillation Method

The oscillation method is a somewhat more sophisticated technique for recording the x-ray diffraction patterns from single crystals. Reflections are produced, in accordance with the Bragg equation, by varying the angle θ for a given

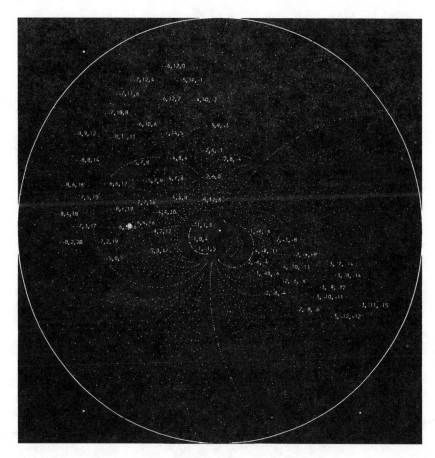

FIGURE 4.18.   Computer simulation diagram of a Laue diffraction pattern for ribonuclease-3$'$ cytidine monophosphate complex (after C. D. Reynolds and S. Prince).

wavelength λ. The variation of θ is brought about by oscillating or rotating the crystal about a crystallographic axis, and λ is "fixed" by the use of an appropriate filter placed in the path of the incident x-ray beam or selected at a synchrotron installation.

The basic arrangement used in the oscillation method is illustrated schematically in Figure 4.19. X-ray reflections produced by the moving crystal are recorded on a cylindrical film coaxial with the axis of oscillation of the crystal. The general appearance of an oscillation photograph is illustrated in Figure 4.20. For the moment, we shall concentrate on the periodicity of the crystal parallel to the oscillation axis. Thinking of the crystal as a (vertical) row of scattering centers, and following (3.39), we see that, for normal incidence ($\phi = 90°$), the diffracted beams will lie on cones that are coaxial with the oscillation axis and intersect the film in

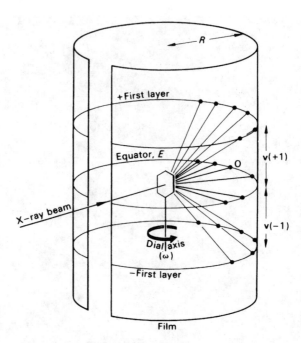

FIGURE 4.19.   Geometry of the oscillation method. The diffraction spots on any layer line are formed by the intersections with the cylindrical film of the diffracted beams that lie on the surface of a cone, the axis of which coincides with the rotation axis. The locus of these intersections is a circle, which becomes a straight line when the film is laid flat.

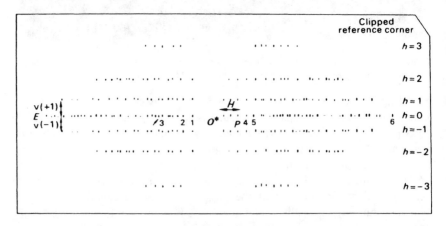

FIGURE 4.20.   Sketch of a 15° oscillation photograph of an orthorhombic crystal mounted on the x axis ($a = 6.167$ Å); the camera radius R is 30.0 mm and λ(Cu $K\alpha$) = 1.5418 Å. The film is flattened out and the right-hand corner, looking toward the x-ray source, is clipped in order to provide a reference mark. P represents any equatorial reflection at a distance $OP(= H\text{mm})$ from the center O. Reflections numbered 1–6 on the zero level are indexed by the method given on page $242\,ff$. The linear scale of the diagram is (1/1.96).

circular traces. When the film is flattened out for inspection, the spots are found to lie on parallel straight lines, known as *layer lines*.

Axial Spacings from Oscillation Photographs. The equatorial layer line, or zero-layer line E, passes through the origin O where the direct x-ray beam intersects the film (Figures 4.19 and 4.20). The layer-line spacings, measured with respect to the zero layer, are denoted by $v(\pm n)$, where $v(+n) = v(-n)$, and are related to the repeat distance in the crystal parallel to the oscillation axis, as is shown in the following treatment.

Let the oscillation axis be a, and let R be the radius of the film, measured in the same units as v. Consider the crystal to be acting as a one-dimensional diffraction grating, with respect to the direction of the a axis, and giving rise to spectra of order $\pm 1, \pm 2, \ldots, \pm n$; $\psi(n)$ is the scattering angle for the nth-order maximum, measured with respect to the direct beam. Normal beam diffraction uses the geometry of Figure 4.21. The path difference for rays scattered at an angle $\psi(n)$ by successive elements of the grating is $a \sin \psi(n)$ which, for maximum intensity,

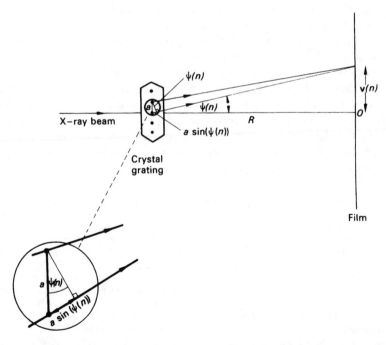

FIGURE 4.21. Diffraction-grating analogy explaining the layer-line spacings on oscillation photographs. Monochromatic x-rays are incident normal to a (oscillation axis) in the crystal. The size of any spot at height such as $v(n)$ depends upon the experimental conditions.

is equal to $n\lambda$. Hence, for layer n,

$$a \sin \psi(n) = n\lambda \qquad (4.5)$$

where $\psi(n)$ is measured experimentally as

$$\tan \psi(n) = v(n)/R \qquad (4.6)$$

Hence

$$a = \frac{n\lambda}{\sin\{\tan^{-1}[v(n)/R]\}} \qquad (4.7)$$

For a known wavelength, this equation provides a convenient and reasonably accurate method for determining unit-cell spacings. In practice, we measure the double spacing between the $\pm n$th orders so as to enhance the precision of the result.

Symmetry in Oscillation Photographs. Oscillation photographs have several useful symmetry properties. A horizontal mirror line along the equator E of a general oscillation photograph indicates a mirror plane perpendicular to the oscillation axis in the corresponding Laue group of the crystal (Table 4.2).

Further observations on the symmetry of the Laue group can be made by arranging for a particular crystal symmetry direction to be parallel to the x-ray beam at the midpoint of the oscillation range (symmetric oscillation photograph). The situations that can arise are summarized in Table 4.2. Note that the highest symmetry observable by the symmetric oscillation method is $2mm$, because mmm is the symmetry of an oscillation movement. However, if an R-fold rotation axis ($R > 2$) is parallel to the beam at the midpoint of the oscillation, then the central portion of the photograph will reveal an *approximate* R-fold symmetry pattern, particularly where the reciprocal unit cell is small. The true symmetry will not appear exactly, as it is degraded by the symmetry of the oscillation movement; the exact symmetries are subgroups of $2mm$.

TABLE 4.2.   Symmetry Indications from Oscillation Photographs

Feature of photograph	Interpretation(s)
Horizontal m line	Horizontal m plane in the corresponding Laue group
Vertical m line[a]	m plane in Laue group of crystal, parallel to the plane defined by the oscillation axis and the beam
2-fold symmetry about the center of the photograph[a]	2-fold axis in the Laue group of the crystal, and parallel to the x-ray beam
Approximate R-fold symmetry around the central portion of the photograph[a]	R-fold axis in Laue group of crystal, and parallel to the x-ray beam

[a] Symmetric oscillation photographs.

Detection of 3-fold, 4-fold, or 6-fold rotational symmetry parallel to the oscillation axis may be effected by taking a series of photographs, the first of which is taken with the crystal oscillating about an arbitrary setting (ω_1) of the dial axis (Figure 4.19). The identical appearance of succeeding photographs with the dial axis set at $\omega_1 + 60°, \omega_1 + 90°$, or $\omega_1 + 120°$ indicates 6-fold, 4-fold, or 3-fold (and 6-fold) symmetry, respectively. A 2-fold axis cannot be detected by this method, because of Friedel's law.

Indexing the Zero Level of a Crystal with an Orthogonal Lattice. We discuss next the relatively small portion of the x-ray diffraction pattern produced in an oscillation photograph. It is of great importance in structure analysis to assign the correct indices *hkl* to each observed reflection. This process is known as indexing, and is reasonably straightforward. In this discussion, we shall consider the indexing of an orthogonal reciprocal lattice, using, as an example, an orthorhombic crystal mounted with *a* as the oscillation axis. It may be noted in passing that monoclinic *b*-axis and hexagonal and trigonal *c*-axis photographs can be indexed in a similar manner. Two prerequisites to indexing are the reciprocal unit-cell dimensions, and the orientation of the reciprocal lattice axes perpendicular to the oscillation axis (b^* and c^* in the example) with respect to the incident x-ray beam. The reciprocal unit-cell dimensions may be derived from the corresponding direct space values (see Section 2.4).

4.4.3 Crystal Setting

We consider here the problem of bringing a given reciprocal lattice plane (equatorial plane) to a position normal to the crystal rotation axis, prior to taking an oscillation photograph. We shall assume that our crystal has a well-developed morphology, such as a prismatic (needle-shaped) habit (Figure 4.4).

The crystal is set up on a goniometer head (see Figure 8.2), with its prism axis along the axis of rotation. Two arc adjustments, A and B, and two sledges, C and D enable the crystal to be set, initially to better than 5°, and arranged so as to rotate within its own volume.

Setting Technique

A method of Weisz and Cole, as modified by Davis, will be considered: it has the advantage that each arc can be treated independently of the other.

A 15° oscillation photograph is taken with the arcs A and B at 45° to the x-ray beam (Figure 4.22a) at the midpoint of the oscillation range, using unfiltered radiation and an exposure time sufficient to produce intense reflections. The goniometer head is then turned through exactly 180° and a second oscillation photograph taken on the same film, but with an exposure time of about one third that of the first. The form of the double oscillation zero layer-line curve is shown in Figure 4.22b.

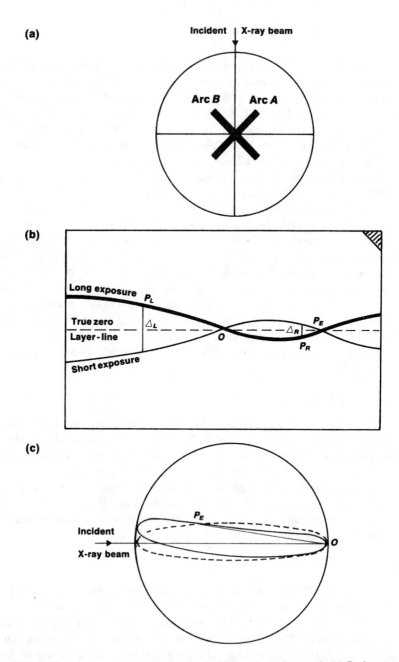

FIGURE 4.22. Crystal setting for an oscillation (or Weissenberg) photograph. (a) Goniometer arcs are set at 45° to the x-ray beam. (b) Appearance of a double oscillation photograph: the traces are outlined by the spots from the $K\alpha$ and $K\beta$ radiation and the Laue 'streaks'; the top right-hand corner of the film (x-ray beam directed toward the observer) is clipped for identification of the orientation. The distances Δ_L and Δ_R are measured at 45° from the centre O of the film. (c) Ewald sphere: the dashed line is the correct position of the equatorial circle, and the full line represents the longer of the two exposures, related to the correct position by a rotation about the line OP_E.

From this figure, the distance of the curve above (P_L) or below the true zero layer-line position at $\theta = 45°$ is $\tan^{-1}(\Delta_L/D)$, where D is the diameter of the film. This value is also that of the ζ reciprocal lattice coordinate of a possible reflection at P_L. Thus, the angle of elevation of the reciprocal lattice vector OP_L is $\delta_L = \tan^{-1}(\zeta/OP_L) = \tan^{-1}(\Delta_L/D\sqrt{2})$, taking $\kappa = \lambda$, as in Section 4.4.4. Similarly, $\delta_R = \tan^{-1}(\Delta_R/D\sqrt{2})$. For values of $\delta < 4°$, $\tan\delta = \delta$ to 0.1%. Hence, for $D = 57.3$ mm, $\delta = 0.707\Delta°$, with Δ expressed in mm.

In order to apply the corrections, we return the goniometer head to a reading at or near the center of the range for the longer exposure. With the photograph marked as shown in Figure 4.22b, consider the more intense curve. The correction δ_R is applied to arc A, lying in the NE–SW direction (N toward the x-ray source). The direction of movement of the arc is such that a reciprocal lattice vector at $\theta=45°$, imagined to be protruding from the crystal in the NE direction (the reciprocal lattice origin is transferred to the crystal at this point), will be brought to the equatorial plane. The correction δ_L is applied to arc B in a similar manner.

4.4.4 Ewald's Construction

The geometric interpretation of x-ray diffraction photographs is greatly facilitated by means of a device due to Ewald, and known as the Ewald sphere, or sphere of reflection. The sphere is centered on the crystal (C) and drawn with a radius of one reciprocal space unit (RU) on the x-ray beam (AQ) as diameter (Figure 4.23). The Bragg construction for reflection is superimposed, and a reflected beam hkl cuts the sphere in P. The points A, P, and Q lie on a circular section of the sphere, which passes through the center C.

From the construction

$$AQ = 2 \quad \text{(by construction)} \tag{4.8}$$

$$\widehat{APQ} = 90° \quad \text{(angle in a semicircle)} \tag{4.9}$$

Hence,

$$QP = AQ \sin\theta(hkl) = 2\sin\theta(hkl) \tag{4.10}$$

From Bragg's equation (3.55),

$$2\sin\theta(hkl) = \lambda/d(hkl) \tag{4.11}$$

and from the definition of the reciprocal lattice (see Section 2.4), we identify the point P with the reciprocal lattice point hkl; hence

$$QP = d^*(hkl) \tag{4.12}$$

with $\kappa = \lambda$ and, from (2.11),

$$d^*(hkl) = 2\sin\theta(hkl) \tag{4.13}$$

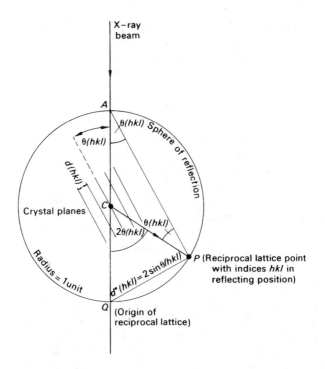

FIGURE 4.23.   Ewald construction illustrating how an x-ray reflection may be considered to arise when a reciprocal lattice point P passes through the sphere of reflection. AP is parallel to the (hkl) planes, and the reciprocal lattice vector QP forms a right angle at P. If the rotation is about a, normal to the circular section shown, this circle becomes the zero-layer circle and reflected beams such as CP will all be denoted $0kl$.

We now have a mechanism for predicting the occurrence of x-ray reflections and their directions in terms of the sphere of reflection and the reciprocal lattice. The origin of the reciprocal lattice is taken at Q and, although the crystal is at C, it may be helpful to imagine a conceptual crystal at Q identical to the real crystal and moving about a parallel oscillation axis in a synchronous manner.

The condition that the crystal is in the correct orientation for a Bragg reflection hkl to take place is that the corresponding reciprocal lattice point P is on the sphere of reflection. As the crystal oscillates, an x-ray reflection flashes out each time a reciprocal lattice point cuts the sphere of reflection, and the direction of reflection is given by CP.

Ewald's construction provides an elegant illustration of the formation of layer lines on an oscillation photograph. Figure 4.24 shows an Ewald sphere and portions of several layers of an orthogonal reciprocal lattice. As the sphere and the x-ray beam oscillate about the x axis, reciprocal lattice points cut the sphere

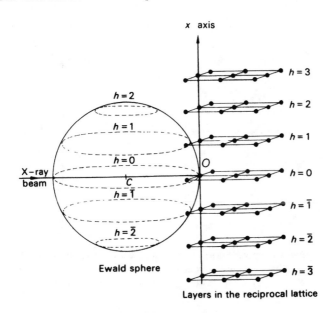

FIGURE 4.24. Formation of layer lines in terms of the Ewald construction. Layers with $|h| \geq 3$ lie outside the range of recording in this illustration. Note that except for $h = 0$, the circles labeled $\pm h$ ($h = 1, 2, 3, \ldots$) cannot be identified with the circular section of Figure 4.23.

of reflection in circles because the axis of the cylindrical film is arranged to be parallel to the oscillation axis x.

The relationship between the reciprocal lattice spacing of the layers and the corresponding repeat distance a in direct space is shown in Figure 4.25. From the similar triangles AOC and BCD,

$$AO/OC = BC/BD \tag{4.14}$$

or

$$v_1/R = \zeta_1/(1 - \zeta_1^2)^{1/2} \tag{4.15}$$

where v_1 is the distance between the zero layer and the first layer line and R is the radius of the film. If the lattice is orthogonal in the aspect illustrated (x^* coinciding with x), then from (4.11) and (4.13), since ζ_1 is equivalent to $d^*(100)$,

$$a = \lambda/\zeta_1 \tag{4.16}$$

Although (4.16) holds generally, if the lattice is not orthogonal, $\zeta_1 \neq d^*(100)$ and the appropriate expressions are a little more complicated (see Table 2.4), requiring a knowledge also of the interaxial angles.

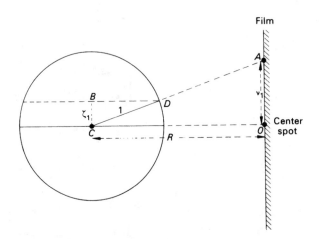

FIGURE 4.25.   Relationship between a reciprocal lattice spacing ζ_n and the corresponding layer-line spacing ν_n on the film; the repeat distance t along the rotation axis is given by $t = n\lambda/\zeta_n$. From the figure, $\zeta_n^2 = (\nu_n/R)^2/[1 + (\nu_n/R)^2]^{1/2}$, which permits an expression for a repeat distance alternative to (4.7).

Indexing Procedure for an a-Axis Oscillation Photograph.   On the zero level of an a-axis oscillation photograph, reflections are of the type $0kl$. The relevant portion of the reciprocal lattice is the y^*z^* net which, for an orthorhombic crystal, is determined by b^* ($= \lambda/b$), c^* ($= \lambda/c$), and α^* ($= 90°$). Without going into further detail, we note that the simplest method of determining b^* and c^* would be from the values of b and c, through (4.7) in the appropriate forms.

A drawing of the reciprocal net is prepared carefully, using a convenient scale, for example, 1 RU = 50 mm. The values of $d^*(0kl)$ are obtained from measurements of $H(0kl)$ on the zero-layer line (Figure 4.20), noting also whether the spot lies to the left or the right of the center O. Since the angular deviation of the x-ray beam is 2θ (Figure 4.23),

$$2\theta = H/R \qquad (4.17)$$

in radian measure. Using degree measure, we find

$$d^*(0kl) = 2 \sin\theta(hkl) = 2 \sin[180H(0kl)/2\pi R] \qquad (4.18)$$

H and R are, conveniently, measured in millimeters.

Worked Example of Indexing.   The a-axis oscillation photograph of an orthorhombic crystal (Figure 4.20) was taken with $+b^*$ pointing toward the x-ray source at the start of a $15°$ anticlockwise oscillation. A sample of reflections recorded on the zero level of this photograph at H mm from the center is listed

TABLE 4.3.   Measurements on the Zero-Layer a-Axis
Photograph

Reflection number	H (mm LHS of center)	Reflection number	H (mm RHS of center)
1	11.2	4	14.1
2	15.7	5	17.4
3	25.3	6	83.3

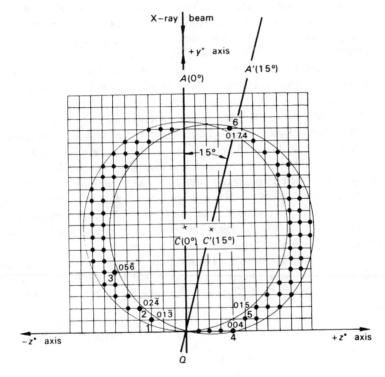

FIGURE 4.26.   Indexing the 15° oscillation photograph in Figure 4.20: orthorhombic crystal, a parallel to the rotation axis, $-y^*$ along the direction of the incident x-ray beam at the start of the oscillation, scale 1 RU $\approx$ 27.7 mm. *Possible* reflections are shown as points within the lunes. Reflections 1–6 indexed in Table 4.4 are indicated.

in Table 4.3. For the x-ray wavelength used, $b^* = 0.114$ and $c^* = 0.113$ RU. A y^*z^* reciprocal net was constructed and used to index these reflections.

The scheme for carrying out this indexing is illustrated in Figure 4.26, with the help of Table 4.4. Instead of thinking in terms of the crystal oscillating (first anticlockwise), we imagine that the sphere of reflection oscillates (first clockwise)

TABLE 4.4.   Indexed Reflections for the $0kl$ Layer Line

Reflection number	d^*	hkl	Reflection number	d^*	hkl
1	0.371	$01\bar{3}$	4	0.464	004
2	0.516	$02\bar{4}$	5	0.573	015
3	0.818	$05\bar{6}$	6	1.967	017,4

about an axis through Q, normal to the x-ray beam, from $QA(0°)$ to $QA'(15°)$, taking the incident x-ray beam with it. The reciprocal lattice points that would intersect the sphere of reflection are those within the lunes swept out during this motion. Points lying within these lunes are shown in Figure 4.26, and they correspond to possible $0kl$ reflections for this oscillation movement. By measuring the d^* values from Q to reciprocal lattice points within the lunes on the left- or right-hand side, as appropriate, the required indices may be determined for the reflections that *do* occur for this crystal.

The construction of the lunes in the correct orientation greatly reduces the number of reciprocal lattice points to be considered. The indexed points are shown in both Table 4.4 and Figure 4.26. Use the diagram to determine which other reflections might have been recorded on this photograph. It should be clear that no reflection for which d^* is greater than 2 RU can be observed; this number is the radius of another sphere, the *limiting sphere*, which is that sphere swept out in reciprocal space by a complete rotation of the Ewald sphere. Problem 4.4 is based on this example. It is sometimes necessary to consider a point just outside the lunes, depending on the probable experimental errors.

Some authors introduce the reciprocal space coordinate ξ, as well as ζ, where ξ is equivalent to d^*. In terms of a horizontal distance H of a spot on the film from the centre, Figure 4.20, and its vertical distance V from the zero layer, we have:

$$\xi^2 = 1 + (1 - \zeta^2) - 2(1 - \zeta^2)^{1/2} \cos(2\theta)$$

$$= 2 - \zeta^2 - 2(1 - \zeta^2)^{1/2} \cos(2\theta)$$

Since H is a circular arc on the film of radius R subtending the angle 2θ at its centre, $2\theta = H/R$. Thus,

$$\xi^2 = 2 - \zeta^2 - 2(1 - \zeta^2)^{1/2} \cos(H/R)$$

which may be expressed as

$$\xi^2 = 1 + [R^2/(V^2 + R^2)] - 2[R^2/(V^2 + R^2)]^{1/2} \cos(H/R)$$

since we have already shown the relationship for ζ. For the zero layer,

$$\xi^2 = 2 - 2\cos(H/R) = 4\sin^2[H/(2R)]$$

But on the zero layer, $H = 2R\theta$, so that

$$\xi = 2\sin\theta$$

which is equivalent to the Bragg equation where $\kappa=1$. Generally, however, it is simpler to work in terms of the Cartesian coordinates of a diffraction spot on the film.

Flat-Plate Oscillation Photography

In this specialized technique, for crystals having large unit cells, it is necessary to employ a large crystal-film distance (at least 60 mm) in order to effect resolution of the diffraction spots on the film. The Arndt–Wonnacott camera, specifically designed for this purpose, employs a flat-plate perpendicular to the x-ray beam and incorporates a mechanism for automatically changing the cassette (a carousel device) to enable several exposures to be set concurrently.

Figure 4.27 indicates the geometry necessary to define the (X, Y) coordinates of an upper-level spot on the film. In this diagram CO' is perpendicular to the film plane. Triangle $CO'P'$ is right-angled at O', and $PP'O'$ at P'; triangles $DD'C$

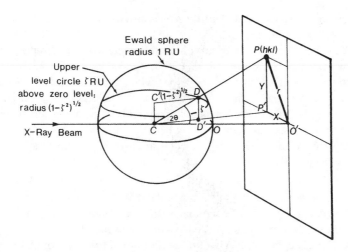

FIGURE 4.27. Coordinate pair X, Y of diffraction spot P on a flat-plate oscillation photograph for a general upper level reflection: C, crystal; O, reciprocal lattice origin; O', origin of coordinates on the film; $OD = d^*(hkl); CP$, reflected beam, intersecting the Ewald sphere at $D; DD'$, perpendicular distance above the zero level. Since the radius of the Ewald sphere is 1 RU, the radius of the upper level circle at level ζ is $(1 - \zeta^2)^{\frac{1}{2}}; CO'$ is the crystal-to-film distance.

and $PP'C$ are similar:

$$Y/CP' = \zeta/CD' = \zeta/(1 - \zeta^2)^{1/2}$$
$$Y = CP'\zeta/(1 - \zeta^2)^{1/2}$$
$$CP'^2 = R^2 + X^2$$
$$Y^2 = (R^2 + X^2)\zeta/(1 - \zeta^2) \qquad (4.19)$$

where R is the distance CO'. Further,

$$r/R = \tan 2\theta \quad \text{and} \quad r^2 = X^2 + Y^2$$

where r is the distance PO'. Therefore,

$$R^2 \tan^2 2\theta = X^2 + Y^2 \qquad (4.20)$$

Combining these results, it follows that

$$X = R[\tan^2 2\theta(1 - \zeta^2) - \zeta^2]^{1/2} \qquad (4.21)$$
$$Y = R\zeta[1 + \tan^2 2\theta]^{1/2} \qquad (4.22)$$

If a reciprocal axis is parallel to the rotation axis, ζ RU corresponds to the appropriate level along the axis.

We will illustrate these ideas with the following worked example. A protein crystal with a monoclinic unit cell, $a = 30.0$, $b = 50.0$, $c = 40.0$ Å, $\beta = 100°$, is mounted on an oscillation camera equipped with a flat-plate cassette placed with its plane perpendicular to the x-ray beam and at a distance $R = 60$ mm from the center of oscillation. The b axis of the crystal is vertical and perpendicular to the x-ray beam, with $-a$ parallel to the beam at the start of a counterclockwise rotation; x-rays with wavelength $\lambda = 1.2$ Å are used in the experiment. Consider the following problems:

1. At what point in the rotation will the 004 reflection take place?
2. Calculate the (X, Y) coordinates (in mm) of the 004 spot on the film, X being the horizontal coordinate and Y the vertical.
3. Calculate the (X, Y) coordinates (in mm) of the 014 spot on the film.

1. Refer to Figure 4.28 to see that the required rotation is $\theta(004)$, calculated as follows:

$$d^*(004) = 4c^*$$

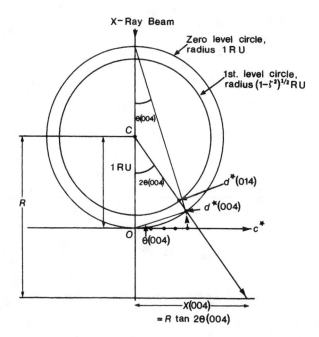

FIGURE 4.28.   Plan view of flat-plate oscillation geometry, looking down the oscillation axis at C. In the accompanying example, the $y(b)$ axis of the crystal (C) is perpendicular to the diagram, and coincides with the rotation axis. Because the crystal is monoclinic, the a^*c^* levels superimpose in this orientation. The diagram shows c^* at the beginning of an anticlockwise rotation perpendicular to the x-ray beam. The crystal rotates through $\theta(004)$ and, at the reciprocal lattice point 004, $d^*(004)$ intersects the Ewald sphere and the 004 reflection is produced. For the 014 reflection, a further small rotation is necessary.

where $c^* = \lambda/c \sin\beta$ (for a monoclinic crystal, see Section 2.4)

$\qquad = 1.2/40 \sin 100 = 0.0305$

$d^*(004) = 0.1220 = 2\sin\theta(004)$

$\sin\theta(004) = 0.0610$

$\qquad \theta(004) = 3.497°$ (rotation angle from starting position)

    2. Also from Figure 4.28 we see that

$$X(004) = R\tan 2\theta(004)$$

$$= 60\tan 6.994$$

$$= 7.361 \text{ mm}$$

$$Y(004) = 0 \text{ mm}$$

3. For the 014 reflection:

$$\zeta = b^* = \lambda/b = 0.024$$

Using (2.16), it follows that

$$d^{*2}(014) = b^{*2} + 16c^{*2}$$
$$= 0.024^2 + 16(0.0305)^2$$
$$= 0.01546$$

so

$$d^*(014) = 0.1243 = 2 \sin\theta(014)$$

Hence,

$$\sin\theta(014) = 0.06215$$

and

$$\theta(014) = 3.563° \quad \text{and} \quad 2\theta(014) = 7.126°$$

Using equation (4.21)

$$X(014) = 60[\tan^2 7.128(1 - 0.024^2) - 0.024^2]^{1/2} = 7.357 \text{ mm}$$

and from equation (4.22)

$$Y(014) = 60 \times 0.024[1 + \tan^2 7.126]^{1/2} = 1.451 \text{ mm}$$

Hence, the 004 reflection has coordinates (7.361, 0.0) mm; the 014, (7.359, 1.451) mm.

4.4.5   Weissenberg Method

The oscillation method suffers from the main disadvantage that it presents three-dimensional information on a two-dimensional film. It is both a distorted and collapsed diagram of the reciprocal lattice. This situation leads, in turn, to overlapping spots, particularly if the reciprocal lattice dimensions are small, with consequent ambiguity in indexing. More advanced x-ray photographic methods permit rapid, unequivocal indexing without graphical construction.

Figure 4.29 is an extension of Figure 4.19. Each cone corresponds to possible reflections of constant h index. In the Weissenberg method, all cones but one are excluded by means of adjustable metal screens, shown in the figure in the position which permits only the $0kl$ reflections to pass through and reach the film. If the

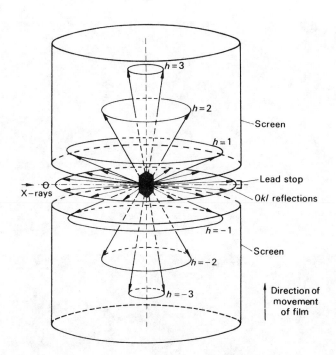

FIGURE 4.29.   Cones representing possible directions of diffracted x-rays from a crystal rotating on its *a* axis. In this diagram, the screens are set to exclude all but the aero layer, $0kl$.

film were kept stationary, the exposed record would still look like the zero-layer line in Figure 4.20. However, in the Weissenberg technique, the film is translated parallel to the oscillation axis, synchronously with the oscillatory motion of the crystal. The result of this procedure is a spreading of the spots over the surface of the film on characteristic straight lines and curves (Figure 4.30). Each spot has a particular position on the film, governed, in this figure, by the values of h and l. With practice, the film can be indexed by inspection.

If the layer-line screens are moved by the appropriate distances parallel to the oscillation axis, higher-order layers can be recorded on the film. Generally, nonzero layers are recorded by the equi-inclination technique,[a] which ensures that their interpretation is very similar to that for the corresponding zero layer.

The Weissenberg photograph is clearly much simpler to interpret than the oscillation photograph, but the apparatus required is more complex. Although the photograph of the reciprocal lattice is not collapsed, it is clear from the figure that it is still distorted: the orthorhombic reciprocal net does not appear as rectangular.

[a] See Bibliography (Jeffery).

(a)

(b)

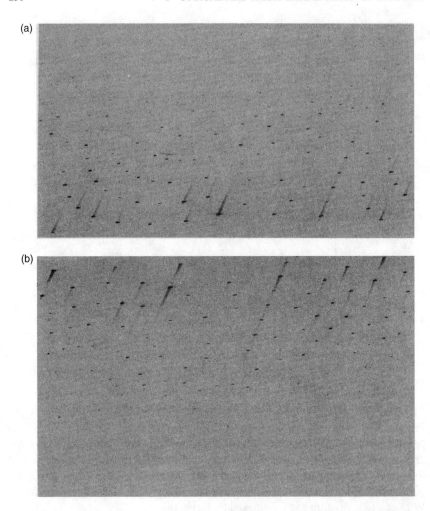

FIGURE 4.30.   Typical Weissenberg x-ray photograph: zero-layer, $h0l$ reflections for a monoclinic crystal, taken with Cu $K\alpha$ radiation. For the more intense reflections, we can see (a) less intense reflections, for the same plane but at a smaller θ value, which arise from the Cu $K\beta$ radiation that is not completely filtered out (Section 3.1.4); (b) Laue streaks that arise from the "white" radiation in the beam (Sections 4.4.1 and 3.1.1) as the crystal selects each wavelength over a small, continuous range of θ.

Unit-Cell Dimensions from a Weissenberg Photograph

We shall use Figure 4.31, which is a scaled-down reproduction of a Weissenberg photograph, as an example. The principal reciprocal lattice axes, y^* and z^*, are straight lines of spots on the photograph, and have a slope of

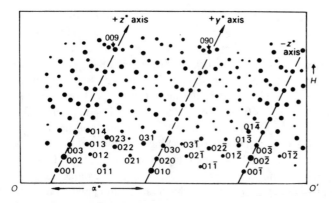

FIGURE 4.31.   Sketch of one half of a partly indexed zero-layer Weissenberg photograph of an orthorhombic crystal mounted about a. The horizontal travel of the film is 1 mm per $2°$ rotation of the crystal, and α^* is $90°$, which is correct for an orthorhombic crystal. The reflections shown are confined to a single line (zero layer) on the corresponding oscillation photograph. However, the $0kl$ reciprocal net is still presented in a distorted manner on the Weissenberg photograph, although it is no longer collapsed. The linear scale of the diagram is 1/1.8, λ is 1.542 Å, and $2R$ is 57.30 mm. A sketch of a precession photograph for the same crystal is shown in Figure 4.33.

2 $(\tan 63.435°)$ and a separation of 45.0 mm; the crystal rotates through $2°$ for every 1 mm of travel of the cassette, so that α^* $(\widehat{y^*z^*})$ is $90°$.

Prior to making measurements on the film, the spacing along the rotation axis of the crystal (a in this example) would have been measured from an oscillation photograph taken with the same camera (see Section 4.4.2). We must now obtain the values of b^*, c^*, and α^* from the Weissenberg photograph. Measurement of unit-cell dimensions from the zero-level Weissenberg photograph makes use of the fact that, for a given reflection, 2θ is calculated according to (4.17), where H is the perpendicular distance from a spot to the central line $O\,O'$, and R is the film radius, as before. A standard Weissenberg camera has a diameter $(2R)$ of $180/\pi$, or 57.30 mm.

In practice, the line $O\,O'$ may be difficult to locate, and the value of a given H may be best obtained by measuring $2H'$ along the actual reciprocal lattice row, across the film between corresponding spots on the top and bottom halves. (A zero-layer Weissenberg photograph has inherent twofold symmetry, and each row and curve shown repeats on the other side of $O\,O'$.) Then we can calculate H from

$$H = \tfrac{1}{2}(2H') \sin(\tan^{-1} 2)$$ (4.23)

From (4.18) we have, as before,

$$d^* = 2\sin[180H/2\pi R] = 2\sin H$$ (4.24)

where H is measured in millimeters. The angle α^* is determined by measuring the separation between y^* and z^* along the OO' direction and converting to degrees by multiplying by the camera constant of 2 deg mm^{-1}.

In Figure 4.31, the value of $H(090)$, for example, is 61.2 mm, using (4.23) with H' measured as 38.0 mm and multiplied by the scale factor of 1.8. Hence, $d^*(090)$ is 1.753 and b^* is 0.195. Similarly, $H(009)$ is $(38.5 \times 1.8) \sin \tan^{-1}(2)$, or 62.0 mm, whence c^* is 0.196. The distance labeled α^* in Figure 4.31 is 25.0 mm, whence α^* is 25.0×1.8 (scale factor) $\times 2$ (camera constant), or 90°, which is correct for the orthorhombic system. Since the photograph was taken with Cu $K\alpha$ radiation ($\lambda = 1.542$ Å), it follows that $b = 7.91$ Å and $c = 7.87$ Å. In other crystal systems, similar relationships apply. A monoclinic crystal is very conveniently studied in this way. If the crystal is mounted along b, an oscillation photograph (in the Weissenberg camera, without the screens) enables b to be determined. Then, from the corresponding zero-level Weissenberg photograph, we can obtain a, c, and β.

Further examples of Weissenberg photographs can be found in Chapter 8.

4.4.6   Precession Method

The precession method produces an undistorted picture of the reciprocal lattice. This is achieved, in principle, by ensuring that a crystal axis t precesses about the x-ray beam and that the film follows the precession motion in such a way that the film is always perpendicular to the crystal axis (Figure 4.32). This is much easier to say than to carry out, and apparatus of appreciable mechanical complexity is required.[a] However, the precession photograph is symmetry-true and readily indexed, as shown by Figure 4.33. It is a sort of "contact print" of a layer of the reciprocal lattice.

Unit-Cell Dimensions from a Precession Photograph

The setting of a crystal for precession photography is discussed in the next section.

A reduced version of the precession photograph of the same crystal as that used for the Weissenberg photograph is shown in Figure 4.33. It was taken in a precession camera with a crystal-to-film distance of 60.0 mm,[b] with the crystal mounted such that a is along the x-ray beam when the camera is in the stationary position, that is, when the precession angle $\bar{\mu}$, defined as the angle between t and the x-ray beam in Figure 4.32, is zero.

---

[a] See Bibliography (Buerger).

[b] Distances in mm measured on the actual film are divided by 60.0 to convert to dimensionless reciprocal lattice units.

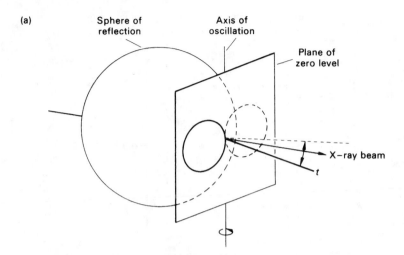

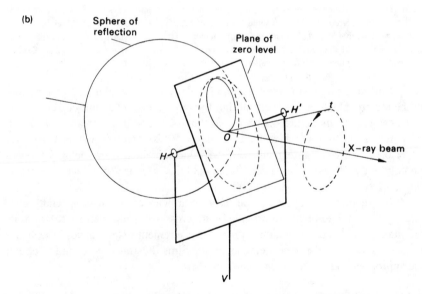

FIGURE 4.32. Oscillation and precession geometry compared: (a) reciprocal lattice zero-level plane whose normal t is oscillating to equal limits on each side of the x-ray beam. Maximum symmetry information about the direction t is $2mm$, the projected symmetry of an oscillation movement; (b) reciprocal lattice zero-level plane whose normal t is precessing about the x-ray beam. HOH' and OV are horizontal and vertical axes. The symmetry about t is now the true symmetry for that direction. [Reproduced from *The Precession Method* by M. J. Buerger, with the permission of John Wiley and Sons Inc., New York.]

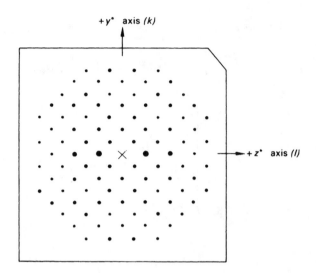

FIGURE 4.33. Sketch of a precession photograph of an orthorhombic crystal precessing about a; an undistorted $0kl$ reciprocal net is represented, permitting b^*, c^*, and α^* to be obtained directly from the film. Remember that b^* and c^* are magnified by a factor equal to the crystal-to-film distance, 60.0 mm in this example. The diagram has been scaled by a factor of 0.267, and λ(Cu $K\alpha$) is 1.542 Å. A sketch of a Weissenberg photograph for the same crystal is shown in Figure 4.31.

Since the precession photograph provides a completely undistorted picture of the reciprocal lattice, it is very straightforward to obtain the unit-cell constants. For example, from Figure 4.33 we find that $14b^*$ and $14c^*$ measure 43.7 and 44.1 mm, respectively, which leads to $b^*(43.7/0.267)/(60 \times 14) = 0.195$ and $c^* = 0.197$, in good agreement with the values obtained from the Weissenberg photograph. It is easy to see from Figure 4.33 that $\widehat{y^*z^*}$ is 90°, so we may report $b = 7.91$ Å, $c = 7.83$ Å, and $\alpha = 90°$.

In practice, measurements on the film are carried out with specially constructed film holders that are provided with cross-wires and vernier scales, thus facilitating the process of obtaining precise measurements. Nevertheless, the reader is encouraged to make measurements on the films illustrated here, and to obtain the appropriate unit-cell dimensions from them.

4.4.7   Setting a Crystal for Precession Photography

As with other methods of single-crystal x-ray photography, production of a good precession photograph requires the crystal to be aligned, accurately with respect to the mechanical parts of the camera and with the x-ray beam direction. There are two principal aims of this setting procedure:

(a) to set the crystal so that an axis of the direct lattice (usually a, b, c, or a reasonably simple $[UVW]$ direction) is parallel to the x-ray beam when the precession angle $\bar{\mu}$ is set at zero;

(b) to align a reciprocal lattice axis parallel to the horizontal, or dial, axis of the camera.

Of these two conditions, (a) is absolutely essential to the setting, and (b), although not essential, is highly desirable, as it allows for subsequent rotation of the crystal in order to pick up a second axis to satisfy condition (a) while maintaining the constant horizontal reciprocal lattice row common to both films. This condition will usually facilitate a complete survey of the reciprocal lattice of the crystal in the minimum of settings.

Setting a Crystal Axis Parallel to the X-ray Beam

Optical examination of the crystal will usually suggest a possible crystallographic axial direction (Section 4.3ff). A useful tip to remember is that a principal axis will often be found perpendicular to a broad face of the crystal. Stable crystals are usually mounted, using adhesive, on to a glass fiber, while unstable crystals, such as proteins, are mounted in the presence of mother liquor inside sealed capillary tubes. For precession photography a crystallographic axis should be perpendicular or closely perpendicular to the fiber or glass capillary. The final alignment of the axis can be undertaken with the crystal mounted on the camera using the goniometer arcs (Figure 8.2). The camera is fitted with a telescope by means of which the crystal can be centered, that is, adjusted on the axes such that it rotates within its own volume, and the desired axis aligned approximately along the x-ray beam. On setting the precession angle to a given $\bar{\mu}$ value, the axis will describe a cone defined by this angle (see Section 4.4.6ff). A zero-level precession photograph taken with unfiltered x-radiation (usually Cu or Mo) will be characterized by Laue streaks radiating from the point d (Figure 4.34), where the direct x-ray beam would strike the film when $\bar{\mu} = 0$. The ends of the Laue streaks define a circle, center c. When the crystal axis is perfectly aligned, points d and c coincide. Otherwise the situation is as shown in Figure 4.34. Two adjustments are required, defined in this diagram by a horizontal component δ_A and a vertical component δ_D. The values of δ are measured as shown from d to the circumference of the circle, taking the *larger* of the two distances in each case.

The Vertical Correction ε_D^0, Defined by δ_D on the Film

This correction is applied to the *dial axis* (D) of the precession camera. The conventional direction for the correction to be made (clockwise or anticlockwise) is explained in Figure 4.34. The reading of δ_D mm is converted into ε_D^0 by use of the chart[a] shown. Values of ε_D are given for selected $\bar{\mu}$ values, other $\bar{\mu}$ values being accessible by interpolation.

---

[a] D. J. Fischer, *Amer. Mineral.* **37** (1952).

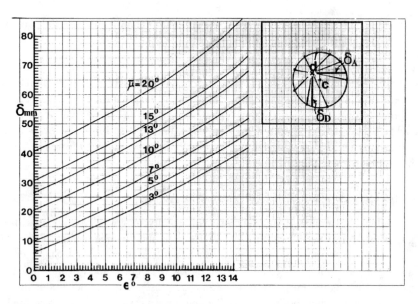

FIGURE 4.34.   Chart and setting instructions for precession photography. The graph shows the variation of δ/mm with ε/deg for various values of $\bar{\mu}$ (indicated for each curve), and with a crystal-to-film distance M of 60.00 mm. The horizontal arc A and dial D corrections are both clockwise for the situation given: d, the direct beam spot may be found by extending the Laue streaks toward the centre of the photograph; c, the centre of the precession circle, radius $2M \tan(\bar{\mu})$. Assume that the face of the arc is horizontal, with the inscribed angular scale pointing upward toward the observer. (i) *Dial correction (d above c)*: for the situation in the diagram, the dial correction ε_D/deg derived from δ_D/mm using the chart is *clockwise*. (ii) *Arc correction (d left of c)*: Similarly, ε_A/deg, the horizontal arc correction derived from δ_A/mm, is clockwise for the given situation. (iii) Corrections corresponding to other relative positions of d and c are made by analogy with (i) and (ii). (iv) The diagram and explanation assume that the observer is looking at the x-ray film, with the beam coming towards him, and with the dial on the right-hand side, as with most cameras. (v) The desired setting can usually be achieved by means of a fairly small $\bar{\mu}$ setting. A value of $10°$ for $\bar{\mu}$ is recommended normally, with $2r = 10.6$ mm (hole) for $s = 30$ mm.

The Horizontal Correction ε_A^0, Defined by δ_A on the Film

This correction applies strictly to a goniometer arc (A) which is *horizontal* (or parallel to the x-ray beam direction) when $\bar{\mu} = 0$. The conventional direction for the correction to be made is explained in Figure 4.34. The reading δ_A is converted to ε_A^0 by use of the same chart. If the goniometer arcs are significantly off-parallel and off-perpendicular to the x-ray beam, ε_A should be resolved into appropriate components depending on the cosine and sine of the offset angular value.

Application of the ε_D and ε_A corrections as explained should result in coincidence of points d and c. Further correction may be required and implemented by taking a second photograph.

Setting a Reciprocal Lattice Row Horizontally

A reciprocal lattice row which is offset by a few degrees (usually no more than $\pm5°$) from the horizontal may be leveled quite satisfactorily. It is achieved simply by adjusting the *vertical* goniometer arc (perpendicular to the beam) by the required angle. Again, if the arcs are not horizontal and vertical, the procedure becomes more difficult. Corrections larger than about 5° may be worth attempting so as to ensure a good sequence of precession photographs, but can usually be carried out only by physically pushing over the crystal on its mount (glass fiber or capillary) or by completely remounting the specimen.

Screen Setting

A zero level precession photograph, with any given zone axis precessing, may be isolated by placing a metal screen with an annular opening of radius r mm at a distance s mm from the crystal. For a given precession angle $\bar{\mu}$, $s=r\cot\bar{\mu}$. Typical values of r, s, and $\bar{\mu}$ are $r=5.3$ mm (in this case the annular opening is simply a hole), $s=30$ mm, $\bar{\mu}=10°$; for a final photograph, $r=20$ mm, $s=40$ mm, $\bar{\mu}=26.6°$.

4.5 Recognition of Crystal System

We mentioned previously several pointers to the system of a crystal under investigation. They include the results of an optical examination under polarized light (Section 4.3ff) and x-ray photographs that may have been obtained by one or more of the several commonly available techniques (Section 4.4ff).

Table 4.1 summarizes the main crystallographic features that can be expected from the preliminary optical examination. It indicates that a cubic crystal should be recognized from the optics alone (isotropic). If the crystal is optically uniaxial the optical examination should reveal the direction of the z axis, and the rotational symmetry (3-, 4-, or 6-fold) could be inferred from a symmetric oscillation photograph (Table 4.2) for a crystal mounted perpendicular to this direction, from a sequence of oscillation photographs taken with the crystal mounted along z (see Section 4.4.2), from zero-level and the first-level Weissenberg photographs from a crystal mounted parallel to the z axis, or from zero- and first-level precession photographs from a crystal mounted with z along the x-ray beam when the equi-inclination angle μ is reset to zero (see Sections 4.4.5 and 4.4.6). Any zero-level photograph has at least 2-fold symmetry, which would modify the actual symmetry about the 3-fold axis of a trigonal crystal, giving it the appearance of a 6-fold axis; it is therefore necessary to obtain at least one upper-level photograph so as to reveal the true symmetry. For a monoclinic crystal (optically biaxial, Table 4.1), the optical examination should reveal the direction of the unique y axis. Since this direction could also be one of the three principal axes of an orthorhombic crystal,

TABLE 4.5.   Scheme for Recognition of Crystal System and Axes

Classification from optical examination under crossed polars	Possible crystal system	Axis usually recognizable from optical and/or morphological examination	Type of x-ray photograph and crystal mounting recommended, minimal features to look for in photograph[a]	
			Weissenberg	Precession
Isotropic	Cubic only	[100] and/or [010] and/or [001]	0 level: 4-fold[b] (for 432, $\bar{4}3m$, or $m3m$); 2-fold[c] (for 23 or $m3$) [100]; [010]; or [001], mounting	0 level: 4-fold[b] (for 432, $\bar{4}3m$, or $m3m$); 2-fold[c] (for 23 or $m3$) [100], [010], or [001] precessing, i.e., mounting perpendicular to one of the axes.
Anisotropic/uniaxial	Hexagonal	[0001]	0 level ⎫ Upper level ⎬ 6-fold [01$\bar{1}$0] mounting or equivalent	0 level ⎫ Upper level ⎬ 6-fold [10$\bar{1}$0] mounting or equivalent, [0001] precessing
	Trigonal (indexed on hexagonal axes)	[0001]	0 level: 6-fold Upper level: 3-fold [0001] mounting	0 level: 6-fold Upper level: 3-fold [10$\bar{1}$0] mounting or equivalent [0001] precessing
	Tetragonal	[001]	0 level ⎫ Upper level ⎬ 4-fold [001] mounting	0 level ⎫ Upper level ⎬ 4-fold [100] mounting [001] precessing

Anisotropic/biaxial	Orthorhombic	[100] and/or [010] and/or [001]	0 level Upper level } mm2 [100], [010], or [001] mounting	0 level Upper level } mm2 [100], [010], or [001] precessing, i.e., mounting perpendicular to one of the axes recognized.
	Monoclinic	[010]	0 level Upper level } 2-fold [010] mounting	0 level mm2 Upper level: m (⊥ [010]) [010] mounting [100] or [001] precessing
		[100] or [001]	0 level: mm2 Upper level: m (⊥ [010]) [100] or [001] mounting	0 level Upper level } 2-fold [100] or [001] mounting [010] precessing
	Triclinic	Axes not necessarily revealed by optics or by morphology	0 level: 2-fold Upper level: 1-fold Any axis mounting	0 level: 2-fold Upper level: 1-fold Any axis precessing

^{a} See also Table 1.6.
^{b} Full symmetry is actually 4mm.
^{c} Full symmetry is actually 2mm.

Notes:

1. Optical examination alone should identify a transparent crystal as cubic. For a cubic crystal, the conventional (Bravais) unit cell is either P, I, or F. A nonstandard axial assignment would give rise to a nonstandard unit cell, which should be transformed accordingly.

2. For a triclinic crystal, a unit cell may be defined (see text) with respect to a reciprocal axis parallel to the dial axis of a precession camera. Precession photographs about two real axes perpendicular to this reciprocal axis are required. A nonprimitive triclinic unit cell may be transformed to the standard P form.

3. Preliminary x-ray studies of a single crystal require at least two zero-level and one upper-level Weissenberg or precession photographs for establishing the unit cell and space group.

x-ray photographs would be required to resolve this ambiguity. The symmetry on any oscillation photograph taken with this direction as the oscillation axis would be m perpendicular to that axis. A zero-level Weissenberg photograph would show symmetry 2 for a monoclinic crystal and symmetry $2mm$ for an orthorhombic crystal. Similar results would be obtained from precession photographs of the crystal mounted perpendicular to y with this axis along the beam when $\bar{\mu} = 0$. For the crystal mounted about y, a zero- and first-level photograph would be required to establish the system as monoclinic (one m plane, parallel to y, disappears for the upper level). If the crystal is orthorhombic and mounted about one axis (say y) each of two zero-level precession photographs, separated by dial axis rotation of 90°, would show $2mm$ symmetry, thus identifying all three crystallographic axes.

Consider finally the two extreme cases, the cubic and triclinic systems. Having established the crystal as cubic (optically isotropic), the direction of x, y, or z may be found from x-ray photographs, which would show the symmetry summarized in Table 4.5. For a triclinic crystal, the choice of axes is arbitrary, but unit-cell definition is best achieved with precession photographs. For this purpose a reciprocal unit-cell axis must be lined up along the dial direction. A series of zone axes along the beam may then be identified for different settings of the dial axis while maintaining the selected reciprocal axis along the dial axis. One such photograph will give two reciprocal axial spacings and the angle between them. Selection of another zone (ideally ≈90° away) would then give the third reciprocal cell spacing and second angle; the third angle would be derived from the difference in dial settings. A mixture of real and reciprocal unit-cell parameters is obtained, requiring use of Table 2.4 to separate them into six real and six reciprocal cell parameters. With a diffractometer, three non-coplanar axes would be established, and the conventional unit cell determined finally by reduction (q.v.).

II Measurement of Intensity Data from Single Crystals

A variety of options are open to the crystallographer for obtaining measurements of x-ray intensities. These include the photographic method, the use of a single counter or serial diffractometer, and the latest area detectors.

Photographic Method

X-ray intensities are measured on photographs from the blackening of the photographic emulsion. This method is now rarely used but may serve in some teaching laboratories as an introductory technique and so is retained in this edition.

Single Counter or Serial Diffractometers

Traditional diffractometers that employ scintillation counters[a] to detect and measure x-ray reflections from single crystals, one by one, incorporate a mechanical goniometer to orientate the crystal into the correct reflecting position for each hkl reflection and to rotate the counter to receive the scattered radiation from this single reflection. The energy is transformed electronically into a form suitable for conversion to $I(hkl)$. Because each reflection is measured individually, with a count time typically of around 60 s, the process is very slow, particularly for proteins, which routinely involve the measurement of tens of thousands of reflections. The dynamic range of the instrument can be enhanced to enable measurement of very strong reflections, through the use of a calibrated attenuator, such as a strip of Ni foil in the case of Cu radiation (see Section 3.1.4). For unstable crystals, such as proteins and macromolecules in general, several crystals may be required in the production of a complete data set. However, the accuracy attainable is possibly better than for most other methods.

Area Detectors

The disadvantages of single counter diffractometry that apply mainly to large molecule structures are a slow data-collection rate and the requirement of several crystals for collection of a complete data set, with the attendant errors associated with scaling and crystal deterioration. However, they have been largely overcome by the use of "electronic film" *area detectors* and *image plates* that have enjoyed rapid development in recent years. In this method, during each exposure, a series of x-ray reflections is produced by oscillating the crystal through a small angle and received by the detector, which is effectively a flat surface with a uniform sensitivity to x-rays. In many laboratories, this method of data collection is used for both large and small molecule crystallography, and is especially suitable for carrying out experiments at low temperature.

In the following sections we review the types of equipment available for data collection. Some practical details are given here but this important aspect is dealt with in more depth in Chapter 10.

4.6 Intensity Measurements on Photographs

X-ray intensities are measured on photographs from the blackening of the photographic film emulsion.

---

[a] Similar to a Geiger counter.

The optical density D of a uniformly blackened area of an x-ray diffraction spot on a photographic film is given by

$$D = \log(I_0/I) \tag{4.25}$$

where I_0 is the intensity of light hitting the spot and I is the intensity of light transmitted by it: D is proportional to the intensity of the x-ray beam I_0 for values of D less than about 1. In practice, this means spots which range from just visible to those of a medium-dark gray on the film.

An intensity scale can be prepared by allowing a reflected beam from a crystal to strike a film for different numbers of times and according each spot a value in proportion to this number; Figure 4.35 shows one such scale. Intensities may be measured by visual comparison with the scale, and, with care, the average deviation of intensity from the true value would be about 15%.

In place of the scale and the human eye, a photometric device may be used to estimate the blackening. In this method, the background intensity is measured and subtracted from the peak intensity. This process is carried out automatically in the visual method. Carefully photometered intensities would have an average deviation of less than 10%.

The accuracy of film measurements can be enhanced if an integrating mechanism is used in conjunction with either a Weissenberg or a precession camera in recording intensities. In this method, a diffraction spot (Figure 4.36a) is allowed to strike the film successively over a grid of points (Figure 4.36b). Each point acts as a center for building up the spot. The results of this process are a central plateau of uniform intensity in each spot and a series of spots of similar, regular shape: Figure 4.37 illustrates, diagrammatically, the building up of the plateau, and Figure 4.38 shows a Weissenberg photograph comparing the normal and integrating methods with the same crystal.

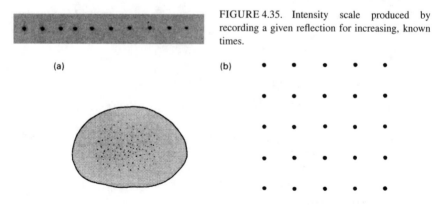

FIGURE 4.35. Intensity scale produced by recording a given reflection for increasing, known times.

(a)

(b)

FIGURE 4.36. Spot integration 1. (a) Typical diffraction spot, as received on an x-ray film. (b) A 5×5 grid of points.

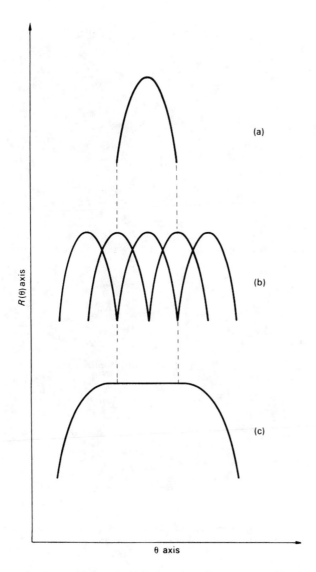

FIGURE 4.37.   Spot integration 2. (a) Ideal single peak profile. (b) Superposition, by translation, of five profiles. (c) Integrated profile, showing a central plateau.

The average deviation in intensity measurements from carefully photometered, integrated Weissenberg photographs is about 5%. The general subject of accuracy in photographic measurements has been discussed exhaustively by Jeffery.[a]

[a] J. W. Jeffery, *loc. cit.*

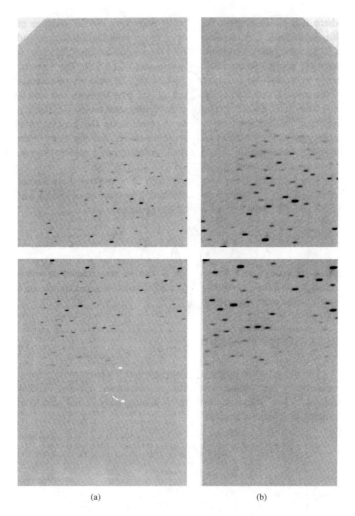

(a)   (b)

FIGURE 4.38. Weissenberg photographs. (a) Normal. (b) Integrated: the spots present the central plateau shown in Figure 4.37c from integration over the 5 × 5 grid.

4.7 Single-Crystal X-ray Diffractometry

It has become commonplace for x-ray diffraction data to be collected by means of a diffractometer. We give here a brief description of the Enraf–Nonius CAD4 instrument.

4.7.1 Instrument Geometry

The CAD4 diffractometer is characterized by its κ-goniometer, which differs in geometry from other four-circle diffractometers. The κ-goniometer carries the goniometer head (arcs) and keeps the crystal at the center of the diffractometer throughout the data collection. The κ-goniometer is a combination of three parts, which carry the rotation axes. All axes intersect in the center of the diffractometer.

The arcs are mounted on the ϕ axis, on the κ block, as shown in Figure 4.39; the angle of rotation about this axis is ϕ_κ. The κ block is rotated about the κ axis, being itself carried by the ω block. In turn, the ω block is rotated about the axis (ω_κ) and is carried on the diffractometer base. The angle χ between the ω and X

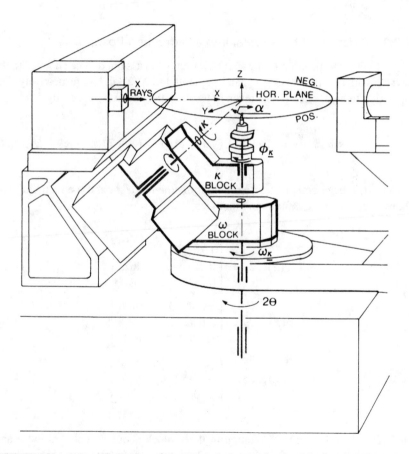

FIGURE 4.39. Schematic diagram of the CAD4 single crystal x-ray diffractometer, showing the X, Y, Z instrumental coordinate system and the rotation directions. (Reproduced with the permission of Enraf–Nonius, Delft.)

axes is ca 50°, and that between the κ and ϕ axes is also ca 50°. Thus, the-goniometer can access all directions of χ within about 100° of the zero setting of the instrument. This suspension gives the CAD4 an enhanced flexibility over the traditional Eulerian cradle. There is also a 2θ axis, coinciding with the ω axis, which carries the scintillation counter detector. The coincidence of the ϕ and ω axes corresponds to $\kappa=0$; $\omega_\kappa=0$ for κ in the plane of the diffractometer axes X and Z, with the κ block opposite $+X$, and $2\theta=0$ when the centre of the detector lies in the plane of X and Z and opposite $+X$. The definition of ϕ_κ is arbitrary, and a suitable working procedure is set up. Starting from $\kappa=\omega_\kappa=0$, positive rotations of θ, ω, and ϕ move a vector from Y toward X, and a positive rotation about κ moves a vector from Y to a position below the horizontal plane.

4.7.2 Rotation of the Crystal into a Diffracting Position

In the zero position of the CAD4, a vector $\mathbf{c}$ is assumed to be attached to the crystal, with components c_1, c_2, and c_3 parallel to X, Y, and Z, respectively. The operation to be applied to $\mathbf{c}$ is given in terms of the angles $\omega_\kappa, \kappa, \phi_\kappa$ by

$$c(\omega, \kappa, \phi) = Z(\omega_\kappa) \cdot Y(-\alpha) \cdot Z(\kappa) \cdot Y(\alpha) \cdot Z(\phi_\kappa) \cdot |\mathbf{c}| \qquad (4.26)$$

We can define general clockwise rotations of amounts α, β, and γ about X, Y, and Z, respectively, by the following matrices:

$$X(\alpha)_{Z \to Y} = \begin{bmatrix} 1 & 0 & 0 \\ 0 & \cos\alpha & \sin\alpha \\ 0 & -\sin\alpha & \cos\alpha \end{bmatrix} \qquad (4.27)$$

$$Y(\beta)_{X \to Z} = \begin{bmatrix} \cos\beta & 0 & -\sin\beta \\ 0 & 1 & 0 \\ \sin\beta & 0 & \cos\beta \end{bmatrix} \qquad (4.28)$$

$$Z(\gamma)_{Y \to X} = \begin{bmatrix} \cos\gamma & \sin\gamma & 0 \\ -\sin\gamma & \cos\gamma & 0 \\ 0 & 0 & 1 \end{bmatrix} \qquad (4.29)$$

In (4.26), α is the angle of rotation of the κ block about Y such that the κ and Z axes coincide. The term $Y(-\alpha)$ in (4.26) returns the goniometer to its original position.

4.7.3 Transformation from Miller Indices to Diffractometer Angles

To set a reflecting plane in the diffracting position, its Miller indices are transformed to a scattering vector $\mathbf{c}$ by an orientation matrix $\mathbf{R}$:

$$\mathbf{R} = \begin{bmatrix} a_X^* & b_X^* & c_X^* \\ a_Y^* & b_Y^* & c_Y^* \\ a_Z^* & b_Z^* & c_Z^* \end{bmatrix} \tag{4.30}$$

where the reciprocal unit-cell vectors $\mathbf{a}^*$, $\mathbf{b}^*$, and $\mathbf{c}^*$ are resolved into components along X, Y, and Z respectively. Then

$$\begin{bmatrix} c_1 \\ c_2 \\ c_3 \end{bmatrix} = \mathbf{R} \cdot \begin{bmatrix} h \\ k \\ l \end{bmatrix} \tag{4.31}$$

or

$$\mathbf{c} = \mathbf{R} \cdot \mathbf{h} \tag{4.32}$$

Other similar transformations are built into the software of the diffractometer so that the scattering vector $\mathbf{c}$ is brought into the horizontal plane and, via Eulerian forward and inverse transformations (for mathematical convenience), the values of the angles ω_κ, κ and ϕ_κ, by which the diffractometer circles must be moved, are determined. Then the intensity of the reflection that corresponds to $\mathbf{c}(\omega_\kappa, \kappa, \phi_\kappa)$ can be measured and recorded.

4.7.4 Data Collection

Flexible routines built into the CAD4 permit selection of the hkl reflection in differing ways, including Friedel pairs (hkl and $\bar{h}\bar{k}\bar{l}$, Sections 3.6.1 and 6.5). The azimuthal angle, the angle between the vertical and the normal to the ϕ circle, can be variously specified, including multiple measurements of reflections at different azimuth values so as to derive an empirical absorption correction curve as a function of the angle ϕ (see Section 3.9.3).

Intensity and background measurements are carried out, and the crystal may be monitored for decay and movement, and reoriented as necessary, during the data collection. If the space group is known, only the geometrically permitted reflections need be scanned. This procedure is of great significance with centered unit cells. In addition, more than the unique portion of reciprocal space can be explored. By means of the CAD4 Structure Determination Package, the symmetry equivalent reflections can be merged to give the best set of unique data, Lorentz and polarization corrections applied, and an absorption correction included if deemed desirable. The data set can be truncated so as to exclude weak reflections of lower accuracy, for example, those for which $|F_0|^2$ is less than $3\sigma(|F_0|^2)$, or they may be remeasured over a longer time period so as to increase their precision. The CAD4 is provided with a means of taking x-ray photographs of the crystal on the diffractometer, using Polaroid film. This procedure may be very helpful where there is a "difficult" space group, or where twinning is suspected, but it is often not a substitute for a thorough initial examination of the crystal by conventional x-ray photography.

The automatic search and indexing routines for determining the unit cell are not always definitive. This situation may arise for various reasons, such as twinning, poor crystal quality (leading to weak reflections), or insufficient spread of reflections in reciprocal space. Several options are available for operator intervention. One of the most useful is to take a Polaroid rotation photograph about the randomly oriented ϕ axis and use it to determine the x, y coordinates of reflections on the film. About 10 such pairs of values will often suffice to determine an approximate unit cell. Then the unit cell can be refined by locating more reflections, well distributed in reciprocal space, and applying the normal least-squares procedure (Section 7.4.1). The ϕ-axis photograph may detect poor crystal quality but not twinning, and it will not help with space-group determination; normal rotation photographs about crystallographic axes are needed in these situations.

The techniques available with the CAD4 have been well tested in laboratories throughout the world, and their careful application can lead to a data set of high quality, capable of solving and refining a crystal structure to high precision, with R [equation (6.30)] approximately 5% on average, and possibly as low as 1% in the best cases.

4.7.5   Scanning Over a Peak: ω/θ versus ω Scans

The most widely used, and arguably the best, scanning option employed with the CAD4 and similar types of diffractometer involves a coupled rotation of the crystal (ω-axis) and the counter (2θ-axis), Figure 4.39, both rotation axes being vertical during the scan. For a given reflection, the centre of the scan coincides with the calculated position of the maximum intensity, based on the orientation matrix and unit-cell parameters determined previously. Figure 4.40 shows the variation of intensity recorded for such a scan, using a good quality crystal. In order to obtain a measure of the intensity it is necessary to evaluate the counts recorded in the area P on the diagram. For most reflections there is a general background intensity, shown in the diagram as being slightly asymmetrical. The background levels B1 and B2 are established by short scans on either side of the peak. The scans are not smooth, and consist of, say, n short steps in both B1 and B2 and m similar steps in P. If the total counts are $b1$ for region B1, $b2$ for region B2, and N for the peak P, it can be shown that the number of counts in P is $N - m(b1 + b2)/2n$. For

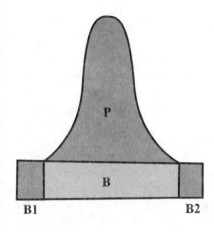

FIGURE 4.40.   Profile of a reflection peak scanned in the ω/θ mode. The background level to be subtracted on either side of the peak is indicated. It is typically slightly asymmetrical.

most small molecule crystals the peak width would be about 40–60 min of arc and the total scan time about 60 s. In this method the two backgrounds are usually fairly similar. The facility for inspecting individual scans is extremely useful for establishing the correct scan width to use for a given crystal, and to check for crystal splitting, which would produce a double (or even a multiple) peak.

For crystals with very large unit cells, which is mainly the situation with macromolecules, neighbouring peaks are usually too close together (the reciprocal lattice constants are very small) to allow the use of ω/θ scans. Instead the alternative ω-scan method is used, involving a stationary counter, fixed at the correct value of 2θ. This reduces the possibility of the diffraction maxima for neighbouring reflections overlapping, but does not eliminate it altogether.

4.8   Area Detectors (Position-Sensitive Detectors)

The use of an *area detector* or *image plate* allows many nonoverlapping reflections to be recorded in a single exposure. Each exposure is produced by oscillating the crystal through a small angle, $\Delta\varphi$. The reflection is received by a detector, which is an effectively flat surface with a uniform sensitivity to x-rays. A series of sequential exposures is recorded, during which the crystal may be rotated through a total of 180°, for a triclinic crystal, or as little as 30°, for a hexagonal or cubic crystal. The X, Y position of each reflection (Figure 4.27) enables its *hkl* index to be computed, and the signals received at the detector are converted electronically into intensity data. Area detectors are therefore also position-sensitive detectors, because both the intensity of a diffracted beam and the exact position where it strikes the detector are determined. Several designs are commercially available, as discussed next.

270 4 II MEASUREMENT OF INTENSITY DATA

4.8.1 Multiwire Proportional Counter

The multiwire proportional counter (MWPC) is a digital detector, working on the principle of detection of x-rays by ionization. This type of detector was used by protein crystallographers in the 1980s and 90s, and may still be found in some laboratories. The design of the MWPC is based on a xenon-filled chamber with an anode lying between two cathodes (Figure 4.41). Each anode and each cathode consists of a plane of parallel wires, with the direction of the wires in the first cathode and the anode perpendicular to those in the second cathode (Figure 4.42). During exposure to x-radiation, x-ray photons enter the chamber. For each photon that is absorbed, a xenon molecule is ionized into a positive xenon ion and an inner shell electron having a kinetic energy that is virtually all of the energy of the absorbed photon. The liberated electron ionizes further xenon molecules. In the case of Cu $K\alpha$ radiation, a single 8-keV photon (the energy of a Cu $K\alpha$, photon) produces about 300 ions and primary electrons. An electric field accelerates the positive ions toward the first cathode and the cloud of electrons toward the anode. The electron cloud passes the first cathode to hit the anode plane, producing an avalanche of secondary ionization, with an amplification factor of the order of 10^4 or more for each primary electron. While the electrons give a negative pulse

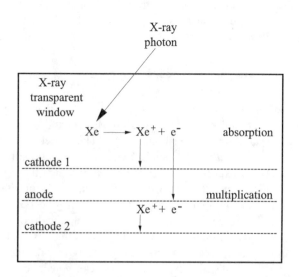

FIGURE 4.41. Principle of the Multiwire Proportional Counter.

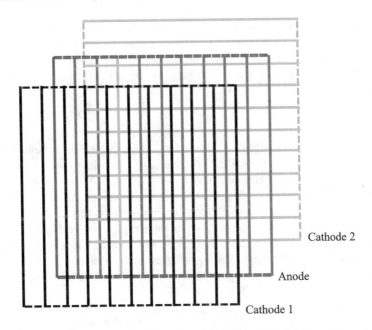

Cathode 2

Anode

Cathode 1

FIGURE 4.42.   Design of the Multiwire Proportional Counter.

on the nearest anode wire, the positive ions move toward the second cathode giving a positive pulse on the nearest cathode wire. To prevent ultraviolet photons, produced in the secondary ionization, from restarting ionization in another region of the chamber, a small amount of quenching gas is added to the xenon. Counting is accomplished through the orthogonal array of wires that samples the centroid of the charge distribution. The analog signals from the detector are digitized to produce the corresponding reading. This type of counter has a low noise level, is extremely sensitive, and covers a wide dynamic range of intensities. Unfortunately MWPCs are severely hampered by size limitations (which reduce the resolution range achievable with Cu $K\alpha$ radiation), suffer from errors due to parallax at high angles of incidence, and require a helium path in order to limit absorption, because of the large crystal-to-detector distances employed. Commercially available MWPCs include the Bruker (formerly Siemens) AXS HI-STAR (11.5 cm diameter) and the larger Xuong–Hamlin (also called San Diego Multiwire System or SDMS) (30 cm × 30 cm).

4.8.2   FAST Area Detector (Enraf–Nonius FAST)

In this method the x-ray reflections strike a phosphor coated fibre optics screen that converts the signals into light photons, which are then intensified, integrated, and digitized. The recording device incorporates a television scanning system and the goniometer is essentially that used in the CAD4 (Section 4.7ff). The disadvantages of this method include high electronic noise and consequent low dynamic range $(1:10^3)$, requiring remeasurement of strong intensities at decreased camera voltage.

4.8.3 Image Plate

In this method the detector[a,b] consists of a barium halide phosphor doped with europium, Eu^{2+}, Figure 4.43a. On exposure to x-rays, Eu^{2+} is excited into the metastable Eu^{3+} state, Figure 4.43b. After completion of each exposure, during which time the crystal is oscillated in the x-ray beam for a few minutes, the image plate is scanned with a fine helium–neon laser beam (Figure 4.43c), which causes the regions converted to Eu^{3+} (where the x-ray spots would be located on a photographic film) to emit violet light ($\lambda = 3900$ Å with an intensity proportional to the absorbed x-ray energy). This light is then detected with a photomultiplier system, integrated, and digitized. After reading the stored data the plate is cleaned

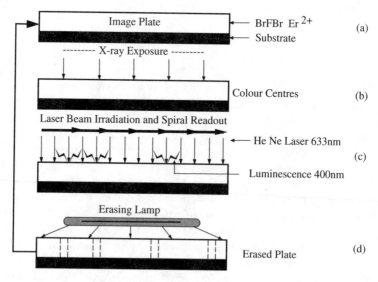

FIGURE 4.43. (a) Resting state of the image plate. (b) Exposure to x-ray diffraction from the oscillating crystal. (c) Scanning with laser light and measurement of the emitted light from each spot with a photomultiplier. (d) Cleaning up the plate ready for reuse.

[a] Mar Research (MAR IP), Norderstedt, Germany.
[b] J. Miyahara, K. Takahashi, Y. Amemiya, N. Kamiya, and Y. Satow, *Nuclear Instrumentation and Methods* A**246**, 572 (1986).

by exposure to bright yellow light (Figure 4.43d), and the "film" is then ready to record the next image.

Image plates up to 30 cm diameter are available (Mar Image Plate or MAR IP), which enable data to be collected to a resolution of about 1.4 Å using Cu $K\alpha$ radiation, and better with synchrotron radiation at smaller wavelengths. The resolution restriction normally precludes the choice of this device for routine small molecule analysis, but it is very popular for large molecule work. A very wide range of intensities, approximately $1 : 10^5$, can be recorded compared to approximately $1 : 200$ for x-ray film.

4.8.4 Charge-Coupled Device Area Detectors

In this type of system, which is rapidly becoming the method of choice in both small molecule and macromolecular crystallography, the video tube of the *FAST* area detector (Section 4.8.2) is replaced by a charge-coupled device (CCD). CCDs have been developed for a multitude of applications where extremely efficient detection of photons is required, as in astronomy and other branches of physics as well as x-ray crystallography. We could say that the ideal characteristics of a detector are:

- 100% quantum efficiency (independent of wavelength);
- Perfectly uniform response and unlimited dynamic range (both energy-wise);
- Electronically noiseless;
- Completely characterized components.

Since the conception of CCDs in 1970,[a] the state-of-the-art in 2002 is about 98% of the way to achieving these criteria. In theory, a CCD performs the following tasks:

(1) Generation of charge. When a photon strikes a CCD, it generates electron-hole pairs by the photoelectric effect.[b]

(2) Collection of the resulting charge or charges in the pixels formed by an array of electrodes or gates.

(3) Transference of the charge on each pixel by application of a differential voltage across the gates. This information is conveyed pixel by pixel for counting.

[a] W. Boyle and G. Smith, Charge coupled semiconductor devices, *Bell Systems Technical Journal* **49**, 587 (1970).
[b] A. Einstein, *Annalen der Physik* **17**, 132 (1905).

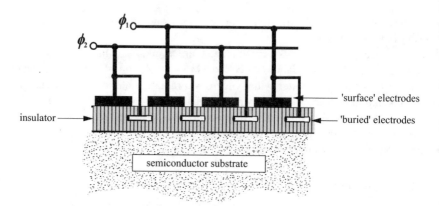

FIGURE 4.44.   Design of a CCD chip.

(4) Detection of the individual charges and conversion to an output voltage, followed by digital encoding prior to computer processing.[a,b]
Figure 4.44 shows a possible design for a CCD chip.[c]

The detector used with the Nonius Kappa CCD diffractometer (Figure 4.45)
achieves high efficiency through the use of a special grade Gd_2O_2S phosphor
with directly bonded high quality fiber-optics, in turn also directly bonded to the
CCD chip. The CCD chip requires to be actively cooled to between $-20°C$ and
$-50°C$, and this is achieved through the incorporation of a stack of four Peltier
elements again bonded directly to the chip. This results in a stable low temperature
at the chip, which is a requirement for low noise and consequent high sensitivity
with low background. Figure 4.46 shows a composite representation of the novel
molecule 1,2-ethene-3,3′-di-*tert*-butyl diimidazole-2,2′-diylidene indium(III) trihydride, $C_{16}H_{32}In_2N_4$, superimposed on part of its diffraction pattern, which was
recorded on a Nonius Kappa CCD diffractometer.[d]

4.8.5   Charge-Coupled Device versus Image Plate

One great advantage of the CCD over the image plate is its short readout time: a millisecond order readout is possible. The disadvantages are
that the detector area of the CCD is smaller than those available in other

[a] S. M. Gruner, *Current Opinion on Structural Biology* **4**, 765 (1994).
[b] E. M. Westbrook and I. Naday, *Methods in Enzymology* **276**, 244 (1997).
[c] The Electronic Universe Project (University of Oregon) nuts@moo.uoregon.edu
[d] C. Jones, *Chemical Communications* **2293** (2001).

area detectors, such as the MAR IP, and also that it needs a cooling system. However, with their high sensitivity, high resolution, and low noise, CCDs are now in regular use for high-resolution macromolecular structure determinations.[a,b] CCD diffractometers, including software packages, are available from Nonius (www.nonius.com), Bruker (www.bruker-axs.com), and Oxford-diffraction (www.oxford-diffraction.com), a system which includes optional liquid nitrogen or liquid helium cooling of the crystal; other systems are run routinely

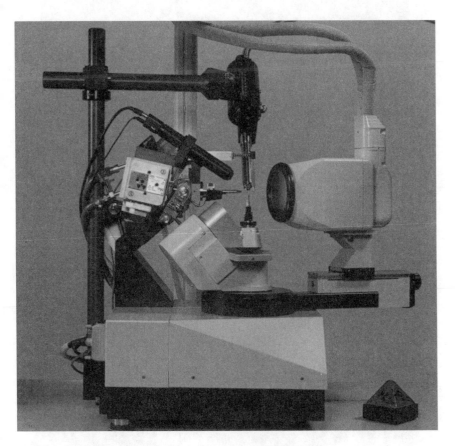

FIGURE 4.45.   A view of the Nonius Kappa CCD diffractometer.

[a] A. Deacon, T. Gleichmann, A. J. Kalb, H. Price, J. Raftery, G. Bradbrook, J. Yariv, and J. R. Helliwell, *Journal of Chemical Society, Faraday Transactions* **93**, 4305 (1997).

[b] R. L. Walter, D. J. Thiel, S. L. Barna, M. W. Tate, M. E. Wall, E. F. Eikenberry, S. M. Gruner, and S. E. Ealick, *Structure* **3**, 835 (1995).

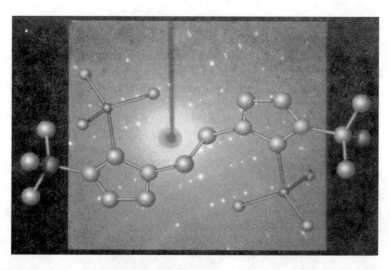

FIGURE 4.46.   Composite of the molecule 1,2-ethene-3, 3′-di-*tert*-butyl diimidazole-2, 2′-diylidene indium(II) trihydride, $C_{16}H_{32}In_2N_4$, superimposed on part of its diffraction pattern, as recorded on a Nonius Kappa CCD diffractometer.

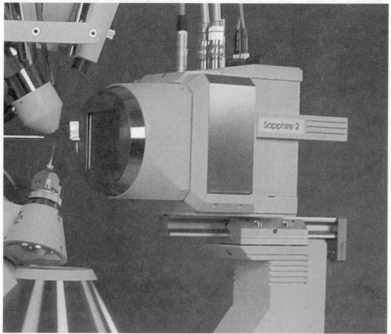

FIGURE 4.47.   Close up of the CCD detector on the Oxford-diffraction Xcalibur diffractometer. [Reproduced by permission of Oxford Cryostream.]

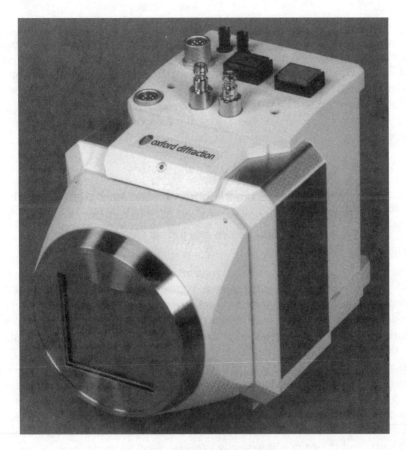

FIGURE 4.47. (Continued)

with liquid nitrogen cooling. Figure 4.47 shows a close up of the Oxford-diffraction CCD detector.

4.9 Monochromators

In this section, we consider monochromators for both the traditional x-ray tube and the synchrotron.

4.9.1 Sealed or Rotating Anode Tube Radiation

In all aspects of x-ray crystallography except for Laue diffraction (see Section 4.4.1), where a continuous spectrum is employed, intensity data are measured with x-rays that are as close as possible to being monochromatic. We have seen in Section 3.1.4 that characteristic x-radiation from a sealed or rotating anode tube can be effectively monochromatized by means of an absorption filter. In the case of a sealed or rotating anode tube with a copper target, a nickel filter will eliminate most (99.4%) of the $K\beta$ radiation, thus providing an effectively monochromatic x-ray beam. The continuous background radiation is also considerably reduced by this method but not removed altogether. Unfortunately, approximately 75% of the intensity of the $K\alpha$ radiation is also lost.

Single and Double Type Crystal Monochromators

The crystal structure of graphite consists of layers of covalently-bonded, planar, hexagonal carbon rings stacked perpendicular to the c-axis. The stacking distance of $c/2 = 3.41$ Å approximates to the van der Waals distance for carbon (Section 7.7.) This layered structure maximizes the carbon content of successive (002) planes which thus produces an extremely strong 002 reflection. The single type crystal graphite monochromator operates by using this reflection as the primary beam. For a copper target there is a greater loss of intensity (about 20%) compared to that suffered with a nickel filter. However, the range of selected wavelengths is very narrow and effectively monochromatic. This results in a dramatic improvement in the peak to background ratio, owing to a virtual elimination of the general background radiation from a sealed tube source. Reduced heating effects resulting in improved crystal life have also been reported for some proteins. For a monochromator tuned to wavelength λ, the harmonically related wavelengths $\lambda/2, \lambda/3, \ldots, \lambda/n$ will be reflected at the same angle. This could, of course, present problems with the measured intensities, which would be from a composite of wavelengths. However, in practice the harmonic wavelengths tend to be very weak in intensity, and choice of a structurally suitable monochromator material, such as graphite, will further ensure that the corresponding $I(h/n, k/n, l/n)$ values are also negligibly small.

One disadvantage with the single type crystal monochromator is that once tuned to select a given wavelength, the instrument cannot easily be adapted for use with a different wavelength source, because the reflecting angle will be different. This can be overcome by employing a double type crystal monochromator (Figure 4.48) in which the incident and emergent beam directions are the same. Tuning of the instrument is achieved by pivots fitted to both halves of the monochromator. Obviously, a further loss of intensity will be suffered with this arrangement, from the second crystal, which offsets the above advantage to some extent.

4.9.2   Monochromators for Synchrotron Radiation

The divergence of the x-ray beam from a synchrotron is small but undesirable (at Daresbury, U.K., the beam divergence is ca 1 min of arc). This effect can be

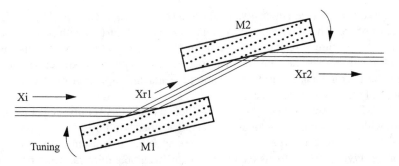

FIGURE 4.48.   Double type crystal monochromator. The incident X-ray beam Xi is reflected as an essentially monochromatic component Xr1 from the first crystal M1. The second crystal M2 is set at the same angle and is therefore tuned to the same wavelength and reflects the emergent component Xr2. Tuning to a different wavelength is achieved by coupled rotation of the two crystals (curved arrows) and maintains the direction of incident and emergent beams, thus requiring minimal adjustment of the diffractometer. The second crystal M2 is absent in a single type crystal monochromator.

corrected by using a focusing monochromator, which may be either single or double. A single crystal of germanium or silicon is used for this purpose, because it has the additional advantage of a much lower bandwidth selection (wavelength range) compared to graphite. The plane of the crystal monochromator is carefully bent in order to produce the focusing effect, which is in one direction only (line focus). Further focusing may be achieved by the use of highly polished quartz or glass toroidal mirrors. Thin diamond plates, which have a high thermal tolerance, also provide excellent mirrors but suffer from the disadvantage of being extremely small.[a]

4.10   Focusing Mirrors

As we have seen, important features of the x-ray beam used in data collection include monochromatization, small beam divergence, high intensity, and optimal focusing. In macromolecular crystallography, a small divergence of the beam enables the use of a larger oscillation range for individual exposures without spot overlap and consequently fewer exposures in total. It will also improve the resolution of individual spots that are close together in a crystal of large unit-cell dimensions.

X-rays can be reflected by mirrors when the angle of incidence is smaller than a critical angle (about 0.1°), and the use of the Franks' double focusing mirror

[a] A. K. Freund, *Structure* **4**, 155 (1996).

to improve x-ray beam intensity, first introduced nearly 50 years ago,[a] has recently been revived. This device consists of a pair of curved mirrors with perpendicular axes of curvature, which produces a point-focussed x-ray beam, with consequent high intensity. The first mirror thus focuses the beam in one direction and the second mirror focuses the beam in the perpendicular direction to give a small, highly concentrated spot size.

More recently the design and construction of confocal ellipsoidal mirrors for use with microfocus x-ray tubes has been described.[b] In this device, two mirrors are glued together in perpendicular arrangement at the same distance from the x-ray source, giving a fixed focus beam. Alternatively, Göbel mirrors, using parabolic focusing in a sequential double-mirror set up, can be used to produce a collimated parallel beam.

The advantages of the use of mirrors include an increased flux, by a factor of at least 3 or 4 compared to a graphite monochromator; a narrow angular divergence and higher brilliance, a small spot size and higher brightness, a low background and some degree of monochromatization. The disadvantages are a degree of competition between monochromatization and beam flux, leading to the use of a nickel filter (for copper radiation) placed in front of the mirror, associated problems with alignment of the x-ray generator and filament (rotating anode), and the requirement of a helium path to counter air-absorption, because of increased path lengths (around 200 mm). Commercially available mirrors include versions by MAR, MAC Science, Nonius XOS, and Charles Supper (Franks' mirrors); Bruker AXS (Göbel mirrors); Osmic's Max Flux[TM] by Bruker, MSC, MAR, and Nonius (AXS Confocal MaxFlux optics) (all confocal mirrors).

4.11   Twinning

During formation, many crystals may undergo some sort of growth stress that causes them to continue crystallizing in other directions. The resultant material is called a *twin crystal*. The principal mechanisms of twinning may be regarded as rotation, reflection, or glide deformation. The relationship of a crystal and its twin counterpart may be expressed in terms of the symmetry operation needed to bring it about. The most common type of twin is formed by rotation of 180° about a zone axis, leading to pseudo-2-fold symmetry. It is evident that such a crystallographic direction could not already be a crystallographic rotation axis of even degree. A well-known example of such twinning is found in gypsum (point group $2/m$), shown in Figure 4.49 twinned by a rotation of 180° about the x axis. The plane of contact of the two parts of the twin, (100) in this example, is called

[a] A. Franks, *Proceedings of Physical Society of London* **B68**, 1054 (1955).

[b] A. C. Bloomer and U. W. Arndt., *Acta Crystallographica* **D55**, 1672 (199).

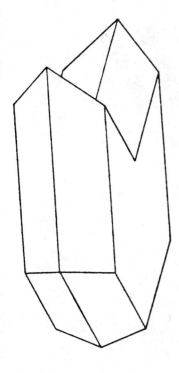

FIGURE 4.49. Gypsum crystal twinned by 180° rotation about x; (100) is the composition plane. (Reproduced from *An Introduction to Crystallography*, 4th Edition, by F. C. Phillips, with the permission of Longmans Group.)

the *composition plane*. Characteristic features of twin crystals are the existence of reentrant angles, shown by Figure 4.49, and pseudosymmetry, $mm2$ for the gypsum twin.

In addition to these types of twinning, multiple twins can occur. The components may take the form of lathlike fragments, parallel to the composition plane, the whole crystal being called a *lamellar* twin.

The existence of twinning can complicate the x-ray diffraction record from such a crystal. When the crystals are large enough, it may be possible to take an appropriate slice of the twin so as to give a single crystal before carrying out further examination. More frequently, twinning is not visible directly, but it may be revealed by microscopic examination under polarized light. Figure 4.50 shows two possible situations where adjacent parts of a specimen are not simultaneously in extinction. Before proceeding to an x-ray examination, one must obtain a single crystal. It may often be accomplished by cutting a single-crystal fragment under the microscope with a sharp, thin blade, and with a "collar" (Figure 4.51) around the crystal to retain the fragments as they are cut. The same apparatus may be used to cut crystals generally, as the occasion demands.

(a)   (b)

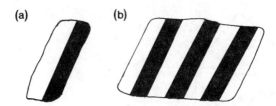

FIGURE 4.50.   Thin sections of twinned crystals under the microscope, in plane-polarized light: (a) simple contact twin, (b) lamellar twin. In each case, adjacent parts of the crystal are not in extinction simultaneously.

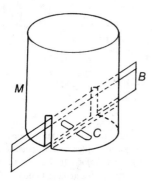

FIGURE 4.51.   Half razor blade B and metal (or plastic) collar M, for cutting a crystal C under the microscope.

The structure refinement program SHEL-X (see Appendix A10) has facilities (BASF and TWIN), which enable tests for twinning to be carried out at the end of the least squares refinement.

Bibliography

Crystal Optics

GAY, P., *An Introduction to Crystal Optics*, London, Longmans (1967).
HARTSHORNE, N. H., and STUART, A., *Crystals and the Polarising Microscope*, London, Arnold (1970).

X-ray Scattering and Reciprocal Lattice

ARNDT, U. W., and WONACOTT, A. J. (Editors), *The Rotation Method in Crystallography: Data Collection from Macromolecular Crystals*, Amsterdam, North-Holland (1977).
BUERGER, M. J., *X-Ray Crystallography*, New York, Wiley (1942).
JEFFREY, J. W., *Methods in X-Ray Crystallography*, London, Academic Press (1971).
WOOLFSON, M. M., *An Introduction to X-Ray Crystallography*, 2nd ed., Cambridge, Cambridge University Press (1997).

Interpretation of X-ray Diffraction Photographs

HENRY, N. F. M., LIPSON, H., and WOOSTER, W. A., *The Interpretation of X-ray Diffraction Photographs*, London, Macmillan (1960).
JEFFERY, J. W., *Methods in X-ray Crystallography*, London, Academic Press (1971).

Precession Method

BUERGER, M. J., *The Precession Method*, New York, Wiley (1964).

Diffractometry

ARNDT, U. W., and WILLIS, B. T. M., *Single Crystal Diffractometry*, Cambridge, Cambridge University Press (1966).

Problems

4.1. Crystals of KH_2PO_4 are needle shaped and show straight extinction parallel to the needle axis. A Laue photograph taken with the x-rays parallel to the needle axis shows symmetry 4*mm*.

(a) What is the crystal system and Laue group, and how is the optic axis oriented?

(b) Describe and explain the appearance between crossed Polaroids of a section cut perpendicular to the needle axis.

(c) What minimum symmetry, would be observed on both general and symmetric oscillation photographs taken with the crystal mounted on the needle axis?

4.2. Crystals of acetanilide (C_8H_9NO) are brick-shaped parallelepipeda, showing straight extinction for sections cut normal to each of the three edges of the "brick."

(a) What system would you assign to the crystals?

(b) Allocate suitable crystallographic axes.

(c) What minimum symmetry would be shown by general oscillation photographs taken, in turn, about each of the three crystallographic axes?

(d) What symmetry would an oscillation photograph exhibit where the crystal is oscillating about the *a* axis such that *b* is parallel to the x-ray beam at the center of the oscillation range?

4.3. Crystals of sucrose show the extinction directions indicated on the schematic crystal drawing of Figure P4.1; the arrows indicate the directions of the cross-wires at extinction.

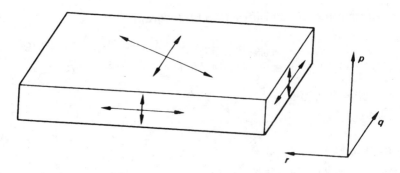

FIGURE P4.1. Crystal section of sucrose, showing extinction directions.

(a) To what crystal system does sucrose belong?
(b) How are the morphological directions, p, q, and r related to the crystallographic axes?
(c) How would you mount the crystal in order to test your conclusions with (i) Laue photographs, in a single mounting of the crystal, and (ii) oscillation photographs? In each case, indicate the symmetry you would expect the photographs to exhibit in the orientations you have chosen.

4.4. General oscillation photographs of an orthorhombic crystal mounted, in turn, about its a, b, and c axes had layer-line spacings, measured between the zero and first levels, of 5.07, 7.74, and 9.43 mm, respectively. If $\lambda = 1.50$ Å and $R=30.0$ mm, calculate a, b, and c and a^*, b^*, and c^* for the given wavelength.

Explain why the 146 reflection for this crystal could not be recorded with x-rays of the given wavelength. What symmetry would be observed on the above oscillation photographs? Is this evidence alone conclusive that the crystal is orthorhombic?

4.5. (a) A tetragonal crystal is oscillated through 15° (i) about [110] and (ii) about the c axis. The layer-line spacings, measured between the $+2$ and -2 layers in each case, were 3.5 and 2.7 cm, respectively. If $\lambda = 1.54$ Å and $R = 3.00$ cm, calculate the a and c dimensions of the unit cell.
(b) How many layers could be recorded in position (ii) if the overall film dimension parallel to the oscillation axis is 8.0 cm?
(c) What is the minimum symmetry obtainable on films such as (i) and (ii)?
(d) How would a third photograph taken after rotating the crystal in (ii) through 90° compare with photograph (ii)?

4.6. Euphenyl iodoacetate is monoclinic, with $a=7.260$, $b=11.55$, and $c=19.22$ Å. Figure P4.2 is the Weissenberg zero level ($h0l$). If the film translation constant is 1 mm per $2°$ rotation, determine β^*, β, and the unit-cell volume.

4.7. A cubic crystal ($a = 5.0$ Å) is mounted on a flat-plate Laue camera so that one axis (b or c) lies along the rotation axis and a is inclined to the x-ray beam at an angle $\phi = 70°$. If the crystal is irradiated with an x-ray beam possessing the wavelength range 0.2–2.5 Å, determine the highest order ($h00$) reflection that can be recorded. If the cassette is placed at a distance $R = 60$ mm from the crystal and the film plate is 125 mm square, determine whether this reflection will be recorded and, if so, find its coordinate in mm on the film.

4.8. A Laue photograph of a protein crystal is recorded on a pack of two films separated by black paper. A particular reflection (hkl) for a wavelength λ in the 'white' radiation is overlapped on the films by the reflection ($2h, 2k, 2l$) for the wavelength $\lambda/2$. The black paper transmits 65% of the radiation of shorter wavelength and 35% of the longer wavelength radiation. If the relative intensity of the composite reflection measured on the film first to receive the reflected beam is 300 but only 130 on the second film, what are the relative intensities of the two reflections on the first film?

4.9. An oscillation photograph of a crystal is taken with Mo $K\alpha$ x-radiation ($\lambda=0.7107$ Å) about an axis of repeat distance 8.642 Å. The camera diameter is 60.00 mm and the x-ray beam is perpendicular to the rotation axis. Calculate the height of the first, second, and third layer lines above the zero-layer line.

4.10. Oscillation photographs are taken of a cubic crystal set successively on [100], [110], and [111], using Cu $K\alpha$ radiation ($\lambda = 1.5418$ Å), the camera diameter was 57.3 mm, and the double separations ($2h_n$) of the layer lines were measured:

Axis	[100]	[110]	[111]
n	3	2	1
$2h_n/mm$	41.0	37.6	52.0

Determine the unit-cell dimension and type.

4.11. β-Zinc sulphide, ZnS, crystallizes in space group $F\bar{4}3\,m$, with $a = 5.41$ Å. A flat-plate Laue photograph, taken with a crystal-to-film distance of 30.00 mm, exhibits symmetry $2mm$. (a) What was the direction of the incident beam in the crystal?

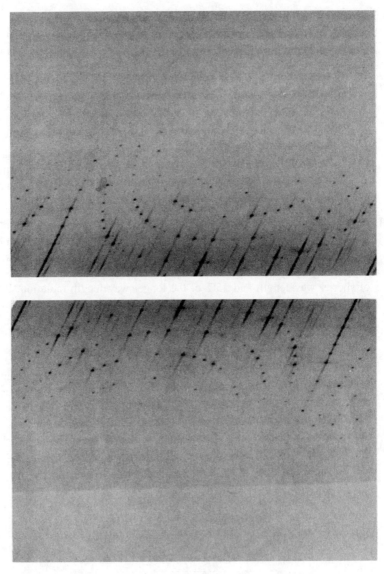

FIGURE P4.2.   Weissenberg photograph of the $h0l$ layer of euphenyl iodoacetate taken with $CuK\alpha$ radiation. There is evidence of a very small piece of crystal attached in nearly the same orientation; it is not a twin.

The horizontal m line on the film displays two pairs of reflections, symmetrically disposed about the centre of the film. The two reflections of the outer pair are 77.5 mm apart, and the inner pair 43.5 mm apart. For each reflection in the two pairs, find (b) the Bragg angle θ, (c) the indices hkl and (d) the wavelength producing each spot.

4.12. Oscillation photographs of manganese carbonate, $MnCO_3$, taken about three mutually perpendicular axes, and Laue photographs along the same three axes gave the following results ($l = 1.5418$ Å):

Axis	1	2	3
Layer line	8	4	2
Distance/mm from zero layer	38.40	33.63	25.40
Laue symmetry	$3m$	m	2

Determine the unit-cell dimensions and the Laue group of the substance; the diameter of the oscillation camera was 60.00 mm.

4.13. The 0002 reflection from a single crystal of graphite can be used to obtain monochromatic x-radiation from the (0001) plane. If the c dimension of graphite is 6.696 Å, calculate the appropriate Bragg angle for reflection of (a) Cu $K\alpha$ (λ=1.5418 Å) and (b) Mo $K\alpha$ (λ=0.7107 Å) radiation at which the crystal monochromator must be set.

4.14. An orthorhombic crystal was set in a random orientation on an automatic four-circle single-crystal x-ray diffractometer. A peak search led to a primitive reciprocal unit cell of dimensions $a^*=b^*=0.239$, c=0.184, $\alpha^*=\beta^*=90°$, $\gamma^*=152.4°$, using Mo $K\alpha$ radiation (λ=0.7107 Å). Determine the transformation matrix to convert the 'diffractometer' cell to the true cell, which was known from photographs to have the approximate dimensions $a = 3.1$, $b = 12.5$, and $c = 3.9$ Å (a) in real space and (b) in reciprocal space.

5

Fourier Series and Fourier Transforms

In Chapter 3, we touched upon the analogy between the diffraction of x-rays and that of visible light. Here, we extend that discussion and consider some aspects of Fourier series and Fourier transforms.

5.1   Image Formation and Focusing

The formation of the optical image of an object involves first, a scattering of light from it, and then a recombination of the scattered light rays. We could, for example, prepare a transparency of one half of Figure 1.1 and project it on to a screen, focusing correctly, thus revealing an enlarged image of the model of the unit cell of sodium chloride.

If we now take the lens out of the projector, there will be just a patch of light on the screen, even though the object, the transparency, is still in the same position. All the information provided by the transparency is still present in the patch of light, but it is not immediately decipherable. The lens has no information about the transparency, but once it is in position in the projector the image becomes clear. The lens rearranges the scattered light so as to be understandable to us.

The process of focusing is complex, but we perform it by adjusting the position of the lens until the image assumes our expectation of the object. Evidently, some foreknowledge of the object is needed, and we assume that the appearance of sharply defined boundaries in the image is a condition of being in focus. If we do not have the necessary foreknowledge, we must determine the relative positions of transparency, lens and screen by the methods of geometrical optics. It is possible to show that the scattered radiation is everywhere within the scattered light by moving through it with a hand lens and a piece of white card, although the resulting image will not be as satisfactory as that obtained with the projector lens system. If we cannot obtain sharp boundaries in the image, we need an aid to focusing. One simple practical method, sometime used in microscopy, would be to have a minute speck of dust adhering to the surface of the transparency. When the speck is in focus, it can be assumed that the whole image is also in focus.

Visible light can resolve separations in an object down to approximately 2500 Å, if a high-quality microscope is employed. Resolution and radiation wavelength go hand in hand. The human eye can observe two objects as separate entities provided that they are no less than approximately 0.15 mm apart. Rayleigh's formula for the limit of resolution, or resolving power, R of a microscope is given as

$$R = 0.61\,\lambda/(n\sin\alpha) = 0.61\,\lambda/N \qquad (5.1)$$

where 2α is the angle of scatter, n is the refractive index of the medium and N is the numerical aperture of the objective. For a microscope working dry N can reach 0.95, and in oil immersion up to 1.45. Hence, R is approximately 3850 Å (dry) or 2520 Å (oil), corresponding to magnifications of 400 and 600, respectively. More recently, higher resolution has been demonstrated by *scanning near-field optical microscopy* (SNOM),[a] and resolution ranges between 0.1λ and 0.01λ have been reported.

The refractive index of materials for x-rays is approximately unity, so that from (5.1) wavelengths in the range 1/0.61, or approximately 1.6 Å, are required in order to resolve an atomic separations of 1 Å; such wavelengths are obtained with x-rays or neutrons. X-rays and neutrons of wavelength 1–2 Å can be used to resolve atomic detail, but they can be focused only by special systems of curved mirrors that lead to impracticably low magnification. With x-rays or neutrons, the scattered radiation must be recombined by calculation.

The electron microscope can provide resolution of atomic detail, provided that the structure is not too complex, but the resolution is limited by spherical aberration of the focusing system.

Certain simplifications exist in the applications of these radiations compared to visible light. The regular packing of atoms and molecules in crystals and a restriction of the radiation to a monochromatic source together give rise to a spot diffraction pattern rather than a diffuse patch.

A close analogy to x-ray diffraction, albeit in two dimensions, can be seen by viewing a sodium street-lamp through a fine, stretched gauze, such as a net curtain or a handkerchief. The spot pattern that is obtained is invariant under translation of the object, but rotates as the object is rotated; we shall return to these two properties later. To form an image from the diffraction pattern, the scattered radiation must be recombined in both amplitude and phase. The lens system used with visible light enables this process to be carried out directly. With x-rays or neutrons, however, not only can the focusing not be done directly, but the important phase information required is not obtained explicitly from the experimental

[a] V. Sandoghdar, in *Proceedings of the International School of Physics, Enrico Fermi*, IOS Press, Amsterdam (2001).

procedure: we record $I(hkl)$, but from Figure 3.28, we see that $|F(hkl)|^2 = A'(hkl)^2 + B'(hkl)^2 = |F(hkl)|^2 \cos^2 \phi(hkl) + |F(hkl)|^2 \sin^2 \phi(hkl) = I(hkl)$, so that information about $\phi(hkl)$ is not given directly.

In crystal structure analysis, we use Fourier series to carry out the focusing process. In the next section, we shall assume that we have the necessary phase information; and the acquisition of this phase information will be the subject of much of the remainder of this book.

5.2 Fourier Series

We consider the function $\psi(X)$ in Figure 5.1: it is continuous, single-valued, and periodic in the repeat distance a. According to Fourier's theorem, it can be represented by a series of cosine and sine terms that may be written conveniently as

$$\psi(X) = \sum_{h=-\infty}^{\infty} \{C(h)\cos(2\pi hX/a) + S(h)\sin(2\pi hX/a)\}$$

which can also be expressed as

$$\psi(X) = C(0) + 2\sum_{h=1}^{\infty}\{C(h)\cos(2\pi hX/a) + S(h)\sin(2\pi hX/a)\} \qquad (5.2)$$

The index representing the hth term in this series is a frequency, or wavenumber, that is, the number of times its own wavelength fits into the repeat period. In order to find the coefficients $C(h)$ and $S(h)$, we form the integral

$$\mathcal{I} = \int_{0}^{a} \psi(X)\cos(2\pi HX/a)\,\mathrm{d}x \qquad (5.3)$$

for a general value H of the frequency variable h. Substituting (5.2) into (5.3), and noting that the integral of a sum is equal to the sum of the integrals of the separate

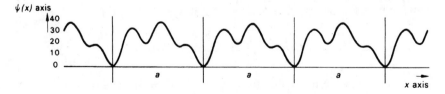

FIGURE 5.1. One-dimensional periodic function $\psi(X)$ of repeat a.

parts, we have

$$\mathcal{I} = C(0) \int_0^a \cos(2\pi H X/a) \, dx$$

$$+ \sum_{h=1}^{\infty} \left\{ C(h) \int_0^a 2[\cos(2\pi h X/a) \cos(2\pi H X/a)] \, dx \right.$$

$$\left. + S(h) \int_0^a 2[\sin(2\pi h X/a) \cos(2\pi H X/a)] \, dx \right\} \tag{5.4}$$

Equation (5.4) contains a number of integrals in x of the forms

$$\int_0^a 2 \cos(2\pi h X/a) \cos(2\pi H X/a) \, dx \tag{5.5}$$

and

$$\int_0^a 2 \sin(2\pi h X/a) \cos(2\pi H X/a) \, dx \tag{5.6}$$

It is straightforward to show, using identities given in Appendix A6, that these integrals are orthogonal, that is, (5.5) $= 0$ except for $H = h$, when it is equal to a, whereas (5.6) $= 0$ for all values of H. Thus, $C(h)$ and $S(h)$ may be expressed by the equations

$$C(h) = \frac{1}{a} \int_0^a \psi(X) \cos(2\pi h X/a) \, dx \tag{5.7}$$

and

$$S(h) = \frac{1}{a} \int_0^a \psi(X) \sin(2\pi h X/a) \, dx \tag{5.8}$$

If the form of the function $\psi(X)$ is known, $C(h)$ and $S(h)$ can be evaluated. We shall carry out this process for the square wave shown in Figure 5.2.

5.2.1   Analysis of the Square Wave

Let the square-wave function $\psi(X)$ be defined in the range $-\pi \le X \le \pi$, with a repeat of 2π. For $X < 0$, $\psi(X) = 0$, and for $0 < X \le \pi$, $\psi(X) = \pi$ (we shall see from the analysis that $\psi(X) = \pi/2$ at $X = 0$). Hence,

$$C(h) = \frac{1}{2\pi} \int_0^{\pi} \pi \cos(hX) \, dx$$

Integration gives the result that for $h \ne 0$, $C(h) = 0$, whereas for $h = 0$, $C(h) = \frac{1}{2} \int_0^{\pi} dx = \pi/2$. In a similar manner, we find that for $h = 0$, $S(h) = 0$,

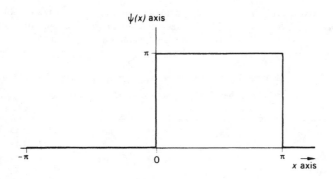

FIGURE 5.2.   Square wave $\psi(X)$, defined in the range $-\pi$ to π and with a repeat of 2π.

but for $h \neq 0$, $S(h) = (1/2h)[1 - \cos(h\pi)]$, Substituting these results in (5.2), we find

$$\psi(X) = \pi/2 + 2 \sum_{h=1}^{\infty} (1/2h)[1 - \cos(\pi h)] \sin(hX)$$

or

$$\psi(X) = \pi/2 + 2 \sum_{\substack{h=1 \\ (h=2n+1)}}^{\infty} (1/h) \sin(hX) \qquad (5.9)$$

since the term $[1 - \cos(\pi h)]$ is zero for even values of h. Hence, finite values of $\psi(X)$ arise for $h = 2n + 1$ ($n = 0, 1, 2, \ldots$).

Range of X

In (5.9), the variable X defines a sampling point in any repeat interval between $X = \pm\pi$. For convenience, we will choose a zero arbitrarily at $X = 0$ and sample the function at intervals of $2\pi/50$, that is, we shall calculate the function from 0 to $50(2\pi/50)$. The results will show that we could, and in general would, make use of reflection symmetry in the function at $X = (2m + 1)\pi/2$ ($m = 0, 1, 2, \ldots, \infty$) to decrease the amount of calculation.

Range of h

The summations of a Fourier series extend, theoretically, from $-\infty$ to ∞. In practice, however, the range becomes $h_{min} \leq h \leq h_{max}$, where the limits of h are preset, normally by experimental conditions. In this example h_{min} is unity, and h_{max} values of 3 and 7 are used in the results presented in Table 5.1 and Figure 5.3. It is notable, even with these few numbers of terms, that increasing h_{max} has a dramatic effect on the series. As h_{max} increases so the series (5.9) approaches more

TABLE 5.1.   Values of the function $(\pi/2) + 2 \sum_{h=1}^{h_{max}} (1/h) \sin(hX)$ (h odd) for $h_{max} = 3$ and 7. The true value $\psi(X)$, for all h, appears in the extreme right-hand column

$X/2\pi$	$\psi(X) \simeq \frac{1}{2}\pi + 2\sin X$ $+ \frac{2}{3}\sin 3X$ $h_{max} = 3$	$\psi(X) \simeq \frac{1}{2}\pi$ $+2\sin X + \frac{2}{3}\sin 3X$ $+\frac{2}{5}\sin 5X + \frac{2}{7}\sin 7X$ $h_{max} = 7$	$\psi(X)$
0/50	1.571	1.571	3.142
1/50	2.067	2.522	3.142
2/50	2.525	3.186	3.142
3/50	2.910	3.428	3.142
4/50	3.200	3.330	3.142
5/50	3.380	3.109	3.142
6/50	3.454	2.977	3.142
7/50	3.433	3.017	3.142
8/50	3.343	3.158	3.142
9/50	3.215	3.265	3.142
10/50	3.081	3.249	3.142
11/50	2.972	3.137	3.142
12/50	2.912	3.034	3.142
m—			—m
13/50	2.912	3.034	3.142
14/50	2.972	3.137	3.142
15/50	3.081	3.249	3.142
16/50	3.215	3.265	3.142
17/50	3.343	3.158	3.142
18/50	3.433	3.017	3.142
19/50	3.454	2.977	3.142
20/50	3.380	3.109	3.142
21/50	3.200	3.330	3.142
22/50	2.910	3.428	3.142
23/50	2.525	3.186	3.142
24/50	2.067	2.522	3.142
25/50	1.571	1.571	3.142
26/50	1.075	0.619	0
27/50	0.617	−0.044	0
28/50	0.231	−0.287	0
29/50	−0.058	−0.188	0
30/50	−0.239	0.033	0
31/50	−0.312	0.164	0
32/50	−0.291	0.125	0
33/50	−0.201	−0.017	0
34/50	−0.073	−0.123	0
35/50	0.061	−0.107	0
36/50	0.169	0.005	0
37/50	0.230	0.108	0
m—			—m
38/50	0.230	0.108	0
39/50	0.169	0.005	0

TABLE 5.1. (Continued)

$X/2\pi$	$\psi(X) \simeq \frac{1}{2}\pi + 2\sin X$ $+ \frac{2}{3}\sin 3X$ $h_{max} = 3$	$\psi(X) \simeq \frac{1}{2}\pi$ $+2\sin X + \frac{2}{3}\sin 3X$ $+\frac{2}{5}\sin 5X + \frac{2}{7}\sin 7X$ $h_{max} = 7$	$\psi(X)$
40/50	0.061	−0.107	0
41/50	−0.073	−0.123	0
42/50	−0.201	−0.017	0
43/50	−0.291	0.125	0
44/50	−0.312	0.164	0
45/50	−0.239	0.033	0
46/50	−0.058	−0.188	0
47/50	0.231	−0.287	0
48/50	0.617	−0.044	0
49/50	1.075	0.619	0
50/50	1.571	1.571	3.142

[a] The corresponding curves are shown in Figure 5.3.

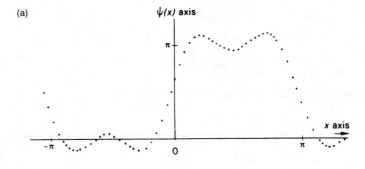

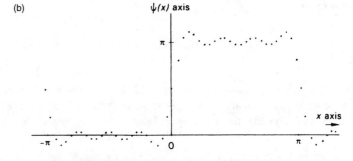

FIGURE 5.3. Square waves calculated from (5.9): (a) $h_{max} = 3$, (b) $h_{max} = 7$. The positive and negative fluctuations of the calculated function arise because there are insufficient terms to provide good convergence of the Fourier series; they are known as series termination errors.

closely the square-wave function in Figure 5.2. In general, the more independent terms that can be included in a Fourier series, the better it represents the periodic function under investigation, from which the terms have been derived.

The process of determining the coefficients of a Fourier series is called Fourier analysis, and the process of reconstructing the function by the summation of a series such as (5.9) is Fourier synthesis. A microscope, in forming an image of an object, effectively performs a Fourier synthesis of the scattered light, a process that we shall have occasion to refer to again later.

5.2.2   Exponential Forms of Fourier Series

Let a function $G(h)$ be represented as

$$G(h) = C(h) + iS(h) \tag{5.10}$$

Using de Moivre's theorem, $\exp(iX) = \cos(X) + i\sin(X)$, it follows that

$$
\begin{aligned}
G(h) &\exp(-i2\pi hX/a) + G(-h)\exp(i2\pi hX/a) \\
&= [C(h) + iS(h)][\cos(2\pi hX/a) - i\sin(2\pi hX/a)] \\
&\quad + [C(h) - iS(h)][\cos(2\pi hX/a) + i\sin(2\pi hX/a)] \\
&= 2C(h)\cos(2\pi hX/a) + 2S(h)\sin(2\pi hX/a)
\end{aligned}
\tag{5.11}
$$

Comparison with (5.2) shows that $\psi(X)$ may be cast as

$$\psi(X) = \frac{1}{2}\sum_{h=0}^{\infty} G(h)\exp(-i2\pi hX/a) + G(-h)\exp(i2\pi hX/a)$$

which reduces to

$$\psi(X) = \sum_{h=-\infty}^{\infty} G(h)\exp(-i2\pi hX/a) \tag{5.12}$$

The multipliers $1/2$ in (5.12a) disappear in (5.12) because the summation limits are now $-\infty$ to ∞, whereas in (5.11) the positive and negative values of h appear explicitly; we shall consider this point again later. Following Section 5.2, we can show

$$G(h) = \frac{1}{a}\int_{0}^{a} \psi(X)\exp(i2\pi hX/a)\,\mathrm{d}x \tag{5.13}$$

Equations (5.12) and (5.13) are Fourier transforms of each other, a topic that we shall consider in more detail shortly; the signs of the exponents should be noted.

5.3 Fourier Series in X-ray Crystallography

The lattice basis of crystal structures introduces a three-dimensional periodicity which pervades the properties of crystals, including the electron density distribution. The square wave that we have just analyzed may be likened to a one-dimensional crystal, or the projection of a crystal structure on to a single axis. The first applications of Fourier series in crystallography were with one-dimensional series, and for good reason. We must consider first the significance of the functions $\psi(X)$ and $G(h)$.

We have shown in Chapter 3 how x-rays are scattered by the electrons associated with atoms in a crystal. The concentration of electrons and its distribution around an atom is called the electron density, and it is measured in electrons per unit volume (usually $\mathring{A}^{-3}$ or nm^{-3}). At any point X, Y, Z the electron density is written normally as $\rho(XYZ)$, and we may identify $\rho(XYZ)$ with the function $\psi(X)$ in (5.12), so that we must now determine the meaning of $G(h)$.

We consider electrons in an atom as though they were concentrated at a point, but specify their distribution by a shape factor, the atomic scattering factor f, which is equivalent to an amplitude. The exponential term in (5.12) represents the phase of the wave scattered by an atom at X with respect to the origin of the unit cell.

5.3.1 One-Dimensional Function

The one-dimensional electron density function $\rho(X)$ (e $\mathring{A}^{-1}$) shown in Figure 5.4 was calculated from a small number of experimental terms. In a small interval dX along the X axis, the electron density may be regarded as being constant, so that the associated electron count is $\rho(X)\, dX$. Its contribution to the hth

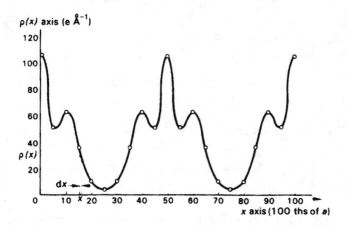

FIGURE 5.4. One-dimensional electron density projection $\rho(X)$ for pyrite, FeS_2.

structure factor $F(h)$ is given, following Section 3.5.3, by $\rho(X)\exp(\mathrm{i}2\pi hX)\,\mathrm{d}X$, where $\exp(\mathrm{i}2\pi hX)$ is the phase associated with $\rho(X)\,\mathrm{d}X$ with respect to the origin. The contribution from the whole repeat period, the structure factor $F(h)$, is now given by

$$F(h) = \int_0^a \rho(X)\exp(\mathrm{i}2\pi hX/a)\,\mathrm{d}X \qquad (5.14)$$

where we choose the integration limits 0 to a instead of $-a/2$ to $a/2$. Equation (5.14) is a generalized one-dimensional structure factor; analogous expressions exist in two and three dimensions. Substituting for $\rho(X)$, equivalent to $\psi(X)$, from (5.12), we have

$$F(h) = \int_0^a \sum_{h'=-\infty}^{\infty} G(h')\exp(-\mathrm{i}2\pi h'X/a)\exp(\mathrm{i}2\pi hX/a)\,\mathrm{d}X \qquad (5.15)$$

where h' indicates the range of values of h under the summation sign. Since the integral of a sum is equal to the sum of the integrals of the separate terms, we write

$$F(h) = \sum_{h'=-\infty}^{\infty} G(h') \int_0^a \exp[\mathrm{i}2\pi(h-h')X/a]\,\mathrm{d}X \qquad (5.16)$$

The integral evaluates to $\exp[\mathrm{i}2\pi(h-h')X/a]/(\mathrm{i}2\pi(h-h')/a)$. Since both h and h' are integers, the numerator of this expression is zero, except when $h' = h$. In this special case, we can see from (5.16) that the integral becomes $\int_0^a \mathrm{d}x$ which has the value a. It follows immediately that $G(h) = F(h)/a$, so that from (5.12) we'write

$$\rho(X) = \frac{1}{a}\sum_{h=-\infty}^{\infty} F(h)\exp(-\mathrm{i}2\pi hX/a) \qquad (5.17)$$

which is the Fourier transform of (5.14).

From Figure 3.28, we see that we can write

$$\begin{aligned} F(h) &= A(h) + \mathrm{i}B(h) \\ F(-h) &= A(h) - \mathrm{i}B(h) \end{aligned} \qquad (5.18)$$

so that (5.17) may be written as

$$\rho(X) = \frac{1}{a}\left\{ F(0) + \sum_{h=1}^{\infty}[A(h)+\mathrm{i}B(h)][\cos(2\pi hX/a) - \mathrm{i}\sin(2\pi hX/a)] \right.$$

$$\left. + \sum_{h=1}^{\infty}[A(h)-\mathrm{i}B(h)][\cos(2\pi hX/a) + \mathrm{i}\sin(-2\pi hX/a)] \right\} \qquad (5.19)$$

which reduces to

$$\rho(X) = \frac{1}{a} \left\{ F(0) + 2 \sum_{h=1}^{\infty} [A(h)\cos(2\pi h X/a) + B(h)\sin(2\pi h X/a)] \right\} \quad (5.20)$$

which may be compared with (5.2).

5.3.2 Two- and Three-Dimensional Functions

Analogous expressions can be formulated for two and three dimensions; we will consider the two-dimensional case in detail as the program system XRAY* (see Chapter 11) uses this form of the Fourier series. First, we state, by analogy with (5.17) and using fractional coordinates, the three-dimensional electron density equation as

$$\rho(xyz) = \frac{1}{V_c} \sum_{h=-\infty}^{\infty} \sum_{k=-\infty}^{\infty} \sum_{l=-\infty}^{\infty} F(hkl)\exp[-i2\pi(hx + ky + lz)] \quad (5.21)$$

Since, from Figure 3.28, $A(hkl) = |F(hkl)|\cos\phi(hkl)$ and $B(hkl) = |F(hkl)|\sin\phi(hkl)$, it follows from the form of (5.20) that $\rho(xyz)$ may be written as

$$\rho(xyz) = \frac{1}{V_c} \sum_{h=-\infty}^{\infty} \sum_{k=-\infty}^{\infty} \sum_{l=-\infty}^{\infty} |F(hkl)|\cos[2\pi(hx + ky + lz) - \phi(hkl)]$$

$$(5.22)$$

which serves to show how the electron density depends upon the phase angles: only $|F(hkl)|$ is measured by experiment, $\phi(hkl)$ must be determined before (5.22) can be summed. This situation constitutes the *phase problem* in crystallography, of which we shall hear more.

Equation (5.21) can be rewritten in the form of (5.20), which is most convenient for calculation by again making use of Friedel's law, namely that $|F(\bar{h}\bar{k}\bar{l})| = |F(hkl)|$, or $A(\bar{h}\bar{k}\bar{l}) = A(hkl)$ and $B(\bar{h}\bar{k}\bar{l}) = -B(hkl)$. Thus, we obtain

$$\rho(xyz) = \frac{1}{V_c} \left\{ F(000) + 2 \sum_{h=1}^{\infty} \sum_{k=-\infty}^{\infty} \sum_{l=-\infty}^{\infty} A(hkl)\cos[2\pi(hx + ky + lz)] \right.$$

$$\left. + B(hkl)\sin[2\pi(hx + ky + lz)] \right\} \quad (5.23)$$

Again, this equation could be put in the form of (5.22) if desired, and variations can be developed according to space-group symmetry, as discussed in Chapter 3 for structure factors.

We consider the two-dimensional equation. The generalized structure factor $F(hkl)$ is given by

$$F(hkl) = V_c \int_0^1 \int_0^1 \int_0^1 \rho(xyz)\exp[i2\pi(hx + ky + lz)]\,dx\,dy\,dz \quad (5.24)$$

we are using the limits 0 to 1 because we are now using fractional coordinates x, y, and z. For a projection along the z axis, we need the $F(hk0)$ reflections, where

$$F(hk0) = V_c \int_0^1 \int_0^1 \left\{ \int_0^1 \rho(xyz)\,dz \right\} \exp[i2\pi(hx + ky + lz)]\,dx\,dy \quad (5.25)$$

To interpret the integral over z, consider an element of structure of cross-sectional area $dx\,dy$ and length c along the z axis. In an element of length dz, the electron content is $\rho(xyz)c\,dz$, so that the total electron content in the element of length c is $c\,dx\,dy \int_0^1 \rho(xyz)\,dz$. Hence, the projected electron density at a point x, y is given by

$$\rho(xy) = c \int_0^1 \rho(xyz)\,dz\,dx\,dy$$

from which $F(hk0)$, or $F(hk)$ becomes

$$F(hk) = \mathcal{A} \int_0^1 \int_0^1 \rho(xy) \exp[i2\pi(hx + ky)] \quad (5.26)$$

since $\int_0^1 \exp(-i2\pi/z)\,dz$ is zero unless $l = 0$.
where $\mathcal{A}$ is the area of the a, b face of the unit cell. It follows that

$$\rho(xy) = \frac{1}{\mathcal{A}} \sum_{h=-\infty}^{\infty} \sum_{k=-\infty}^{\infty} F(hk) \exp[-i2\pi(hx + ky)] \quad (5.27)$$

which may be written more conveniently as

$$\rho(xy) = \frac{1}{\mathcal{A}} \left\{ F(000) + 2 \sum_{h=1}^{\infty} \sum_{k=-\infty}^{\infty} [A(hk) \cos(2\pi(hx + ky)) \right.$$

$$\left. + B(hk) \sin(2\pi(hx + ky))] \right\} \quad (5.28)$$

Unless one is concerned with absolute values of electron density, the $F(000)$ term can be omitted and $\rho(xy)$ scaled to a convenient maximum value.

We conclude this section by a consideration of the one-dimensional electron function which we discussed in Section 5.2. We draw Figure 5.1 again as Figure 5.5, now with the amplitudes and phases of the waves shown. Following the arguments above, we can write the electron density function in terms of amplitude and phase as

$$\rho(x) = \tfrac{1}{5}\{20 + 5\cos[2\pi(x) - 3.023] + 2.5\cos[2\pi(2x) - 2.400]$$

$$+ 4\cos[2\pi(3x) - 4.060]\}$$

This equation would then be evaluated at a suitable interval x, 20th of a would be appropriate, and the function plotted. Practice with Fourier series follows from the Problems section.

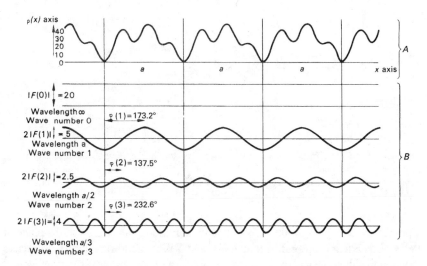

FIGURE 5.5. The periodic function of Figure 5.1, with amplitudes and phases for waves of indices 0, 1, 2, and 3; the amplitudes for the indices 1, 2, and 3 have been given twice their weight according to (5.20) for $\rho(X)$. The portion A may be regarded as the electron density $\rho(X)$ with a repeat distance a of 5 Å. The portion B corresponds to the structure factors, or harmonics, with their phases. X-ray diffraction may be regarded as the path $A \rightarrow B$. Structure determination is the path $B \rightarrow A$, and the need for correct phases is clear. The values for the phases given are based on their own wavelengths, 0, a, $2a$, and $3a$ in order to comply with the definition of phase (Section 3.2.3).

5.4 Holes and Atoms

When visible light is incident upon a circular hole in an otherwise opaque card, its diffraction pattern is somewhat diffuse. The theory of this scattering is complex, but it can be simplified by considering the pattern at an ideally infinite distance—the Fraunhofer diffraction pattern of the hole. Figure 5.6 shows a simple schematic experimental arrangement for viewing a Fraunhofer diffraction pattern. Parallel light from a laser source S is incident upon the object hole at O, which must be smaller than the diameter of the laser beam. The diffraction pattern can be viewed or photographed at F, the back focal plane of the lens L. The whole of the radiation is not completely in phase at F, and its intensity falls off with the distance from F. The larger the hole, the more rapid is the fall off, since the addition of waves at a given distance from F is then less complete.

The diffraction pattern of the hole has circular symmetry, and it may be represented by a radial distribution function of the form

$$A(\theta) = J_1(X)/X \tag{5.29}$$

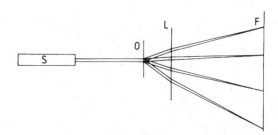

FIGURE 5.6. Simple experimental arrangement for producing Fraunhofer diffraction: *S*, helium–neon laser source; *O*, object; *L*, lens (focal length *ca* 1 m); *F*, back focal plane of lens *L*, where the diffraction pattern may be recorded. [Reproduced from *Diffraction*, by C. A. Taylor, with the permission of the Institute of Physics.]

where $A(\theta)$ is the amplitude of the scattered radiation at an angle θ to the plane of the hole, J_1 is a first-order Bessel function, and X is a function that is proportional to the radius of the hole and to $(\sin\theta)/\lambda$. Figure 5.7a illustrates the diffraction pattern of a hole; Figure 5.7b is a plot of (5.29), normalized to $A(\theta) = 6$ at $(\sin\theta)/\lambda = 0$, together with the atomic scattering factor curve for carbon, atomic number 6. Whereas $A(\theta)$ alternates in sign, the atomic scattering factor f does not, but at low values of $(\sin\theta)/\lambda$ the two curves are closely similar. The diffraction pattern of the hole is its Fourier transform, and (5.29) is a mathematical representation of the transform. Similarly, f is the transform of an atom, and the f-curve shows its variation with $(\sin\theta)/\lambda$.

5.5   Generalized Fourier Transform

We refer back to Figure 3.8, in Chapter 3, and some development thereof. Let the three-dimensional electron density for the body at the point A be $\rho(\mathbf{r})$ with respect to a single electron at the origin O. An element of volume δV around O has an electron content of $\rho(\mathbf{r})\delta V$, and its phase with respect to O is $2\pi\mathbf{r}\cdot\mathbf{S}$. Hence, the contribution of the quantity $\rho(\mathbf{r})\delta V$ to scattering in the direction θ is

$$\rho(\mathbf{r})\exp[\mathrm{i}2\pi(\mathbf{r}\cdot\mathbf{S})]\delta V \qquad (5.30)$$

Then, the total scattering for the body is

$$\int_V \rho(\mathbf{r})\exp[\mathrm{i}2\pi(\mathbf{r}\cdot\mathbf{S})]\,\mathrm{d}V \qquad (5.31)$$

where the integral extends over the volume V of the body. This expression is the Fourier transform of the body, and may be written

$$G(\mathbf{S}) = \int_V \rho(\mathbf{r})\exp[\mathrm{i}2\pi(\mathbf{r}\cdot\mathbf{S})]\,\mathrm{d}V \qquad (5.32)$$

(a)

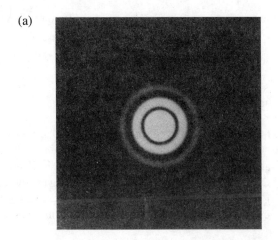

(b)

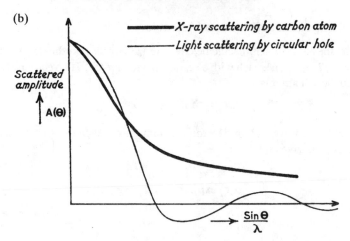

FIGURE 5.7.   Scattering from a circular hole. (a) Diffraction pattern for a hole of 5 mm diameter. [Reproduced from *Diffraction*, by C. A. Taylor, with the permission of the Institute of Physics.] (b) The scattered amplitude function $A(\theta)$ for a given hole radius, normalized to equal the value of f (carbon) at $(\sin \theta)/\lambda = 0$. [Reproduced from *Optical Transforms*, by C. A. Taylor and H. Lipson, with the permission of Routledge.]

Following Section 3.6.1, we have

$$I(\mathbf{S}) = G(\mathbf{S})G^*(\mathbf{S}) \tag{5.33}$$

So that $I(\mathbf{S})$ is the intensity of the transform. For an atom, we may usually assume spherical symmetry, so that the Fourier transform for an atom is just $f(\mathbf{S})$, where

f is the atomic scattering factor. Since centrosymmetry is a subgroup of spherical symmetry

$$f(\mathbf{S}) = f^*(\mathbf{S}) \qquad (5.34)$$

so that

$$f(\mathbf{S}) = [\, f(\mathbf{S}) f^*(\mathbf{S})]^{1/2} = [\, f^2(\mathbf{S})]^{1/2} \qquad (5.35)$$

In practice, $f(\mathbf{S})$ is defined to be positive for all values of $\sin \theta$. Strictly, there is a phase change of π when x-rays are scattered, as we discussed in Chapter 3, but since it is true for all atoms it is usually ignored, and $f(\mathbf{S})$ is given by the Fourier transform of (5.32), where $\rho(\mathbf{r}) = 4\pi r^2 \psi(r)$, $\psi(r)$ being an appropriate radial wave function, or a combinations of wave functions, for the atom. Evidently, we evaluated the Fourier transform $f(\mathbf{S})$ surreptitiously in Section 3.2.5 and Problem 3.3.

5.5.1   Fourier Transform of a Molecule

Let the atoms of a molecule be characterized by the coordinates X_j, Y_j, and Z_j ($j = 1, 2, 3, \ldots, n$), each with a scattering factor $f(\mathbf{S})$. The vector $\mathbf{r}_j$ from the origin to the jth atom is

$$\mathbf{r}_j = \mathbf{X}_j + \mathbf{Y}_j + \mathbf{Z}_j \qquad (5.36)$$

From (5.31), the wave scattered by the jth atom at a distance $\mathbf{r}$ from it, with respect to the origin, is given by

$$\int_V \rho(\mathbf{r}) \exp[\mathrm{i}2\pi(\mathbf{r} + \mathbf{r}_j) \cdot \mathbf{S}]\, \mathrm{d}V \qquad (5.37)$$

or

$$\int_V \rho(\mathbf{r}) \exp[\mathrm{i}2\pi(\mathbf{r} \cdot \mathbf{S})]\, \mathrm{d}V \, \exp[\mathrm{i}2\pi(\mathbf{r}_j \cdot \mathbf{S})] \qquad (5.38)$$

which becomes

$$f_j(\mathbf{S}) \exp[\mathrm{i}2\pi(\mathbf{r}_j \cdot \mathbf{S})] \qquad (5.39)$$

Thus, the total wave from all n discrete atoms in the molecule is its Fourier transform $G(\mathbf{S})$, given by

$$G(\mathbf{S}) = \sum_{j=1}^{n} f_j \exp[\mathrm{i}2\pi(\mathbf{r}_j \cdot \mathbf{S})] \qquad (5.40)$$

We write f_j for $f_j(\mathbf{S})$, because we have already decided that f_j is spherically symmetrical, and its variation with θ is inherent in f itself.

5.5.2   Fourier Transform of a Unit Cell

Let x_j, y_j, z_j now be the fractional coordinates of the jth atom in a unit cell, so that

$$\mathbf{r}_j = x_j\mathbf{a} + y_j\mathbf{b} + z_j\mathbf{c} \tag{5.41}$$

From the Bragg equation, $(2 \sin \theta)/\lambda = 1/d$, and the fact that

$$\mathbf{S} = \mathbf{d}^* = h\mathbf{a}^* + k\mathbf{b}^* + l\mathbf{c}^* \tag{5.42}$$

we have

$$\mathbf{r}_j \cdot \mathbf{S} = hx_j + ky_j + lz_j \tag{5.43}$$

Then, the total transform for the unit cell is

$$G(\mathbf{S}) = \sum_{j=1}^{n} f_j \, \exp[\text{i}2\pi(hx_j + ky_j + lz_j)] \tag{5.44}$$

which is identical to the structure factor equation (3.107) for the reflection hkl, where hkl is related to $\mathbf{S}$ through (5.42). The Fourier transform (5.44) is valid for all values of h, k, and l. In a crystal, however, because of interference, the transform can be observed only at those points where scattering is reinforced, that is, at the reciprocal lattice points hkl, where h, k, and l are necessarily integral. Thus, the structure factor equation for a crystal is its Fourier transform sampled at the reciprocal lattice points. We may imagine the Fourier transform for a molecule overlaid by the reciprocal lattice in the correct orientation: only those points that satisfy the limiting conditions for the space group could give rise to x-ray reflections.

5.6   Practice with Transforms

We can calculate transforms by (5.44), and we can prepare them experimentally by means of the optical diffractometer. Both techniques can provide useful results in developing Fourier transform theory.

5.6.1   Optical Diffractometer

The optical diffractometer permits the preparation of diffraction patterns in a relatively straightforward manner. In Figure 5.8, S is a helium–neon laser of ca 50 mW intensity, M_1 and M_2 are mirrors, E is an expander that extends the laser beam without loss of spatial coherence, L_1 and L_2 are lenses, and O is the object. The diffraction pattern is brought to a focus at F, the back focal plane of lens L_2, where it may be viewed with an eyepiece, photographed or input to a television camera; Figures 5.7a and 5.9 were produced with such an instrument.

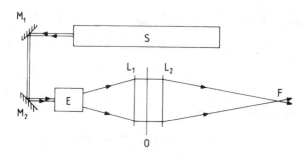

FIGURE 5.8.   Schematic arrangement of an optical diffractometer: S, helium–neon laser source; M_1 and M_2, mirrors; E, beam expander; O, object; L_1 and L_2, lenses; F, back focal plane of lens L_2, where the diffraction pattern may be recorded. [Reproduced from *Diffraction*, by C. A. Taylor, with the permission of the Institute of Physics.]

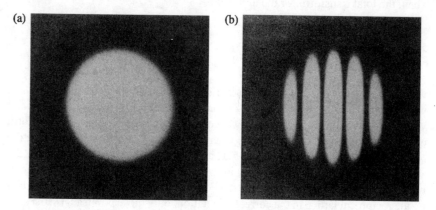

FIGURE 5.9.   Diffraction patterns from circular holes: (a) one hole 1 mm diameter, (b) two holes each of 1 mm diameter, set 3 mm apart in the horizontal direction. [Reproduced from *Diffraction*, by C. A. Taylor, with the permission of the Institute of Physics.]

5.6.2   Single Hole

The diffraction pattern for a single hole has been illustrated in Figure 5.7a, and equation (5.29), which represents the diffraction function, or transform, is plotted in Figure 5.7b. The diffraction pattern of a circular hole is often called the Airy disk, after its discoverer.

5.6.3   Two or More Holes

The Fourier transform of a single hole may be represented generally by (5.32), as well as by (5.29). A second hole, displaced by a vector distance **a** from the first, has an identical transform function but with a vector distance **r** + **a** from the origin. The total transform is the sum of the displaced and undisturbed transforms:

$$
\begin{aligned}
G_T(\mathbf{S}) &= \int_V \rho(\mathbf{r}) \exp[i2\pi(\mathbf{r} \cdot \mathbf{S})] \, dV + \int_V \rho(\mathbf{r}) \exp[i2\pi(\mathbf{r} + \mathbf{a}) \cdot \mathbf{S}] \, dV \\
&= \int_V \rho(\mathbf{r}) \exp[i2\pi(\mathbf{r} \cdot \mathbf{S})] \, dV \, \{1 + \exp[i2\pi(\mathbf{a} \cdot \mathbf{S})]\} \\
&= G_0(\mathbf{S})\{1 + \exp[i2\pi(\mathbf{a} \cdot \mathbf{S})]\}
\end{aligned}
\tag{5.45}
$$

where $G_0(\mathbf{S})$ is the transform of the undisturbed function. The term $\{1 + \exp[i2\pi(\mathbf{a} \cdot \mathbf{S})]\}$ is a fringe function modifying $G_0(\mathbf{S})$. It has the value 2 when $\mathbf{a} \cdot \mathbf{S}$ is integral, and is zero halfway between, as can be shown from de Moivre's theorem. Thus, the total transform is that of a single hole (Figures 5.7a or 5.9a) crossed by a system of planar fringes (Young's fringes), as shown by Figure 5.9b. The fringe system lies perpendicular to the direction joining the two holes. As the distance a is increased, the distance between the individual fringes becomes smaller, an example of the reciprocal nature of the diffraction process. The amplitudes of the fringes vary sinusoidally, alternate fringes having a relative phase difference of π.

The addition of further pairs of holes of different spacings and differing orientations gives rise to more fringe systems, all with the same reciprocal property discussed above. The complete diffraction pattern of a molecule may be thought of as a superposition of many sets of fringes. The sequence of optical transforms in Figure 5.10 shows not only an increase in the number of pairs of holes, as with the benzene ring itself, but also the effect of increasing the numbers of benzene ring entities in both one and two dimensions.

5.6.4 Change of Origin

If the origin to which a scattering species is referred be changed by the addition of a fixed vector $\mathbf{p}$ to all $\mathbf{r}_j$ vectors, then from (5.40)

$$
\begin{aligned}
G_p(\mathbf{S}) &= \sum_{j=1}^{n} f_j \exp[i2\pi(\mathbf{r}_j + \mathbf{p}) \cdot \mathbf{S}] \\
&= \sum_{j=1}^{n} f_j \exp[i2\pi(\mathbf{r}_j \cdot \mathbf{S})] \exp[i2\pi(\mathbf{p} \cdot \mathbf{S})] \\
&= G_0(\mathbf{S}) \exp[i2\pi(\mathbf{p} \cdot \mathbf{S})]
\end{aligned}
\tag{5.46}
$$

In (5.46), $G_0(\mathbf{S})$ is modified by the fringe function $\exp[i2\pi(\mathbf{p} \cdot \mathbf{S})]$. Since, in practice $\mathbf{p} \cdot \mathbf{S}$ is integral at points where the transform can be observed, it follows that $\exp[i2\pi(\mathbf{p} \cdot \mathbf{S})]$ has a magnitude of unity. Thus, the amplitude (and intensity) of

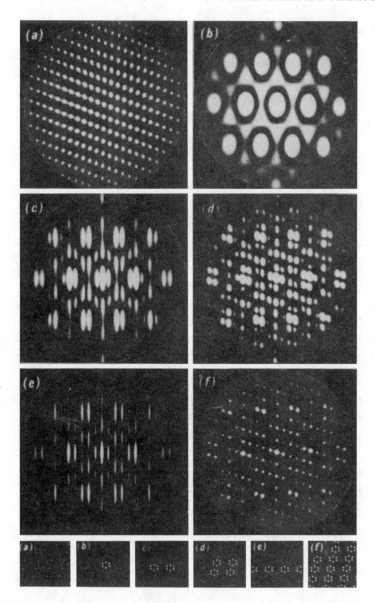

FIGURE 5.10.   Optical diffraction patterns illustrating scattering: (a) two-dimensional lattice (portion only); (b) single molecule (simulated benzene ring); (c) two such molecules; (d) four molecules; (e) a row (portion only) of six molecules; (f) a net (portion only) of molecules. [Reproduced from *Optical Transforms*, by C. A. Taylor and H. Lipson, with the permission of Routledge.]

the transform is invariant under translation, as can be demonstrated as described in Section 5.1. The phase, however, is dependent upon position, from which it is apparent that a true position cannot be determined without a knowledge of relative phases—the central problem in crystal structure determination, as we have remarked already.

5.6.5 Systematic Absences

At this point, we can show how the Fourier transform of a crystal can predict systematic absences. Suppose that the vector displacement $\mathbf{a}$ in (5.45) is replaced by $(\mathbf{a}/2 + \mathbf{b}/2)$, consistent with a C-centered unit cell. Then, we have

$$G_C(\mathbf{S}) = G_0(\mathbf{S})\left\{1 + \exp\left[i2\pi\tfrac{1}{2}(\mathbf{a} + \mathbf{b}) \cdot \mathbf{S}\right]\right\} \qquad (5.47)$$

which, from (5.42), becomes

$$G_C(\mathbf{S}) = G_0(\mathbf{S})\{1 + \exp[i\pi(h + k)]\} \qquad (5.48)$$

For $(h + k)$ odd, $G_C(\mathbf{S})$ is identically zero, but it equals $2G_0(\mathbf{S})$ for $(h + k)$ even, which we recognize as characteristic for a C-centered unit cell (see Section 3.7). Equations like (5.48) can be developed for all translational symmetries. Such results show clearly those reciprocal lattice points that cannot be sampled, whatever the nature of the contents of the unit cell.

5.6.6 Reconstruction of the Image

Consider again the diffraction pattern of the hypothetical two-dimensional crystal in Figure 5.10f. The lattice in this example is defined by the separations of the simulated benzene molecules. The weighted reciprocal lattice is the transform of a single molecule (Figure 5.10b) crossed by fringes, the principal sets of which are governed by the basic translations of the lattice (Figure 5.10a). Again, we see the continuous transform decomposed into, or sampled as, a spot pattern.

In reconstructing an image from its diffraction pattern, our object is to attain, say, the structure of Figure 5.10f given only its diffraction pattern, or transform. The fringes themselves are rather like a diffraction grating. A set of true sinusoidal fringes used as a mask would produce a diffraction pattern that is the original two-hole object. Thus, in reconstructing the object from its diffraction image, we are really seeking the diffraction pattern of the diffraction pattern, or the transform of the transform. We may express this result mathematically in the following manner.

The transform of (5.32) will, for a crystal, give an expression for the distribution of the electron density $\rho(\mathbf{r})$ in the form (note the change of sign of the exponent)

$$\rho(\mathbf{r}) = \int_v G(\mathbf{S}) \exp[-i2\pi(\mathbf{r} \cdot \mathbf{S})]\, dv \qquad (5.49)$$

where the integral extends over a volume v in reciprocal space. Using (5.40) in the context of the unit cell of a crystal, we write

$$\rho(\mathbf{r}) = \int_v F(hkl) \exp[-i2\pi(\mathbf{r} \cdot \mathbf{S})] \, dv \qquad (5.50)$$

Equation (5.49), or the equivalent (5.50), is the Fourier transform of (5.32). Thus, the transform in Figure 5.10f corresponds to the calculation in (5.32), whereas the structure in Figure 5.10f is the result of evaluating (5.49) or (5.50).

From the $F(h0l)$ weighted reciprocal lattice section of the crystalline platinum derivative of phthalocyanine, a mask (Figure 5.11a) was prepared in which the relative amplitudes of the structure factors were indicated by the sizes of the holes. When this mask was used in the optical diffractometer, the transform obtained was that shown by Figure 5.11b. Comparison with the electron density map from the fully solved structure (Figure 5.11h) shows that the recombination obtained furnishes a good reconstruction of the molecule that, in the crystal, gave the spot pattern from which the mask was prepared. One will ask immediately how the phase problem has been overcome in this reconstruction, as it appears not to have been even considered.

We have, in this example, a rather special case. In the crystal of platinum phthalocyanine, the space group is $P2_1/c$ and there are two molecules in the unit cell. Thus, the platinum atom occupies the origin ($\bar{1}$) in the asymmetric unit, and makes a positive contribution to all structure factors. It is such a heavy atom (high atomic number) that it dominates the contributions from all other atoms in the structure so causing all $F(h0l)$ structure factors to have a positive sign (zero phase angle). Thus, all $h0l$ structure factors have the same relative phase, and the transform of the diffraction pattern gives a true representation of the structure without further reference to phase.

Figure 5.11c–g and c′–g′ shows different portions of the $h0l$ reciprocal lattice section and the corresponding transforms. The effect of the cut-off on the resolution is well illustrated.

Representation of Fourier Transforms

If an object is centrosymmetric and the origin is taken at a center of symmetry, then the Fourier transform is real (see Section 3.6.2) and only one diagram is needed to represent it. This situation exists for the $h0l$ section of the transform of platinum phthalocyanine (Figure 5.11a). If the object is not centrosymmetric, then it is necessary to use two diagrams to display the transform, either amplitude and phase or real and imaginary parts. The phase, or the real and imaginary components, will vary according to the choice of origin, but the amplitudes remain invariant under change of origin. (See Section 5.6.4.)

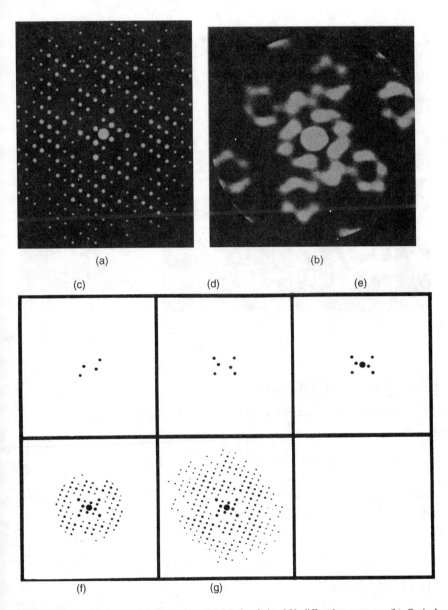

(a)   (b)

(c)   (d)   (e)

(f)   (g)

FIGURE 5.11.   Platinum phthalocyanine. (a) Mask of the $h0l$ diffraction pattern; (b) Optical transform of (a), showing a complete molecule and portions of neighboring molecules in the (projected) crystal structure. [After Dunkerley and Lipson, *Nature*, **176**, 81 (1955).] (c)–(g) Increasingly large portions of the diffraction pattern; (c′)–(g′) Corresponding transforms—the effect of the cut-off of the pattern on the resolution in the transform is evident; (h) Electron density contour map. All the diffraction patterns relate to the $h0l$ data, so that the corresponding transforms are x, z projections in real space [after Taylor and Lipson, *loc. cit.*].

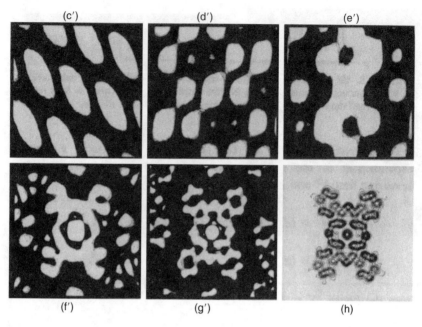

FIGURE 5.11.   (Continued)

5.6.7   Transforms and Inverse Transforms

Consider a one-dimensional square-wave function defined by

$$f(x) = \begin{cases} 2 & \text{for } -a/2 \le x \le a/2 \\ 0 & \text{for } -a/2 > x > a/2 \end{cases}$$

which has the form shown in Figure 5.12. The Fourier transform in the one dimension x is, from (5.32),

$$G(S) = \int_{-a/2}^{a/2} 2 \exp(\mathrm{i}2\pi S x)\, dx \tag{5.51}$$

which is solved readily to give

$$G(S) = 2 \sin(\pi S a)/(\pi S) = 2a \sin(\pi S a)/(\pi S a) \tag{5.52}$$

This transform, in which the function $\sin(\pi S a)/(\pi S a)$ is typical for the transform of a pulse waveform, has the form shown in Figure 5.13, where $G(S)$ is plotted as a function of S in units of $1/a$. It may be noted that, characteristically, the length $2/a$

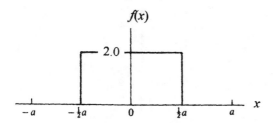

FIGURE 5.12.   Square wave of amplitude 2.0, defined for the period $-a/2$ to $a/2$; $f(x) = 2.0$ for $-a/2 \leq x \leq a/2$; $f(x) = 0$ for $|x| > a/2$.

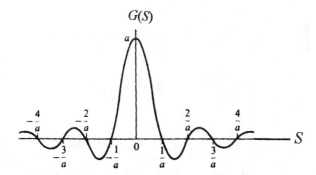

FIGURE 5.13.   Fourier transform of the square wave in Figure 5.12. Subsidiary, decreasing maxima arise at intervals $(2n + 1)a/2$ $(n = 1, 2, \ldots)$.

between the first two nodes on each side of the central maximum is the reciprocal of the width of the function $f(x)$. We would have obtained the same result if we had used the real part of $\exp(i2\pi Sx)$ in (5.51), that is, $2 \int_0^{a/2} 2\cos(2\pi Sx)\,dx$, because Figure 5.12 indicates an *even* function: $f(-x) = f(x)$.

Consider next the function

$$f(x) = \begin{cases} 1 - 2|x| & \text{for } -\tfrac{1}{2} \leq x \leq \tfrac{1}{2} \\ 0 & \text{for } |x| > \tfrac{1}{2} \end{cases}$$

which has a saw-tooth wave form, Figure 5.14. Because of the symmetry of this function about the point $x = 0$, the Fourier transform of this function has the general form

$$G(S) = 2 \int_{-1/2}^{0} (1 + 2x)\cos(2\pi Sx)\,dx + 2 \int_{0}^{1/2} (1 - 2x)\cos(2\pi Sx)\,dx$$

$$(5.53)$$

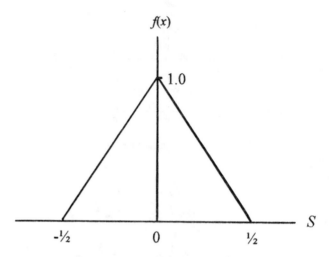

FIGURE 5.14.   Saw-tooth wave form defined for the period $-\frac{1}{2}$ to $\frac{1}{2}$; $f(x) = 1 - 2|x|$ for $|x| \leq 1/2$; $f(x) = 0$ for $|x| > \frac{1}{2}$.

which may be simplified to $2 \int_0^{1/2} \cos(2\pi S x) \, dx - 4 \int_0^{1/2} x \cos(2\pi S x) \, dx$. The first of these two integrals solves to $\sin(\pi S)/(\pi S)$. Integrating by parts, the second integral gives $(1/\pi^2 S^2)[1 - \cos(\pi S)] - (1/\pi S) \sin(\pi S)$, so that the total result is

$$G(S) = (1/\pi^2 S^2)[1 - \cos(\pi S)] \tag{5.54}$$

It is left as an exercise to the reader to plot this transform, as a function of S, ($S = 0, 0.05, 0.10, \ldots$), and to show that (5.54) is obtained also if $\exp(i2\pi S x)$ is used in place of $\cos(2\pi S x)$ in (5.53). The plot of the transform should have a maximum at $S = 0$, zeroes at $S = 2n$ ($n = 1, 2, 3, \ldots$), and small decreasing maxima at $S = 2n + 1$ ($n = 1, 2, 3, \ldots$).

A function and its Fourier transform are reciprocally related; we noted this feature in studying Fourier series in Section 5.3. We can illustrate this property by means of two programs, TRANS1* and FOUR1D*, that are part of the suite of programs that is included with this book.

Consider the one-dimensional, periodic function $f(X)$ listed in Table 5.2 at 30ths of the repeat distance along the X axis, from $n = 0$ to 30/30. It can be represented by a series of sine and cosine terms with coefficients determined by the program TRANS1* (see Chapter 11), which calculates the Fourier transform of the function. When this program is executed on the data shown in Table 5.2, with $h_{max} = 10$, the coefficients listed in Table 5.3 are obtained. *Note that only the values of the function f(x) are entered as data.* If these coefficients are used with the program FOUR1D*, which calculates a one-dimensional Fourier series, the initial function is regained (Table 5.4): the transform of the transform is the

TABLE 5.2. Periodic function $\rho(X)$, listed at 30ths (n) of the repeat distance along the X axis

n	$f(x)$	n	$f(x)$	n	$f(x)$	n	$f(x)$
0	50	1	52	2	57	3	63
4	69	5	76	6	81	7	84
8	85	9	83	10	79	11	73
12	66	13	60	14	50	15	42
16	37	17	32	18	41	19	62
20	90	21	117	22	126	23	123
24	113	25	95	26	78	27	65
28	56	29	51	30	50		

TABLE 5.3. Coefficients $A(h)$ and $B(h)$, for $h = 0$ to 10, transformed from $\rho(X)$ data in Table 5.2 using TRANS1*

h	$A(h)$	$B(h)$	h	$A(h)$	$B(h)$	h	$A(h)$	$B(h)$	h	$A(h)$	$B(h)$
0	71.87	0.00	1	2.22	−4.61	2	−15.08	−2.53	3	0.55	5.84
4	2.14	−1.54	5	−0.99	−0.21	6	0.19	0.36	7	0.15	−0.25
8	−0.26	−0.04	9	−0.07	0.13	10	0.43	0.06			

original function. Further details on the manipulation of these programs are given in Chapter 11. However, we mention here that the value of h should be below the critical value of $\Delta/2$, where Δ is the sampling interval, 30 in this example. The *sampling theorem* states that if the Fourier transform of a function is zero for all frequencies greater than a critical frequency f_c, that is, its bandwidth is limited to frequencies smaller than f_c, then the continuous function can be determined from a knowledge of its sampled values. The function in Table 5.2 is not of this character, so that the maximum frequency h used should be $\frac{1}{4}$ to $\frac{1}{3}$ of the sampling interval, 30. The reader may care to investigate this property by using the programs with different values of h. Further reading about sampling may be found in the literature.[a] This problem is rarely manifested in the normal Fourier syntheses of x-ray crystallography. The experimental $|F_o|$ data fall towards zero at the higher values of h, k, and l because of the attenuation arising from the temperature factor effects on the atomic scattering factors (see Section 3.9.8). In working with the sharpened Patterson function and the E-map (q.v.), the sharpening inherent in the coefficients of the former reduces the tendency to zero of the higher order spectra, whereas in the latter the coefficients in the Fourier transformation to an E-map

[a] See Bibliography (Brigham).

TABLE 5.4.   Transformation of the Coefficients $A(h)$ and $B(h)$ from Table 5.3, using FOUR1D*, to give the Original Function $f(x)$ of Table 5.2

n	$f(x)$	n	$f(x)$	n	$f(x)$	n	$f(x)$
0	50.43	1	51.95	2	56.47	3	63.52
4	69.15	5	75.50	6	81.72	7	83.67
8	84.48	9	83.91	10	78.39	11	72.55
12	67.43	13	59.02	14	50.57	15	43.00
16	34.66	17	32.62	18	41.49	19	60.90
20	90.76	21	117.64	22	126.27	23	122.34
24	112.70	25	95.35	26	77.50	27	65.48
28	55.99	29	50.54	30	50.43		

are both sharpened and significantly decreased in number. We consider in later chapters how these cases are treated in practice.

5.6.8   Delta Function

Another important function, considered in Chapter 3, is the Gaussian distribution

$$f(x) = (1/k) \exp(-\pi x^2 / k^2) \tag{5.55}$$

where k may be regarded as the width of the function. In finding the Fourier transform of this function, we can make use of the fact that it is an even function, so that we use the cosine part of $\exp(i2\pi Sx)$ and determine twice the sum from zero to infinity, that is,

$$\frac{1}{k} \int_{-\infty}^{\infty} \exp(-\pi x^2 / k^2) \exp(i2\pi Sx) \, dx$$

$$= \frac{2}{k} \int_{0}^{\infty} \exp(-\pi x^2 / k^2) \cos(2\pi Sx) \, dx$$

From tables of standard integrals, or otherwise,[a] $\int_{0}^{\infty} \exp(-a^2 x^2) \cos(bx) \, dx = (\pi/4a^2)^{1/2} \exp(-b^2/4a^2)$, so that the required Fourier transform becomes

$$G(S) = \exp(-\pi k^2 S^2) \tag{5.56}$$

Thus, the transform of a Gaussian function is another Gaussian, in reciprocal space, of a width $1/k$, the reciprocal of the width k in real space.

The integral

$$\frac{1}{k} \int_{-\infty}^{\infty} \exp(-\pi x^2 / k^2) \, dx$$

[a] *Handbook of Chemistry and Physics*, edited by R. C. Weast, Chemical Rubber Co. Cleveland, Ohio (1974–1975).

evaluates to $(1/\pi)^{1/2}\Gamma\left(\frac{1}{2}\right)$, which is unity (normalized) for all values of k. Consider next the function

$$\delta(x) = \lim_{k \to 0} (1/k) \exp(-\pi x^2/k^2) \qquad (5.57)$$

This function has the following properties:

$$\delta(x) = \begin{cases} 0 & \text{for } x \neq 0 \\ \infty & \text{for } x = 0 \end{cases}$$

$$\int_{-\infty}^{\infty} \delta(x)\,dx = 1$$

and is known as the (Dirac) δ-function; it corresponds to an infinitely sharp line of unit weight at the origin. As the width k in (5.57) tends to zero, so the transform of the δ-function tends to unity, and in the limit where $k = 0$ reaches it.

This result follows, since

$$\int_{-\infty}^{\infty} f(x)\delta(x)\,dx = f(0)$$

and if we let $f(x) = \exp(\mathrm{i}2\pi Sx)$, then

$$\int_{-\infty}^{\infty} \exp(\mathrm{i}2\pi Sx)\delta(x)\,dx = \exp(0) = 1$$

If the δ-function is located at $x = x_0$, we have $f(x) = \delta(x - x_0)$. The Fourier transform of $f(x)$ is then

$$G(S) = \int_{-\infty}^{\infty} \delta(x - x_0) \exp(\mathrm{i}2\pi Sx)\,dx$$

$$= \int_{-\infty}^{\infty} \delta(x) \exp[\mathrm{i}2\pi S(x + x_0)\,dx$$

$$= \exp(\mathrm{i}2\pi Sx_0)$$

Thus, the Fourier transform of a δ-function at a point x_0 is equal to $\exp(\mathrm{i}2\pi Sx_0)$; when the δ-function is at the origin, its Fourier transform is at unity for all values of S.

Of particular interest is a set of δ-functions that define a one-dimensional lattice of spacing a. Its Fourier transform will be another set of δ-functions of spacing $1/a$ which define the corresponding one-dimensional reciprocal lattice. While the above discussions have been confined, for convenience, to one-dimensional space, the results are equally true in higher dimensions, and we shall consider such applications in later sections.

5.6.9 Weighted Reciprocal Lattice

As a final, practical example, we illustrate the power of the Fourier transform with the crystal structure of euphenyl iodoacetate, Figures 1.2 and 1.3. The crystal is monoclinic, with unit-cell dimensions $a = 7.260, b = 11.547, c = 19.217$ Å, $\beta = 94.10°$. There are two molecules in the unit cell, and systematic absences indicated space group $P2_1$ or $P2_1/m$. The latter is not possible, because the two molecules in the unit cell would have to lie on special positions of symmetry either m or $\bar{1}$. The chiral nature of the molecule, a tetracyclic triterpene, precludes both of these symmetries, so that the two molecules lie in general positions in space group $P2_1$.

Figure 5.15 is a reconstruction of the x-ray photograph, or weighted reciprocal lattice of the $h0l$ section for this crystal; this section is centric, symmetry $p2$. Since the iodine atom is very much heavier that the other atoms present in the

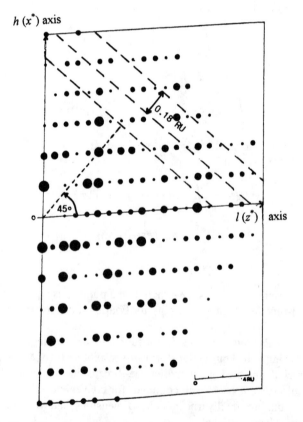

FIGURE 5.15. The $h0l$ section of the weighted reciprocal lattice of euphenyl iodoacetate; the I–I fringe spacing (three lines of zero amplitude are indicated) is 0.18 reciprocal lattice units, and the I–I vector makes an angle of *ca* 45° with the z^* axis (It may help to inspect the diagram edgewise).

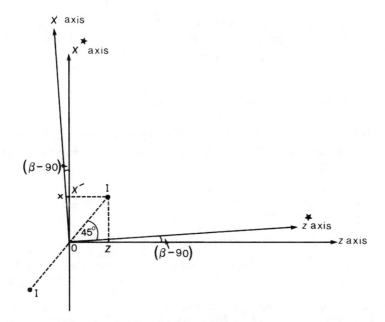

FIGURE 5.16.   Geometry of the reconstruction of the coordinates of the iodine atoms in euphenyl iodoacetate. The β angle is $94.1°$; X, X', and Z are measured along the x, x^*, and z axes, respectively. Note that since the space group of euphenyl iodoacetate is $P2_1$, the y coordinates of the iodine atoms may be set at $\pm\frac{1}{4}$, so as to fix the position of the origin with respect to the symmetry elements.

crystal, the unit cell may be regarded, to a first approximation, as a pair of iodine atoms distant d apart.

From Section 5.6.3, we would expect to find a set of fringe systems running through the transform, arising from the heavy iodine atoms. In Figure 5.15, the two most obvious fringe systems at an angle of $85.90°$ to each other arise from the unit-cell translations a and c. In addition, the dashed lines indicate another prominent fringe system, that of the iodine atoms. Note that the system is more prominent at higher angles, because the scattering from heavy atoms falls of relatively much less than that from the lighter atoms present in the structure.

The perpendicular spacing of the fringes is 0.18 reciprocal units which, because Cu $K\alpha$ x-radiation ($\lambda = 1.5418$ Å) was used in the experiment, corresponds to an I–I spacing of 8.565 Å (in projection) lying in a direction normal to the fringe system, that is, at approximately $45°$ to z^*, Figure 5.16. Measuring X' and z along the x^* and z axes respectively, we have

$$X' = (d/2)\sin(45 + \beta - 90)° = 3.237 \text{ Å}$$

$$Z = (d/2)\cos(45 + \beta - 90)° = 2.804 \text{ Å}$$

From the figure, $X = X' \cos(\beta-90)°$, whence $X = 3.229$ Å. Hence, the fractional coordinates are $x = 0.445$ and $z = 0.146$. After the structure was fully solved, the refined values of these coordinates were $x = 0.4274$ and $z = 0.1431$.

5.7 Some General Properties of Transforms

We summarize here some of the more important properties of Fourier transforms in their applications to x-ray crystallography; for each property the spaces may be interchanged.

Operation in Bravais space	Result in reciprocal space
Rotation about an axis	Rotation about a parallel axis at the same speed
Change of scale in a given direction	Reciprocal change of scale in the same direction
Translation	Modulus unchanged; phase modified by a fringe function
Addition of n units	Vector summation of n transforms referred to a common origin
(a) Two parallel units	Transform for one unit crossed by parallel, planar fringes; maximum amplitude doubled
(b) Two units related by a center of symmetry	Transform for one unit crossed by wavy fringes which may be approximately planar in limited regions
Convolution of two functions	Transform is the product of the individual transforms

The discussion of the convolution of two functions is our next task.

5.8 Convolution

We consider here the last of the properties of transforms listed in the previous section. The convolution integral is often called the "folding" integral, for a reason that will become clear as we continue the discussion.

5.8.1 Convolution and Diffraction

The Fourier transform of a slit diffraction grating of a given width can be considered in terms of the diffraction at a single slit of that width, together with that of an ideal, infinite grating. We have, first, the diffraction pattern, or transform, of the single slit, shown by Figure 5.17a, and then the transform of the ideal grating, Figure 5.17b. These two transforms, in reciprocal space, are multiplied, point by point, to give the product in real space (Figure 5.17c), which is called the *convolution* of the two functions. In a crystal, we have the contents of the unit cell and a point function, the lattice, in Bravais space. The convolution of these two functions is the diffraction pattern in reciprocal space. The transform of a lattice is another set of points, the reciprocal lattice points, and it has unit

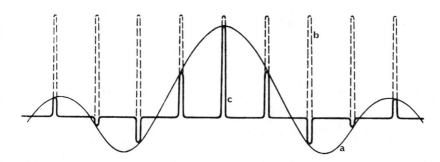

FIGURE 5.17. Relationship between amplitudes of diffraction. (a) Single slit of finite width (thin line); (b) Ideal slit grating (dashed line); (c) Grating of finite-width slits (bold line). The transform of the grating of finite-width slits is the convolution of the transforms of the single slit and the ideal grating. [Reproduced from *Optical Transforms*, by C. A. Taylor and H. Lipson, with the permission of Routledge.]

value at each point and zero value elsewhere. The product of the two transforms is the transform of the contents of the unit cell sampled at the reciprocal lattice points.

The image of a diffraction spectrum is influenced also by the divergence of the incident x-ray beam, the crystal shape and the mosaic structure of the crystal, because all of these factors have a bearing on the Fourier transform of the crystal.

5.8.2 Convolution Integral

The convolution of two functions is a physical concept of significance in divers scientific fields. We consider the process of convolution now in a little more detail. Consider two functions $f(\mathbf{r})$ and $g(\mathbf{r})$ in Bravais (real) space, and let their transforms in reciprocal space be $F(\mathbf{S})$ and $G(\mathbf{S})$, respectively. The Fourier transform of the product of these two transforms may be written as

$$T(\mathbf{r}) = \int F(\mathbf{S})G(\mathbf{S})\exp(-\mathrm{i}2\pi\mathbf{r}\cdot\mathbf{S})\,\mathrm{d}V_{\mathbf{S}} \qquad (5.58)$$

and is a function in Bravais space. From earlier sections we know that $F(\mathbf{S})$ is given by

$$F(\mathbf{S}) = \int f(\mathbf{r}')\exp(\mathrm{i}2\pi\mathbf{r}'\cdot\mathbf{S})\,\mathrm{d}V_{\mathbf{r}} \qquad (5.59)$$

where $\mathbf{r}'$ is a vector different from $\mathbf{r}$ in general, but ranging in the same Bravais space, and the change of sign arises because we are considering the reverse transformation. Thus, we have

$$T(\mathbf{r}) = \int G(\mathbf{S}) \left\{ \int f(\mathbf{r}') \exp(\mathrm{i}2\pi \mathbf{r}' \cdot \mathbf{S}) \, dV_{\mathbf{r}'} \right\} \exp(-\mathrm{i}2\pi \mathbf{r} \cdot \mathbf{S}) \, dV_{\mathbf{S}}$$

$$= \int f(\mathbf{r}') \left\{ \int G(\mathbf{S}) \exp[-\mathrm{i}2\pi (\mathbf{r} - \mathbf{r}') \cdot \mathbf{S}] \, dV_{\mathbf{S}} \right\} dV_{\mathbf{r}'}$$

$$= \int f(\mathbf{r}') g(\mathbf{r} - \mathbf{r}') \, dV_{\mathbf{r}'} \tag{5.60}$$

which is the convolution $f(\mathbf{r}) * g(\mathbf{r})$, or $c(\mathbf{r})$; $g(\mathbf{r} - \mathbf{r}')$ is the function $g(\mathbf{r})$ with its origin moved from $\mathbf{r} = 0$ to $\mathbf{r} = \mathbf{r}'$.

We can give a physical interpretation to this process. In Figure 5.18a, an element $\delta \mathbf{r}'$ of the function $f(\mathbf{r}')$ is defined. According to (5.60) each such element of $f(\mathbf{r}')$ must be multiplied by the sharp function $g(\mathbf{r} - \mathbf{r}')$ in Figure 5.18b before integration. In Figure 5.18(c), this process is shown for three elements, with the result in Figure 5.18(d). The completed convolution is shown by Figure 5.18e; it is evident that the function $f(\mathbf{r}')$ has been repeated at each value

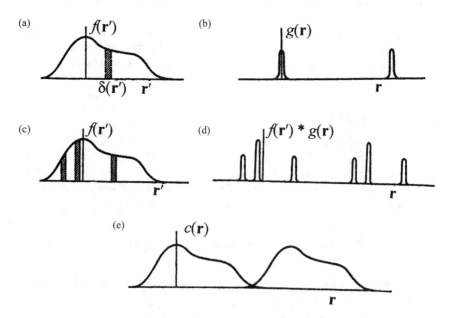

FIGURE 5.18. Convolution of two functions in the one-dimension space of $\mathbf{r}$. (a) Function $f(\mathbf{r}')$, showing an element of width $\delta \mathbf{r}'$; (b) Function $g(\mathbf{r})$, showing two sharp peaks; (c) Function $f(\mathbf{r}')$ with three selected elements $\delta \mathbf{r}'$; (d) Convolution of (b) and (c); the three elements are reproduced by $g(\mathbf{r})$, modified by the values of that function; (e) the completed convolution $f(\mathbf{r}') * g(\mathbf{r})$; because the function $g(\mathbf{r})$ does not have infinitesimal width, $f(\mathbf{r}')$ is reproduced in a slightly modified form.

of $g(\mathbf{r})$. Note that $g(\mathbf{r} - \mathbf{r}')$ is $g(\mathbf{r}')$ mirrored, or *folded*, across the ordinate axis and shifted by an amount $\mathbf{r}$.

The converse of (5.60) is equally true, that is,

$$T[c(\mathbf{r})] = F(\mathbf{S})G(\mathbf{S}) = \int \left\{ \int f(\mathbf{r}')g(\mathbf{r} - \mathbf{r}')\,dV_{\mathbf{r}'} \right\} \exp(-i2\pi\mathbf{r} \cdot \mathbf{S})\,dV_{\mathbf{r}}$$

$$= \int f(\mathbf{r}') \left\{ \int g(\mathbf{r} - \mathbf{r}')\,dV_{\mathbf{r}} \right\} \exp(-i2\pi\mathbf{r} \cdot \mathbf{S})\,dV_{\mathbf{r}'} \qquad (5.61)$$

If we now let $\mathbf{r} - \mathbf{r}' = \mathbf{r}''$, where $\mathbf{r}''$ is another independent variable in the same space as $\mathbf{r}$ and $\mathbf{r}''$. Hence,

$$T[c(\mathbf{r})] = \int f(\mathbf{r}')\exp(-i2\pi\mathbf{r}' \cdot \mathbf{S})\,dV_{\mathbf{r}'} \int g(\mathbf{r}'')\exp(-i2\pi\mathbf{r}'' \cdot \mathbf{S})\,dV_{\mathbf{r}'}$$

$$= F(\mathbf{S})G(\mathbf{S}) \qquad (5.62)$$

so that the transform of the convolution is the product of the individual transforms.

5.8.3 Crystal Structure and Convolution

We have shown in the previous section, and illustrated in Figure 5.18, that the convolution, in one-dimensional space, of a function $f(\mathbf{r})$ with a function $g(\mathbf{r})$ that consists of two sharp peaks, results in $f(\mathbf{r})$ being repeated, modified slightly in accordance with the width of $g(\mathbf{r})$, at the two locations in $g(\mathbf{r})$. A pair of δ-functions can be derived from $g(\mathbf{r})$ by allowing the width of the sharp peaks, normalized to unit area, to be reduced while maintaining the area at unity. In the limit as the width becomes infinitely small, the function becomes infinitely high at each location. We now have a pair of δ-functions that have the value unity at each location but zero otherwise. If these δ-functions were to be convolved with $f(\mathbf{r})$, then $f(\mathbf{r})$ would be repeated exactly at the locations of the $\mathbf{r}$ vectors. We may write this process in mathematical terms in the following way. Initially, we have the convolution expressed as

$$c(\mathbf{r}) = \int f(\mathbf{r}' - \mathbf{r})g(\mathbf{r})\,dV_{\mathbf{r}}$$

where $g(\mathbf{r})$ is normalized to unit area. As the width tends to infinitesimal size, so we have

$$c(\mathbf{r}) = \int f(\mathbf{r}' - \mathbf{r})\delta(\mathbf{r})\,dV_{\mathbf{r}} = f(\mathbf{r}') \qquad (5.63)$$

from Section 5.6.8. If we translate the δ-function by a vector $\mathbf{r}''$ along the positive axis, so that the δ-function is defined as $\delta(\mathbf{r} - \mathbf{r}'')$, then the convolution becomes

$$c(\mathbf{r}) = \int f(\mathbf{r}' - \mathbf{r})\delta(\mathbf{r} - \mathbf{r}'')\,dV_{\mathbf{r}} \qquad (5.64)$$

so that the function $f(\mathbf{r}')$ is reproduced at $\mathbf{r}''$ from the origin, that is, it is the function $f(\mathbf{r} - \mathbf{r}'')$. It is straightforward to extend the argument to a one-dimensional lattice. The convolution of $f(\mathbf{r}')$ with an infinite array of δ-functions, $g(\mathbf{r}) = \sum_{j=-\infty}^{\infty} \delta(\mathbf{r} - j\mathbf{r}'')$, is given by

$$c(\mathbf{r}) = \sum_{j=-\infty}^{\infty} f(\mathbf{r} - j\mathbf{r}'') \tag{5.65}$$

which is periodic in $\mathbf{r}''$, Figure 5.19.

Thus, a crystal structure may be regarded as the convolution of a unit cell, defined by vectors $\mathbf{a}$, $\mathbf{b}$, and $\mathbf{c}$, and a three-dimensional set of δ-functions that have the value unity at each Bravais lattice points but zero elsewhere.

The transform of a unit cell contents may involve one or more molecules, and is a continuous function in reciprocal space. However, the conditions for interference lead to the fact that this transform may be sampled, experimentally, only at the reciprocal lattice points, so that the convolution of the unit cell and the lattice of δ-functions is the product of the transform of the unit cell contents and the reciprocal lattice, and is manifested in the experimental diffraction pattern.

The essence of this process may be grasped from Figure 5.20, in which the transform of the contents of the centrosymmetrical unit cell of naphthalene, projected on to the ac plane, has been overlaid with a drawing of the x^*, z^* reciprocal lattice in the correct orientation. The weights of the transform at the reciprocal lattice points $hk0$ correspond to the values of $|F(hk0)|$.

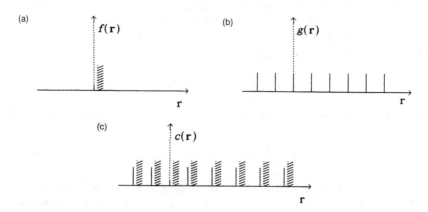

FIGURE 5.19.   Convolution: (a) Structural entity $f(\mathbf{r})$; (b) The function $g(\mathbf{r})$ is an infinite set of δ-functions $\delta(\mathbf{r})$; (c) Convolution $c(\mathbf{r}) = f(\mathbf{r}) * \delta(\mathbf{r})$.

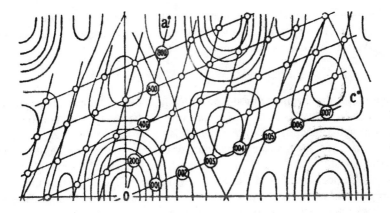

FIGURE 5.20.   Fourier transform of the $h0l$ projection of the centrosymmetric structure of naphthalene, showing the a^* and c^* directions. The transform is overlaid with a drawing of the reciprocal lattice in the correct orientation. The amplitudes $|F(h0l)|$ are the amplitudes of the transform at the reciprocal lattice points. Reflections such as 201, 202, and 801 are sensitive to small changes in the orientation of the reciprocal lattice, whereas reflections such as 203, 603, and 802 are relatively insensitive under rotation.

5.9   Structure Solution in Brief

An x-ray diffraction pattern can be recombined to form an image of the object from which it was derived provided that we know the indices h, k, l, the amplitude $|F|$ and the relative phase ϕ for each reflection. The absence of direct measurements of the phases means that they must be obtained indirectly. If it can be done satisfactorily, a Fourier transform (Fourier series) can be used to convert the reciprocal-space diffraction pattern into a real-space object, the crystal structure. This section is just a brief overview of some of the problems in crystal structure analysis, and how they may be overcome.

5.9.1   Use of Heavy Atoms

We considered a special case of this method in Section 5.6.6, where all $F(h0l)$ data for platinum phthalocyanine were positive in sign. In a more general case, the heavy atom (which acts like the speck of dust on the transparency, Section 5.1) will lie at a general position in the unit cell, probably related to one or more atoms by symmetry. Assuming for the moment we can determine the coordinates of the heavy atoms in the unit cell (see, for example, Section 5.6.9), we can use them to calculate approximate phases for the reflections. Then, the Fourier transform (5.50) becomes a summation, or series, with coefficients of the form

$|F_o|\exp(-i\phi_c)$, where $|F_o|$ is an experimentally determined structure amplitude and ϕ_c is the relative phase calculated from the positions of the heavy atoms alone. We may write these coefficients in the form

$$|F_o|\exp(-i\phi_c) = (|F_o|/|F_c|)|F_c|\exp(-i\phi_c) \qquad (5.66)$$

where $|F_c|$ is the amplitude produced by the heavy atoms alone. The right-hand side of (5.66) is the product of the transform of the heavy-atom portion of the structure and a function $(|F_o|/|F_c|)$ which is of zero phase for every reflection. The result in real space is the convolution of the heavy-atom portion of structure with the transform of $(|F_o|/|F_c|)$. This transform, because of its zero phase, will have a large peak at the origin, and the convolution will tend to make the heavy-atom part of the structure dominant. However, because $|F_o| \neq |F_c|$, there will be a background effect imposed upon the heavy-atom part of the structure that will modify it in the direction of $|F_o|$. If the heavy-atom positions are correct, or nearly so, the convolution will lead to an improved model for the structure. It is important to realize, however, that because of (5.66) such a convolution is always biased toward the heavy-atom part of the structure, and success with this method depends strongly on the degree of correctness of the heavy-atom positions.

Series Termination Effect

If a Fourier series is calculated with data only up to a certain value $\sin\theta_{max}$, it is equivalent to multiplying the transform of the structure by a function that is unity up to $\sin\theta_{max}$ and zero above it. The result is the convolution of the complete transform with that of the exclusion function. In two dimensions, the exclusion function simulates a circular hole, and its transform is a Bessel function (Figure 5.7b) which has a central maximum surrounded by maxima of alternating sign and decreasing magnitude. The transform of the limited data set is, thus, the complete transform convoluted with the Bessel function. The result can be seen in electron density maps as contours around the atomic positions that decrease to zero from the central maximum, and then are surrounded by *ripples*, or rings of decreasing magnitude and alternating sign. The effect is most noticeable in the regions around the heavy atoms in a structure (see, for example, Figure 6.47a).

5.9.2   General Phase-Free Transform

Although a diffraction pattern is invariant under the operation of translation, a representation of translation is contained within the pattern, because when the pattern is recombined the translational property of the object crystal is revealed. The relative phases at different parts of a transform do change under translation, but the changes are not evident because we observe the intensity of the transform

$$I = |F|^2 \qquad (5.67)$$

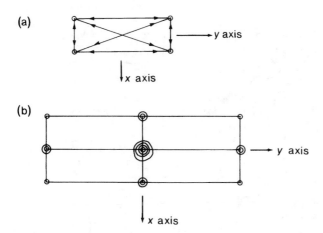

FIGURE 5.21.   Convolution with a rectangle of scattering centers (atoms). (a) Hypothetical four-atom structure of coordinates $\pm(0, 1, 0.2; 0.1, -0.2)$; interatomic vectors are shown by arrows. (b) Schematic convolution of the four-atom structure with its inversion in the origin. Multiweight peaks arise from a superposition of identical vectors.

Equation (5.67) is the counterpart of (5.33) for observations at reciprocal lattice points.

Recombination based on intensity alone contains no information about the lateral position of the object. All possible pairs of scattering points in the object will be reproduced with the correct orientation and separation, but symmetrically disposed about the center of the image. Hence, the image from the transform of intensities is that of vector positions between pairs of scattering points (atoms) in the object, all taken to a common origin. It is the Patterson function (see Section 6.4.2), and may be thought of as the result of the superposition of numerous fringe systems, all of which have a positive sign at the origin of the unit cell. Mathematically, the transform of intensity, $G(S)G^*(S)$, is the convolution of the transform of $G(S)$, which is the electron density, with the transform of $G^*(S)$, which is the electron density inverted in the origin.

In the hypothetical four-atom rectangle structure (Figure 5.21a), there are 16 (4^2) interatomic vectors. Four of them are of zero length and coincide at the origin (Figure 5.21b), and 12 are arranged in centrosymmetric pairs. Those formed by the m symmetries of the rectangle (Figure 5.21a) are of double weight, whereas those formed by symmetry 2, are of single weight. Figure 5.21b can be considered in terms of Figure 5.21a by transferring all interatomic vectors (arrowed) to a common origin.

Alternatively, we can think of Figure 5.21b as Figure 5.21a convoluted with its inversion in the origin (which is the same as itself in this centrosymmetric

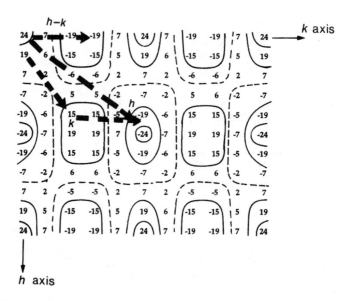

FIGURE 5.22. Calculated transform for the four-atom structure (Figure 5.21a) convolved with its inversion in the origin; contours are shown at -20, -10, 10, and 20. Fringe systems are apparent, corresponding to the normals to the directions between all pairs of atoms in the four-atom structure. (The labelled vectors **h**, **k**, and **h** $-$ **k** refer to the discussion Section 5.9.3.)

arrangement). Thus, the structure in Figure 5.21a is drawn four times, with its center of inversion at each of the atoms of the inverted structure in turn, and in the same orientation to give Figure 5.21b.

We can arrive at the same results by calculating the appropriate transforms. Using (5.44) with $f_j = 6$ (carbon), and neglecting the change in f_j with θ, Figure 5.22 was plotted for $h = 0$ to 10 and $k = 0$ to 10. The contours (disregard the three vectors highlighted for this application) show clearly the fringe systems that are perpendicular to each of the six pairs of points in Figure 5.21a. In Figure 5.23, the phase-free transform of the intensities (the squares of the values shown in Figure 5.22, suitably scaled), we see the 16 peaks and 4 rectangles, just as in Figure 5.21b. The orientations and separations are correct, but there is a four-fold ambiguity with respect to the lateral positions. If, by some stratagem, we were able to determine the correct relative phases, that those shown in Figure 5.22 were in fact correct, then the transform of these *amplitudes*, with those phases, would lead to the one correct result (Figure 5.24), from which we determine the atomic positions $\pm(0.1, 0.2; 0.1, -0.2)$. We shall investigate the required cunning in the next two chapters.

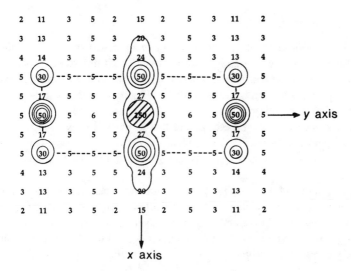

FIGURE 5.23.   Calculated transform from Figure 5.22 for h and $k = 0$ to 10, using the squares of the given amplitudes (phase-free) suitably scaled; contours are shown at 20, 30, 40, and 50, with the origin peak shaded. The rectangles are correct in size and orientation, but there is a 4-fold ambiguity in the lateral positions in the x,y plane. The ambiguity is related to the symmetry and not to the number of atoms in the structure.

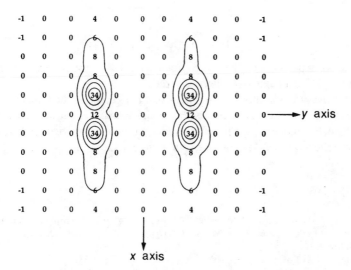

FIGURE 5.24.   Calculated transform from Figure 5.22 for h and $k = 0$ to 10, using the given amplitudes and signs; contours are shown at 6, 14, 22, and 30. A single, correct structure is now revealed.

5.9.3   Sign Relationships

Sign relationships (phases in the most general case) form a basis of the direct method of structure determination (see Chapter 7). At this stage, it is interesting to show how an important sign relationship may be deduced from the transform of a centrosymmetric arrangement of atoms (Figure 5.22) that will be considered to exist within a crystal unit cell.

Three vectors of the type $\mathbf{h}$, $\mathbf{k}$, and $\mathbf{h} - \mathbf{k}$, where we use $\mathbf{h}$ to represent the triplet hkl and $\mathbf{k}$ another triplet $h'k'l'$, form a triangle in reciprocal space, with one vertex of the triangle at the origin; these vector terminations correspond to x-ray reflections.

If, as in this chosen example, the vectors terminate in regions of relatively *high magnitude* in the transform (strong reflections), then by counting the number of times that the zero-boundary contour is traversed we can arrive at the signs of the transform at the reciprocal lattice points $\mathbf{h}$, $\mathbf{k}$, and $\mathbf{h} - \mathbf{k}$, since the transform must be positive at the origin:

Vector	hkl	Times zero-boundary crossed	Sign
$\mathbf{h}$	450	3	−
$\mathbf{k}$	420	2	+
$\mathbf{h} - \mathbf{k}$	030	1	−

Hence,

$$s(\mathbf{h})\, s(\mathbf{k})\, s(\mathbf{h} - \mathbf{k}) = +1 \tag{5.68}$$

which is an expression of the triple product (Σ_2) sign relationship, or Sayre's equation (7.7) for a centrosymmetric crystal.

If a vector terminates in a region of low magnitude in the transform (weak reflection), then no certain conclusion can be drawn about its phase, because of the difficulty in locating exactly the zero boundary contour in practice. In this context, consider reflections such as 840 or 810 in Figure 5.22.

Bibliography

HARBURN, G., TAYLOR, C. A., and WELBERRY, T. R., *An Atlas of Optical Transforms*, Bell (1975).

JAMES, R. W., *The Crystalline State*, Vol. II: *The Optical Principles of the Diffraction of X-rays*, Bell (1950).

LIPSON, H., *Optical Transforms*, Academic Press (1972).

ORAN BRIGHAM, E., *The Fast Fourier Transform and its Applications*, Prentice Hall (1988).

TAYLOR, C. A., *Diffraction*, Institute of Physics (1975).

TAYLOR, C. A., and LIPSON, H., *Optical Transformations*, Bell (1964).

WOOLFSON, M. M., *An Introduction to Crystallography*, 2nd ed., Cambridge University Press (1997).

Problems

5.1. Show that $\int_{-c/2}^{c/2} \sin(2\pi mx/c)\cos(2\pi nx/c)\,dx$ is zero for all x, except when $m = \pm n$; m and n are both integers. Hence, state the single finite value of the integral. One or more of the identities in Appendix A6 may be useful.

5.2. Magnesium fluoride is tetragonal, with space $P4_2/mnm$, $a = 4.625$, $c = 3.052$ Å and two species MgF_2 per unit cell at positions

$$2Mg \quad 0,0,0; \tfrac{1}{2},\tfrac{1}{2},\tfrac{1}{2}$$

$$4F \quad \pm\left(x,x,0; \tfrac{1}{2}+x, \tfrac{1}{2}-x, \tfrac{1}{2}\right)$$

The $F(h00)$ data are listed below:

h	2	4	6	8	10	12
$F(h00)$	−2.7	12.0	7.2	0.1	3.2	0.1

Calculate $\rho(x)$ using the program FOUR1D* with a subdivision of 40, and plot the function. Determine x. What length of the repeat is sufficient to define the complete function. Investigate the effects of (a) using terms up to $h = 6$, and (b) changing the sign of $F(600)$.

5.3. Find the Fourier transform of the function

$$f(x) = \begin{cases} a & \text{for } |x| < p \\ a/2 & \text{for } x = \pm p \\ 0 & \text{for } |x| > p \end{cases}$$

Then, transform the transform, and show that it equals the original function.

5.4. Find the Fourier transform and its inversion for the periodic function

$$f(t) = A\cos(2\pi f_0 t).$$

5.5. Using Equation (5.40), deduce the Fourier transform for two asymmetric scattering units related by a center of symmetry; the center may be taken as the origin. What general feature might be expected in the resulting transform?

5.6. Show, from Fourier transform theory, the nature of the systematic absences that would arise from a 2_1 screw axis along the line $\left[0,\ y,\ \tfrac{1}{4}\right]$ in an orthorhombic unit cell.

5.7. Tetraethyldiphosphine disulfide, $(C_2H_5)_4P_2S_2$, crystallizes in the triclinic system, with space group $P\bar{1}$ and one molecule in the unit cell. Figure P5.1 is the $hk0$ section of the weighted reciprocal lattice for this crystal. Make a photocopy of the diagram, identify, and draw lines to indicate, the fringe system for the (double weight) P–S vector. Hence, allocate signs to the more intense reflections (about 25) in the asymmetric portion of the reciprocal space diagram.

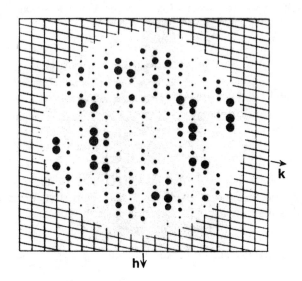

FIGURE P5.1. The $hk0$ section of the weighted reciprocal lattice of tetraethyldiphosphine disulphide. [Reproduced from *Optical Transforms*, by C. A. Taylor and H. Lipson, with the permission of Routledge.]

5.8. A hypothetical, two-dimensional, three-atom structure has the atomic coordinates $0.1, 0.1$; $0.1, -0.1$; $0.2, 0.3$, with respect to the origin of rectangular axes. Draw a diagram to represent the convolution of this structure with its inversion in the origin.

5.9. The figure field of Figure P5.2 represents the phase-free Fourier transform of the intensities of the diffraction pattern of the three-atom structure in Problem 5.8. How many images of the structure are present, and what are the sets of possible atomic coordinates?

5.10. Refer to Figure 5.20. What are the signs of $|F|$ for the reflections $00l$, $l = 1\text{–}6$?

5	10	7	14	6	21	6	14	7	10	5
6	12	8	20	7	22	7	16	7	10	6
7	14	9	68	7	22	9	22	8	10	7
7	20	9	24	8	31	10	90	9	15	7
9	82	9	17	9	32	11	23	9	16	9
11	23	9	23	11	322	11	23	9	23	11 ⟶ y axis
9	16	9	23	11	32	9	17	9	82	9
7	15	9	90	10	31	8	24	9	20	7
7	10	8	22	9	22	7	68	9	14	7
6	10	7	16	7	22	7	20	8	12	6
5	10	7	14	6	21	6	14	7	10	5

x axis

FIGURE P5.2.   Figure field of the phase-free Fourier transform of the intensities of the hypothetical structure in Problem 5.8.

5.11. If $f(x) = 1/\sqrt{(2\pi)}\exp(-x^2/2)$ and $g(x) = \delta(x-2)$, find the convolution $c(x) = f(x)^*g(x)$ by multiplying the transforms of $f(x)$ and $g(x)$.

6

Fourier Techniques in X-ray Structure Determination

6.1 Introduction

We have reached the stage where we can consider how to attack the solving of a crystal structure. After the earliest trial and error determinations in the 1920s with very simple and highly symmetrical structures, it was found that the application of Fourier series, initially in one dimension, led to the electron density function, in which peaks of density corresponded to atomic positions. As we have seen in the previous chapters, it is necessary to have the phases of the structure factors for this synthesis to be carried out meaningfully. One way in which phase information is obtained is through the Patterson function of vector density, a function of interatomic vectors in the crystal structure.

In this chapter, we shall examine the application of Fourier series to crystal structure analysis, together with some of its developments. However, in any structure analysis there are certain preliminary investigations that can throw light on the problem in hand. One such investigation, which we have already discussed, leads to the unit-cell dimensions and space group. We study next some example crystal structures in order to show how other, readily available information can be used to assist in the structure solution process.

6.2 Analysis of the Unit-Cell Contents

The density D_m of the crystals under examination may be measured by suspending them in a liquid or liquid mixture. The composition of the liquid is altered until the crystals neither rise nor fall; then the density of the liquid, equal to D_m, is measured with a pyknometer. Many organic materials can be suspended in aqueous sodium bromide. At flotation equilibrium, the refractive index of the solution may be measured, and the density determined by interpolation of a graph of the density of the solution as a function of its refractive index. The flotation procedure is best carried out in a thermostat. It may still happen, however, that the demarcation between sinking and floating is a little ill defined. Inclusion of air or solvent in the crystal will lead to a smaller apparent density, and the flotation position corresponding to a maximum value for the density measured should be appropriate.

If the crystal contains the number Z of chemical species, each of relative molar mass M_r, in the unit cell, then the following relationship holds for the density D_m of the crystal:

$$D_m = ZM_r u/V_c \qquad (6.1)$$

where u is the atomic mass unit. If the volume of the unit cell is in Å^3 and the density in g cm^{-3}, (6.1) may be written as

$$D_m = 1.6605 Z M_r/V_c \qquad (6.2)$$

If several measurements are made, the standard deviation $\sigma(D_m)$ can be deduced. It is useful to calculate the density D_c from the unit cell volume and the (integral) value of Z. A significant discrepancy between D_m and D_c should be examined, as it might point to an error in the unit-cell dimensions or to solvent of crystallization not included in M_r at that stage.

6.2.1   Papaverine Hydrochloride, $C_{20}H_{21}NO_4 \cdot HCl$

Crystal Data

> System: monoclinic.
> Unit-cell dimensions: $a = 13.059, b = 15.620, c = 9.130$ Å, $\beta = 92.13°$.
> V_c: 1861.1 Å^3.
> D_m: 1.33 g cm^{-3}.
> M_x: 375.85.
> Z: 4 to the nearest integer [3.97 from (6.2)].
> Unit-cell contents: 80C, 88H, 4N, 16O, 4Cl atoms.
> Absent spectra: $h0l$: l odd; $0k0$: k odd.
> Space group: $P2_1/c$.

All atoms are in general equivalent positions. The molecular conformation, obtained by a complete structural analysis,[a] is shown in Figure 6.1.

6.2.2   Naphthalene, $C_{10}H_8$

Crystal Data

> System: monoclinic.
> Unit-cell dimensions: $a = 8.658, b = 6.003, c = 8.235$ Å, $\beta = 122.92°$.

[a] C. D. Reynolds et al., *Journal of Crystal and Molecular Structure* **4**, 213 (1974).

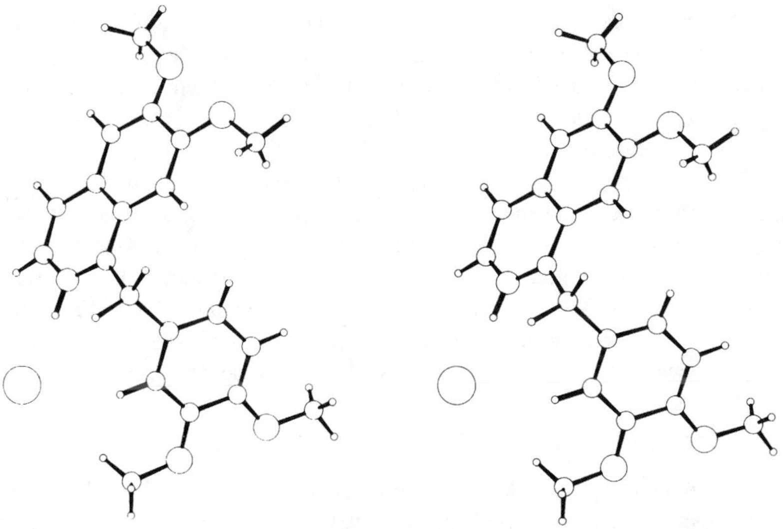

FIGURE 6.1.   Stereoview of the molecular conformation of papaverine hydrochloride; the circles, in order of decreasing size, represent Cl, N, C, and H.

V_c: 359.28 Å^3.
D_m: 1.152 g cm^{-3}.
M_r: 128.17.
Z: 2 to the nearest integer [1.94 from (6.2)].
Unit-cell contents: 20C, 16H atoms.
Absent spectra: $h0l$: $l = 2n + 1$; $0k0$: $l = 2n + 1$.
Space group: $P2_1/c$.

6.2.3   Molecular Symmetry

In papaverine hydrochloride, the four molecules in the unit cell occupy a set of general positions; each atom at coordinates x_j, y_j, z_j ($j = 1, 2, \ldots, 48$) is repeated by the space-group symmetry so as to build up the crystal structure. There are, therefore, 48 atoms, including hydrogen, in the asymmetric unit to be located by the structure analysis.

Naphthalene is not quite so straightforward. With two molecules per unit cell, there are 20 carbon atoms and 16 hydrogen atoms that may be distributed in four equivalent-position sets of five and four atoms, respectively, in the unit cell. This means that in order to solve the structure, we have to locate five carbon atoms and four hydrogen atoms. This number is only half that expected: since Z is 2, each atom is related by one of the symmetry elements of the space group to a second atom of the same type in the same molecule, so as to generate $C_{10}H_8$ from C_5H_4. There are three different symmetry elements to consider: the 2_1 axis, the *c*-glide plane, and the center of symmetry. The screw axis and glide plane are discounted because they involve translational symmetry, which would generate an infinite molecule with translational repeats. We must, therefore, conclude that the atom pairs are related by a center of symmetry, which in turn means that the molecule of naphthalene is centrosymmetric.

The symmetry analysis for naphthalene has served two very useful purposes: it has halved the work of the subsequent structure analysis, and shown that the molecules in the crystal exhibit a certain minimum symmetry ($\overline{1}$). This result is, of course, in agreement with chemical knowledge, which ordinarily we are quite entitled to use. The conventional notion that naphthalene should have *mmm* symmetry (Figure 6.2) is not supported directly, although the crystal structure analysis shows that this symmetry holds within experimental error.

6.2.4   Special Positions

The molecules of naphthalene lie on special positions in $P2_1/c$ (Figure 6.3). Special position sites correspond in symmetry to one of the 32 crystallographic

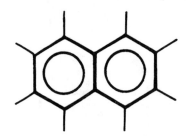

FIGURE 6.2.   Naphthalene molecular structure; the 9–10 carbon–carbon bond lies on a center of symmetry.

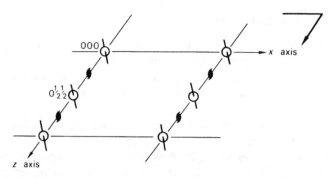

FIGURE 6.3.   Grouping of one of the special positions sets in $P2_1/c$; the arrangement of molecules (symmetry $\overline{1}$) with their centers at 0, 0, 0 and 0, $\frac{1}{2}$, $\frac{1}{2}$ is shown.

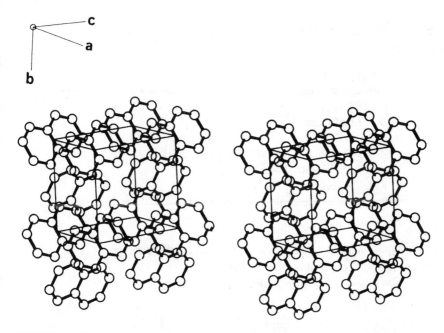

FIGURE 6.4. Stereoview of the crystal structure of naphthalene (for clarity, H atoms are not shown).

point groups and, in subsequent examples, we shall see that both atoms and molecules can occupy special positions.

Glide planes and screw axes do not usually accommodate atoms or molecules; an atom lying exactly on a translational symmetry element would introduce a pseudo-half-axial translation, thus creating special reflection conditions which, depending on the atomic number, may be observable among the x-ray data (see Problem 6.1).

Although they are in special positions, the molecules of naphthalene are subject to the space-group symmetry inherent in the general positions: if one molecule is located at 0, 0, 0, then the second molecule is at 0, $\frac{1}{2}$, $\frac{1}{2}$ (see Section 2.7.5). This set may be determined by substituting $x = y = z = 0$ into the set of general positions. The structure of naphthalene[a] is shown in Figure 6.4. The reader may like to consider the three other possible sets of special positions that could be used to represent this structure, and then show from the structure factor equation that $|F(hkl)|$ is invariant with respect to each set of special positions.

[a] D. W. J. Cruickshank, *Acta Crystallographica* **10**, 504 (1957), used the (nonstandard) space group $P2_1/a$, equivalent to $P2_1/c$, with **a** and **c** interchanged and **b** changed to $-\mathbf{b}$.

6.2.5   Nickel Tungstate, NiWO₄

Crystal Data

System: monoclinic.
Unit-cell dimensions: $a = 4.60, b = 5.66, c = 4.91$ Å, $\beta = 90.1°$.
V_c: 127.84 Å^3.
D_m: 7.964 g cm^{-3}.
M_x: 306.81.
Z: 2 to the nearest integer [2.00 from (6.2)].
Unit-cell contents: 2 Ni, 2 W, 8 O atoms.
Absent spectra: $h0l$: $l = 2n + 1$.
Possible space groups; Pc or $P2/c$.

We shall use space group $P2/c$, since the structure was determined successfully only with this space group.[a]
The general equivalent positions in $P2/c$ are

$$\pm\left\{x, y, z; \quad x, \bar{y}, \tfrac{1}{2} + z\right\}$$

but in order to study NiWO₄ further, we must consider the possible special positions for this space group; they are located on either the twofold axes or the centers of symmetry. The reader should make a drawing for space group $P2/c$, using the coordinates listed above and inserting the symmetry elements.

Special Positions on 2-fold Axes

The 2-fold axes lie along the lines $\left[0, y, \tfrac{1}{4}\right], \left[\tfrac{1}{2}, y, \tfrac{1}{4}\right], \left[0, y, \tfrac{3}{4}\right]$, and $\left[\tfrac{1}{2}, y, \tfrac{3}{4}\right]$. The equivalent positions generated by the space-group symmetry show that the special position sets are

$$\pm\left\{0, y, \tfrac{1}{4}\right\} \quad \text{or} \quad \pm\left\{\tfrac{1}{2}, y, \tfrac{1}{4}\right\}$$

and each set satisfies $P2/c$ symmetry by accommodating in the unit cell two structural entities with symmetry 2.

[a] R. O. Keeling, *Acta Crystallographica* **10**, 209 (1957).

Special Positions on Centers of Symmetry

If we repeat the above analysis for the eight centers of symmetry in the space group, we will develop four special position sets:

$$0,0,0; \quad 0,0,\tfrac{1}{2}$$

$$\tfrac{1}{2},0,0; \quad \tfrac{1}{2},0,\tfrac{1}{2}$$

$$0,\tfrac{1}{2},0; \quad 0,\tfrac{1}{2},\tfrac{1}{2}$$

$$\tfrac{1}{2},\tfrac{1}{2},0; \quad \tfrac{1}{2},\tfrac{1}{2},\tfrac{1}{2}$$

The Ni and W atoms must lie on special positions, with either 2 or $\bar{1}$ symmetry. Nothing can be said about the position of the oxygen atoms, and without further detailed analysis we cannot define this structure further. However, to complete the picture, we list the atomic parameters for this structure, and illustrate it in Figure 6.5:

$$2\text{Ni} \quad \pm\left\{\tfrac{1}{2},0.653,\tfrac{1}{4}\right\}$$

$$2\text{W} \quad \pm\left\{0,0.180,\tfrac{1}{4}\right\}$$

$$4\text{O} \quad \pm\{0.22,0.11,0.96; \quad 0.22,0.89,0.46\}$$

$$4\text{O}' \quad \pm\{0.26,0.38,0.39; \quad 0.26,0.62,0.89\}$$

The heavy atoms (W and Ni) were found to occupy the four 2-fold axes in pairs. This conclusion, although not uniquely derivable from the symmetry

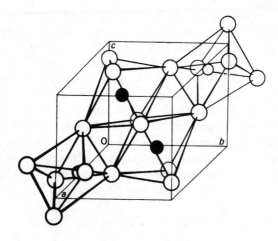

FIGURE 6.5.   Structure of $NiWO_4$, showing the WO_6 ad NiO_6, octahedra: large open circles O, small open circles W, small black circles Ni.

analysis alone, was at least partially indicated by it. Once again, a pencil and paper operation saved considerable effort in the subsequent detailed structure analysis by pointing to the proper course of action.

In these few examples, we have shown the value of a symmetry analysis in the early stages of a structure determination. The procedure may be regarded as a routine, to be carried out before the more detailed calculations required in the elucidation of the atomic parameters.

6.3   Interpretation of Electron Density Distributions

We have discussed different forms of the electron density equation in Chapter 5, and we now make use of that theory in studying the distribution of electrons density in crystal structures.

Electron density is concentrated in the vicinity of atoms, rising to peaks at the electron density maxima, corresponding to atomic positions, and falling to relatively low values between them. The wavelengths of x-rays used in crystal structure analysis are too long to reveal the intimate electronic structure of atoms themselves, which are seen, therefore, somewhat blurred in the calculated electron density function. Atoms appear as peaks in this function, and the peak position of a given atom is assumed to correspond to its atomic center, within the limit of experimental error. In general, the more complete and accurate the experimental $|F|$ data, the better will be the atomic resolution and the more precise the final structure model.

Peak Heights and Weights

To a first approximation, the heights of the peaks in an electron density distribution of a crystal are proportional to the corresponding atomic numbers. The hydrogen atom, at the extreme low end of the atomic numbers, is resolved with difficulty in electron density maps; its small electron density merges into the background density that arises from errors in both the data and the structure model. However, hydrogen atoms can be detected by a difference-Fourier technique, as discussed later (Section 6.4.5), and by neutron diffraction (see Section 6.6).

A better measure of the electron content of a given atom may be obtained from an integrated peak weight, in which the absolute values of $\rho(x, y, z)$ are summed over the volume occupied by the atom. This technique makes some allowance for the variation of individual atomic temperature factors, high values of which tend to decrease peak heights for a given electron content.

Computation and Display of Electron Density Distributions

Assuming for the moment that phases are available, the electron density function may be calculated over a grid of chosen values of x, y, and z. For this purpose, the unit cell is divided into a selected number of equal divisions, in a manner similar to that employed in the synthesis of the square-wave function (Section 5.2.1). Intervals corresponding to about 0.3 Å are satisfactory for most electron density maps. The symmetry of $\rho(x, y, z)$ is that of the space group of the crystal under investigation. Consequently, a summation over a volume equal to, or just greater than, that of the asymmetric unit is adequate.

In order to facilitate the interpretation of $\rho(x, y, z)$, it is essential to present the distribution of the numerical values in such a way that the geometric relationships between the peaks are easily inspected. This feature is afforded by first calculating the electron density in sections, each corresponding to a constant value of x, y, or z using (5.23). Each section consists of a field of figures arranged on a grid, closely true to scale for preference, which may be contoured by lines passing through points of equal electron density, interpolating as necessary (Figure 6.6). The grading of the contour intervals is selected to produce a reasonable number of contours around the higher-density areas. The contouring should be carried out with care; this exercise leads to fairly precise peak positions and a desirable familiarity with the problem. Sophisticated map-plotting and peak-searching facilities are available, but they should be treated with caution by the beginner.

The contoured sections are finally transferred to a transparent medium, such as thin perspex or clear acetate sheets, which are then stacked at the requisite spatial intervals and viewed over a diffuse light source. Figure 1.2 is a photograph of such a display, extending through 17 sections.

An alternative method of displaying the results of an electron density calculation is by means of a ball-and-stick model. An example of this form of representation is shown in Figure 6.7.

In the analysis of small molecules it is not usually considered necessary always to plot and contour the electron density function, although it can be done through some program packages, such as WinGX. There are many graphics programs available that recognize the highest peaks and carry a geometrical interpretation in terms of their coordinates: Platon for small molecules, and O or Turbo Frodo for proteins (see Appendix A10 for these programs).

Projections

The use of two-dimensional studies in crystallography is fairly restrictive but, nevertheless, worthy of mention because of the relative ease of calculation and preparation of Fourier maps. For example, the function

$$\rho(x, z) = \frac{2}{A_b} \sum_h \sum_l |F(h0l)| \cos[2\pi(hx + lz) - \phi(h0l)] \qquad (6.3)$$

is calculated with the data from only one level of the reciprocal lattice, the zero level, perpendicular to b, and plotted over the area A_b of the a,c plane, or the

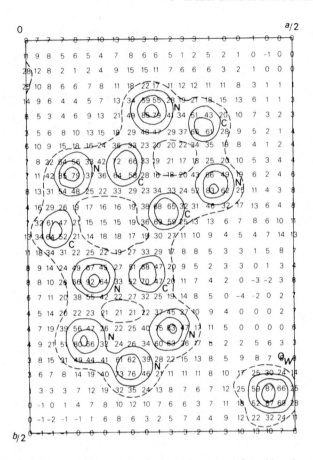

FIGURE 6.6.   Two-dimensional electron density projection $\rho(x, y)$ for azidopurine monohydrate, $C_5H_3N_7 \cdot H_2O$, calculated from the data of Glusker et al., *Acta Crystallographica B* **24**, 359 (1968). The isolated peak (O_w) in the lower right-hand region of the map represents the oxygen atom of the water molecule. Hydrogen atom positions are not obtained in this electron density synthesis. The field figures are $10\rho(x, y)$ in electrons per Å^2 contoured at intervals of 20 units.

asymmetric portion thereof. The simplification in the calculations is offset, however, by a corresponding complexity in the interpretation of the maps, arising from the superposition of peaks in projection on to the given plane, although this effect is not as severe as in one dimension. Equation (6.3) corresponds to the projection of the electron density along the *b* axis: it is essential to appreciate the difference between the meaning of $\rho(xz)$ and $\rho(x0z)$; the latter represents the section of the three-dimensional electron density function at $y = 0$. Equations for projections along other principal axes may be written by analogy with (6.3).

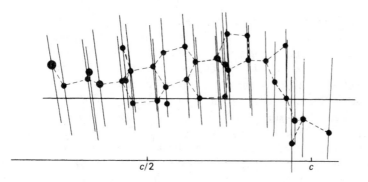

FIGURE 6.7.   Three-dimensional model of euphenyl iodoacetate (see also Figures 1.2 and 1.3).

Even simple atomic arrangements may appear distorted in projections, with individual molecules overlapping to some degree, but we would not wish to discourage their consideration. We shall restrict their use to examples illustrating various aspects of structure analysis. Practice in the calculation and interpretation of Fourier series is afforded by Problems 6.11 and 6.12, and also the exercises with the $XRAY^*$ system (see Chapters 8 and 11).

6.4   Methods of Solving the Phase Problem

At this point, it may be convenient for the reader to revise some of the ideas presented in Chapter 5, since a general overview of the present topic was presented there.

The set of $|F_o(hkl)|$ data constitutes the starting point of an x-ray structure determination. The approximate number of symmetry-independent reflections measurable may be calculated in the following manner.

6.4.1   Number of Reflections in the Data Set

The radius of the limiting sphere is 2 RU,[a] and its volume is therefore 33.51 RU^3. The number of reciprocal lattice points within the limiting sphere is approximately equal to the number of times the reciprocal unit cell volume V^* will fit into 33.51; since, for this application, $V^* = \lambda/V$, this number is 33.51 V_c/λ^3. The number of symmetry-independent reflections observable, N_{max}, in a given experiment in which θ_{max} represents the practical upper limit of θ is given by

$$N_{max} = 33.51\ V_c \sin^3 \theta_{max}/\lambda^3 Gm \qquad (6.4)$$

[a] Reciprocal lattice units.

where G is the unit-cell translation constant (Table 3.2) for nonzero reflections and m is the number of symmetry-equivalent reflections, or the number of general equivalent points in the appropriate Laue group. For zones and rows, m may take different values from that for hkl, and a number of systematic absences within the sphere of radius $2 \sin \theta_{max}$ may have to be subtracted.

As an example, consider an orthorhombic crystal of space group $Cmm2$, with unit cell dimensions $a = 9.00, b = 10.00$, and $c = 11.00$ Å. For Cu $K\alpha$ radiation ($\bar{\lambda} = 1.542$ Å) and θ_{max} of $85°$ ($d_{min} = 0.77$ Å), N_{max} is ($33.51 \times 9 \times 10 \times 11 \times \sin^3 85°)/(1.542^3 \times 2 \times 8) = 559$. If Mo $K\alpha$ radiation[a] ($\bar{\lambda} = 0.7107$ Å) had been used instead of Cu $K\alpha$, the number would have been 5710. Such a structure might contain, say, 15 atoms in the asymmetric unit. In the structure analysis, each atom would be determined by three positional parameters (x_j, y_j, z_j) and, say, one isotropic thermal vibration parameter, which, with an overall scale factor, totals 61 variables. Even with Cu $K\alpha$ radiation, there are nine reflections per variable, a situation which, from a mathematical point of view, is considerably overdetermined. This feature is important, since the experimental intensity measurements contain random errors which cannot be eliminated, and a preponderance of data is needed to ensure good precision in the structural parameters. We shall consider this situation again in Chapter 7 but, as will be explained in Chapter 10, such a degree of over-determination is not usually possible with macromolecules.

6.4.2   The Patterson Function

Although the connection between Fourier theory and x-ray diffraction was recorded first in 1915, it was not until about 1930 that very much practical use was made of it. Before the advent of computing facilities, the calculation of even a Fourier projection, involved considerable time and effort. Add to this the phase problem, which necessitated many such calculations, and it is easy to understand that x-ray analysts were not anxious to become involved with extensive Fourier calculations; many early structure analyses were based on two projections.

In 1934, Patterson reported a new Fourier series which could be calculated directly from the experimental intensity data. However, because phase information is not required in the Patterson series, the result cannot be interpreted as a set of atomic positions, but rather as a collection of interatomic vectors all taken to a common origin (Section 5.9.2). Patterson was led to the formulation of his series from considerations of an earlier theory of Debye on the scattering of x-rays by liquids—a much more difficult problem.

Patterson functions are of considerable importance in x-ray structure analysis, and their application will be considered in some detail. We will study first a one-dimensional function.

---

[a] $\theta_{max} ca 27°$.

One-Dimensional Patterson Function

The electron density at any fractional coordinate x is $\rho(x)$, and that at the point $(x + u)$ is $\rho(x + u)$. The average product of these two electron densities in a repeat of length a, for a given value of u, is

$$A(u) = \int_0^1 \rho(x)\rho(x + u)\,dx \tag{6.5}$$

where the upper limit of integration corresponds to the use of fractional coordinates. Using (5.21) in a form appropriate to a one-dimensional unit repeat, we obtain

$$A(u) = \int_0^1 \frac{1}{a^2} \sum_h F(h)e^{-i2\pi hx} \sum_{h'} F(h')e^{-i2\pi h'(x+u)}\,dx \tag{6.6}$$

The index h' lies within the same range as h, but is used to effect distinction between the Fourier series for $\rho(x)$ and $\rho(x + u)$. Separating the parts dependent upon x, and remembering that the integral of a sum is the sum of the integrals of the separate terms, we may write

$$A(u) = \frac{1}{a^2} \sum_h \sum_{h'} F(h)F(h')e^{-i2\pi h'u} \int_0^1 e^{-i2\pi(h+h')x}\,dx \tag{6.7}$$

Considering the integral

$$\int_0^1 e^{-i2\pi(h+h')x}\,dx = \left. \frac{e^{-i2\pi(h+h')x}}{-i2\pi(h + h')} \right]_0^1 \tag{6.8}$$

$e^{-i2\pi(h+h')}$ is unity, since h and h' are integral, from De Moivre's theorem (Section 3.2.3), and the integral is, in general, zero. However, for the particular value of h' equal to $-h$, it becomes indeterminate and we must consider making this substitution before integration. Thus,

$$\int_0^1 dx = 1 \tag{6.9}$$

Hence, from (6.7), for nonzero values of $A(u)$, where $h' = -h$,

$$A(u) = \frac{1}{a^2} \sum_h \sum_{-h} F(h)F(-h)e^{i2\pi hu} \tag{6.10}$$

Equation (6.10) is not really a double summation, since h and $-h$ cover the same field of the function. Furthermore, $F(-h)$ is really the conjugate $F^*(h)$, and using (3.102), we obtain

$$A(u) = \frac{1}{a^2} \sum_h |F(h)|^2 e^{i2\pi hu} \qquad (6.11)$$

where the index h ranges from $-\infty$ to ∞. Now using Friedel's law (3.108), we find

$$A(u) = \frac{1}{a^2} \sum_h (|F(h)|^2 e^{i2\pi hu} + |F(h)|^2 e^{-i2\pi hu}) \qquad (6.12)$$

and from De Moivre's theorem,

$$A(u) = \frac{2}{a^2} \sum_h |F(h)|^2 \cos 2\pi hu \qquad (6.13)$$

where h now ranges from 0 to ∞. The corresponding Patterson function $P(u)$ is usually defined as

$$P(u) = \frac{2}{a} \sum_h |F(h)|^2 \cos 2\pi hu \qquad (6.14)$$

a small difference from the averaging function $A(u)$.

The practical evaluation of $P(u)$ proceeds through (6.14), but its physical interpretation is best considered in terms of (6.5), neglecting the small difference between $P(u)$ and $A(u)$.

Figure 6.8a shows one unit cell of a one-dimensional structure containing two different atoms A and B situated at fractional coordinates x_A and x_B, respectively. Equation (6.5) represents the value of the electron density product $\rho(x)\rho(x+u)$, for any constant value of u, averaged over the repeat period of the unit cell. The average will be zero if one end of the vector u always lies in zero regions of electron density, small if both ends of the vector encounter low electron densities, large if the electron density products are large, and a *maximum* where u is of such a length that it spans two atomic positions in the unit cell.

For values of u less than u_{min} in Figure 6.8a, no peak will arise from the pair of atoms. As u is increased, however, both ends of the vector will come simultaneously under the electron density peaks, and from (6.5) a finite value of $A(u)$, or $P(u)$, will be obtained. The integration can be simulated by sliding a vector of a given magnitude u along the x axis, evaluating the product $\rho(x)\rho(x+u)$ for all sampling intervals between zero and unit fractional repeat; this process is carried out for all fractional values of u between zero and one. The graph of $P(u)$ as a function of u is similar in appearance to an electron density function, but we must be careful not to interpret it in this way.

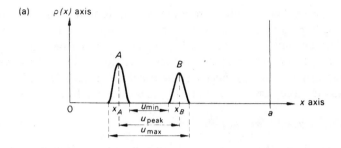

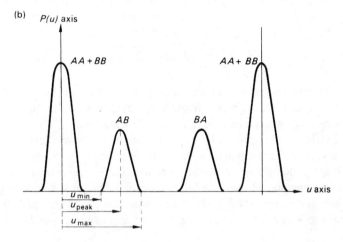

FIGURE 6.8.   Development of a one-dimensional Patterson function (b) for a two-atom structure (a). Note the centrosymmetry of the Patterson function that is lacking in the $\rho(x)$ function.

As we proceed through the values of u, we encounter u_{peak}, the interatomic vector $A - B$, which gives rise to the maximum value of $P(u)$, labeled AB in Figure 6.8b. As u increases to u_{max}, the electron density product falls to zero and $P(u)$ decreases correspondingly. Since we are concerned with interatomic *vectors*, negative values of u are equally important; $-AB$ is marked off on the negative side of the origin, or at BA within the given unit cell.

If we consider next *very* small values of u, both ends of such vectors will lie inside one and the same electron density peak, and $P(u)$ will be large. In the limit as $u \rightarrow 0$, the product involves that of the electron density maximum with itself, which is a local maximum for each atom, and a very large peak at the origin ($u = 0$) is to be expected. Thus the Patterson function is represented as a map of interatomic vectors, including null vectors, all taken to the origin (see also Section 5.9.2).

The reader should confirm from Figure 6.8, using tracing paper, that the positions of the peaks in Patterson space can be plotted graphically by placing each atom of the structure $\rho(x)$ in turn at the origin of the Patterson map, in parallel orientation, and marking the positions of the other atoms on to the Patterson unit cell. Because of the centrosymmetry of the Patterson function, implicit in (6.14), it is not strictly necessary to plot vectors lying outside one-half of the unit cell.

Three-Dimensional Patterson Function

If we replace $\rho(x)$ and $\rho(x + u)$ in (6.5) by the three-dimensional analogs $\rho(x, y, z)$ and $\rho(x + u, y + v, z + w)$ and integrate over a unit fractional volume, we can derive the three-dimensional Patterson function:

$$P(u, v, w) = \frac{2}{V_c} \sum_h \sum_k \sum_l |F(hkl)|^2 \cos 2\pi(hu + kv + lw) \qquad (6.15)$$

where the summations range, in the most general case, over one half of experimental reciprocal space. This equation should be compared with (5.22) and (6.14): it is a Fourier series with zero phases and $|F|^2$ as coefficients. Since $|F|^2$ is $F \cdot F^*$, we see from (5.21) that (6.15) represents the convolution of the electron density $\rho(\mathbf{r})$ with its inversion in the origin, that is, with $\rho(-\mathbf{r})$. In practice, (6.15) may be handled like the corresponding electron density equation, with u, v, w replacing x, y, z, but it should be remembered that both functions explore the same field, the unit cell. The roving vector is now specified by three coordinates, $u, v,$ and w, and $P(u, v, w)$ is a maximum where the corresponding vector spans two atoms in the crystal.

Positions and Weights of Peaks in the Patterson Function

The positions of the peaks in $P(u, v, w)$ may be plotted in three dimensions by placing each atom of the unit cell of a structure in turn at the origin of Patterson space, in parallel orientation, and mapping the positions of all other atoms on to the Patterson unit cell. Examples of this process are illustrated graphically in Figure 6.9; for simplicity the origin peak is not shown in Figure 6.9d. In Figure 6.9a, all atoms and their translation equivalents produce vector peaks lying on the points of a lattice that is identical in shape and size to the crystal lattice. For example, atom 1 at x, y, z and its translation equivalent, $1'$, at $x, 1 + y, z$ give rise to a vector ending at 0, 1, 0 on the Patterson map. Peaks of this nature accumulate at the corners of the Patterson unit cell in exactly the same way as those of the origin peak, $P(0, 0, 0)$. From (6.15), we can derive the height of the origin peak;

$$P(0, 0, 0) = \frac{2}{V_c} \sum_{h=0}^{\infty} \sum_{k,} \sum_{l=-\infty} |F_o(hkl)|^2 \qquad (6.16)$$

In general, (6.16) is equivalent to a superposition at the origin of all N products like $\rho(x_j, y_j, z_j)\rho(x_j, y_j, z_j)$, where N is the number of atoms in the unit

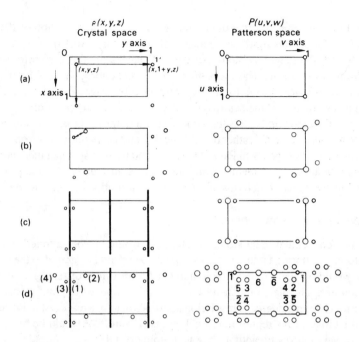

FIGURE 6.9.   Effects of symmetry-related and symmetry-independent atoms on the Patterson function. The weights of the peaks are approximately proportional to the diameters of the circles: (a) $P1(N = 1)$; (b) $P1(N = 2)$—two atoms per unit cell produce $(2^2 - 2)$ nonorigin peaks; (c) $Pm(N = 2)$—two nonorigin peaks, but with coordinates $\pm\{0, 2y, 0\}$; (d) $Pm(N = 4)$—12 nonorigin peaks per unit cell; for clarity the origin peak has not been drawn. The *Patterson* space group is $P\bar{1}$ in (a) and (b) and $P2/m$ in (c) and (d).

cell. Since $\rho(x_j, y_j, z_j)$, is proportional to the atomic number Z_j of the jth atom (Section 6.3), we have

$$P(0, 0, 0) \propto \sum_{j=1}^{N} Z_j^2 \tag{6.17}$$

A single vector interaction between two atoms j and k (Figure 6.9b) will have a Patterson peak of height proportional to $Z_j Z_k$. Hence, the height $H(j, k)$ of this peak will be given by

$$H(j, k) \approx P(0, 0, 0) Z_j Z_k \Big/ \sum_{j=1}^{N} Z_j^2 \tag{6.18}$$

where $P(0, 0, 0)$ is calculated from (6.16). This equation can serve as a useful guide, but overlapping vectors may give rise to misleading indications. The reservations on peak heights already mentioned (Section 6.3) apply also to Patterson peaks.

It should be remembered that the correct geometrical interpretation of Patterson peaks is of far greater significance than is an adherence to (6.18).

In a structure with N atoms per unit cell, each atom forms a vector with the remaining $N - 1$ atoms. There are, thus, $N(N - 1)$ nonorigin peaks. From (6.15), substitution of $-u, -v, -w$ for u, v, w, respectively, leaves $P(u, v, w)$ unaltered, which is a statement of the centrosymmetry of the Patterson function.

The Patterson unit cell is the same size and shape as the crystal unit cell, but it has to accommodate N^2 rather than N "peaks" and is, therefore, correspondingly overcrowded. Thus, peaks in Patterson space tend to overlap when there are many atoms in the unit cell, a feature which introduces difficulties into the process of unraveling the function in terms of the correct distribution of atoms in the crystal.

Sharpened Patterson Function

In a conceptual point atom, the electrons would be concentrated at a point. The atomic scattering factor curves (Figure 3.39) would be parallel to the abscissa and f would be equal to the atomic number for all values of $(\sin \theta)/\lambda$ and at all temperatures. The electron density for a crystal composed of point atoms would show a much higher degree of resolution than does that for a real crystal. Put another way, the broad peaks representing real atoms (Figure 5.4) would be replaced by peaks of very narrow breadth in the point-atom crystal.

A plot of the mean value of $|F_0|^2$ against $(\sin \theta)/\lambda$ for a typical set of data is shown in Figure 6.10. The radial decrease in $\overline{|F_0|^2}$ can be reduced by modifying $|F_0|^2$ by a function which increases as $(\sin \theta)/\lambda$ increases. The coefficients for a sharpened Patterson synthesis may be calculated by the following equation. Sharpening can be effected also through the use of $|E|$ values (see Chapter 7).

$$|F_{\text{mod}}(hkl)|^2 = \frac{|F_0(hkl)|^2}{\exp[-2B(\sin^2 \theta)/\lambda^2]\{\sum_{j=1}^{N} f_j\}^2} \qquad (6.19)$$

where N is the number of atoms in the unit cell and B is an overall isotropic temperature factor (Section 3.9.8).

The effect of sharpening on a Patterson synthesis is illustrated in Figure 6.17d, the Harker section $(u, \frac{1}{2}, w)$ for papaverine hydrochloride. It should be compared with Figure 6.17b; the increased resolution is very apparent.

Oversharpening of Patterson coefficients may lead to spurious peaks because of series termination errors (Section 5.9.1), particularly where heavy atoms are present, and the technique should not be applied without care. Sometimes the coefficients can be further modified to advantage by multiplication by a function such as $\exp(-m \sin^3 \theta)$, where m is chosen by trial, but might be about 5. This function has the effect of decreasing the magnitude of the $\overline{|F_0|^2}$ curve at high θ values. Many other sharpening functions have been proposed, but we shall not

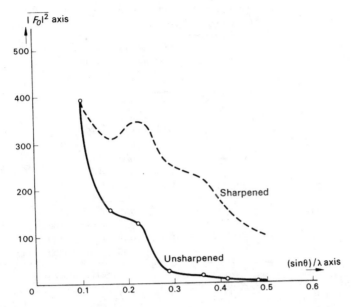

FIGURE 6.10.   Effect of sharpening on the radial decrease of the local average intensity $\overline{|F_o|^2}$.

dwell on this subject. It is often helpful to calculate both the normal and sharpened Patterson functions for comparison. Practice can be gained through exercises with the *XRAY** program system.

Symmetry of the Patterson Function for a Crystal of Space Group *Pm*

An inspection of Figures 6.9c and 6.9d shows that the peaks on the line $[0, v, 0]$ arise from atom pairs related by the m planes. The vector interactions for case (d) are listed in Table 6.1, and may be easily verified by the reader; the values $z_1 = z_2 = 0.0$ were chosen for convenience only.

The m planes in Pm are carried over into Patterson space, and relate the following pairs of peaks in the vector set:

$$1,\bar{1}; \quad 2,5; \quad \bar{2},\bar{5}; \quad 3,4; \quad \bar{3},\bar{4}; \quad 6,\bar{6} \qquad (6.20)$$

Furthermore, the presence of a center of symmetry in the diffraction pattern generates $2/m$ symmetry in the Patterson map, which corresponds to the Laue symmetry of all monoclinic crystals. Evidently, the symmetry of the diffraction pattern is impressed on to the Patterson function by the use of $|F|^2$ coefficients in the Patterson–Fourier series. As a consequence, the Patterson synthesis is computed

TABLE 6.1.   Vectors Generated by Two Independent Atoms and their Symmetry
Equivalents in Space Group Pm^a

Atom pair	Analytical form of vector	Subtraction of coordinates		Reduced to one unit cell		Point in Figure 6.9d
		u	v	u	v	
(1), (3)	$\pm\{0, 2y_1, 0\}$	0	0.10	0	0.10	1
		0	-0.10	0	0.90	$\bar{1}$
(1), (2)	$\pm\{x_1 - x_2, y_1 - y_2, z_1 - z_2\}$	0.15	-0.15	0.15	0.85	2
		-0.15	0.15	0.85	0.15	$\bar{2}$
(1), (4)	$\pm\{x_1 - x_2, y_1 + y_2, z_1 - z_2\}$	0.15	0.25	0.15	0.25	3
		-0.15	-0.25	0.85	0.75	$\bar{3}$
(2), (3)	$\pm\{x_1 - x_2, -y_1 - y_2, z_1 - z_2\}$	0.15	-0.25	0.15	0.75	4
		-0.15	0.25	0.85	0.25	$\bar{4}$
(3), (4)	$\pm\{x_1 - x_2, -y_1 + y_2, z_1 + z_2\}$	0.15	0.15	0.15	0.15	5
		-0.15	-0.15	0.85	0.85	$\bar{5}$
(2), (4)	$\pm\{0, 2y_2, 0\}$	0	0.40	0	0.40	6
		0	-0.40	0	0.60	$\bar{6}$

[a] The coordinates of the four atoms in two sets of general positions are $x, y, z; x, \bar{y}, z$ with $x_1 = 0.20$, $y_1 = 0.05$, $x_2 = 0.05$, $y_2 = 0.20$, and $z_1 = x_2 = 0.00$.

in the primitive space group corresponding to the Laue symmetry of a crystal, and this situation is similar for all space groups.

We can detect the presence of the 2-fold axis parallel to b in Figure 6.9d through vector peaks such as $5, \bar{2}$ and $3, \bar{4}$. Finally, the symmetry-related pairs of atoms in the crystal, 1, 3 and 2, 4, give rise to vectors along the line $[0, v, 0]$—the peaks $1, 6, \bar{6}$, and $\bar{1}$ in Patterson space. The presence of a large number of peaks along an axis in a three-dimensional Patterson map may be used as evidence for a mirror plane perpendicular to that axis in the crystal. This feature is important because an m plane does not give rise to systematic absences in the diffraction pattern (Table 3.7). The existence of peaks, arising from symmetry-related atoms, in certain regions of Patterson space was noted first by Harker in 1936. The line $[0, v, 0]$ for Pm is called a Harker line; *planes* containing peaks arising from pairs of symmetry-related atoms are called Harker sections. We shall consider some examples below.

Vector Interactions in Other Space Groups

We shall consider atoms in general positions in a number of space groups which should be now familiar.

Space Group $P\bar{1}$
General positions: $x, y, z; \bar{x}, \bar{y}, \bar{z}$.
Vectors: $\pm\{2x, 2y, 2z\}$.

Harker peaks lie in general positions in Patterson space.

Space Group P2
General positions: $x, y, z; \bar{x}, y, \bar{z}$.
Vectors: $\pm\{2x, 0, 2z\}$.
Harker section: $(u, 0, w)$.

It may be noted that for complex structures, not all of the peaks on Harker sections are necessarily true Harker peaks. If in this structure there are two atoms not related by symmetry, which, by chance, have the same or nearly the same y coordinates, the vector between them will produce a peak on the Harker section.

Space Group P2/m

Vectors:			
	$\pm\{2x, 0, 2z\}$	double weight	type 1
	$\pm\{0, 2y, 0\}$	double weight	type 2
	$\pm\{2x, 2y, 2z\}$	single weight	type 3
	$\pm\{2x, 2\bar{y}, 2z\}$	single weight	type 4.
Harker section:	$(u, 0, w)$.		
Harker line:	$[0, v, 0]$.		

Vector type 1 arises in two ways, once from the pair $x, y, z; \bar{x}, y, \bar{z}$ and once from the pair $x, \bar{y}, z; \bar{x}, \bar{y}, \bar{z}$. These two interactions give rise to identical vectors, which therefore superimpose in Patterson space and form a double-weight peak. Similar comments apply to type 2, but the centrosymmetrically related atoms give rise to single-weight peaks, types 3 and 4. Figure 6.11 illustrates these vectors, as seen along the z axis. The reader may now consider how the Patterson function might be used to differentiate among space groups $P2$, Pm, and $P2/m$. Statistical methods, discussed in Chapter 3, are often employed to verify the results obtained from a study of the vector distribution.

6.4.3   Examples of the Use of the Patterson Function in Solving the Phase Problem

In this section, we shall consider how the Patterson function was used in the solution of three quite different structures.

Bisdiphenylmethyldiselenide, $(C_6H_5)_2CHSe_2CH(C_6H_5)_2$

Crystals of this compound form yellow needles, with straight extinction under crossed Polaroids for all directions parallel to the needle axis, and oblique extinction on the section normal to the needle axis. Photographs taken with the crystal oscillating about its needle axis show only a horizontal m line, while

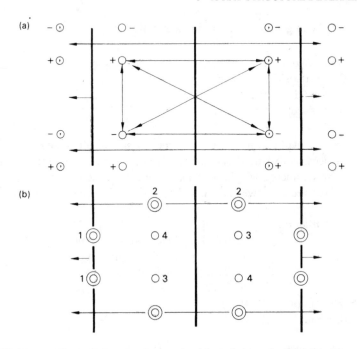

FIGURE 6.11.   (a) Vectors between symmetry-related atoms in general equivalent positions in space group $P2/m$. Coordinates like $\bar{x}$ have been treated as $(1 - x)$ in drawing the vectors. (b) One unit cell of the Patterson function: the twofold axes intersect the m planes in centers of symmetry. Note the single-weight and double-weight peaks, and their relation to the space-group symmetry.

zero- and upper-layer Weissenberg photographs show only symmetry 2. The crystals are therefore monoclinic, with b along the needle direction.

Crystal Data

System: monoclinic.

Unit-cell dimensions: $a = 18.72, b = 5.773, c = 12.594$ Å$, \beta = 125.47°$.

V_c: 1108.5 Å^3.

D_m: 1.49 g cm^{-3}.

M_x: 492.38.

Z: 2.02, or 2 to the nearest integer.

Unit-cell contents: 4Se, 52C, and 44H atoms.

Absent spectra: hkl: $h + k = 2n$.

Possible space groups: $C2, Cm, C2/m$.

Symmetry Analysis.   Where the space group is not determined uniquely by the x-ray diffraction pattern, it may be possible to eliminate certain alternatives at the outset of the structure determination by other means.

Space groups $C2$ and Cm each require four general positions:

$$C2: \left(0,0,0; \tfrac{1}{2},\tfrac{1}{2},0\right) + \{x,y,z; \bar{x},y,\bar{z}\}$$

$$Cm: \left(0,0,0; \tfrac{1}{2},\tfrac{1}{2},0\right) + \{x,y,z; x,\bar{y},z\}$$

Since Z is 2, the molecular symmetry is either 2, in $C2$, or m, in Cm. In both $C2$ and Cm, all atoms could satisfy general position requirements, and neither arrangement would be stereochemically unreasonable.

Space group $C2/m$ requires eight general equivalent positions per unit cell. Only special position sets, such as 0, 0, 0 and $\tfrac{1}{2},\tfrac{1}{2},0$ correspond with $Z = 2$. These positions have symmetry $2/m$, but it is not possible to construct the molecule in this symmetry without contradicting known chemical facts. Consequently, we shall regard this space group as highly improbable for the compound under investigation.

Patterson Studies. Whatever the answer to the questions remaining from this symmetry analysis, we expect, from chemical knowledge (Tables 7.24 and 7.25, and Bibliography, Sutton), that the two selenium atoms will be covalently bonded at a distance of about 2.3 Å. This Se–Se interaction will produce a strong peak in the Patterson function at about 2.3 Å from the origin.

The atomic numbers of Se, C, and H are 34, 6, and 1, respectively. Hence, the important vectors in the Patterson function would have approximate single-weight peak heights, from (6.18), as follows:

(a) Se–Se: 1156.
(b) Se–C: 204.
(c) C–C: 36.

Because of the presence of identical vectors arising from the C unit cell, all vectors will be double these values.

Figure 6.12 is the Patterson section $P(u,0,w)$, calculated with 1053 data of $|F_o(hkl)|^2$, with grid intervals of 50ths along u,v, and w. The origin peak $P(0,0,0)$ was scaled to 100 and, from (6.17), $\sum_{j=1}^{N} Z_j^2 = 6540$. Hence, the vector interactions (a), (b), and (c) should have peak heights in the approximate ratio $32:5.7:1$.

The section is dominated by a large peak of height 39 at a distance of about 2.3 Å from the origin. Making the reasonable assumption that it represents the Se–Se vector, and since there are no significant peaks on the v axis, the Harker line in Cm, it follows that the space group cannot be Cm, thus leaving $C2$ as the most logical choice.

By measurement on the section, the Patterson coordinates are $u = 6.7/50$ and $w = 2.2/50$, and from the study of space group $P2$, so that $x_{Se} = 0.067$ and $z_{Se} = 0.022$.

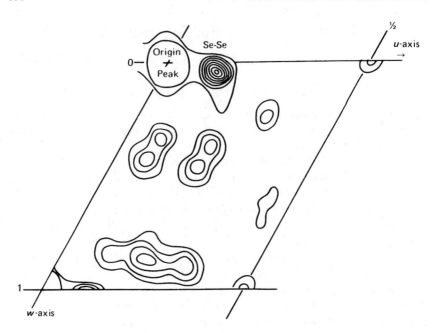

FIGURE 6.12.   Patterson section, $P(u, 0, w)$; the origin peak (height $= 100$) has not been contoured. Contours around the Se–Se peaks are at intervals of 4; elsewhere at intervals of 2.

In space group $C2$, the unit-cell origin is fixed in the xz plane by the 2-fold axis. There is no symmetry element that defines the origin in the y direction, which must be fixed by specifying the y coordinate for a selected atom. For convenience, we may set $y_{Se} = 0$, and our analysis so far may be given as the positions

$$Se: \quad 0.067, 0, 0.022$$

$$Se': \quad -0.067, 0, -0.022$$

A space-group ambiguity is not always resolved in this manner. Sometimes it is necessary to proceed further with the structure analysis, even to refinement stages, before confirmation is obtained.

What of the atoms other than selenium? Is it possible to determine the positions of the carbon and hydrogen atoms? We shall find that we can locate the carbon atoms in this structure from the Patterson synthesis. To explain the procedure, we consider first only part of the structure, including one phenyl ring of the asymmetric unit (Figure 6.13a), and neglect all but the C–Se vectors. The vector set generated by the two Se atoms and 14 C atoms in this hypothetical arrangement contains two images of the structure fragment (one per Se atom), which are displaced from each other by the Se–Se vector. The idealized vector

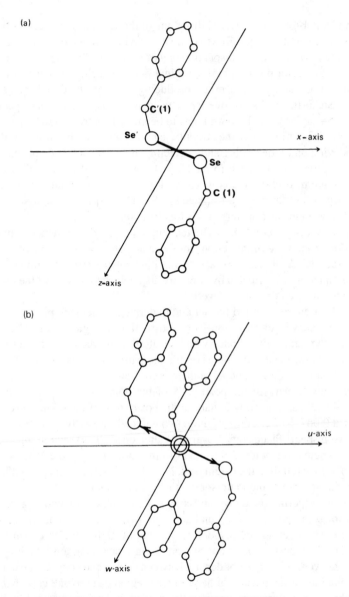

FIGURE 6.13.   (a) Hypothetical structure fragment $C_6H_5CHSe_2CHC_6H_5$ in real space; (b) idealized set of Se–Se and Se–C vectors in Patterson space.

set is shown in Figure 6.13b. By shifting one of these images by a *reverse*[a] Se–Se vector displacement, it is possible to bring the two images into coincidence. Verify this statement by making a transparent copy of Figure 6.13b and placing its origin over an Se–Se vector position in the original figure, keeping the pairs of u and w axes parallel. Certain peaks overlap, producing a single, displaced image of the structure. Shade the peaks that overlap. This image is displaced with respect to the true space-group origin, which we know to be midway between the two Se atoms. A correctly placed image of the structure can be recovered by inserting the true origin position on to the tracing and neglecting any peaks that are not shaded.

The partial vector set was formed from the image of all atoms of the fragment in each Se atom; each image is weighted by Z_j, the atomic number of the jth atom, (carbon, in this example), imaged in Se. The displacement arises because the Patterson synthesis transfers all vectors to a common origin.

Patterson Superposition. The technique just described depends upon the recognition of the vector interaction from a given pair of atoms, the two Se atoms in this example. At least a partial unscrambling of the structure images in the Patterson function was effected by correctly displacing two copies of the Patterson map and noting the positions of overlap.

To illustrate the method further and to derive a systematic procedure for its implementation, we return to the Patterson section in Figure 6.12. The two Se atoms have the same y coordinate, which means that the vector shift takes place in this section. Now, make two copies on tracing paper of the half unit-cell *outline*, $x = 0$ to $\frac{1}{2}$ and $z = 0$ to 1, and label them copy 1 and copy 2.

On copy 1 mark in the position S of the point, $-(x_{Se}, z_{Se})$, which is at $-0.067, -0.022$, and on copy 2 mark in the position S' of the point $-(x_{Se'}, z_{Se'})$, which is at $0.067, 0.022$. Think of these two unit cells as existing in crystal space, not Patterson space. Place copy 1 over the Patterson $(u, 0, w)$ section, maintaining a parallel orientation, with S over the origin, and trace out the Patterson map (Figure 6.14a), excluding the origin peak in each case. Repeat this procedure with copy 2, placing S' over the Patterson section origin (Figure 6.14b).

Finally, superimpose copy 1 and copy 2. As in the exercise with Figures 6.13a and 6.13b, some peaks overlap and some lie over blank regions in one or the other map. The overlaps correspond to regions of high electron density in the crystal. They are best mapped out by compiling a new diagram which contains the *minimum* value of the vector density between copy 1 and copy 2 for each point, thus eliminating or decreasing in height those regions where one copy has no or only slight overlap. A map prepared in this way is shown in Figure 6.14c.

[a] If the forward direction of this vector is used, the structure obtained would, in general, be inverted through the origin. This does not happen with the example under study because the molecule possesses 2-fold symmetry.

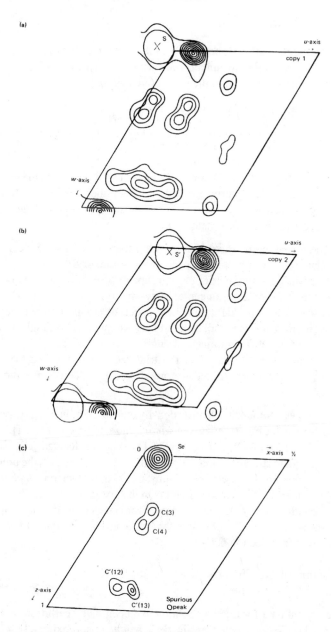

FIGURE 6.14. Bisdiphenylmethyldiselenide: (a,b) Shifted copies 1 and 2 prepared from the $(u, 0, w)$ section; (c) minimum-function M_2 section at $y = 0.0$; C′(12) and C′(13) are symmetry-related to C(12) and C(13) in Figure 6.15.

Minimum Function

The technique outlined above follows the method of Buerger.[a] An analytical expression for the minimum function $M_n(x, y, z)$ is given by (6.21); it may be regarded as an approximation to the electron density $\rho(x, y, z)$.

$$M_n(x, y, z) = \text{Min}[P(u - x_1, v - y_1, w - z_1),$$
$$P(u - x_2, v - y_2, w - z_2), \ldots, P(u - x_n, v - y_n, w - z_n)]$$
$$(6.21)$$

where $\text{Min}(P_1, P_2, \ldots, P_n)$ is the lowest value at the point x, y, z in the set of superpositions $P_1, P_2, \ldots, P_n$; n corresponds with the number of known or trial atomic positions. The following general comments on the application of the minimum function procedure should be noted:

1. The n trial atoms should form within themselves a set or sets of points related by the appropriate space-group symmetry.
2. In a noncentrosymmetric space group, n should be three or more in order to remove the Patterson center of symmetry.
3. If the various n trial atoms have different atomic numbers, the corresponding Patterson copies should be weighted accordingly in order to even out the different image strengths.
4. Incorrectly placed atoms in the trial set tend to confuse the structure image. New atom sites therefore should be added to the model with caution.

Figure 6.15 shows a composite electron density map of the atoms in the asymmetric unit that were revealed by a *three-dimensional* minimum function M_2. This result is quite satisfactory; only C(9), C(10), and C(11) are not yet located. The composite map of the complete structure[b] and the packing of the molecules in the unit cell are shown in Figure 6.16. In favorable circumstances, the Patterson function can be solved for the majority of the heavier atoms in the crystal structure. The atoms not located by M_2 in this example were obtained from an electron density map phased on those atoms that were found, a standard method for attempting to complete a partial structure (see Section 6.4.4).

Determination of the Chlorine Atom Positions in Papaverine Hydrochloride

The crystal data for this compound have been given earlier in this chapter (Section 6.2.1). The calculated origin peak height is approximately 4700, and

[a] See Bibliography.
[b] H. T. Palmer and R. A. Palmer, *Acta Crystallographica* B**25**, 1090 (1969).

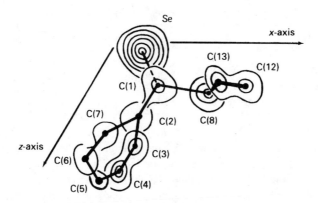

FIGURE 6.15.   Composite map of the three-dimensional minimum function $M_2(x, y, z)$ for bisdiphenylmethyldiselenide.

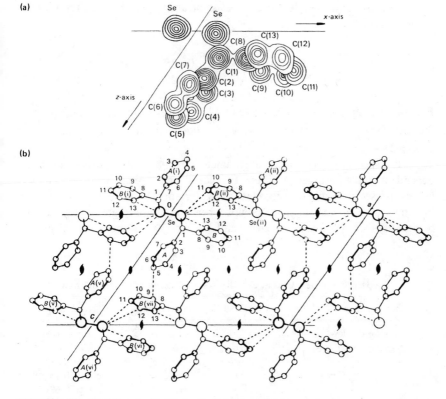

FIGURE 6.16.   Bisdiphenylmethyldiselenide: (a) Composite electron density map as seen along b; (b) crystal structure as seen along b; the dashed lines indicate the closest intermolecular contacts.

TABLE 6.2.   Patterson Peaks in Space group $P2_1/c$

Label	Vector	Peak strength	Harker region
A	$\pm\{0, \frac{1}{2} + 2y, \frac{1}{2}\}$	Double weight	Line:$[0, v, \frac{1}{2}]$
B	$\pm\{2x, \frac{1}{2}, \frac{1}{2} + 2z\}$	Double weight	Section: $(u, \frac{1}{2}, w)$
C	$\pm\{2x, 2y, 2z\}$	Single weight	General region
D	$\pm\{2x, 2\bar{y}, 2x\}$	Single weight	General region

a single-weight Cl–Cl vector would have a height of about 6% of that of the origin peak. The Cl–Cl vector may not be located as easily as that of Se–Se in the previous example. The general equivalent positions in $P2_1/c$ give rise to the vectors shown in Table 6.2. The assignment of coordinates to the chlorine atoms follows the recognition of peaks A, B, and C as Cl–Cl vectors on the Patterson maps (Figure 6.17a–c). Figure 6.17d is the sharpened section, $(u, \frac{1}{2}, w)$. The steps in the solution of the problem are set out in Table 6.3.

The results are completely self-consistent, and we may list the Cl coordinates in the unit cell:

$$4Cl: \quad 0.025, \quad 0.169, \quad 0.038; \quad 0.025, \quad 0.331, \quad 0.538$$
$$-0.025, \quad -0.169, \quad -0.038; \quad -0.025, \quad -0.331, \quad -0.538$$

For simplicity, peak A was assigned as $-\left(\frac{1}{2} + 2y\right)$, which is crystallographically the same as $\frac{1}{2} - 2y$, in order to obtain $y \leqslant \frac{1}{2}$. For a similar reason, B was retained as $\frac{1}{2} + 2z$.

The specification of the peak parameters in this manner is, to some extent, dependent on the observer. A different choice, for example, $\frac{1}{2} + 2y$ in A, merely results in a set of atomic positions located with respect to one of the other centers of symmetry as origin. In space groups where the origin might be defined with respect to other symmetry elements, similar arbitrary peak specifications may be possible.

TABLE 6.3.   Heavy-atom Coordinates for Papaverine Hydrochloride

Patterson map	Label	Vector coordinates[a]	Cl coordinates
Figure 6.17a, level $v = 8.4/52$	A	$\frac{1}{2} - 2y = 8.4/52$	$y = 0.169$
Figure 6.17b, level $v = \frac{1}{2}$	B	$2x = 2.2/44,$	$x = 0.025$
		$\frac{1}{2} + 2z = 17.3/30$	$z = 0.038$
Figure 6.17c, level $v = 17.6/52$	C	$2x = 2.2/44$	$x = 0.025$
		$2y = 17.6/52$	$y = 0.169$
		$2z = 2.3/30$	$z = 0.038$

[a] The Patterson synthesis was computed with the intervals of subdivision 44, 52, and 30 and along u, v, and w, respectively.

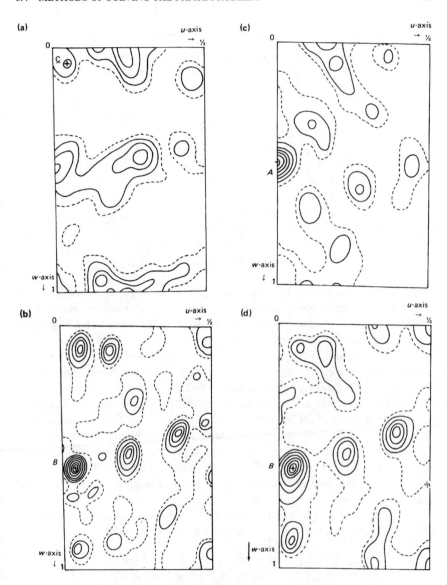

FIGURE 6.17.   Three-dimensional Patterson sections for papaverine hydrochloride; the Cl–Cl vectors are labeled A, B, and C: (a) $v = 8.4/52$, (b) $v = \frac{1}{2}$, (c) $v = 17.6/52$, (d) $v = \frac{1}{2}$ (sharpened section).

Determination of the Mercury Atom Positions in KHg_2

This example illustrates the application of the Patterson function to the determination of the coordinates of atoms in special positions of space group *Imma*.

> *Crystal Data*[a]
> System: orthorhombic.
> Unit-cell dimensions: $a = 8.10, b = 5.16, c = 8.77$ Å.
> V_c: 366.55 Å^3.
> D_m: 7.95 g cm^{-3}.
> M_r: 440.28.
> Z: 3.99, or 4 to the nearest integer.
> Unit-cell contents: 4K and 8Hg atoms.
> Absent spectra: hkl: $h + k + l = 2n + 1$
> $\qquad\qquad\quad hk0$: $h = 2n + 1, (k = 2n + 1)$.

From the diffraction data, possible space groups are *Im2a*, *I2ma*, or *Imma*. In the absence of further information on the space group, we shall proceed with the analysis in *Imma* (Figure 6.18a,b). The reader may like to consider how easily these diagrams may be derived from *Pmma* (origin on $\bar{1}$) $+ I$.

Symmetry and Packing Analyses. Since Z is 4 and there are 16 general equivalent positions in *Imma*, all atoms must lie in special positions. Table 6.4 lists these positions for this space group, with a center of symmetry $(2/m)$ as origin.[b]

This list presents a quite formidable number of alternatives for examination. The eight Hg atoms could lie in (f), (g), (h), or (i). However, further consideration of sets (f), (g), and (i) and sets (c) and (d) shows that they would all involve pairs of Hg atoms being separated by distances less than $b/2$ (2.58 Å). This value is much shorter than known Hg–Hg bond distances in other structures, and we shall reject these sets. The positions in these sets may be plotted to scale in order to verify the spatial limitations.

Of the remaining sets, (a) and (b) together would again place neighboring Hg atoms too close to one another. There are three likely models:

> Model I: four Hg in (a) + four Hg in (e).
> Model II: four Hg in (b) + four Hg in (e).
> Model III: eight Hg in (h).

The Patterson function enables us to differentiate among these alternative models.

---

[a] In the original paper, the origin in *Imma* was chosen on a center of symmetry displaced by $\frac{1}{4}, \frac{1}{4}, \frac{1}{4}$ from this origin.

[b] E. J. Duwell and N. C. Baenziger, *Acta Crystallographica* **8**, 705 (1995).

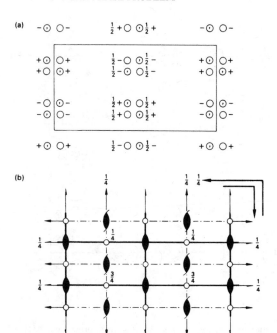

FIGURE 6.18. Space group *Imma* (rotated by 90° from the standard setting): (a) general equivalent positions, (b) symmetry elements.

TABLE 6.4. Special Positions in *Imma*

4	(a)	$2/m$	$0,0,0; \frac{1}{2},0,0; \frac{1}{2},\frac{1}{2},\frac{1}{2}; 0,\frac{1}{2},\frac{1}{2}$
4	(b)	$2/m$	$0,\frac{1}{2},0; \frac{1}{2},\frac{1}{2},0; \frac{1}{2},0,\frac{1}{2}; 0,0,\frac{1}{2}$
4	(c)	$2/m$	$\frac{1}{4},\frac{1}{4},\frac{1}{4}; \frac{1}{4},\frac{3}{4},\frac{1}{4}; \frac{3}{4},\frac{3}{4},\frac{3}{4}; \frac{3}{4},\frac{1}{4},\frac{3}{4}$
4	(d)	$2/m$	$\frac{1}{4},\frac{1}{4},\frac{3}{4}; \frac{1}{4},\frac{3}{4},\frac{3}{4}; \frac{3}{4},\frac{3}{4},\frac{1}{4}; \frac{3}{4},\frac{1}{4},\frac{1}{4}$
4	(e)	$mm2$	$\frac{1}{4},0,z; \frac{3}{4},0,\bar{z}; \frac{3}{4},\frac{1}{2},\frac{1}{2}+z; \frac{1}{4},\frac{1}{2},\frac{1}{2}-z$
8	(f)	2	$\pm\{0,y,0; \frac{1}{2},y,0; \frac{1}{2},\frac{1}{2}+y,\frac{1}{2}; 0,\frac{1}{2}+y,\frac{1}{2}\}$
8	(g)	2	$\pm\{x,\frac{1}{4},\frac{1}{4}; x,\frac{3}{4},\frac{1}{4}; \frac{1}{2}+x,\frac{3}{4},\frac{3}{4}; \frac{1}{2}+x,\frac{1}{4},\frac{3}{4}\}$
8	(h)	m	$\pm\{x,0,z; \frac{1}{2}-x,0,z; \frac{1}{2}+x,\frac{1}{2},\frac{1}{2}+z; \bar{x},\frac{1}{2},\frac{1}{2}+z\}$
8	(i)	m	$\pm\{\frac{1}{4},y,z; \frac{1}{4},\bar{y},z; \frac{3}{4},\frac{1}{2}+y,\frac{1}{2}+z; \frac{3}{4},\frac{1}{2}-y,\frac{1}{2}+z\}$

Vector Analysis of the Alternative Hg Positions. Model I would produce, among others, an Hg–Hg vector at $u = \frac{1}{2}$, $w = 0$, from the atoms in set (a). The *b* axis Patterson projection (Figure 6.19) shows no peak at that position, and we eliminate model I. For a similar reason, with the atoms of set (b), model II is

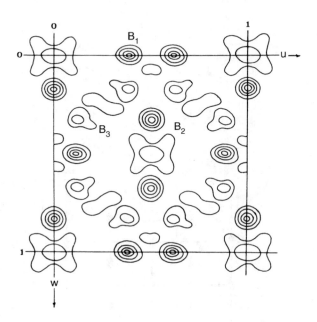

FIGURE 6.19.   Patterson projection $P(u, w)$ for KHg$_2$; the origin peak has not been contoured and the labelled peaks are Hg–Hg vectors.

rejected. It is necessary to show next that model III is consistent with the Patterson function. The a-axis projection is shown in Figure 6.20.

Interpretation of $P(u, w)$.   In this projection, no reference is made to the y coordinates, and we look for vectors of the type $\pm\{\frac{1}{2}+2x, 0\}$ and $\pm\{\frac{1}{2}, 2z\}$, and four vectors related by $2mm$ symmetry $\pm\{2x, 2z\}$ and $\pm\{2\bar{x}, 2z\}$. The double-weight peak labeled $B(1)$ is on the line $w = 0$, and $B(2)$ is on the line $u = \frac{1}{2}$. Hence, $x_{Hg} = 0.064$ and $z_{Hg} = 0.161$. These values are corroborated by measurements from the single weight peak $B(3)$.

Interpretation of $P(v, w)$.   Vectors like A (Figure 6.20) are of the type $\pm\{0, 2z\}$. We deduce $z_{Hg} = 0.161$, in excellent agreement with the value obtained from the b-axis projection.

Superposition techniques applied to the a-axis projection indicate that the K atoms are in special positions (b), but this result is not supported by the b-axis projection. Evidently, the Patterson results can give only a partial structure, and supplementary methods are needed to carry the analysis to completion. In summary, we have determined the positions of the mercury atoms to be in set (h),[a] Table 6.4, with $x = 0.064, z = 0.161$.

---

[a] In the work of Duwell and Baenziger (*loc. cit.*), the positions listed are 8 (i), with $x = 0.186$ and $z = 0.089$, each being $\frac{1}{4}$ *minus* the value given here (see footnote to page 366).

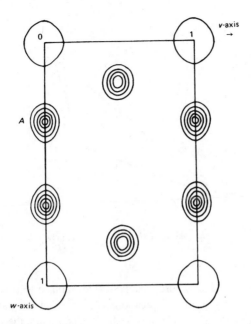

FIGURE 6.20. Patterson projection $P(v, w)$ for KHg_2; the origin peak has not been contoured.

6.4.4 Heavy-Atom Method and Partial Fourier Synthesis

The heavy-atom method was conceived originally as a method for determining the positions of light atoms in a structure containing a relatively small number of heavier atoms. However, the technique can be applied to most situations where a partial structure analysis has been effected, provided that certain condition are met.

Imagine a situation where N_k, of the N atoms in a unit cell have been located; N_k may be only one atom, if it is a heavy atom. There will be N_u atoms remaining to be located, and we may express the structure factor in terms of known (k) and unknown (u) atoms:

$$F(hkl) = \sum_{j=1}^{N_k} g_j \exp[\mathrm{i}2\pi(hx_j + ky_j + lz_j)]$$

$$+ \sum_{u=1}^{N_u} g_u \exp[\mathrm{i}2\pi(hx_u + ky_u + lz_u)] \qquad (6.22)$$

or

$$F(hkl) = F_c(hkl) + F_u(hkl) \qquad (6.23)$$

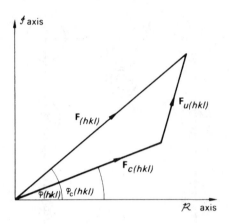

FIGURE 6.21.   Partial-structure phasing; $F(hkl)$ is the true structure factor of modulus $|F_0(hkl)|$ and phase $\phi(hkl)$.

In practice, $|F_0|$ data, appropriately scaled, replace $|F(hkl)|$, and $F_c(hkl)$ refers to the known (N_k) atomic positions. As more of the structure becomes known, the values of $|F_c(hkl)|$ approach $|F_0(hkl)|$ and the phase angle ϕ_c approaches the unobservable but required value $\phi(hkl)$. Figure 6.21 illustrates this argument for any given reflection. The values of ϕ_c may provide sufficiently reasonable approximations to $\phi(hkl)$ for an electron density map to be calculated with some confidence. The nearer F_c is to $F(hkl)$, the better the value of the phase angle, and this is clearly dependent upon the percentage of the scattering power which is known. As a guide to the effective phasing power of a partial structure, the quantity r may be calculated:

$$r = \sum_{j=1}^{N_k} Z_j^2 \bigg/ \sum_{u=1}^{N_u} Z_u^2 \qquad (6.24)$$

where Z refers to the atomic number of a species. A value of r near unity is considered to provide a useful basis for application of the heavy-atom method. However, values of r quite different from unity have produced successful results, because for a given reflection the important quantity is really r', the scattering ratio, given by

$$r' = \sum_{j=1}^{N_k} g_j^2 \bigg/ \sum_{u=1}^{N_u} g_u^2 \qquad (6.25)$$

If r is large, however, the heavy-atom contributions tend to swamp those from the lighter atoms, which may then not be located very precisely from electron density maps. On the other hand, if r is small, the calculated phase may deviate widely from the desired value, and the resulting electron density map could be very difficult to interpret. These extreme situations are found in two of the structures just studied, bisdiphenylmethyldiselenide ($r = 2.4$) and papaverine hydrochloride ($r = 0.28$), based, in each case, on the heavy atoms alone in N_k.

The underlying philosophy of the heavy-atom method depends on the acceptance of calculated phases, even if they contain errors, for the computation of the electron density synthesis. Large phase errors give rise to high background features, which mask the image of the correct structure. The calculated phases ϕ_c contain errors arising from inadequacies in the model, but the $|F_o|$ data, although subject to experimental errors, hold information on the complete structure. Phase errors may be counteracted to some extent by weighting the Fourier coefficients according to the degree of confidence in a particular phase. For centrosymmetric structures, the weight $w(hkl)$ by which $|F_o(hkl)|$ is multiplied is given by[a]

$$w(hkl) = \tanh(\chi/2) \qquad (6.26)$$

where χ is given by

$$\chi = 2|F_o||F_c|\Big/ \sum g_u^2 \qquad (6.27)$$

The subscripts c and u refer, respectively, to the known and unknown parts of the structure. We can show, in a simplified manner, how one may reasonably expect the heavy-atom procedure to be successful. In a centrosymmetric structure, the two terms on the right-hand side of (6.22) would be cosine expressions. The sum over the N_k atoms would have a magnitude $M1$ and be either $+$ or $-$ in sign. Similarly, the sum over the N_u atoms would have a magnitude $M2$ together with a $+$ or a $-$ sign. Over a number of reflections, we may say that there is a 50% chance that the true signs are those given by the heavy atom. For the other 50%, there is a 25% chance that $M1 > M2$, so that again the sign given by the heavy atom is correct. Thus, there is a good chance that a large percentage of the reflections will be given the correct sign in a favorable heavy-atom application. In noncentrosymmetric structures, $w(hkl)$ can be obtained from the graph in Figure 6.22.[b] Weighting factors should be applied to $|F_o|$ values that have been placed on an absolute, or approximately absolute, scale.

Bearing these points in mind, it follows that the best electron density map one can calculate with phases determined from a partial structure is given by

$$\rho(x, y, z) = \frac{2}{V_c} \sum_{h=0}^{\infty} \sum_{k=-\infty}^{\infty} \sum_{l=-\infty}^{\infty} w(hkl)|F_o(hkl)|$$
$$\times \cos[2\pi(hx + ky + lz) - \phi_c(hkl)] \qquad (6.28)$$

[a] M. M. Woolfson, *Acta Crystallographica* **9**, 804 (1956).
[b] G. A. Sim, *Acta Crystallographica* **13**, 511 (1960).

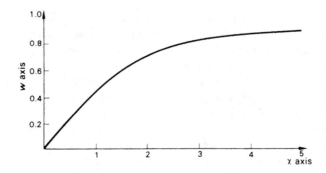

FIGURE 6.22.   Weighting factor $w(hkl)$ as a function of χ in noncentrosymmetric crystals.

where

$$\phi_c(hkl) = \tan^{-1}[B'_c(hkl)/A'_c(hkl)] \qquad (6.29)$$

$A'_c(hkl)$ and $B'_c(hkl)$ are the real and imaginary components, respectively, of the calculated structure factor, which is included in the right-hand side of (6.23).

Electron density maps calculated from partial-structure phasing contain features which characterize both the true structure and the partial, or trial, structure. We have considered this situation in Section 5.9.1. Now, we may let each observation in (5.66) be multiplied by the weight w, so as to give a better statistical significance to each term in the calculation.

If the model includes atoms in reasonably accurate positions, we can expect two important features in the electron density map: (a) atoms of the trial structure should appear, possibly in corrected positions, and (b) additional atoms should be revealed by the presence of peaks in stereochemically sensible positions.

If neither of these features is observed in the electron density synthesis, it may be concluded that the trial structure contains very serious errors, and we would be on a false trail. Correspondingly, there would be poor agreement in the *pattern* of relationship between $|F_o|$ and $|F_c|$ as we describe next.

Reliability Factor

The differences between the *scaled*-observed and the calculated structure-factor amplitudes are a measure of the quality of the trial structure. Large differences correspond to poor reliability, and vice versa. An overall reliability factor (R factor) is defined as

$$R = \sum_{hkl} |K|F_o| - |F_c|| \Big/ \sum_{hkl} K|F_o| \qquad (6.30)$$

For a well refined structure model, the value of R approaches a small value (about 1% at best), corresponding to the residual errors in both the experimental data and the model. In the early stages of the analysis, however, R may lie between 0.4 and 0.5. It expresses the first criterion of correctness, namely, good agreement between $|F_o|$ and $|F_c|$. It should be noted that trial structures with an R factor of more than 50% have been known to be capable of refinement—R is only a rough guide at that stage of the analysis. A better basis for judgment is a comparison of the *pattern* of $|F_o|$ and $|F_c|$, which requires care and experience.

Pseudosymmetry in Electron Density Maps

The electron density map calculated with phases derived from the heavy-atom positions may not exhibit the true space-group symmetry. Suppose space group $P2_1$, for example, has one heavy atom per asymmetric unit. The origin is defined with respect to the x and z axes by the 2_1 axis along $[0, y, 0]$, but the y coordinate of the origin is determined with respect to an arbitrarily assigned y coordinate for *one* of the atoms. Consider the heavy atoms at x, y, z and, symmetry-related, at $\bar{x}, \frac{1}{2} + y, \bar{z}$. This arrangement has the symmetry of $P2_1/m$, with the m planes cutting the y axis at whatever y coordinate is chosen for the heavy atom, say y_H, and at $\frac{1}{2} + y_H$. If $y_H = \frac{1}{4}$, a center of symmetry is at the origin, and the calculated phases will be 0 or π. This situation is illustrated in Figure 6.23, which indicates that an unscrambling of the images must be carried out. If the heavy atom is given any general value for y_H, $B'(hkl)$ will not be zero and the phase angles will not be 0 or π, but the pseudosymmetry will still exist.

Successive Fourier Refinement

A single application of the Fourier method described above does not usually produce a complete set of atomic coordinates. It should lead to the inclusion of more atoms into subsequent structure factor calculations and so to a better electron density map, and so on. This iterative process of Fourier refinement should, after several cycles, result in the identification of all nonhydrogen atoms in the structure to within about 0.1 Å of their true positions. Further improvement of the structure would normally be carried out by the method of least squares, which is described in Chapter 7.

6.4.5 Difference-Fourier Synthesis

Some errors present in the trial structure may not be revealed by Fourier synthesis. In particular, the following situations are important.

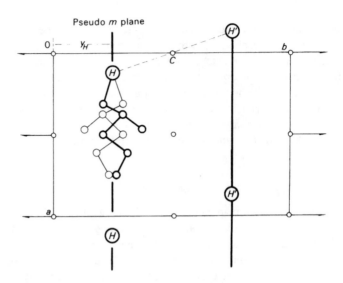

FIGURE 6.23.   Introduction of pseudosymmetry into space group $P2_1$ by single heavy-atom phasing. H is the heavy atom and C is a center of symmetry introduced between H and its $P2_1$ equivalent H'. The space group (for the heavy atoms alone) thus appears as $P2_1/m$ with mirror (and $\bar{1}$) pseudosymmetry. The electron density map, phased on the H and H' species, will contain two mirror-related images in the asymmetric unit, with a certain degree of confusion between them.

1. Atoms in completely wrong positions tend to be returned by the Fourier process with the same fractional coordinates, but sometimes with a comparatively low electron density.
2. Correctly placed atoms, may have been assigned either the wrong atomic number, for example, C for N, or an incorrectly estimated temperature factor.
3. Small corrections to the fractional coordinates may be difficult to assess from the Fourier map.

In these circumstances, a difference-Fourier synthesis is valuable. We shall symbolize the Fourier series with $|F_o|$ coefficients as $\rho_o(x, y, z)$ and the corresponding synthesis with $|F_c|$ instead as $\rho_c(x, y, z)$; the difference-Fourier synthesis $\Delta\rho(x, y, z)$ may be obtained in a single-stage calculation from the equation

$$\Delta\rho(x, y, z) = \frac{2}{V_c} \sum_h \sum_k \sum_l (|F_o| - |F_c|) \cos[2\pi(hx + ky + lz) - \phi_c] \quad (6.31)$$

Since the phases are substantially correct at this stage, it is in effect, a subtraction, point by point, of the "calculated," or trial, Fourier synthesis from that of the "observed," or experimentally based, synthesis. The difference synthesis has the following useful properties.

1. Incorrectly placed atoms correspond to regions of high electron density in $\rho_c(x, y, z)$ and low density in $\rho_o(x, y, z)$; $\Delta\rho(x, y, z)$ is therefore negative in these regions.
2. A correctly placed atom with either too small an atomic number or too high a temperature factor shows up as a small positive area in $\Delta\rho$. The converse situations produce negative peaks in $\Delta\rho$.
3. An atom requiring a small positional correction tends to lie in a negative area at the side of a small positive peak. The correction is applied by moving the atom into the positive area.
4. Very light atoms, such as hydrogen, may be revealed by a $\Delta\rho$ synthesis when the phases are essentially correct, after least-squares refinement has been carried out.
5. As one final test of the validity of a refined structure, the $\Delta\rho$ synthesis should be effectively featureless within 2–3 times the standard deviation of the electron density (see Section 7.7).

6.4.6 Limitations of the Heavy-Atom Method

The Patterson and heavy-atom techniques are effective for structures containing up to about 100 atoms in the asymmetric unit. It is sometimes necessary to introduce heavy atoms artificially into structures. This process may not be desirable because a possible structural interference may arise, and there will be a loss in the accuracy of the light-atom positions. An introduction to "direct methods," capable of solving the phase problem for such structures, is given in the next chapter.

Patterson Selection

It is possible that the Patterson function for the crystal of a heavy-atom compound may not reveal the heavy-atom vector unambiguously. Figure 6.24a is the Patterson projection on to the xz plane for euphenyl iodoacetate, $C_{32}H_{53}O_2I$, and Figure 6.24b is the Harker section for the same material. There are two high peaks, A and B, in the asymmetric unit where only one was expected. For the Harker section, the coefficients were sharpened, and further modified by the multiplicative function $\exp(-9\sin^3\theta)$ so as to smooth out any undesirable fluctuations, caused by the sharpening which enhances the high-order reflections relative to those of low order.

The $|F_o|^2$ data, averaged in zones of $2\sin\theta$, are plotted as a function of $2\sin\theta$ in Figure 6.24c(i). Between the values for $2\sin\theta$ of 0.5 and 0.7, the average values of $|F_o|^2$ are enhanced, owing to the multiplicity of similar distances in the structure (see Figure 1.2), compared with the corresponding smoothed curve,

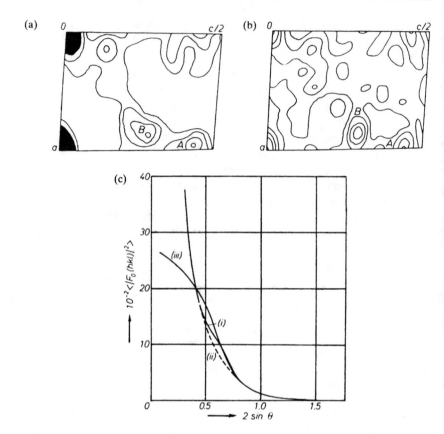

FIGURE 6.24.   Patterson studies on the heavy-atom compound $C_{32}H_{53}O_2I$ ($P2_1$; $Z = 2$): (a) Projection $P(uw)$, (b) Sharpened Patterson–Harker section $P(v\frac{1}{2}w)$, (c) plots of average $|F_o|^2$ against $2\sin\theta$: (i) experimental data, (ii) "smooth" curve through the experimental points, (iii) hypothetical (random) structure $C_{32}O_2I$, with the same crystal geometry as $C_{32}H_{53}O_2I$.

Figure 6.24c(ii). The many equal, or nearly equal, vectors between atoms in the molecule are superimposed, in the Patterson function and lead to additional large peaks.

The $|F_o|^2$ data were "selected" by excluding from the next Patterson synthesis all those data lying within the range $0.5 < 2\sin\theta < 0.7$, and the resulting sharpened Harker section is shown in Figure 6.25; clearly, the heavy-atom vector is at B. The selection process has effectively removed the "structure" that was giving rise to the additional peak A, so that the heavy-atom vector was then sought among the vectors from a more random array of atoms.[a]

[a] M. F. C. Ladd, *Zeitschrift fur Kristallographie* **124**, 64–68 (1967).

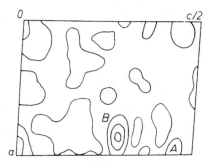

FIGURE 6.25. Sharpened Patterson–Harker section $P(u, \frac{1}{2}, w)$, with the $|F_o|^2$ data selected as described in the text. The I–I vector is clearly at B, whereas in Figures 6.24a,b, there were *two* peaks of equal height in the asymmetric unit.

Figure 6.24c(iii) is the average $|F_o|^2$ curve for a random arrangement corresponding to $C_{32}O_2I$, with the same unit-cell geometry as that of euphenyl iodoacetate, and with no two atoms closer than 1.6 Å. The R factor against euphenyl iodoacetate itself was 0.58; the value for a completely random noncentrosymmetric structure is 0.586.[a] The curve evolves in a manner similar to an f^2 curve and, as would be expected, shows no structural effects of the nature of Figure 6.24c(i).

6.4.7 Isomorphous Replacement

A common feature of biologically important substances is their high molecular weight. Proteins and enzymes, for example, are polymers built up from various amino acid residues and forming very large atomic assemblies with molecular weights greater than 5000. The study of the conformations of these giant molecules is necessary for an understanding of their biological functions, and the principal method of obtaining structural detail is by x-ray analysis.

Because of their high molecular weight, protein structures do not yield to analysis by the heavy-atom method. The value of r, from (6.24), is typically 0.03 for a protein molecule of molecular weight 5000 containing one (added) mercury atom. This value of r is too small to be useful. Another difficulty is that most proteins and enzymes contain neither very heavy atoms nor easily replaceable groups to facilitate the introduction of heavy atoms. In spite of these difficulties, if a heavy-atom derivative of a large molecule can be prepared, it may be possible to induce it to crystallize in a similar size of unit cell and with the same space group as the native compound. Such pairs of compounds are said to be isomorphous.

The structure factor of the heavy-atom derivative may be expressed vectorially as F_{PH}, where

$$F_{PH} = F_P + F_H \qquad (6.32)$$

[a] A. J. C. Wilson, *Acta Crystallographica* **3**, 397–398 (1950).

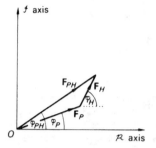

FIGURE 6.26. Graphical interpretation of the isomorphous replacement equation. In practice, the phase ϕ_P and ϕ_{PH} are unknown initially. F_H may be known with a fair degree of accuracy if the heavy-atom position in the isomorphous derivative are known. This enables a solution, as illustrated in Figure 6.27, to be obtained.

F_P and F_H are the structure factors for the parent protein and the heavy atoms alone, respectively, for the same reflection. This relationship is shown in Figure 6.26.

Assuming that the positions of the N_H heavy atoms in the unit cell can be determined, their contribution can be calculated:

$$F_H = \sum_{j=1}^{N_H} g'_j \exp[\mathrm{i}2\pi(hx_j + ky_j + lz_j)] \qquad (6.33)$$

where $g'_j = f'_j \exp(-B_j \sin^2\theta/\lambda^2)$, and $f'_j = K_j f_j$; K_j is a site occupation factor, less than or equal to unity[a] depending on the degree of substitution at the heavy-atom site j; some heavy-atom binding sites of the protein molecules in the crystal may not be fully substituted.

To obtain an idea of the effect of a heavy atom on the intensities of x-ray reflections from a protein, we shall carry out a simple calculation for a crystal containing one protein molecule per unit cell in space group $P1$. Assuming that it has a molecular weight of about 13,000, about 1000 nonhydrogen atoms would comprise the molecule; we shall assume that they are all carbon ($Z_C = 6$). Accepting Wilson's approximation (3.173) and replacing g_j by f_C, we have

$$\overline{|F_P|^2} \approx \sum_{j=1}^{1000} f_C^2 \qquad (6.34)$$

At $\sin\theta = 0, \overline{|F_P|^2}$ is 36,000. If the derivative contains one mercury atom ($Z_{\mathrm{Hg}} = 80$),

$$\overline{|F_{PH}|^2} \approx \sum_{j=1}^{1000} f_C^2 + \sum_{j=1}^{N_H} f'^2_j \qquad (6.35)$$

---

[a] Assuming $|F_P|$ and $|F_{PH}|$ to be on an absolute scale.

which has the value 42,400 at $\sin \theta = 0 (f_j = 80)$. Hence, the maximum change in intensity is about 18%, which is a surprisingly high value.

Experimentally, two sets of data $|F_P(hkl)|$ and $|F_{PH}(hkl)|$ are measured and, because of the comparative nature of the phase-determining procedure with isomorphous compounds, they must be placed on the same relative scale, which can be achieved by Wilson's method.

Rewriting (6.32), we have

$$|F_P| \exp(\mathrm{i}\phi_P) = |F_{PH}| \exp(\mathrm{i}\phi_{PH}) - F_H \qquad (6.36)$$

Assuming that F_H can be determined, this equation involves two unknown quantities, ϕ_P and ϕ_{PH}, and cannot yield a unique solution. However, Figure 6.27 shows that ideally only two solutions for ϕ_P are real, corresponding to the vectors OP_1 and OP_2, one of which is the true F_P vector. A second isomorphous derivative with a *different set* of heavy-atom positions, will also have two solutions for F_P. The derivatives are denoted 1 and 2, and the solutions for F_P are OP_{11} and OP_{12} (derivative 1) and OP_{21} ad OP_{22} (derivative 2), as shown in Figure 6.28. Two of the solutions should agree with either OP_1 or OP_2 within experimental error, thus resolving the ambiguity (Figure 6.28). With a more extensive series of isomorphous derivatives, it is possible to obtain phases capable of yielding interpretable

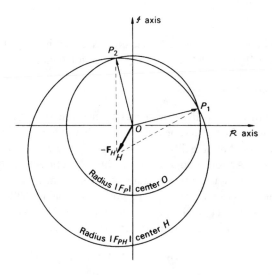

FIGURE 6.27. Single isomorphous replacement (SIR) phase–amplitude diagram in the single isomorphous replacement method. $OH(-F_H)$ is the known reversed heavy-atom vector. The triangles OHP_1 and OHP_2 both satisfy (6.32), giving a twofold ambiguity with either OP_1 or OP_2 as the solution for F_P.

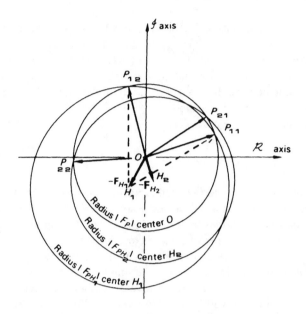

FIGURE 6.28.   Multiple isomorphous replacement (MIR) resolution of the phase ambiguity with a second isomorphous heavy-atom derivative PH_2. The determined direction of F_P is near OP_{11} and OP_{21}. In practice, P_{11} and P_{21} rarely coincide, due to inaccuracies in the heavy-atom parameters and lack of complete isomorphism.

electron density maps. Many protein structures have been investigated successfully by this technique. Further details are given in Section 6.4.8.

Centrosymmetric Projections

Proteins always crystallize in noncentrosymmetric space groups because the amino acid residues in the polypeptide have "left-handed" configurations about the α-carbon atoms. Amino acid residues with "right-handed" configurations are very rare in nature. Although noncentrosymmetric structures usually present more difficulties than centrosymmetric structures, there is a compensation in the relative ease of determination of the space group; ambiguities such as $P2_1$ and $P2_1/m$ do not exist for the protein crystallographer. Most noncentrosymmetric space groups have at least one centric zone. In such a case, (6.32) becomes

$$s_{PH}|F_{PH}| = s_P|F_P| + s_H|F_H| \qquad (6.37)$$

where s refers to the sign of the structure factor, and is ± 1.

Unless both $|F_{PH}|$ and $|F_P|$ are very small compared with $|F_H|$, it is unlikely that s_{PH} will differ from s_P. Generally F_P and F_{PH} are pointing in the same direction. Accepting this statement, we may substitute, s_P for s_{PH} in (6.37):

$$s_P(|F_{PH}| - |F_P|) = s_H|F_H| \qquad (6.38)$$

or

$$s_P = s_H|F_H|/\Delta F \qquad (6.39)$$

where $\Delta F = |F_{PH}| - |F_P|$. Since we are interested only in the signs ($\pm$), (6.39) may be rewritten as

$$s_P = s_H s_\Delta \qquad (6.40)$$

where s_Δ is $+1$ if $|F_{PH}| > |F_P|$ and -1 if $|F_{PH}| < |F_P|$. In this way, signs can often be determined for centric reflections in a protein crystal with only a single isomorphous derivative, and we shall illustrate the method by the following example.

Sign Determination for Centric Reflections in Protein Structures

We shall consider data for both the enzyme ribonuclease and a heavy-atom derivative prepared by soaking pregrown crystals of the enzyme in $K_2[PtCl_6]$ solution.

Crystal Data for Ribonuclease
System: monoclinic.
Unit-cell dimensions: $a = 30.31, b = 38.26, c = 52.91$ Å, $\beta = 105.9°$.
M_X: 13,500 (ribonuclease).
Z_P: two molecules of ribonuclease plus an unknown number of water molecules.
Z_{PH}: as for $Z_P + N_H[PtCl_6]^{2-}$ groups per unit cell.
Absent spectra: $0k0: k = 2n + 1$.
Space group: $P2_1$. The $h0l$ zone is centrosymmetric.

Given the heavy-atom positions, Table 6.5 shows how the signs for some $h0l$ reflections have been determined. Notice that experimental errors in $|F_P|$ and $|F_{PH}|$, together with errors in the calculated $|F_H|$ arising from inaccuracies in the heavy-atom model, are reflected in the inequality of ΔF and $|F_H|$. The validity of (6.40) is upheld by these data.

Location of Heavy-Atom Positions in Proteins

In a centrosymmetric zone, it follows from (6.39), since s_P and s_H are ± 1, that

$$|F_H| = |\Delta F| \qquad (6.41)$$

TABLE 6.5. $h0l$ Data for Ribonuclease

hkl	$\|F_P\|$	$\|F_{PH}\|$	$\|\Delta F\|$	s_Δ	$\|F_H\|$	s_H	$s_P = s_H s_\Delta$
	Observed data				Calculated data		Deduced sign
003	437	326	111	-1	50	$+1$	-1
006	59	48	11	-1	27	-1	$+1$
007	182	109	73	-1	90	-1	$+1$
$10\overline{17}$	144	196	52	$+1$	31	-1	-1
1013	146	82	64	-1	52	$+1$	-1
$10\overline{9}$	97	165	68	$+1$	55	-1	-1
106	183	242	59	$+1$	45	$+1$	$+1$
$30\overline{4}$	746	861	115	$+1$	72	$+1$	$+1$
405	103	57	46	-1	56	$+1$	-1

where $|\Delta F| = ||F_{PH}| - |F_P||$. A Patterson function calculated with $|F_H|^2$ as coefficients would give the vector set of the substituted heavy atoms in the protein molecule. Since $|F_H|$ cannot be observed directly, the next best procedure is to calculate a difference Patterson map with $(\Delta F)^2$ as coefficients. If the experimental errors in $|F_P|$ and $|F_{PH}|$ are not significant, and not too many sign "cross-overs" with s_P and s_{PH} occur, then the $(\Delta F)^2$ Patterson projection would be expected to reveal the heavy-atom vectors. In the case of general noncentrosymmetric reflections, we note in Figure 6.29 that since $OQ = |F_{PH}|$, $OP = |F_P|$, and $OR = |F_P| \cos(\phi_{PH} - \phi_P)$, we have

$$RQ = |F_H| \cos(\phi_H - \phi_{PH}) \quad \text{and} \quad OQ = OR + RQ \qquad (6.42)$$

Hence,

$$|F_{PH}| - |F_P| \cos(\phi_{PH} - \phi_P) = |F_H| \cos(\phi_H - \phi_{PH}) \qquad (6.43)$$

If $\phi_{PH} - \phi_P$ is small it follows that $\cos(\phi_{PH} - \phi_P) \approx 1.0$ and

$$|F_{PH}| - |F_p| = \Delta F \approx |F_H| \cos(\phi_H - \phi_{PH}) \qquad (6.44)$$

In practice, since the angle $\phi_H - \phi_{PH}$ is undeterminable at this stage, the best one can do is to calculate a Patterson function with $(\Delta F)^2$ coefficients as for centrosymmetric reflections, but as an added precaution to ensure that $(\phi_{PH} - \phi_P)$ is small, use only those terms for which both $|F_P|$ and $|F_{PH}|$ are large. Although the noncentrosymmetric $(\Delta F)^2$ synthesis is not a true Patterson function, it has been used successfully to determine the heavy-atom distribution in proteins.

The most useful derivatives contain a small number of highly substituted sites. Unlike the structure analysis of smaller molecules, it is not known initially how many heavy-atom sites have been incorporated into the molecule.

As an example, we shall consider the $(\Delta F)^2$ Patterson map for the Pt derivative of ribonuclease, space group $P2_1$. The vectors between symmetry-related

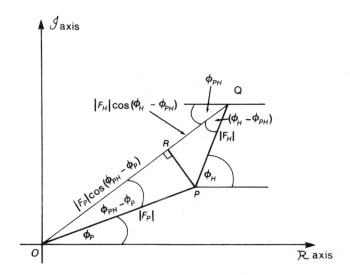

FIGURE 6.29. Location of heavy-atom positions in proteins.

atoms occur on the Harker section $\left(u, \frac{1}{2}, w\right)$. Eight peaks occur on the Harker section and four at $v = 0$ (Figure 6.30a and 6.30b). This result suggests that there is more than one heavy-atom site per protein molecule. The most obvious choice is two, since four heavy atoms per unit cell would give rise to 12 nonorigin peaks. If the two sites are labeled 1 and 2, their Harker peaks will be of the form $\pm\left\{2x_1, \frac{1}{2}, 2z_1\right\}$ and $\pm\left\{2x_2, \frac{1}{2}, 2z_2\right\}$.

Interpretation of the Patterson function is best undertaken in terms of the Harker section, assuming that the peaks represent nonoverlapping vectors and ignoring the possibility that some peaks could be non-Harker peaks. Since the true Harker peaks are of the form $2x$, $2z$, values of x and z can be obtained by dividing by 2 the fractional coordinates on the Harker section.

This analysis may be carried out graphically. The peak positions from the Harker section are replotted, on tracing paper, on a unit cell projection in which the a and c dimensions are each reduced by a factor of $\frac{1}{2}$. This procedure results in one quadrant of Figure 6.31a. The diagram is completed by operating on the first quadrant with the translation of $a/2$, and then on both quadrants by $c/2$, thus completing an area the size of the true unit-cell projection.

All points marked on this map locate potential (x, z) coordinates for the heavy atoms. In fact, it contains four equivalent solutions with respect to the four unique 2_1 axes in the unit cell. Cross-vector peaks are found by moving this *implication diagram*[a] to other sections of the Patterson function, using pairs of potential sites

[a] See Bibliography (Buerger).

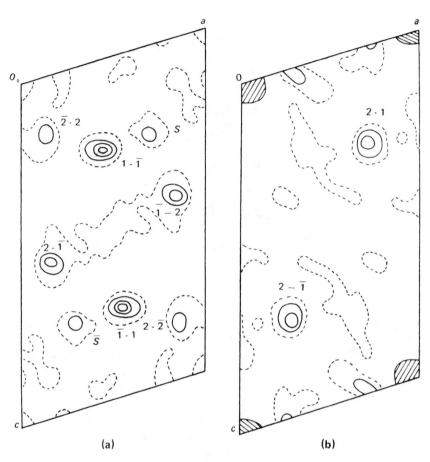

FIGURE 6.30. $(\Delta F)^2$ Patterson sections for the Pt derivative of ribonuclease: (a) $P\left(u, \frac{1}{2}, w\right)$, (b) $P(u, 0, w)$.

to generate potential vectors. To see how this mechanism operates, place the site marked 2 on the tracing paper over the origin of the section $v = 0$ and note the coincidence of site 1 with the peak 2-1. Similarly, the peak 2-$\bar{1}$ and others on the section $v = \frac{1}{2}$ can be generated from the sites 1, 2, $\bar{1}$, and $\bar{2}$ on the implication diagram. Peaks S and $\bar{S}$ are not explained in this way; they may be assumed to be spurious (remember $(\Delta F)^2$ is not a true representation of $|F_H|^2$).

Figure 6.31b shows a composite electron density map of the Pt atom sites which were prepared by an independent method, and confirm the Patterson analysis. The y coordinates of the two heavy-atom sites are almost equal, which accounts for the presence of the non-Harker peaks $\pm(2\text{-}1)$ on the Harker section. Figure 6.32

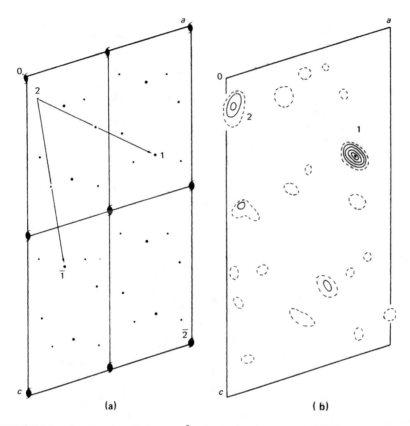

FIGURE 6.31. Interpretation of the $(\Delta F)^2$ Patterson sections for ribonuclease (a) implication diagram, (b) electron density map showing the Pt atom sites.

is a stereo-pair showing the course of the polypeptide chain in ribonuclease and the position of the main site in the Pt derivative.

6.4.8 Further Details of the Isomorphous Replacement Phasing Procedure

In single isomorphous replacement (SIR), the ambiguity in ϕ_P is best resolved (see Figure 6.33) by taking $\phi_P = \phi_H$ with F_P along the median OM between P_1 and P_2; $|F_P|$ should be weighted[a] by $m = \cos\psi$ where ψ is the

---

[a] M. G. Rossman and D. M. Blow, *Acta Crystallographica* **14**, 641–646 (1961).

FIGURE 6.32.   Stereoviews of the polypeptide chain in ribonuclease; the main site in the Pt derivative is shown as a simulated octahedrally coordinated group.

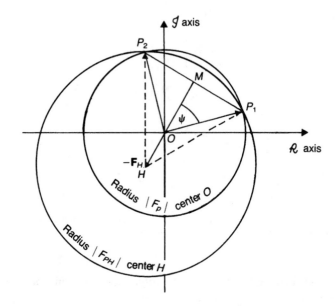

FIGURE 6.33.   Phase-amplitude diagram in the single isomorphous replacement method; $\mathbf{OH}(-F_H)$ is the known reversed heavy-atom vector. The triangles OHP_1 and OHP_2 both satisfy (6.81), giving twofold ambiguity with either $\mathbf{OP}_1$ or $\mathbf{OP}_2$ as the solution for F_P; OM defines the weighted SIR solution for F_P.

semi-angle between P_1 and P_2. By the cosine rule

$$m = \cos \psi = \left||F_{PH}|^2 - |F_P|^2 - |F_H|^2\right|/2|F_P||F_H| \qquad (6.45)$$

we see that m would have a maximum value of unity in the special case for which $|F_H| = ||F_{PH}| - |F_P||$, where the two circles are tangential. The coefficients in the SIR electron density map would be composed of $m|F_P|$ and ϕ_H. The electron density of such a map would be subject to the pseudosymmetry effects discussed in Section 6.4.4. Taking $\phi_P = \phi_H$ is thus the SIR equivalent of the initial stage of the heavy-atom method, in which we take $\phi = \phi_H$.

Analytical Calculation of Phases in SIR and MIR

The geometrical determination of phases by the isomorphous replacement method using Harker's construction is impractical for several reasons:

(i) In MIR (multiple isomorphous replacement), phase-circle intersections, due to accumulated errors, do not usually give absolutely clear indications of ϕ_P, as shown in Figure 6.34 by P_{11} and P_{21}. Actual phase determination in MIR, exemplified by Figure 6.35, contains a complexity of multiple-derivative phase indications, the phase circles intersecting in rather ill-defined regions.

(ii) The size of the task of estimating thousands of $\phi_P(hkl)$ values in a typical protein analysis necessitated the development of an analytical formula suitable for computer programming, as outlined below.

The basis for a computational algorithm, alternative to the Harker construction for SIR, is shown in Figure 6.36, in which the inner circle represents $|F_P|$ and

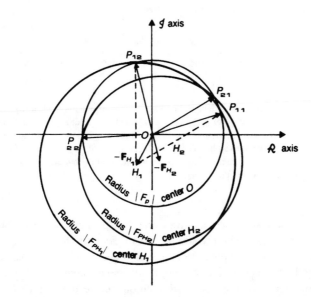

FIGURE 6.34. Multiple isomorphous replacement (MIR) resolution of the phase ambiguity with a second isomorphous derivative PH_2. The determined direction is near $\mathbf{OP}_{11}$ and $\mathbf{OP}_{21}$. In practice P_{11} and P_{21} rarely coincide, owing to inaccuracies in the heavy-atom parameters and a lack of true isomorphism.

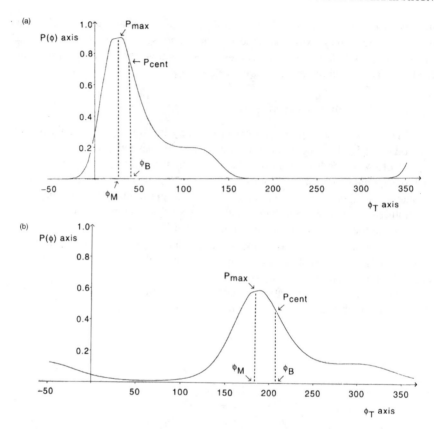

FIGURE 6.35. Examples of MIR where the phase circles do not intersect at a point. The most probable value of the phase $\phi\,(\phi_M$ at $P_{max})$ and the centroid phase (ϕ_B at $P_{cent})$ are indicated: (a) $P(\phi)$ calculated for the two-derivative case in Figure 6.34, with $\phi_B = 43°, \phi_M = 31°$, and $m = 0.80$; and (b) an example of three-derivative phasing (see Problem 6.14) with $\phi_B = 204°, \phi_M = 185°$, and $m = 0.59$. Although based on three derivatives, the probability distribution in (b) is not as sharp as that in (a), resulting in a lower figure of merit, m, and larger $\phi_B \to \phi_M$ difference.

the spokes represent ϕ_T, a series of trial values of ϕ_P for $\phi_T = 0$ to 360° in steps of 30°. The vector F_H, which would be *calculated* from (6.33), using the known heavy-atom parameters, is plotted at the end of each spoke. In order to simplify the drawing, the third side of the isomorphous replacement triangle, representing F_{PH}, has not been joined up. The SIR solutions (corresponding to P_1 and P_2 in Figure 6.33) would occur when F_H just touches the F_{PH} circle, which is plotted concentric with the F_P circle and is the outer circle in Figure 6.36. These two positions are indicated in the diagram, which should be compared with Figure 6.33. Because of the method of selecting ϕ_T, neither is generated exactly in this method.

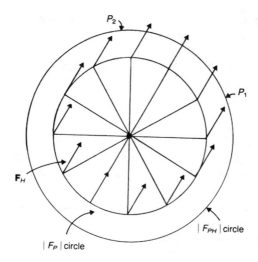

FIGURE 6.36.   The concept of lack of closure; the inner circle represents $|F_P|$ and the outer circle $|F_{PH}|$. Trial values of ϕ_P are plotted at 30° intervals, each carrying the known F_H vector. At P_1 and P_2, F_H ends exactly on the F_{PH} circle; otherwise it fails to close, being too long for the smaller region spanning P_1–P_2 and too short for the rest (see Figure 6.27 for Harker's construction of this SIR case).

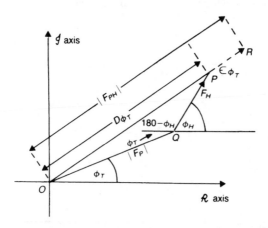

FIGURE 6.37.   Calculation of the lack-of-closure error $\epsilon_{\phi T}$.

Now consider Figure 6.37, which shows a more detailed representation of the case where $\phi_T = 30°$ (Figure 6.36). This is one of the general cases (not P_1 or P_2) where the F_{PH} trial vector would not close the third side of the phase triangle properly. In the case shown F_{PH} is too short; other situations evident in Figure 6.36 would correspond to F_{PH} being too long. In these situations there is a *lack-of-closure error* denoted by ε_{ϕ_T} (Figure 6.37). For the SIR solutions, $\varepsilon_{\phi_T} = 0$. In general, ε_{ϕ_T} may be calculated as follows:

$$D^2_{\phi_T} = |F_P|^2 + |F_H|^2 + 2|F_P||F_H|\cos(\phi_T - \phi_H) \tag{6.46}$$

and

$$\varepsilon^2_{\phi_T} = (|F_{PH}| - D_{\phi_T})^2 \tag{6.47}$$

The SIR solutions could be determined to a satisfactory degree of precision by plotting ε_{ϕ_T} against ϕ_T and locating the two ϕ_T values for which $\varepsilon_{\phi_T} = 0$. This is shown for the example in Figures 6.33 and 6.36 by the graph of Figure 6.38. Both solutions P_1 and P_2 are of course equally probable in the SIR method. In the theory of phase analysis by the MIR method, errors may be assumed to reside in $|F_{PH}|$,[a] which simplifies the calculations. For a given trial value of ϕ_T, the probability that ϕ_T is the correct value is

$$P(\phi_T) = \exp(-\varepsilon^2_{\phi_T}/2E^2) \tag{6.48}$$

where E is the root-mean-square error in $|F_{PH}|$ arising from data errors.

In MIR there would be one value of ε_{ϕ_T} per derivative. Let $\varepsilon_i(\phi_T)$ be the value for derivative i, where $i = 1, 2, \ldots$ to the total number of derivatives. Then

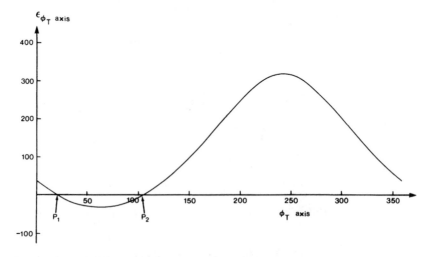

FIGURE 6.38.   Values of ε_{ϕ_T} plotted against ϕ_T; P_1 and P_2 are the two positions for which $\varepsilon_{\phi_T} = 0$, corresponding to the SIR solutions (see Figures 6.33 and 6.36).

[a] D. M. Blow and F. H. C. Crick, *Acta Crystallographica* **12**, 794–799 (1959).

the probability for the ith derivative is

$$P_i(\phi_T) = \exp[-\varepsilon_i^2(\phi_T)/2E_i^2] \tag{6.49}$$

and the combined or joint probability over all derivatives is

$$P(\phi_T) = P_1(\phi_T) \cdot P_2(\phi_T) \cdots$$

or

$$P(\phi_T) = \exp\left[-\sum \varepsilon_i^2(\phi_T)/2E_i^2\right] \tag{6.50}$$

Typical examples of probability distributions met in practice are given in Figure 6.35. Generally the distributions are bimodal, indicating a stronger preference for one maximum over the other. The most probable electron density map uses coefficients $\{|F_P|, \phi_M\}$, where ϕ_M is the phase angle corresponding to the maximum probability in the range 0–360°. However, the electron density map with the least overall root-mean-square error uses coefficients $\{|F_P|, \phi_B\}$, where ϕ_B is the "best" phase angle, corresponding to the centroid (center of gravity) of the probability distribution, and m is a weighting function (or figure of merit), given by

$$m \cos \phi_B = \sum_{\phi_T} P(\phi_T) \cos \phi_T \Big/ \sum P(\phi_T) \tag{6.51}$$

$$m \sin \phi_B = \sum_{\phi_T} P(\phi_T) \sin \phi_T \Big/ \sum P(\phi_T) \tag{6.52}$$

It is convenient in practice to evaluate these expressions by stepping from 0 to 360° in regular intervals of 5° or 10°. The probability distributions and corresponding phases may be readily evaluated by suitable programming.[a]

For each derivative the root-mean-square estimate of error may be taken initially as the average

$$E_j^2 = \langle(|\Delta F_i| - |F_{H_i}|)^2\rangle_{hkl} \tag{6.53}$$

where

$$\Delta F_i = |F_{PH_i}| - |F_P| \tag{6.54}$$

evaluated for centric reflections only.

The error in a phase angle may be defined as $\Delta\phi = \phi_B - \phi_M$ and $m = \cos\Delta\phi$. A value of $m = 0.7$ corresponds to $\Delta\phi \cong 45°$. The average value of m is a measure of the average of $\cos\Delta\phi$. In a typical protein analysis at resolution 2 Å[b] ($\sin\theta_{max} = \lambda/4$), an average m of 0.6–0.7 would be acceptable. Further practical details of MIR are to be found in Chapter 10.

[a] R. E. Dickerson, J. E. Weintzierl, and R. A. Palmer, *Acta Crystallographica* B24, 997–1003 (1968).
[b] The resolution of a protein x-ray analysis is loosely defined as d_{min}, where $d_{min} = \lambda/2\sin\theta_{max}$, θ_{max} being the maximum Bragg angle associated with the analysis: θ_{max} may be restricted in order to limit the labor required, at the expense of the quality of the electron density image.

Electron Density Maps Used in Large-Molecule Analysis

The correlation of heavy-atom sites between derivatives requires one to establish the coordinates of heavy atoms in derivative i with respect to those of another derivative or combination of derivatives for which phases $\phi_{P \neq i}$ have been determined. A difference electron density map may be calculated as

$$\rho \Delta_i (xyz) = \frac{1}{V} \sum_h \sum_k \sum_l (|F_{PH_i}| - |F_P|) \cos[2\pi(hx + ky + lz) - \phi_{P \neq i}]$$

(6.55)

This should reveal the heavy atoms in derivative i with respect to the same origin as in the other heavy-atom derivatives. Derivative i can then be added into the MIR procedure.

For a trial structure in which phases have been calculated, as in small-molecule analysis, a difference electron density map may be used in order to effect corrections to the structure:

$$\Delta \rho(xyz) = \frac{1}{V} \sum_j \sum_k \sum_l (|F_P| - |F_c| \cos[2\pi(hx + ky + lz) - \phi_c]$$ (6.56)

Alternatively a double-difference map

$$\rho'(xyz) = \frac{1}{V} \sum_h \sum_k \sum_l (2|F_P| - |F_c|) \cos[2\pi(hx + ky + lz) - \phi_c]$$ (6.57)

where $\rho'(xyz)[= \rho_c(xyz) + \Delta\rho(xyz)]$ may be used since new features may be more easily recognized in $\Delta\rho(xyz)$ against the background of the known $\rho_c(xyz)$ structure. This map is very useful in computer graphics analysis.

In MIR the most error-free electron density is calculated as

$$\rho_P(xyz) = \frac{1}{V} \sum_h \sum_k \sum_l m|F_P| \cos[2\pi(hx + ky + lz) - \phi_B]$$ (6.58)

where ϕ_B is the MIR phase corresponding to the centroid of the phase probability distribution (the best phase) and m is the figure of merit; see Equations (6.51) and (6.52).

6.5 Anomalous Scattering

Friedel's law is not an exact relationship, and becomes less so as the atomic numbers of the constituent atoms in a crystal increase. The law breaks down severely if x-rays are used that have a wavelength just less than that of an absorption edge (see Section 3.1.3) of an atom in the crystal. However, this criterion is not essential for anomalous scattering to be used in two important aspects of crystal structure analysis, namely, the determination of absolute stereochemical configurations and the phasing of reflections.

Anomalous scattering introduces a phase change into a given atomic scattering factor, which becomes complex:

$$f = f_0 + \Delta f' + i\Delta f'' = f' + i\Delta f'' \tag{6.59}$$

$\Delta f'$ is a real correction, usually negative, and $\Delta f''$ is an imaginary component which is rotated anticlockwise through $90°$ in the complex plane with respect to f_0 and $\Delta f'$, that is, to f'.

A possible situation is illustrated in Figure 6.39. In Figure 6.39a, atom A is assumed to be scattering in accordance with Friedel's law, and it is clear that $|F(\mathbf{h})| = |F(\bar{\mathbf{h}})|$, where $\mathbf{h}$ stands for hkl. In Figure 6.39b, atom A is represented as an anomalous scatterer, with its three components, according to (6.59). In this situation, $|F(\mathbf{h})| \neq |F(\bar{\mathbf{h}})|$, and intensity measurements of Friedel pairs of reflections produce different values.

We can safely assume that procedures for measuring $I(hkl)$ correctly differentiate between hkl and $\bar{h}\bar{k}\bar{l}$. In any non-centrosymmetric space group a structure model can be inverted, as if through a center of symmetry, and used to recalculate

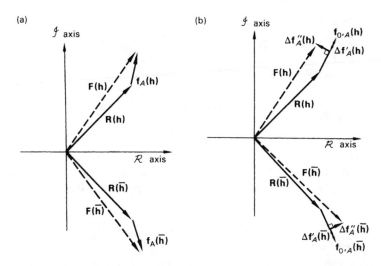

FIGURE 6.39.   Anomalous scattering of atom A with respect to the rest of the structure R: (a) normal case—$|F(\mathbf{h})| = |F(\bar{\mathbf{h}})|$; (b) anomalous case—$|F(\mathbf{h})| \neq |F(\bar{\mathbf{h}})|$. The general non-centrosymmetric case is illustrated. For centrosymmetric crystals, $|F_{\mathbf{h}}|$ always equals $|F_{\bar{\mathbf{h}}}|$, but $\phi_{\mathbf{h}}$ differs to a small degree from 0 or π if anomalous scattering is significant.

TABLE 6.6.   Example of Some Friedel
Pairs and the Corresponding $|F|$ Values

hkl	$|F_o|$	$|F_c|x,y,z$	$|F_c|\bar{x},\bar{y},\bar{z}$
121	17.0	19.1	18.3
122	21.2	22.9	21.9
123	41.4	44.4	42.8
341	36.7	38.7	35.5
342	7.8	9.5	8.2
413	14.2	15.3	13.5

structure factors. Because structure factor formulae involve $(hx + ky + lz)$, the two models will produce different values for $|F_c(hkl)|$ and $|F_c(\bar{h}\bar{k}\bar{l})|$. The correct enantiomorph is expected to produce better agreement between $|F_o|$ and $|F_c|$ and, thus, a lower R-factor. Some typical results are listed in Table 6.6, from which it may be deduced that the structure giving $|F_c|\bar{x},\bar{y},\bar{z}$ corresponds to the absolute configuration. An equivalent procedure would be to measure the values of both $|F(hkl)|$ and $|F(\bar{h}\bar{k}\bar{l})|$ and compare them with $|F_c|x,y,z$. The technique can be used only with crystals which are noncentrosymmetric, because $F(\mathbf{h}) = F(\bar{\mathbf{h}})$ in centrosymmetric crystals, but this limitation is not important because molecules which crystallize with a single enantiomorph cannot do so in a space group containing any form of inversion symmetry.

6.5.1   Hamilton's R Ratio and Flack's Parameter

The reliability of the determination of the absolute configuration by the technique just described may be checked by means of Hamilton's R ratio test.[a] The generalized weighted R factor is defined by

$$R_g = \left(\frac{\sum_{hkl} w(hkl)[|F_o(hkl)| - K^{-1}|F_c(hkl)|]^2}{\sum_{hkl} w(hkl)|F_o(hkl)|^2} \right)^{1/2} \tag{6.60}$$

where the summations are taken over the unique set of reflections. Let the generalized weighted R factors for a structure and its enantiomorph be R_1, and R_2, respectively. The R factor ratio is compared with $R'_{p,n-p,\alpha}$ where

$$R = \frac{R_2}{R_1}; \quad R' = \left| \frac{p}{n-p} F_{p,\,n-p,\,\alpha} + 1 \right|^{1/2} \tag{6.61}$$

p is the number of parameters and n is the number of reflections, so that $n - p$ is the number of degrees of freedom. The null hypothesis for parameter sets r_1 and

[a] W. C. Hamilton, *Ac lu Crystallographica* **18**, 502 (1965).

r_2, corresponding to R_1 and R_2,

$$H_0 = r_2 - r_1 \tag{6.62}$$

can be tested at the $100\alpha\%$ significance level, where α defines a chosen level of significance. If $R > R'$, then we can reject the hypothesis at the $100\alpha\%$ level. When the number of degrees of freedom $(n - p)$ is large,

$$F_{p,\,n-p} = \chi_p^2/p \tag{6.63}$$

and R may be tested against

$$R' = R'_{p,\,n-p,\,\alpha} = \left(\frac{\chi_p^2}{n - p} + 1 \right)^{1/2} \tag{6.64}$$

The absolute configuration of stercuronium iodide[a] was confirmed by applying Hamilton's ratio test at the 0.01 level. The data are as follows:

$$n = 2420 \quad p = 304$$

$$n - p = 2116 \quad \alpha = 0.01$$

Following Hamilton, we obtain

$$R' = R'_{304,\,2116,\,0.01} = \left(\frac{367.5}{2116} + 1 \right)^{1/2} = 1.083$$

$$R = R_2/R_1 = 0.1230/0.1037 = 1.186$$

Since $R > R'$, the structure with $R_g = R_2$ was rejected at the 1% level (99% certain).

In order to overcome the complexities and uncertainties in applying Hamilton's test for enantiomorph definition, other, more automated methods have been sought.[b,c] The method currently employed in program packages such as SHELX-97 defines a parameter x that is refined by least squares, together with the other structural parameters, to a final value and corresponding standard deviation. The Flack parameter x is defined in terms of $|F_c|$ by the equation

$$|F_c(hkl)|^2 = (1 - x)|F_c(hkl)|^2 + x|F_c(\bar{h}\bar{k}\bar{l})|^2 \tag{6.65}$$

[a] J. Husain, R. A. Palmer, and I. J. Tickle, *Journal of Crystal and Molecular Structure* **11**, 87 (1981).
[b] D. Rogers, *Acta Crystallographica* **A37**, 734 (1981).
[c] H. D. Flack. *Acta Crystallographica* **A39**, 876 (1983).

When the atomic coordinate set and the crystal have the same chirality, x takes the value zero; if they are different, x is equal to unity. A result is considered acceptable if x lies within 3 standard deviations of either zero or unity. If a value of unity is returned, the coordinates of the model should be inverted (data collection routines ensure that the hkl data have the correct polarity). After further refinement a value of zero for x should result. This method has withstood innumerable tests and can be considered reliable in most circumstances. A further advantage of this method is that although, as with all analyses, the measured intensity data set should cover as large a volume of reciprocal space as possible, it is not absolutely necessary to measure Friedel or Bijvoet pairs. In practice, for the test to be reliable, the standard deviation in x should be 0.05 or less. Since the value of x may be related to twinning (Section 4.11), it is advisable to carry out further tests for this effect as prescribed, for example, in the SHELX system.

6.5.2 Effect of Anomalous Scattering on the Symmetry of Diffraction Patterns

We have seen that when Friedel's law holds, the x-ray diffraction pattern, considered as a three-dimensional weighted reciprocal lattice, exhibits symmetry equivalent to that of the point group of the crystal, with an additional center of symmetry, if not already present. There are, thus, 11 diffraction symmetry groups (Laue groups) corresponding to the centrosymmetric point groups listed in Table 1.6. However, for a structure in which some of the atoms scatter anomalously Friedel's law breaks down, and the symmetry of the diffraction pattern then reverts to that of the point group of the crystal. For a centrosymmetric crystal $|F(hkl)|$ still equals $|F(\bar{h}\bar{k}\bar{l})|$, although the phase angle is no longer 0 or π. As an example to illustrate the effects of anomalous scattering let us consider a crystal in space group $P2_1$.

Diffraction Symmetry for a Crystal in Space Group $P2_1$ with No Anomalous Scattering

The crystal is monoclinic, belonging to point group 2, and the diffraction symmetry if Friedel's law holds is $2/m$. The $|F(hkl)|$ equivalents are thus $|F(hkl)| = |F(\bar{h}k\bar{l})| = |F(h\bar{k}l)| = |F(\bar{h}\bar{k}\bar{l})|$. On x-ray photographs, for example, zero-level and upper-level a-axis precession photographs, this symmetry will manifest itself as indicated in Figure 6.40. The zero level, Figure 6.40a(i), clearly demonstrates symmetry $mm2$ with $|F(0kl)| = |F(0\bar{k}l)| = |F(0k\bar{l})| = |F(0\bar{k}\bar{l})|$, while upper-level photographs exhibit symmetry m (perpendicular to b^*) with $|F(1kl)| = |F(1\bar{k}l)$ and $|F(1k\bar{l})| = |F(1\bar{k}\bar{l})|$ respectively, but $|F(1kl)| \neq |F(1k\bar{l})|$ [Figure 6.40a(ii)].

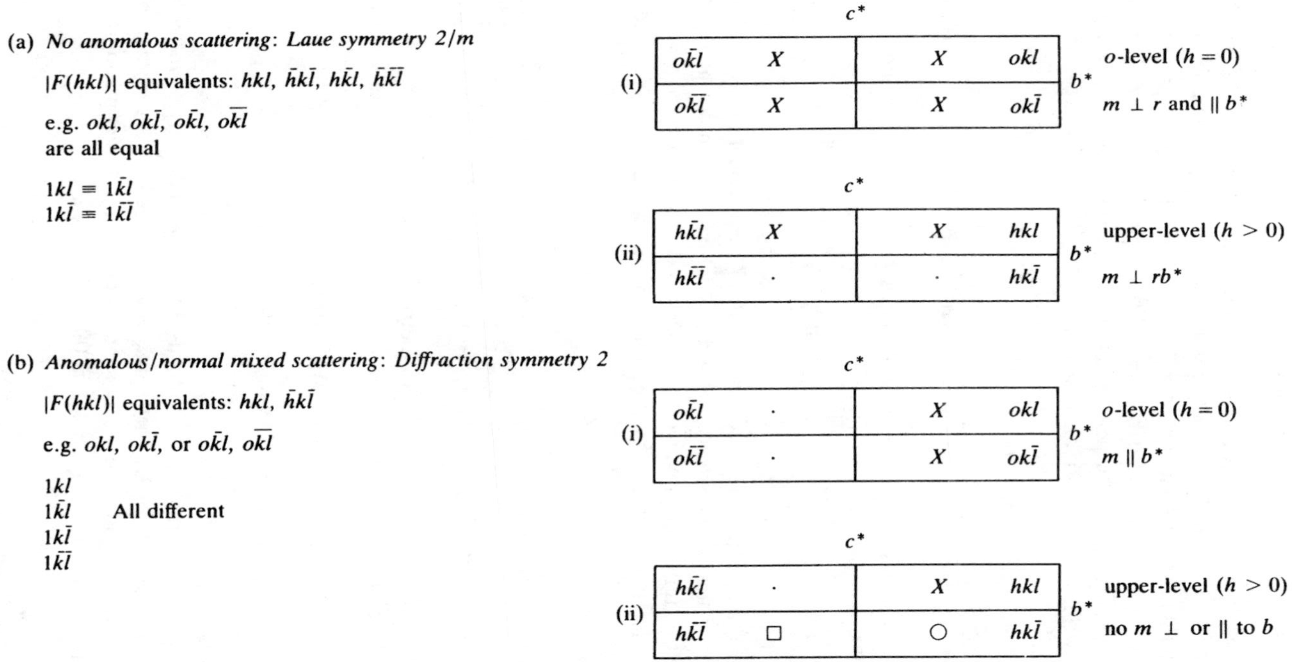

(a) No anomalous scattering: Laue symmetry $2/m$

$|F(hkl)|$ equivalents: hkl, $\bar{h}k\bar{l}$, $\bar{h}kl$, $\bar{h}\bar{k}\bar{l}$

e.g. okl, $ok\bar{l}$, $\bar{o}kl$, $\bar{o}k\bar{l}$
are all equal

$1kl \equiv 1\bar{k}\bar{l}$
$1k\bar{l} \equiv 1\bar{k}l$

(i)

| $o\bar{k}l$ | X | X | okl |
| $o\bar{k}\bar{l}$ | X | X | $ok\bar{l}$ |

o-level ($h=0$)

$m \perp r$ and $\parallel b^*$

(ii)

| $h\bar{k}l$ | X | X | hkl |
| $h\bar{k}\bar{l}$ | . | . | $hk\bar{l}$ |

upper-level ($h>0$)

$m \perp rb^*$

(b) Anomalous/normal mixed scattering: Diffraction symmetry 2

$|F(hkl)|$ equivalents: hkl, $\bar{h}\bar{k}l$

e.g. okl, $\bar{o}k\bar{l}$, or $\bar{o}kl$, $ok\bar{l}$

$1kl$
$1\bar{k}l$ All different
$1k\bar{l}$
$1\bar{k}\bar{l}$

(i)

| $o\bar{k}l$ | . | . | okl |
| $o\bar{k}\bar{l}$ | . | . | $ok\bar{l}$ |

o-level ($h=0$)

$m \parallel b^*$

(ii)

| $h\bar{k}l$ | . | X | hkl |
| $h\bar{k}\bar{l}$ | □ | O | $hk\bar{l}$ |

upper-level ($h>0$)

no $m \perp$ or $\parallel$ to b

FIGURE 6.40.   Schematic representation of symmetry exhibited in b^*c^* sections of the reciprocal lattice for a crystal with space group $P2_1$, (or any other monoclinic space group in crystal class 2). (a) No anomalous scattering. The diffraction symmetry is $2/m$. (i) Symmetry of $0kl$ section is $2/m$. (ii) Symmetry of $1kl,2kl,\ldots$ sections is $mm2$. Equivalent hkl are represented by the same symbol X etc., (b) Anomalous scattering case. The diffraction symmetry is 2. (i) Symmetry of $0kl$ section is $m\parallel b^*$. (ii) Symmetry of $1kl,2kl,\ldots$ sections has reduced to 1 (no m present).

Diffraction Symmetry for a Crystal in Space Group $P2_1$ for a Structure Containing Some Anomalous Scatterers

The diffraction symmetry is now that of point group 2, for which $|F(hkl)| = |F(\bar{h}k\bar{l})|$ and $|F(\bar{h}\bar{k}\bar{l})| = |F(h\bar{k}l)|$. On the a-axis precession photographs shown schematically in Figure 6.40b(i) we observe, on the zero level, $|F(0kl)| = |F(0\bar{k}\bar{l})|$ and $|F(0\bar{k}l)| = |F(0k\bar{l})|$, but $|F(0kl)| \neq |F(0\bar{k}l)|$ and $|F(0k\bar{l})| \neq |F(0\bar{k}\bar{l})|$ because there is now no m plane perpendicular to b^*. On the upper-level photographs, no symmetry is observable and all four $|F|$ values are different, Figure 6.40b(ii).

We may define the anomalous difference as

$$\Delta F_{\text{ANO}} = (|F(hkl)| - |F(\bar{h}\bar{k}\bar{l})|) \quad \text{or} \quad (|F(+)| - |F(-)| \tag{6.66}$$

From the above discussion, for the case of space group $P2_1$ with anomalous scatterers present, ΔF_{ANO} is also given by

$$\Delta F_{\text{ANO}} = (|F(hkl)| - |F(h\bar{k}l)|) \tag{6.67}$$

since $|F(\bar{h}\bar{k}\bar{l})| = |F(h\bar{k}l)|$. Differences such as (6.67), equivalent by point-group symmetry to the difference between $|F|$ for a Friedel pair, are known as *Bijvoet differences*, the two reflections involved being denoted as a *Bijvoet pair*.[a]

Bijvoet differences can be observed on x-ray photographs represented in Figure 6.40a,b as follows:

(a) $h = 0$

$$(|F(0kl)| - |F(0\bar{k}l)|) \quad \text{Bijvoet pair}$$

$$(|F(0k\bar{l})| - |F(0\bar{k}\bar{l})|) \quad \text{Bijvoet pair}$$

These differences will be equivalent by symmetry.

(b) upper level

$$(|F(hkl)| - |F(h\bar{k}l)|) \quad \text{Bijvoet pair}$$

$$(|F(hk\bar{l})| - |F(h\bar{k}\bar{l})|) \quad \text{Bijvoet pair}$$

These differences are not equivalent.

It is, thus, possible to monitor anomalous differences indirectly on *the same photograph* through the use of Bijvoet pairs, whereas Friedel pairs, in the true sense of the definition would necessarily always occur on different photographs (except for axial reflections, such as $00l$ and $00\bar{l}$).

[a] A. F. Peerdeman and J. M. Bijvoet, *Acta Crystallographica* **9**, 1012 (1956).

6.5.3 Form of the Structure Factor for a Structure Composed of Heavy-Atom Anomalous Scatterers

For a structure composed of N_H heavy-atom anomalous scatterers the structure factor becomes

$$F(hkl) = \sum_{j=1}^{N_H} (f'_j + i\Delta f''_j) \exp 2\pi i (hx_j + ky_j + lz_j) \qquad (6.68)$$

This can be written as

$$F_H(+) = F'_H(+) + iF''_H(+)$$

where

$$F'_H(+) = \sum_{j=1}^{N_H} f'_j \exp 2\pi i (hx_j + ky_j + lz_j) \qquad (6.69)$$

and

$$F''_H(+) = \sum_{j=1}^{N_H} \Delta f''_j \exp 2\pi i (hx_j + ky_j + lz_j) \qquad (6.70)$$

Similarly the structure factor $F(\bar{h}\bar{k}\bar{l})$ can be written as

$$F_H(-) = F'_H(-) + iF''_H(-)$$

where

$$F'_H(-) = \sum_{j=1}^{N_H} f'_j \exp\{-2\pi i (hx_j + ky_j + lz_j)\} \qquad (6.71)$$

and

$$F''_H(-) = \sum_{j=1}^{N_H} \Delta f''_j \exp\{-2\pi i (hx_j + ky_j + lz_j)\} \qquad (6.72)$$

The form of the structure factor for a protein heavy-atom derivative crystal composed of protein atoms P (light atom, negligible-anomalous scatterers, mainly C, N, O) and heavy atoms H (anomalous scatterers) thus becomes

$$F_{PH}(+) = F_P(+) + F'_H(+) + iF''_H(+) \quad \text{for } hkl \qquad (6.73)$$

$$F_{PH}(-) = F_P(-) + F'_H(-) + iF''_H(-) \quad \text{for } \bar{h}\bar{k}\bar{l} \qquad (6.74)$$

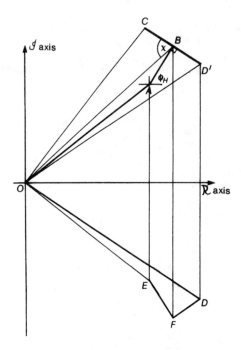

FIGURE 6.41. Structure factors $F_{PH}(+)$ and $F_{PH}(-)$ for a protein crystal containing heavy-atom anomalous scatterers: $\widehat{ROC} = \phi_{PH}(+)$; $\widehat{ROD} = \widehat{ROD}'$; $\widehat{ROB} = \phi_{PH}$; $\widehat{ROA} = \phi_P$; $\widehat{OAB} = \phi_P + (180° - \phi_H)$; $OA = |F_P(+)|$; $AB = |F_H(+)|$; $BC = |F_H''(+)|$; $OB = |F_{PH}|$; $OC = |F_{PH}(+)|$; $OD = |F_{PH}(-)|$; OD' = mirror reflection of OD; $OE = |F_P(-)|$; $EF = |F_H(-)|$; $FD = |F_H''(-)|$; $\widehat{OBC} = \chi = \phi_{PH} - \phi_H + 90°$; $AC = AD' = ED = |F_H|$.

F_P is the normal protein structure factor and $F_P(+)$ is clearly equal to $F_P(-)$ (Figure 6.39a).

The structure factors $F_{PH}(+)$ and $F_{PH}(-)$ are represented in Figure 6.41, where $OC = F_{PH}(+)$ and $OD = F_{PH}(-)$. Clearly $|F_{PH}(+)| \neq |F_{PH}(-)|$. Figure 6.41 also shows the effect of reflecting $F_{PH}(-)$ across the real axis OR. This device simplifies the following calculation. Using the cosine rule in triangles OBC and OBD', we obtain

$$|F_{PH}(+)|^2 = |F_{PH}|^2 + |F_H''|^2 - 2|F_{PH}||F_H''|\cos(\phi_{PH} - \phi_H + 90°) \quad (6.75)$$

$$|F_{PH}(-)|^2 = |F_{PH}|^2 + |F_H''|^2 - 2|F_{PH}||F_H''|\cos(90° - \phi_{PH} - \phi_H) \quad (6.76)$$

Subtracting (6.76) from (6.75) gives

$$|F_{PH}(+)|^2 - |F_{PH}(-)|^2 = [|F_{PH}(+)| + |F_{PH}(-)|][|F_{PH}(+)| - |F_{PH}(-)|]$$
$$= -2|F_{PH}||F_H''|[\cos(\phi_{PH} - \phi_H + 90°)$$
$$- \cos(90° - \phi_{PH} - \phi_H)]$$
$$= 4|F_{PH}||F_H''|\sin(\phi_{PH} - \phi_H) \qquad (6.77)$$

But, intuitively,

$$|F_{PH}(+)| + |F_{PH}(-)| = OC + OD' \approx 2OB \approx 2|F_{PH}| \qquad (6.78)$$

Thus, it follows that the anomalous difference

$$\Delta F_{\text{ANO}} = |F_{PH}(+)| - |F_{PH}(-)| \approx 2|F_H''|\sin(\phi_{PH} - \phi_H) \qquad (6.79)$$

For a given heavy-atom type, it is known that the ratio $|F_H'|/|F_H''| = f'/\Delta f''$, and is approximately constant, κ, say. Thus,

$$\Delta F_{\text{ANO}} \approx (2|F_H'|/\kappa)\sin(\phi_{PH} - \phi_H)$$

or

$$|F_H'|\sin(\phi_{PH} - \phi_H) \approx (\kappa/2)\Delta F_{\text{ANO}} \qquad (6.80)$$

From (6.44), we see that $\Delta F_{\text{ISO}} \approx |F_H|\cos(\phi_H - \phi_{PH})$, and in the case of anomalous scattering $|F_H|$ becomes $|F_H'|$. It follows that $|F_H'| \approx [(\Delta F_{\text{ISO}})^2 + ((\kappa/2)\,\Delta F_{\text{ANO}})^2]^{1/2}$.

This provides a possible method for estimating (F_H') for calculation of a difference Patterson map when anomalous scattering measurements are available.

6.5.4 Phasing by Use of Anomalous Scattering

Anomalous scattering can be used in phasing reflections. We saw in the previous section that the isomorphous replacement technique in noncentrosymmetric crystals leads to an ambiguity in phase determination (Figure 6.27). The ambiguity cannot be resolved unless the replaceable site is changed. Merely using a third derivative with the same replaceable site would lead to a situation comparable with that in Figure 6.27. The heavy atom vector would still be directed along OH, and its different length would be just balanced by the change in $|F_o|$, so that *three* circles would intersect at P_1 and P_2.

A vector change in the heavy atom contribution can be brought about through anomalous scattering in a given derivative, instead of invoking a different replaceable site. Following Figure 6.39b we see that two different $|F_o|$ values can arise for $\mathbf{h}$ and $\bar{\mathbf{h}}$. Consequently, $\phi(\mathbf{h}) \neq \phi(\bar{\mathbf{h}})$, and the ambiguity can be resolved by the experimental data. This technique is particularly important with synchrotron radiation, where the wavelength can be tuned to the absorption edge of a relatively heavy atom in the structure so as to obtain the maximum difference between $|F_o(\mathbf{h})|$ and $|F_o(\bar{\mathbf{h}})|$.

6.5.5 Resolution of the Phase Problem for Proteins Using Anomalous Scattering Measurements (SIRAS[a] Method)

A Harker diagram for any hkl reflection $F_P(+)$ may be constructed as discussed earlier by first drawing a circle of center O and radius $|F_P(+)|$. From C^+, the end of the vector $-F_H(+)$, as center, a second circle of radius $|F_{PH}(+)|$ is drawn, as shown in Figure 6.42. It intersects the $F_P(+)$ circle in points $P_1(+)$ and $P_2(+)$; $OP_1(+)$ and $OP_2(+)$ represent the SIR phase ambiguity noted previously (see Section 6.4.8)

The Harker diagram for the corresponding $\bar{h}\bar{k}\bar{l}$ reflection $F_P(-)$ is constructed by drawing a circle of radius $|F_P(-)|$, center O, and finally a circle of radius $|F_{PH}(-)|$, centered at C^-, the end of the vector $-F_H(-)$, as shown

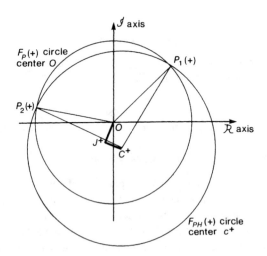

FIGURE 6.42.   Harker diagram for $F_P(+)(hkl)$ for a protein crystal with heavy atoms scattering anomalously. OJ^+ is the vector $-F'_H(+)$, and J^+C^+ is the imaginary component of OC^+, the $-F_H(+)$ vector. The $F_P(+)$ and $F_{PH}(+)$ circles intersect at $P_1(+)$ and $P_2(+)$. $OP_1(+)$ and $OP_2(+)$ are the ambiguous solutions for $F_P(+)$.

[a] Single Isomorphous Replacement with Anomalous Scattering.

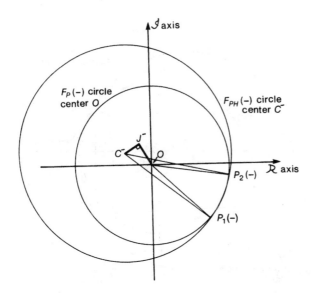

FIGURE 6.43. Harker diagram for $F_P(-)(hkl)$ for a protein crystal with heavy atoms scattering anomalously. OJ^- is the vector $-F_H(-)$, and J^-C^- is the imaginary component of OC^-, the $-F_H(-)$ vector. The $F_P(-)$ and $F_{PH}(-)$ circles intersect at $P_1(-)$ and $P_2(-)$. $OP_1(-)$ and $OP_2(-)$ are the ambiguous solutions for $F_P(-)$.

in Figure 6.43. The two circles this time intersect in points $P_1(-)$ and $P_2(-)$, representing the ambiguous SIR solution for $F_P(-)$.

Since $F_P(+)$ and $F_P(-)$ are related by reflection across the real axis OR (Figures 6.42 and 6.43) the *correct* solutions for $F_P(+)$ and $F_P(-)$ will be mirror-related in this way, while the incorrect pair will not. In the present case the corresponding correct solutions are $P_1(+)$ and $P_1(-)$; $P_2(+)$ and $P_2(-)$ are thus the unwanted erroneous solutions. In order to rationalize this process a combined Harker diagram can be conceived as in Figure 6.44. This involves plotting the $F_{PH}(-)$ circle at the end of the *mirrored* $F_H(-)$ vector, thus enabling solutions $P_1(+)$ and $P_1(-)$ mirrored to coalesce. In practice, the phasing process is carried out by calculating a probability distribution in a similar manner to that used in the MIR technique. Algorithms for carrying out these computations have been developed.

6.5.6 Protein Phasing Using the Multiple-Wavelength Anomalous Dispersion Technique (MAD) with Synchrotron Radiation (SR)

The above treatment shows that, in principle, the phase ambiguity associated with the single isomorphous replacement technique can be resolved by

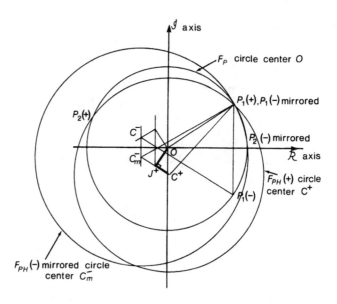

FIGURE 6.44.   Combined Harker diagram for $F_P(+)$ and $F_P(-)$, with $F_P(-)$ mirrored across the real axis of the Argand diagram. C_m^- is the mirrored point of C^-, where OC^- is the $-F_H(-)$ vector, as in Figure 6.43. The correct solution for $F_P(+)$ is $OP_1(+)$, coinciding with $OP_1(-)$ mirrored.

incorporating anomalous dispersion measurements. Conventional laboratories are usually equipped with either sealed-tube or rotating-anode x-ray sources generating x-radiation from a Cu target. Anomalous scattering effects for Cu $K\alpha$ radiation are quite small (Table 6.7), the actual differences between $|F(hkl)|$ and $|F(\bar{h}\bar{k}\bar{l})|$ being difficult to detect without extremely careful measurements. This practical limitation to the method may be overcome by the use of synchrotron radiation. Optimization of anomalous scattering information can be achieved by selection of a wavelength close to an absorption edge of the heavy atom, where $\Delta f''$ is maximized. It should be remembered, however, that anomalous differences are still small, even for measurements made for wavelengths tuned in this way. There is another advantage to be gained by the availability of SR radiation, namely that measurements can be made at different wavelengths, possibly even on the same crystal specimen. The second wavelength should be selected such that f' is large and f'' is small for the anomalous scatterer. Measurements for this wavelength would be made only for hkl reflections. This technique is known as the *multiple-wavelength anomalous dispersion* (MAD) method for phasing. The measurements required are $|F_P(hkl)|$, $|F_{PH}(+)|_{\lambda_1}$, $|F_{PH}(-)|_{\lambda_2}$, and $|F_{PH}(+)|_{\lambda_2}$, and Figure 6.45 shows an idealized phase diagram for a typical hkl reflection using the MAD technique.

TABLE 6.7. Values of the Real (Dispersion) $\Delta f'$ and Imaginary (Absorption) $\Delta f''$ Components of Anomalous Scattering for Cu $K\alpha$ x-rays. The Atoms Selected are Those Found in Proteins and Those Used Frequently for Heavy-Atom Derivatives

| | | Cu $K\alpha$ radiation, $\lambda = 1.542$ Å | | | | |
| | | $\Delta f'$ | | $\Delta f''$ | | |
	Atomic number	$(\sin\theta)/\lambda = 0$	$= 0.6$	$(\sin\theta)/\lambda = 0$	$= 0.4$	μ
C	6	0	0	0	0	4.6
N	7	0	0	0	0	7.52
O	8	0	0	0.1	0.1	11.5
S	16	0.3	0.3	0.6	0.6	89.1
Fe	26	−1.1	−1.1	3.4	3.3	308
Zn	30	−1.7	−1.7	0.8	0.7	60.3
Pd	46	−0.5	−0.6	4.3	4.1	206
Ag	47	−0.5	−0.6	4.7	4.5	218
I	53	−1.1	−1.3	7.2	6.9	294
Sm	62	−6.6	−6.7	13.3	12.8	397
Gd	64	−12	−12	12.0	11.6	439
Lu	71	−7	−7	5	5	153
Pt	78	−5	−5	8	7	200
Au	79	−5	−5	8	8	208
Hg	80	−5	−5	9	8	216
Pb	82	−4	−5	10	9	232
U	92	−4	−5	16	16	306

[a] μ is the absorption coefficient (see Section 3.1.3).

6.6 Neutron Diffraction

Neutrons are scattered by crystalline material, and the resultant radiation contains both coherent and incoherent components. We are concerned here with only the coherent scattering, that is, where the incident and diffracted beams have a distinct phase relationship with each other. In this case, the diffraction of neutrons by crystals is, normally, similar to x-ray diffraction; that is, the phase difference between incident and scattered beams is π.

Neutrons from an atomic reactor, slowed by collisions with a moderator such as graphite, typically have a speed v of about 4 km s^{-1}. From the de Broglie equation

$$\lambda = h/(m_n v) \qquad (6.81)$$

where h is the Planck constant and m_n is the mass of the neutron. It follows that the corresponding neutron wavelength would be approximately 1 Å, which is commensurate with the wavelengths of x-rays used in diffraction. In practice, a neutron beam from a reactor contains a range of velocities, according to the

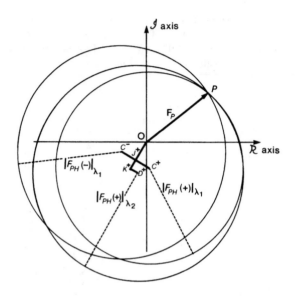

FIGURE 6.45.   MAD phasing for wavelengths λ_1 and λ_2. The circles based on λ_1 for $|F_{PH}(+)|$ and $|F_{PH}(-)|$ are drawn as before. Adding the measurement at λ_2 of $|F_{PH}(+)|_{\lambda_2}$ is comparable to the information from a second heavy-atom derivative in the MIR technique. OK^+ is the real component and K^+D^+ the imaginary component of OD^+, the vector $F_H(+)_{\lambda_2}$. *Note.* For clarity the F_P circle is not shown; F_P gives the MAD solution for the protein structure factor. The measurements are sometimes made only for the derivative crystal, using first λ_1 and then λ_2, during the same experimental session at the synchrotron radiation station. In such a case, the $|F_P|$ data may not be available, whereupon the phasing would be carried out for $|F_{PH}|$ instead. The ensuing electron density map would then apply to the derivative crystal.

equation

$$\tfrac{1}{2}m_n v^2 = \tfrac{2}{3}\kappa T \qquad\qquad (6.82)$$

where k is the Boltzmann constant and T is the temperature of the reactor. A range of wavelengths is thus implied, and the temperature of a reactor can be adjusted so that the neutrons emitted have wavelengths from 1.4 to 1.7 Å, which values bracket the wavelength of Cu $k\alpha$ x-radiation. The beam is collimated and reflected from a single crystal to give a monochromatic beam. Perfection cannot be achieved and, in practice, the neutron beam will have a wavelength spread of up to 0.03 Å.

The intensity of a neutron beam is much less than that of an x-ray beam. Consequently, a beam of large cross-sectional area must be employed, with a correspondingly larger crystal specimen than is used with x-rays. Neutron diffraction follows the same physical principles as do x-rays, but the scattering mechanisms differ. X-rays are scattered by the electrons in an atom, and their distribution leads to interference such that the scattering intensity varies with the scattering angle.

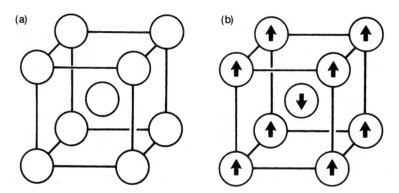

FIGURE 6.46. Unit cell of the crystal structure of chromium: (a) by x-rays—body-centered cubic, (b) by neutron diffraction—primitive cubic. The arrows represent the magnetic moment vectors.

When neutrons are scattered by the *nucleus* of the atom of a nonmagnetic species, there is no recoil of the nucleus and the scattering amplitude at distance r from the nucleus is given by β/r, where β is the *scattering length* for the given species. There is no polarization of the scattered beam and, since the nuclear diameter is small compared with the neutron wavelength, β is almost independent of the scattering angle and of the wavelength of the neutron beam. However, unlike with x-rays, there is no simple relationship between β and atomic number. Indeed, different isotopes of one and the same element have differing scattering lengths. Thus, β must be determined by experiment. For some species, such as hydrogen and titanium, β is negative, so that the phase change on scattering is zero. Where β is positive the phase change on scattering is π, as with x-rays. If a species in a crystal contains unpaired electrons, there will also be an interaction between the magnetic moment of the neutron and that of the atom. This scattering is a form of neutron–electron scattering, and its amplitude decreases as a function of $(\sin\theta)/\lambda$. In this context, we may recall our earlier statement that symmetry may depend upon the nature of the examining probe (Section 1.4). For example, elemental chromium $(Ar\,3d^5 4s^1)$ is body-centered cubic with respect to x-ray diffraction (Figure 6.46a), but magnetically primitive cubic with respect to neutron diffraction (Figure 6.46b).

6.6.1 Complementary Nature of Neutron Diffraction

We can see now that neutron diffraction experiments can produce important results that complement those from x-ray diffraction. They can be used to distinguish between isotopes, to study ordering in certain alloy systems, such as the Cu–Zn phases, to examine magnetically ordered structures, and to determine the positions of light atoms in the presence of heavy atoms—a difficult matter with x-rays.

Of particular note is hydrogen, which has a very small scattering for x-rays. For example, the x-ray examination of sodium hydride, NaH, showed that the Na^+ species formed a face-centered cubic structure, but failed to locate the H^- species. Sodium and hydrogen scatter neutrons to about the same extent, but opposite in phase. A study of the neutron diffraction data on sodium hydride showed conclusively that it has the sodium chloride structure type.

Hydrogen atoms produce a high intensity of incoherent neutron scattering, which enhances the background. Deuterium, however, has a large, positive coherent scattering with only small background. Thus, it is common practice to replace hydrogen by deuterium in a study of hydrogen atom positions in crystal structures. Notable examples of this application are the precise location of the hydrogen–deuterium atom positions in potassium dihydrogen phosphate, KH_2PO_4 (Figure 6.47), and in sucrose, $C_{12}H_{22}O_{11}$. More recently, neutron scattering has been used in protein crystallography to distinguish between N–H and C–H on histidine rings; the proper positioning of these groups is important for understanding the biological action of proteins.

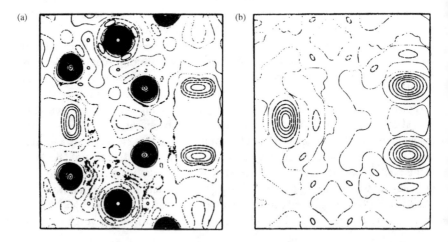

FIGURE 6.47.   Projections on to the xy plane of the neutron scattering density for potassium dihydrogen phosphate, KH_2PO_4, in the room-temperature tetragonal phase: (a) direct synthesis, showing all atoms in the structure, the most intense being due to K and P superimposed in this projection. The other peaks are O (full lines) and H (dashed lines indicating the negative scattering amplitude of this species), (b) corresponding difference synthesis, in which only the H atoms appear. It is noteworthy that the diffraction ripples (series termination errors) around the heavy-atom positions in (a) are subtracted out in (b), and that the contours of H are free from the distorting effects of the ripples. [Reproduced from *Neutron Diffraction*, 2nd ed., by G. E. Bacon, with the permission of Oxford University Press.]

6.6.2 Refinement of Hydrogen Atom Positions

The neutron diffraction results for sucrose give the value 1.09 Å for the C–H bond length, whereas by x-rays it is shown to be 0.98 Å; the corresponding O–H bond length values are 0.98 Å and 0.80 Å. These results are of interest in connection with the refinement of a structure. If the hydrogen atoms are included in the calculations but are not themselves refined, a not uncommon procedure, then with x-ray data it is reasonable to constrain the C–H bond length to be about 0.98 Å rather than 1.09 Å. The polarity of the C–H bond leads to a displacement of the electron density of hydrogen toward the carbon atom. Thus, the apparent electron density maximum for the hydrogen atom will not coincide with that given by neutron diffraction.

Neutron diffraction investigations can also be carried out on powdered crystalline samples, and the information contained in the pattern extracted by the Rietveld technique (see Chapter 9).

Bibliography

Introductory Structure Analysis

BRAGG, W. L., *A General Survey* (*The Crystalline State*, Vol. I), London, Bell (1949).
HAHN, T. (Editor), *International Tables for Crystallography*, Vol. A, 5th Ed., Kluwer Academic (2002).
HENRY, N. F. M., and LONSDALE, K. (Editors), *International Tables for X-Ray Crystallography*, Vol. I, Birmingham, Kynoch Press (1965).

General Structure Analysis

BUERGER, M. J., *Vector Space*, New York, Wiley (1959).
DUNITZ, J. D. *X-ray Analysis and the Structure of Organic Molecules*, Verlag Helvetica Chimica Acta (1995).
WOOLFSON, M. M., *An Introduction to X-ray Crystallography*, 2nd Edition, Cambridge, Cambridge University Press (1997).

Protein Crystallography

BLUNDELL, T. L., and JOHNSON, L. N., *Protein Crystallography*, New York, Academic Press (1977).
EISENBERG, D., in *The Enzymes*, Vol. I, *X-ray Crystallography and Enzyme Structure*, New York, Academic Press (1970).
PHILLIPS, D. C., *Advances in Structure Research by Diffraction Methods*, **2**, 75 (1966).

Neutron Diffraction

BACON, G. E., *Neutron Diffraction*, 3rd ed., Oxford (1975).

Chemical Data

SUTTON, L. E. (Editor), *Tables of Interatomic Distances and Configuration in Molecules and Ions*, London, The Chemical Society (1958; supplement, 1965).

Problems

6.1. A structure with the apparent space group $P2_1/c$ consists of atoms at $0.2, \frac{1}{4}, 0.1$ and the symmetry-related positions; the center of symmetry is at the origin. Evaluate the geometric structure factor for the four given positions in the unit cell, and, hence, determine the systematic absences among the hkl reflections. What are the consequences of these absences as far as the true structure is concerned? Sketch the structure in projection along b. What is the true space group?

6.2. Rh_2B crystallizes in space group $Pnma$ with $a = 5.42, b = 3.98, c = 7.44$ Å, and $Z = 4$. Consider Figure 2.36. Show that if no two Rh atoms may approach within 2.5 Å of each other, they cannot lie in general positions. Where might the Rh atoms be placed? Illustrate your answer with a sketch showing possible positions for these atoms in projection on (010).

6.3. Trimethylammonium chloride,

$$\begin{bmatrix} H_3C \diagdown \\ H_3C \text{-}-N\text{-}H \\ H_3C \diagup \end{bmatrix}^+ \quad Cl^-$$

crystallizes in a monoclinic, centrosymmetric space group, with $a = 6.09, b = 7.03, c = 7.03$ Å, $\beta = 95.73°$, and $Z = 2$. The only limiting condition is $0k0: k = 2n$. What is the space group? Comment on the probable positions of (a) Cl, (b) C, (c) N, and (d) H atoms.

6.4. Potassium hexachloroplatinate (IV), $K_2[PtCl_6]$, is cubic, with $a = 9.755$ Å. The atomic positions are as follows ($Z = 4$):

$$\left(0, 0, 0; \quad 0, \tfrac{1}{2}, \tfrac{1}{2}; \quad \tfrac{1}{2}, 0, \tfrac{1}{2}; \quad \tfrac{1}{2}, \tfrac{1}{2}, 0 \right) +$$

4	Pt:	$0,0,0$
8	K:	$\tfrac{1}{4}, \tfrac{1}{4}, \tfrac{1}{4}; \quad \tfrac{3}{4}, \tfrac{3}{4}, \tfrac{3}{4}$
24	Cl:	$\pm\{x, 0, 0; \quad 0, x, 0; \quad 0, 0, x\}$

Show that $F_c(hhh) = A'(hhh)$, where

$$A'(hhh) = 4g_{Pt} + 8g_K \cos(3\pi h/2) + 24g_{Cl} \cos 2\pi h x_{Cl}$$

Calculate $|F_c(hhh)|$ for the values of h tabulated below, with $x_{Cl} = 0.23$ and 0.24. Obtain R factors for the scaled $|F_o|$ data for the two values of x_{Cl}, and indicate which value of x_{Cl} is the more acceptable. Calculate the Pt–Cl distance, and sketch the $[PtCl_6]^{2-}$ ion. What is the point group of this species?

hkl	111	222	333
$|F_o|$	491	223	281
g_{Pt}	73.5	66.5	59.5
g_K	17.5	14.5	12.0
g_{Cl}	15.5	13.0	10.5

Atomic scattering factors g_j may be taken to be temperature-corrected values.

6.5. USi crystallizes in space group $Pbnm$, with $a = 5.65, b = 7.65, c = 3.90$ Å, and $Z = 4$. The U atoms lie at the positions

$$\pm\left\{x, y, \tfrac{1}{4}; \quad \tfrac{1}{2} - x, \tfrac{1}{2} + y, \tfrac{1}{4}\right\}$$

Obtain a reduced expression for the geometric structure factor ($\bar{1}$ at $0, 0, 0$) for the U atoms. From the data below, determine approximate values for x_U and y_U; the Si contributions may be neglected.

hkl	200	111	210	231	040	101	021	310
$I_o(hkl)$	0	236	251	200	0	170	177	0

Proceed by using 200 to find a probable value for x_U. Then find y_U from 111, 231, and 040.

6.6. Methylamine forms a complex with boron trifluoride of composition $CH_3NH_2BF_3$.

Crystal Data
System: monoclinic.
Unit-cell dimensions: $a = 5.06, b = 7.28, c = 5.81$ Å, $\beta = 101.5°$.
V_c: 209.7 Å^3.
D_m: 1.54 g cm^{-3}.
M_x: 98.86.
Z: 1.97, or 2 to the nearest integer.
Unit-cell contents: 2C, 10H, 2N, 2B, and 6F atoms.
Absent spectra: $0k0 : k = 2n + 1$.
Possible space groups: $P2_1$ or $P2_1/m$ ($P2_1/m$ may be assumed).
Determine what you can about the crystal structure.

6.7. Write the symmetry-equivalent amplitudes of $|F(hkl)|, |F(0,k,l)|$, and $|F(h0l)|$ in (a) the triclinic, (b) the monoclinic, and (c) the orthorhombic crystal systems. Friedel's law may be assumed.

6.8. (a) Determine the orientations of the Harker lines and sections in Pa, $P2/a$, and $P222_1$.

(b) A monoclinic, noncentrosymmetric crystal with a primitive space group shows concentrations of peaks on $(u, 0, w)$ and $[0, v, 0]$. How might this situation arise?

6.9. Diphenyl sulfoxide, $(C_6H_5)_2SO$, is monoclinic, with $a = 8.90, b = 14.08, c = 8.32$ Å, $\beta = 101.12°$, and $Z = 4$. The conditions limiting possible x-ray reflections are as follows.

$$hkl: \quad \text{none;} \quad h0l: \quad h+l = 2n; \quad 0k0: \quad k = 2n$$

(a) What is the space group?

(b) Figures P6.1a–c are the Patterson sections at $v = \frac{1}{2}, v = 0.092$, and $v = 0.408$, respectively, and contain S–S vector peaks. Write the coordinates of the nonorigin S–S vectors (in terms of x, y, and z), and from the sections provided determine the best values for the S atoms in the unit cell. Plot these atomic positions as seen along the b axis, with an indication of the heights of the atoms with respect to the plane of the diagram.

6.10. Figure P6.2 shows an idealized vector set for a hypothetical structure C_6H_5S in space group $P2$ with $Z = 2$, projected down the b axis. Only the S–S and S–C vector interactions are considered.

(a) Determine the x and z coordinates for the S atoms, and plot them to the scale of this projection.

(b) Use the Patterson superposition method to locate the carbon atom positions on a map of the same projection.

6.11. Hafnium disilicide, $HfSi_2$, is orthorhombic, with $a = 3.677, b = 14.55, c = 3.649$ Å, and $Z = 4$. The space group is $Cmcm$, and the Hf and Si atoms occupy three sets of special positions of the type

$$\pm\left\{0, y, \tfrac{1}{4}; \quad \tfrac{1}{2}, \tfrac{1}{2} + y, \tfrac{1}{4}\right\}$$

The contributions from the Hf atoms dominate the structure factors. By combining the terms $\cos 2\pi ky$ and $\cos 2\pi(ky + k/2)$, show that the geometric structure factor $A(0k0)$ is approximately proportional to $\cos 2\pi y_{Hf}$. The

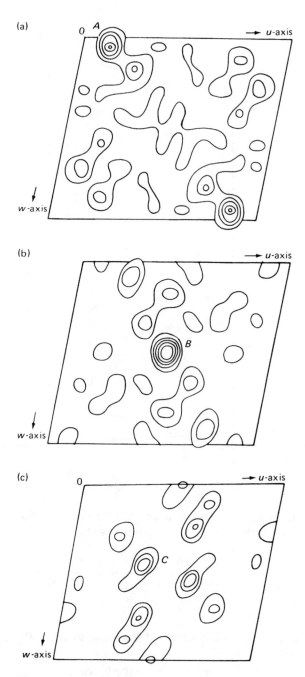

FIGURE P6.1. Patterson sections at (a) $v = \frac{1}{2}$, (b) $v = 0.092$, (c) $v = 0.408$.

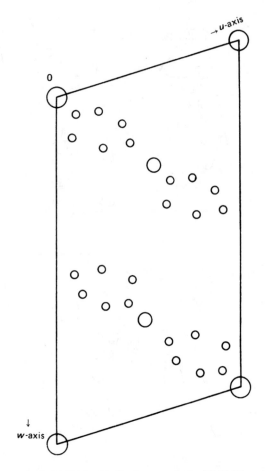

FIGURE P6.2.   Idealized vector map for "C_6H_4S."

$|F(0k0)|$ data are listed below, from which the values of $|F(0k0)|^2$, divided by 10 and rounded to the nearest integer, have been derived.

$0k0$	020	040	060	080	010,0	012,0	014,0	016,0
$|F(0k0)|$	7	14	18	13	12	<1	20	<1
$|F(0k0)|^2$	5	20	32	17	14	0	40	0

(a) Calculate the one-dimensional Patterson function $P(v)$, using the equation

$$P(v) \propto \sum_k |F(0k0)|^2 \cos 2\pi k v$$

The multiplying factor $2/b$ and the $F(000)$ term have been omitted to simplify the calculation; they can never change the form of the synthesized function, although the neglect of the term involving $F(000)$ gives rise to negative values in the calculated $P(v)$.

The Fourier summation here can be carried out readily by means of the program FOUR1D*, which also gives a plot of the function at the on-line printer.

However, for many years following their introduction in 1936, the summation aid known as Beevers-Lipson[a] strips was used for Fourier summations in one and two dimensions, and even Harker sections in three dimensions. For historical interest, Table P6.1 shows the strips that would be used for this summation. Each line contains the value of $(+)\frac{|F|^2}{10}\cos 2\pi h(n/60)$ for $n = 0$ to 15/60 (for $(-)|F|$ the sign of all terms for $n = 0$–15 are changed). For reflection symmetry at $\frac{1}{4}$, only the values of n from 0–15 are needed. The columns are added vertically to give the sum over h at each value of n. The range of n can be increased by making use of the properties of the cosine function. A Table such as P6.2 shows clearly how an error is of greater consequence in a phase (sign) than in an amplitude. Consider changing the sign of 080 in the calculation of $\rho(y)$ and then the amplitude by, say, 20%.

Plot the function, extend it to one repeat unit, interpret the four highest nonorigin peaks, and determine y_{Hf}.

(b) Using the value of y_{Hf} and the form of the geometric structure factor $A(0k0)$, determine the signs for the $0k0$ reflections. Hence, compute the electron density:

$$\rho(y) \propto \sum_k \pm |F(0k0)| \cos 2\pi ky$$

Again the $2/b$ factor and $F(000)$ have been omitted.[b] Plot the function and determine y_{Hf}. What can be deduced about the positions of the Si atoms? In the light of your results, study $P(v)$ again. (Table P6.2 contains the simulated Beevers-Lipson strips, with *positive* values of $|F|$.)

6.12. The alums, $MAl(NO_4)_2 \cdot 12H_2O$, where $M = NH_4$, K, Rb, Tl, and $N = S$, Se, are isomorphous. They crystallize in the cubic centro-symmetric space group $Pa3$, with the unit-cell side a in the range 12.2–12.4 Å and $Z = 4$.

---

[a] H. Lipson and C. A. Beevers, *Proceedings of the Physical Society* **48**, 772 (1936).
[b] These omissions give rise to the proportionality signs.

TABLE P6.1.　Simulated Beevers-Lipson "strips"

Amplitude	Index	$\frac{0}{60}$	$\frac{1}{60}$	$\frac{2}{60}$	$\frac{3}{60}$	$\frac{4}{60}$	$\frac{5}{60}$	$\frac{6}{60}$	$\frac{7}{60}$	$\frac{8}{60}$	$\frac{9}{60}$	$\frac{10}{60}$	$\frac{11}{60}$	$\frac{12}{60}$	$\frac{13}{60}$	$\frac{14}{60}$	$\frac{15}{60}$
5	2	5	5	5	4	3	2	2	1	$\bar{1}$	$\bar{2}$	$\bar{2}$	$\bar{3}$	$\bar{4}$	$\bar{5}$	$\bar{5}$	$\bar{5}$
20	4	20	18	13	6	$\bar{2}$	$\bar{10}$	$\bar{16}$	$\bar{20}$	$\bar{20}$	$\bar{16}$	$\bar{10}$	$\bar{2}$	6	13	18	20
32	6	32	26	10	$\bar{10}$	$\bar{26}$	$\bar{32}$	$\bar{26}$	$\bar{10}$	10	26	32	26	10	$\bar{10}$	$\bar{26}$	$\bar{32}$
17	8	17	11	$\bar{2}$	$\bar{14}$	$\bar{17}$	$\bar{8}$	5	16	16	5	$\bar{8}$	$\bar{17}$	$\bar{14}$	$\bar{2}$	11	17
14	10	14	7	$\bar{7}$	$\bar{14}$	$\bar{7}$	7	14	7	$\bar{7}$	$\bar{14}$	$\bar{7}$	7	14	7	$\bar{7}$	$\bar{14}$
40	14	40	4	$\bar{39}$	$\bar{12}$	37	20	$\bar{32}$	$\bar{27}$	27	32	$\bar{20}$	$\bar{37}$	12	39	$\bar{4}$	$\bar{40}$
Add columns for $\sum_k$		128	71	$\bar{20}\cdots$													

TABLE P6.2. Simulated Beevers-Lipson "strips"

7	2	7	7	6	6	5	3	2	1	$\bar{1}$	$\bar{2}$	$\bar{3}$	$\bar{5}$	$\bar{6}$	$\bar{6}$	$\bar{7}$	$\bar{7}$
14	4	14	13	9	4	$\bar{1}$	$\bar{7}$	$\bar{11}$	$\bar{14}$	$\bar{14}$	$\bar{11}$	$\bar{7}$	$\bar{1}$	4	9	13	14
18	6	18	15	6	$\bar{6}$	$\bar{15}$	$\bar{18}$	$\bar{15}$	$\bar{6}$	6	15	18	15	6	$\bar{6}$	$\bar{15}$	$\bar{18}$
13	8	13	9	$\bar{1}$	$\bar{11}$	$\bar{13}$	$\bar{6}$	4	12	12	4	$\bar{6}$	$\bar{13}$	$\bar{11}$	$\bar{1}$	9	13
12	10	12	6	$\bar{6}$	$\bar{12}$	$\bar{6}$	6	12	6	$\bar{6}$	$\bar{12}$	$\bar{6}$	6	12	6	$\bar{6}$	$\bar{12}$
20	14	20	2	$\bar{20}$	$\bar{6}$	18	10	$\bar{16}$	$\bar{13}$	13	16	$\bar{10}$	$\bar{18}$	6	20	$\bar{2}$	$\bar{20}$

TABLE P6.3. $|F(hhh)|$ for Isomorphous Alums

hkl	NH_4^+ (10 electrons)	K^+ (18 electrons)	Rb^+ (36 electrons)	Tl^+ (80 electrons)
111	86	38	19	113
222	0	19	79	195
333	111	125	158	236
444	25	6	55	125
555	24	49	64	131
666	86	86	122	164
777	53	34	0	18
888	0	16	22	56

A symmetry analysis leads to the following atomic positions:

$4M$: $\quad 0,0,0; \quad 0,\frac{1}{2},\frac{1}{2}; \quad \frac{1}{2},0,\frac{1}{2}; \quad \frac{1}{2},\frac{1}{2},0$

$4Al$: $\quad \frac{1}{2},\frac{1}{2},\frac{1}{2}; \quad \frac{1}{2},0,0; \quad 0,\frac{1}{2},0; \quad 0,0,\frac{1}{2}$

$8N$: $\quad \pm\{x,x,x; \quad \frac{1}{2}+x,\frac{1}{2}-x,x; \quad \bar{x},\frac{1}{2}+x,\frac{1}{2}-x; \quad \frac{1}{2}-x,\bar{x},\frac{1}{2}+x\}$

The N atoms lie on cube diagonals, and x_N may be obtained by a one-dimensional Fourier synthesis along the line [111], using $F(hhh)$ data. Table P6.3 lists these data for four alums ($N = S$). Tl may be assumed to be sufficiently heavy to make all F values positive in this derivative. The same sites in each crystal are occupied by the replaceable atoms.

(a) Use the isomorphous replacement technique to determine the signs of the reflections in Table P6.3.

(b) Compute $\rho[111]$ for K alum, using the following equation:

$$\rho(D) \propto \sum_h \pm|F(hhh)| \cos 2\pi hD$$

where D is the sampling interval along [111]. Plot the function and determine a probable value for d_s.

(c) The corresponding hhh data for the isomorphous K–Se alum are listed below. The signs have been allocated by a similar isomorphous replacement procedure. Calculate and plot $\rho(D)$ for these data. Compare the two electron density plots and comment upon the results.

hkl	111	222	333	444	555	666	777	888
$\pm\|F\|$	−48	−52	64	0	116	100	−16	0

6.13. A crystal contains five atoms per unit cell. Four of them contribute together $100e^{i\phi}$ to F(010). The fifth atom has fractional coordinates 0.00, 0.10, 0.00, and its atomic scattering factor components f_o, $\Delta f'$, and $\Delta f''$ are 52.2, −2.7, and 8.0, respectively. If $\phi = 60°$, determine, graphically or otherwise, $|F(010)|$, $|F(0\bar{1}0)|$, $\phi(010)$, and $\phi(0\bar{1}0)$.

6.14. A protein crystal structure is to be solved using MIR. Three isomorphous derivatives are prepared using platinum, uranium, and iodine compounds. For the reflection 060 the following measurements were recorded:

For protein $|F_p| = 858$

For Pt derivative $|F_{PH_1}| = 756$, $|F_{H_1}| = 141$, $\phi_{H_1} = 78°$
U derivative $|F_{PH_2}| = 856$, $|F_{H_2}| = 154$, $\phi_{H_2} = 63°$
I derivative $|F_{PH_3}| = 940$, $|F_{H_3}| = 100$, $\phi_{H_3} = 146°$

Use Harker's construction to obtain an estimate for ϕ_P for this reflection from the native protein crystal.

6.15. Consider Equation (6.19)

$$\psi(x) = \frac{\pi}{2} + 2\sum_{h=1}^{\infty} \frac{1}{h}\sin hx$$

Show that $\sin hx$ can be replaced by $\cos(hx - \phi)$, where $\phi = \pi/2$.

6.16. A protein with molecular weight 18,000 Daltons crystallizes in space group C2 with unit-cell dimensions $a = 40, b = 50, c = 60$ Å, $\beta = 100°$.

(a) Estimate the number of protein molecules per unit cell if there are equal masses of protein and solvent in the unit cell ($M_H = 1.66 \times 10^{-24}$ g).

(b) What symmetry would the protein molecule be required to adopt in the crystalline state?

6.17. A noncentrosymmetric structure is composed entirely of N_H identical heavy atom anomalous scatterers per unit cell (no normal scatterers). Show, graphically or otherwise, that $|F(hkl)| = |F(\bar{h}\bar{k}\bar{l})|$ but that $\phi(hkl) \neq -\phi(\bar{h}\bar{k}\bar{l})$.

6.18. A centrosymmetric structure contains a mixture of anomalously scattering atoms and normal scatterers. Show that $|F(hkl)|^2 = |F(\bar{h}\bar{k}\bar{l})|^2$ and $\phi(hkl) = \phi(\bar{h}\bar{k}\bar{l}) \neq 0$ or π.

6.19. Given the $|F(hkl)|$ equivalents for each of the following space groups, list corresponding Bijvoet pairs:

(a) $C2$: $|F(hkl)| = |F(\bar{h}k\bar{l})| = |F(h\bar{k}l)| = |F(\bar{h}\bar{k}\bar{l})|$

(b) Pm: $|F(hkl)| = |F(\bar{h}k\bar{l})| = |F(h\bar{k}l)| = |F(\bar{h}\bar{k}\bar{l})|$

(c) $P2_12_12_1$: $|F(hkl)| = |F(\bar{h}kl)| = |F(h\bar{k}l)| = |F(hk\bar{l})|$

$= |F(\bar{h}k\bar{l})| = |F(h\bar{k}\bar{l})| = |F(\bar{h}\bar{k}l)| = |F(\bar{h}\bar{k}\bar{l})|$

(d) $P4$: $|F(hkl)| = |F(\bar{k}hl)| = |F(\bar{h}\bar{k}l)| = |F(k\bar{h}l)|$

$= |F(k\bar{h}\bar{l})| = |F(hk\bar{l})| = |F(\bar{k}h\bar{l})| = |F(\bar{h}\bar{k}\bar{l})|$

6.20. A neutron beam has a wavelength spread of 1.0 to 2.0 Å. The beam is not quite parallel, the angular divergence being ± 0.25 Å from the ideal path. The beam is monochromatized with a single crystal of lead (cubic, $a = 4.954$ Å) by reflection from the (111) face. At what angle should the (111) planes be set so as to give a wavelength of 1.25 Å? What would be the approximate spread of wavelength in the monochromatized beam?

6.21. Sodium hydride has the sodium chloride structure type, with $a = 4.88$ Å, Na^+ at 0, 0, 0, and H^- at $0, 0, \frac{1}{2}$. Using the data below, calculate $F(111)$ and $F(220)$ for NaH and NaD, for both x-rays and neutrons; the temperature factor may be neglected for the x-ray case. First, formulate a simplified structure factor equation for the NaH–NaD structure.

$$f_{Na^+}(111) = 8.1 \qquad \beta_{Na^+} = 0.35$$
$$f_{H^-/D^-}(111) = 0.38 \qquad \beta_{H^-} = -0.37$$
$$f_{Na^+}(220) = 6.7 \qquad \beta_{D^-} = 0.67$$
$$f_{H^-/D^-}(220) = 0.21$$

6.22. X-ray intensity data are to be measured for an orthorhombic crystalline protein in three stages:

(a) For $0° < \theta < 10°$, both $I(hkl)$ and $I(\bar{h}\bar{k}\bar{l})$ symmetry-equivalent reflections are measured;

(b) For $10° < \theta < 20°$, $I(hkl)$ alone are measured;

(c) For $20° < \theta < 25°$, $I(hkl)$ alone are measured.
If the unit cell is primitive, with $a = 30, b = 50$ and $c = 40$ Å, and the x-ray wavelength is 1.5 Å, estimate the number of reflections measured and the corresponding resolution for the three data sets.

7

Direct Methods and Refinement

7.1 Introduction

In this chapter, we consider *direct methods*, also known as phase probability methods, of solving the phase problem, together with *Patterson search* techniques, *least-squares refinement*, and other important procedures that are usually involved in the overall investigation of crystal and molecular structure are also discussed.

7.2 Direct Methods of Phase Determination

Direct methods of solving the phase problem are an important technique, particularly in their ability to yield good phase information for structures containing no heavy atoms.

One feature common to the structure-determining methods that we have encountered so far is that values for the phases of x-ray reflections are derived initially by structure factor calculations, albeit on only part of the structure. Since the data from which the best phases are ultimately derived are the $|F_o|$ values, we may imagine that the phases are encoded in these quantities, even though their actual values are not recorded experimentally. This philosophy led to the search for analytical methods of phase determination, which are independent of trial structures, and initiated the development of direct methods, or phase probability techniques.

7.2.1 Normalized Structure Factors

A simplification in direct phase-determining formulae results by replacing $|F(hkl)|$ by the corresponding normalized structure factor $|E_o(hkl)|$, which is

TABLE 7.1. Some Theoretical and Experimental Values Related to $|E|$ Statistics

Mean values	Theoretical values $P\bar{1}(C)$	$P1(A)$	Experimental values and conclusions Crystal 1	Crystal 2
$|E|^2$	1.00	1.00	0.99	0.98
$|E|$	0.80	0.89	0.85 A/C	0.84 A/C
$||E|^2 - 1|$	0.97	0.74	0.91 C	0.82 A
Distribution	%	%	%	%
$|E| > 3.0$	0.30	0.01	0.20 C	0.05 A
$|E| > 2.5$	1.24	0.19	0.90 C	0.98 C
$|E| > 2.0$	4.60	1.80	2.70 A/C	2.84 A/C
$|E| > 1.75$	8.00	4.71	7.14 C	6.21 A/C
$|E| > 1.5$	13.4	10.5	12.9 C	10.5 A
$|E| > 1.0$	32.0	36.8	33.7 C	37.1 A

given by the equation[a]

$$|E_o(hkl)|^2 = \frac{K^2|F_o(hkl)|^2}{\varepsilon \sum_{j=1}^{N} g_j^2} \tag{7.1}$$

The summation in the denominator of (7.1), which is θ dependent through g_j, may be obtained through a curve (K curve) similar to that in Figure 6.24c(ii). The data must be on an absolute scale ($K = 1$), and ε is incorporated in setting up the K curve.[b]

The E values have properties similar to those of the sharpened F values derived for a point-atom model (Section 6.4.2); they are compensated for the fall-off of f with $\sin \theta$. High-order reflections with comparatively small $|F|$ values can have quite large $|E|$ values, an important fact in the application of direct methods. We may note in passing that $|E^2|$ values (or $|E||F|$) can be used as coefficients in a sharpened Patterson function, and since $\overline{|E|^2} = 1$ (see Table 7.1), the coefficients $(|E|^2 - 1)$ produce a sharpened Patterson function with the origin peak removed. This technique is useful because, in addition to the general sharpening effect, vectors of small magnitude which are swamped by the origin peak may be revealed.

ε Factor

Because of the importance of individual reflections in direct phasing methods, care must be taken to obtain the best possible $|E|$ values. The factor ε in the

[a] We shall generally omit the subscript o in the $|E|$ symbol here, since such terms normally refer to observed data.
[b] See also Sections 3.10.1, 11.4.6, and 11.4.10.

denominator of (7.1) takes account of the fact that reflections in certain reciprocal lattice zones or rows may have an average intensity greater than that for the general reflections. The ε factor depends upon the crystal class, and its values for all crystal classes are listed in Table 3.10. Detailed considerations of ε and of the $|E|$ statistics in the next section will be found in Section 3.10.

$|E|$ Statistics

The distribution of $|E|$ values holds useful information about the space group of a crystal. Theoretical quantities derived for equal-atom structures[a] in space groups $P1$ and $P\bar{1}$ are listed in Table 7.1, together with the experimental results for two crystals.

Crystal 1 is pyridoxal phosphate oxime dihydrate, $C_8H_{11}N_2O_6P \cdot 2H_2O$, which is triclinic. The values in Table 7.1 favor the centric distribution C, and the structure analysis[b] confirmed the assignment of space group $P\bar{1}$. Crystal 2 is a pento-uloside sugar; the results correspond, on the whole, to an acentric distribution A, as expected for a crystal of space group $P2_12_12_1$.[c]

It should be noted that the experimentally derived quantities do not always have a completely one-to-one correspondence with the theoretical values, and care should be exercised in using these statistics to select a space group.

7.2.2 Structure Invariants and Origin-Fixing Reflections

The formulae used in direct phasing require, initially, the use of a few reflections with phases known, either uniquely or symbolically. In centrosymmetric crystals, the origin is taken on one of the eight centers of symmetry in the unit cell, and we speak of the sign $s(hkl)$ of the reflection; $s(hkl)$ is $F(hkl)/|F(hkl)|$ and is either $+$ or $-$. We shall show next that, in any primitive, centrosymmetric space group in the triclinic, monoclinic, or orthorhombic systems, arbitrary signs can be allocated to three reflections in order to specify the origin at one of the centers of symmetry. These signs form a basic set, or "fountainhead," from which more and more signed reflections emerge as the analysis proceeds.

From (3.112), we have

$$F(hkl)_{0,0,0} = \sum_{j=1}^{N} g_j \cos 2\pi (hx_j + ky_j + lz_j) \qquad (7.2)$$

---

[a] Theoretical arguments are simplified by assuming that the N atoms in the unit cell are of the same species.

[b] A. N. Barrett and R. A. Palmer, *Acta Crystallographica* B**25**, 688 (1969).

[c] H. T. Palmer, personal communication (1973).

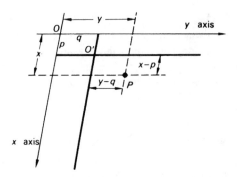

FIGURE 7.1.   Transformation of the point $P(x, y)$, with respect to two-dimensional axes, by moving the origin from O to $O'(p, q)$; the transformed coordinates of P are $(x - p, y - q)$.

where $F(hkl)_{0,0,0}$ indicates an origin of coordinates at the point 0, 0, 0. If this origin is moved to a center of symmetry at $\frac{1}{2}, \frac{1}{2}, 0$, the point that was originally x_j, y_j, z_j becomes $x_j - \frac{1}{2}, y_j - \frac{1}{2}, z_j$ (Figure 7.1, with $p = q = \frac{1}{2}$). The structure factor equation is now

$$F(hkl)_{1/2,1/2,0} = \sum_{j=1}^{N} g_j \cos 2\pi [(hx_j + ky_j + lz_j) - (h + k)/2] \qquad (7.3)$$

Expanding the cosine term, and remembering that $\sin[2\pi(h + k)/2]$ is zero, we obtain

$$F(hkl)_{1/2,1/2,0} = (-1)^{h+k} F(hkl)_{0,0,0} \qquad (7.4)$$

Equation (7.4) demonstrates that $|F(hkl)|$ is invariant under change of origin, as would be expected, but that a change of sign may occur, depending on the parity of the indices hkl. The complete results are listed in Table 7.2. The use of this table will be illustrated by the following examples.

Reflection 312 belongs to parity group 7 (ooe, in short). If $s(312)$ is given a plus sign, the origin could be regarded as being restricted to one from the following list:

$$0, 0, 0; \quad 0, 0, \tfrac{1}{2}; \quad \tfrac{1}{2}, \tfrac{1}{2}, 0; \quad \tfrac{1}{2}, \tfrac{1}{2}, \tfrac{1}{2}$$

Similarly, if $s(322)$, parity group 2(oee), is also given a plus sign, the possible origins are

$$0, 0, 0; \quad 0, \tfrac{1}{2}, 0; \quad 0, 0, \tfrac{1}{2}; \quad 0, \tfrac{1}{2}, \tfrac{1}{2}$$

Combining these two sign allocations, the common origins are

$$0, 0, 0; \quad 0, 0, \tfrac{1}{2}$$

TABLE 7.2. Effect of a Change of Origin of Coordinates, among Centers of Symmetry, on the Sign of a Structure Factor

Centers of symmetry	Parity group							
	1	2	3	4	5	6	7	8
	h even k even l even	h odd k even l even	h even k odd l even	h even k even l odd	h even k odd l odd	h odd k even l odd	h odd k odd l even	h odd k odd l odd
$0, 0, 0$	+	+	+	+	+	+	+	+
$\frac{1}{2}, 0, 0$	+	−	+	+	+	−	−	−
$0, \frac{1}{2}, 0$	+	+	−	+	−	+	−	−
$0, 0, \frac{1}{2}$	+	+	+	−	−	−	+	−
$0, \frac{1}{2}, \frac{1}{2}$	+	+	−	−	+	−	−	+
$\frac{1}{2}, 0, \frac{1}{2}$	+	−	+	−	−	+	−	+
$\frac{1}{2}, \frac{1}{2}, 0$	+	−	−	+	−	−	+	+
$\frac{1}{2}, \frac{1}{2}, \frac{1}{2}$	+	−	−	−	+	+	+	−

In order to fix the origin uniquely at, say, $0, 0, 0$, we select another reflection with a plus sign with respect to $0, 0, 0$. Reference to Table 7.2 shows that parity groups 4, 5, 6, and 8 each meet this requirement.

Parity groups 1 and 3 are excluded from the choice as the third origin-specifying reflection. Group 1 is a special case discussed below. Group 3 (eoe) is related to groups 2 and 7 through an addition of indices:

$$312 + 322 \rightarrow 634 \tag{7.5}$$

or, more generally,

$$ooe + oee \rightarrow eoe \tag{7.6}$$

since $o + o$ or $e + e \rightarrow e$, and $e + o \rightarrow o$.

Parity groups 2, 3, and 7 are said to be linearly related, and cannot be used together in defining the choice or origin.

Structure factors belonging to parity group 1 do not change sign on change of origin, as is evident from both the development of (7.4) and Table 7.2. Reflections in this group are called structure invariants; their signs depend on the actual structure and cannot be chosen at will.

7.2.3 Sign Determination: Centrosymmetric Crystals

Many equations have been proposed that are capable of providing sign information for centrosymmetric crystals. Two of these expressions have proved to be outstandingly useful, and it is to them that we first turn our attention.

Triple-Product Sign Relationship

In 1952, Sayre[a] derived a general formula for hypothetical structures containing identical resolved atoms. For centrosymmetric crystals, it may be given in the form

$$s(hkl)s(h'k'l')s(h - h', k - k', l - l') \approx +1 \qquad (7.7)$$

where $\approx$ means "is probably equal to." The vectors associated with these reflections, $d^*(hkl), d^*(h'k'l')$, and $d^*(h - h', k - k', l - l')$ form a closed triangle, or vector triplet, in reciprocal space. In practice, it may be possible to form several such vector triplets for a given hkl; Figure 7.2a shows two triplets for the vector 300. If two of the signs in (7.7) are known, the third can be deduced, and we can extend the sign information beyond that given in the starting set.

A physical meaning can be given to Equation (7.7) by drawing the traces, in real space, of the three planes that form a vector triplet in reciprocal space (Figure 7.2a,b). For a centrosymmetric crystal, we write

$$\rho(x, y, z) = \frac{2}{V_c} \sum_h \sum_k \sum_l \pm |F(hkl)| \cos 2\pi(hx + ky + lz) \qquad (7.8)$$

The $|F(hkl)|$ terms in this equation take a *positive* sign if the traces of the corresponding planes pass through the origin, like the full lines in Figure 7.2b, and a *negative* sign if they lie midway between these positions, like the dashed lines in the figure. The combined contributions from the three planes in question will thus have maxima at the points of their mutual intersections, which are therefore potential atomic sites, and correspond to regions of high electron density.

This argument is particularly strong if the three planes have large $|E|$ values, because in $|E|$ the damping effect of f has been significantly reduced, leaving a term that is governed by the structure itself. It may be seen from the diagram that triple intersections occur only at points where either three full lines $(+ + +)$ meet, or two dashed lines and one full line meet (some combination of $+ - -$). This result is in direct agreement with (7.7). It is interesting to note that the structure of hexamethylbenzene was solved in 1929 by Lonsdale[b] through drawing the traces of three high-order, high-intensity reflection planes, $7\bar{3}0$, 340, and $4\bar{7}0$, and placing atoms at their intersections. These planes form a vector triplet, and this structure determination contained, therefore, the first but apparently inadvertent use of direct methods.

[a] D. Sayre, *Acta Crystallographica* **5**, 60 (1952).
[b] K. Lonsdale, *Proceedings of Royal Society* A**123**, 494(1929).

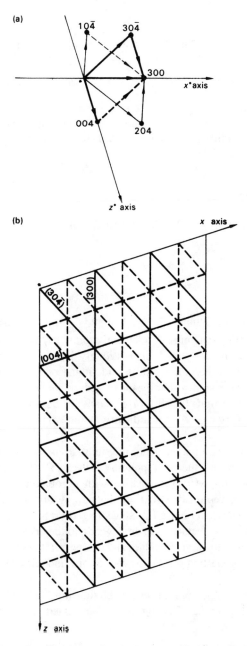

FIGURE 7.2.   (a) Vector triplets 300, 204, 10$\bar{4}$ and 300, 30$\bar{4}$, 004; (b) physical interpretation of (7.7); the lines are traces of the families (300), (004), and (30$\bar{4}$), and the points of triple intersection correspond with (7.7).

Σ_2 Formula

Hauptman and Karle[a] have given the more general form of (7.7) as

$$s[E(hkl)] \approx s\left[\sum_{h'k'l'} E(h'k'l')E(h-h',k-k',l-l')\right] \tag{7.9}$$

the summation being over all vector pairs with known signs which form a triplet with hkl. The probability associated with (7.9) is given by[b]

$$P_+(hkl) = \tfrac{1}{2} + \tfrac{1}{2}\tanh[(\sigma_3/\sigma_2^{3/2})\alpha'] \tag{7.10}$$

where α' is given by

$$\alpha' = |E(hkl)|\sum_{h'k'l'} E(h'k'l')E(h-h',k-k',l-l') \tag{7.11}$$

and σ_n by

$$\sigma_n = \sum_j Z_j^n \tag{7.12}$$

where Z_j is the atomic number of the jth atom.

For a structure containing N identical atoms, $\sigma_3/\sigma_2^{3/2}$ is equal to $N^{-1/2}$. From (7.11), we see that the probability is strongly dependent upon the magnitudes of the $|E|$ values. Furthermore, unless glide-plane or screw-axis symmetry is present, or there exists some other means of generating negative signs, (7.9) will produce only positive signs for all $E(hkl)$. Such a situation would correspond to a structure with a very heavy atom at the origin and would, in general, lead to an incorrect solution.

If the combination of signs under the summation in (7.11) produces a large and negative value for α', the corresponding value of $P_+(hkl)$ may tend to zero. This result indicates that $s(hkl)$ is negative, with a probability that tends to unity.

Probability curves for different numbers N of atoms in the unit cell as a function of α' are shown in Figure 7.3. Since the most reliable signs from (7.9) are associated with large $|E|$ values, we can now add to the origin-specifying criteria the requirements of both large $|E|$ values and a large number of Σ_2 interactions for each reflection in the starting set. In this way, strong and reliable sign propagation is encouraged.

To illustrate the operation of the Σ_2 relationship, we shall consider the two vector triplets in Figure 7.2. The sign to be determined is $s(300)$, the others are

[a] H. Hauptman and J. Karle, *Solution of the Phase Problem, 1. The Centrosymmetric Crystal*, American Crystallographic Association Monograph No. 3 (1953).
[b] See Bibliography (Woolfson).

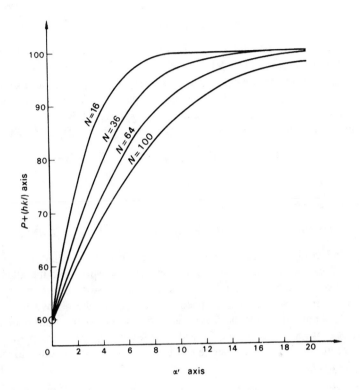

FIGURE 7.3.   Percentage probability of a single triple-product (Σ_2) sign relationship as a function of α' for different numbers N of atoms in a unit cell, according to (7.10).

assumed to be known. It may be noted that sometimes we speak of a sign as $+$ or $-$ and at other times as $+1$ or -1. The latter formulation is clearly more appropriate to computational methods. The data are tabulated as follows.

hkl	$|E(hkl)|$					
300	2.40					
$h'k'l'$	$E(h'k'l')$	$h-h', k-k', l-l'$	$E(h-h', k-k', l-l')$	α'	$s(hkl)$	$P_+(hkl), \%$
$10\bar{4}$	$+2.03$	204	-2.22	19.3	-1	0.8
004	-1.95	$30\bar{4}$	$+1.81$		-1	

Assuming that N is 64, the indication given is that $s(300)$ is negative with a probability of 99.2%.

7.2.4   Amplitude Symmetry and Phase Symmetry

In space group $P\bar{1}$, the only symmetry-related structure factors are $F(hkl)$ and $F(\bar{h}\bar{k}\bar{l})$. According to Friedel's law the intensities and, hence the amplitudes, of these structure factors are equal, and in centrosymmetric space groups $s(hkl) = s(\bar{h}\bar{k}\bar{l})$. Thus, the amplitude symmetry and the phase symmetry follow the same law, but this will not necessarily be true in other space groups.

From the geometric structure factor for space group $P2_1/c$, (3.130),

$$|F(hkl)| = |F(\bar{h}\bar{k}\bar{l})| = |F(h\bar{k}l)| = |F(\bar{h}k\bar{l})| \tag{7.13}$$

and for the signs there are two possibilities:

$$k+l = 2n: \qquad s(hkl) = s(\bar{h}\bar{k}\bar{l}) = s(h\bar{k}l) = s(\bar{h}k\bar{l}) \tag{7.14}$$

$$k+l = 2n+1: \qquad s(hkl) = s(\bar{h}\bar{k}\bar{l}) = -s(h\bar{k}l) = -s(\bar{h}k\bar{l}) \tag{7.15}$$

These relationships provide enhanced opportunities for Σ_2 relationships to be developed, and in this way space-group symmetry can improve the chances of successful phase determination. When considering phase relationships in a non-centrosymmetric space group, we need to take note of the change in sign and magnitude of both the A and the B components of the geometrical structure factor, as h, k, or l take a negative sign. Consider, for example, space group Pc (b axis unique; origin on c). We can separate (3.128) into two parts, according to whether l is odd or even:

$l = 2n$

$$A = 2\cos 2\pi(hx + lz)\cos 2\pi ky$$
$$B = 2\sin 2\pi(hx + lz)\cos 2\pi ky$$

$l = 2n + 1$

$$A = -2\sin 2\pi(hx + lz)\sin 2\pi ky$$
$$B = 2\cos 2\pi(hx + lz)\sin 2\pi ky$$

It is clear that, in all cases, $\phi(hkl) = -\phi(\bar{h}\bar{k}\bar{l}) \neq \phi(\bar{h}k\bar{l})$. However, for the two parities of l, the following expressions hold.

$$l = 2n: \qquad \phi(hkl) = \phi(h\bar{k}l) \text{ and } \phi(\bar{h}kl) = -\phi(hk\bar{l})$$

$$l = 2n+1: \qquad \phi(hkl) = \pi + \phi(h\bar{k}l) \text{ and } \phi(\bar{h}kl) = \pi - \phi(hk\bar{l})$$

The amplitude symmetry and phase symmetry for all space groups are contained in the *International Tables for X-ray Crystallography*, Volume A (or Volume 1).[a]

[a] See Bibliography, Chapter 1.

7.2.5 Σ_2 Listing

Because of both the increased probability in relationships developed for reflections with high $|E|$ values and the existence of many vector triplets in a complete set of data, the initial application of direct methods is limited to reflections with large $|E|$ values, say, greater than 1.5.

A Σ_2 listing is prepared by considering each value of $|E(hkl)|$ greater than the preset limit in order of decreasing magnitude, as a basic hkl vector, and searching the data for all possible interactions with $h'k'l'$ and $h - h', k - k', l - l'$. Some reflections will enter into many such interactions, while others will produce only a small number.

7.2.6   Symbolic-Addition Procedure

Karle and Karle[a] have described a technique for the systematic application of the Σ_2 formula for building up a self-consistent sign set. The various steps involved are outlined below, using results obtained with pyridoxal phosphate oxime dihydrate.[b]

Crystal Data

Formula: $C_8H_{11}N_2O_6P \cdot 2H_2O$
System: triclinic
Unit-cell dimensions: $a = 10.94, b = 8.06, c = 9.44$ Å, $\alpha = 57.18°$,
  $\beta = 107.68°, \gamma = 116.53°$
V_c: 623.75 Å^3
D_m: 1.57 g cm^{-3}
M_r: 298.19
Z: 1.98, or 2 to the nearest integer.
Absent spectra: none.
Possible space groups: $P1$ or $P\bar{1}$. $P\bar{1}$ was chosen on the basis of intensity statistics (Table 7.1).
All atoms are in general positions.

Sign Determination

1. A total of 163 reflections for which $|E| \geqslant 1.5$ was arranged in descending order of magnitude, and a Σ_2 listing was obtained using a computer program.

---

[a] J. Karle and I. L. Karle, *Acta Crystallographica* **21**, 849 (1966).
[b] Barrett and Palmer, *loc. cit.*

2. From a study of the Σ_2 listing, three reflections were allocated $+$ signs (Table 7.3); they are the origin-fixing reflections, selected according to the procedures already discussed.

3. Equation (7.9) was used by searching, initially, between members of the origin-fixing set and other reflections. To maintain a high probability, only the highest $|E|$ values were used. For example, $9\bar{5}\bar{5}(|E| = 2.31)$ is generated by the combination of $8\bar{1}5$ and $\bar{1}40$:

$$s(9\bar{5}\bar{5}) \approx s(8\bar{1}5)s(\bar{1}40) = (+1)(+1) = +1 \qquad (7.16)$$

From (7.11), α' is 16.5, and Figure 7.3 shows ($N = 38$, excluding hydrogen) that the probability of this indication is about 99.7%. The new sign was accepted and used to generate more signs. This process was continued until no new signs could be developed with high probability.

4. At this stage, it is often found that the number of signs developed with confidence is small. This situation arose with pyridoxal phosphate oxime dihydrate, and the Σ_2 formula was then applied to reflections with symbolic signs. In this technique, a reflection was selected, again by virtue of its high $|E|$ value and long Σ_2 listing, and allocated a letter symbol (Table 7.3). Generally, less than five symbolic phases are sufficient, and there are no necessary restrictions on the parities of these reflections. However, it is desirable that there are no redundancies in the complete starting set, that is, no three reflections in the set should themselves be related by a triple-product relationship.

As a symbol became involved in a sign of a reflection, it was written into the Σ_2 listing. The example in Table 7.4 shows a Σ_2 entry for $9\bar{8}6$. Reading across the table, sign combinations are seen to be generated by multiplying $s(h'k'l')$ by $s(h-h',k-k',l-l')$, which are then written as $s(9\bar{8}6)$ in the penultimate column.

TABLE 7.3. Starting Set for the Symbolic-Addition Procedure

hkl	$|E|$	Sign	
$9\bar{1}\bar{4}$	2.97	$+$	$\left.\right\}a$
$8\bar{1}5$	3.00	$+$	
$\bar{1}40$	2.38	$+$	
020	4.50	A	$\left.\right\}b$
253	2.24	B	
822	2.71	C	
303	2.69	D	
023	2.28	E	

^{a} Origin-fixing reflections.
^{b} Letter symbols, each representing $+$ or $-$.

TABLE 7.4. Σ_2 Listing for the Reflection $98\bar{6}$ of Pyridoxal Oxime Phosphate with Appropriate Phase Symbols Added[a]

$h'k'l'$	$s(h'k'l')$	$|E_2|$ $|E(h'k'l')|$	$h - h'$, $k - k'$, $l - l'$	$s(h - h'$, $k - k'$, $l - l')$	$|E_3|$ $E(h - h'$, $k - k'$, $l - l')$	$|E_1| \times$ $|E_2| \times$ $|E_3|$	$s(98\bar{6})$	$P_+(98\bar{6})$, %
$1\,\bar{5}\,0$	BD	2.16	$8\ \bar{3}\ \bar{6}$	A	1.63	6.40	ABD	90
$10,\bar{2}\,\bar{2}$	AB	2.04	$1\ 6\ 4$	D	1.88	6.97	ABD	91
$10,\bar{7}\,\bar{1}$	D	1.87	$1\ 1\ 5$	AB	1.63	5.54	ABD	87
$4\,\bar{8}\,\bar{3}$	D	1.83	$5\ 0\ \bar{3}$	ECD	1.58	5.25	EC	85
$3\,\bar{9}\,\bar{4}$		1.76	$6\ 1\ \bar{2}$		1.58	5.03		
$3\,\bar{5}\,\bar{6}$		1.70	$6\ \bar{3}\ 0$		1.51	4.66		
$6\,\bar{7}\,\bar{2}$		1.68	$3\ \bar{1}\ \bar{4}$		1.63	4.98		
$10,\bar{4}\,\bar{2}$	B	1.62	$1\ 4\ 4$	AD	1.67	4.93	ABD	84
$0\,\bar{2}\,0$	$-A$	4.50	$9\ 6\ \bar{6}$		1.73	14.08		
$0\,\bar{8}\,0$	$+$	2.48	$9\ 0\ \bar{6}$		1.85	8.30		

[a] $|E(98\bar{6})| = |E_1| = 1.89$. We can use both $\mathbf{h} - \mathbf{k}$ and $\mathbf{h} + \mathbf{k}$ in these triple products.

Recurring combinations, such as ABD, gave rise to consistent indications. If the probability that $s(98\bar{6}) = s(ABD)$ is sufficiently large, this sign value is entered for $s(98\bar{6})$ wherever these indices occur. In the final column of the table, the probability of each sign indication is listed. Although they are small individually, the combined probability from (7.10) that $s(98\bar{6})$ was ABD is 100%.

5. When this process had been exhausted, the results were examined for agreement among sign relationships. For example, in Table 7.4 there is a weak indication that $ABD = EC$. The most significant relationships found overall were $AC = E, C = EB, B = ED, AD = E$, and $AB = CD$. Multiplying the first by the second, and the first by the fourth, and remembering that products such as A^2 equal $+1$, reduces this list to $A = B, C = D$, and $E = AC$.

The five symbols were reduced, effectively, to two, A and C. The sign determination was rewritten in terms of signs and the symbols A and C; reflections with either uncertain or undetermined signs were rejected from the first electron density calculation.

7.2.7 Calculation of E Maps

The result of the above analysis meant that four possible sign sets could be generated by the substitutions $A = \pm 1, C = \pm 1$. The set with $A = C = +1$ was rejected because this phase assignment implies a very heavy atom at the origin of the unit cell. The three other sign combinations were used to calculate E maps.

These maps are obtained by Fourier syntheses, using (7.8), but with $|E|$ replacing $|F|$ as the coefficients. The sharp nature of E implicit in (7.1) is advantageous when using a limited number of data to resolve atomic peaks in the electron density map, although normally about eight reflections per atom in the asymmetric unit are desirable. Spurious peaks can arise, like that next to O(8), because of the limited number of coefficients in the Fourier series for an E map.

The sign combination for pyridoxal phosphate oxime dihydrate that led to an interpretable E map was $A = C = -1$. The atomic positions from this map (Figure 7.4a) were used in a successful refinement of the structure, and Figure 7.4b shows the conformation of this molecule.

If there are n symbolic signs in the final centrosymmetric phase solution, there will be 2^n combinations, each of which can give rise to an E map, and it is desirable to set up criteria that will seek the most probable set. We shall consider such criteria during our discussion of the noncentrosymmetric case, where they are of even greater importance.

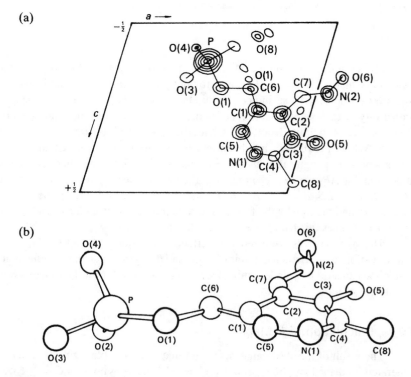

FIGURE 7.4.   Pyridoxal phosphate oxime dihydrate: (a) composite three-dimensional E map as seen along b; (b) molecular conformation viewed at about 10° to the plane of the six-membered ring.

7.2.8 Phase Determination: Noncentrosymmetric Crystals

The noncentrosymmetric case is more difficult, both in theory and in practice. Much of this difficulty stems from the fact that the phase angle can take on any value between 0 and 2π, with a consequent imprecision in its determination. Nevertheless, direct methods are used regularly to solve such structures, with *ca* 100 atoms in the asymmetric unit. The limiting factor is not actually the size of the molecule being studied, but rather the number of atoms N in the unit cell, which is space-group dependent. Structures where N is 400 or more can be expected to be solved without great difficulty using current versions of the various programs now available.

Equations for a noncentrosymmetric crystal are, not surprisingly, more generalized expressions of those such as (7.9), for a centrosymmetric crystal. Using the fact that the electron density distribution is a non-negative function, Karle and Hauptman derived a set of inequality relationships.[a] Here, we stress the vector nature of the structure factor by writing it in bold type. The first three inequalities may be written as non-negative functions.

$$F_{000} \geqslant 0 \tag{7.17}$$

$$|F_{\mathbf{h}}| \leqslant F_{000} \tag{7.18}$$

$$\mathbf{F_h} - \delta_{\mathbf{h},\mathbf{k}} \leqslant \mathbf{r} \tag{7.19}$$

where

$$\delta_{\mathbf{h},\mathbf{k}} = \mathbf{F_{h-k}}\mathbf{F_k^b}/F_{000} \tag{7.20}$$

and

$$|\mathbf{r}| = \frac{\begin{vmatrix} F_{000} & \mathbf{F_{h-k}^*} \\ \mathbf{F_{h-k}} & F_{000} \end{vmatrix}\begin{vmatrix} F_{000} & \mathbf{F_k^*} \\ \mathbf{F_k} & F_{000} \end{vmatrix}}{F_{000}} \tag{7.21}$$

We shall use here the convenient notation $\mathbf{h}$ for hkl, $\mathbf{k}$ for $h'k'l'$, $\mathbf{h} - \mathbf{k}$ for a third reflection that forms a vector triplet with $\mathbf{h}$ and $\mathbf{k}$, and, in order to avoid excessive parentheses, $\mathbf{F_h}$ for $\mathbf{F(h)}$. Equations (7.17) and (7.18) are immediately acceptable in terms of earlier discussions in this book.

The structure factor $\mathbf{F_h}$, given only $|F_{\mathbf{h}}|$, must lie on a circle of that radius on an Argand diagram (Figure 7.5). Equation (7.19) then indicates that $\mathbf{F_h}$ lies within a circle, center $\delta_{\mathbf{h},\mathbf{k}}$ and radius $|\mathbf{r}|$, between the points P and Q. Expansions of the determinants in (7.21), remembering that $\mathbf{F} \cdot \mathbf{F^*} = |F|^2$, shows that the larger the

---

[a] J. Karle and H. Hauptman, *Acta Crystallographica* **3**, 181 (1950).
[b] $\mathbf{F_k}\mathbf{F_{h-k}}$ is a multiplication of two complex numbers and may be interpreted as $|F_{\mathbf{k}}||\mathbf{F_{h-k}}|\exp\{i(\phi_{\mathbf{k}}+\phi_{\mathbf{h-k}})\}$.

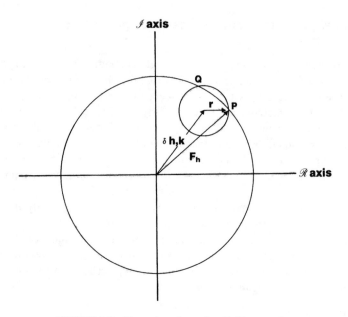

FIGURE 7.5.   Illustration of equation (7.19) at equality.

values of $|F_k|^2$ and $|F_{h-k}|^2$, the closer $\mathbf{F_h}$ approaches $\delta_{h,k}$. For a given $\mathbf{h}$, as $\mathbf{k}$ is varied, $\mathbf{F_h}$ is proportional to the average over $\mathbf{k}$:

$$\mathbf{F_h} \propto \langle \mathbf{F_k} \mathbf{F_{h-k}} \rangle_k \tag{7.22}$$

the proportionality constant being F_{000}. We can see how equation (7.22) can give rise to (7.9), or to (7.7) for a single interaction.

Using the general relation

$$\mathbf{F_h} = |F_h| \exp(i\phi_h) \tag{7.23}$$

we obtain the phase addition formula

$$\phi_h \approx \phi_k + \phi_{h-k} \tag{7.24}$$

The sign $\approx$ indicates an approximation which is better the larger the values of the corresponding structure factors. Where several triplets are involved with a given $\mathbf{h}$, (7.24) becomes

$$\phi_h \approx \langle \phi_k + \phi_{h-k} \rangle_k \tag{7.25}$$

where $\langle \; \rangle_k$ implies an average, taken over a number of triple product relationships (TPRs) common to $\mathbf{h}$.

The $|F_o|$ data derived experimentally are converted to $|E|$ values. Again, we commence phase determination with $|E|$ values greater than about 1.5, in order to maintain acceptable probability limits. Equation (7.25) is illustrated by an Argand diagram in Figure 7.6 for four values of $\mathbf{k}$; $\phi_{\mathbf{h}}$ is the estimated phase angle associated with the resultant vector $\mathbf{R_h}$. Each vector labeled κ depends on a product $|E_{\mathbf{k}}||E_{\mathbf{h-k}}|$ and may be resolved into components A and B along the real and imaginary axes, respectively, such that

$$A = |E_{\mathbf{k}}||E_{\mathbf{h-k}}|\cos(\phi_{\mathbf{k}} + \phi_{\mathbf{h-k}}) \tag{7.26}$$

and

$$B = |E_{\mathbf{k}}||E_{\mathbf{h-k}}|\sin(\phi_{\mathbf{k}} + \phi_{\mathbf{h-k}}) \tag{7.27}$$

It follows from (7.25)–(7.27), with the enhancement $w_{\mathbf{h}}$, that

$$\tan \phi_{\mathbf{h}} \approx \frac{\sum_{\mathbf{h}} w_{\mathbf{h}}|E_{\mathbf{k}}||E_{\mathbf{h-k}}|\sin(\phi_{\mathbf{k}} + \phi_{\mathbf{h-k}})}{\sum_{\mathbf{h}} w_{\mathbf{h}}|E_{\mathbf{k}}||E_{\mathbf{h-k}}|\cos(\phi_{\mathbf{k}} + \phi_{\mathbf{h-k}})} \tag{7.28}$$

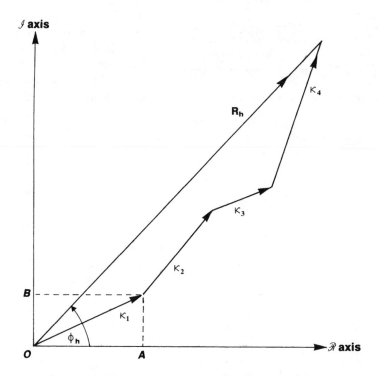

FIGURE 7.6.   Summation of four vectors $\kappa_1 - \kappa_4$ on an Argand diagram; the resultant vector is $\mathbf{R_h}$, with a phase angle $\phi_{\mathbf{h}}$.

Equation (7.28) is a weighted tangent formula, where weights w_h may be either unity or given values as explained in Section 7.2.12. Current phase-determining procedures are based largely on (7.28). The reliability of (7.28) can be measured by the variance $V(\phi_h)$. Figure 7.7 shows $V(\phi_h)$ as a function of α_h, where

$$\alpha_h^2 = \left[\sum_h \kappa_{hk} \cos(\phi_k + \phi_{h-k}) \right]^2 + \left[\sum_k \kappa_{hk} \sin(\phi_k + \phi_{h-k}) \right]^2 \quad (7.29)$$

and

$$\kappa_{hk} = 2\sigma_3 \sigma_2^{-3/2} |E_h||E_k||E_{h-k}| \quad (7.30)$$

with

$$\sigma_n = \sum_{j=1}^{N} Z_j^n \quad (7.31)$$

as before, and the sum taken over the N atoms in the unit cell. The parameter α_h gives a measure of the reliability with which ϕ_h is determined by the tangent formula. When (7.29) contains only one term, as it may in the initial stages of phase determination, then $\alpha_h = \kappa_{hk}$ and is strongly dependent on the product $|E_h||E_k||E_{h-k}|$. Figure 7.7 shows clearly that $V(\phi_h)$ has acceptably small values when α_h is greater than about 4 ($V^{1/2} < 30°$), but increases rapidly for α_h decreasing below about 3 ($V^{1/2} > 40°$); α_h depends also on $\sigma_3 \sigma_2^{-3/2}$, which

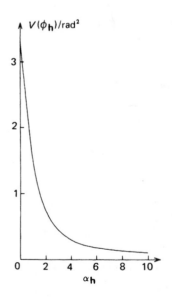

FIGURE 7.7. Variance $V(\phi_h)$ as a function of α_h.

TABLE 7.5. Values of
$|E_{\min}|$ for $\alpha_h = 3.0$ in
Structures Containing N
Identical Atoms Per Unit Cell

N	$|E_{\min}|$
25	1.96
36	2.08
49	2.19
64	2.29
81	2.38
100	2.47

value depends on the number and types of atoms in the unit cell. This dependence may be illustrated by hypothetical structures containing different numbers N of identical atoms; α_h $(= \kappa_{hk})$ is then given by

$$\alpha_h = \frac{2}{N^{1/2}}|E_h||E_k||E_{h-k}| \qquad (7.32)$$

Table 7.5 lists the values of $|E_{\min}|$ needed to obtain $\alpha_h = 3$ for selected values of N from 25 to 100. The table illustrates clearly an important limitation of direct methods: the required $|E_{\min}|$ increases dramatically as a function of N, whereas, as indicated earlier, the distribution of $|E|$ values is largely independent of structural complexity. Therefore it becomes more and more difficult to form a good starting set as N becomes larger and larger.

Calculation of α_h from (7.29) is possible only when phases are available. In the initial stages of phase determination this is not practicable, and the following formula for the expectation value (α_E^2) of α_h^2, which uses only the values of κ_{hk}, has been developed:

$$\alpha_E^2 = \sum_k \kappa_{hk}^2 + \sum_k \sum_{\substack{k' \\ k \neq k'}} \kappa_{hk}\kappa_{hk'} \frac{I_1(\kappa_{hk})I_1(\kappa_{hk'})}{I_0(\kappa_{hk})I_0(\kappa_{hk'})} \qquad (7.33)$$

where I_0 and I_1 are modified Bessel functions of the zero and first orders, respectively. The function $I_1(\kappa)/I_0(\kappa)$ has the form shown in Figure 7.8 and may be expressed as the polynomial

$$I_1(\kappa)/I_0(\kappa) \approx 0.5658\kappa - 0.1304\kappa^2 + 0.0106\kappa^3$$

in the range $0 \leqslant \kappa \leqslant 6$; for $\kappa > 5$ the value of the function is essentially unity. These principles, used in conjunction with those discussed earlier for selecting the origin-determining reflections, may help a direct methods analysis to be established

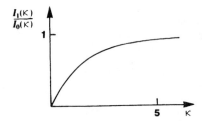

FIGURE 7.8.   Variation of $I_1(\kappa)/I_0(\kappa)$ with κ.

on a sound basis right from the beginning, and so lead to a number of sufficiently accurate phases to give an interpretable E map. Experience shows, however, that even with great care, the development of phases may not always be successful. In such an event the remedy is often to try again with a different starting set of reflections.

Σ_1 Relationships

Let $-\mathbf{k} = \mathbf{h}$ in (7.24). Then

$$\phi_{\mathbf{h}} \approx \phi_{-\mathbf{h}} + \phi_{2\mathbf{h}} \approx \phi_{2\mathbf{h}}/2 \qquad (7.34)$$

If the structure is centrosymmetric then, in (7.7), $S_{\mathbf{h}}$ is ± 1; since $S_{\mathbf{h}}S_{-\mathbf{h}} = 1$, it follows that

$$S_{2\mathbf{h}} \approx 1 \qquad (7.35)$$

These two Σ_1 relationships, like (7.7) and (7.24) themselves, require large values of both $|E_{\mathbf{h}}|$ and $|E_{2\mathbf{h}}|$ for a high probability of their validity.

7.2.9   Enantiomorph Selection

In those noncentrosymmetric space groups, such as $P2_1$ and $P2_12_12_1$, that contain no inversion symmetry (enantiomorphous space groups), it is always possible to specify two enantiomorphic arrangements of the atoms in the structure that will lead to the same values of $|F|$. For example, in the structure in Figure 1.3, which has two molecules per unit cell in space group $P2_1$, the two arrangements would be related by inversion through the origin, and will be referred to as the structure (S) and its inverse (I). From the structure factor theory discussed earlier, we can write

$$F(\mathbf{h})_S = A(\mathbf{h})_S + iB(\mathbf{h})_S$$

for the structure, and   (7.36)

$$F(\mathbf{h})_I = A(\mathbf{h})_I + iB(\mathbf{h})_I$$

for its inverse. From the inversion relationship, we know that $F(\mathbf{h})_S$ and $F(\mathbf{h})_I$ are complex conjugates; hence,

$$A(\mathbf{h})_S = A(\mathbf{h})_I$$

and   (7.37)

$$B(\mathbf{h})_S = -B(\mathbf{h})_I$$

For either the structure or its inverse, we can choose $B(\mathbf{h})$ to be positive, so that the corresponding phase angle $\phi(\mathbf{h})$ lies in the range $0 \leqslant \phi(h) \leqslant \pi$. This procedure was followed in the structure analysis of tubercidin (Section 7.2.10), where the phase of symbolic reflection $a(13\bar{8})$ was restricted to a value between 0 and π, specifically $3\pi/4$.

In $P2_12_12_1$, another noncentrosymmetric space group of frequent occurence in practice, the zonal reflections $0kl, h0l$, and $hk0$ are centric, and may be given phases equal to $m\pi/2$. The value of m takes the same parity as the index following zero, working in a cyclic manner. Thus, an origin and an enantiomorph could be specified in this space group by the selection

$$
\left.
\begin{array}{cccc}
5 & 2 & 0 & +\pi/2 \\
0 & 1 & 1 & +\pi/2 \\
11 & 3 & 0 & +\pi/2
\end{array}
\right\} \quad \text{Origin}
$$

$$
\begin{array}{cccc}
11 & 0 & 0 & +\pi/2 \quad \text{Enantiomorph}
\end{array}
$$

A detailed practical treatment on the origin and enantiomorph for all space groups has been given by Rogers[a].

It is important not to confuse the specifying of the enantiomorph with the selection of the absolute configuration of a structure: in both cases, the same type of space group is involved. Selection of the enantiomorph is essential to a correct application of direct methods to a structure with an enantiomorphous space group. However, the derived solution of the structure may correspond to either the absolute configuration or its inverse. This dilemma has to be resolved by further tests, usually involving anomalous scattering (see Section 6.5).

7.2.10 Phase Determination in Space Group $P2_1$

The method of symbolic addition can be used for phase determination in noncentrosymmetric crystals, but it is extremely laborious because of the general nature of $\phi(hkl)$, which can have any value in the range $0 - 2\pi$. The structure

---

[a] D. Rogers in *Theory and Practice of Direct Methods in Crystallography*, edited M. F. C. Ladd and R. A. Palmer, p. 147, Plenum Press (1980).

TABLE 7.6.   Crystal Data for
Tubercidin

Formula	$C_{11}H_{14}N_4O_4$
M_r	266.26
Space group	$P2_1$
a	9.724(9) Å
b	9.346(11)
c	6.762(10)
β	94.64(10)°
V_c	612.52Å^3
D_m	1.449gcm^{-3}
D_x	1.444
Z	2
$F(000)$	280

of tubercidin was determined by Stroud[a] using this method: Table 7.6 lists the crystal data for this compound.

In space group $P2_1$, $|E(hkl)|$ has the following symmetry equivalence:

$$|E(hkl)| = |E(\bar{h}k\bar{l})| = |E(h\bar{k}l)| = |E(\bar{h}\bar{k}\bar{l})| \qquad (7.38)$$

The phases of the symmetry-related reflections in this space group are also linked, but in a different way, according to the parity of k:

$$k = 2n: \qquad \phi(hkl) = \phi(\bar{h}k\bar{l}) = -\phi(h\bar{k}l) = -\phi(\bar{h}\bar{k}\bar{l}) \qquad (7.39)$$

$$k = 2n + 1: \qquad \phi(hkl) = \pi + \phi(\bar{h}k\bar{l}) = \pi - \phi(h\bar{k}l) = -\phi(\bar{h}\bar{k}\bar{l}) \qquad (7.40)$$

Although ϕ_h can, in general, have a value anywhere in the range $0 - 2\pi$, the $h0l$ reflections are restricted to the values 0 or π in this space group; in other words the $h0l$ zone is centric.

The origin was specified by assigning phases to three reflections, according to the rules discussed above, as shown by Table 7.7. Next, new phases were determined according to (7.24) or (7.25). In order to maintain an expected variance $V(\phi_h)$ (Figure 7.7) of no more than $0.5\,rad^2$, the product $|E_h||E_k||E_{h-k}|$ must be greater than 8.5 for this structure. Two new phases $\phi(80\bar{2})$ and $\phi(612)$ were thus determined from the origin set (Table 7.8), further phases being determined in terms of symbols (Table 7.9). Eleven phases were generated in terms of the origin phases and symbol a, 20 after adding letter b, and 47 after adding the third symbolic phase c. Table 7.9 illustrates the initial stages of this process. The criteria

[a] R. M. Stroud, *Acta Crystallographica* B**29**, 60 (1973).

TABLE 7.7.
Origin-Specifying Phases for
Tubercidin

khl	$|E_h|$	$\phi_{\mathbf{h}}$
$10\bar{6}$	1.95	0
$40\bar{1}$	2.09	0
$71\bar{4}$	2.45	0

TABLE 7.8.   Course of the Phase Determination
Procedure for Tubercidin

hkl	$\phi_{\mathbf{h}}$	$|E_{\mathbf{h}}|$	Number of numerical or symbolic phases
Origin set	See Tables 7.7 and 7.9		5
$13\bar{8}$	a	2.99	11
206	b	2.20	20
790	c	2.76	47

for accepting a phase were as follows:

1. that $V(\phi_{\mathbf{h}})$, irrespective of the actual choice for phases c and a (b is a structure invariant with phase 0 or π), should be less than $0.5\,\text{rad}^2$, no matter how many contributors there were to the sum in (7.25);
2. that where there were two or more different indications for a phase, the phase would be accepted only when indications of one type predominated strongly.

Manual phase determination using a Σ_2 listing indicated relationships such as

$$\phi(63\bar{3}) = c - 2a - b \tag{7.41}$$

and

$$b = 0 \tag{7.42}$$

which are strong because they both come from multiple indications, 6 and 3, respectively. By reiteration of the phase addition procedure described above, the results in Table 7.10 indicate relationships between a and c.

Bearing in mind that the objective is to obtain a self-consistent set of phases, it is well to consider how this might now be achieved. Refinement of phases could in principle be achieved by application of (7.28). However, this would be possible only if numerical values for a and c (taking b as zero) were available. Alternatively,

TABLE 7.9.   Initial Development of
Phases for Tubercidin

$\mathbf{h}^a$	$|E_{\mathbf{h}}|$	$\phi_{\mathbf{h}}$	$|E_{\mathbf{h}}||E_{\mathbf{k}}||E_{\mathbf{h}-\mathbf{k}}|$
Origin set			
$40\bar{1}$	2.09	0	
$10\bar{6}$	1.95	0	
$71\bar{4}$	2.45	0	
New phases			
$40\bar{1}$		0	
$40\bar{1}$		0	
$*80\bar{2}$	2.33		10.2
$\bar{1}06$		0^b	
$71\bar{4}$		0	
$*612$	1.83	0	8.7
First letter			
$13\bar{8}$	2.99	a	
612		0	
$*74\bar{6}$	2.20	a	12.0
$7\bar{1}4$		π^c	
$\bar{1}38$		$\pi + a^c$	
$*624$	2.19	a	16.0

^{a} An asterisk denotes a new phase in the list.
^{b} Symmetry-related phase used from (7.39).
^{c} Symmetry-related phase used from (7.40).

TABLE 7.10.   Relationships between
Letter Symbols

Form of relationship	Number of indications
$c = \pi + 2a$	7
$c = \pi + 3a$	15
$c = \pi + 4a$	19
$c = 3a$	5
$c = 4a$	2
$c = -3a$	4 or 5
$a = 0$	2
$a = \pi$	2
$b = 0$	Many
$b = \pi$	None

if a working formula relating a and c could be found, (7.28) could be implemented by substitution of values for one symbol only. Table 7.10 shows that there were 41 indications that

$$c = \pi + pa \tag{7.43}$$

where the best numerical value for p was found to be 3.29. Hence,

$$c = \pi + 3.29a \qquad (7.44)$$

The symbol a was then limited to the range

$$0 < a < \pi \qquad (7.45)$$

in order to fix the enantiomorph (Section 7.2.9). Values for a were chosen such that

$$a = n\pi/8 \quad (n = 1, 2, \ldots, 8) \qquad (7.46)$$

and converted into phases by (7.44); each set was expanded and refined by (7.28) (taking $w_{\mathbf{h}} = 1$) for up to 419 reflections with $|E_{\min}| \geqslant 1.0$. Some phases were rejected because of inconsistencies in their phase indications. An interpretable E map was obtained using the refined phase set with $a = 6\pi/8$; a composite diagram is given in Figure 7.9.

7.2.11   Advantages and Disadvantages of Symbolic Addition

Symbolic addition has several advantages and disadvantages, summarized as follows:

Advantages:

1. The user is in control throughout the analysis. He has the responsibility of making sure that all formulae, including symmetry relationships, are applied correctly.
2. The user can make decisions regarding criteria of acceptance of phase indications, the number of $|E|$ values to include, the number of symbolic phases, the choice of starting set, and so on.

Disadvantages:

1. The analysis can be carried out only by a specialist in crystallography.
2. The procedure is slow, requiring many hours of preparation before meaningful results emerge.
3. If a large number of symbols is required, many phase sets will be produced, each of which requires refinement by the tangent formula.

Not surprisingly, alternative rapid and more automatic methods of applying direct methods formulas were sought in the late 1960s, leading to development of the multisolution methods.[a]

---

[a] M. M. Woolfson and G. Germain, *Acta Crystallographica* B**24**, 91 (1968); G. Germain, P. Main and M. M. Woolfson, *Acta Crystallographica* B**26**, 274 (1970).

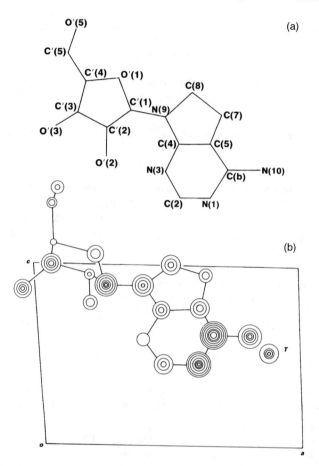

FIGURE 7.9. Tubercidin, $C_{11}H_{14}N_4O_4$: (a) structural formula in approximately the same orientation as in the E map; (b) composite E map; contours (idealized) are drawn at arbitrary equal intervals. Some peaks are heavier than others because of the limited data set used; peak T was the only significant spurious peak.

7.2.12   Multisolution Philosophy and Brief Description of the Program MULTAN

In symbolic addition, we saw that a new phase may be indicated several times by the same combination of symbols. Then, the individual indications reinforce one another to produce an improved joint probability, since α_h (7.29) is then given by

$$\alpha_h = \sum_{k_r} \kappa_{h,k} \tag{7.47}$$

taking $\kappa_{h,k}$ from (7.30).

The combination of indications involving entirely different symbols presents a problem. For instance, suppose two separate indications for $\phi_\mathbf{h}$ are $a + b$ and $c+d$, where $a = \pi/4, b = 3\pi/4, c = -\pi/4$, and $d = -3\pi/4$, say. The individual indications $a + b$ and $c + d$ both predict $\phi_\mathbf{h} = \pi$. Symbolic combination would yield an indication $\frac{1}{2}(a + b + c + d)$, which results in the false value of $\phi_\mathbf{h} = 0$. Even if values of $c = -\pi/4+2\pi = 7\pi/4, d = -3\pi/4+2\pi = 5\pi/4$ are used the combination indication would again be false, since $\frac{1}{2}(a+b+c+d) \equiv 0$ modulo 2π. In such a situation, the usual practice with symbolic addition is to accept the strongest indication and put other indications aside for possible future use.

The multisolution philosophy gets around this problem by introducing numerical phases rather than symbols at an early stage. It is then always possible to combine individual phase indications by means of the tangent formula. This strategy was implemented in the program MULTAN.[a] This phase-determining procedure is based on a starting set of phases, formed in a similar way to that used in symbolic addition, that may be categorized mainly as follows:

1. origin and enantiomorph definition, requiring up to four phase assignments according to the rules already discussed, and
2. further phases required to initiate a continuous phase-determining process by tangent formula expansion; these phases are variables.

It is in category (2) where MULTAN differs fundamentally from the symbolic addition technique. Instead of introducing letter phases for these reflections, they are given numerical values. Specifically, according to space-group symmetry, values such as $0, \pi$, or $\pm\pi/2$ may be assigned. General noncentrosymmetric phases are assigned the values $\pm\pi/4$ and $\pm3\pi/4$, or for enantiomorph specification merely $\pm\pi/4$. In this way, a total of p variable phases, including enantiomorph specification, would therefore yield $2 \times 4^{\,p-1}$ possible phase sets for a noncentrosymmetric crystal. The method is justified numerically in that it gives a maximum error of $45°$ for any of the initial variable phases with a mean error of only $22.5°$. The tangent formula is used to determine probable values of new phases, and the number of possible phase sets rises rapidly with p, as Table 7.11 shows.

TABLE 7.11.   Number of Phase Sets Generated by the Multisolution Method

	Number of phase sets generated	
Number of variable phases p	For a noncentrosymmetric crystal $2 \times 4^{p-1}$	For a centrosymmetric crystal 2^p
1	2	2
2	8	4
3	32	8
4	128	32
5	512	64

[a] P. Main et al., Department of Physics, University of York, York, England (1974).

Magic Integer Phase Assignment

A more efficient strategy is to represent several unknown variable phases in terms of *magic integers*.[a] The phases are expressed by equations of the type

$$\phi_1 = m_1 x$$
$$\phi_2 = m_2 x$$
$$\phi_3 = m_2 x \qquad\qquad (7.48)$$
$$\vdots$$
$$\phi_n = m_n x$$

where $\phi_1, \phi_2, \ldots, \phi$ are the phases $m_1, m_2, \ldots, m_n$ that form a *magic integer sequence*, and the various phase sets are generated by assigning a suitable number of different, equally spaced values between 0 and 2π to x.[b] A simple illustration of such a phase permutation scheme may be devised for the magic integer sequence $m_1 = 2, m_2 = 3$, representing two variable phases ϕ_1 and ϕ_2. Thus,

$$\phi_1 = 2x \quad \phi_2 = 3x \qquad\qquad (7.49)$$

or

$$\phi_2 = \tfrac{3}{2}\phi_1 \qquad\qquad (7.50)$$

Equation (7.50) defines the line shown in Figure 7.10a, and the values of ϕ_1 and ϕ_2 satisfying this equation for a series of values of x at intervals of $\pi/6$ are in Table 7.12a. Since a phase value of $\phi + 2\pi$ or $\phi + 4\pi$, and so on, is equivalent to a phase value of ϕ itself, the values of ϕ in Table 7.12c may be reduced to within the range $0-2\pi$. This process is defined mathematically as expressing phases according to *modulo* 2π, or simply *mod* 2π. The above values of ϕ_1, ϕ_2 reduced mod 2π are given in Table 7.12b, and it can be seen that, through this process, the line *OD* in Figure 7.10a now reduces to a series of equispaced lines in phase space in a box bounded by $\phi_1 = 0$ to $2\pi, \phi_2 = 0$ to 2π (Figure 7.10b). This diagram also shows that phase space can be efficiently covered for ϕ_1 and ϕ_2 by 12 values on a close-packed grid. These 12 values replace the 16 values (4^2) that would have been required by assigning ϕ_1 and ϕ_2 values of $\pm\pi/4$ and $\pm3\pi/4$, as in our previous considerations. The phase values 1–12 are given in Table 7.12c. The root-mean-square error of a typical phase represented in the above magic integer strategy can be shown to be 27.4°, which is only marginally greater than the previous value of 22.5°. This is a small price to pay for the effected reduction of 25% in the number

[a] P. S. White and M. M. Woolfson, *Acta Crystallographica* A**31**, 53–56 (1975).
[b] P. Main, *Acta Crystallographica* A**33**, 750–759 (1977).

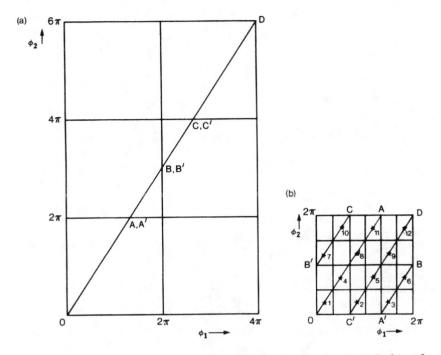

FIGURE 7.10.   Magic integers: (a) the line represented by $\phi_1 = 2x$, $\phi_2 = 3x$; (b) graph of $\phi_1 = 2x$, $\phi_2 = 3x$, mod 2π in the range $0 < x < 2\pi$. The 12 phase sets (Table 7.12c) generated for two unknown phases by magic integer permutation are represented on a close-packed grid.

of phase permutations to be explored. In practice it is frequently necessary to use three or more variable phases in order to obtain a useful phase expansion, and a correspondingly more extensive magic integer sequence can then be used to define an efficient starting set. Such sequences include, for example, $3, 4, 5$; $5, 7, 8, 9$; $7, 9, 10, 11, 12, 13$. Magic integers can also be used to express Σ_2 relationships. For example, in the case where the origin and enantiomorph are defined by the four phases ϕ_1, ϕ_2, ϕ_3, and ϕ_4, respectively, the starting set could be expanded in terms of the magic integer sequence $3, 4, 5$ for three variables x, y, and z for phases $\phi_5, \ldots, \phi_{13}$ as

$$\phi_5 = 3x \quad \text{mod } 1 \qquad \phi_8 = 3y \quad \text{mod } 1 \qquad \phi_{11} = 3z \quad \text{mod } 1$$

$$\phi_6 = 4x \quad \text{mod } 1 \qquad \phi_9 = 4y \quad \text{mod } 1 \qquad \phi_{12} = 4z \quad \text{mod } 1$$

$$\phi_7 = 5x \quad \text{mod } 1 \qquad \phi_{10} = 5y \quad \text{mod } 1 \qquad \phi_{13} = 5z \quad \text{mod } 1$$

TABLE 7.12a. Values Satisfying $\phi_1 = 2x$, $\phi_2 = 3x$ at Intervals of $x = \pi/6$ (Figure 7.10a)

x	0	$\frac{\pi}{6}$	$\frac{2\pi}{6}$	$\frac{3\pi}{6}$	$\frac{4\pi}{6}$	$\frac{5\pi}{6}$	$\frac{6\pi}{6}$	$\frac{7\pi}{6}$	$\frac{8\pi}{6}$	$\frac{9\pi}{6}$	$\frac{10\pi}{6}$	$\frac{11\pi}{6}$	$\frac{12\pi}{6}$
ϕ_1	0	$\frac{2\pi}{6}$	$\frac{4\pi}{6}$	$\frac{6\pi}{6}$	$\frac{8\pi}{6}$	$\frac{10\pi}{6}$	$\frac{12\pi}{6}$	$\frac{14\pi}{6}$	$\frac{16\pi}{6}$	$\frac{18\pi}{6}$	$\frac{20\pi}{6}$	$\frac{22\pi}{6}$	$\frac{24\pi}{6}$
					AA'				BB'				D
ϕ_2	0	$\frac{3\pi}{6}$	$\frac{6\pi}{6}$	$\frac{9\pi}{6}$	$\frac{12\pi}{6}$	$\frac{15\pi}{6}$	$\frac{18\pi}{6}$	$\frac{21\pi}{6}$	$\frac{24\pi}{6}$	$\frac{27\pi}{6}$	$\frac{30\pi}{6}$	$\frac{33\pi}{6}$	$\frac{36\pi}{6}$

TABLE 7.12b. Corresponding Values of ϕ_1, ϕ_2 Reduced mod(2π) (Figure 7.10b)

ϕ_1	0	$\frac{2\pi}{6}$	$\frac{4\pi}{6}$	$\frac{6\pi}{6}$	$\frac{8\pi}{6}$	$\frac{10\pi}{6}$	2π i.e. 0	$\frac{2\pi}{6}$	$\frac{4\pi}{6}$	$\frac{6\pi}{6}$	$\frac{8\pi}{6}$	$\frac{10\pi}{6}$	2π i.e. 0
					AA'		BB'		CC'				D
ϕ_2	0	$\frac{3\pi}{6}$	$\frac{6\pi}{6}$	$\frac{9\pi}{6}$	2π i.e. 0	$\frac{3\pi}{6}$	$\frac{6\pi}{6}$	$\frac{9\pi}{6}$	2π i.e. 0	$\frac{3\pi}{6}$	$\frac{6\pi}{6}$	$\frac{9\pi}{6}$	2π i.e. 0

TABLE 7.12c. Close-Packed Network of $12\phi_1$, ϕ_2 Values Derived by Magic Integer Sequence 2, 3 (Figure 7.10b)

Point number	1	2	3	4	5	6	7	8	9	10	11	12
ϕ_1	$\frac{\pi}{6}$	$\frac{5\pi}{6}$	$\frac{9\pi}{6}$	$\frac{3\pi}{6}$	$\frac{7\pi}{6}$	$\frac{11\pi}{6}$	$\frac{11\pi}{6}$	$\frac{5\pi}{6}$	$\frac{9\pi}{6}$	$\frac{3\pi}{6}$	$\frac{\pi}{6}$	$\frac{11\pi}{6}$
ϕ_2	$\frac{\pi}{4}$	$\frac{\pi}{4}$	$\frac{\pi}{4}$	$\frac{3\pi}{4}$	$\frac{3\pi}{4}$	$\frac{3\pi}{4}$	$\frac{5\pi}{4}$	$\frac{5\pi}{4}$	$\frac{5\pi}{4}$	$\frac{7\pi}{6}$	$\frac{7\pi}{4}$	$\frac{7\pi}{4}$

Here x, y, z are expressed in cycles, where one cycle $= 2\pi$. Thus, if an unknown phase ϕ_{14} is involved in the triplet

$$\phi_{14} + \phi_3 + \phi_9 \approx 0 \qquad (7.51)$$

ϕ_{14} can be represented as

$$\phi_{14} \approx -(\phi_3 + \phi_9) \quad \text{or} \quad \phi_{14} \approx -(3x + 4y) \quad \text{mod } 1$$

In general, any phase relationship can be formed in terms of an expression involving magic integers as an equation of the type

$$2\pi(Hx + Ky + Lz) + b \approx 0 \quad \text{mod } 2\pi \qquad (7.52)$$

where b is a phase constant and H, K, L are the representative magic integers. The MULTAN program employs a modified tangent formula given by

$$\tan \phi_{\mathbf{h}} = \frac{\Sigma_k Q_{\mathbf{h},k} \sin(\phi_k + \phi_{\mathbf{h}-k})}{\Sigma_k Q_{\mathbf{h},k} \cos(\phi_k + \phi_{\mathbf{h}-k})} = \frac{T_{\mathbf{h}}}{B_{\mathbf{h}}} \qquad (7.53)$$

where

$$Q_{\mathbf{h},k} = w_k w_{\mathbf{h}-k} |E_k||E_{\mathbf{h}-k}|/(1 - |U_{\mathbf{h}}|^2) \qquad (7.54)$$

with

$$w_{\mathbf{h}} = \tan[\sigma_3 \sigma_2^{-3/2} |E_{\mathbf{h}}| (T_{\mathbf{h}}^2 + B_{\mathbf{h}}^2)^{1/2}] \qquad (7.55)$$

and

$$|U_{\mathbf{h}}| = |F_{\mathbf{h}}|/\varepsilon^{1/2} \sum_{j=1}^{N} g_j \quad \text{(the unitary structure factor)} \qquad (7.56)$$

Thus, each phase assignment carries a weight designed such that poorly determined phases have little effect in the generation of new phases, while the fact that all phases are included leads to efficient propagation of phase information throughout the data set.

The choice of starting-set reflections for MULTAN is made automatically by a subroutine called CONVERGE. In the case of an unsatisfactory choice by CONVERGE, facilities are available for the user to make his own origin selection. CONVERGE forms an ordered map of reflections such that, from the starting set, each reflection may be determined in terms of all those preceding it. In the initial stages of phase determination, the first 60 phases from the convergence map are used and only those phase relationships determined with $\alpha_{\mathbf{h}}$ (7.29) greater than 5 are accepted. Several passes are made through the tangent formula, lowering this limit each time. As phases are developed the $\alpha_{\mathbf{h}}$s increase, self-consistency being defined as a change of less than 2% in $\Sigma_{\mathbf{h}} \alpha_{\mathbf{h}}$ from one cycle to the next. Up to this stage the phases of the starting reflections have been kept constant. They are now allowed to vary and refine to produce a trial phase set.

7.2.13 Figures of Merit

It is helpful to be able to choose the most probable set of phases prior to the calculation of an E map. In the centrosymmetric case, the following quantities may be determined:

$$M_1 = \sum s_{\mathbf{h}} s_{\mathbf{k}} s_{\mathbf{h}-\mathbf{k}} \qquad (7.57)$$

$$M_2 = |E_{\mathbf{h}} E_{\mathbf{k}} E_{\mathbf{h}-\mathbf{k}}| s_{\mathbf{h}} s_{\mathbf{k}} s_{\mathbf{h}-\mathbf{k}} \qquad (7.58)$$

$$M_3 = \sum P_{\mathbf{hk}} s_{\mathbf{h}} s_{\mathbf{k}} s_{\mathbf{h}-\mathbf{k}} \qquad (7.59)$$

where $P_{\mathbf{hk}}$ is given by (7.10), written for a single TPR, that is, without the summation in (7.11). It is easy to see that M_1, M_2, and M_3 should all be large for a set of reliable and consistent sign relationships. If the E map produced by the most clearly indicated set of signs does not contain a correct structure, the next set in the order of merit would be tried.

In MULTAN, phase sets are assessed by three figures of merit, ABS FOM, PSI ZERO, and RESID, computed for each phase set. In addition, a fourth combined figure of merit, COMBINED FOM, is produced in order to provide an overall picture of the three individual quantities. ABS FOM is a measure of the internal consistency among the Σ_2 relationships and is given by Z, where

$$Z = \sum_{h}(\alpha_h - \alpha_{R_h}) \bigg/ \sum_{h}(\alpha_{E_h} - \alpha_{R_h})$$ (7.60)

α_{R_h} is the value expected from random phases given by

$$\alpha_R = \left\{ \sum_k \kappa^2 \right\}^{1/2}$$ (7.61)

with κ given by (7.30); α_{E_h} is the estimated value of α_h calculated from (7.33) during the convergence procedure. Thus, ABS FOM is zero for random phases and unity if α_h is equal to its expectation value. For crystal structures containing translational symmetry elements, the correct set of phases should correspond to one of the higher values of ABS FOM, because the tangent formula tends to maximize phase relation consistencies. In practice, a correct set of phases has been found usually to correspond to values of ABS FOM in the range of 1.0–1.4, but phase sets with values as low as 0.7 have yielded interpretable E maps; there are also many instances of values of ABS FOM much larger than 1.5 leading to correct structures.

PSI ZERO is defined as

$$\psi_0 = \sum_h \sum_k |E_k||E_{h-k}|$$ (7.62)

where the $|E|$ values in this summation are either very small or zero. For small $|E_h|$, ψ_0 should have a small value for the correct phase set. It is independent of the tangent formula and therefore may be useful as a discriminator when ABS FOM yields similar values for different phase sets.[a] RESID corresponds to the Karle R_K parameter, and is calculated in a way similar to the familiar crystallographic R factor:

$$\text{RESID} = R_K = \frac{\Sigma_h ||E_h| - |E_h|_{calc}|}{\Sigma_h |E_h|}$$ (7.63)

where

$$|E_h|_{calc} = K \langle |E_k||E_{h-k}| \rangle_h$$ (7.64)

and K is a scale factor given by

$$K = \sum_h |E_h|^2 \bigg/ \sum_h \langle E_k||E_{h-k} \rangle^2$$ (7.65)

The correct set of phases should correspond to that with the lowest RESID.

[a] Often with space groups containing no translational symmetry elements.

Experience has shown that discrimination between phase sets is often possible in terms of either Z or R_K. However, sometimes both may fail to enable the correct phase set to be selected easily. A further useful indicator is the combined figure of merit C, given by

$$C = W_1 \frac{Z - Z_{min}}{Z_{max} - Z_{min}} + W_2 \frac{(\psi_0)_{max} - \psi_0}{(\psi_0)_{max} - (\psi_0)_{min}}$$
$$+ W_3 \frac{(R_K)_{max} - R_K}{(R_K)_{max} - (R_K)_{min}} \tag{7.66}$$

where W_1, W_2, and W_3 are weights, often unity, which may be changed to give more emphasis to ψ_0 and less to Z for space groups without translational symmetry elements.

A more complete description of MULTAN is not possible here.[a] Recent features include the automatic production of E maps and their interpretation in terms of molecular geometry, the use of known atomic positions, and subtraction of contributions from heavy atoms in special positions. The success of the program in numerous structure determinations speaks for itself. Those structures that have failed to yield provide the incentive for further developments.

7.2.14   Example of the Use of MULTAN: Methyl Warifteine (MEW)

Crystal data for MEW (Figure 7.11) are listed in Table 7.13. The starting-set reflections, generated automatically by CONVERGE, are given in Table 7.14. Using 228 data with $|E| > 1.70$, the number of Σ_2 interactions was limited (by the user) to 2000, and 16 phase sets were generated. The phase set having the second highest value of Z (0.96) and lowest R_K (0.23) produced an E map from which 36 of the 45 nonhydrogen atoms were identified in geometrically acceptable positions. These 36 sites were drawn from the highest 60 peaks in the map. The remaining nine atoms of MEW were located by an electron density synthesis using all reflections, the initial R factor being 0.35. After refinement the final R factor was 0.057 for 2508 observed reflections.[b] A stereoview of the molecule is shown in Figure 7.12.

7.2.15   Some Experiences

Direct methods have now been developed to the extent where many, even very complex, structures can be solved in a straightforward manner (see Section 8.5). This is the result of the experience and application of many workers in this

---

[a] See CCP4 Program Suite RANTAN; www.yorvic.york.ac.uk/~yao/rantan.html
[b] N. Borkakoti and R. A. Palmer, *Acta Crystallographica* B**34**, 482 (1978).

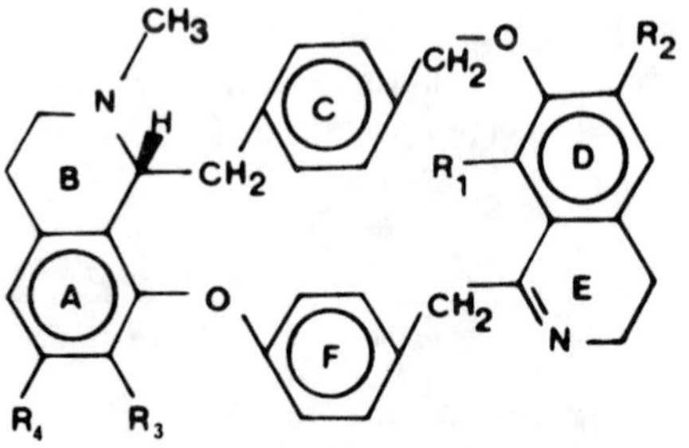

FIGURE 7.11.   Structural formula of methyl warifteine, $R_1 = OH$; $R_2 = R_3 = R_4 = OCH_3$.

TABLE 7.13.   Crystal Data for MEW

Formula	$C_{37}H_{38}N_2O_6$	V_c	$3085.8\ \text{Å}^3$
M_r	606.72	D_x	$1.306\ \text{g cm}^{-3}$
Space group	$P2_12_12_1$	Z	4
a	17.539(4) Å	$F(000)$	1288
b	12.224(3)	$\mu(\text{Cu}\ K\alpha)$	$5.6\ \text{cm}^{-1}$
c	14.393(3)	Crystal size	0.2, 0.3, 0.3 mm

TABLE 7.14.   Starting-Set
Reflections for MEW

hkl	$\lvert E\rvert$	$\phi/\deg$
0 1 3	2.10	90 ⎫
1 0 3	2.30	90 ⎬ origin
14 3 0	4.73	360 ⎭
0 11 1	4.40	90, 270
1 1 7	2.36	45, 135, 225, 315
9 7 2	3.07	45, 315 (enantiomorph)

field, enabling the drawing up of a number of strategies which have often proved significant in structure determinations.

Some Prerequisites for Success in Using Direct Methods

Some of the rules that have emerged can be summarized as follows:

1. As complete and accurate a set of $\lvert F_o\rvert$ data as possible must be used. Sometimes a small crystal size results in many weak-intensity measurements and consequent difficulty in fulfilling this condition.
2. If automatic solution of the phase problem fails, we must be prepared to intervene (a) by selecting origin and starting phases, (b) by varying the value of $\lvert E_{\min}\rvert$, (c) by varying the number of interactions present, and

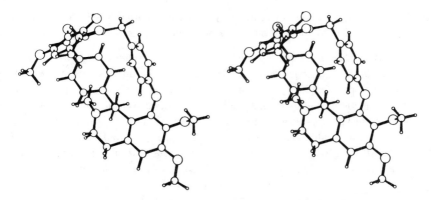

FIGURE 7.12. Stereoscopic view of MEW looking along c; the circles, in increasing order of size, represent H, C, N, and O.

(d) by employing special techniques, possibly involving tangent-formula recycling and so on.

3. Remember that a computer program may contain mistakes, even after considerable length of use. Always check the validity of results as far as possible, especially assignments of basic origin- and enantiomorph-fixing reflections and values of special phases.

4. Do not give in!

Figures of Merit: A Practical Guide

We have mentioned the four criteria (Section 7.2.13) calculated in MULTAN for assessment and ordering of phase sets. This facility is necessary in order to overcome the basic disadvantage inherent in the multisolution method, namely, that many possible phase sets may have to be considered. In principle, the phase sets are explored in order until the structure is found; normally, one does not investigate them further.

Experience suggests that two of the figures of merit currently calculated in MULTAN, Z and R_K, are often sufficient to enable the most likely E maps to be explored; C appears also to be very useful, and ψ_0 can be used in difficult cases. As a guide, and not as an absolute measure, we may use the following values in selecting the order of calculation of E maps: $Z > 1.0$, $R_K < 0.20$, and $C > 2.5$. Other programs may calculate different figures of merit, and one should check with the program author for the corresponding recommended values.

Signs of Trouble, and Past Remedies when the Structure Failed to Solve

This section summarizes some of our own experiences, and records factors involved when a structure solution failed to emerge from direct phase-determining methods. Although these comments apply primarily to the use of MULTAN, similar considerations would apply to other like program systems.

1. *All values of Z are too low ($<<1.0$) or all values of R_K are too high ($>>2.0$)—both conditions can arise*. The criterion R_K applies strictly speaking only when all possible Σ_2 interactions have been used. The usual maximum number of Σ_2 interactions allowed in MULTAN is 2000, but a cutoff can be applied by the user; some versions allow up to 4000 Σ_2 interactions. If a cutoff is applied, the correct phase set could have a much higher value of R_K (up to 0.3 has been noted).

2. *All phase sets and their figures of merit are similar*. This situation may occur if too few Σ_2 interactions are being used. Try more, by lowering $|E_{min}|$.

3. *The origin-defining set is poor or incorrect*. Try another one, choosing it with the aid of the rules given.

4. *The E map contains one very large peak*. The phases are probably very inaccurate; the heavy peak may be located in the center of a closed ring. Start again; do not waste time trying to interpret the E map.

5. *The E map is not interpretable or chemically sensible*. The phases are incorrect. Try again.

6. *If heavy atoms are present in the structure, they alone may show up*. Proceed to Fourier methods using interpretable heavy-atom sites. A check against the Patterson function might prove useful here.

7. *Only a small molecular fragment is discernible from the E map*. Try recycling, basing phases on the fragment found, or try to obtain more phases by increasing the initial data set.

8. *The program selects an incorrect or poor starting set (e.g., too few Σ_2 interactions)*. Select your own starting set. If you suspect that the program may contain a fault, inform the author; do not attempt to correct it.

9. *The solution still fails to emerge*. Review the calculation of the $|E|$ values; perhaps omit reflections that appear to have a bad influence on the phase-determining pathway.

10. *All fails*. Go back to fundamentals. Check the space group, data collection, processing, and any other factor that might be at fault.

If you exhaust these possibilities without achieving success, try another method for determining the structure. Or give it a rest and try again later. Or study *recent* Bibliography on direct methods.

7.2.16 Structure Invariants: Triplets, Quartets, and the SHELX Program Strategy[a]

We discussed in Sections 7.2.2 and 7.2.9 certain entities known as *structure invariants*. They have values or properties that are completely independent of the

[a] G. M. Sheldrick, *Acta Crystallogr* A**46**, 467 (1990).

coordinate origin, and depend only on the structure in question. In this section, we examine how this concept can be used to enhance direct methods of phase determination. The structure factor equation has been be expressed as

$$F(hkl) = \sum_{j=1}^{N} g_j \exp[i2\pi(hx_j + ky_j + lz_j)] = |F(hkl)| \exp i\phi(hkl) \quad (7.67)$$

where N is the number of atoms in one unit cell. In the following discussions it will generally be assumed, for simplicity, that we are dealing with an equal-atom structure.

Referring to Section 7.2.2 and Figure 7.1, it can be seen that changing the origin to any point $(\Delta x, \Delta y, \Delta z)$, changes each atom coordinate to $(x_j - \Delta x, y_j - \Delta y, z_j - \Delta z)$, and that the structure factor in (7.67) will change to a new structure factor $F'(hkl)$, given as

$$F'(hkl) = \sum_{j=1}^{N} \{g_j \exp[i2\pi(hx_j + ky_j + lz_j)]\}\{\exp[-i2\pi(h\Delta x + k\Delta y + l\Delta z)]\}$$

$$(7.68)$$

where $\{\exp[-i2\pi(h\Delta x + k\Delta y + l\Delta z)]\}$ is a term external to the summation, so that

$$F'(hkl) = F(hkl)\{\exp[-i2\pi(h\Delta x + k\Delta y + l\Delta z)]\} \quad (7.69)$$

Thus, changing the origin to the point $(\Delta x, \Delta y, \Delta z)$ causes a change in $\phi(hkl)$ given by

$$\Delta\phi(hkl) = -2\pi(h\Delta x + k\Delta y + l\Delta z) \quad (7.70)$$

$|F(hkl)|$ is, of course, invariant, as is the intensity of the reflection $I(hkl)$.

If we now consider the structure factors for *two reflections* we see that changing the origin to the point $(\Delta x, \Delta y, \Delta z)$ will produce a *combined phase change* $\Delta\phi_{12}$ given by

$$\Delta\phi_{12} = -2\pi[(h_1 + h_2)\Delta x + (k_1 + k_2)\Delta y + (l_1 + l_2)\Delta z] \quad (7.71)$$

where the two reflections have indices $h_1k_1l_1$ and $h_2k_2l_2$. Using vector notation for the indices, that is, $\mathbf{h}_1$ for $h_1k_1l_1$ and $\mathbf{h}_2$ for $h_2k_2l_2$ we see that if $\mathbf{h}_1 + \mathbf{h}_2 = 0$, then $\Delta\phi_{12} = 0$. It follows that if $\mathbf{h}_1 + \mathbf{h}_2 = 0$, the phase sum $(\phi_1 + \phi_2)$ is a *structure invariant*. This result can be generalized for three and four (or more) reflections to give the following expressions:

$$(\phi_1 + \phi_2 + \phi_3) \text{ is a structure invariant if } \mathbf{h}_1 + \mathbf{h}_2 + \mathbf{h}_3 = 0 \quad (7.72)$$

$$(\phi_1 + \phi_2 + \phi_3 + \phi_4) \text{ is a structure invariant if } \mathbf{h}_1 + \mathbf{h}_2 + \mathbf{h}_3 + \mathbf{h}_4 = 0 \quad (7.73)$$

Notice that the *individual* phases in these expressions are not invariant, and that nothing has been said so far about the actual values of these phase sums. For (7.72), we can show that we have assumed tacitly (Section 7.2) that, for sufficiently large values of α, that is, of $2N^{-1/2}|E_{\mathbf{h}_1}||E_{\mathbf{h}_2}||E_{\mathbf{h}_3}|$,

$$\phi_1 + \phi_2 + \phi_3 \approx 0 \qquad (7.74)$$

Rearranging (7.74) gives:

$$\phi_1 \approx -\phi_2 - \phi_3 \qquad (7.75)$$

We make a small change in notation, so as to accord with that used in SHELX-97 (see below) and to focus on determining $\phi_{\mathbf{h}}$. Let $\mathbf{h}_1 = \mathbf{h}, -\mathbf{h}_2 = \mathbf{k}, -\mathbf{h}_3 = \mathbf{h} - \mathbf{k}$; then if Freidel's law holds, that is $\phi(hkl) = -\phi(\bar{h}\bar{k}\bar{l})$, then (7.75) becomes

$$\phi_{\mathbf{h}} \approx \phi_{\mathbf{k}} + \phi_{\mathbf{h}-\mathbf{k}} \qquad (7.76)$$

which is the phase addition formula (7.24) from which the tangent formula (7.28) was developed as a general means of exploiting TPRs for expanding and refining phases by direct methods. Equation (7.73) represents a four-phase structure invariant, or quartet, and, like three-phase triplets, may be a source of phase information, since it can be shown[a,b] that for a sufficiently large value of the expression $2N^{-1}|E_{\mathbf{h}_1}||E_{\mathbf{h}_2}||E_{\mathbf{h}_3}||E_{\mathbf{h}_4}|$

$$\cos(\phi_1 + \phi_2 + \phi_3 + \phi_4) \approx +1 \qquad (7.77)$$

provided that $|E_{\mathbf{h}_1+\mathbf{h}_2}|, |E_{\mathbf{h}_1+\mathbf{h}_3}|$, and $|E_{\mathbf{h}_1+\mathbf{h}_4}|$ are all large. Alternatively,

$$\cos(\phi_1 + \phi_2 + \phi_3 + \phi_4) \approx -1 \qquad (7.78)$$

if $|E_{\mathbf{h}_1+\mathbf{h}_2}|, |E_{\mathbf{h}_1+\mathbf{h}_3}|$, and $|E_{\mathbf{h}_1+\mathbf{h}_4}|$ are all small. Equations (7.77) and (7.78) are *positive quartets* and *negative quartets* (NQRs), respectively. Thus, the sum of four phases is dependent not only on the intensities of the four corresponding reflections, the *primary* terms, but also on the intensities of three other index-related reflections, the *cross-terms*. As in the treatment for triplets, we need to cast the indices of the phase-quartet in terms of a target reflection $\mathbf{h}$. The quartet phase sum then becomes

$$(\phi_{\mathbf{h}} + \phi_{-\mathbf{k}} + \phi_{-\mathbf{l}} + \phi_{-\mathbf{h}+\mathbf{k}+\mathbf{l}}) \qquad (7.79)$$

where $\mathbf{h} = \mathbf{h}_1, -\mathbf{k} = \mathbf{h}_2, -\mathbf{l} = \mathbf{h}_3$, **and** $-\mathbf{h} + \mathbf{k} + \mathbf{l} = \mathbf{h}_4$. The primary terms are $|E_{\mathbf{h}}|, |E_{\mathbf{k}}|, |E_{\mathbf{l}}|$, and $|E_{-\mathbf{h}+\mathbf{k}+\mathbf{l}}|$, which should all be large; the cross terms are $|E_{\mathbf{h}-\mathbf{k}}|, |E_{\mathbf{h}-\mathbf{l}}|$, and $|E_{\mathbf{k}+\mathbf{l}}|$, which should all be large for a positive quartet and small for a negative quartet.

[a] H. Hauptman, *Crystal Structure Determination: The Role of the Cosine Seminvariants*, Plenum Press (1972).
[b] H. Hauptman, *Acta Crystallographica* B**28**, 2337 (1972).

We have seen (Section 7.2.11) that multiple solution techniques for phase determination employ a number of initial phase assignments for a relatively small number of reflections. Each phase set is then expanded *via* the tangent formula, which is based on the use of strong triplets. Only a very small number of such assignments can be expected to converge to produce a phase set that is somewhere near the correct one. To avoid the use of the time-consuming inspection of a large number of E-maps, figures of merit (Section 7.2.12) have been devised as a means of recognizing the phase set which is most likely to produce the correct structure, particularly ABSFOM, equation (7.56). Experience has shown[a] that, with large structures, figures of merit such as ABSFOM that are based only on the consistency of triplet relationships, may not discriminate the correct phase set. This failure has led to the use of NQRs, as well as triple phase relations, for both phase determination and calculation of combined figures of merit.

Direct Methods in the Program SHELX-97

The structure solving program system SHELX, of which the most recent version is SHELXL-97, is currently used extensively in the solution of a wide range of crystal structures. It can be accessed from the web (see Appendix A10), and users are advised to read the operation manual[a] very carefully.

The system is capable of handling both small molecules and proteins, and contains a number of executable programs, such as those that solve structures by both Patterson and direct methods, carry out detailed least-squares refinement, and locate water molecules of crystallization. The heavy-atom method can be invoked, and the Patterson search method used is effectively PATSEE, which we describe in Section 7.3. The least-squares procedures encompass *inter alia* dispersion, absorption, and extinction, together with a wide range of constraints and restraints, and routines for fixing and refining hydrogen atom positions. The direct methods routine is based on a random start multisolution strategy, or more accurately a multi-permutation single-solution procedure, since it endeavours to identify the correct solution and then to improve on it by $|E|$-map partial structure extension, or successive refinement. Geometry routines calculate bond lengths, bond angles, and torsion angles and identify possible hydrogen-bonds, all of the results being tabulated for publication purposes. The system is strongly recommended to the serious structure analyst.

The triple phase and NQR relationships discussed above are employed in the SHELX-97 to generate phases by a modified tangent formula. We write

$$\alpha = 2N^{-1/2}|E_{\mathbf{h}}|E_{\mathbf{k}}E_{\mathbf{h}-\mathbf{k}} \tag{7.80}$$

$$\eta = gN^{-1}|E_{\mathbf{h}}|E_{\mathbf{k}}E_{\mathbf{l}}E_{\mathbf{h}-\mathbf{k}-\mathbf{l}} \tag{7.81}$$

[a] G. M. Sheldrick, *SHELX-97 Program Manual*, University of Gottingen (1997).

where $E_j = |E_j|(\cos \phi_j + i \sin \phi_j)(j = k, l)$, and g is a positive constant set by the program to account for the cross-term $|E|$-values; (7.80) and (7.81) are subject to the same conditions as (7.79)

It has been found that computer time is optimized for cases where the number of NQRs is restricted to between 1000 and 8000, only the most reliable relationships being retained in this process, strictly where *all three* cross-terms have actually been measured and found to be weak. However, all interconnecting triple phase relationships are used, except for Σ_1 terms (Section 7.2.8) and those which all involve restricted phases that prevent the resultant phase from being zero[a]. A process of *phase annealing*, based on a principle similar to that of simulated annealing used in the refinement of macromolecular structures (Section 10.9.1), is employed in the next stage of phase refinement. The results from these two stages are then processed with full tangent formula refinement for the best retained reflections. The total number of different attempts using these procedures can be set by the user and may be as many as 5000 for really difficult structures.

Initial Stages.   In the initial stages of SHELX-97, a number of cycles of weighted tangent formula (7.28) are performed, starting with a selected number of randomly generated phases. The best phase sets as judged by NQR and triplet consistency are retained and the process repeated to give a number of parallel-generated phase sets. After each iteration, the total number of phase sets processed in parallel is reduced until only 25% of the original number of phase sets is generated. Typically 8 or 16 best phase sets are retained from each cycle for a total run of 128 parallel permutations. The best reflections and strongest triple product relationships are retained and passed to the next stage of the procedure. The program uses a TPR figure of merit similar to ABSFOM, combined with a new indicator NQUAL based on the NQRs, where

$$NQUAL = \Sigma|\alpha \cdot \eta|/\Sigma|\alpha||\eta| \qquad (7.82)$$

The two figures of merit together combine to give an indicator CFOM that is *smallest* for the best phase set, NQUAL being a negative quantity.

Phase Annealing.   This method is used to refine the phase sets retained after the initial stages. Phase annealing supplies a correction to the phase produced by the tangent formula. As only a limited number of reflections is involved the process uses computer time efficiently, employing only the strongest triplets and quartets.

Final Stages of Phase Determination.   The total number of direct methods attempts (*np* on the TREF instruction in SHELX-97) to be employed may have to be given a very high value for difficult structures. The routines in SHELX-97 are written with efficiency in mind and the program requires only a few minutes on a PC for quite large structures, even when TREF is as high as 5000. The program will

[a] C. Giacovazzo, *Acta Cryst.* A**30**, 631 (1974).

select the number of reflections nE to be involved at this stage if not stipulated, but may have to be reset to a higher value if the program fails to produce an interpretable E-map. An example of the use of this program for solving a difficult structure (cyclosporin H) is given in Chapter 8.

7.3 Patterson Search Methods

We saw from our earlier discussion in this chapter, that a Patterson synthesis must contain a complete set of peaks, that is, $N^2 - N$ peaks for a crystal containing N atoms in the unit cell, excluding the origin peak. Since we can always calculate the vector set for a model structure, or just a part of it, the Patterson function may be unscrambled, wholly or partially, in terms of a set of atomic coordinates. This idea has led to a technique in structure analysis called Patterson search methods.

7.3.1 General Comments for Small Molecules and Macromolecules

We introduce here, a method for structure determination that is useful and applicable either to small molecules, for which the method would normally be used in case of failure of direct methods, or to macromolecules, for which the method is now frequently used as a first choice where possible. The method may be designated Patterson Search for small molecules and Molecular Replacement (MR) in the case of macromolecules. In order to apply the method to solve a structure the initial requirements are:

1. For the crystal under investigation, the *target structure*, a set of $|F_o(hkl)|$ values to as high a resolution as possible (see Section 7.9). This resolution would normally be atomic resolution for small molecules, but in the case of a macromolecular crystal, generally far short of atomic resolution, because large molecule crystals tend to be poor diffractors (see Chapter 10);
2. The availability of the coordinates of a good quality structure, the *search structure*, which forms a relatively small fragment of the target structure for small molecules, whereas in the macromolecular case the search structure should ideally be similar in size and structure to the target molecule;
3. A sound understanding of the principles and practices involved;
4. State-of-the-art software and hardware.

Conceptually the basic principle of Patterson Search or MR is quite straightforward. Remember, we know that the intensity data $|F(hkl)|^2$ as a whole inherently contain phase information. The MR method as proposed by Rossman and Blow,[a] involves a critical and quantitative comparison of the Patterson functions of the target and search models. Similar comments apply to Patterson Search, a method implemented successfully some years ago, for example by Braun *et al.* in the Vector Verification Method.[b] Although quite successful and relatively easy to use, this method has now been superseded by the program PATSEE, which will be described in detail. In contemporary software for carrying out MR applications, for example, AmoRe and X-PLOR, the method is strengthened through the use of a variety of other crystallographic techniques, which are now within the capabilities of modern computers; PATSEE is similarly strengthened by the use of direct methods.

The Patterson function has peaks of high density at locations corresponding to the ends of atom–atom vector pairs, with one atom of each pair occupying a common origin. For complex structures like proteins, the interatomic vectors are densely packed in the unit cell, and most of them will not be resolved in the Patterson map; lack of atomic resolution in the X-ray data will also cause a further blurring of the vector distribution density. The two atoms forming the Patterson vector can be in either the same molecule, *intramolecular*, or symmetry related molecules, *intermolecular*.

7.3.2   Intramolecular Interatomic Vectors and Molecular Orientation

The vectors in this set arise through interactions *within* one molecule and therefore tend to be shorter than intermolecular vectors, which span structurally related molecules. This *self-vector set*, as it is also known, is consequently situated around the origin of the Patterson function, the longest vectors arising between atoms at extreme ends of the molecule. Each atom, in theory, images[c] both the structure and its inverse, that is, vector types AB and BA. Because of the lack of resolution in most protein structures, this will be in the form of a *blurred molecular envelope* of density, whereas for small molecules at atomic resolution the individual images will be much more clearly defined. There will be one such image of the structure, plus its inverse, per atom, forming a centrosymmetric distribution. Figure 7.13 shows two similar (homologous) two-dimensional molecules, in which *S* represents a suitable search molecule, and *T* the target molecule whose structure is being determined. In the case of a small molecule we can consider the four or five discs to represent resolvable atoms. For macromolecules, the surrounding sheaths in these diagrams represent the overall molecular shape that MR is seeking.

[a] M. G. Rossman and D. M. Blow, *Acta Crystallographica* 14, 641 (1961).
[b] P. B. Braun, J. Hornstra, and J. I. Leenhouts, *Philips Research Reports* 42, 85 (1969); and the 3rd edition of this book.
[c] See Bibliography, Chapter 6

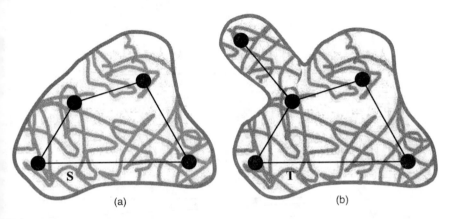

FIGURE 7.13.   (a) Known, search model S to be used in Patterson Search or Molecular Replacement. (b) The unknown target molecule T, that is similar to the search model but has an extra structural feature.

Figure 7.14a,b shows the two actual structures in their different unit cells. (In small molecule structures, where the search and target molecules include a relatively small part, maybe 50%, of the structure in common, we would expect the two unit cells and space groups to be different. For macromolecular structures, which are more similar to each other, this may not be so, although it frequently is.) As can be seen in this example, the known search structure S is based on an inclined unit cell, whilst the target structure T has an orthogonal cell. Both cells incorporate 2-fold symmetry; Figure 7.14(b) shows some of the intermolecular interatomic vectors as discussed in Section 7.3.3. For the two molecules S and T, Figure 7.15 shows the corresponding resolved *self-vector set* peak positions in the Patterson functions for small molecules and the corresponding simulated vector sheaths representing the envelope of the Patterson function, for macromolecules; for clarity only those peaks *not* related by the Patterson centre of symmetry are shown.

Rotation Stage of Patterson Search or Molecular Replacement

In the above simulated example the vectors defining the *orientation* of the unknown structure T in its unit cell (Figure 7.15b) can be seen to occur in the Patterson of the known, search structure S, Figure 7.15a. To demonstrate this fact the reader should make a *transparent* copy of Figure 7.14a, which shows the Patterson peaks of the known molecule S, and place it over Figure 7.14b, the Patterson function of the unknown structure T. Locate the two diagrams such that their *origins are in register*, and show that maximum correspondence occurs for an *anticlockwise rotation* of your copy by 67°.

In a real structure determination this orientation angle would be calculated by one of the available computer programs, as described below. Note that because

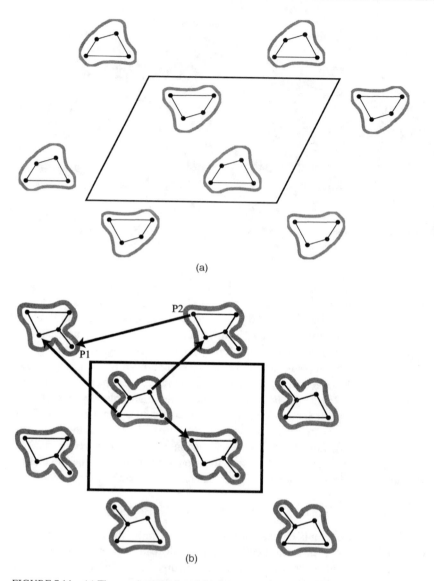

FIGURE 7.14.   (a) The search model S, and (b) the target molecule T packed in their respective unit cells. It is necessary that the structure of S is known and that a set of atomic coordinates is available. The structure (b) is to be determined by Patterson Search or Molecular Replacement. Some of the intermolecular Patterson vectors are shown in (b).

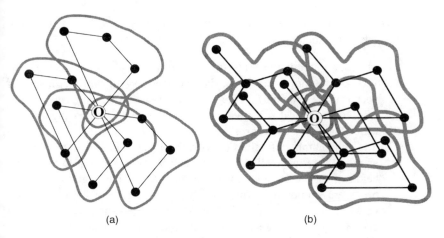

(a)   (b)

FIGURE 7.15. Interatomic vectors for (a) the search model S, and (b) the target molecule T. Only those vectors independent of the Patterson centre of symmetry are shown. Both diagrams contain multiple images, or part images, of structures. There is a rotational relationship between the two Patterson functions that the reader should try to determine (see text). Note that the sets of vectors shown here are arranged around the origin of the Patterson function at distances corresponding to the distances between atoms within a given molecule. The high vector density which always occurs at the Patterson origin has been omitted.

the unknown structure, Figure 7.13b contains an additional moiety, vectors for this part of the structure are missing in the search Patterson, Figure 7.14a. For small molecule analysis the missing vectors may form a large fraction of the target molecule, whereas for macromolecules only a small fraction would normally be involved. Subsequent Fourier and least-squares refinement of the target structure, once the search model has been located, will serve both to locate any missing atoms or groups and to eliminate others which may have been included, but which do not form part of the target structure.

In practice the process of matching the orientations of the two Patterson functions is carried out, by computation, the *rotation stage*, by testing over a series of orientation angles. For three-dimensional structures, it is necessary to perform these rotations about three independent axes, normally called α, β and γ and not to be confused with the unit-cell angles. Axial systems used for the rotation procedure differ among different programs, and it is advisable to read the program manual in each case for information on this point (see Chapter 10 and Appendix A7).

The angular rotation ranges and intervals have to be chosen carefully in order to cover a sufficient number of possibilities, thus ensuring that the correct angular triplet is not overlooked. Although computationally wasteful it is better to include too many trials rather than too few. In PATSEE it is not unusual to use several thousand random angle triplets in the rotation stage of the analysis.

On account of the complexity of protein structures, the following two conditions apply to the rotation function:

1. In order to limit the Patterson vectors to lengths that include self-vectors (intermolecular vectors) but exclude cross vectors (intermolecular vectors), the rotation function $R(\alpha, \beta, \gamma)$ should be calculated over a restricted spherical volume U centred at the origin, and having a radius called the *radius of integration*.

2. The large number of values of $R(\alpha, \beta, \gamma)$, calculated over the required angular range, generally contains many peaks in addition to those that belong to the correct solution, and having comparable magnitudes. These peaks are simply signifying that there is a degree of correspondence between the two Patterson functions in this orientation, albeit a wrong one. Because of this uncertainty, some programs, such as AmoRe, retain all peaks greater than 50% of the highest peak (even more in some programs) for transference to the *translation stage* of MR. The problem is less significant for small molecules but nevertheless must not be overlooked. *Figures of merit* will be described that are used to discriminate between the true and false solutions.

7.3.3   Intermolecular Interatomic Vectors: Translation Stage of MR

Assuming that the correct orientation of the search fragment or molecule has been determined in the rotation stage, the correctly oriented structure must then be *correctly located spatially* in position in the unit cell of the target crystal by means of a *translation stage*. The origin with respect to this translation process is usually governed by the space group. Translation is carried out rigorously by placing the oriented search fragment in a *large number of test positions* located on a fine grid. This process must cover a sufficient number of finely selected translational increments in three dimensions, designated as either Δt_1, Δt_2, Δt_3 or t_x, t_y, t_z, in order to ensure that the correct location of the molecule is scanned. In essence, the correct structure is recognized initially as corresponding to the highest degree of overlap between the calculated and observed Patterson functions, when superimposed after applying the given translation vector. During the above process all of the lattice and symmetry operations for the target crystal are fully applied. It is therefore absolutely essential that the space group of the target crystal has been correctly assigned. If this is not the case there will be no outstanding solution at the translation stage, and further refinement of the structure will not be possible. Figure 7.14b, which represents the correct solution for the target structure, includes some of the intermolecular interatomic vectors. For comparison the corresponding vectors are shown in Figure 7.16, in which the search molecules S is correctly oriented but *incorrectly* translated into the target unit cell. Very large changes in

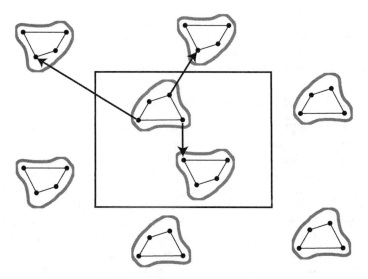

FIGURE 7.16.   A possible solution for Patterson Search or Molecular Replacement, where the *correct rotation* (or orientation) has been applied to the search model S but the subsequent translation is incorrect (compare with Figure 7.14b). Three intermolecular vectors are shown for this situation which, because the translation is incorrect, will not be present in the Patterson function for the target molecule. Consequently, there will be a lack of correspondence between the S and T Patterson functions, which will result in poor figures of merit for this solution; these solutions will therefore be rejected.

these vectors are evident, which would have a profound effect on the quality of this particular translation when tested computationally. The changes in how the molecules are packed (see below) while retaining the 2-fold symmetry and unit-cell translations in the structure should be noted by comparing Figures 7.14b and 7.16.

7.3.4   Crystal Packing and Refinement of the Structure

We consider next the packing of the rotated–translated (RT) molecules in the unit cell. An incorrect RT solution will produce incorrect crystal packing, which can be easily detected from a thorough calculation of interatomic distances. What usually happens is that some atoms in different molecules will be unacceptably close to each other (see Tables 7.24 and 7.25). Evidence of the correctness of the combined RT search results can thus be easily achieved by inspecting the crystal packing generated through each promising set of rotation and translation parameters that has been retained, usually because the figures of merit are encouraging. Crystal packing can be easily examined with the use of a suitable molecular graphics program. For small molecules this can be carried out, for example, in the program package WinGX using the routine PLATON99 while for macromolecules the program MOLPACK has been found to be useful and reliable (see

Appendix A10). These programs provide not only graphical representations of the packed molecules but also quantitative values for the intermolecular contacts; in particular any violent clashes between the symmetry-related molecules, which would exclude a physically viable solution, are flagged.

Expansion and Refinement of the Structure

The final stage of this procedure, as with other methods of structure analysis, attempts to locate atoms that are missing from the model and to refine the positions of all atoms to produce a value for the R-factor (see Section 6.4.4) that is compatible with the accuracy of the intensity data. As we have seen, missing atoms can be located by calculating the electron density, based on phasing from the partial model. Initially the structure factors will be of low accuracy, but with a reasonably sized fragment; as will be seen in the following examples, the electron density for the correct solution can be expected to reveal at least some of the missing atoms, which are then added to the phasing process and passed on to the next cycle. Program suites like WinGX have facilities for checking the molecular geometry at each stage so as to reduce the possibility of adding incorrectly placed atoms to the model. The partial model may be subjected to a few cycles of least-squares refinement, as this can accelerate the procedure toward convergence. However, it should be noted that only when all atoms are present in the model will the R-factor be meaningful. This refinement procedure will be designated loosely as "Fourier-least-squares".[a]

7.3.5 Patterson Search Methods for Small Molecules

In describing the heavy-atom method (Section 6.4.4) we showed that, for a structure containing a small number of relatively heavy atoms, the Patterson function can lead to successful determination of the atomic coordinates of the heavy atoms, the remaining atoms being located by subsequent calculation of electron density maps prior to refinement. Direct methods can be used to solve light-atom structures, for which there are no predominant interatomic vectors as would be necessary for application of the heavy-atom method. We have seen that the Patterson function necessarily contains the complete set of interatomic vectors for the given crystal structure, in the form of peaks located at the ends of such vectors, one end of each vector being located at the origin. Reference to Figure 7.15 shows that in effect each atom forms an image of the structure with itself at the origin. When the Patterson centre of symmetry is added this results in $2N$ displaced images of the structure being present around the Patterson origin. A *single weight* peak, that is, one generated by a given pair of atoms 1 and 2, will

---

[a] The method of least-squares that is central to structure refinement is described in detail in Section 7.4.2.

have a value or peak height roughly proportional to $Z_1 Z_2$, where Z_j is the atomic number of the jth atom.

The total of $N^2 - N$ non-origin Patterson peaks per unit cell containing N atoms increases rapidly with N; the peaks may become overcrowded and overlapping, and unresolvable for larger structures. In this section we will describe applications of Patterson methods to approximately equal-atom structures. These structures are usually of relatively low molecular weight, containing 30–40 carbon-like atoms, with crystals that diffract to atomic resolution, leading to well-resolved interatomic vectors in the Patterson function. Macromolecules, which generally diffract, far short of atomic resolution will be treated separately in Chapter 10.

In the case of small molecules we know, assuming peak overlap is not a major problem, that the Patterson function necessarily contains the complete set of interatomic vectors for the crystal structure. Thinking of the logistics in reverse, for a *known* light-atom structure we could use the values of the atomic coordinates to generate the set of interatomic vector coordinates, and these would match the Patterson function $P(uvw)$ within limits of error that are dictated by the quality of the $|F_o|$ data. We have seen that each atom in the structure forms, in the Patterson function, an image of the structure in itself (see Section 6.4.2) and there are, therefore, N such images scrambled, or convoluted together, in $P(uvw)$. The complexity of the Patterson function generally makes deconvoluting it, in terms of the individual atomic coordinates, almost impossible unless some additional information can be used initially. This information is usually in the form of a reasonably precise model of the geometry of a fragment of the molecule whose three-dimensional structure is known. The method thus depends on a good knowledge of the chemical identity of the molecule in question. Knowing the geometry of such a fragment of the structure, interatomic vectors for the fragment may be calculated and searched for in the Patterson function; hence the term Patterson Search methods. Coordinates for the molecular fragment to be used, may be derived from the library of known crystal structures in the Cambridge Crystallographic Data Base, or generated from graphics programs, such as Chem-X,[a] which incorporate tables of standard bond lengths and angles (see Section 7.7). For example, a suitable molecular fragment may simply be a benzene ring, which comprises a flat regular hexagon with sides of 1.40 Å and internal angles of 120°. The coordinates for the search model in this case could be obtained by drawing (see Problem 7.10). If the structure contains a cyclohexane ring, this structure is truly three-dimensional, and one of the other two methods mentioned above must be used to derive its coordinates (see also Chapter 11). The examples discussed below both involve somewhat more complex search fragments.

---

[a] Oxford Molecular 1999: http://www.oxmol.co.uk

7.3.6   The Program PATSEE

In the discussion so far, we have established that a Patterson Search in vector space consists of the following stages:

1. Acquisition of a set of accurate atom coordinates for a suitable search model;
2. Calculation and storage of the Patterson function (or self-vector set) for the model;
3. Calculation and storage of the Patterson function for the target crystal, based on the $|F_o(hkl)|$ data;
4. Rotation search, which provides several possible orientations for the search model to occupy in the target unit cell, listed in order of likely feasibility or precedence;
5. Translation search which attempts to place each of the rotation solutions in turn into the correction position in the target unit cell;
6. Refinement and expansion of the rotated and translated models, again in order of precedence, to finally converge on the correct solution for the target structure.

The program PATSEE follows stages (1)–(5) in the above list, using some important and powerful new algorithms in its implementation. Stage (6) reverts to the standard procedures for structure refinement described in the next section. The PATSEE program is highly recommended, is easy to use, and readily available to crystallographers, being incorporated into the structure analysis suite WinGX. As with other methods that utilize Patterson Search, the program determines the orientation of the search fragment from the rotation function. However, instead of then using the conventional translational function, direct methods are applied in order to determine the position of the oriented fragment in the unit cell. This stage involves the use of several triple phase invariants (see Section 7.2.16) selected by the program as being sensitive to the location of the fragment within the unit cell. The weighted sum of the cosines of these phases is maximized with respect to the position of the search fragment in order to determine its most probable location.

As a very large number of possible solutions for the orientation and position of the search fragment in the unit cell is explored, it is necessary to try to pin-point the correct solution; this is done through the use of a *figure of merit* based upon:

1. the agreement with the Patterson function;
2. the triple-phase consistency;
3. an R-index between the observed $|E_o|$ (see Section 7.2.1) and calculated $|E_c|$ (see Section 11.4.10) for the partial structure.

The Patterson function is calculated using $|E_o||F_o|$ as coefficients instead of $|F_o|^2$ in order to sharpen or improve the resolution of peaks (see also Chapter 11). The complete Patterson function is then stored in the computer, each grid value being represented by a digit between 0 and 7 so that it can be stored in three bits of computer memory. This is one step better than the method employed in the Vector Verification, which employed only a two-bit representation. In order to make efficient use of computer memory, the Patterson values are encoded according to seven test levels, with level 2 equal to the median of the cumulative Patterson distribution, and the difference between two successive test levels being about half the expected height of the highest single vector. The user can, in fact, supply different test levels from those noted above but it is probably not necessary.

Rotation Search Strategy used in PATSEE

The rotation search procedure used in PATSEE is summarized in Table 7.15.

TABLE 7.15.   Rotation Search in PATSEE

---

1. Compilation of the intramolecular self-vector set from the model coordinates with distances between 2 Å (too short to provide orientation data) and 6 Å (the point at which errors in distance affect the accuracy with which vectors superimpose).
2. Generation of random orientations, typically between 10,000 and 60,000 angle triplets (α, β, γ) at about 7° intervals.
3. For each orientation, calculation of the correlation between the rotated intramolecular vector set and the Patterson function. This is computed as a sum function to give a figure of merit

$$\text{RFOM} = (1/n) \sum_i P_i/w_i \qquad (7.83)$$

where $n \approx 0.3n_{\text{total}}$, w_i is the calculated vector weight, P_i is the nearest Patterson grid value and n_{total} is the number of worst-fitting vectors, that is, those with the lowest P_i/w_i value; the summation is carried out over the range 1 to n.
4. An overlap or packing test ensures that when symmetry and lattice translations are applied there are no serious interatomic clashes.
5. An equivalence test which excludes similar solutions from being retained.
6. Sorting of the solutions in descending order of RFOM.
7. Refinement of the best solutions, carried out by testing up to 1000 additional random rotations around each retained solution, at approximately 2° intervals.

Translation Search Strategy Used in PATSEE

The translation search procedure used in PATSEE is summarized in Table 7.16.

TABLE 7.16.   Translation Search in PATSEE

---

1. Search for the most probable direct methods TPRs;
2. Calculation of $|E_0(hkl)|$ for a given orientation of the search model;
3. Selection of suitable phase relationships from a *relatively small number* of large $|E_0(hkl)|$ values. This strategy considerably speeds up the procedure and especially enhances the efficiency for larger structures;

   Note that in small molecule work, the search fragment is of course usually very incomplete and possibly inaccurate. Nevertheless, if its scattering power ($\approx \Sigma_j Z_j^2$) is large enough, the TPRs used here should hold, at least approximately, for the correct solution, and be nonrandom in character. Hence the importance of having a sufficiently large fragment of the whole structure as the search model, as discussed further in the next section. (The reader should refer to Section 7.2 to aid the appreciation of these three steps.);
4. For each oriented search model from the rotation stage, sets of atom coordinates are generated, the oriented search model being placed at a position in the unit cell generated in a random manner. Each of these sets of coordinates is used for calculation of trial phases and assessment through their agreement with the selected TPRs from stage 1;
5. A packing test is carried out on rotated-translated fragments from stage 4 and a model is eliminated if short intermolecular distances occur;
6. Initial refinement of an RT fragment: this procedure involves optimizing a figure of merit TPSRSUM by fine tuning the position of the fragment:

$$\text{TPSRSUM} = [\Sigma |E_\mathbf{h}||E_\mathbf{k}||E_{-\mathbf{h}-\mathbf{k}}| \cos(\phi_\mathbf{h} + \phi_\mathbf{k} + \phi_{-\mathbf{h}-\mathbf{k}})]/[\Sigma |E_\mathbf{h}||E_\mathbf{k}||E_{-\mathbf{h}-\mathbf{k}}|] \quad (7.84)$$

   the two sumations being taken over all selected three-phase structure invariants. TPSRSUM is expected to be large and positive for the correct solution, up to a maximum value of 1.0. During this process the step sizes are reduced from around 0.2Å to 0.05Å;
7. A further distance test is then carried out to check the packing of the fragment;
8. Solutions that have survived all of these rigorous tests are further tested against the Patterson function of the target crystal. In earlier Patterson Search programs, such as Vector Verification, this stage was carried out immediately after the rotation stage,[a] and consequently more time consuming and less likely to succeed. In PATSEE the correlation between the Patterson function and the fragment-derived intermolecular vector set is examined by comparing the weight of each vector with the nearest grid value. The fit is measured by calculating a further figure of merit:

$$\text{TFOM} = (1/n) \sum_i (P_i/w_i) \quad (7.85)$$

   where $n \approx 0.2 n_{\text{total}}$; as before, w_i is a calculated vector weight, P_i is the nearest Patterson grid value and n_{total} is the number of worst fitting vectors, those with the lowest P_i/w_i value, the summation being carried out from 1 to n;
9. Final selection and ordering of the possible solutions. At this stage a small number of the most promising solutions according to TPSRSUM and TFOM will have been stored in the computer. An R_E index is calculated as

$$R_E = [\Sigma(|E_0| - |E_c|)/p]/\Sigma |E_0| \quad (7.86)$$

   where p^2, equal to $\Sigma Z_{\text{frag}}^2 / \Sigma Z_{\text{molecule}}^2$, is the fractional scattering power of the search model (frag) compared to the whole molecule. Only selected positive terms are considered, as it is assumed that negative terms indicate complete agreement. The solutions are then sorted according to a combined figure of merit:

$$\text{CFOM} = 0.1 \, (\text{RFOM} + \text{TFOM})(\text{TPRSUM}^{1/2})/R_E \quad (7.87)$$

---

[a] M. F. C. Ladd and R. A. Palmer, *Structure Determination by X-ray Crystallography*, 3rd ed., 406ff (1993).

7.3.7 Examples of Structure Solution using PATSEE

The following examples indicate various other features of the PATSEE program. In particular, the use of the figures of merit RFOM, TPSRSUM, and CFOM in practice to pin-point the correct Patterson Search solution.

Structure of 5,7-methoxy-8-(3-methyl-1-buten-3-ol)-coumarin

The crystal and molecular structure of the antimalarial compound 5,7-methoxy-8-(3-methyl-1-buten-3-ol)-coumarin, $C_{16}H_{18}O_5$, $M_r = 290.3$ Da was reported recently.[a] This molecule, Figure 7.17a, was selected for investigation of the features of the PATSEE program, because the coumarin moiety, Fig 7.17b, is known to be fairly rigid and planar (rings A and B) and as such is an ideal search molecule. The presence of potentially more flexible side groups provides an element of challenge to the method. The atomic coordinates of the 11 atoms in the coumarin moiety are readily available from either the published structure[b] or the Cambridge Crystallographic Data Base. In fact, we chose the method of molecular graphics employing the Chem-X package to generate coumarin search model 1.

Search Model 1 from Chem-X. The material crystallizes in the monoclinic space group $P2_1/c$ with four molecules per unit cell of dimensions $a = 8.9044(9)$, $b = 17.623(1)$, $c = 10.175(1)$Å, $\beta = 113.97(1)°$. The x-ray intensity data were collected on a Nonius CAD4 diffractometer (see Chapter 4) using Cu $K\alpha$ radiation, $\lambda = 1.5418$ Å. A total of 3193 reflections was collected, to a θ_{max} of 74.22°, of which 2972 are independent with R_{int} (see Section 10.4.7)= 0.0175. Using the coumarin ring system as search model provides a fragment consisting of 11 out

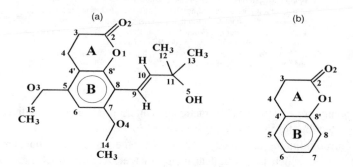

FIGURE 7.17. Chemical formulae: (a) the Coumarin derivative (see text); (b) Coumarin. Rings A and B form a planar group for use in Patterson Search. [Diagrams produced by Chemwindow (Softshell International Limited).]

[a] J. N. Lisgarten, B. Potter, R. A. Palmer, H. Oketch-Rabah, and J. Aymami, *Journal of Chemical Crystallography* to be published.

[b] E. Gavuzzo, F. Mazza, and E. Giglio, *Acta Crystallographica* **B30**, 1351 (1974).

TABLE 7.17.   Input Data for Coumarin Model 1
PATSEE Search

TITLE	Coumarin derivative Chem-X model for PATSEE
CELL	1.5418 8.9044 17.6236 10.1757 90.0 113.97 90.0
ZERR	4 0.0009 0.001 0.001 0 0.01 0
LATT	1
SYMM	$X, 0.5 - Y, 0.5 + Z$
SFAC	C H O
UNIT	64 72 20
ROTS	10000 20
TRAN	
FRAG	1 100.00 100.00 100.00 90.0 90.0 90.0
O1	3 −0.01015 0.00495 0.000
C2	1 −0.02219 −0.00188 0.000
O2	3 −0.03271 0.00429 0.000
C3	1 −0.02228 −0.01572 0.000
C4	1 −0.01034 −0.02273 0.000
C4'	1 0.00169 −0.01589 0.000
C5	1 0.01363 −0.02289 0.000
C6	1 0.02566 −0.01605 0.000
C7	1 0.02576 −0.00221 0.000
C8	1 0.01382 0.00479 0.000
C8'	1 0.00179 −0.00205 0.000

Notes: The above data have the following interpretation: CELL: wavelength and target unit-cell parameters; ZERR: number of molecules in target cell and errors in experimental unit cell parameters $a, b, c, \alpha\ \beta, \gamma$; LATT 1: primitive unit cell and centrosymmetric space group; SYMM: other space-group symmetry operations; SFAC: atom types in molecule; UNIT: number of each atom type in unit cell; ROTS: 10000 20 (use 10,000 random test orientations and retain the best 20); TRAN: initiate translation search using n_t random positions calculated by the program [the user also has the option of supplying a value for this number here]; FRAG: 1 = fragment 1 (a second fragment can be supplied in addition as in the example below). The next six values are the unit-cell parameters for the fragment. This is not a real crystal cell as this fragment was model built in Chem-X.

   The final 11 lines of data here represent the model atom names, types, and coordinates in the FRAG unit cell.

of 21 nonhydrogen atoms. The fractional scattering power (p^2), defined in the previous section, for this model, is very high (50.4%).

   The coordinates generated by Chem-X for the first search model are shown in Table 7.17, which is a complete listing of the PATSEE input data, whereas Table 7.18 summarizes the search results.

   Discussion of Results and Expansion and Refinement of the Model. From the above extract from the PATSEE output we can see:

1. The top 15 values of RFOM, after the rotation stage, are quite similar.

TABLE 7.18. Summary of Results for Coumarin Model 1

		Rotation stage			Translation stage
Solution	RFOM	$\alpha°$	$\beta°$	$\gamma°$	CFOM
1	0.514	5.103	5.844	2.295	0.559
2	0.514	5.109	5.887	0.527	0.678
3	0.514	1.261	3.580	0.844	0.540
4	0.514	1.317	2.667	0.593	0.665
5	0.514	3.524	4.462	1.948	0.542
6	0.514	3.185	1.821	0.953	0.854
7	0.514	1.325	3.594	2.521	0.698
8	**0.514**	**6.163**	**5.078**	**0.884**	**1.053 ←**
9	0.514	5.153	5.841	0.462	0.909
10	0.514	1.832	0.475	0.891	0.881
11	0.514	2.868	4.993	1.146	0.554
12	0.514	5.136	0.437	2.633	0.716
13	0.491	1.939	5.704	2.240	0.470
14	0.491	1.862	5.706	2.253	0.594
15	0.490	1.315	2.543	2.291	0.547
16	0.364	5.305	0.977	1.172	0.103
17	0.356	0.438	1.273	2.666	0.409
18	0.329	4.078	4.008	1.198	0.094
19	0.299	5.091	0.103	2.478	0.089
20	0.298	1.162	2.826	2.258	0.132

2. The values of CFOM, after the translation stage, are more widely spread and only solution 8 has a value greater than 1.0.
3. Solution 9 has CFOM = 0.909 and similar rotation solution angles.

Using the coordinates of the coumarin fragment corresponding to solution 8, all of the atoms in the structure were located and refined within two iterations of Fourier-least-squares in the program WinGX. Figure 7.18 shows how the structure developed at these three stages. At stage (a) (Figure 7.18a) the R-factor was, as expected, very high with a value of 57.3%, reducing to 42.3% after the stage of Figure 7.18b, and dramatically to 15.5% at Figure 7.18c, when all nonhydrogen atoms were included. Experience tells us that a structure which refines isotropically, that is, with isotropic temperature factors, to this sort of R-value is probably correct. Further refinement of the model leads to the published structure for which $R = 3.9\%$ (Figure 7.18d). It has been shown that the use of model 1 with PATSEE has led to a successful determination of the structure with the RT solution corresponding to the highest CFOM value. The strategy of carrying 20 rotation stage solutions through to the translation stage was necessary in view of the lack of discrimination in the RFOM values.

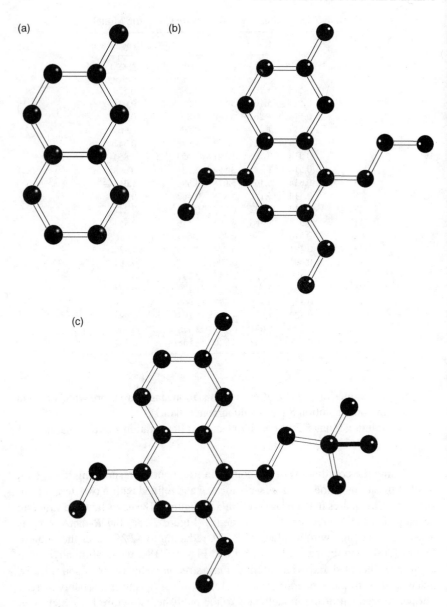

FIGURE 7.18. Stages in determination of the structure of the coumarin derivative: (a) the search model; (b) the partial structure developed after one cycle of refinement; (c) after two cycles of refinement; (d) after further refinement. [Diagrams by POV-Ray^{tm} VERSION 3.1 (www.povray.org), as implemented in WinGX and generated by Ortep-3 for Windows.]

(d)

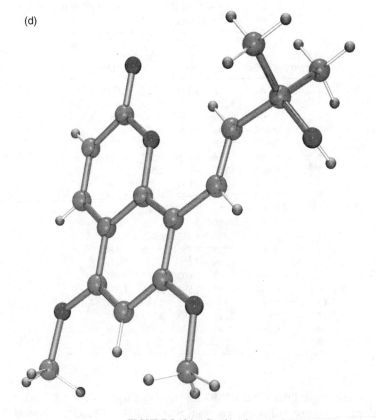

FIGURE 7.18d.   Continued.

Search Model 2 from Chem-X.   In order to understand further the working
and effectiveness of the PATSEE program, the above calculations were repeated
with a smaller search model. A smaller, six-atom fragment, model comprising ring
B (Figure 7.17b) was generated again using Chem-X. This is essentially a benzene
ring with six equivalent bonds and angles. The fractional scattering power p^2 is
now much lower, 24.1% instead of 50.4% for model 1, and is therefore expected to
be less effective in the Patterson search. In other words even if PATSEE can locate
the model correctly in the unit cell, this may not be the top solution and may be
more difficult to expand to the full molecule.

The program was again instructed to retain the top 20 rotation solutions,
which were then passed on to the translation procedure, with the results as
summarized in Table 7.19.

Expansion and refinement of the model.   The coordinates for the six atoms
corresponding to the best RT solution (solution 16) were input to the WinGX

TABLE 7.19.   Summary of RT Results for Coumarin
Model 2

Solution	RFOM	Rotation stage			Translation stage
		$\alpha°$	$\beta°$	$\gamma°$	CFOM
1	0.604	0.515	1.328	2.756	0.960
2	0.604	5.015	5.907	2.162	0.661
3	0.604	5.088	5.846	0.637	0.881
4	0.604	3.746	1.826	1.492	0.437
5	0.604	5.088	5.284	2.035	0.264
6	0.604	1.973	5.816	0.615	0.983
7	0.604	5.956	4.351	2.546	1.335
8	0.604	1.278	2.682	2.257	1.234
9	0.604	1.895	5.873	0.623	1.071
10	0.604	0.353	1.714	1.263	0.783
11	0.604	3.158	4.603	2.316	0.787
12	0.604	3.353	4.417	2.025	0.879
13	0.604	2.724	1.756	0.429	0.830
14	0.604	4.548	3.606	0.925	0.715
15	0.604	1.468	3.602	0.781	0.691
16	**0.604**	**6.289**	**4.886**	**0.936**	**1.478** ←
17	0.604	2.741	4.932	1.130	0.697
18	0.604	4.301	3.585	2.717	1.055
19	0.604	1.843	5.736	2.330	1.174
20	0.604	1.217	3.681	2.402	0.650

package and refined with SHELX-97 for two least-squares cycles. Inspection of the electron density, using the program SXGRAPH in WinGX, allowed all of the missing structure to be built in and refined with two iterations of Fourier and isotropic least squares to an R-factor of 15.5%. The above result is quite pleasing, but may be somewhat surprising in view of the relatively few atoms used in the search.

Search Model 3 from Chem-X.   The search model was further modified and reduced by removing atoms C4$'$ and C8$'$ (Figure 7.17b) and the PATSEE procedure was repeated. A correctly rotated and translated search model was found in the PATSEE RT listing. It was expanded and refined using WinGX to the same isotropic R-factor as in the previous cases, but with a little more difficulty.

The reason this model did not produce the top RT solution is undoubtedly due to its size. Compared to models 1 and 2, the fractional scattering power for model 3 is only 16.1%. It is therefore a tribute to the method that the correct crystal structure can be generated with PATSEE without too much difficulty; remember though that it is often much easier to solve a problem when the answer is known.

The above experiments with a second and third model emphasize the power of the method and it should be remembered that some structures do still fail to be

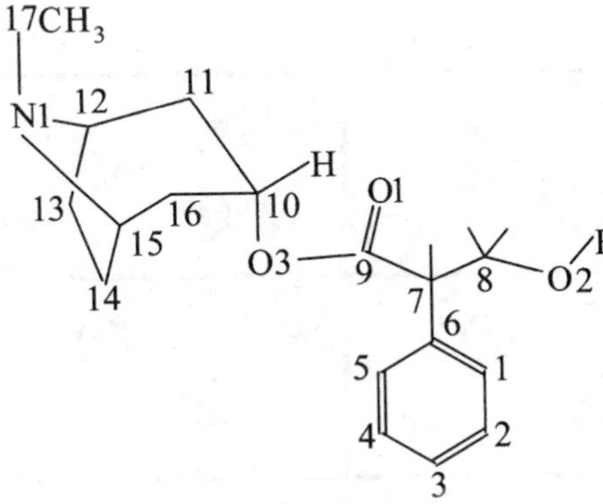

FIGURE 7.19. Chemical formula of atropine. [Diagram produced by Chemwindow (Softshell International Limited).]

determined by direct methods. We know that a 21-atom structure can be solved with a 4-atom search model. Similar models may be useful for solving small polypeptide structures, for example, using the well-defined planar peptide group in the initial search. A dipeptide search model with one degree of rotational freedom may enable a tetrapeptide structure to be solved. This procedure will become clearer after studying the next section.

Structure of Atropine: α-[Hydroxymethyl]benzeneacetic Acid 8-methyl-8-azabicyclo[3.2.1]oct-3-yl Ester

The chemical formula of the atropine molecule is shown in Figure 7.19. Until recently[a] there were no reports of the structure of this classic molecule in the literature. Atropine is a competitive antagonist at central and peripheral synapses and has been used, somewhat unadvisedly, as a beautifying agent; hence its alternative, better known name *Belladonna*.

Atropine, $C_{17}H_{23}NO_3$, $M_r = 289.4$ Da, is α-[hydroxymethyl]benzeneacetic acid 8-methyl-8-azabicyclo[3.2.1]oct-3-yl ester, and is known also as tropine tropate.

Practical details. Unit-cell determination and refinement, data collection, and data reduction were carried out on a Nonius Kappa CCD diffractometer[b] (see Chapter 4) using Mo $K\alpha$ radiation $\lambda = 0.71073$ Å, cooled to $-173°C$ (liquid nitrogen temperature) with an Oxford Cryostreams cooler. At this temperature

[a] J. W. Saldanha, B. S. Potter, B. Howlin, A. Tanczos, and R. A. Palmer, *Journal of Chemical Crystallography* **,** (2003).
[b] EPSRC Data Collection Service, University of Southampton, UK.

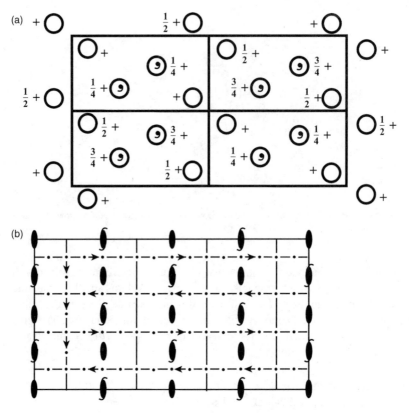

FIGURE 7.20. Space group diagrams for $Fdd2$: (a) general equivalent positions; (b) symmetry elements.

the crystal diffracted strongly over the 90-min period of data collection. Data frames were processed using the Nonius software. The material crystallizes in the orthorhombic system, with the unusual space group, $Fdd2$ (Figure 7.20), with 16 molecules per unit cell of dimensions $a = 24.291(5)$, $b = 39.538(8)$, $c = 6.472(1)$ Å. A total of 2701 independent reflections was collected, to θ_{max} ($MoK\alpha$) $25.02°$.

Patterson Search Models. The molecule contains a benzene ring, as in the previous example, which is an obvious fragment to use for Patterson Search. This can be easily extended by adding atom C7 to form a slightly larger fragment comprising 7 of the 21 nonhydrogen atoms, with a fractional scattering power $p^2 = 0.30$. However with this model, trials with PATSEE fail to identify an RT solution that could be expanded and refined into the complete molecule.

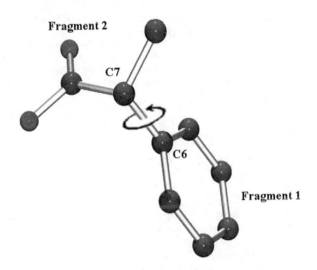

FIGURE 7.21.   The search model for atropine; torsional flexibility between the two rigid groups occurs about the C6–C7 bond indicated. [Diagram produced by POV-Ray[tm] VERSION 3.1 (www.povray.org), as implemented in WinORTEP (www.chem.gla.ac.uk/~louis/ortep3/).]

It can be demonstrated, by applying the usual Fourier-least-squares proce-dure to the *known* atom coordinates for this fragment, that phasing on the 7 atoms of this model leads to a complete and refinable structure. We conclude therefore, that if the correct solution is in fact produced by PATSEE for this small fragment, it does not appear as one of the best solutions, as judged by the figures of merit.

To overcome this problem and further test the features of PATSEE, a model was built with Chem-X, which included fragment 1 (atoms 1–6) and five other atoms, C7, C9, O1, O3, and C8, as fragment 2 (Figure 7.21). Fragment 2 is itself structurally rigid, but for the combined search model of fragments 1 and 2 there is *rotational freedom* about the bond C6–C7. PATSEE allows the linkage torsion angle (see Section 7.5) between two such fragments to be systematically varied during the Patterson Search to produce a series of test models. Each new set of coordinates produced by a change of this torsion angle is then treated as an independent search model. The computer time required to complete the rotation search is thus multiplied by the number of individual models explored. The best models are again passed to the translation stage.

Patterson Search for Atropine.   The input data are shown in Table 7.20. The reader should identify the differences between the data in this table and those in Table 7.17 used for coumarin. Obvious changes arise from the differences in unit-cell dimensions and space group and molecular formula. The reader should study Figure 7.20 so as to identify the entries from this which appear in Table 7.20; LATT-4 is the code for a noncentrosymmetric space group with an *F* unit cell. The

TABLE 7.20.   Input Data for PATSEE Search for
Atropine Fragment

TITL	Atropine PATSEE using dual model
CELL	0.71073 24.2913 39.5380 6.4727 90.000 90.000 90.000
ZERR	16.00 0.0049 0.0079 0.0013 0.000 0.000 0.000
LATT	-4
SYMM	$-X, -Y, Z$
SYMM	$0.25 + X, 0.25 - Y, 0.25 + Z$
SYMM	$0.25 - X, 0.25 + Y, 0.25 + Z$
SFAC	C H N O
UNIT	272 368 16 48
ROTS	10000 20
TRAN	
	(*Fragment from crystal via Chem-X*)
FRAG	1 100.00 100.00 100.00 90.0 90.0 90.0
C1 1	0.12068 0.04724 0.06165
C2 1	0.12991 0.05601 0.06751
C3 1	0.13407 0.05393 0.08046
C4 1	0.12927 0.04327 0.08761
C5 1	0.12003 0.03461 0.08187
C6 1	0.11568 0.03661 0.06881
TWIS	0 2 360
C7 1	0.10546 0.02722 0.06245
C8 1	0.07132 0.04096 0.06734
C9 1	0.09194 0.02896 0.06908
O3 3	0.08480 0.03817 0.06248
O1 3	0.08835 0.02332 0.07902

rotation (ROTS) and translation (TRAN) instructions have been retained without change. The next major change occurs in the FRAG listing of atom positions: the instruction TWIS 0 2 360 causes the program to systematically vary the torsion angle about C6–C7 from 0° to 360° in steps of 2°, thus involving 181 different search models in the rotation procedure. The final application of the torsional change at 360° gives the same result as for the first at 0° and merely acts as a check.

Rotation stage. The results of the rotation analysis are summarized below as Table 7.21. In this table and Table 7.22, the torsion angle refers to the relative orientation, about the C6–C7 bond, of the two rigid fragments of the search model (Figure 7.21).

Translation stage. The possible rotation function solutions are then each transferred to the translation algorithm in turn. The best twenty solutions (Table 7.21) were subjected to the translation algorithms as before. Values of CFOM after translation and optimization range from 0.929 to 2.640, as shown in Table 7.22.

TABLE 7.21. Atropine Fragment Rotation Search

Solution	RFOM	$\alpha°$	$\beta°$	$\gamma°$	Torsion angle°
1	1.742	2.942	2.912	0.174	182
2	1.732	5.892	3.527	1.736	2
3	1.701	6.036	3.359	2.980	180
4	1.676	6.052	3.352	2.984	182
5	1.563	2.932	2.916	0.176	184
6	1.544	5.827	3.474	1.766	358
7	1.535	5.933	3.540	1.737	360
8	1.535	5.933	3.540	1.737	0
9	1.522	6.101	3.295	2.999	184
10	1.438	5.938	3.567	1.705	4
11	1.403	2.883	2.927	0.158	176
12	1.353	2.690	5.981	1.726	356
13	1.348	2.912	2.876	0.179	180
14	1.334	6.105	3.408	2.969	186
15	1.332	2.702	5.949	1.783	354
16	1.312	5.862	3.549	1.745	6
17	1.285	2.946	2.883	0.176	178
18	1.272	6.114	3.360	3.008	188
19	1.246	2.594	6.027	1.757	352
20	1.222	2.588	5.984	1.748	356

Note: Each entry in this table is the result of 10,000 test rotations.

In this analysis the order of precedence has changed when using CFOM as the final discriminator. As would be expected the best solutions (arrowed) all have similar rotational and torsional parameters. Although not shown here, the corresponding translation components derived in the second stage of PATSEE are also quite similar, as would be expected.

Expansion and refinement of the model. The overall best solution was taken to be number 2 in Table 7.22, and the model coordinates corresponding to this solution were subjected to Fourier-least-squares expansion and refinement as before. After two iterations of Fourier-least squares all noncarbon atoms were located and isotropically refined to an *R* factor of 0.105. In the published structure *anisotropic* refinement led to final *R* factor of 0.0453; see Figure 7.22.

Conclusions. These examples illustrate many of the features of the powerful Patterson Search technique. However, for small molecule analysis, because direct methods are so much more easy to apply, if not to understand, Patterson search will remain as a reserve technique, albeit a very useful and powerful one, in the crystallographer's armoury.

We have discussed two small molecule analyses where Patterson Search, with PATSEE, has led successfully to refinable models of the crystal structure. Provided that a suitable search model is available there is no reason to doubt that most small molecule structures could be solved in a similar way. The most

TABLE 7.22. Atropine Fragment
Translation Search Results

Solution	RFOM	Torsion angle°	CFOM
1	1.742	182	1.408
2	**1.732**	**2**	**2.640** ←
3	1.701	180	1.373
4	1.676	182	1.250
5	1.563	184	1.424
6	1.544	358	2.284 ←
7	1.535	360	1.232
8	1.535	0	1.232
9	1.522	184	1.029
10	1.438	4	2.342 ←
11	1.403	176	1.081
12	1.353	356	1.342
13	1.348	180	1.039
14	1.334	186	1.114
15	1.332	354	1.307
16	1.312	6	1.793
17	1.285	178	1.077
18	1.272	188	0.929
19	1.246	352	1.763
20	1.222	356	1.694

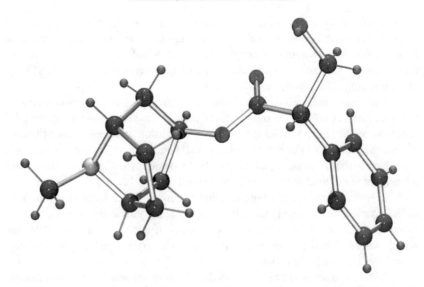

FIGURE 7.22. The completed structure of atropine. [Diagram produced by POV-Ray^tm
VERSION 3.1 (www.povray.org), as implemented in WinORTEP (www.chem.gla.ac.uk/~louis/
ortep3/).]

useful discriminator provided by PATSEE is the combined figure of merit CFOM. For all but the smallest search models employed in the examples discussed, the RT solution with the largest value of CFOM has proved to be correct.

A search model with a fractional scattering power as low as 16.1% and including only 6 out a total of 21 nonhydrogen atoms can provide the correct answer, but it requires more sifting of coordinate sets, using Fourier and least squares, to be selected. We shall see in Chapter 10 that the related method of MR is routinely used in macromolecular structure analysis, and for this reason the examples discussed in the present chapter will be invaluable for a sound understanding of the principles of the method when we come to discuss the large molecule case.

7.3.8   Shake 'n' Bake

We mention here, very briefly and with no technical details, another procedure that is a stage further toward solving really complicated structures. Like PATSEE the program is readily available and is quite fun to use, as the name *Shake 'n' Bake* suggests.

In outline, the Shake 'n' Bake technique alternates phase refinement in reciprocal space with electron density refinement in real space. Numerous sets of phases are generated and evaluated but, unlike traditional methods, the reciprocal-space refinement requires reducing the value of a minimal function $R(\Phi)$ to a global minimum. The function $R(\Phi)$ is defined by

$$
R(\Phi) = \left\{ \sum_{H,K} A_{HK} [\cos T_{HK} - I_1(A_{HK})/I_0(A_{HK})]^2 \right.
$$

$$
\left. + \sum_{L,M,N} |B_{LMN}| [\cos Q_{LMN} - I_1(B_{LMN})/I_0(B_{LMN})]^2 \right\}
$$

$$
\times \left[\sum_{H,K} A_{HK} + \sum_{L,M,N} |B_{LMN}| \right]^{-1}
$$

where T_{HK} is a structure invariant triple defined by $T_{HK} = \Phi_H + \Phi_K + \Phi_{-H-K}$, and Q_{LMN} is the structure invariant quartet $Q_{LMN} = \Phi_L + \Phi_M + \Phi_N + \Phi_{-L-M-N}$; $A_{HK} = (2/N^{1/2})|E_H E_K E_{H+K}|$ and $B_{LMN} = (2/N)|E_L E_M E_N E_{L+M+N}| \times (|E_{L+M}|^2 + |E_{M+N}|^2 + |E_{N+L}|^2 - 2)$, and N is the number of atoms, assumed equal, in the unit cell.

The technique has solved a number of structures, including one with over 600 atoms in the asymmetric unit. For further details the reader is referred to the reference in Appendix A10.

7.4   Least-Squares Refinement

In Chapter 1, we used the equation of a line in two-dimensional space

$$Y = mX + b \qquad (7.88)$$

If we have two pairs of values of X and Y for measurements which are related by this equation, we can obtain a unique answer for the constants m and b. Sometimes, as in the Wilson plot, we have several pairs of values, which contain random errors, and we need to obtain those values of m and b that best fit the complete set of observations.

In practical problems, we have often a situation in which the errors in the X values are negligible compared with those in Y. Let the best estimates of m and b under these conditions be m_0 and b_0. Then, the error of fit in the ith observation is

$$e_i = m_0 X_i + b_0 - Y_i \qquad (7.89)$$

The principle of least squares states that the best-fit parameters are those that minimize the sum of the squares of the errors. Thus,

$$\sum_i e_i^2 = \sum_i (m_0 X_i + b_0 - Y_i)^2 \quad (i = 2, \dots, N) \qquad (7.90)$$

has to be minimized over the number N observations. This condition corresponds to differentiating partially with respect to m_0 and b_0, in turn, and equating the derivatives to zero. Hence

$$m_0 \sum_i X_i^2 + b_0 \sum_i X_i = \sum_i X_i Y_i \qquad (7.91)$$

$$m_0 \sum_i X_i + b_0 N = \sum_i Y_i \qquad (7.92)$$

which constitute a pair of simultaneous equations (normal equations) easily solved for m_0 and b_0.

In a crystal structure analysis, we are always manipulating more observations than there are unknown quantities; the system is said to be overdetermined. We shall consider some applications of the method of least squares.

7.4.1   Unit-Cell Dimensions

In Chapter 4, we considered methods for obtaining unit-cell dimensions with moderate accuracy. Generally, we need to enhance the precision of these measurements, which may be achieved by a least-squares analysis. Consider, for example, a monoclinic crystal for which the θ values of a number of reflections, preferably high-order, of known indices, have been measured to the nearest $0.01°$. In the monoclinic system, $\sin\theta$ is given [Table 2.4 with $\kappa = \lambda$, and (3.55)] by

$$4\sin^2\theta = h^2a^{*2} + k^2b^{*2} + l^2c^{*2} + 2hla^*c^* \cos\beta^* \tag{7.93}$$

In order to obtain the best values of a^*, b^*, c^*, and $\cos\beta^*$, we write, following (7.90),

$$\sum_i (h_i^2 a^{*2} + k_i^2 b^{*2} + l_i^2 c^{*2} + 2h_i l_i a^* c^* \cos\beta^* - 4\sin^2\theta_i)^2 \tag{7.94}$$

and then minimize this expression, with respect to a^*, b^*, c^*, and $\cos\beta^*$, over the number of observations i. The procedure is a little more involved numerically; we obtain four simultaneous equations to be solved for the four variables, but the principles are the same as those involved with the straight line, (7.88)–(7.92).

7.4.2 Least-Squares Parameters

Correct trial structures are refined by the least-squares method. In essence, this process involves, adjusting a scale factor and the positional and temperature parameters of the atoms in the unit cell so as to obtain the best agreement between the experimental $|F_o|$ values and the $|F_c|$ quantities derived from the structure model. In its most usual application, the technique minimizes the function.

$$R' = \sum_{\mathbf{h}} w(|F_o| - G(|F_c|)^2 \tag{7.95}$$

where the sum is taken over the set of crystallographically independent terms $\mathbf{h}$, w is a weight for each term, and G is the reciprocal of the scale factor K for $|F_o|$. Let $p_j(j = 1, 2, \ldots, n)$ be the variables in $|F_c|$ whose values are to be refined. Then

$$\frac{\partial R'}{\partial p_j} = 0 \tag{7.96}$$

or

$$\sum_{\mathbf{h}} w\Delta \frac{\partial|F_c|}{\partial p_j} = 0 \tag{7.97}$$

where Δ is $|F_o| - |F_c|$. For a trial set of parameters not too different from the correct values, Δ is expanded as a Taylor series to the first order:

$$\Delta(\mathbf{p} + \xi) = \Delta(\mathbf{p}) - \sum_{i=1}^{n} \xi_i \frac{\partial|F_c|}{\partial p_j} \tag{7.98}$$

where the shift ξ_i is the correction to be applied to parameter p_i; $\mathbf{p}$ and ξ represent the complete sets of variables and corrections. Substituting (7.98) in (7.97) leads to the normal equations

$$\sum_{i=1}^{n}\left[\sum_{\mathbf{h}} w\frac{\partial|F_c|}{\partial p_i}\frac{\partial|F_c|}{\partial p_j}\right]\xi_i = \sum_{\mathbf{h}} w\Delta\frac{\partial|F_c|}{\partial p_j} \tag{7.99}$$

The n normal equations may be expressed neatly in matrix form:

$$\mathbf{A}\xi = \mathbf{b} \quad \text{or} \quad \sum_i a_{ij}\xi_i = b_j \tag{7.100}$$

where

$$a_{ij} = \sum_{\mathbf{h}} w\frac{\partial|F_c|}{\partial p_i}\frac{\partial|F_c|}{\partial p_j} \tag{7.101}$$

and

$$b_j = \sum_{\mathbf{h}} w\Delta\frac{\partial|F_c|}{\partial p_j} \tag{7.102}$$

The solution of the normal equations is a well-documented mathematical procedure that we shall not dwell upon. Instead, we draw attention to a few features of least-squares refinement. It is important to remember that least squares provides the best fit for the parameters that have been put into the model. Hence, it is essential to examine a final difference Fourier map at the completion of a least-squares refinement, after several cycles of calculations have led to negligible difference ξ_i. The techniques of least squares have been reported fully at Crystallographic Computing Conferences (see Bibliography).

Temperature Factors

An overall isotropic temperature factor T, as obtained from a Wilson plot, is the simplest approximation (see Section 3.10.1). A better procedure allots B parameters to atoms and refines them as least-squares parameters. We can write

$$T_i = \exp[-(B_i\lambda^{-2}\sin^2\theta)] \tag{7.103}$$

and the equation

$$B_i = 8\pi^2\overline{U_i^2} \tag{7.104}$$

relates B_i to the mean square amplitude $\overline{U_i^2}$ of the ith atom; the surface of vibration is a sphere.

A more sophisticated treatment describes the vibrations of each atom by a symmetrical tensor $\mathbf{U}$ having six independent components in the general case. Now, we have

$$T_i = \exp[-2\pi^2(U_{11}h^2a^{*2} + U_{22}k^2b^{*2} + U_{33}l^2c^{*2}$$
$$+ 2U_{23}klb^*c^* + 2U_{31}lhc^*a^* + 2U_{12}hka^*b^*)] \qquad (7.105)$$

and the (anisotropic) U_{ij} parameters are refined as part of the model. The surface of vibration is now a biaxial (thermal) ellipsoid, and the mean-square amplitude of vibration in the direction of a unit vector $\mathbf{L} = (L_1, L_2, L_3)$ is given by

$$\overline{U^2} = \sum_{i=1}^{3} \sum_{j=1}^{3} U_{ij}L_iL_j \qquad (7.106)$$

Biaxial ellipsoids for the molecule of lamotrigine are illustrated by Figure 7.23. Since $\mathbf{L}$ is defined with respect to the reciprocal lattice, the component of $\mathbf{U}$ with $\mathbf{l} = (1, 0, 0)$, parallel to a^*, is

$$\overline{U^2} = U_{11} \qquad (7.107)$$

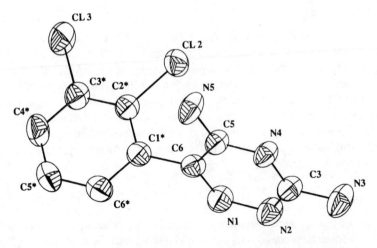

FIGURE 7.23. Anisotropic thermal ellipsoids for lamotrigine, $C_9H_7N_5Cl_2$, an anticonvulsant drug. The ellipsoids are plotted at the 50% probability level, which means that at the maximum radius drawn the exponential expression of (7.105) is 0.5. In terms of an isotropic thermal surface, the radius would be almost 0.6 Å corresponding to $\sin^2\theta/\lambda^2$, or $(1/2d)^2$, of 0.694 Å^{-2}. (After R. W. Janes, J. N. Lisgarten, and R. A. Palmer.]

In an orthorhombic crystal, for example, a direction $30°$ from a^* in the a^*b^* plane has $\mathbf{l} = (\sqrt{3}/2, 1/2, 0)$, and the component of $\mathbf{U}$ in that direction is

$$\overline{U^2} = U_{11}(\sqrt{3}/2)^2 + U_{22}(1/2)^2 + 2U_{12}(\sqrt{3}/2)(1/2) \qquad (7.108)$$

The following relationships among the values of B, $\overline{U^2}$, and the root mean square (rms) amplitude are often useful:

B (Å^2)	$\overline{U^2}$ (Å^2)	rms amplitude (Å)
0.10	0.0013	0.036
0.50	0.0063	0.080
1.0	0.013	0.11
5.0	0.063	0.25
10	0.13	0.36

The smallest rms amplitudes encountered are ca 0.05 Å. Values of B between 3 and 10 Å^2 are found in organic structures at ambient temperatures. The larger rms amplitudes require caution in interpreting them in terms of bond lengths and their precision. For example, static or dynamic disorder (Section 7.9), not included in the model, may be manifested as abnormally large temperature factors.

Scale Factor

In least-squares refinement, the $|F_o|$ data must not be adjusted, and so the parameter G in (7.95) is introduced. The inverse of the refined value of G may be applied to $|F_o|$ at the end of a refinement cycle. Several cycles of refinement may be needed before the parameters reach a sensibly constant value. Generally full-matrix least-squares refinement is to be preferred. However, where the number of parameters is very large or where computer availability is limited, an approximation may be used. One such method is the block-diagonal refinement, in which certain off-diagonal a_{ij} terms in (7.100) are neglected. Generally, more cycles are necessary in this procedure.

Weights

In the initial stages of refinement, weights may be set at unity or chosen so as to accelerate the process, such as downweighting reflections of small $|F_o|$ or of high order, or both. In the final stages, weights should be related to the precision of $|F_o|$, which can be achieved in two ways:

1.
$$w(hkl) = 1/\sigma^2(|F_o(hkl)|) \qquad (7.109)$$

where the estimated standard deviation, $\sigma|F_o(hkl)|$, is obtained from counting statistics in diffractometer data by the relationship

$$\sigma = \sqrt{N} \tag{7.110}$$

N being related to the total counts, peak, and background, for the given reflection. Sometimes $p|F_o|^2$ is added to the right-hand side of (7.109), where p is adjusted so that $w\Delta^2$ is constant over ranges of $|F_o|$.

2.
$$w(hkl) = \Phi(|F_o|) \tag{7.111}$$

where the function Φ is chosen so that $w\Delta^2$ is again sensibly constant. One weighting scheme is given by

$$w = (A + |F_o| + B|F_o|^2 + C|F_o|^3)^{-1} \tag{7.112}$$

where the constants A, B, and C may be obtained by a least-squares fit of mean values of Δ, in ranges of $|F_o|$, to the inverse of the right-hand side of (7.112).

Precision

The choice of absolute weights (7.109) should yield parameters of lowest variances:

$$\sigma^2(p_j) = (a_{jj})^{-1} \tag{7.113}$$

where $(a_{jj})^{-1}$ is an element of the matrix inverse to that of the a_{ij} elements (7.100). With weights related to $|F_o|$,

$$\sigma^2(p_j) = (a_{jj})^{-1}\frac{\sum_h w\Delta^2}{m-n} \tag{7.114}$$

where m is the number of reflections and n the number of parameters to be refined. Generally, $m \geqslant 5n$ is a satisfactory relationship. In a block-diagonal approximation, the standard deviations are usually underestimated by 15–20%.

Atoms in Special Positions

If any symmetry operation of the space group of a structure leaves an atom invariant, the atom is on a special position. Consider one molecule of formula AB_2 in the unit cell of space group $P2$. The A atoms occupy, say, the special positions $0, y, 0$, and the B atoms occupy the general positions x, y, z and $\bar{x}, y, \bar{z}$. Several important points arise:

1. The x and z coordinates of atoms A remain invariant at zero during refinement.

2. In this space group, the origin is along the 2-fold axis y and must be specified by fixing the y coordinate of an atom; the heavier the atom the better. It could be $y_B = 0$, in which case this parameter also remains invariant.
3. With respect to the symmetry operations of the space group, atoms A must be given an atom multiplicity factor of $\frac{1}{2}$ so that a total of one A atom per unit cell obtains.
4. In an isotropic refinement, one of the three 2-fold axes of the biaxial thermal ellipsoid must lie along the 2-fold axis y of the space group. In this case, the U_{12} and U_{23} elements of $\mathbf{U}$ remain invariant at zero value.

7.4.3 Strategy in Least-Squares Refinement

The calculations of least squares, although lengthy, have been implemented in now well tested program systems, such as WinGX and SHELX, that are available to workers in structure determination. In a number of instances, the application of least-squares calculations is straightforward, and results of sufficient precision are obtained. In other cases, and perhaps more generally, a consideration of the strategy to be employed, within the constraints of the program system, is necessary and desirable. In the structure determination process, a model is proposed on the basis of the diffraction data, and then subjected to adjustments (refinement) of the parameters of that model so as to reach a solution with a defined precision.

Model

For a least-squares refinement to be successful, the model must be sufficiently close to the truth and contain all the elements of the structure. The multidimensional surface that corresponds to an n-dimensional refinement is complex and contains many false minima, into any of which an insufficiently correct model may converge. Hence, the correctness of a structure should be based on several criteria (see Section 7.7).

Data Errors

Experimental data are subject to systematic and random errors. In x-ray diffraction, a systematic error may be, for example, the lack of an absorption correction: the values of $|F_o|$ will often be less than those of $|F_c|$ for the partially refined structure, noticeably for the low-order reflexions of high intensity. The minimization of the difference between $|F_o|$ and $|F_c|$ will lead, in this case, to unwarranted discrepancies in the temperature factors of the atoms and the scale factor of the data.

Increasing the temperature factors and decreasing G, the scale factor applied inversely to $|F_c|$ during refinement, will tend to decrease the $|F_c|$ values so as to compensate for the effect of absorption. Random errors are unavoidable in any physical experiment, but they can be minimized by careful attention to the experimental procedure. For example, reflections of low intensity and consequent high probable error can be improved in reliability by increasing the measurement time. This procedure may be costly, but is very straightforward with a diffractometer.

The data set must be sufficient in size for the work in hand. The success of least-squares refinement is partially dependent on the fact that the calculations are appreciably overdetermined. This excess of data over variables is needed in order to average out discrepancies in individual measurements. As an approximate guide, the ratio of data to variables should be at least 8. The ratio can be improved by decreasing the number of variables. There are instances, such as with phenyl rings, where the refinement of the parameters of the hydrogen atoms might be regarded as exaggerated. The hydrogen atoms can, and should, be allowed to contribute to the calculated structure factors, but unless there is reason to suppose otherwise they would be expected to display sp^2 geometry, with C–H = 1.00 Å. Their isotropic temperature factors may be refined, or assumed to be, say, 1.3 times that of the carbon atoms to which they are attached. In a structure like euphenyl iodoacetate (see Section 1.1), $C_{32}H_{53}O_2I$, the number of variables may be very significantly reduced by this type of approach. The strategy should be determined by the nature of the problem in hand, and computer programs that handle least-squares refinement have been designed to accommodate a range of constraints.

If the data-to-variables ratio is too low, the final structure may contain an inherent lack of resolution that may not be immediately apparent from the numerical least-squares results, although all the expected atomic positions will be represented, provided that the model itself was complete. Least squares, unlike the Fourier process, cannot find anything that is not present in the model. It will obtain a best fit, under the given strategy, for the model supplied. A lack of resolution may arise because of an unsatisfactory termination of the data set. If the cut-off criterion for acceptable data leads to too few reflections, those that have been excised could be subjected to re-measurement for longer times so that they can be accepted under the same criterion. If this is not done, then imperfections will exist in the refinement, and may be manifested in large estimated standard deviations for some parameters or in unsatisfactory temperature factors. Incorrect B values may lead to highly improbable root mean square atomic displacements or to U_{ij} values that are nonpositive definite, that is, they do not define an ellipsoid.

There is some concern with published papers on structure determination that report 50% or more of the unique experimental $|F_o|$ data rejected by some criterion, such as $I \leqslant 3\sigma(I)$, apparently to the satisfaction of both the worker and the journal editor. The cosmetic effect of this practice serves merely to reduce the R factor rather than to improve the esds of the structural parameters. The weak intensity data do, in fact, contain structural information. Protein crystallographers, who of necessity work with poorly diffracting crystal specimens, are not in a position to discard data in this fashion.

Least-Squares Refinement Procedure

The least-squares refinement of a model leads to those parameters that minimize $(|F_o| - |F_c|)^2$, as in equation (7.90), over the whole data set. The first application of the calculations will generally not lead to the best-adjusted parameters, and it will be necessary to cycle through the calculations until the results converge; that is, until the calculated shifts are less than the estimated standard deviations of the corresponding parameters. The refinement is continued until the shifts in the parameters are some small fraction (0.1 is often quoted) of the estimated standard deviations.

A good procedure will begin by ensuring the maximum number of good observations, and by minimizing the number of variables consistent with the requirements of the problem. The starting model must be sufficiently good so that the minimization does not fall into a local, false position. This is unlikely if Fourier synthesis or difference-Fourier synthesis has been used in establishing the model and approximate scale and temperature factors have been obtained by statistical methods. Although Fourier methods of refinement are less convenient than those of least squares, they do have the power of revealing necessary information that may not be contained in an initial model. From the above, it should be clear that such information is vital to a good least-squares refinement.

Another constraint that may be considered, as well as the fixed C–H geometry, is the rigid-body specification of a group of atoms. For example, unless there is any reason to suppose otherwise, a phenyl ring can be considered to obey the geometry of Tables 7.24 and 7.25 (or other preferable compilation) and thus be refined as a single, rigid entity.

The least-squares equations outlined in Section 7.4.2 include a weight for each observation. They may be unity or calculated from (7.104) for the early stages of refinement. Subsequently, weighting schemes based on $|F_o|$ and/or $\sin \theta$ may be used. The validity of any weighting scheme should be checked for the given problem, as already indicated. Attention to detail in least-squares refinement is generally rewarding, and it is worth remembering that it consumes about three quarters of the total calculation time of a structure analysis.

7.5 Molecular Geometry

7.5.1 Bond Lengths and Angles

When the formal structure analysis is complete, we need to express our results in terms of molecular geometry and packing. This part of the analysis

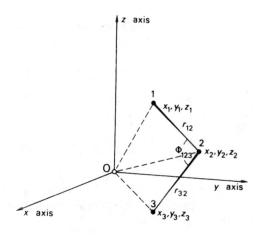

FIGURE 7.24. Geometry of the calculation of interatomic distances and angles; points 1, 2, and 3 represent atomic positions.

includes the determination of bond lengths, bond angles, and intermolecular contact distances, with measures of their precision.

Consider three atoms with fractional coordinates x_1, y_1, z_1; x_2, y_2, z_2, and x_3, y_3, z_3 in a unit cell of sides a, b, and c (Figure 7.24). The vector $\mathbf{r}_j$ from the origin O to any atom j is given by

$$\mathbf{r}_j = x_j\mathbf{a} + y_j\mathbf{b} + z_j\mathbf{c} \tag{7.115}$$

The vector $\mathbf{r}_{12}$ is given by

$$\mathbf{r}_{12} = \mathbf{r}_2 - \mathbf{r}_1 \tag{7.116}$$

or, using (7.115),

$$\mathbf{r}_{12} = (x_2 - x_1)\mathbf{a} + (y_2 - y_1)\mathbf{b} + (z_2 - z_1)\mathbf{c} \tag{7.117}$$

Forming the dot product of each side with itself, remembering that

$$\mathbf{p} \cdot \mathbf{q} = pq \cos \widehat{\mathbf{pq}} \tag{7.118}$$

we obtain

$$
\begin{aligned}
r_{12}^2 = {} & (x_2 - x_1)^2 a^2 + (y_2 - y_1)^2 b^2 + (z_2 - z_1)^2 c^2 \\
& + 2(y_2 - y_1)(z_2 - z_1)bc \cos \alpha + 2(z_2 - z_1)(x_2 - x_1)ca \cos \beta \\
& + 2(x_2 - x_1)(y_2 - y_1)ab \cos \gamma
\end{aligned}
\tag{7.119}
$$

This equation may be simplified for crystal systems other than triclinic. Thus, if the atoms exist in a tetragonal unit cell,

$$r_{12}^2 = [(x_2 - x_1)^2 + (y_2 - y_1)^2]a^2 + (z_2 - z_1)^2 c^2 \qquad (7.120)$$

In a similar manner, we can evaluate r_{32} (Figure 7.24).

Using (7.118), for the tetragonal system,

$$\cos \Phi_{123} = \frac{[(x_2 - x_1)(x_2 - x_3) + (y_2 - y_1)(y_2 - y_3)]a^2 + (z_2 - z_1)(z_2 - z_3)c^2}{r_{12}r_{32}}$$

$$(7.121)$$

where r_{12} and r_{32} are evaluated following (7.120). Similar equations enable any distance or angle to be calculated, in any crystal system, in terms of the atomic coordinates and the unit-cell dimensions.

When the asymmetric unit of a crystal contains more than one copy of a given molecule, or when similar molecules occur in different crystals, the question arises of whether or no the several sets of molecular dimensions are significantly different. The statistical test applicable in this situation is the chi-square $(\chi)^2$ test. It involves calculation of $\sum_{i=1}^{n}[\Delta_i/\sigma\Delta_i]^2$, which is distributed as χ^2 with n degrees of freedom; Δ_i is the difference between one measured property, such as a bond length, in a pair of molecules and $\sigma\Delta_i$ is the standard deviation in Δ_i, estimated typically as $[\sigma^2(d_{i1}) - \sigma^2(d_{i2})]^{1/2}$, assuming no correlation between d_{i1} and d_{i2}, and n is the number of pairs of measurements.

The significance of the result can be tested in the usual way[a] by making the null hypothesis that all of the differences can be accounted for by random errors in the experimental procedures, and then obtaining from statistical tables the significance level of the test, that is, the probability of incorrectly rejecting a good hypothesis. Normally, the test is not regarded as significant unless $\alpha \leqslant 0.05$ (see Section 6.5.1).

7.5.2 Torsion Angles

Torsion angles are useful conformational parameters with which to compare different, related molecules or, indeed, different conformations of one and the same molecule. In a freely rotating moiety, a torsion angle may be a function of the environment of the molecule. Consider an arrangement of four atoms 1, 2, 3, 4 (Figure 7.25). The torsion angle $\chi(1, 2, 3, 4)$ is defined by the angle between the planes 1, 2, 3 and 2, 3, 4, and lies in the range $-180° < \chi \leqslant 180°$; the sign is an important property of the parameter.

In the planar, eclipsed conformation shown in Figure 7.25, χ is zero. The torsion angle is the amount of rotation of 1, 2 about 2, 3 and, looking along the

[a] Hussain, Palmer, and Tickle, loc. cit.

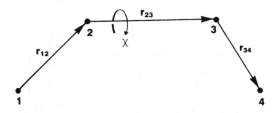

FIGURE 7.25. Torsion angle $\chi(1, 2, 3, 4)$: the torsion angle has a positive sign for a rotation of 1, 2, about 2, 3 as shown, that is, it is positive if, looking along $\mathbf{r}_{23}$, a clockwise rotation is required to bring atom 1 into atom 4.

direction $2 \rightarrow 3$; a positive value of χ corresponds to the clockwise rotation that brings atom 1 into atom 4. Let

$$\mathbf{p}_1 = \mathbf{r}_{23} \times -\mathbf{r}_{12} \tag{7.122}$$

and

$$\mathbf{p}_2 = \mathbf{r}_{23} \times \mathbf{r}_{34} \tag{7.123}$$

Then

$$\chi(1, 2, 3, 4) = \cos^{-1}\left(\frac{\mathbf{p}_1 \cdot \mathbf{p}_2}{p_1 p_2}\right) \tag{7.124}$$

and the sign of χ is that of $\mathbf{p}_1 \times \mathbf{p}_2 \cdot \mathbf{r}_{24}$. Torsion angle calculation is provided by the program MOLGOM* in the CD program suite (see Chapter 11) and by programs such as SHELX-97.

7.5.3 Conformational Analysis

Confusion has arisen in the literature over the use of torsion angles in conformational analysis. It is often convenient to quote values of torsion angles as lying within certain ranges. For example, $\chi \approx 0°$ may be called *cis*, $\chi = 180°$ is *trans* (*t*), and $\chi \approx \pm60°$ is $\pm gauche$ ($\pm g$). However, because of changing conventions, it is best to quote the actual value of χ, and to state how it is defined (Figure 7.25). This procedure will minimize ambiguities in the future.

Ring Conformations

Two types of symmetry (or pseudosymmetry) must be considered in order to define ring conformations[a] namely mirror planes perpendicular to the dominant ring plane and 2-fold axes lying in the ring plane. If there is an odd number (usually 5 or 7) of atoms in the ring, all symmetry elements pass through one of

---

[a] W. L. Duax et al., *Atlas of Steroid Structure*, Vols. 1 and 2, New York, IFI/Plenum Data Company (1975, 1984).

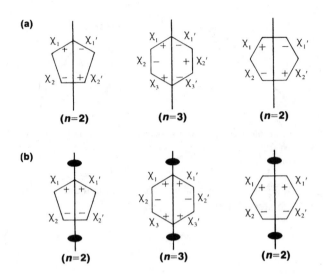

FIGURE 7.26.   Conformations in five- and six-membered rings: (a) Torsion angles related by mirror planes (—) have opposite signs, and (b) angles related by 2-fold rotation axes (Φ) have the same sign.

the ring atoms and bisect the opposite bond (Figure 7.26). In rings containing an even number of atoms (usually 6), symmetry elements may pass through two ring atoms located directly across the ring, or else bisect two opposite ring bonds.

Ten symmetry elements are possible in five-membered rings. The *planar* five-membered ring possesses all 10, five mirror planes and five 2-fold axes. The ideal *envelope* conformation has only a single *m* plane, and it passes through the out-of-plane atom. The ideal *half-chair* has one 2-fold axis bisecting the bond between the two out-of-plane atoms.

Six-membered rings possess 12 locations for symmetry elements. In determining the ring conformation, we can ignore the 2-, 3-, and 6-fold collinear rotation axes perpendicular to the ring plane. Figure 7.27 illustrates the symmetry elements that define the ideal forms of commonly observed conformations. The planar ring, such as in benzene, has one *m* plane and one 2-fold axis at each of six locations ($6/m\ mm$). The *chair* form of cyclohexane has three *m* planes and three 2-fold axes ($\bar{3}m$). The *boat* and *twist-boat* have symmetry $mm2$ and 222 respectively, while the *sofa* has symmetry *m* and the *half-chair* symmetry 2.

Asymmetry Parameters

Once the atom coordinates are available, torsion angles can be calculated. Because of errors in the data and for stereochemical reasons, a particular cyclic structure will often depart from its ideal symmetry. The degree of this departure,

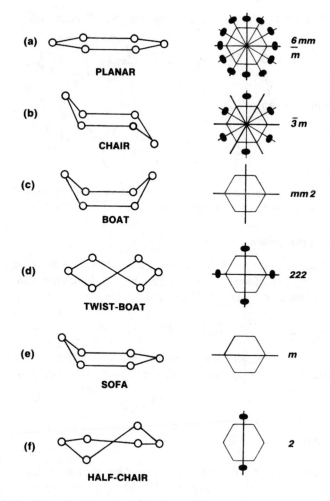

FIGURE 7.27. Commonly observed conformations of six-membered rings. The mirror and 2-fold rotational symmetries are indicated on the right.

its asymmetry, may be calculated in terms of asymmetry parameters. For this purpose, related or possibly related torsion angles are compared in a way that will result in a value of zero for a parameter if the corresponding symmetry is realized in the molecule. Mirror-related torsion angles have equal magnitude but opposite sign, and such torsion angles are compared by addition. The torsion angles related by 2-fold symmetry are equal in both magnitude and sign, and are compared by subtraction. The rms value of each discrepancy yields a measure of the deviation from ideal symmetry at the location in question. We calculate

$$\Delta C_s = \left(\frac{\Sigma_{i=1}^{n} (\chi_i + \chi_i')^2}{n} \right)^{1/2} \tag{7.125}$$

in respect of m symmetry, and

$$\Delta C_2 = \left(\frac{\Sigma_{i=1}^{n} (\chi_i - \chi_i')^2}{n} \right)^{1/2} \tag{7.126}$$

in respect of 2-fold symmetry; n is the number of individual comparisons, and χ_i and χ_i' are the related torsion angles in question.

7.5.4 Mean Planes

In discussing the geometry of a molecule, it may be desirable to test the planarity of a group of atoms. For a number n ($n > 3$) of atoms, the best plane may be obtained by the method of least squares. Let the plane be given by

$$PX + QY + RZ + S = 0 \tag{7.127}$$

The constants P, Q, R, and S are obtained through a procedure similar to that expressed by (7.88)–(7.92). It is convenient to work with Cartesian coordinates (see Appendix A7).

7.6 Precision

Closely related to the calculations of bond lengths and angles is the expression of the precision of these quantities. The least-squares refinement procedure establishes values for an estimated standard deviation (esd) in each of the variables used in these calculations. Thus, a fractional coordinate of 0.3712 might have a standard deviation of 0.0003, written as 0.3712(3).

We need to know further how errors are propagated in a quantity which is a function of several variables, each of which contains some uncertainty arising from random errors. The answer is provided by the statistical principle of superposition of errors.

Let q be a function of several variables $p_i (i = 1, 2, 3, \dots, N)$, with known standard deviations $\sigma(p_i)$. Then the esd in q is given by

$$\sigma^2(q) = \sum_{i=1}^{N} \left[\frac{\partial q}{\partial p_i} \sigma(p_i) \right]^2 \tag{7.128}$$

A simple example may be given, through (7.117) and (7.128), for a bond between two atoms lying along the c edge of a tetragonal unit cell. Let c be 10.06(1) Å, z_1 be 0.3712(3), and z_2 be 0.5418(2). From (7.117),

$$r_{12} = (z_2 - z_1)c = (0.5418 - 0.3712)10.06 = 1.716 \text{ Å} \qquad (7.129)$$

and from (7.128),

$$\sigma^2(r_{12}) = (0.5418 - 0.3712)^2(0.01)^2$$
$$+ (10.06)^2(0.0002)^2 + (10.06)^2(0.0003)^2 \qquad (7.130)$$

Thus, $\sigma(r_{12})$ is 0.004 Å and we write $r_{12} = 1.716(4)$ Å. Similar calculations may be used for all distance and angle calculations in all crystal systems, but the general equations are quite involved numerically and best handled by computer methods.

The esd of a torsion angle can be calculated along the lines given in this section; the significances of differences between torsion angles may be as important as the differences themsleves. Again, in a discussion of best-plane results it is essential to evaluate the perpendicular distances (deviations) of atoms from the plane, and their esds. Let the jth atom have the coordinates X_j, Y_j, Z_j. Then it is a simple exercise in coordinate geometry to show that the deviation of this atom from the best plane (7.127) is given by

$$\Delta_j = (PX_j + QY_j + RZ_j + S)/K \qquad (7.131)$$

where K is given by

$$K = (P^2 + Q^2 + R^2)^{1/2} \qquad (7.132)$$

To obtain the esd in Δ_j the esds in X_j, Y_j, and Z_j are first obtained, from Appendix A7 and (7.128). Then

$$\sigma(\Delta_j) = \{[P\sigma(X_j)]^2 + [Q\sigma(Y_j)]^2 + [R\sigma(Z_j)]^2\}^{1/2}K \qquad (7.133)$$

7.7 Correctness of a Structure Analysis

At this stage we may summarize four criteria of correctness of a good structure analysis. If we can satisfy these conditions in one and the same structure model, we shall have a high degree of confidence in it.

1. There should be good agreement between $|F_o|$ and $|F_c|$, expressed through the R factor. Ultimately, R depends upon the quality of the experimental data. At best, it will probably be about 1% greater than the average standard deviation in $|F_o|$. Assuming the desirable situation that two or more asymmetric units of data have been collected, a value R close to that for R_{int} for the data is acceptable.

2. The electron density map should show neither positive nor negative density regions that are unaccountable, other than Fourier series termination errors.

3. The difference-Fourier map should be relatively flat. This map eliminates series termination errors as they are present in both ρ_o and ρ_c. Random errors produce small fluctuations on a difference map, but they should be less than 2.5–3 times the standard deviation of the electron density $\sigma(\rho_o)$:

$$\sigma(\rho_o) = \frac{1}{V_c} \left[\sum_{hkl} (\Delta F)^2 \right]^{1/2} \tag{7.134}$$

where $\Delta F = |F_o| - |F_c|$ and the sum extends over all symmetry-independent reflections.

4. The molecular geometry should be chemically sensible, within the limits of current structural knowledge. Abnormal bond lengths and angles may be correct but they must be supported by strong evidence of their validity in order to gain acceptance. Normally a deviation of less than three times the corresponding standard deviation is not considered to be statistically significant.

As a guide to the interpretation of acceptable stereochemistry, we include selections of ionic radii, bond lengths, and bond angles in Tables 7.23, 7.24, and 7.25, respectively (see also Table 11.4).

7.7.1   Data Bases

TABLE 7.23.   Selected Ionic Radii (Å) Referred
to Coordination Number 6[a]

Ag^+	1.27	Hg^{2+}	1.02	Sr^{2+}	1.32
Al^{3+}	0.54	K^+	1.44	Th^{4+}	0.94
Ba^{2+}	1.49	La^{3+}	1.03	Ti^{2+}	0.86
Be^{2+}	0.48				
Ca^{2+}	1.18	Li^+	0.86	Ti^{4+}	0.61
Cd^{2+}	0.95	Mg^{2+}	0.87	Tl^+	1.54
Ce^{3+}	1.01	Mn^{2+}	0.83	Zn^{2+}	0.74
Ce^{4+}	0.87	Na^+	1.12	NH_4^+	1.66
				H	1.39
Co^{2+}	0.75	Ni^{2+}	0.69	Br^-	1.87
Co^{3+}	0.61	Pb^{2+}	1.19	Cl^-	1.70
Cr^{3+}	0.62	Pd^{2+}	0.86	F^-	1.19
Cs^+	1.84	Pt^{2+}	0.80	I^-	2.20
Cu^+	0.77	Pt^{4+}	0.63	O^{2-}	1.40
Cu^{2+}	0.73	Ra^{2+}	1.43	S^{2-}	1.70
Fe^{2+}	0.78	Rb^+	1.58	Se^{2-}	1.81
Fe^{3+}	0.65	Sn^{2+}	0.93	Te^{2-}	1.97

[a] The change in an ionic radius from coordination number 6 to coordination numbers 8, 4, 3, and 2 is approximately 1.5%, −1.5%, −3.0%, and −3.5%, respectively.

TABLE 7.24.   Selected Bond Lengths $(\text{Å})^a$

Formal single bonds				Formal double bonds			
C4–H	1.09	C3–C2	1.45	C3–C3	1.34	C2–O1	1.16
C3–H	1.08	C3–N3	1.40	C3–C2	1.31	N3–O1	1.24
C2–H	1.06	C3–N2	1.40	C3–N2	1.32	N2–N2	1.25
N3–H	1.01	C3–O2	1.36	C3–O1	1.22	N2–O1	1.22
N2–H	0.99	C2–C2	1.38	C2–C2	1.28	O1–O1	1.21
O2–H	0.96	C2–N3	1.33	C2–N2	1.32		
				Formal triple bonds			
C4–C4	1.54	C2–N2	1.33	C2–C2	1.20	N1–N1	1.10
C4–C3	1.52	C2–O2	1.36	C2–N1	1.16		
C4–C2	1.46	N3–N3	1.45				
C4–N3	1.47	N3–N2	1.45				
				Aromatic bonds			
C4–N2	1.47	N3–O2	1.36	C2–C3	1.40	N2–N2	1.35
C4–O2	1.43	N2–N2	1.45	C2–N2	1.34		
C3–C3	1.46	N2–O2	1.41				

^{a} The notation in the table indicates the connectivity of the atoms.

TABLE 7.25.   Selected Bond Angles

Atom	Geometry	Angle (degrees)
C4	Tetrahedral	109.47
C3	Planar	120
C2	Bent	109.47
C2	Linear	180
N4	Tetrahedral	109.47
N3	Pyramidal	109.47
N3	Planar	120
N2	Bent	109.47
N2	Linear	180
O3	Pyramidal	109.47
O2	Bent	109.47

Tables of standard (average) bond lengths and angles (with esds) are useful aids to structure determination. In any research or advanced study, it is necessary to take cognizance of all work in the given field that has already been published. The number of crystal structures that has been solved and published is now vast, and data files have been constructed that can be interrogated by computer. The best known of these is the Cambridge Structural Database (CSD).

This database contains the results of both x-ray and neutron diffraction studies on organic and organometallic compounds. About 250,000 structures are now filed therein, and the database is available in about 25 countries. (see also Appendix 10).

Not every structure in the above classes that has been ever published will be found in the CSD. Each entry has to pass a scrutiny that involves such checks as the consistency between the published coordinates and bond lengths. The information in the CSD falls into three categories, namely, bibliographic, connective, and crystallographic. The bibliographic file contains information such as the chemical name, the type of structure analysis, the chemical class, the molecular formula, and relevant literature for a given compound. The connective file contains chemical structural formulae encoded as atom and bond parameters, and the crystallographic file contains parameters relevant to the crystal structure data and its solving.

Retrieval from the database is flexible, and the software permits many different types of search, such as on chemical name, formula, or class, and compounds containing specific chemical fragments can be sought. The results of searches can be recorded by printing and plotting techniques.

Similar databases have been organized for proteins by the Brookhaven National Laboratory, for inorganic structures by the University of Bonn, and for metals by the National Research Council of Canada. These, too, are available in several countries.

7.8   Limitations of X-ray Structure Analysis

There are certain things which x-ray analysis cannot do well, and it is prudent to consider the more important of them.

Liquids and gases lack three-dimensional order, and cannot be used in diffraction experiments in the same way as are crystals. Certain information about the radial distribution of electron density can be obtained, but it lacks the distinctive detail of crystal analysis.

It is not easy to locate light atoms in the presence of heavy atoms. Difference-Fourier maps alleviate the situation to some extent, but the atomic positions are not precise. Least-squares refinement of light-atom parameters is not always successful, because the contributions to the structure factor from the light atoms are relatively small.

Hydrogen atoms are particularly difficult to locate with precision because of their small scattering power and the fact that the center of the hydrogen atom does not, in general, coincide with the maximum of its electron density. Terminal hydrogen atoms have a more aspherical electron density distribution than do hydrogen-bonded hydrogen atoms, and their bond distances, from x-ray studies, often appear short when compared with spectroscopic or neutron diffraction values. For similar reasons, refinement of hydrogen atom parameters in a structure analysis may be imprecise, and the standard deviations in their coordinate values may be as much as 10 times greater than those for a carbon atom in the same structure. It is, nevertheless, desirable to include hydrogen atom positions in the final structure model. They lead to a best fit, and are useful when comparing the results of x-ray structure determination with those of other techniques, notably nuclear magnetic resonance.

In general, bond lengths determined by x-ray methods represent distances between the centers of gravity of the electron clouds, which may not be the same as the internuclear separations. Internuclear distances can be found from neutron diffraction data, because neutrons are scattered by the atomic nuclei. If, for a given crystal, the synthesized neutron scattering density is subtracted from that of the x-ray scattering density, a much truer picture of the electron density can be obtained (see Section 6.6.2).

7.9   Disorder in Single Crystals

A typical small-molecule analysis involves less than about 100 nonhydrogen atoms in the asymmetric unit. With Cu $K\alpha$ radiation and to $\theta \leqslant 70°$, it would be

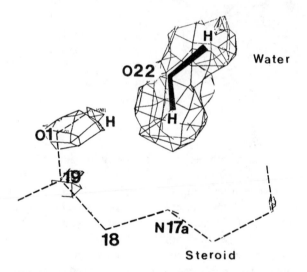

FIGURE 7.28.   Difference electron density map for the azasteroid HS626, showing the hydrogen atom on O1 (hydroxyl) and the two water-hydrogen atoms, none of which was included in the structure factor calculation. The steroid molecule, part of which is shown by the dashed line, has been subtracted out in the difference synthesis.

Note: This and the next five figures are electron density and difference electron density maps photographed from the screen of an Evans and Sutherland Picture System 2 cathode ray tube display unit coupled to a computer that holds the electron density data. The interactive computer graphics system is programmed such that the user can simulate a three-dimentsional effect by rotating the map about one or more of three mutually perpendicular axes. The contouring of the maps encloses the electron density in a cage of "chicken wire" hoops running in several directions. Unlike the sectional contour maps used elsewhere in this book, only one contour level is used, selected so as to optimize the desired features of electron density.

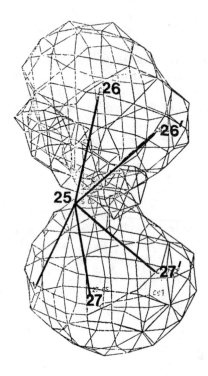

FIGURE 7.29.   Electron density at the end of the cholesteryl side chain of HS650 (molecule B). The density is smeared out, and at least two stereochemically sensible positions for $C25\genfrac{}{}{0pt}{}{C26}{C27}$ can be fitted to the density, as indicated.

expected to lead to the determination of bond lengths with esds of about 0.005 Å and of bond angles with esds of about 0.2°. Isotropic thermal parameters (U_{iso})[a] for nonhydrogen atoms usually range from 0.050 to 0.090 Å^2 and may have esds from 0.003 to 0.007 Å^2. However, it is sometimes found that the refined thermal parameters for certain atoms in a structure have atypical values. For example, U_{iso} may increase progressively and significantly toward the end of a chainlike moiety compared to the more rigid areas of the structure. The obvious and reasonable physical interpretation is simply that the atoms near the end of the chain experience greater thermal motion than do the atoms in the bulk of the molecule. For example, the hydroxyethyl side-chain atoms of an azasteroid derivative[b] have the following U_{iso} parameters:

$$\text{>N(0.058)–C(0.073)H}_3\text{–C(0.190)H}_2\text{–O(0.176)H}$$

[a] $U_{\text{iso}} = 8\pi^2\overline{U^2}\sin\theta/\lambda^2$, where $\overline{U^2}$ is the mean-square amplitude of vibration of an atom.

[b] A. I. El-Shora, R. A. Palmer, H. Singh, T. R. Bhardwaj, and D. Paul, *Journal of Crystallographic and Spectroscopic Research* **14**, 89 (1984).

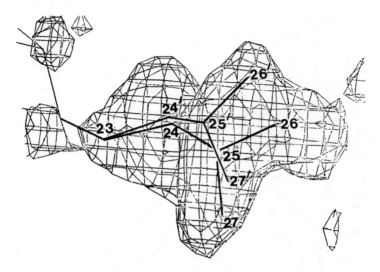

FIGURE 7.30. As for Figure 7.29, but with molecule A of HS650, in which the disorder in the side chain is more extensive and encompasses C23 to C27. In this structure the side chains are loosely held, having little contact with neighbouring molecules in the crystal.

In this analysis all atoms, including those in the side chain, were resolved and refined successfully by least squares.

Atoms in solvent of crystallization molecules may exhibit high thermal parameters, and for similar reasons. Exceptions occur from time to time, and in the above example a well-resolved solvent water-oxygen atom had a refined U_{iso} value of 0.088 Å^2 and was so well ordered that its hydrogen atoms were clearly located in a ΔF map (Figure 7.28). In this particular case, the clarity of definition in the electron density is associated with the formation of two strong hydrogen bonds donated by each of the water hydrogen atoms holding it firmly in position. However, in the structure of another steroid derivative,[a] the carbon atoms of the side chains were so badly disordered that some atoms were not resolved in the difference electron density and appeared as diffuse patches (Figures 7.29 and 7.30). Such disorder is probably of a statistical nature, with the atoms taking up slightly different positions from one unit cell to another. The effect can be compensated, albeit somewhat artificially, by the refinement of the isotropic thermal parameters assigned to the atoms concerned. In the example, the U_{iso} values are 3–6 times greater than those of the ordered atoms in the structure.

[a] J. Husain, I. J. Tickle, R. A. Palmer, H. Singh, and K. K. Bhutani, *Acta Crystallographica* B**38**, 2845–2851 (1982).

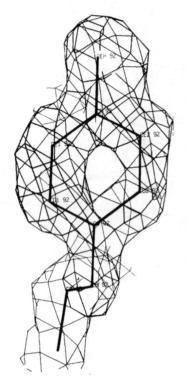

FIGURE 7.31. Figures 7.31–7.33 are extracts from the electron density map of ribonuclease-A at 1.45 Å resolution. The maps are calculated with the coefficients $|F_o| + (|F_o| - |F_c|) = 2|F_o| - |F_c|$ and thus show features of both the electron density and the difference electron density. This figure shows a tyrosyl residue $(HOC_6H_4CH_2CH\diagdown)$, with the hole of the phenyl ring and the −OH group (on the top) clearly indicated.

Disorder may also arise by groups of atoms either in free rotation in the solid state (dynamic disorder) or in more than one position of similar energy (static disorder). Methyl groups in large organic molecules often show this type of behavior. It may be possible to distinguish between dynamic and static disorder by a complete re-examination of the structure at a much reduced temperature.

Protein structures are of particular topical interest, and recent innovations in this field include the development of techniques for refining structures.[a] The molecules involved in protein analysis are very large, typically with more than 1000 nonhydrogen atoms in the asymmetric unit. Consequently, the crystals have large unit cells, and many possible x-ray reflections occur within a given θ range compared to small-molecule crystals. It is customary to limit severely the maximum θ value during the course of a protein structure analysis, depending on the particular stage reached. Corresponding to a given maximum θ value (θ_{max}) there is a minimum d value (d_{min}) (given by $\lambda/2 \sin \theta_{max}$), and it is customary to

[a] D. S. Moss and A. Morffew, *Computational Chemistry* **6**, 1–3 (1981).

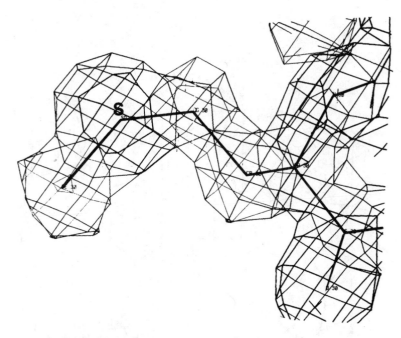

FIGURE 7.32. Methionyl residue ($CH_3SCH_2CH_2CH\langle$) showing the outstanding electron density region around the sulfur atom.

speak of the d_{min} of an analysis as the (nominal) resolution. The protein analysis may proceed through stages of progressively higher resolution—for example, 6 Å, 3 Å, 2.5 Å, 2 Å, and 1.5 Å—the electron density image undergoing gradually improved mathematical focusing in the process. In addition to the large quantity of data associated with protein analysis, there is a further technical problem, which limits the quality of most studies. During the crystallization process from solution, solvent molecules (typically 40–60% by weight) are trapped in the structure: the protein molecules almost float in a solvated crystalline state (see for example Blundell and Johnson),[a] and consequently many regions of electron density in the protein structure may be subject to the type of disorder described above. Even in a good high-resolution analysis, protein data rarely extend beyond 1.5 Å, and individual atoms in the protein molecule will not be revealed. A protein structure refinement involves the use of both least-squares analysis and geometrically constrained positioning of groups in order to produce a plausible model.

[a] Bibliography, Chapter 6.

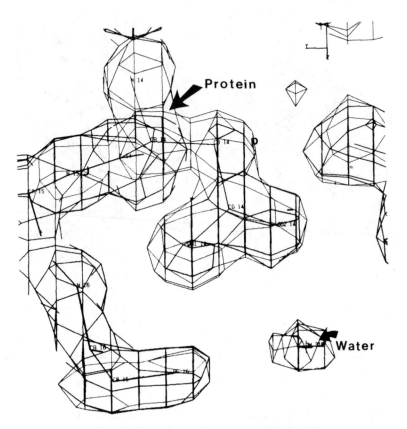

FIGURE 7.33.   This electron density portion shows a clearly resolved solvent molecule (water), not included in the structure factor calculation.

We conclude this section with an example of electron density determined in the high-resolution x-ray analysis of the enzyme ribonuclease,[a] a small protein of molecular weight 13,700. In the first example, a tyrosine residue (Figure 7.31) is seen at 1.45 Å resolution as a hollow ring of density and, although the individual atoms are not resolved, the shape of the density image is strikingly good. As would be expected, the sulfur atom of a methionine residue (Figure 7.32) is quite outstanding, but it does not swamp the rest of this slender aliphatic side chain. At this resolution, the high quality of the refined analysis is evident in the appearance of resolved, solvent (water) molecules, as shown in Figure 7.33. Further practical details of protein analysis are to be found in Chapter 10.

[a] N. Borkakoti, D. S. Moss, M. J. Stanford, and R. A. Palmer, *Journal of Crystallographic and Spectroscopic Research* **14**, 467 (1984).

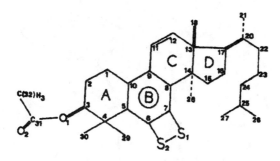

FIGURE 7.34.

7.10   Computer Prediction of Crystal Structures

An early use of theoretical calculations in crystal structure studies was carried out on 3β-acetoxy-6,7-epidithio-19-norlanosta-5,7,9,11-tetraene; the dithiete structure (Figure 7.34) was unusual[a,b]. CNDO/2 calculations were carried out on a model consisting of ring B (C_6S_2), for a range of C_7–S_1, C_6–S_2, C_7–S_1–S_2 and C_6–S_2–S_1 with the following results for the minimum energy conformation[c]:

	Calculated	Experimental
C_7–S_1	1.74 Å	1.767(4) Å
C_6–S_2	1.76 Å	1.782(4) Å
C_7–S_1–S_2	100°	103.3(3)°
C_6–S_2–S_1	99°	100.5(3)°

A compound corresponding to the (C_6S_2) fragment cannot be isolated, as it dimerizes through S—S bonding[d]. The dimer, dibenzo[c,g]tetrathiocin, could exist in a rigid *trans* form (Figure 7.35), or in a mobile form with a range of possible conformations: CNDO/2 energy calculations as a function of the dihedral angle between the two rings of the dimer showed the *trans* form to be the most stable (minimum energy) conformation.

7.10.1   Crystal Structure of 5-Azauracil

Recent work involves computer programs that fit structures to a given molecular conformation and minimise the lattice energy of the chosen structure model

[a] R. B. Boar, D. W. Hawkins, J. F. McGhie, S. C. Misra, D. H. R. Barton, M. F. C. Ladd, and D. C. Povey, *J. C.. S. Chem. Comm.* 756 (1975).
[b] M. F. C. Ladd and D. C. Povey, *Acta Crystallogr.* B32, 1311 (1976).
[c] M. F. C. Ladd *Unpublished results* (1976).
[d] W. D. Stephens, *PhD Thesis*, Vanderbilt University (1960).

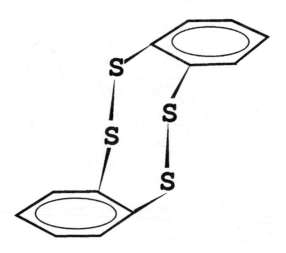

FIGURE 7.35.

so as to obtain an energetic "best fit". As an example, we consider the crystal structure of 5-azauracil, $C_3H_3N_3O_2$[a], a rigid molecular species (at 0 K).

A successful modelling procedure requires an accurate model of the molecule, a formula for the intermolecular force function, and a method for generating close-packed structures. In this example, the molecular structure model was obtained through an energy optimisation of an SCF 6-31G** wavefunction, using the program CADPAC[b]. The model for the electrostatic contribution to the lattice energy involved calculating sets of atomic multipoles derived by a distributed multipole analysis[c] of the MP2 6-31G** wavefunction for 5-azauracil. Other intermolecular forces were represented by an isotropic atom–atom repulsion and attraction (dispersion) potential. Thus the lattice energy U was written as

$$U = \sum_{i \in 1, j \in 2} (A_{pp}A_{qq})^{1/2}\exp[-(B_{pp}B_{qq})R_{ij}/2] - (C_{pp}C_{qq})/R_{ij}^6 \qquad (7.135)$$

where atom i in molecule 1 is of type p, and atom j in molecule 2 is of type q; the best known values[d,e] of the parameters A, B and C for the C, H, O and N atoms

[a] B. S. Potter, R. A. Palmer, R. Withnall, B. Z. Chowdhry and S. L. Price, *J. Mol. Struct.* **485–486**, 349 (1999).

[b] R. D. Amos *et al*, The Cambridge Analytical Derivatives Package, 5th Edition (1992).

[c] D. E. Williams and S. R. Cox, *Acta Crystallogr.* B**40**, 404 (1984).

[d] D. E. Williams and S. R. Cox, *Acta Crystallogr.* B**40**, 404 (1984).

[e] S. R. Cox, L. Y. Hsu and D. E. Williams, *Acta Crystallogr.* A**37**, 293 (1981).

TABLE 7.26.   Crystal Structure Parameter for Modelled Structures and the X-ray Structure

Space group	$-U/\text{kJ mol}^{-1}$	$a/\text{Å}$	$b/\text{Å}$	$c/\text{Å}$	β/deg	$(V_c/Z)/\text{Å}^3$	NH...O/Å	NH...N/Å
$Pbca$	109.4	13.777	9.197	6.814	90	107.9	2.17	1.87
$P2_1/c$	108.2	9.545	7.293	9.710	140.3	107.9	2.51	1.83
$Pbca$	108.1	7.277	9.730	12.243	90	108.3	2.48	1.82
$P2_1$	106.8	4.767	9.717	4.767	80.3	108.8	2.49	1.81
$Pbca$	106.4	7.302	9.694	12.296	90	108.8	2.51	1.81
$Pca2_1$	105.2	13.517	3.682	9.129	90	113.6	2.09	1.86
X-ray	—	6.5135	13.5217	9.5824	90	105.5	2.03	1.99

TABLE 7.27.   Programs for Organic Crystal Structure Prediction [Courtesy S. L. Price, *loc. cit.*, (2003.)]

Program	Type of molecule used in development	Search type
Chin	Crystal engineering – diketopiperazines	Monte Carlo simulated annealing with hydrogen-bonding bias
CRYSTALG	Rigid organics – amides, bases	Self-consistent basin-to-deformed-basin mapping global optimization
CRYSCA	Pigments, organometallics	Random search with steepest descent
ICE9	Rigid hydrocarbons	Systematic grid search to generate close-packed structures
MDCP	Small rigid	Constant pressure molecular dynamics to find crude structures
MOLPAK	Energetic materials, rigid	Systematic search for high density
DMAREL	Rigid polar and hydrogen bonded	structures in common coordination types
MPA, extended to Mpg	Small rigid	Lattman systematic, or random generation of expanded trial unit cell
Perlstein	Moderate sized, semi-flexible organics	Aufbau search for low energy 1D and 2D aggregates, primarily for monolayer predictions
PMC	Hydrocarbons	Symmetry adapted grid systematic
Polymorph Predictor	Flexible organics, including pharmaceuticals	Monte Carlo simulated annealing with intermediate clustering
PROMET	Rigid hydrocarbons	Selecting cohesive dimer, ribbons and layer sub-structures of partial spacegroup
SySe and PP	Pigments	Grid based systematic
UPACK	Sugars and alcohols	Systematic grid or random search, with intermediate clustering

were used. This form of the potential function has been found to be successful in representing other small, rigid C, H, O, N molecular species.

The program MOLPAK[a] was used to select close-packed molecular structures, using a simple hard-sphere repulsion potential function. The lattice energies were calculated with the program DMAREL[b], and minimization of the lattice energy was carried out by a modified Newton-Raphson method that optimised the unit cell dimensions and the rotations and translations of each molecule in the unit cell; the symmetry was generally maintained, although the minimization procedure did not enforce it. The 10 structures with the lower initial lattice energies corresponded to space groups $P\bar{1}$, $P2_1$, $P2_1/c$ and $P2_12_12_1$. Structures in these space groups were minimized, and on the basis of the results MOLPAK generated crystal structures with the less common space groups $P1$, $C2/c$, $Pna2_1$, $Pca2_1$ and $Pbca$. Table 7.26 lists the unit cell dimensions and hydrogen-bond distances, which are very structure-sensitive, for the 6 lower energy-minimized structures, together with the corresponding results from the x-ray study.

The x-ray determination[c] and the best modelled structure (**cab** setting) of 5-azauracil are in good agreement. However, there is only a small energy difference, of the order of $1\text{--}2\,\text{kJ mol}^{-1}$, between the best solution and other close agreements, so that it may be necessary to obtain final confirmation by x-ray methods.

7.10.2 Developments in Computer Crystal Structure Prediction

The state of the art in the computer prediction of organic crystal structures has been discussed in recent authoritative reviews[d,e]. Many programs have been written that attempt to predict structures, and a list of them is given in Table 7.27. The assumption is made that the true crystal structure will correspond to the global minimum lattice energy among the predicted structures. Since the structure is calculated for atoms at rest, that is, at 0 K, the lattice energy will differ from the lattice free energy at ambient temperature by an entropic component, which could affect the choice of structure when several results of similar lattice energy are calculated.

In structures that can exist in polymorphic modifications, the minimum energy conformation should correspond to the thermodynamically most stable form. In practice, however, various factors, such as temperature, rate of crystallization and nature of solvent, could affect the form that is observed. It seems probable that improvements in the precision of the prediction method will involve

[a] J. R. Holden, Z. Y. Dy and H. L. Ammon, *J. Comp. Chem.* **14**, 422 (1993).

[b] D. J. Willcock, S. L. Price, M. Leslie and C. R. A. Catlow, *J. Comp. Chem.* **16**, 628 (1995).

[c] B. S. Potter *loc. cit.* (1999).

[d] S. L. Price in *Encyclopedia of Supramolecular Chemistry*, edited J. Atwood and J. Steed, Marcel Dekker Inc. (2003).

[e] T. Beyer, T. Lewis and S. L. Price, *Cryst. Eng. Comm.* **3**, 178 (2001).

incorporating more accurate forms of the intermolecular potential into the program, as well as allowing for the possibility of kinetic control of crystallization.

While computer prediction clearly has a part to play in crystal structure determination, it will probably remain that x-ray diffraction will be needed for confirmation of the structure and particularly for obtaining accurate molecular geometry.

Bibliography

Crystallographic Computing

AHMED, F. R., HALL, S. R., and HUBER, C. P. (Editors), *Crystallographic Computing*, Copenhagen, Munksgaard (1970).

DIAMOND, R., RAMASESHAN, S., and VENKATESAN, K. (Editors), *Computing in Crystallography*, Bangalore, Indian Academy of Sciences (1980).

FLACK, H. D., PÁRKÁNYI, L., and SIMON, K. (Editors). *International Union of Crystallography symposia on Crystallographic Computing*, No. 6, A Window on Modern Crystallography, Oxford University Press/International Union of Crystallography, Oxford (1993).

MITCHELL, D. J., and LIPPERT, E. L., *Acta Crystallogr.* **18**, 559 (1965).

MITCHELL, E. J., *PhD Thesis*, Vanderbilt University (1965).

MORAS, D., PODJARNY, A. D., and THIERRY, J. C., (Editors). *International Union of Crystallography Symposia on Crystallographic Computing*, No. 5, From Chemistry to Biology, Oxford University Press/International Union of Crystallography, Oxford (1991).

PEPINSKY, R., ROBERTSON, J. M., and SPEAKMAN, J. C. (Editors), *Computing Methods and the Phase Problem in X-Ray Crystal Analysis*, Oxford, Pergamon Press (1961).

ROLLETT, J. S. (Editor), *Computing Methods in Crystallography*, Oxford, Pergamon Press (1965).

Direct Methods

GIACOVAZZO, C., *Direct Methods in Crystallography*, New York, Academic Press (1980).

LADD, M. F. C., and PALMER, R. A. (Editors), *Theory and Practice of Direct Methods in Crystallography*, New York, Plenum Press (1980).

WOOLFSON, M. M., *An Introduction to X-Ray Crystallography*, 2nd ed., Cambridge, Cambridge University Press (1997).

WOOLFSON, M. M., *Direct Methods in Crystallography*, Oxford, Clarendon Press (1961).

Chemical Data

ALLEN, F. H. et al., *Journal of the Chemical Society Perkin Transactions II*, London (1987), pages S1 to S19.

KITAIGORODSKII, A. I., Molecular Crystals and Molecules, Academic Press (1973).

MEGAW, H. D., *Crystal Structures*, Philadelphia, Saunders (1973).

SUTTON, L. E. (Editor), *Tables of Interatomic Distances and Configurations in Molecules and Ions*, London, The Chemical Society (1958; supplement, 1965).

Problems

7.1. Choose three of the following reflections to fix an origin in space group $P\bar{1}$, giving reasons for your choice.

hkl	$|E|$	hkl	$|E|$
705	2.2	$6\bar{1}\bar{7}$	3.2
$42\bar{6}$	2.7	203	2.3
$4\bar{3}2$	1.1	$8\bar{1}\bar{4}$	2.1

Are there any triplets which meet the vector requirements of the Σ_2 formula?

7.2. The geometric structure factor formulae for space group $P2_1$ are

$$A = 2\cos 2\pi(hx + lz + k/4)\cos 2\pi(ky - k/4)$$
$$B = 2\cos 2\pi(hx + lz + k/4)\sin 2\pi(ky - k/4)$$

Deduce the amplitude symmetry and the phase symmetry for this space group according to the two conditions $k = 2n$ and $k = 2n + 1$.

7.3. In space group $P2_1/c$, two starting sets of reflections for the application of the Σ_2 formula are proposed:

	Origin-fixing	Symbols
(a)	041, 117, $\bar{1}23$	242, $\bar{1}62$
(b)	223, 012, $13\bar{7}$	111, 162

Which starting set would be chosen in practice? Give reasons. What modification would have to be made to the starting set if the space group is $C2/c$?

7.4. The following values of $\ln[\Sigma_j f_j^2(hkl)/|F_o(hkl)|^2]$ and $(\sin^2\theta)/\lambda^2$ were obtained from a set of three-dimensional data for a monoclinic crystal. Use the method of least squares (program LSLI*) to obtain values for the scale K (of $|F_o|$) and temperature factor B by Wilson's method.

$\log_e[\Sigma_j f_j^2(hkl)/|F_o(hkl)|^2]$	$(\sin^2\theta)/\lambda^2$
4.0	0.10
5.6	0.20
6.5	0.30
7.9	0.40
9.4	0.50

What is the value of the root mean square atomic displacement corresponding to the derived value of B?

7.5. An orthorhombic crystal contain four molecules of a chloro-compound in a unit cell of dimensions $a = 7.210(4)$, $b = 10.43(1)$, $c = 15.22(2)$ Å. The coordinates of the Cl atoms are

$$\tfrac{1}{4}, y, z; \quad \tfrac{3}{4}, \bar{y}, z; \quad \tfrac{1}{4}, \tfrac{1}{2} + y, \tfrac{1}{2} + z; \quad \tfrac{3}{4}, \tfrac{1}{2} - y, \tfrac{1}{2} + z$$

with $y = 0.140(2)$ and $z = 0.000(2)$. Calculate the shortest Cl $\cdots$ Cl contact distance and its estimated standard deviation.

7.6. The following data give phase indications for the reflection 771 ($|E_h| = 2.2$, $\phi_{\text{calc}} = -14°$) in a crystal of space group $P2_12_12_1$. Determine ϕ_h by both (7.25) and (7.28). For simplicity, let w_h in (7.28) be taken as unity.

k			ϕ_k deg	h − k			ϕ_{h-k} deg	$|E_k||E_{h-k}|$
12	1	0	0	$\bar{5}$	6	1	−37	4.4
7	$\bar{1}$	4	177	0	8	$\bar{3}$	180	5.1
12	0	$\bar{1}$	90	$\bar{5}$	7	2	−144	4.5
12	0	1	90	$\bar{5}$	7	0	−90	3.3
6	1	3	102	1	6	$\bar{2}$	−64	2.7
$\bar{1}$	4	5	−79	8	3	$\bar{4}$	92	3.7

7.7. When employing Patterson Search methods for structure analysis, under what circumstances would you expect the search molecule to be (a) very similar in size to and (b) much smaller than the target molecule? Discuss your answer in some detail.

7.8. Figures 7.14a and 7.14b show unit cells for a hypothetical search model (S) and a target structure (T) respectively. Assuming these structures to be correct, which of the intermolecular vectors indicated in Figure 7.14b will not actually occur in the Patterson for the search molecule? Explain your answer.

7.9. (a) Why would you not use a molecular graphics package alone to generate coordinates for all atoms in the coumarin derivative shown in Figure 7.17a?

(b) The z coordinates of all atoms in the coumarin model built with Chem-X in Table 7.18 are all 0.0000. Why is this so?

(c) The unit cell for the coumarin model in Table 7.18 has three sides each of 100 Å and all angles 90°. Why do you think this is so? [Hint: apparently this is not a real unit cell.]

7.10. Determine graphically the Å coordinates, and hence fractional coordinates of the atoms of a (planar) benzene ring to be used in Patterson Search. The bond lengths are all 1.40 Å and angles 120°. Hints: Construct a regular hexagon with its base parallel to the x (horizontal) axis and its centre at the origin. The Å coordinates of 2 atoms are then quite obvious. The remaining 4 atom positions are generated by symmetry. All z coodinates are of course 0.0. Check your results with the program INTXYZ* (see Chapter 11).

7.11. From the definition of $|E|$, show how a Patterson function with $(|E|^2 - 1)$ values as coefficients leads to a sharpened Patterson function with the origin peak removed.

8

Examples of Crystal Structure Determination

8.1 Introduction

In this chapter we wish to draw together, by means of actual examples, material presented earlier in the book. It may be desirable for the reader to refer back to the previous chapters for descriptions of the techniques used, since we shall present here mainly the results obtained at each stage.

The first two examples can be solved by either the heavy-atom method or direct methods. Nowadays, it is quite commonplace to attempt the solving by direct methods of those structures which, at one time, would have been treated by the heavy-atom method. Where a powerful and sophisticated computer package is available, direct methods frequently provide the most expeditious route to the solution of a structure. However, in order that we may illustrate the methods described, we shall use the heavy-atom technique for the first structure and direct methods for the others.

8.2 Crystal Structure of 2-Bromobenzo[b] indeno[1,2-e] pyran[a]

2-Bromobenzo[b]indeno[1,2-e]pyran (BBIP) is an organic compound which is prepared by heating a solution in ethanol of equimolar amounts of 3-bromo-6-hydroxybenzaldehyde (I) and 2-oxoindane (II) under reflux in the presence of piperidine acetate. The two molecules condense with the elimination of two molecules of water. Upon recrystallization of the product from toluene, it has a m.p. of 176.5–177.0°C. Its molecular formula is $C_{16}H_9BrO$, and its classical structural formula is shown by III.

---

[a] M. F. C. Ladd and D. C. Povey, *Journal of Crystal and Molecular Structure* **2**, 243 (1972).

8.2.1   Preliminary Physical and X-ray Measurements

The compound was recrystallized from toluene by slow, isothermal evaporation of the solvent at room temperature. The crystals were red, with an acicular (needle-shapped) habit, with the forms (subsequently named) {100}, {110}, {001}, and {011} predominant (Figure 8.1). The red color is characteristic of the chromophoric nature of a conjugated double-bond system.

The density of the crystals was measured by suspending them in aqueous sodium bromide solution in a stoppered measuring cylinder in a thermostat bath at 20°C. Water or concentrated sodium bromide solution, as necessary, was added to the suspension until the crystals neither settled to the bottom of the cylinder nor

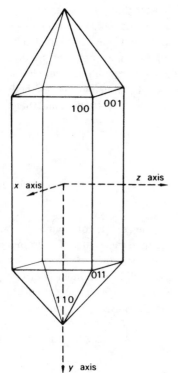

FIGURE 8.1. Crystal habit of BBIP with the crystallographic axes drawn in; the forms shown are {100}, {110}, {001} and {011}.

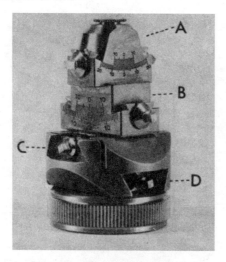

FIGURE 8.2.   Standard goniometer head (by courtesy of Stoe et Cie); *A* and *B* are two arcs for angular adjustments; *C* and *D* are two sledges for horizontal adjustments.

floated to the surface of the solution. Then, the crystals and solution were of the same density, and the density of the solution was measured with a pyknometer. A convenient variant here is to measure the refractive index of the final solution with an Abbe refractometer, having first prepared a graph of refractive index against density from data in the literature.[a]

Under a polarizing microscope, the crystals showed straight extinction on (100) and (001), and oblique extinction (about 3° to a crystal edge) on a section cut normal to the needle axis (y). These observations suggested that the crystals were probably monoclinic.

The crystals chosen for x-ray studies had the approximate dimensions 0.2, 0.4, and 0.3 mm parallel to a, b, and c, respectively. A crystal was mounted on the end of an annealed quartz fiber with "Araldite" (or "Eastman 910") adhesive and the fiber attached to an x-ray goniometer head (arcs) (Figure 8.2) with dental wax. The arcs were affixed to a single-crystal oscillation camera, and the crystal was set with the needle axis accurately parallel to the axis of oscillation, first by eye and finally by x-ray methods. Copper $K\alpha$ radiation ($\bar{\lambda} = 1.5418$ Å) was used throughout the work.

A symmetrical oscillation photograph taken about the b direction is shown in Figure 8.3. The horizontal mirror symmetry line indicates that the Laue group of the crystal has an m plane normal to the needle axis. Further x-ray photographs, for

---

[a] *International Critical Tables* **3**, 80 (1928), **7**, 73 (1930); *Mellor's Treatise on Inorganic and Theoretical Chemistry* **2**, 941 (Supplement, 1961).

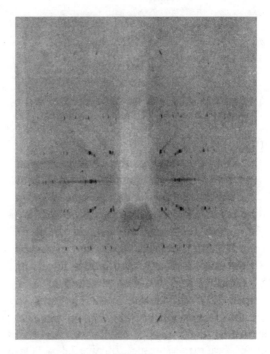

FIGURE 8.3. Symmetrical oscillation photograph taken with the x-ray beam normal to b; the "shadow" arises from the beam stop and holder. The horizontal m line indicates an m plane in the *Laue group* of the crystal, normal to the axis of oscillation.

example, the Laue photograph in Figure 8.4, showed that the only axial symmetry was 2 parallel to b, thus confirming the monoclinic system for BBIP.

Weissenberg photographs are shown in Figures 8.5–8.7. The indexing of the reflections can be understood with reference to Figures 8.8 and 8.9. There are no systematic absences for the hkl reflections, so that the unit cell is primitive (P), but systematic absences do arise for $h0l$ with l odd and for $0k0$ with k odd. These observations confirm the monoclinic symmetry, and the systematic absences lead unambiguously to space group $P2_1/c$.

Measurements on the x-ray photographs gave the approximate unit cell dimensions $a = 7.51$, $b = 5.96$, $c = 26.2$ Å, and $\beta = 92.5°$.

The Bragg angles θ of 20 high-order reflections of known indices, distributed evenly in reciprocal space, were measured to the nearest $0.01°$ on a four-circle diffractometer. From these data, the unit-cell dimensions were calculated accurately by the method of least squares (see Section 7.4ff). The complete crystal data are listed in Table 8.1. The calculated density D_x, for $Z = 4$, is in good agreement with the measured value D_m, which indicates a high degree of self-consistency in the parameters involved. The estimated standard deviations of the measured and

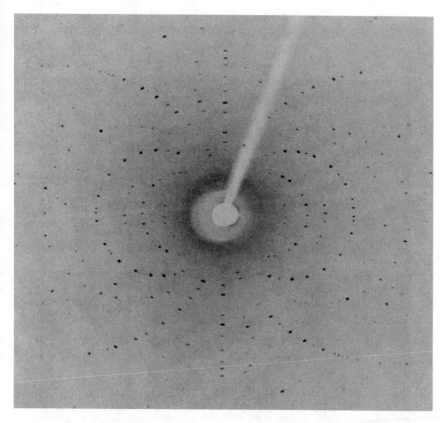

FIGURE 8.4.   Laue photograph taken with the x-ray beam along b, showing clearly the 2-fold symmetry axis along this direction.

calculated density values are important, because a significant difference ($>3\sigma$) may indicate the presence of unsuspected solvent of crystallization.

8.2.2   Intensity Measurement and Correction

Intensity data were collected up to $\sin\theta = 65°$. The number N of data to be expected may be calculated from the formula $N = (4\pi/3)[2(\sin\theta_{max})/\lambda]^3 V_c/mG$, where mG is the number of general positions in the Laue group of the crystal, and the other symbols have their usual meanings. Since $mG = 4$ in the present example, $N = 1990$. From this number we subtract the number of systematic absences, 165, to give 1825. In practice, 1724 data were collected, indicating 101 accidental absences. Thus, a total of 1623 reflections was used for the structure analysis, giving 10 reflections per parameter, assuming anisotropic temperature factors for the non-hydrogen atoms and a single scale factor in the least-squares refinement.

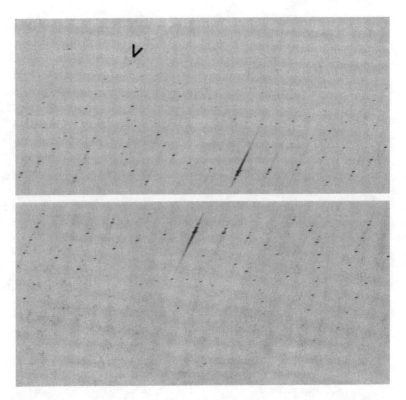

FIGURE 8.5.   Weissenberg photograph of the $h0l$ layer. The more intense reflections show spots arising from both Cu $K\alpha(\lambda = 1.542$ Å) and Cu $K\beta(\lambda = 1.392$ Å) radiations. In some areas, spots from W $L\alpha$ radiation ($\lambda = 1.48$ Å) arise due to sputtering of the copper target in the x-ray tube with tungsten from the filament (see Section 3.1.1). In each case, the spots for a given $h0l$ reflection lie along the directions of axial rows at decreasing values of $\sin\theta$, in proportion to the change in λ. The continuous streaks (Laue streaks) arise from the "white" radiation; filtering (Section 3.1.4) is never perfect.

The data were corrected for Lorentz and polarization effects, but not for absorption. Approximate scale K and overall isotropic temperature B factors were obtained by Wilson's method (Section 3.10.1). The parameters, (3.174)–(3.176), were fitted by least squares, and the line obtained had the equation

$$\ln\left\{\frac{\sum_j f_j^2}{\overline{|F_o|^2}}\right\} = -1.759 + 3.480\,\overline{\sin^2\theta} \tag{8.1}$$

From the slope $(2B/\lambda^2)$ and intercept $(2\ln K)$, $B = 4.1$ Å^2 and $K = 0.41$, where K is the scale factor for $\overline{|F_o|^2}$. The graphical Wilson plot is shown in Figure 8.10. We are now ready to proceed with the structure analysis.

FIGURE 8.6. Weissenberg photograph of the, $h1l$ layer. The $01l$ reciprocal lattice row, indicated by an arrowhead, illustrates clearly the effect of slight mis-setting of the crystal.

8.2.3 Structure Analysis in the xz Projection

This projection of the unit cell has the largest area and thus would be expected to show good resolution of the molecule. It is uncommon for a normal three-dimensional study to be preceded by an analysis in projection. However, from the standpoint of introductory study, carefully chosen two-dimensional examples have much to offer.

Using Figure 2.32 (see Section 2.7.5), we can associate the coordinates of the general equivalent positions

$$\pm \left\{ x, y, z; \; x, \tfrac{1}{2} - y, \tfrac{1}{2} + z \right\}$$

with the four bromine atoms in the unit cell. In the xz projection, these coordinates give rise to two repeats within the length c, so we may consider this projection in

FIGURE 8.7. Upper part of a Weissenberg photograph of the 0*kl* layer, Cu *K β* spots can be seen for the more intense reflections. The 00*l* (z^*) reciprocal lattice row, indicated by an arrowhead, is common to this photograph and that of the *h*0*l* layer.

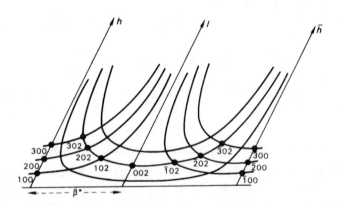

FIGURE 8.8. Sample of indexed reflections on an *h*0*l* Weissenberg photograph diagram.

terms of plane group *p*2 and compute the Patterson function from 0 to $a/2$ and 0 to $c/2$, which is equivalent to one-half of the unit cell in *p*2. This portion of the projection $P(u, w)$ would be expected to show one Br–Br vector at $(2x, 2z)$, as in Figure 8.11.

The two rows of peaks indicate that, in this projection, the molecules lie closely parallel to the *z* axis; this conclusion is supported by the large magnitude

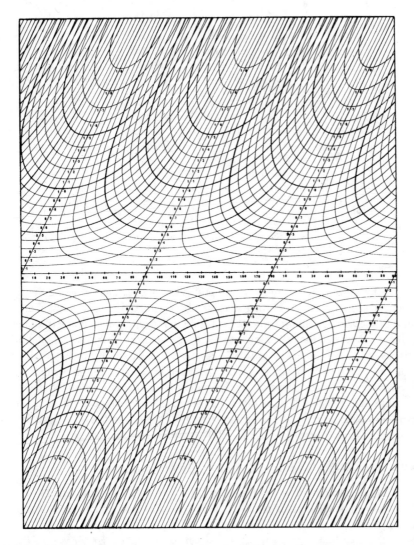

FIGURE 8.9.   Weissenberg chart: camera diameter 57.30 mm, 2° rotation per mm travel.
(Reproduced with the permission of the Institute of Physics and the Physical Society, London.)

of $|F(200)|$, equal to 336. This value may be compared with $F(000)$ in Table 8.1,
and, more significantly, with $\sum_j f_{j,\theta_{200}}$, which is 474. The peak arising from the
Br–Br vector is marked A, and by direct measurement we obtain the fractional
coordinates $x = 0.25$, $z = 0.015$ for the Br atom in the asymmetric unit.

TABLE 8.1.   Crystal Data for
BBIP at 20°C^a

Molecular formula	$C_{16}H_9BrO$
Molecular weight	297.16
Space group	$P2_1/c$
a (Å)	7.508(4)
b (Å)	5.959(5)
c (Å)	26.172(6)
β (deg)	92.55(2)
Unit-cell-volume (Å^3)	1169(2)
Radiation (Cu$K\alpha$)	1.5418Å
D_m (g cm^{-3})	1.68(1)
D_x (g cm^{-3})	1.688(3)
Z	4
$F(000)$	592

^{a} The numbers in parentheses are estimated
standard deviations, to be applied to the
least significant figure.

An electron density map, in this projection, was calculated using the signs given by F_{Br} with the experimental values of $|F_o(h0l)|$. If F_{Br} was less than one third of the corresponding value of $|F_o|$ for any reflection, the sign was assumed to be uncertain and the reflection omitted from the electron density calculation at this stage of the analysis. Figure 8.12 shows the electron density map with the molecule, fitted with the aid of a model, marked in. The resolution is moderately good, and we can see that we are working along the right lines. From the shapes of the rings, it is apparent that the molecule is inclined to the plane of this projection, and there will be a limit to the improvement of the resolution attainable in this projection. Consequently, we shall begin three-dimensional studies.

8.2.4   Three-Dimensional Structure Determination

In order to obtain values for all spatial coordinates, we proceeded first to a three-dimensional Patterson map $P(u, v, w)$, calculated section by section normal to the b axis.

The coordinates of the general positions show that the Br–Br vectors in the asymmetric unit will be found at $2x, 2y, 2z$ (single-weight peak), $0, \frac{1}{2} - 2y, \frac{1}{2}$ (double-weight peak), and $2x, \frac{1}{2}, \frac{1}{2} - 2z$ (double-weight peak). Hence, we must study the Patterson map carefully, particularly the Harker line $[0, v, \frac{1}{2}]$ and section $(u, \frac{1}{2}, w)$—why? Figures 8.13 and 8.14 show these two regions of Patterson space, and Figure 8.15 illustrates a general section calculated close to the single-weight peak B. From the peaks B, C, and D, the coordinates for the bromine atom in the asymmetric unit were found to be 0.248, 0.188, 0.016. Repeating the phasing procedure, but now for hkl reflections, and calculating a three-dimensional electron

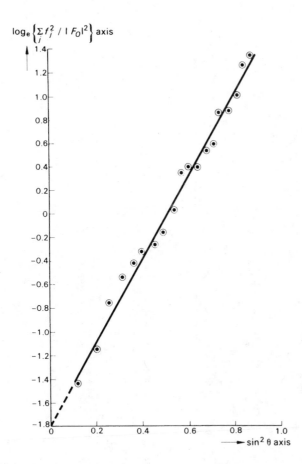

FIGURE 8.10.   Wilson plot for BBIP; the slope is 3.48 and the intercept is -1.76.

density map produced a good resolution of the complete structure, with the exception of the hydrogen atoms. Figure 8.16 illustrates a composite electron density map, which consists of superimposed sections calculated at intervals along a.

The scattering of x-rays by hydrogen atoms is small in magnitude, and these atoms are not normally resolved by the direct summation of the electron density. If the data are of good quality and all other atoms in the structure have been found, a difference Fourier synthesis (see Section 6.4.5) will generally result in the hydrogen atoms being located, provided the other atoms are not themselves too large in scattering power. It is possible also to calculate the positions of hydrogen atoms from the geometry of the structure, if the positions of sufficient surrounding groups are known. Nowadays, most program systems for x-ray structure analysis include routines for calculating the coordinates of hydrogen atoms, according to

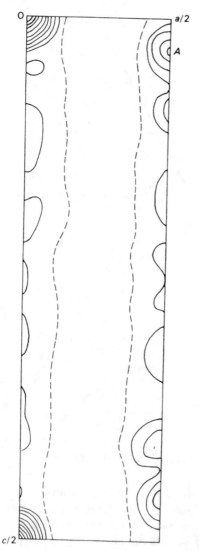

FIGURE 8.11.   Asymmetric unit of $P(u, w)$. Since we are concerned here mainly with the Br–Br vector (A), the slight distortion arising from drawing β as $90°$ is inconsequential.

their nature, for example, $R_1 R_2 R_3$CH and $R_1 R_2$CH$_2$; the hydrogen atoms in a CH$_3$ group can be calculated if a position with respect to rotation about the R–CH$_3$ bond can be postulated.

Finally, we arrive at the complete structure for BBIP, as shown in Figure 8.17 with a convenient numbering scheme.

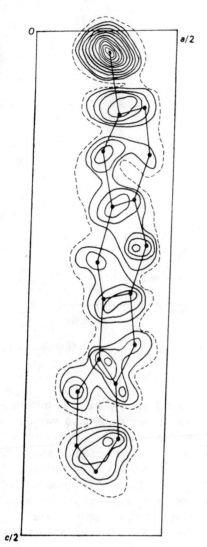

FIGURE 8.12.   Asymmetric   unit   of $\rho(x, z)$ phased on the bromine atoms; the probable atomic positions are marked in.

8.2.5   Refinement

During the final stages of refinement of the structure, the hydrogen atoms were included in the evaluation of the structure factors F_c, but no attempt was made to refine the parameters of the hydrogen atoms because the main interest in the problem lay in determining the molecular conformation. We note also that we

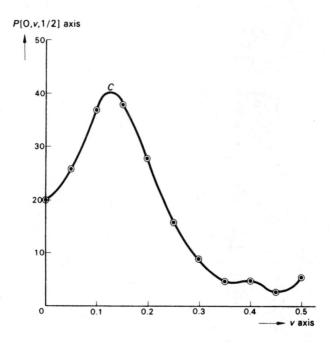

FIGURE 8.13.   Patterson function along the Harker line $[0, v, \frac{1}{2}]$, showing a double-weight Br–Br vector at C.

have a favorable ratio of 10.4 for data to variables. Had we included the hydrogen atoms in the refinement, with isotropic temperature factors, the ratio would have been decreased to 8.5 (see Section 7.4.3). The final adjustments of the structural parameters of the Br, O, and C atoms $(x, y, z,$ and anisotropic temperature factors) and the scale factor were carried out by the method of least squares (see Section 7.4.2). The refinement converged with an R factor of 0.070, and a final difference-Fourier synthesis showed no fluctuations in density greater than about twice $\sigma(\rho_o)$, and then only around the position of the bromine atom. The analysis was considered to be satisfactory, and the refinement was terminated at this stage.

8.2.6   Molecular Geometry

It remained to determine the bond lengths, bond angles, and other features of the geometry of the molecule and its relationship with other molecules in the unit cell.

From the coordinates of the atomic positions (Table 8.2) and using equations such as (7.119) and (7.121), bond lengths and angles were calculated. They are

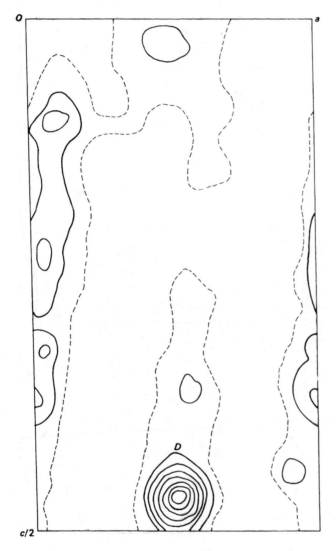

FIGURE 8.14.   Patterson section $\left(u, \frac{1}{2}, w\right)$ showing a double-weight Br–Br vector at D.

shown on the drawings of the molecule in Figures 8.18 and 8.19. Figure 8.20 illustrates the packing of the molecules in the unit cell, as seen along b; the average intermolecular contact distance is 3.7 Å, a typical intermolecular contact distance in organic compounds, in which van der Waals forces link the molecules in the solid state.

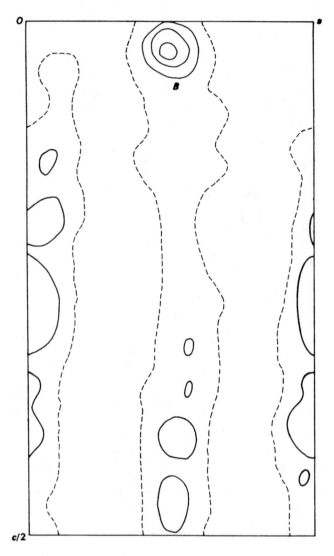

FIGURE 8.15.   Patterson section $(u, 0.375, w)$, showing a single-weight Br–Br vector B.

In a molecule of this nature, the planarity of the ring system is of stereo-chemical interest. The equation of a plane, $Ax + By + Cz = D$, can be solved by three triplets x, y, z. Hence, the best molecular plane is obtained by a least-squares procedure that minimizes the sum of the squares of the deviations d of all of the

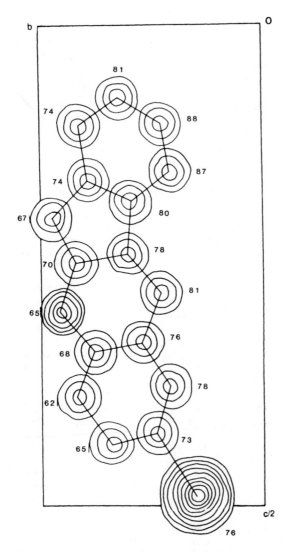

FIGURE 8.16.   Composite three-dimensional electron density map with the molecule (excluding H atoms) marked in, as seen along a. The contour of zero electron density is not shown, and the numbers represent $100x$ for each atom. A symmetry-related position to that chosen in Section 8.2.4 has been selected here for the Br atom. (What is the symmetry operation?)

atoms from the plane. The results are listed in Table 8.3. It can be seen that the deviations of the atoms from the best plane are not significant, and it is possible to conclude, therefore, that the introduction of the heteroatom has but little effect on the planarity of the benzofluorene moiety.

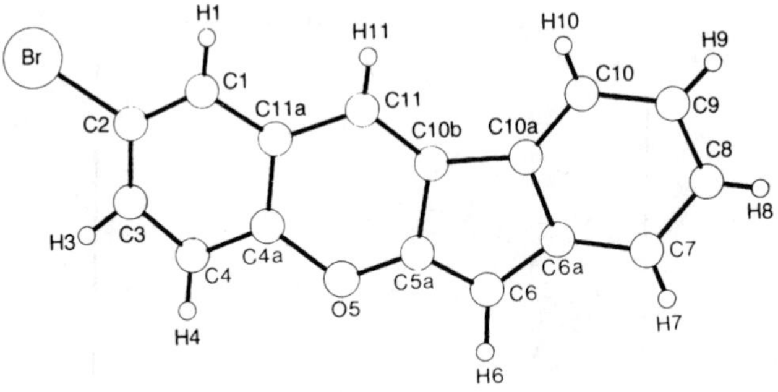

FIGURE 8.17.   Structural formula for BBIP.

TABLE 8.2.   Fractional Atomic Coordinates in
BBIP, with esd's in Parentheses[a]

Atom	x	y	z
Br	0.7602(2)	0.3152(3)	0.4848(0)
C(1)	0.7820(16)	0.4187(22)	0.3789(4)
C(2)	0.7310(16)	0.4951(24)	0.4252(4)
C(3)	0.6524(16)	0.7075(25)	0.4297(5)
C(4)	0.6214(16)	0.8413(23)	0.3871(5)
C(4a)	0.6794(16)	0.7619(22)	0.3406(4)
O(5)	0.6520(11)	0.9051(14)	0.2995(3)
C(5a)	0.6973(14)	0.8329(21)	0.2526(4)
C(6)	0.6714(14)	0.9397(19)	0.2077(5)
C(6a)	0.7384(15)	0.7990(19)	0.1678(4)
C(7)	0.7401(17)	0.8230(24)	0.1150(4)
C(8)	0.8078(18)	0.6526(24)	0.0858(4)
C(9)	0.8766(17)	0.4574(24)	0.1079(5)
C(10)	0.8731(16)	0.4268(21)	0.1605(4)
C(10a)	0.8035(16)	0.5954(20)	0.1908(4)
C(10b)	0.7767(14)	0.6076(21)	0.2454(5)
C(11)	0.8064(15)	0.4734(21)	0.2850(4)
C(11a)	0.7593(14)	0.5475(20)	0.3359(5)
H(1)	0.809	0.239	0.375
H(3)	0.622	0.789	0.460
H(4)	0.565	0.999	0.389
H(6)	0.630	0.121	0.208
H(7)	0.674	0.944	0.097
H(8)	0.804	0.667	0.043
H(9)	0.886	0.361	0.076
H(10)	0.809	0.253	0.375
H(11)	0.870	0.311	0.285

[a] There are no esd's for the hydrogen atom coordinates because these
parameters were not included in the least-squares refinement.

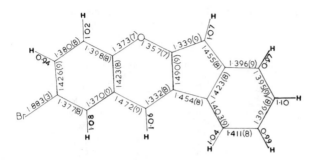

FIGURE 8.18.   Bond lengths in BBIP, with their estimated standard deviations in parentheses.

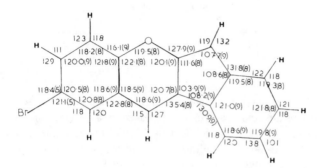

FIGURE 8.19.   Bond angles in BBIP, with their estimated standard deviations in parentheses.

FIGURE 8.20.   Stereoview of the molecular packing in the structure of BBIP, as seen along a.

8.3   Crystal Structure of Potassium 2-Hydroxy-3,4-dioxocyclobut-1-ene-1-olate Monohydrate (KHSQ)[a]

1,2-Dihydroxy-3,4-dioxocyclobut-1-ene (IV) may be prepared by the acid-catalyzed hydrolysis of 1,2-diethoxy-3,3,4,4-tetrafluorocyclobut-1-ene (V). On recrystallization from water, it has a melting point of 293°C, at which temperature it decomposes.

[a] R. J. Bull, M. F. C. Ladd, D. C. Povey, and R. Shirley, *Crystal Structure Communications* **2**, 625 (1973).

TABLE 8.3.   Deviations of Atoms from
the Least-Squares Plane through the
Molecule[a]

Atom	Deviation (Å)	Atom	Deviation (Å)
Br	0.03	C(6a)	0.02
C(1)	0.01	C(7)	−0.06
C(2)	−0.04	C(8)	−0.08
C(3)	−0.05	C(9)	−0.03
C(4)	−0.06	C(10)	0.03
C(4a)	0.01	C(10a)	0.04
O(5)	0.05	C(10b)	0.04
C(5a)	0.06	C(11)	0.03
C(6)	0.02	C(11a)	0.03

[a] The mean estimated standard deviation of the deviations
is 0.02, so that hardly any atoms deviate significantly
from the best plane at a 3σ level.

It has been called by the trivial name, squaric acid; the hydrogen atoms in the hydroxyl groups are acidic, and can be replaced by a metal. Potassium hydrogen squarate monohydrate (VI), which is the subject of this example, can be obtained by mixing hot, concentrated, equimolar aqueous solutions of potassium hydroxide and squaric acid and then cooling the reaction mixture.

8.3.1   Preliminary X-ray and Physical Measurements

The compound was recrystallized from water as colorless, prismatic crystals with the forms {001}, {110}, and {100} predominant (Figure 8.21). Under a polarizing microscope, straight extinction was observed on {001} and {100}, and an extinction angle of about 2° was obtained on a section cut normal to b. These results suggest strongly that the crystals belong to the monoclinic system.

The crystal specimen chosen for x-ray work had the dimensions 0.5, 0.5, and 0.3 mm parallel to a, b, and c, respectively. The details of the preliminary measurements are similar to those described for the previous example, and we list the crystal data immediately (Table 8.4). Copper $K\alpha$ radiation ($\bar{\lambda} = 1.5418$ Å) was used throughout this work.

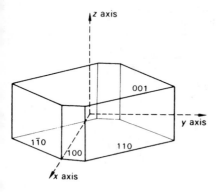

FIGURE 8.21. Crystal habit of potassium hydrogen squarate monohydrate (KHSQ) with the crystallographic axes drawn in.

TABLE 8.4.   Crystal Data for KHSQ
at 20°C

Molecular formula	$C_4HO_4^-, K^+, H_2O$
Molecular weight	170.17
Space group	$P2_1/c$
a (Å)	8.641(1)
b (Å)	10.909(1)
c (Å)	6.563(2)
β (deg)	99.81(1)
Unit-cell volume (Å^3)	609.6(2)
D_m (g cm^{-3})	1.839(7)
D_x (g cm^{-3})	1.854(1)
Z	4
$F(000)$	344

8.3.2   Intensity Measurement and Correction

Nine hundred symmetry-independent intensities with $(\sin\theta)/\lambda \leqslant 0.57\ \text{Å}^{-1}$ were measured. Corrections were applied for polarization and Lorentz effects, but not for absorption. Scale (K) and isotropic temperature (B) factors were deduced by Wilson's method, and the $|F_o|$ data were converted to $|E|$ values through (7.1).

The structure analysis began with 142 $|E|$ values $\geqslant 1.5$, representing 15.8% of the experimental reflection data, and their statistics are shown in Table 8.5. The agreement with the theoretical values for a centric distribution of $|E|$ values is very close, in accord with the chosen centrosymmetric space group (Table 8.4).

8.3.3 Σ_2 Listing

The next stage was the preparation of a Σ_2 listing (see Section 7.2.5). Symmetry-related reflections become very important in generating triplet relationships: 300 and 30$\bar{4}$ can lead to both 004 and 60$\bar{4}$, the latter by replacing 30$\bar{4}$ by

TABLE 8.5. Statistics of $|E|$ Values
in KHSQ

	Acentric	Centric	This structure
$\overline{|E|^2}$	1.00	1.00	1.00
$\overline{|E|}$	0.89	0.80	0.81
$\overline{|E|^2 - 1}$	0.74	0.97	0.95
$\% \geqslant 1.0$	36.8	31.7	33.9
$\% \geqslant 1.5$	10.5	13.4	14.6
$\% \geqslant 1.75$	4.7	8.0	8.4
$\% \geqslant 2.0$	1.8	4.6	4.9
$\% \geqslant 2.5$	0.2	1.2	1.1

$\overline{3}04$, taking note of the phase symmetry. The relevant phase symmetry for space group $P2_1/c$ follows from (3.130), and may be summarized as follows:

$$s(hkl) = s(\overline{h}\overline{k}\overline{l}) \tag{8.2}$$

$$s(hkl) = s(h\overline{k}l)(-1)^{k+1} \tag{8.3}$$

A portion of the Σ_2 listing is shown in Table 8.6; full use of the symmetry relationships (8.2) and (8.3) has been made in setting up the listing. The numbers in parentheses under each $\mathbf{h}$ are the total numbers of Σ_2^a triplets for each of these reflections; $\mathbf{k}$ and $\mathbf{h} - \mathbf{k}$ represent those reflections that form a vector triplet with $\mathbf{h}$.

8.3.4 Specifying the Origin

Following the procedure described in Section 7.2.2 and using the reflections in Table 8.6, three reflections were chosen and allocated positive signs, in order to fix the origin at 0, 0, 0. The symmetry relationships in the space group of this compound allowed, in all, 12 signs in the origin set (Table 8.7). The reader should check the signs, starting from the first one in each group of four, using Equations (8.2) and (8.3).

8.3.5 Sign Determination

The Σ_2 listing was examined with a view to generating new signs, using (7.9), which may be given by

$$s[E(\mathbf{h})] \approx s\left[\sum_{\mathbf{k}} E(\mathbf{k})E(\mathbf{h} - \mathbf{k}) \right] \tag{8.4}$$

---

[a] We use the notation $\mathbf{h}$ for hkl, $\mathbf{k}$ for $h'k'l'$, and $\mathbf{h} - \mathbf{k}$ for $h - h', k - k', l - l'$.

TABLE 8.6.   Part of the Σ_2 Listing for KHSQ

h	$|E(\mathbf{h})|$	k	$|E(\mathbf{k})|$	h − k	$|E(\mathbf{h} - \mathbf{k})|$	$|E(\mathbf{h})||E(\mathbf{k})||E(\mathbf{h} - \mathbf{k})|$
53$\bar{1}$	2.6	010,$\bar{4}$	2.8	5$\bar{7}$3	2.6	18.9
(37)		0$\bar{4}$1	2.2	57$\bar{2}$	3.3	17.2
		0$\bar{4}\bar{1}$	2.0	570	2.7	14.0
		$\bar{1}$14	2.3	62$\bar{5}$	1.7	10.2
		0$\bar{3}$2	1.7	56$\bar{3}$	2.0	8.8
$\bar{1}$14	2.3	5$\bar{7}$2	3.3	482	1.9	14.4
(45)		6$\bar{6}$4	1.8	570	2.7	11.2
		6$\bar{8}$1	1.5	5$\bar{7}$3	2.6	9.0
		$\bar{5}$63	2.0	4$\bar{5}$1	1.5	6.9
		454	1.6	$\bar{5}\bar{4}$0	1.5	5.5
032	1.7	$\bar{5}$31	2.6	56$\bar{3}^a$	2.0	8.8
(54)		5$\bar{7}$2	3.3	5$\bar{4}$0	1.5	8.4
		482	1.9	$\bar{4}$5$\bar{4}$	1.6	5.2
		$\bar{4}\bar{5}$1	1.5	48$\bar{1}^a$	2.0	5.1
11$\bar{2}$	2.5	5$\bar{7}$2	3.3	6$\bar{6}$4	1.8	14.9
(39)		$\bar{4}$8$\bar{2}$	1.9	5$\bar{7}$0	2.7	12.8
		11$\bar{4}$	2.3	002	1.9	10.2
		$\bar{5}$7$\bar{1}$	1.7	68$\bar{1}$	1.5	6.4
010,4	2.8	$\bar{3}$32	2.2	372	1.9	11.7
(35)		$\bar{6}$25	1.7	68$\bar{1}$	1.5	7.1
33$\bar{2}$	2.2	$\bar{3}\bar{3}$2	2.2	66$\bar{4}$	1.8	8.7
(46)		11$\bar{4}$	2.3	242	1.7	8.6
		3$\bar{1}$3	1.8	041	2.0	7.9
		62$\bar{5}$	1.7	$\bar{3}$13	1.8	6.7
002	1.9	041	2.0	0$\bar{4}$1	2.0	7.4
(25)		11$\bar{4}$	2.3	$\bar{1}$16	1.5	6.6
		68$\bar{1}$	1.5	$\bar{6}$8$\bar{1}$	1.6	4.6

^{a} Uses 03$\bar{2}$.

TABLE 8.7.   Origin-Fixing Reflections and Their Symmetry Equivalents

Reflections	Sign	$|E|$	No. of Σ_2 triplets
53$\bar{1}$	+	2.6	37
$\bar{5}\bar{3}$1	+		
5$\bar{3}\bar{1}$	+		
$\bar{5}$31	+		
11$\bar{4}$	+	2.3	45
$\bar{1}\bar{1}$4	+		
1$\bar{1}\bar{4}$	−		
$\bar{1}$14	−		
032	+	1.7	54
0$\bar{3}\bar{2}$	+		
0$\bar{3}$2	−		
03$\bar{2}$	−		

where the sum is taken over the several **k** triplets all involved with the given **h**. The probability of (8.4) is given by (7.10). If only a single Σ_2 interaction is considered, (8.4) becomes

$$s[E(\mathbf{h})] \approx s[E(\mathbf{k})]s[E(\mathbf{h}-\mathbf{k})] \tag{8.5}$$

and the probability calculation omits the summation in (7.11).

We shall assume that the values of the probability $P_+(\mathbf{h})$ are sufficiently high for the signs to be accepted as correct; very small or zero values of $P_+(\mathbf{h})$ indicate strongly a negative sign for **h**. Some examples of the application of (8.5) are given in Table 8.8. It does not matter which reflections in a triplet are labeled **h** and **k**.

Use of Sign Symbols

The above process of sign determination was applied to the entire Σ_2 listing which, although it contained 1276 triple products, was exhausted after only 24 signs had been found. To enable further progress to be made, three reflections were assigned the symbols A, B, and C, where each symbol represented either a plus or minus sign. Twelve symbolic signs (Table 8.9) were thus added to the

TABLE 8.8.   Sign Determination
Starting from the "Origin Set"

k	**h − k**	**h**	Indication for s (**h**)
$53\bar{1}+$	$1\bar{1}4-$	$62\bar{5}$	−
$53\bar{1}+$	$03\bar{2}-$	$56\bar{3}$	−
$56\bar{3}-$	$\bar{1}\bar{1}4+$	451	−
$56\bar{3}-$	$032+$	$59\bar{1}$	−
$451-$	$03\bar{2}-$	$48\bar{1}$	+

TABLE 8.9.   Symbolic Signs

Reflection	Sign	$|E|$	No. of Σ_2 relationships
$11\bar{2}$	A	2.5	39
$\bar{1}\bar{1}2$	A		
$1\bar{1}\bar{2}$	$-A$		
$\bar{1}12$	$-A$		
$010,\bar{4}$	B	2.8	35
$0\bar{1}0,\bar{4}$	B		
$0\bar{1}0,4$	B		
$010,\bar{4}$	B		
$33\bar{2}$	C	2.2	46
$\bar{3}\bar{3}2$	C		
$3\bar{3}\bar{2}$	$-C$		
$\bar{3}32$	$-C$		

TABLE 8.10.   Further Sign Determinations[a]

k	s(k)	(h − k)	s(h − k)	h	Sign indication, sh
010,4	B	62$\bar{5}$	−	68$\bar{1}$	$s(68\bar{1}) = B$
010,4	B	33$\bar{2}$	C	372	$s(372) = -BC$
11$\bar{2}$	A	68$\bar{1}$	B	571	$s(571) = AB$
53$\bar{1}$	+	010,4	B	573	$s(573) = B$
11$\bar{4}$	+	68$\bar{1}$	B	573	$s(573) = B$
33$\bar{2}$	C	33$\bar{2}$	C	66$\bar{4}$	$s(66\bar{4}) = CC = +$
11$\bar{4}$	+	66$\bar{4}$	+	570	$s(570) = -$
11$\bar{2}$	A	570	−	482	$s(482) = A$
11$\bar{2}$	A	66$\bar{4}$	+	57$\bar{2}$	$s(57\bar{2}) = -A$
11$\bar{4}$	+	57$\bar{2}$	−A	482	$s(482) = A$
032	+	57$\bar{2}$	−A	540	$s(540) = A$
032	+	482	A	454	$s(454) = -A$
11$\bar{4}$	+	454	−A	540	$s(540) = A$
33$\bar{2}$	C	11$\bar{4}$	+	242	$s(242) = -C$
11$\bar{2}$	A	11$\bar{4}$	+	002	$s(002) = A$
11$\bar{2}$	A	482	A	570	$s(570) = -AA = -$
570	−	53$\bar{1}$	+	041	$s(041) = -$
62$\bar{5}$	−	33$\bar{2}$	C	31$\bar{3}$	$s(31\bar{3}) = C$
31$\bar{3}$	C	33$\bar{2}$	C	041	$s(041) = -CC = -$
53$\bar{1}$	+	041	−	57$\bar{2}$	$s(57\bar{2}) = +$
002	A	041	−	041	$s(041) = A$
002	A	68$\bar{1}$	B	681	$s(681) = AB$
002	A	11$\bar{4}$	+	11$\bar{6}$	$s(11\bar{6}) = A$

[a] Symmetry relations should be employed as necessary.

set, and the sign determination was continued, now in terms of both signs and symbols. It may be noted that although the symbols are given to reflections with large $|E|$ values and large numbers of Σ_2 interactions, there are not, necessarily, any restrictions on either parity groups or the use of structure invariants.

Some examples of this stage of the process are given in Table 8.10. The values of **h** and **k** are taken from either Tables 8.7 and 8.9, which constitute the "starting set," or as determined through (8.4). The reader is invited to follow through the stages in Table 8.10, working out the correct symmetry-equivalent signs from (8.2) and (8.3) as necessary.

From Table 8.10, we see that six more reflections have been allocated signs, and another 17 are determined in terms of $A, B,$ and C. Multiple indications can now be seen. For example, there are two indications that $s(573) = B$, two indications that $s(570) = -$, and two indications that $s(540) = A$. Three indications for 041 suggest that both $s(041) = -$ and $A = -$.

Continuing in this manner, it was found possible to allot signs and symbols to all 142 $|E|$ values greater than 1.5. The symbols $A, B,$ and C were involved in 65, 72, and 55 relationships, respectively. Consistent indications, such as those mentioned above for $s(041)$, led finally to the sign relationships $A = AC = B = -,$

from which it follows that $C = +$. It does not always turn out that the signs represented by symbols can be allocated from the analysis in this complete and satisfactory manner. If there are n undetermined symbols, then there will be 2^n sets of signs to be examined. In this case, figures of merit, such as those discussed in Section 7.2.13, can be used to indicate that set of signs most likely to be correct. It may not follow that the indicated set *is* correct, and some trials with Fourier syntheses may be needed at this stage in order to elicit the correct result.

8.3.6   The E Map

The signs of the 142 $|E|$ values used in this procedure were obtained with a high probability, and an electron density map was computed using the signed $|E|$ values as coefficients. The sections of this map $\rho(x, y, z)$ at $z = 0.15, 0.20, 0.25,$ and 0.30 are shown in Figures 8.22–8.25. They reveal the K^+ ion and the $C_4O_4^-$ ring system clearly; the oxygen atom O_w of the water molecule was not indicated convincingly at this stage of the analysis. A tilt of the plane of the molecule with respect to (001) can be inferred from Figures 8.23–8.25. Some spurious peaks S may be seen. This is a common feature of E maps. We must remember that a limited data set (142 out of 900) is being used, and that the $|E|$ values are sharpened coefficients corresponding to an approximate point-atom model. The data set is therefore terminated while the coefficients for the Fourier series are relatively large, a procedure that can lead to spurious maxima (see Section 5.9.1). However, such peaks are often of smaller weight than those that correspond to atomic positions.

8.3.7   Completion and Refinement of the Structure

Sometimes, all atomic positions are not contained among the peaks in an E map. Those peaks that do correspond to chemically sensible atomic positions

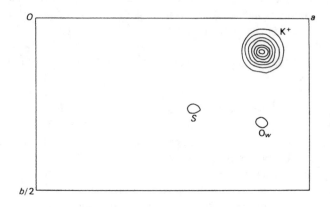

FIGURE 8.22. E map for KHSQ at $z = 0.15$.

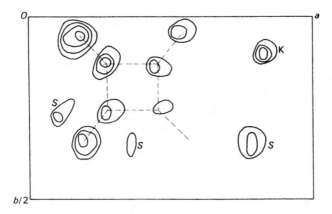

FIGURE 8.23.   *E* map for KHSQ at $z = 0.20$.

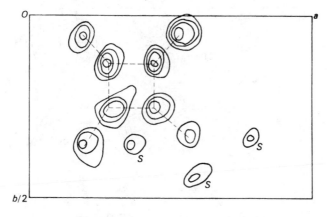

FIGURE 8.24.   *E* map for KHSQ at $z = 0.25$.

may be used to form a trial structure for calculation of structure factors and an $|F_o|$ electron density map. A certain amount of subjective judgment may be required to decide upon the best peaks for the trial structure at such a stage.

This situation was obtained for KHSQ, although it was not difficult to pick out a good trial structure. Coordinates were obtained for all nonhydrogen atoms except the oxygen atom of the water molecule. The R factor for this trial structure was 0.30, and the composite three-dimensional electron density map obtained is shown in Figure 8.26, which now reveals O_w clearly. It may be noted in passing that the small peak labeled O_w in Figure 8.22 corresponds to the position of this atom, but this fact could not be determined conclusively at that stage of the analysis.

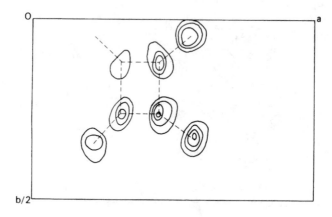

FIGURE 8.25. E map for KHSQ at $z = 0.30$.

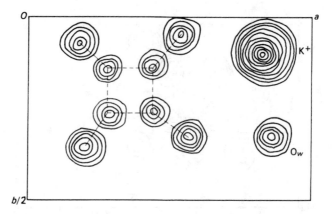

FIGURE 8.26. Composite electron density map for KHSQ (excluding H atoms); the atomic coordinates are listed in Table 8.11.

Further refinement was carried out by the method of least squares, and an R factor of 0.078 was obtained. Figure 8.27 shows a composite three-dimensional difference-Fourier map for KHSQ. Peaks numerically greater than 0.5 electron per Å^3, representing about twice $\sigma(\rho_0)$, are significant, and have been contoured. Some of these peaks indicate areas of small disagreement between the true structure and the model. Three positive peaks, however, are in positions expected for hydrogen atoms. Inclusion of these atoms in the structure factor calculations in the final cycles of least-squares refinement had a small effect on the R factor, bringing it to its final value of 0.077. The fractional atomic coordinates for the atoms in the asymmetric unit are listed in Table 8.11.

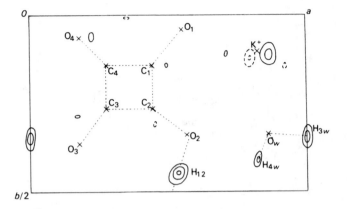

FIGURE 8.27.   Composite difference electron density map for KHSQ. Positive contours are solid lines and negative contours are broken lines. Bonds in the squarate ring and those involving hydrogen atoms are shown in dotted lines. Some spurious, small peaks (unlabeled) we shown by this synthesis.

TABLE 8.11.   Fractional Atomic
Coordinates for KHSQ

	x	y	z
K^+	0.8249(2)	0.1040(2)	0.1295(3)
C(1)	0.4353(9)	0.1295(7)	0.2572(12)
C(2)	0.4495(9)	0.2597(7)	0.2714(12)
C(3)	0.2795(9)	0.2714(8)	0.2462(11)
C(4)	0.2659(9)	0.1345(7)	0.2305(12)
O(1)	0.5399(6)	0.0450(5)	0.2649(10)
O(2)	0.5649(6)	0.3346(5)	0.2920(10)
O(3)	0.1874(7)	0.3582(6)	0.2386(10)
O(4)	0.1578(6)	0.0605(5)	0.2022(10)
O_w	0.8789(7)	0.3429(6)	0.0424(10)
H(12)	0.522	0.413	0.246
H(3w)	1.000	0.346	0.075
H(4w)	0.826	0.400	0.100

Interatomic distances and angles are shown in Figure 8.28, and a molecular packing diagram, as seen along c, is given in Figure 8.29. From the analysis, we find that intermolecular hydrogen bonds exist between O(2) and O(1)' [2.47(1) Å], between O(3)'' and O_w [2.76(1) Å], and between O(4)' and O_w [2.95(1) Å]; they are largely responsible for the cohesion between molecules in the solid state.[a] We cannot determine from this analysis how the electronic charge on the individual species is distributed. With x-rays, a more precise analysis would be needed, followed by a peak integration of an electron density map.

[a] Single and double primes indicate different neighboring molecules.

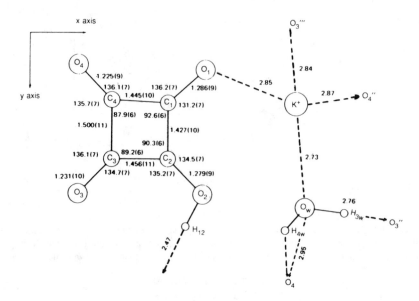

FIGURE 8.28. Bond lengths and bond angles in the asymmetric unit of KHSQ; the OH⋯O distances refer to the overall O⋯O separations. Primes on atom symbols indicate neighboring asymmetric units; this diagram should be studied in conjunction with Figure 8.29.

However, more recently, an *ab initio* calculation with Gaussian 94 on a free KHSQ species resulted in the following electron population parameters p on the individual atomic species.

	p		p		p		p
K	18.111	C_1	5.581	C_2	5.833	C_3	5.605
C_4	5.599	O1	8.808	O2	8.760	O3	8.626
O_4	8.544	0_w	8.855	H_{12}	0.564	H_{3w}	0.560
H_{4w}	0.554	$\Sigma_p = 86 = F(000)/4$					

Although the above results would be modified slightly in the crystal environment, they indicate the drawing of electron density from the less electronegative species, particularly hydrogen, towards the more electronegative oxygen atoms. This effect leads to bond lengths involving hydrogen that measure shorter by x-ray diffraction than with neutrons.

8.4   Structure Analysis by X-ray and Neutron Diffraction

As a further example in this chapter, we discuss the structure determination[a] by both x-ray and neutron diffraction methods of the crown

---

[a] R. A. Palmer, B. N. Potter, J. N. Lisgarten, R. H. Fenn, S. A. Mason, O. S. Mills, P. M. Robinson and C. I. F. Watt, *Acta Crystallographica* B**57**, 339 (2001).

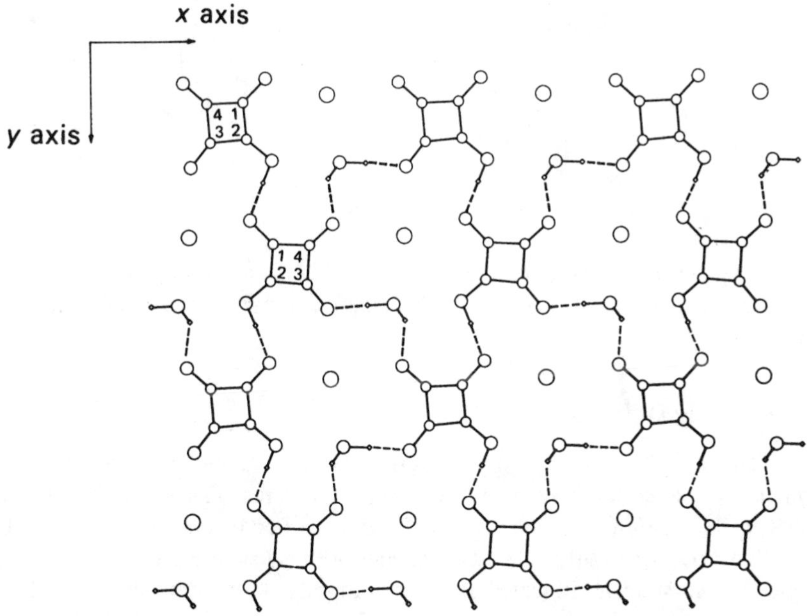

FIGURE 8.29.   Molecular packing diagram of one layer of the KHSQ structure as seen along c. The circles in order of decreasing size represent K, O, C, and H. The hydrogen-bond network is shown by dashed lines.

ether 1,8-(3,6,9-trioxaundecane-1,11-diyldioxy)-9,10-dihydro-10,10-dimethyl-9-ol, $C_{24}H_{30}O_6,H_2O$. Crown ethers are important for their ability to bind ions or neutral species by non-covalent interactions.

A neutron diffraction study was desirable in order to provide accurate positions of the hydrogen atoms; the scattering power of hydrogen for neutrons is comparable to that of other atoms in the structure. As a consequence, the determination of the neutron structure, involving over 60 approximately equal atoms, by *ab initio* direct methods was a formidable task. In this type of study it is customary to determine the structure first by x-rays, and subsequently to marry this structure to that determined with the neutron diffraction data collected in an independent experiment, employing the same unit cell and space group.

8.4.1   Sample Preparation

The subject compound (II) is prepared in high yield by bis-methylation of anthralin (I) at the 10-position, followed by bis-O-alkylation to give a ketone that is reduced to the alcohol by sodium amalgam.

(II)   (I)

Slow evaporation of a solution in ethanol yielded crystals of m.p. 151–153°C. A crystal of approximate dimensions 0.35, 0.25, 0.20 mm was selected for the x-ray study.

8.4.2 X-ray Diffraction Study

The crystal was centered on a Nonius CAD4 diffractometer, and a C-centered unit cell was determined by the diffractometer software package, with the dimensions $a = 12.856(1)$, $b = 24.397(4)$, $c = 14.564(2)$ Å, $\alpha = 89.98(1)$, $\beta = 89.987(9)$, $\gamma = 90.10(1)°$. On account of the close proximity to 90° of all angles, the crystal was allocated to the orthorhombic system; hence, the relevant asymmetric unit of reciprocal space was searched for intensity data. (A later version of the software carries out a check for Laue symmetry, but that was not available at the time.) Data were recorded with Cu $K(\alpha)$ x-radiation for $1° < \theta < 74°$. Inspection of the intensity record led to the assignment of space group $C222_1$, with eight molecules in the unit cell, that is, two per asymmetric unit.

The data were corrected for Lorentz and polarization effects, but not for absorption. Of the 2572 unique intensities collected, 1395 had $|F_o| > 4\sigma|F_o|$, indicating weak diffraction, where $\sigma|F_o|$ is the standard deviation in $|F_o|$ estimated from the counting statistics.

The direct methods program SHELX-S produced no satisfactory solution. Space group $C2_1$ (c unique) was then used, where upon the direct methods program produced a good solution, with an E-map exhibiting the essential features of both molecules in the asymmetric unit, even though the data set was incomplete with respect to monoclinic symmetry. However, the trial structure refined to only 23.8%.

A second, much improved data set was collected using an option that constrained the unit cell to be monoclinic. The unit cell parameters are listed in Table 8.12 (with other relevant data) and the space group was determined as $P2_1$, with four molecules per unit cell, that is, two per asymmetric unit.

The structure was determined by direct methods with the program SHELX-S, and the E-map revealed the positions of all non-hydrogen atoms in both molecules of the asymmetric unit. Difference electron density maps revealed one molecule of water of crystallization bound to each organic molecule. Hydrogen atoms

TABLE 8.12.   X-ray Structure Determination and Refinement Data (Second Set) for 1,8-(3,6,9-Trioxaundecane-1,11- diyldioxy)-9,10-dihydro-10,10-dimethylanthracene-9-ol

Compound	$C_{24}H_{30}O_6.H_2O$
Compound name	1,8-(3,6,9-Trioxaundecane-1,11-diyldioxy)-9,10-dihydro-10,10-dimethylanthracene-9-ol
Colour/shape	Colourless/plates
Formula weight	432.50
Temperature	293(2)K
Crystal system	Monoclinic
Space group	$P2_1$
Unit cell dimensions/Å	a 12.845(5)
(25 reflections	b 14.575(3)
$25° < \theta < 30°$)	c 13.779(2)
	β 117.72(2)°
Unit cell volume/Å^3	V_c 283.6(1)
Z	4
$D_c/\text{g cm}^{-3}$	1.302
μ_c/cm^{-1}	0.754
Diffractometer/scan	Nonius CAD4 $\omega/2\theta$
Radiation	Cu $K\alpha$
Monochromator	Graphite
Wavelength	$\lambda = 1.5418$ Å
$F(000)$	928
Crystal dimensions/mm	0.35, 0.25, 0.20
Reflections measured	17753
Independent/Observed	9092/9092
R_{int}	0.0459
θ range/deg	3.62–72.93
Corrections applied	Lorentz/polarization
Absorption correction	None
Range of relative transmission factors	0.19–0.32
Range of h, k, l	$-15/15, -18/18, -17/17$
Reflections observed	$9092 [I_o > 2\sigma(I_o)]$
Computer programs	
Data processing	CAD4- Express 1988
Structure solution	SHELX-86
Structure refinement	SHELXL-97
Refinement method	Full-matrix least-squares on $|F|^2$
Data/restraints/parameters	9092/217/605
Goodness-of-fit on $|F|^2$	0.908
SHELX-97 weight parameters	0.1791 0.0
Final R indices $[I > 2\sigma(I)]$	$R_1 = 0.0619, wR_2 = 0.1506$
R indices (all data)	$R_1 = 0.1001, wR_2 = 0.1688$
Extinction coefficient	0.000(4)
Flack parameter	0.3(4)
Largest diff. peak and hole/e Å^{-3}	0.232 and -0.260

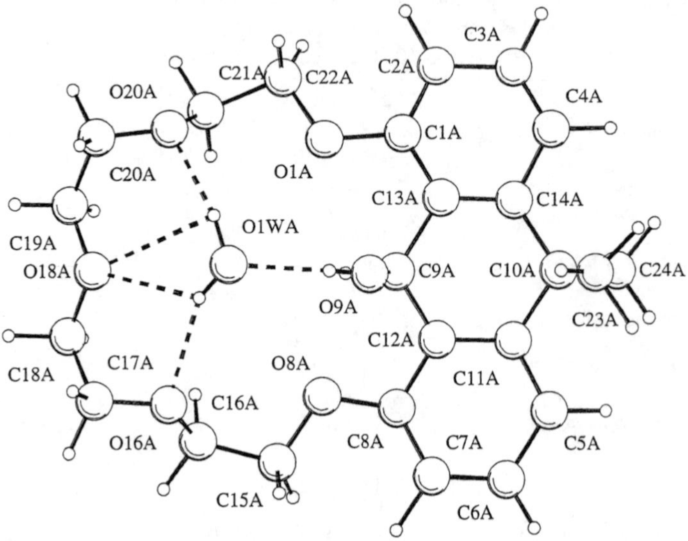

FIGURE 8.30.   One of the molecules of the crown ether in the asymmetric unit; dashed lines indicate intramolecular hydrogen bonds to the molecule of water of crystallization.

were located by geometry and included in the structure factor calculation but not refined. Full-matrix least squares led to convergence at $R = 6.2\%$. The molecular conformation is illustrated by Figure 8.30.

Inspection of various diagrams of the crystal packing strongly suggested an association between the two molecules in the asymmetric unit via a c-glide plane perpendicular to b. If this situation were true, the space group would be $P2_1/c$ rather than $P2_1$. Consequently, when the program PLATON in the WinGX suite was used to check this possibility, it indicated that, with the exception of the water molecules, all atoms exhibited closely the c-glide relationship. Furthermore, manual inspection of the reflection data indicated a number of $|F(h0l)|$ data with $l = 2n + 1$ whose values were significantly greater than zero, thus eliminating the presence of a c-glide operator. The same reflections had significantly non-zero values among the neutron data.

8.4.3   Neutron Diffraction Study

This part of the study was carried out by members of the group working independently at the ILL laboratory in Grenoble. Pressure on booking and using time allocated at this facility accounts for the exchange of information between the x-ray and neutron groups before final confirmation of the symmetry and x-ray structure.

The initial data was the original octant of wrongly assigned C-centered orthorhombic x-ray data. Later the workers were informed that it was actually an incomplete set of C-centered, z-axis unique monoclinic data, with γ very close to 90°, that needed transforming to monoclinic $P2_1$ symmetry; the unit cell, with y unique, was also supplied. Initially, however, the "orthorhombic" unit cell was located on a four-circle neutron diffractometer and data collected to $\theta = 41°$ with a wavelength of 1.538 Å. Later, in order to measure higher-angle data, the wavelength was changed to 1.312 Å. A large crystal of dimensions 2.5, 1.6, 0.2 mm was used in order to provide a sufficient number of intensities measurable at the neutron source. The final neutron data set recorded a complete asymmetric unit of data in space group $P2_1$.

Initially, the refined atomic coordinates from the x-ray study were used. Unexpectedly, there was no satisfactory agreement between the calculated and observed structure factors, although the unit cell parameters from both studies were close (Table 8.12). This situation was traced to idiosyncrasies in the unit cell shape. When the pseudo-orthorhombic unit cell was transformed to monoclinic P, two extremely similar unit cells could be chosen, as shown by Figure 8.31. If the β angle of the C unit cell had been exactly 90°, the transformed parameters for both cells would have been identical. As it was, one cell had been chosen for the x-ray study and the other for the neutron study. The unit cell parameters and hkl indices in the neutron study were modified appropriately. Figure 8.31 shows that only c and β are affected: $c = 13.688$ Å, $\beta = 117.93°$ (neutron experiment) and $c = 13.588$ Å, $\beta = 117.12°$ (neutron transformed). *It is important*, then, not to overlook the value of photographs, or other means, for checking carefully the Laue symmetry, when using automated diffractometer systems.

The neutron diffraction data refined satisfactorily to $R = 4.6\%$. One hydrogen atom from each of the water molecules, undetected by the x-ray study, was found from a difference Fourier map in this study.

8.4.4   Some Structural Features

The two molecules differ slightly but significantly in the patterns of hydrogen bonding, mainly with the water molecules, which feature is responsible for breaking the pseudo-$P2_1/c$ symmetry. In both molecules, the hydrogen atoms on the crown ether oxygen atoms O(9) donate to form intramolecular hydrogen bonds of high (2.57 Å) or medium (2.81 Å) strength, and bifurcated hydrogen bonds, for example to O(16) and O(18). There is no intermolecular hydrogen bonding.

Molecules A and B are geometrically similar and exhibit pseudo-mirror symmetry, as can be seen in Figure 8.30 for molecule A (in projection the m is a line joining O18 and C24 in each molecule; the asymmetry parameters (see Section 7.5.3) derived from torsion angles across these "m" planes are only 5.8° and 6.7°. The molecules possess considerable flexibility, which probably explains

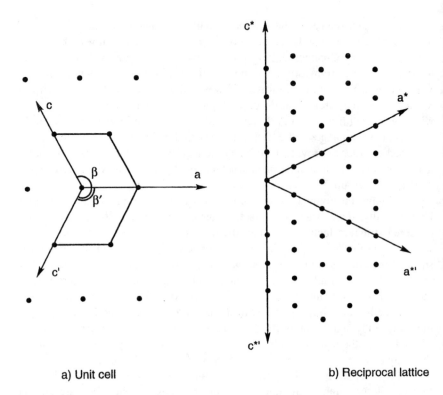

a) Unit cell   b) Reciprocal lattice

FIGURE 8.31.   Alternative choice of unit cell for crystals of the crown ether; the cell **a, c** was used
in the x-ray study, and **a, c′** in the neutron study.

the small differences in molecular geometry between the two molecules in the
asymmetric unit.

8.5   Determination of an Intermediate Size Crystal Structure Using Direct Methods

8.5.1   Introduction

In this section we provide a summary of the studies of a fungal peptide
molecule, cyclosporin H^a (CsH), that have led to the discovery of a previously
unknown conformation for this type of molecule. The unit cell contains 368
non-hydrogen atoms per unit cell (92 per asymmetric unit). The structure was

[a] B. S. Potter, *PhD Thesis*, University of London (2001).

determined in crystal form II to an atomic resolution of 0.80 Å, with 11,512 independent observed values of $I(hkl)$. For comparison, some data are provided below for the related crystal form I, for which the diffraction data are inferior.

8.5.2   What We Expect to Learn from the X-ray Structure of Cyclosporin H

Cyclosporins are cyclic peptides derived from fungi. The original and best known member of this family is cyclosporin A, or CsA,[a,b] which is a leading immunosuppressant drug used to prevent rejection of organ transplants, and also possesses antifungal properties. The importance of CsA can be judged by the fact that, since its discovery in 1976, an average of over 1000 scientific papers per year have appeared, and the total number currently increases by just under 2000 per year. Chemically there are 11 amino acid residues in CsA, some of which rarely, if ever, occur in mammals. These groups are linked, as in a protein, by peptide bonds (see also Chapter 10). In addition, residue 11 forms a peptide link looping back residue 1, generating a closed cyclic peptide. Many of the NH groups are N-methylated, that is, they are replaced by N–CH$_3$, which strongly influences the chemical and pharmacological properties. In addition, many of the amino acid side groups are also hydrophobic, so that they have no affinity with water or other hydrogen-bonding media. CsH[c,d] is a member of the cyclosporin family that is chemically very close to CsA but surprisingly shows no immunosuppressive or antifungal activity. This, in spite of the fact that the *only* difference between CsA and CsH[c] is at peptide 11, which has a valine side chain, $-C\alpha-CH(CH_3)_2$, in both compounds. However, in CsA the absolute configuration at the α-carbon is L, whereas in CsH it is inverted to the D form.

These facts are quite well established from biochemical studies. However, the crystal structure determination of CsH was undertaken in order to verify the configurational features, and to identify structural and surface features important for biological activity and for use in the future design of new cyclosporin derivatives. Prior to this determination of the structure of CsH it had generally been assumed, at least tacitly, in the literature, that the folding of CsH is similar to the open β-sheet structure of CsA.[e]

---

[a]J. F. Borel, Z. L. Kis, and T. Beveridge, in *The Search for Anti-Inflammatory Drugs* (Edited by V. J. Merluzzi and J. Adams) 27–63. Birkhäuser (1985).

[b]J. F. Borel, F. Dipadova, J. Mason, V. Quesniaux, B. Ryffel, and R. Wenger, *Pharmacology Review* **41**, 239 (1989).

[c]K. WenzelSiefert, C. M. Hurt, and R. Seifert, *Journal of Biological Chemistry* **273**, 24181 (1998).

[d]K. Kitegaki, H. Nagai, S. Hayashi, and T. Totsuka, *European Journal of Pharmacology* **337**, 283 (1997).

[e]R. B. Knott, J. Schefer and B. P. Schoenborn, *Acta Crystallographica* **C46**, 1528 (1990).

8.5.3   Crystallization and Data Collection

Crystals of CsH form I were grown over 21 days at $-20°C$ from methanol in partially sealed glass vials. Crystals of form II, which are isomorphous with form I, were grown under identical conditions but in the presence of magnesium perchlorate. This technique led to improvements in the quality of the X-ray diffraction pattern and, once the x-ray structures were completed, was thought to be associated with a stabilising effect on the structure in form II, especially the long MeBmt-1[a] side-chain (Figures 8.32 and 8.33), which is considerably disordered in the absence of magnesium perchlorate (form I). The largest crystals were approximately 1.0, 0.5, 0.3 mm, but were very unstable and disintegrated spontaneously on exposure to air. For collection of x-ray intensity data, crystals were mounted *rapidly* in glass capillary tubes which were immediately sealed (see Section 10.3.1). The unit cell parameters were measured and refined and x-ray data collected with Cu $K\alpha$ radiation on a Nonius CAD4 diffractometer at room temperature. The crystals were stable during the course of data collection, as indicated by regular measurement of three selected standard intensities that did not suffer appreciably from fading. Several ϕ-scan curves were collected for use with the semi-empirical method for absorption correction. Processing of the intensity data was carried out with the Nonius programs CAD77 and CADRAL to correct for fading, absorption, Lorentz and polarization effects, leading to a file of processed $|F(hkl)|$ with $\sigma(F)$ values. These files were checked to appraise the quality of the processed data set. The program CADSHEL then transforms the processed file into a format recognized by the SHELX suite of programs. X-ray data and data relating to the structure determination and refinement are in Table 8.13.

8.5.4   X-ray analysis

The program SHELXL-97 (see Section 7.2.13) as implemented in the suite WinGX, was used to solve and refine the structure of CsH, form II. The direct methods program selected was SHELXS-97. It was run initially with all parameters at default settings, with the following results:

First Application of SHELXS-97

Initially, 357 reflections with $1.5 < |E(hkl)| < 5.0$ were used; 5222 unique triple product relationships were used for phase annealing; subsequent expansion resulted in 655 phases which were refined using 39,357 unique TPRs; further refinement was carried out with 21,250 unique negative quartet relationships; 478 unique NQRs were used in phase annealing, wherein the best CFOM figure of

---

[a] MeBmt is an amino acid, (4R)-4[(E)-2-butenyl]-4-N-dimethyl-L-threonine, not normally found in proteins. It has a methylated amide nitrogen atom, as have five other of the CsH residues.

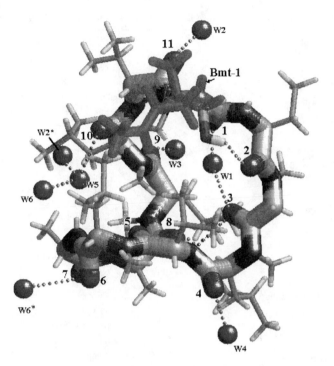

FIGURE 8.32.   Secondary structure of CsH, illustrating the convoluted folding and the water molecules' structure. Water molecule W2* is related by symmetry to water molecule W2 and forms a cluster with water molecules W5 and W6 in CsH form II. Water molecule W6 occurs only in form II, and links two symmetry-related CsH molecules; water molecules W1–W5 are common to both forms. Hydrogen bonds are shown as dashed lines. Main-chain C=O group numbers are in black. Water molecule oxygen atoms are dark spheres labeled W. The bonds in the CsH molecule are represented by "sticks"; the main chain radius is twice that of residue Bmt-1, and three times those of the remaining ten side-chains. Bmt-1 is displayed in a different shade to emphasize its important role in cyclosporin function. Elsewhere C is light grey, H white, N grey, and O dark grey. This diagram was generated using the program RasMol [R. Sayle, Glaxo, Greenford, UK (1994)].

merit was 0.1967; the CPU time was 71.4 s on a PC with a Pentium II processor, 64 MB RAM, and 4GB hard drive. There was no recognizable structure on the resulting E-map which was inspected with the program SXGRAPH. Because default input parameters were used, the number of phase sets refined was 256.

Diagnosis.   The program clearly failed to solve the structure at this attempt. The value of CFOM should be much closer to 0.0 for a correct solution. The program manual for SHELXL-97 advises, as one possibility, an increase in the number np of direct methods attempts, which was 256 in the first run; this parameter is the first entry on the instruction line TREF. The second entry nE on this line is the number

TABLE 8.13.   X-ray Structure Determination Parameters for Cyclosporin H

Molecular formula	$C_{62}H_{111}N_{11}O_{12} \cdot 6H_2O[\cdot 5H_2O]$
Formula weight/Da	1310.72 [1292.71]
Crystal system	Orthorhombic
Unit cell dimensions/Å	$a = 17.358(2)[17.299(1)]$
	$b = 19.515(1)[19.690(1)]$
	$c = 23.225(3)[24.137(2)]$
Cell volume/Å^3	7867(1) [8221(1)]
Space group	$P2_12_12_1$
Number of molecules per cell	4
Calculated density/g cm^{-3}	1.106 [1.052]
Solvent water molecules	Water molecules W1–W6
	[Water molecules W1–W5 and W6[a]]
Solvent clusters	W2 + W5 + W6* [W2 + W5]
Crystallization conditions at $-20°C$	$CH_3OH + Mg(ClO_4)_2$ [CH_3OH]
Crystal mounting	Wet in capillary tube
Diffractometry	$\omega/2\theta$ scans (CAD4)
Data collection time	1 min per reflection
Temperature/°C	20
Total number of data	12,343 [18,238]
Radiation and wavelength/Å	Cu $K\alpha$; 1.5418
Number of observed data with $I > 2\sigma(I)$	11,512 [5,043]
R_{int}	0.0341[0.204]
Maximum θ/deg	74.61 [69.99]
Resolution/Å	0.80[0.82]
Structure solution: direct methods	SHELXS-97
Refinement	Anisotropic
Non-H atoms	
H-atoms (calculated positions)	Isotropic
Structure refinement, full-matrix least-squares	SHELXL-97
Data/restraints/parameters	11,462/0/902
	[9,826/11/959]
Precision/Å (bond lengths)	0.0035[0.0045]
Precision/deg (bond angles)	0.20 [0.35]
Flack parameter	$-0.42(0.27)^b$
R-factor	0.066 [0.088]

[a] Data are for crystal form II, with crystal data for form I given in brackets where there is a significant difference from form II.
[b] Indicates that the correct absolute configuration has been chosen, the Flack parameter being within 3σ of zero.

of reflections used initially. Increasing this number may also result in successful direct methods phasing. On this occasion, however, it proved to be necessary only to increase np to 5000, which means that 5000 random starting sets of phases would be generated instead of 256 and refined using the direct methods protocols.

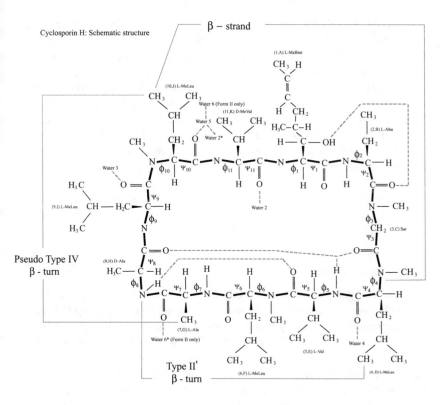

FIGURE 8.33. Schematic structure of cyclosporin H: the symbols *, ----, and IIIII indicate, respectively, a symmetry-related species, a hydrogen bond, and secondary structural features.

Second Application of SHELXS-97

Using 5000 random starting sets resulted in the program carrying out 125 parallel refinements at any one time. The CPU time increased to 1045.5 s and the best CFOM dropped dramatically to 0.0947. This time the E-map for this best solution revealed 88 out of the 95 atoms in the asymmetric unit in excellent stereochemical locations.

Completion and Refinement of the Structure

Isotropic refinement of this model, with all atoms treated as carbon, gave an R-factor of 0.179. The remaining atoms in the structure, including water oxygens, showed up in an electron density map inspected in SXGRAPH. In the next cycle of refinement all atoms were included, still as carbon, and the R-factor dropped to 0.154. The input file was then modified to convert all atoms to their true chemical identity. In WinGX this can be done using the facilities in SXGRAPH. Atom names were changed to be chemically meaningful as part of the same operation. Further refinement, this time with all thermal parameters converted to anisotropic, gave an R-value of 0.084. Hydrogen atoms were then added in geometrically acceptable positions, a procedure again carried out in SXGRAPH. The model then refined to $R = 0.066$, with the coordinates and isotropic thermal parameters for the hydrogen atoms tied to their heavy atoms, carbon or nitrogen. The Flack parameter (see Section 6.5.1) refined to a value of -0.42 (Table 8.13) which is approximately zero within three standard deviations (3×0.27). We conclude that the structure refined has the correct hand, and corresponds to the assignments of chirality made independently from chemical considerations as published in the literature.

Table 8.13 provides details of a similar analysis for the less well resolved crystal structure of CsH, form I. Many items from this table suggest the structure of form I to be inferior (see also Problem 8.4).

Results and Discussion

A principal outcome of this study is the observation that the *local* chemical change between CsA and CsH, the inversion of the chirality at $C\alpha(11)$ from L to D, is associated with a *major structural* transformation from open β-sheet in CsA to a highly convoluted conformation in CsH. Furthermore, the structure of CsH unlike CsA, which has only one associated water molecule, is heavily solvated. All six water molecules in the CsH structure II form hydrogen bonds directly with the CsH molecule, five of them to main chain C=O or N–H groups, and one to a side chain OH group (residue 1). The symmetry generated counterparts of water molecules 2 and 6 form a solvent cluster with water molecule 5. It is possible that the conformational changes in CsH, compared to CsA, are somehow driven by the presence of these solvent molecules. It is, however, somewhat surprising that there are so many waters attached to CsH, as water was excluded at all times during handling of the material; in fact, CsH proved to be extremely difficult to crystallize. The observation that the presence of magnesium perchlorate during crystallization provided a stabilizing effect on the more mobile side chains is also unexplained since, although structural magnesium perchlorate does not occur in any of the crystals, the presence of both Mg^{2+} and Cl^- ions was detected using energy dispersive x-ray fluorescence spectroscopy.

The minor chemical differences between the two cyclosporins invoke unpredictably major differences in their three-dimensional structures. The peptide bond at position 9 is *cis* in CsA, while the peptide bonds in the highly convoluted loop conformation of CsH are all *trans*. CsH forms I and II both adopt the same highly convoluted secondary structure (Figure 8.32). Since the only chemical difference between CsA and CsH is the L to D inversion of MeVal-11, it is surprising that this outwardly minor change produces a significant structural transformation from a fairly open, mainly β-sheet structure in CsA[a,b] to the highly twisted conformation observed in CsH. The observed main chain ϕ, ψ angles (see Section 10.1) exhibit large differences, with an average absolute change in ϕ of $37°$ and in ψ of $56°$.

The observed, and unexpected, conformation of CsH offers a basis for explanation of the significant differences in pharmacological activity compared to the standard cyclosporin drug CsA, and this study illustrates the value of high-resolution crystallography in rational structure-driven drug design strategies. Only very few molecular peptide structures, comparable in size with the cyclosporins, have to date been determined to 0.9 Å resolution or better.[c]

8.6 Concluding Remarks

No description of the process of x-ray crystal structure analysis can be as complete or as satisfying as a practical involvement with the subject. In teaching crystallography and structural chemistry, projects that include crystal structure determinations have become increasingly important. However, in order to attempt to replicate the practical side of structure analysis, insofar as is possible in isolation from the laboratory, problems on this topic additional to those at the end of this chapter are given in Chapter 11; they involve the program system XRAY* and the several sets of data that accompany it. They have been designed to give practice, albeit in two dimensions, with the fundamental techniques of solving crystal structures, given the crystal and reflection data. The programs can be executed on any IBM-type PC. We encourage the reader strongly to tackle these problems and so engage in a practical way with the techniques of crystal structure analysis. We refer the reader also to Appendix A10, wherein are listed the many program systems now freely available to the crystallographer.

Bibliography

Published Structure Analyses

Acta Crystallographica [the early issues (1948–1968) are most suitable for the beginner].
Journal of Crystallographic and Spectroscopic Research, Journal of Chemical Crystallography (formerly *Journal of Crystal and Molecular Structure*).
Zeitschrift für Kristallographie.

[a] H.-R. Loosli, H. Kessler, H. Oschkinat, H. P. Weber, T. J. Petcher, and A. Widmer, *Helvetica Chimica Acta* **68**, 682 (1985).
[b] H. Kessler, M. Kock, M. T. Wein, and M. Gehrke, *Helvetica Chimica Acta* **73**, 1818 (1990).
[c] S. Longhi, M. Czjzek, and C. Cambillau, *Current Opinion in Structural Biology* **8**, 730 (1998).

General Structural Data

KENNARD, O. et al. *Cambridge Structural Database* (CSD)—see page 499 *ff* herein.
WYCKOFF, R. W. G., *Crystal Structures*, Vols 1–6 (1963–1966, 1968, 1971), New York, Wiley.

Problems

8.1. The unit cell of euphenyl iodoacetate, $C_{32}H_{53}O_2I$, has the dimensions $a = 7.26$, $b = 11.55$, $c = 19.22$ Å, and $\beta = 94.07°$. The space group is $P2_1$ and $Z = 2$. Figure P8.1 is the Patterson section $(u, \frac{1}{2}, w)$.

FIGURE P8.1.   Sharpened Harker section, $P(u, \frac{1}{2}, w)$, for euphenyl iodoacetate.

(a) Determine the x and z coordinates for the iodine atoms in the unit cell.

(b) Atomic scattering factor data for iodine are tabulated below; temperature factor corrections may be ignored.

$(\sin\theta)/\lambda$	0.00	0.05	0.10	0.15	0.20	0.25	0.30	0.35	0.40
f_1	53.0	51.7	48.6	45.0	41.6	38.7	36.1	33.7	31.5

   Determine probable signs for the reflections 001 ($|F_o| = 40$), 0014 ($|F_o| = 37$), 106 ($|F_o| = 33$), and 300 ($|F_o| = 35$). Comment upon the likelihood of the correctness of the signs which you have determined.

(c) Calculate the length of the shortest iodine–iodine vector in the structure.

8.2. The following $|E|$ values were determined for the [100] zone of a crystal of space group $P2_1/a$. Prepare a Σ_2 listing, assign an origin, and determine signs for as many reflections as possible, and give reasons for each step that you carry out. In this projection, two reflections for which the indices are not both even may be used to specify the origin.

$0kl$	$|E|$	$0kl$	$|E|$
0018	2.4	0310	1.9
011	1.0	0312	0.1
021	0.1	059	1.9
024	2.8	081	2.2
026	0.3	0817	1.8
035	1.8	011,7	1.3
038	2.1	011,9	2.2

8.3. The chart in Figure P8.2 shows some $|E|$ values taken from the $hk0$ data for potassium hydrogen squarate. Take an origin at the center of a sheet of centimeter graph paper and copy the $|E|$ values on to it, using the top left portion of each appropriate square. For each $|E|$ value plotted, add the corresponding symmetry-related $|E|$ values to the other three portions of the graphical reciprocal space representation. Remember to change the signs of $|E|$ in accordance with the space group. Next, draw an identical chart on transparent paper, but with the $|E|$ values in the bottom right portion of each appropriate square.

(a) Obtain a Σ_2 listing: take each plotted $|E|$ value in turn on the original chart and superimpose the transparency, with the origin of the transparency over the chosen $|E|$ value and keeping the two sets of h, k axes in register. Look for any superimposed $|E|$ values. A Σ_2 triplet is given by the $|E|$ value on the original chart under the origin of the transparency, together with the superimposed values, with h, k read one from the original and the other from the transparency. Thus, with the origin of the transparency on the original $|E(300)|$, we read 840 on the original and 540 on the transparency. Set up the Σ_2 listing as follows:

$\mathbf{h}$	$|E_\mathbf{h}|$	$\mathbf{k}$	$|E_\mathbf{k}|$	$\mathbf{h-k}$	$|E_{\mathbf{h-k}}|$	$|E_\mathbf{h}||E_\mathbf{k}||E_{\mathbf{h-k}}|$
300	1.75	840	1.79	$\overline{5}40$	1.92	6.0
$\vdots$						
700	2.26	...				

The rationale for the graphical procedure may be seen from Figure P8.3.

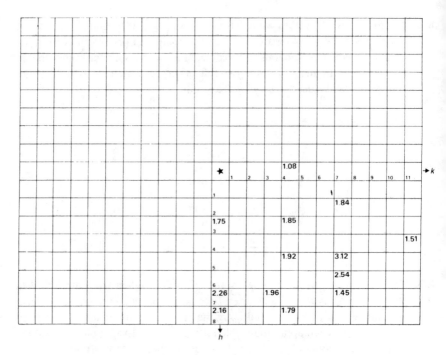

FIGURE P8.2.   Chart of $|E(hk)|$ data for KHSQ.

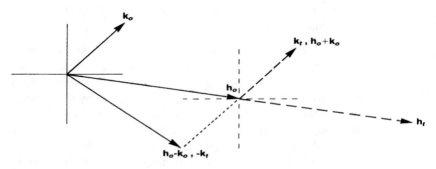

FIGURE P8.3. Σ_2 relationships: subscript o refers to the original chart, and subscript t refers to the transparency. If the origin of the transparency is placed over a chosen $\mathbf{h}_o$, it may be seen that coincidences of $|E|$ values given by $+\mathbf{k}_t, \mathbf{h}_o + \mathbf{k}_o$ or by $-\mathbf{k}_t, \mathbf{h}_o - \mathbf{k}_o$ will represent Σ_2 triplets. It should be noted that this technique applies only to centrosymmetric projections of space groups.

(b) Assign an origin in accordance with the rules discussed in the previous chapter and allocate signs to as many reflections as possible; use symbols if necessary. It may be assumed that the products $|E_\mathbf{h}||E_\mathbf{k}||E_{\mathbf{h}-\mathbf{k}}|$ are all sufficiently large for the indications to be accepted.

8.4. List the factors in Table 8.12 that suggest that CsH crystal form II is better ordered than crystal form I; relevant data for form I are in brackets in this table.

Further problems in structure analysis, involving the interactive program system XRAY* that accompanies this book, are addressed in Chapter 11.

9

X-ray Structure Determination with Powders

9.1 Introduction

Since the earlier editions of this book, developments have taken place with the x-ray examination of polycrystalline materials that have led to complete structure determinations of medium-sized molecules, with up to 60 atoms in the asymmetric unit. It is appropriate, therefore, to include a short account of this topic in the present edition. Further, more detailed discussions of powder methods can be found in the literature.[a,b,c]

The powder method was devised by Hull soon after the discovery of x-ray diffraction. However, until about three decades ago, this technique was used primarily for identification, analysis, and phase-diagram studies (Figures 9.1 and 9.2) in the solid state, particularly among alloys, refractory materials, catalysts, and ferroelectrics.

The modern developments in the powder method have added a new and powerful tool for the determination of the structure of the many substances that could be obtained *only* in microcrystalline form (particle size *ca* 10^{-3} mm).

9.1.1 Structure Determination Scheme

A scheme for structure determination from powder data may be set out in a number of stages:

1. Indexing the powder diffraction pattern, and determining the unit cell;
2. Identifying the space group from the diffraction record, as far as possible;

---

[a] W. I. F. David, K. Shankland, Ch. Baerlocher, and L. B. McCusker (Editors), *Structure Determination from Powder Diffraction Data*, International Union of Crystallography/Oxford University Press, Oxford (2002).

[b] K. D. M. Harris, M. Tremayne, and B. M. Kariuki, *Angewadte Chemie International Edition* **40**, 1626 (2001).

[c] K. D. M. Harris and M. Tremayne, *Angewadte Chem. Mater.* **8**, 2554 (1996).

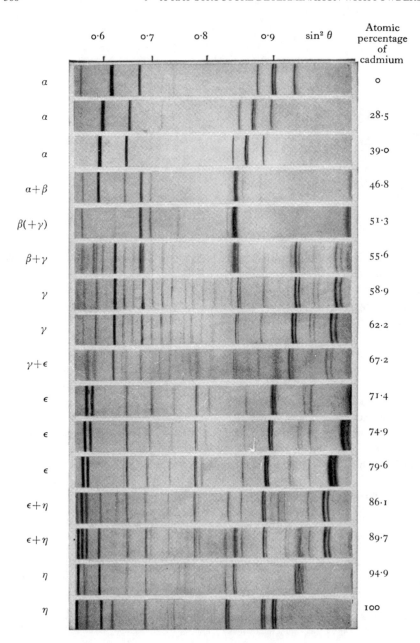

FIGURE 9.1.   X-ray powder photographs of silver–cadmium alloys, taken with Fe $K\alpha$ x-radiation. Silver, the α-phase, takes up cadmium into solid solution to the extent of *ca* 40% Cd. The unit-cell side a of the (cubic) α-phase is proportional to the concentration c of cadmium (Vegard's law). Since the metallic radius of cadmium (1.50 Å) is greater than that of silver (1.44 Å), lines in the α-phase move to lower θ as c_{Cd} increases. Several films show $\alpha_1\alpha_2$ splitting at high θ. [After Westgren and Phragmén, *Metallwirtschaft*, **7**, 700 (1928).]

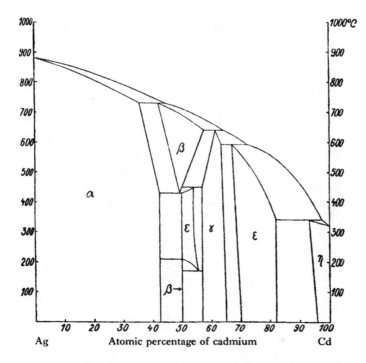

FIGURE 9.2. Phase equilibrium diagram for the silver–cadmium system; silver and cadmium form a continuous range of solid solutions, up to 42.5% Cd. The range at the high cadmium end is much smaller. [After Westgren, *Angewandte Chemie* **45**, 33 (1932).]

3. Decomposing the powder pattern into integrated intensity data, $I(hkl)$;
4. Solving the phase problem;
5. Refining the model structure.

As with single-crystal analysis, the first three stages lead to a set of relative $|F_o(hkl)|$ data. The phase problem may be tackled by Patterson or direct methods, the traditional approach, or with direct-space techniques, such as Monte Carlo or genetic algorithms that make use of the power of modern computers. The refinement of the structure follows the Rietveld technique, developed in the late 1960s, and which has been a driving force for the extension of the powder method to the more complex materials now studied.

The experimental data-collection procedures have progressed greatly since the initial film-based cameras. Nevertheless, these early methods provide a straightforward introduction to powder studies, so we shall consider some of the initial methods that have led up to the techniques of the present day.

9.2   Basis of the Powder Method

Figure 9.3 illustrates the geometrical basis of x-ray powder photography. Monochromatic x-rays enter along a diameter of the cylindrical camera, through a small hole in the film, and exit through a diametrically opposed aperture into a beam trap. In this type of camera, the specimen is a very finely ground powder, either contained in a borosilicate glass capillary or mixed with an adhesive and formed into a cylindrical shape. The mount for the specimen is provided with slides, so that the specimen can be adjusted to rotate on its axis within its own volume. The rotation increases the effective randomness of the orientations of the crystallites in the sample, but is not a necessary feature of the method.

A typical reflection is shown at a scattering angle of 2θ, twice the Bragg angle θ. From the geometry of the camera, it is clear that θ is obtained from

$$\text{arc } A_1 A_2 = 4R\theta \tag{9.1}$$

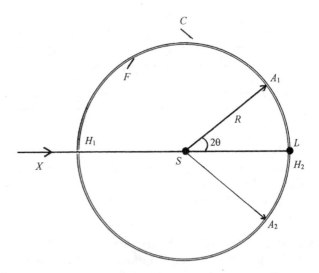

FIGURE 9.3.   Basic, Debye–Scherrer arrangement for x-ray diffraction from polycrystalline specimens: C, camera housing of radius R, with film strip F fitted tightly against its inner surface; X, incident x-ray beam travelling along a diameter of the camera; H_1 and H_2, two small, diametrically opposed holes in the camera and film; S, the powder specimen; L, trap for the transmitted x-ray beam; A_1 and A_2, positions of two arcs of one and the same powder reflection; 2θ, the angle of scatter (twice the Bragg angle). Since the crystallites lie in all possible orientations, the Bragg equation is satisfied by any generator of a cone of semi-vertical angle 2θ and axis along the transmitted x-ray beam. The complete cone of diffraction intercepts the cylindrical film strip in circular arcs (see Figure 9.3); arc $A_1 A_2 = 4R\theta$. If all possible orientations do not exist, as when the crystallites are too large, the arcs are broken to give a 'spotty' powder photograph.

where R is the radius of the film. Since the Bragg equation is satisfied for this reflection, we have

$$\sin^2 \theta = \lambda^2/(4d^2) \tag{9.2}$$

so that d-values are also obtainable. An important limitation of the powder method is immediately evident, namely, that we can measure only one geometrical parameter, the Bragg angle θ for each reflection. For a structure analysis, we know that we must obtain the unit-cell dimensions, the space group, and the hkl indices for each reflection together with its integrated intensity. Until the development of high-speed computing and improvements in data collection, the task was formidable, although some simplification arises for high-symmetry materials.

9.3 Data Collection

In powder diffraction work, many fewer reflections are available than with single-crystal x-ray crystallography. In addition, the problem of determining the unit cell, indices, space group, and intensities of reflections is more difficult.

Powdered specimens contain numerous very small crystallites arranged in completely random orientations, so that there is no explicit information on the location of a reciprocal lattice point other than its distance d^* from the origin. A powder pattern is a collapsed, tangential projection of the weighted reciprocal lattice, in which each diffraction maximum is characterized by its Bragg angle θ and its intensity.

Powder lines have a significant breadth, and tend to merge, even to overlap, with both increasing θ and increasing unit-cell size. In order that powder lines should be sufficiently well resolved for their positions to be measured to better than $0.01°$ in θ, it is necessary that the x-radiation is strictly monochromatic and very finely focussed on to the film, with a low intensity of background radiation.

9.3.1 Guinier-type Cameras

One method of obtaining the desired experimental conditions is with a Guinier-type camera, using a curved crystal monochromator. This x-ray monochromator consists essentially of an asymmetrically ground, curved crystal of quartz, silicon, or germanium. The $(10\bar{1}0)$ diffraction planes of a quartz crystal, or the (111) planes of silicon or germanium, are arranged to lie at a prearranged angle to the crystal surface, Figure 9.4a. The crystal is ground with a cylinder radius of $2R$ and then bent over the fitting area with a radius R. X-rays from the line focus of the x-ray tube strike the crystal as a divergent bundle. In the crystal, a narrow wave band is separated from the polychromatic source by diffraction according to the Bragg equation. The diffracted rays leave the crystal

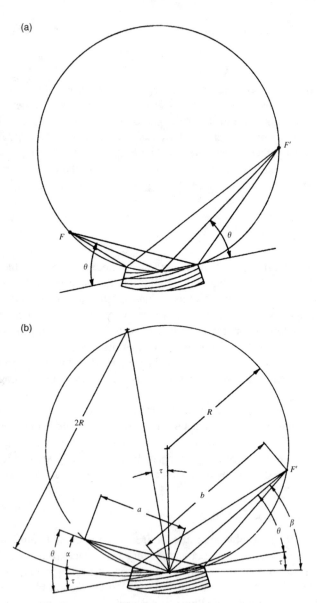

FIGURE 9.4.   (a) Curved crystal monochromator, showing the incident divergent bundle of rays from the x-ray tube at F reflected at the glancing angle θ and the bundle convergent at the point F'. (b) Curved crystal, showing the determinants involved in the setting of the crystal; the constancy of a and b with change of wavelength depends on the correct choice of τ. (c) Principle of the Guinier method, showing the disposition of x-ray tube, monochromator, powder specimen, and focusing cylinder. The focusing cylinder carries a photographic film, or can be replaced by a scintillation counter (diffractometer) or an imaging plate (imaging camera). [Courtesy Huber Diffraktionstechnik GmbH, D-83253 Rimsting, Germany.]

(c)

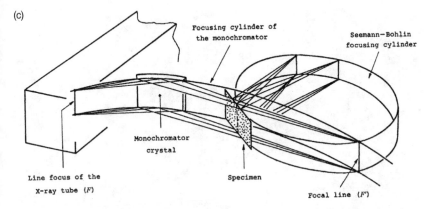

FIGURE 9.4. (Continued)

as a convergent bundle, at the glancing angle θ. The most intense spectral line is the $K\alpha_1$, and the crystal is set to reflect this wavelength, with the elimination of other wavelengths from the diffracted beam.

Figure 9.4b shows the location of the various determinants involved in the setting of the crystal monochromator. The distance a between the line focus F of the x-ray tube and the centre of the crystal, and the distance b between the centre of the crystal and the focal line F' are maintained constant for different wavelengths by judicious choice of τ, the angle between the $(10\bar{1}0)$ or (111) planes and the crystal surface. This feature allows a change of x-ray tube (and wavelength) without major alterations in the experimental arrangement.

The principle of Guinier operation is illustrated by Figure 9.4c. The Guinier-type camera can function in transmission geometry, as with low–medium absorbing specimens, or in reflection geometry, for the case of strongly absorbing materials. In the photographic technique, the x-ray film is mounted on the Seeman–Bohlin focussing cylinder. Figure 9.5 shows a complete Guinier-type film-camera assembly for flat, powdered specimens. If the final film is assessed with a Vernier measuring instrument, the desired accuracy in θ can be achieved.

9.3.2 Image Plate Camera

In a modern development of the Guinier technique, the x-ray film is replaced by an imaging plate, leading to greatly improved speed of data collection without loss of the accuracy that is so important in powder indexing. The x-ray imaging plate is a flexible strip, like an x-ray film, that contains a metastable phosphor. When an x-ray photon hits the image plate its energy is stored, to be released subsequently when a laser beam is scanned over its surface. The advantages

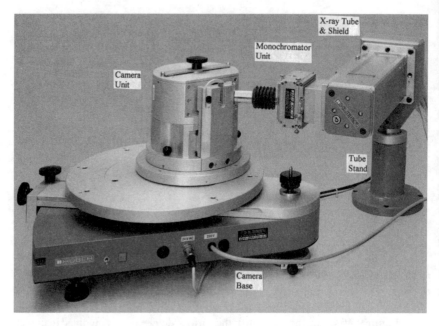

FIGURE 9.5. Huber–Guinier Powder Camera assembly. [Courtesy Huber Diffraktionstechnik GmbH, D-83253 Rimsting, Germany.]

of the imaging plate are parallel data collection over a large area, with high dynamic range, good resolution, and digital readout. The image-recorded intensity tends to decay with time, so that the plates are best handled in darkened conditions.

Figure 9.6 is a schematic diagram of the Huber Imaging-Plate Guinier camera. The incident beam is monochromatic $K\alpha_1$ x-radiation, focused on to a flat powder specimen. The camera enclosure may be evacuated, so as to decrease the background scattering. The imaging plate containing the powder pattern is scanned by a laser, and software handles the scanned data to produce a profile record and data files. Collection times may be reckoned in minutes, particularly where a synchrotron source is available. After use, the imaging plate is restored to its original condition by means of the halogen erasure lamp (see also Section 4.6.3). The data files are provided in the usual format for Rietveld refinement (q.v.). Fuller details of the performance of the camera are available from the manufacturer.[a]

[a] http://www.xhuber.com

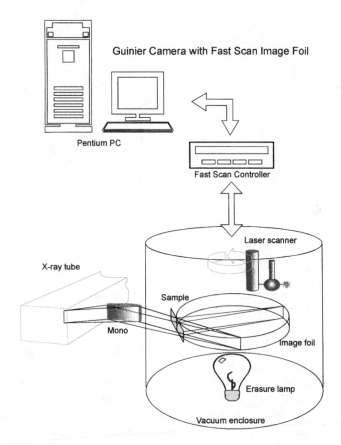

FIGURE 9.6.   Schematic diagram of the Huber Imaging Powder Camera. The powder pattern, collected on the imaging plate, is scanned by a laser and then digitized for computer analysis. [Courtesy Huber Diffraktionstechnik GmbH, D-83253 Rimsting, Germany.]

9.3.3   Powder Diffractometers

Many x-ray powder diffractometers use the Bragg–Brentano parafocusing principle in reflection geometry, Figure 9.7. The $K\alpha$ x-rays impinge on a monochromator (Figure 9.4) and converge at a knife edge where residual $K\alpha_2$ rays are cut off. Scattered x-rays from the flat specimen converge at a receiving slit in front of the detector. The detector rotates about the sample axis synchronously through twice the angular rotation of the sample in $\theta/2\theta$ scans. The effective source, receiving slit and sample lie on the focussing circle, which has a radius dependent on θ. It is important to avoid preferred orientation in the flat powder sample when

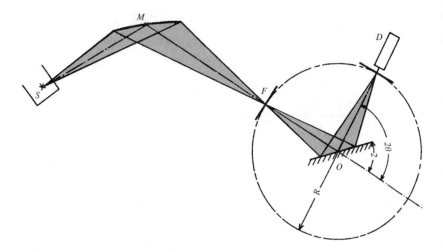

FIGURE 9.7.   Schematic diagram of a powder diffractometer with Bragg–Brentano parafocusing: S, x-ray source focus; M, monochromator; F, adjustable knife edge to cut off residual $K\alpha_2$; O, rotation axis; D detector; R, radius of focussing circle. [After Huber Diffraktionstechnik GmbH, D-83253 Rimsting, Germany.]

preparing it for diffractometer.[a] Figure 9.8 illustrates a modern high resolution powder diffractometer assembly.

The use of synchrotron radiation leads to better resolution in general, but in view of the cost of SR facilities, it is desirable to carry out a preliminary x-ray examination so as to obtain unit-cell data and to index the pattern. Particular advantages of synchrotron radiation are the high intensity and excellent vertical collimation, leading to greatly improved resolution, compared with laboratory x-ray sources. These features decrease the difficulties caused by overlapping reflections, and have enabled structures of considerable complexity to be solved, such as that of the compound $La_3Ti_5Al_{15}O_{37}$, which has 60 atoms in the asymmetric unit.[b]

9.3.4   Diffractometry at a Neutron Source

We mention this topic briefly because of its importance in structural studies. A pulsed-neutron source can be used with a diffractometer for time-of-flight (TOF) studies with powder samples. The resolution is high and the peak shape, though complicated, is well understood and can be modelled well in a refinement process. Because the neutron scattering length (form factor) is more or less constant over

[a] W. Parrish and T. C. Huang, *Advances in X-ray Analysis* **26**, 35 (1983).

[b] R. E. Morris, J. J. Owen, J. K. Stalick, and A. K. Cheetham, *Journal of Solid State Chemistry* **111**, 52 (1994).

(a)

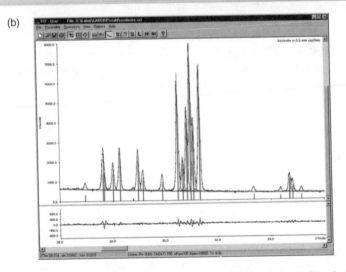

(b)

FIGURE 9.8. (a) Stoe STADI P Powder Diffractometer System: the radiation is $K\alpha_1$, in the range from Fe to Mo, the focusing circle diameter is 130 mm, the 2θ range is 0–140°, and the minimum step size is 0.0005°. Intensities are measured by a scintillation counter, and the whole operation is handled by built-in structure-solving software. (b) Diffractogram of solecite, a Ca-containing zeolite, measured in the Debye–Scherrer transmission mode in a capillary of 0.3 mm diameter: calculated and experimental peaks (with probable errors indicated), positions of main peaks (vertical lines) and, below, the difference curve. [Courtesy Stoe & Cie GmbH, Darmstadt, Germany.]

the whole range of $(\sin\theta)/\lambda$, atomic coordinates and temperature factors can be obtained with high precision. The method can be applied at high pressures and at low temperatures.

The structure of benzene at low temperature has been studied by this technique.[a] Notwithstanding its highly symmetrical shape, benzene crystallizes in the orthorhombic system, with space group *Pbca* with $a = 7.3551(3)$, $b = 9.3712(4)$, $c = 6.6994(3)$ Å, and four molecules in the unit cell; the figures in parentheses are the estimated standard deviations (esds) of the parameters. Thus, the molecular symmetry that is confirmed crystallographically is only $\bar{1}$. In practice, deuterated benzene was used because the incoherent scattering from hydrogen results in a very high background level; deuteration of hydrogen-containing materials is standard practice with neutron experiments. Furthermore, the neutron scattering length for deuterium is positive ($+0.65$) whereas that for hydrogen is -0.38.

The precision of the analysis can be judged from the following bond length parameters, quoted in Å:

$$C_1\text{–}C_2 = 1.3969(7) \qquad C_1\text{–}D_1 = 1.0879(9)$$

$$C_2\text{–}C_3 = 1.3970(8) \qquad C_1\text{–}D_1 = 1.0869(9)$$

$$C_3\text{–}C_1 = 1.3976(7) \qquad C_1\text{–}D_1 = 1.0843(8)$$

The mean values for C–C and C–D are 1.3972(5) Å and 1.0864(7) Å, respectively. The differences between the individual C–C and C–D bond lengths are not significant at the 3σ level, so that there is no evidence to show that the geometry of benzene is other than truly hexagonal. For further discussions on neutron diffractometry, the reader is referred to the literature.[b]

9.4   Indexing Powder Patterns

The simplest powder diffraction pattern derives from cubic crystals. In the cubic system, (9.2) may be written as

$$\sin^2\theta = \frac{\lambda^2}{4a^2}(h^2 + k^2 + l^2) = \frac{\lambda^2}{4a^2}N_c \tag{9.3}$$

where N_c is the integer sum $(h^2 + k^2 + l^2)$. It follows that the values of $\sin^2\theta$ must exhibit integer ratios with cubic crystals, within the limits of experimental error, and that the integer N_c follows one of three clearly defined patterns: $1, 2, 3, 4, \ldots$

---

[a] G. A. Jeffrey, J. R. Ruble, R. K. McMullan, and J. A. S. Pople, *Proceedings of the Royal Society* A **414**, 47 (1987).

[b] See Bibliography (Young and David et al.)

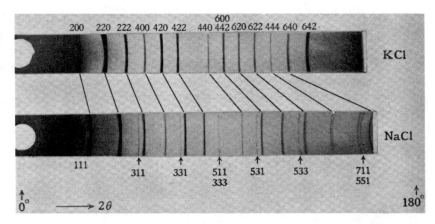

FIGURE 9.9.   Indexed x-ray powder photographs for KCl and NaCl. Both structures have the space group $Fm3m$, but the apparent absence of powder lines with h, k, and l all odd in KCl makes it pseudo-$Pm3m$ with a spacing $a_{apparent} = a_{true}/2$. [After Bragg, *The Crystalline State*, Vol. 1, Bell (1949).]

for a P unit cell; $2, 4, 6, 8, \ldots$ for I; and $3, 4, 8, 11, \ldots$ for F, according to the limiting conditions. Hence, the indexing of a cubic substance is normally a simple matter. But even with this high symmetry, more than one reflection can have the same value of θ. For example, pairs such as 300 and 221 ($N_c = 9$), and 411 and 330 ($N_c = 18$) are superimposed on the x-ray film. Certain values of N_c, such as $7, 15, 23, 28, \ldots$, cannot arise in the cubic system, because they cannot be derived as the sum of the squares of three integers. Although the unit cell can be identified readily, there is already a problem in assigning the correct relative intensities to these superimposed reflections.

It is, perhaps, interesting here to note that Figure 9.9 shows indexed powder photographs of sodium and potassium chlorides, which were the first structures to be analyzed by x-rays. They both have the NaCl-type structure, but the apparent systematic absences in the pattern for potassium chloride caused it to be considered as primitive cubic at first. The explanation here is that the K^+ and Cl^- species are isoelectronic. Hence, the reflections for which h, k, and l are all odd integers, those that depend for their intensity on $f(K^+) - f(Cl^-)$, are absent or only very weak at best; they are accidental absences, not dependent on the symmetry.

Powder patterns may be indexed easily by trial and error for simple crystal structures of the tetragonal and hexagonal systems, by development of (9.3), but as the crystal structure becomes more complex, with lower symmetry and larger unit-cell dimensions, the problem is enhanced: the powder lines become very close, merging into and overlapping one another, so that indexing and intensity measurement become increasingly difficult.

9.4.1   General Indexing

The indexing of powder diffraction patterns is carried out today by computer programs, and success in this process depends to some extent on the complexity of the structure, and to a very large extent on the accuracy of the primary θ-data. An early general procedure for indexing was set out by Ito.[a]

We recall the general equation for $d^*(hkl)$, developed in Chapter 3:

$$d^{*2}(hkl) = h^2 a^{*2} + k^2 b^{*2} + l^2 c^{*2} + 2kl b^* c^* \cos \alpha^*$$
$$+ 2lh c^* a^* \cos \beta^* + 2hk a^* b^* \cos \gamma^* \tag{9.4}$$

Generally, this equation is cast in terms of Q-values, for convenience, such that $Q_{hkl} = d^{*2}(hkl)$, $Q_A = a^{*2}, \dots, Q_D = 2b^* c^* \cos \alpha^*, \dots$, and so on, so that

$$Q_{hkl} = h^2 Q_A + k^2 Q_B + l^2 Q_C + kl Q_D + lh Q_E + hk Q_F \tag{9.5}$$

It follows that

$$Q_{hkl} = 4 \frac{\sin^2 \theta (hkl)}{\lambda^2} \tag{9.6}$$

Magnesium Tungstate

A set of 40 Q-values for magnesium tungstate, $MgWO_4$, collected by a film technique is listed in Table 9.1. Any three noncoplanar reciprocal lattice vectors define a possible unit cell. A unit cell thus chosen may not be the smallest, conventional unit cell, but it may be transformed as we discuss shortly. We will consider Ito's method by means of this hand-solved example.

Normally, there will be no *a priori* knowledge of three suitable vectors that are non-coplanar. Hence, we consider the first three lines in the diffraction pattern, and label them initially as Q_{100}, Q_{010}, and Q_{001}, so that we have $a^* = 0.1761$, $b^* = 0.2131$, and $c^* = 0.2683$ Å. We look through the pattern for multiples of these values, so as to improve the values of the reciprocal parameters by averaging the results. For example, $Q_{200} = 4Q_{100} = 0.1240$, and line number 7 at $Q = 0.1239$ is evidently Q_{200}, within experimental error. Proceeding in this way as far as possible, we obtain the average values $a^* = 0.1761$, $b^* = 0.2131$, and $c^* = 0.2684$ Å.

Next, we consider a zone in which one index is zero, say, $hk0$. If the angle γ^* is not 90°, we would expect to find two lines symmetrically disposed about the position that would correspond to Q_{110} if γ^* was 90°. For example, assuming correctness so far, $Q_{100} + Q_{010} = 0.0764$. In the list of Q-values, we have line 4 at 0.0764, and no pair of lines symmetrically disposed about this position, as

[a] T. Ito, *Nature* **164**, 755 (1949).

TABLE 9.1.   Values of Q from the Powder
Diffraction Pattern of Magnesium Tungstate

Line number	Q	Line number	Q
1	0.0310	21	0.3246
2	0.0454	22	0.3322
3	0.0720	23	0.3364
4	0.0764	24	0.3418
5	0.1164	25	0.3451
6	0.1186	26	0.3505
7	0.1239	27	0.3646
8	0.1649	28	0.3723
9	0.1695	29	0.3824
10	0.1816	30	0.4016
11	0.1957	31	0.4083
12	0.2077	32	0.4432
13	0.2123	33	0.4465
14	0.2386	34	0.4500
15	0.2436	35	0.4608
16	0.2517	36	0.4659
17	0.2563	37	0.4770
18	0.2793	38	0.4863
19	0.2884	39	0.4918
20	0.3055	40	0.4960

TABLE 9.2.   Some Calculated
and Observed Q_{hk0}-values for
Magnesium Tungstate

$hk0$	Q_{calc}	Q_{obs}
110	0.0764	0.0764
210	0.1694	0.1695
310	0.3244	0.3246
120	0.2126	0.2123
220	0.3056	0.3055

there would be if γ^* was not 90°. Thus, we conclude that line 4 is 110, and that $\gamma^* = 90°$.

We expect now to find other Q_{hk0}-values among those listed in Table 9.1. For example, $Q_{210} = 4Q_A + Q_B = 0.1694$, so that the line at 0.1695 corresponds to Q_{210} within a reasonable experimental error in the given data. A few $hk0$ lines are listed in Table 9.2, showing that we are on the right lines.

In a similar manner, $Q_{010} + Q_{001} = 0.1174$. There is no line of this Q-value, but lines 5 and 6 are nearly symmetrically disposed about this position, so that these lines may be taken as $0\bar{1}1$ and 011. Since, $Q_{011} - Q_{0\bar{1}1} = 4b^*c^* \cos \alpha^*$, it follows that $\alpha^* = 89.45°$. As before, we can now check for the presence of Q_{0kl}-values, as desired. For example, $Q_{021} = 4Q_{020} + Q_{001} + 2Q_D (4b^*c^* \cos \alpha^*) = 0.2558$; similarly, $Q_{02\bar{1}} = 0.2514$. These Q-values correspond to lines 17 and 16.

We seek β^* in a similar manner. By trial and error, it is found ultimately that lines 11 and 32 are nearly symmetrically disposed about the calculated value, 0.3190, of Q_{102}. Hence, β^* evaluates to 49.11°. Summarizing the results, we have the reciprocal unit cell

$$a^* = 0.1761 \qquad b^* = 0.2131 \qquad c^* = 0.2684 \text{ Å}$$
$$\alpha^* = 89.45° \qquad \beta^* = 49.11° \qquad \gamma^* = 90.00°$$

Finally, we obtain the direct space unit cell (see Section 3.4 and program RECIP*), and then calculate the Q-values so as to identify and index all lines in the diffraction record, as shown by Table 9.3. The results show good agreement between Q_{obs} and Q_{calc} for most of the lines, although certain fits suggest that some errors exist among the data. We see already in this fairly simple example the importance of the accurate experimental measurement of θ-values.

TABLE 9.3.   Indices and Q-values for Magnesium Tungstate

Line	hkl	Q_{calc}	Q_{obs}	Line	hkl	Q_{calc}	Q_{obs}
1	100	0.03101	0.0310	21	310	0.32444	0.3249
2	010	0.04541	0.0454	22	$41\bar{2}$	0.33250	0.3322
3	001	0.07203	0.0720	23	012	0.33568	0.3364
4	110	0.07644	0.0764	24	$22\bar{2}$	0.34197	0.3418
5	$01\bar{1}$	0.11637	0.1164	25	$32\bar{1}$	0.34491	0.3451
6	011	0.11851	0.1186	26	$3\bar{2}\bar{1}$	0.34937	0.3505
7	200	0.12403	0.1239	27	211	0.36626	0.3646
8	101	0.16491	0.1649	28	$1\bar{2}2$	0.37272	0.3723
9	$2\bar{1}0$	0.16948	0.1695	29	$1\bar{2}\bar{2}$	0.38135	0.3824
10	020	0.18165	0.1816	30	$40\bar{3}$	0.40195	0.4016
11	$30\bar{2}$	0.19597	0.1957	31	030	0.40871	0.4083
12	$21\bar{2}$	0.20791	0.2077	32	$21\bar{3}$	0.44324	0.4432
13	$3\bar{1}\bar{1}$	0.21202	0.2123	33	$41\bar{3}$	0.44409	0.4465
14	$11\bar{2}$	0.23864	0.2386	34	$2\bar{1}\bar{3}$	0.44973	0.4500
15	$3\bar{1}\bar{2}$	0.24357	0.2436	35	320	0.46063	0.4608
16	$02\bar{1}$	0.25177	0.2517	36	$02\bar{2}$	0.46548	0.4659
17	021	0.25582	0.2563	37	$4\bar{2}\bar{2}$	0.47534	0.4770
18	300	0.27908	0.2793	38	$1\bar{1}2$	0.48616	0.4863
19	002	0.28812	0.2884	39	112	0.49041	0.4918
20	220	0.30562	0.3055	40	400	0.49614	0.4960

9.4.2 Reduced and Conventional Unit Cells

From the results so far, the crystal system of magnesium tungstate appears to be triclinic. However, we must determine whether or not this unit cell is the conventional unit cell for the lattice, that is, it is correctly related to the lattice symmetry according to the conventions that we have discussed in earlier chapters. The real space unit cell is derived using the equations developed in Chapter 3, or the program RECIP* by entering the reciprocal unit cell parameters, with the constant κ equal to 1. Thus, we obtain

$$a = 7.512 \qquad b = 4.693 \qquad c = 4.929 \text{ Å}$$
$$\alpha = 90.72° \qquad \beta = 130.89° \qquad \gamma = 89.52°$$

or, in the usual order of increasing unit-cell lengths, as

$$a = 4.693 \qquad b = 4.929 \qquad c = 7.512 \text{ Å}$$
$$\alpha = 130.89° \qquad \beta = 89.52° \qquad \gamma = 90.72°$$

The order makes no difference to the reduced cell; it alters only the transformation matrix between the input and the reduced unit cells. A two-stage process now allows the determination of the reduced and conventional unit cells.

The unit-cell parameters derived from the indexing procedure are transformed to a unique, reduced unit cell based on the three shortest noncoplanar translations in the lattice (Niggli-reduced cell);[a] such a unit cell must always be primitive. In the second stage, the distribution of 2-fold axes is determined, because it will define the lattice symmetry[b] (see also Section 1.4.2 and Table 1.5). We know, for example, that the normal to a plane in a lattice that contains p intersecting 2-fold axes is itself a p-fold axis, and that a mirror plane exists normal to a p-fold axis ($p = 2, 4, 6$) in a lattice.

The two stages have been brought together in the program LEPAGE*, kindly made available to the academic community.[c] We shall describe this program further in Chapter 11, but for the moment we use it with the unit cell derived from indexing the MgWO$_4$ diffraction pattern. Thus, we obtain

	a/Å	b/Å	c/Å	α/deg	β/deg	γ/deg
Input cell, P	4.693	4.929	5.712	130.89	89.52	90.72
Reduced cell, P	4.693	4.930	5.679	90.12	90.01	90.72
Conventional cell, P	4.693	4.929	5.679	90.12	90.01	90.72

[a] I. Křivý and B. Gruber, *Acta Crystallographica* **A32**, 297 (1976); P. Niggli, *Handbuch der Experimentalphysik* Vol. 7, Leipzig, Akademische Verlagsgesellschaft (1928).

[b] Y. Le Page, *Journal of Applied Crystallography* **B15**, 255 (1982).

[c] A. L. Spek, *Journal of Applied Crystallography* **21**, 578 (1988).

We see that the reduced and conventional unit cells are identical in this example. Further interpretation now depends upon the error permitted in the collinearity of the 2-fold axes in the real and reciprocal unit cells, a "2-axis criterion." Ideally, it should be zero. However, there will be experimental errors in the data that are conveyed to the parameters; if we dismiss these errors we may fail to recognize the true symmetry. In the present case, if all angles are regarded as 90° within experimental error, the lattice is P orthorhombic, with a 2-axis criterion of 1°. A more realistic situation could be to set α and β at 90°; then we obtain monoclinic P, under the more stringent 2-axis criterion of 0.5°.

$$a = 4.693 \qquad\qquad b = 5.679 \qquad\qquad c = 4.929 \text{ Å}$$

$$\alpha = 90(89.88)° \qquad\qquad \beta = 90.72° \qquad\qquad \gamma = 90°$$

where the unit cell has been rearranged so that the unique angle is β. Notice that this result would have been obtained immediately from the first reduction by imposing monoclinic symmetry. Some of the difficulty in this indexing analysis for magnesium tungstate arose from the fact that the data was not of the highest quality. The modern camera and, particularly, diffractometer techniques described above ensure data of a sufficiently high quality.

9.4.3   Computer Indexing of the Diffraction Pattern

We have shown that determining the unit cell by hand can be a slow process, and it is not surprising to find that the literature today abounds with computer programs for indexing a powder pattern. Most of these programs that serve to determine the unit cell are stand-alone programs, but a few are part of a structure-solving and refinement package that leads to a complete structure determination, just like the single-crystal methods that we have described.

Indexing a powder pattern, by whatever method, is an inductive process: we must deduce the indices of the diffraction lines from the experimental Q-values. The first 20–30 lines in a pattern, starting from the low θ region, are most important in indexing, because Q_{calc} involves the square of the indices, and so errors in Q_{obs} become more important as θ increases. Among the programs that are in frequent use are ITO, DICVOL, and TREOR; here, we examine the first of them.

ITO Program System

The program ITO12* is the twelfth version of this program, and has been kindly made available by its author (Version 13 is now published; see Appendix A10). It is a deductive program based on the properties of crystal zones, that is, planes of reciprocal lattice points passing through the origin, and is often described as *zone-indexing*. It performs best when given 30–40 accurately measured powder lines.

Any zone is specified by three parameters; for example, the zone $hk0$ may be formulated by

$$Q_{hk0} = h^2 Q_A + k^2 Q_B + hk Q_F \qquad (9.7)$$

Two Q-values are selected and assigned as Q_{100} and Q_{010}, similar to the procedure in Section 9.3. Expanding (9.7), we have

$$Q_{hk0} = h^2 Q_{100} + k^2 Q_{010} + 2hk(Q_{100}Q_{010})^{1/2} \cos \gamma^* \qquad (9.8)$$

Let

$$2hk(Q_{100}Q_{010})^{1/2} \cos \gamma^* = R = \frac{-(h^2 Q_{100} + k^2 Q_{010} - Q_{hk0})}{hk} \qquad (9.9)$$

Values for Q_{hk0} are obtained from the experimental data, and used in the right-hand term of (9.9) so as to obtain a list of $|R|$ values. Agreements in $|R|$, within a permitted error, are then used to find a value for γ^*. Zones that are found are checked and reduced, and the three zone parameters refined by least squares. Zone quality is determined by a parameter $1/P$, where P is the probability that a zone is found by chance. Pairs of zones with a common row are sought and the angle between them calculated. The unit cells found are reduced and transformed to standard form, and the first 20 lines indexed where possible. The fit is assessed by the M_{20} parameter, best judged in a series of results (Table 9.4):

$$M_{20} = \frac{Q_{20}}{2\bar{Q}N_{20}} \qquad (9.10)$$

where Q_{20} is the Q-value for the 20th indexed line, $\bar{Q}$ is the average error between Q_{obs} and Q_{calc} for the first 20 lines , and N_{20} is the number of lines, observed and calculated, up to Q_{20}. The program is optimized for the lower symmetry systems, orthorhombic, monoclinic, and triclinic. High-symmetry lattices may be reported in an orthorhombic setting, with a note that a higher symmetry lattice may exist. Notes on the practical use of the program ITO appear in Chapter 11, and Problems are given at the end of this chapter. We note in passing that the data for magnesium tungstate (Section 9.3.1), when used with ITO12, led to the unit cell dimensions $a = 4.929, b = 5.678, c = 4.693$ Å, $\alpha = 90°, \beta = 90.77°, \gamma = 90°$, in good agreement with the values derived in the example.

CRYSFIRE Program System

The program system CRYSFIRE, which is under further development, provides a detailed set of procedures for indexing powder patterns. In common with all methods, a prerequisite is a set of Q-values with errors less than $ca\ 0.01°$ in θ. As we have seen, this level is achievable experimentally, but the number of lines and their potential overlap increases as d^{*3}. Since powder lines have a finite width, clear resolution exists only at the lower θ-values; at higher values of θ, they merge

TABLE 9.4.   Some Results of an Indexing with CRYSFIRE

I_{20}	M_{20}	$V/\text{Å}^3$	$a/\text{Å}$	$b/\text{Å}$	$c/\text{Å}$	α/deg	β/deg	γ/deg	Link
20	22.3	562.15	7.746	11.482	6.321	90.00	90.00	90.00	ITO12
20	21.2	280.53	5.000	11.475	4.991	90.00	101.6	90.00	KOHL
20	12.6	560.97	9.982	11.475	5.000	90.00	101.6	90.00	KOHL
20	12.1	562.15	7.746	11.481	6.321	90.00	90.00	90.00	ITO12
20	9	561.64	9.995	11.505	4.986	90.00	104.3	90.00	TREOR90
20	6.5	1966.5	16.737	11.507	10.210	90.00	90.00	90.00	DICVOL91
20	6	623.25	13.048	4.894	11.090	90.00	118.3	90.00	TREOR90
19	24.5	561.56	7.742	11.473	5.001	90.00	90.00	90.00	KOHL
19	8	561.90	9.973	11.508	5.001	90.00	101.8	90.00	TREOR90

into semi-continuous profiles of mainly unresolved maxima, each of which may contain 5–50 peaks.

Provided that the average discrepancy between the observed and calculated Q-values is less than about 5%, the true lattice can be extracted from among other approximate solutions. A wholly exhaustive search is prohibitive. We have shown in Chapter 2 that a lattice can be described by any number of alternative unit cells, recognized by having the same reduced unit cell and volume. Less satisfactory solutions will usually occur, but they may be recognized through goodness-of-fit parameters, such as M_{20}, defined in (9.10); the higher the values of these parameters the better is the fit.

The program system CRYSFIRE is actually a master automatic-indexing script, which operates through another program CRYS acting as a front-end 'wizard' to a collection of eight indexing programs written by other workers of lengthy experience in the field of powder indexing, each having its own strategy and its own best applicability. The CRYSFIRE system, now Version 2002, has succeeded in indexing numerous powder patterns of all symmetries; further details of this system may be obtained from its author.[a,b]

As an example of indexing with CRYSFIRE, a set of 40 Q-values for a particular powder sample was input to the program. A series of possible solutions were obtained, listed in Table 9.4 in order of number of lines listed I_{20} out of the first 20, and figure of merit M_{20}, together with the unit-cell volume and other parameters, and the particular link of the program that produced the solution. The program also recorded other possible solutions with I_{20} less than 19. All solutions are indicated as either orthorhombic or monoclinic.

The first solution listed in Table 9.4, with the highest figure of merit and with I_{20} equal to 20, reproduced all 40 observed Q-values, although the experimental error in some lines was greater than the best achievable. This result may be regarded

[a] Shirley, R. *Crysfire 2002; http://www.ccp14.ac.uk.*

[b] Shirley, R., *Powder Indexing with Crysfire 2002*, International Union of Crystallography Commission on Powder Diffraction Newsletter (July 2002)

as the most probable, and a good starting point for further investigations. From a perusal of the indices found by the program, the following limiting conditions were deduced:

$$hkl: h + k = 2n \qquad 0kl: (k = 2n) \qquad h0l: (h = 2n) \qquad hk0: (h + k = 2n)$$

$$h00: (h = 2n) \qquad 0k0: (k = 2n) \qquad 00l: \text{None}$$

Hence, the conclusions $a = 6.321, b = 7.746, c = 11.482$ Å reordered so that $c > b > a$, with the possible space group being one of $Cmmm$, $Cmm2$, and $C222$. However, at this stage we have not considered the problem of overlapping lines, and it may be necessary to review the deduction of the space group after the powder pattern has been decomposed into integrated intensities.

We consider briefly some of the remaining solutions in Table 9.4. The second solution indicates a smaller unit-cell volume, with an apparently monoclinic unit cell. It can be explained by the transformation (from solution 1) $\mathbf{a}' = \mathbf{a}/2 + \mathbf{b}/2, \mathbf{b}' = \mathbf{b}, \mathbf{c}' = -\mathbf{a}/2 + \mathbf{c}/2$, to give a P unit cell, but it would not be chosen as the conventional cell. The penultimate solution in Table 9.4, with $M_{20} = 24.5$, is almost as satisfactory, and would probably make a suitable starting point in the absence of solution 1. Several listed solutions include a dimension of 9.97–10.00 Å: this value is obtainable as $\sqrt{(7.746^2 + 6.321^2)}$ Å, but such solutions also involve an interaxial angle greater than 90°. Finally, the only other orthorhombic unit cell listed in Table 9.4 has a volume of 1966.5 Å^3, where the transformation (from cell 1) $\mathbf{a}' = 2\mathbf{a} + \mathbf{c}$ leads to the value 16.732 Å, and $|\mathbf{a} + \mathbf{c}|$ is approximately 10 Å. Thus, we have a set of unit cells within one and the same lattice, and we have chosen the most probable and conventional one.

9.5   Extracting Integrated Intensities from a Powder Pattern

At this stage, it is prudent to check whether or not the pattern now indexed has, in fact, already been recorded, and to what extent the structure has been determined. The International Centre for Diffraction Data (ICDD),[a] formerly Joint Committee on Powder Diffraction Standards (JCPDS), holds records of over 100,000 powder patterns, for checking with a 'new' pattern. Assuming that the compound is new, we proceed to obtain the intensity data.

The extraction of individual intensities from a powder pattern is complicated by the overlapping of reflections in the pattern. Overlaps may be exact, as

---

[a] http://www.icdd.com

imposed by symmetry, or accidental, arising from near-equivalence of d-values for nonequivalent reflections. Pattern decomposition is usually performed using an adaptation of the Rietveld whole-profile method,[a] whereby the intensities of the reflections, instead of structural parameters, are adjusted so as to give the best match of the calculated diffraction pattern with that obtained experimentally. The intensities thus obtained are proportional to the corresponding values of $|F|^2$. Two algorithms for extracting the integrated intensities in this way have been developed.

One method was proposed by Le Bail *et al.*,[b] who extended a method used by Rietveld to extract intensities for the calculation of Fourier maps during the course of a whole-profile refinement (see Section 9.6) to the case where no initial structural model is available. In the Rietveld procedure, the profile is defined by a number j of digitized points and, in the case of a resolved peak, the background-subtracted points are summed. For overlapping peaks 1 and 2, the integrated intensities $\mathcal{J}_{obs}$ are obtained through the equations

$$\mathcal{J}_{1,\text{obs}} = \sum_j \frac{I_{1_{\text{calc}}} q_{1j}}{I_{1_{\text{calc}}} q_{1j} + I_{2_{\text{calc}}} q_{2j}} (P_j - B_j)$$

$$\mathcal{J}_{2,\text{obs}} = \sum_j \frac{I_{2_{\text{calc}}} q_{2j}}{I_{1_{\text{calc}}} q_{1j} + I_{2_{\text{calc}}} q_{2j}} (P_j - B_j)$$

(9.11)

where $\mathcal{J}_{1,\text{obs}}$ is equal to $m_1 |F_1|^2$, m_1 is the multiplicity factor for $|F_1|$, $I_{1,\text{calc}}$ is the calculated value of $|F_1|^2$ based on an appropriate model, and $q_{1,j}$ contains the Lorentz, polarization, absorption (if necessary), and shape factors associated with peak 1, and similarly for peak 2, $(P_j - B_j)$ is the measured (peak − background) term for the jth point in the pattern. By summing the two equations (9.11), we obtain

$$\mathcal{J}_{1,\text{obs}} + \mathcal{J}_{2,\text{obs}} = \sum_j (P_j - B_j)$$

(9.12)

so that the sum of the peak areas is equal to the background-subtracted area, as in the case of resolved peaks.

The Le Bail procedure is essentially an iterative version of Rietveld's. Generally, rapid convergence is obtained notwithstanding an initial assumption of equal peak areas. The procedures are programmed and available in systems such as FullProf and GSAS.

An alternative approach has been given by Pawley[c] in which the reflection intensities are refined using a least-squares procedure. The function minimized is

[a] H. M. Rietveld, *Journal of Applied Crystallography* **2**, 65 (1969).
[b] A. Le Bail, H. Duroy, and J. L. Fourquet, *Materials Research Bulletin* **23**, 447 (1988).
[c] G. S. Pawley, *Journal of Applied Crystallography* **14**, 357 (1981).

the sum of the squares of the differences between the observed and calculated profiles. The method has been programmed and is available in the system ALLHKL. In both the Le Bail and Pawley algorithms, variables related to peak positions and shapes are the same as in the Rietveld structure refinement method.

9.6   Rietveld Refinement

Refinement appears, logically, as the final stage in the scheme outlined in Section 9.1.1, as indeed it does in single-crystal analysis, since a model is required before it can be applied. However, we discuss it at this point, because Rietveld refinement is invoked in the example structure determinations yet to be discussed, and the solution of the phase problem (Stage 4) emerges from these examples.

In Rietveld refinement, the method of least squares is carried out so as to obtain the best fit between the whole experimental powder diffraction profile and the corresponding pattern calculated from the trial structure. In this way, the explicit decomposition of overlapping reflections can be avoided, because only the points along the observed and calculated profiles, and not the individual reflections, are compared.

The powder pattern is digitized into i steps to give an intensity function y_i. We assume that the digitizing parameter is the scattering angle 2θ for x-rays, or a velocity function for time-of-flight neutron studies.

The experimental diffraction profile $h(x)$ is the convolution of the intrinsic diffraction profile $f(x)$ with an instrumental function $g(x)$, that is,

$$h(x) = \int f(x - x_0) g(x_0)\, dx_0 \qquad (9.13)$$

where x is $2\theta_i$, the value of the function at the ith sampling point, and x_0 is the value corresponding to $2\theta_j$, where θ_j is the Bragg reflection angle; x_0 is a variable in the same field as x. A number of forms have been used for the profile function: one of the simplest, which is appropriate for constant wavelength neutron data, is the Gaussian

$$\sqrt{\frac{C}{\pi H_j^2}} \exp\left[-C\frac{(2\theta_i - 2\theta_j)^2}{H_j^2} \right]$$

where C is $4\ln 2$ and H_j is the full width of the peak at half-maximum height (FWHM) of the jth Bragg reflection. The pseudo-Voigt distribution, which is a combination of a Gaussian and a Lorentzian, is used frequently to model the peak shape in x-ray diffraction patterns:[a] the Gaussian is of the form noted above, while

---

[a] R. A. Young and D. B. Wiles, *Journal of Applied Crystallography* **15**, 430 (1982).

the Lorentzian may be expressed as

$$\frac{2}{\pi H_j} \left[1 + 4\frac{(2\theta_i - 2\theta_j)^2}{H_j^2} \right]^{-1}$$

The quantity minimized by the Rietveld refinement is a residual r, given as

$$r = \sum_j w_j (y_j - y_{c,j})^2 \tag{9.14}$$

and the best fit of the calculated pattern to the observed pattern judged by means of numerical criteria. Some of the functions used are

$$R_F = \frac{\sum_j |\sqrt{I_{obs,j}} - \sqrt{I_{calc,j}}|}{\sum_j \sqrt{I_{obs,j}}} \tag{9.15}$$

which is the conventional R-factor, but written in terms of $\sqrt{I}$ instead of $|F|$;

$$R_B = \frac{\sum_j |I_{obs,j} - I_{calc,j}|}{\sum_j I_{obs,j}} \tag{9.16}$$

which is called the Bragg R-factor;

$$R_p = \frac{\sum_j |y_{obs,j} - y_{calc,j}|}{\sum_j y_{obs,j}} \tag{9.17}$$

which is the profile R-factor; and

$$R_{wp} = \frac{\sum_j w_j (y_{obs,j} - y_{calc,j})^2}{\sum_j w_j (y_{obs,j})^2} \tag{9.18}$$

where w_j is a weighting factor for the jth point, which is the weighted profile R-factor. A goodness-of-fit indicator χ^2 is also used, given by

$$\chi^2 = (R_{wp}/R_e)^2 \tag{9.19}$$

where R_e is the statistically expected R-factor:

$$R_e = \left(\frac{N - P}{\sum_j w_j (y_{obs,j})^2} \right)^{1/2} \tag{9.20}$$

Here, N is the number of data points j in the experimental powder profile and P is the number of parameters refined. We shall indicate satisfactory values for these parameters in the next section. Probably the most meaningful parameter is R_{wp}, since the numerator is the residual (9.14) that is minimized. However, it is sensitive to a small number of poor agreements, which could arise from impurity.

The Rietveld method, now used universally for powder studies, was programmed first by Rietveld,[a] but it is now available in many program packages, combined with other techniques such as profile decomposition in, for example, GSAS (Le Bail decomposition) and WPPF (Pawley decomposition). RIETAN is a system similar to GSAS, but offers additionally a choice of three minimization algorithms that can be introduced under user-control in one and the same minimization process.

A problem with Rietveld refinement, as with other forms of minimization, is the possibility of converging into a false minimum. The risk can be ameliorated by making reasonable variations to the starting model and refining, hopefully, to the same minimum. The three algorithms provided in RIETAN also provide for a possible way out of false minima. Guidelines for structure refinement using the Rietveld method have been published by the International Union of Crystallography (IUCr) Commission on Powder Diffraction.[b]

9.7   Examples of Solved Structures

Once the initial stages indicated in Section 9.1.1 have been completed, attempts can be made to solve the phase problem, so as to obtain a structural model that can be subjected to at least a partial refinement. The attack may take place by means of the so-called traditional approach, that is, by Patterson or direct methods along the lines discussed for single crystals, or newer techniques such as Monte Carlo, simulated annealing, and maximum entropy can be brought to bear on the problem. As with single-crystal structure determination, Patterson methods tend to be applied where heavy atoms are present in the molecules, and direct methods are used mainly for equal-atom structures. Refinement usually involves a combination of difference-Fourier syntheses to locate any missing atoms, and Rietveld whole-profile structure refinement. We shall examine some structures with a view to indicating the stages that have been found necessary in obtaining satisfactory solutions.

---

[a] H. M. Rietveld, *loc. cit.* (1969).

[b] L. B. McCusker, R. B. Von Dreele, D. E. Cox, D. Louër, and P. Scardi, *Journal of Applied Crystallography* **32**, 36 (1999).

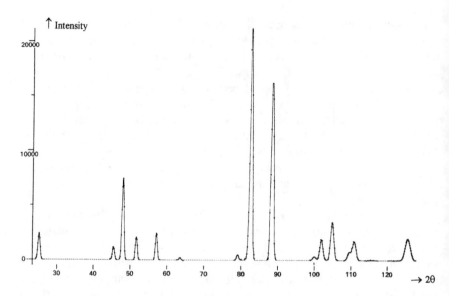

FIGURE 9.10. Neutron diffraction pattern of a powdered sample of $CaUO_4$, measured at $\lambda =$ 2.565 Å: the full line shows the calculated profile and the dots indicate the observed profile. [After Loopstra and Rietveld, *Journal of Applied Crystallography* **2**, 65 (1969).]

9.7.1 Traditional Methods

Calcium Uranate

The early work by Rietveld was carried with data obtained by neutron diffraction, but the use of x-rays, and particularly with a synchrotron source, is now well developed. Figure 9.10 shows an early diffractogram obtained by neutron diffraction on calcium uranate, $CaUO_4$. The fitted and experimental profiles show excellent agreement.

Manganese Phosphate Monohydrate

This structure, originally thought to be a $1\frac{1}{2}$ H_2O hydrate, was solved by Patterson methods,[a] using x-ray powder diffractometer data collected at a synchrotron source. X-rays of wavelength 1.3208 Å were selected by means of a Ge(111) monochromator, and the radiation scattered in the vertical plane was measured.

The pattern was indexed by the program ITO, and the unit cell at an M_{20} of 196 had the dimensions $a = 6.916, b = 7.475, c = 7.361$ Å; $\beta = 112.32°$; Z

---

[a] P. Lightfoot, A. K. Cheetham, and A. W. Sleight, *Inorganic Chemistry* **26**, 3544 (1987).

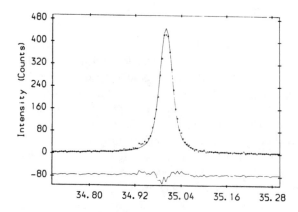

FIGURE 9.11. Manganese phosphate monohydrate, $MnPO_4 \cdot H_2O$, modelled profile: plot of the intensity of the observed (circles) and calculated (solid line) profiles as a function of 2θ for the $31\bar{1}$ peak. The lower difference curve shows the accuracy of the pseudo-Voigt modelling function. [After Lightfoot, Cheetham, and Sleight, *loc. cit.* (1987).]

was 4, and the space group either Cc or $C2/c$ from systematic absences. We may note here that it may be inadvisable to decide the presence or absence of a centre of symmetry by statistical tests, because the distribution of $|F_o|$ values can be affected by the methods of pattern decomposition;[a,b] space group $C2/c$ was confirmed by the structure analysis.

Pattern decomposition was carried out with the pseudo-Voigt function, leading to 61 intensity data; Figure 9.11 shows a typical modelled peak. The intensity data were used to generate a Patterson map. The manganese atom coordinates were determined from this map, and a difference-Fourier synthesis located the phosphorus and three oxygen atoms. Refinement was carried out by minimizing the function $\sum_j w_j (y_{\text{obs},j} - Cy_{\text{calc},j})^2$, where the terms have the meanings already discussed, and C is a scaling constant. The weights for each jth point were calculated from the expression $w_j = [(y_j + b_j) + \sigma^2(b_j)]^{-1}$, where b_j is the background at the jth point and $\sigma(b_j)$ is the esd of b_j. During refinement it was determined that the substance was, in fact, a monohydrate. Satisfactory final refinement was achieved, with the following agreement factors:

$$R_B = 4.74\% \quad R_p = 12.2\% \quad R_{\text{wp}} = 16.1\% \quad R_e = 15.4\% \quad \chi^2 = 1.1$$

[a] G. Cascarano, L. Favia, and C. Giacovazzo, *Journal of Applied Crystallography* **25**, 310 (1992).
[b] M. A. Estermann, L. B. McCusker, and Ch. Baerlocher, *Journal of Applied Crystallography* **25**, 539 (1992).

Cimetidine

The structure of cimetidine, $C_{10}H_{16}N_6S$, has been solved from powder diffraction data.[a] In this example, a diffractometer and a synchrotron source of wavelength 1.4599(1) Å were used to obtain the data. The diffractogram was indexed by the program TREOR ($M_{20} = 176$), and the pattern decomposed by a modified Pawley method. Several attempts were made with direct methods programs, which located only sulfur and three other atoms. Further phase extraction was carried out with the direct methods program SIR88. The structure was completed by iterative Fourier and least-squares procedures. A difference-Fourier synthesis then located all hydrogen atoms. After final Rietveld refinement, the structure converged with the following residuals:

$$R_B = 1.9\% \quad R_{wp} = 8.5\% \quad R_e = 6.9\% \quad \chi^2 = 1.5$$

It is noteworthy that in the absence of the hydrogen atoms R_B and R_{wp} were 10.3 and 16.2%, respectively. Figure 9.12 illustates the profile refinement of cimetidine with and without the inclusion of the hydrogen atoms.

SIR Program System

In the context of direct methods, SIR, as in SIR88, refers to phase determination by the method of *semi-invariants representations*, and should not be confused with SIR as used in Chapters 6 and 10, wherein it means *single isomorphous replacement*.

A structure seminvariant (semi-invariant) is a linear combination of phases, the value of which is uniquely determined by the crystal structure alone, irrespective of the choice of permissible origin. For example, in space group $P2_1$ (y axis unique), $\phi(h0l)$ is a seminvariant if h and l are both even. Again, if $h_1 + h_2 + h_3 = 2n$, $k_1 + k_2 + k_3 = 0$, and $l_1 + l_2 + l_3 = 2n'$, then $\phi(\mathbf{h}_1) + \phi(\mathbf{h}_2) + \phi(\mathbf{h}_3)$ is a seminvariant. Further discussions of the properties of structure seminvariants may be found in the literature.[b]

The SIR method is based on the estimation of 1- and 2-phase structure seminvariants and 3- and 4-phase structure invariants, according to the theory of representations.[c] The program functions in all space groups without user intervention, although a knowledge of partial structure moieties may be exploited with advantage. Later versions of SIR are SIR92, SIR97, and SIR2001 (see Appendix A10).

---

[a] R. J. Cernik, A. K. Cheetham, C. K. Prout, D. J. Watkin, A. P. Wilkinson, and B. T. M. Willis, *Journal of Applied Crystallography* **24**, 222 (1991).

[b] M. F. C. Ladd and R. A. Palmer (Editors), *Theory and Practice of Direct Methods in Crystallography*, Plenum Press, New York (1980), and references therein.

[c] C. Giacovazzo, *Acta Crystallographica* A**33**, 933 (1977; idem. ibid. A**36**, 362 (1980)).

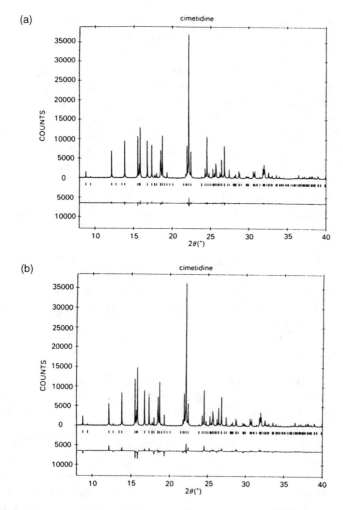

FIGURE 9.12. Rietveld profile refinement of cimetidine: (a) with hydrogen atoms included, $R_{wp} = 8.5\%$; (b) without hydrogen atoms included, $R_{wp} = 16.2\%$. [After Cernik et al., *loc. cit.* (1991).]

Silver–Pyrazole Complex

The structure of the silver–pyrazole complex $[Ag(pz)]_3$ (pz = pyrazole), $C_9H_9N_6Ag_6$, has been solved by direct methods[a] as part of a program investigating complexes between pyrazole, $C_3H_4N_2$, and copper or silver.

[a] N. Masciocchi, M. Moret, P. Cairali, A. Sironi, G. Attilio Ardizzoia, and G. La Monica, *Journal of American Chemical Society* **116**, 7668 (1994).

Powder diffractometer data were collected with graphite-monochromatized x-radiation of wavelength 1.5418 Å, at values of 17–85° in 2θ-steps of 0.02°. The program TREOR was used to index the pattern, and it gave the orthorhombic unit cell $a = 13.13, b = 10.56, c = 8.79$ Å with $M_{20} = 15$. The indexed data suggested the space group *Pbcn*, which was confirmed by the structure analysis. The refined unit cell had the dimensions $a = 13.1469(4), b = 10.5702(10), c = 8.7921(4)$. Since $Z = 4$, there are 12 Ag(pz) moieties in the unit cell.

Pattern decomposition was achieved by the Pawley method through application of the program ALLHKL to 924 reflection data, using 3400 points.

The program SIRPOW92 was used to extract a direct methods trial model comprising the three silver atoms in the asymmetric unit, one of which lay on a 2-fold axis. At this stage, $R_p = 0.29$ and $R_{wp} = 0.37$. The structure was completed and Rietveld refinement, excluding hydrogen atoms, converged with the following set of indicating parameters:

$$R_F = 6.0\% \quad R_p = 11.5\% \quad R_{wp} = 14.9\% \quad R_e = 6.6\% \quad \chi^2 = 5.1$$

The value of χ^2 is larger than normal for a refined structure. It has been noted[a] that the parameters R_e and χ^2 depend on the intensity counting statistics and are, at best, only gross goodness-of-fit parameters.

EXPO Program System

The program SIRPOW92 that was employed in the structure determination just described has been incorporated into the powerful structure-solving program system EXPO that is currently much used. The main addition to the SIRPOW92 system is the routine EXTRA, the purpose of which is to provide for the decomposition of the pattern according to the Le Bail algorithm. Figure 9.13 illustrates the flow diagram for EXPO. Advances in pattern decomposition have made it possible to introduce into EXTRA the positivity of the electron density and Patterson functions, the treatment of pseudo-translational symmetry and of preferred orientation, and the availability of a partial structure. Such information, when detected in SIRPOW, is recycled as shown in Figure 9.13, so as to obtain better extracted values for $|F_o|^2$. Additional user-friendly facilities include contouring of Fourier maps, representation of crystal structures by coordination polyhedra, and automatic preliminary Rietveld refinement, triggered when the structure is incomplete. A more recent innovation is the labelling of peaks in electron density maps, based on chemical information rather than on electron density peak heights[a].

The EXPO system has been made available to the academic community by the courtesy of its authors, and can be obtained from a web site (see Appendix A10).

---

[a] R. J. Hill and I. C. Madsen, *Powder Diffraction* **2**, 146 (1987)

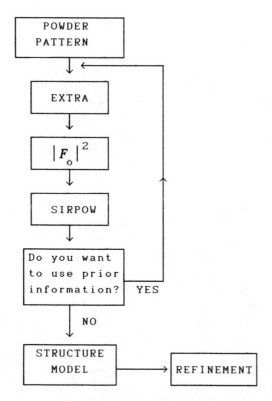

FIGURE 9.13. Flow diagram for the program system EXPO; structural information detected by SIRPOW can be recycled to EXTRA, so as to obtain more reliable intensity data. [After Altomare et al., *loc. cit.* (1995).]

9.7.2 Direct-Space Methods

Direct-space methods, not to be confused with direct methods, have evolved from traditional model building techniques, in which chemical information is used in order to construct a sensible structural model. Such a model can then be used as a basis for calculating a diffraction pattern with which to compare the measured pattern. The inclusion of information on bond lengths, bond angles, connectivities, torsion angles, orientation, and position means that a very large number of chemically sensible models can be produced. Thus, computing power must be brought to bear in order for such an approach to be practicable. Encoding frequently results in algorithms being specific to a given class of compounds.

In the case of molecular compounds, information on bond lengths and bond angles is well documented, so that the number of variables is reduced to that defining orientation and position, and possibly torsion. In the case of zeolite structures, the chemical composition and connectivities of structural units are of significance in designing useful structural models.

An alternative approach involves generating a model by the random placement of atoms of the required number and types in the unit cell, then applying shifts to the atoms in predetermined amounts, and calculating and comparing patterns so as to find a best fit. Constraints in terms of known chemical information can be introduced into the model, so as to increase the plausibility of the model and to reduce the time consumption of computing facilities. We will consider examples of these and other related techniques in the remainder of this section.

Zeolites and the FOCUS Algorithm

Zeolite structures have many important applications as molecular sieves, absorbents, catalysts, and ion-exchange materials. These properties are related to both their unusual structures and, in particular, to their framework topologies, that is, the way in which the tetrahedral structural units are linked in the solid state. As zeolites are microcrystalline, powder diffraction proves to be the only method available for their detailed structural examination.

The structures are complex, often with high symmetry, such as $P\frac{6_3}{m}mc$ or $Fm\bar{3}c$ ($Fm\bar{3}c$), and can have large unit cells, with dimensions up to 40 Å, so that a high degree of overlap of diffraction maxima arises. Generally, direct methods have been used for solving the structures, with difference-Fourier synthesis and Rietveld profile refinement to complete and refine the structure.[a] A different approach is used in the program environment FOCUS[b] indicated by Figure 9.14, where Fourier recycling is combined with a specialized topology search and topology classification scheme. The method makes use of crystal-chemical information such as chemical composition, probable interatomic distances, and the fact that all zeolite structures have three-dimensional four-connected frameworks, in order to aid in the interpretation of electron density maps.

The Q-values of about 20 high-accuracy powder lines serve to determine the unit-cell parameters and index the pattern. The space group follows from the indexed lines, and integrated intensities are extracted.

Random phases consistent with the space group are assigned to the extracted reflection intensities and an electron density map generated. If enough of the phases are (by chance) correct, some of the features of the structure will appear, and an attempt is made to interpret the map using the chemical information outlined above. The resulting model is then used to generate new phases and a new electron density map. This Fourier interpretation and recycling procedure is continued until either

[a] W. I. F. David, K. Shankland, L. B. McCusker, and Ch. Baerlocher, *loc. cit.*

[b] R. W. Groose-Kunstleve, L. B. McCusker, and Ch. Baerlocher, *Journal of Applied Crystallography* **30**, 985 (1997).

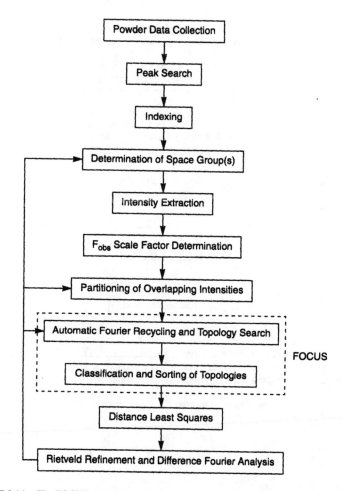

FIGURE 9.14.   The FOCUS structure-solving environment, tailored to the solving of zeolite structures. The essential algorithm for the zeolites is enclosed in dashed lines; the remaining blocks in the
diagram refer to standard procedures. [After R. Grosse-Kuntsleve, *loc. cit.* (1996).]

phase convergence or a maximum number of cycles is reached. Then the process
is started again with a new set of random phases. Each time an electron density
map is generated, a search is also made for a three-dimensional four-connected
net. If one is found it is classified and written to a file. The procedure is terminated
when a sufficient number of such nets have been found. The net that occurs most
frequently is usually the correct framework structure. The model is then used
as a starting point for structure completion and Rietveld refinement. Figure 9.15
illustrates the FOCUS algorithm, the detail of that section of Figure 9.14 that is
enclosed by dashed lines.

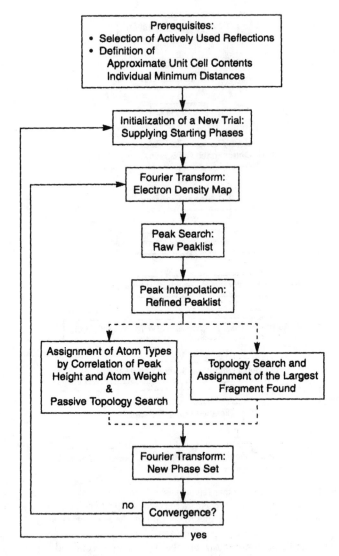

FIGURE 9.15.   The FOCUS algorithm in detail (compare Figure 9.14): automatic Fourier recycling and topology searching is carried out here. [After Gross-Kuntsleve, *loc. cit.* (1996).]

Zincosilicate Complex VIP-9

As an example of a large structure solved only when the FOCUS procedure was applied, we cite the zincosilicate molecular sieve complex

VPI-9.[a] The synthesized material corresponded in chemical composition to $Rb_{38-43}K_{5-10}[Si_{96}Zn_{24}O_{240}]48H_2O$, from chemical analysis and ^{29}Si NMR spectroscopy. Room temperature powder patterns were collected on a Scintag XDS 2000 powder diffractometer, operating in Bragg-Brentano geometry, with a flat-plate sample and Cu $K\alpha$ radiation ($\lambda = 1.54184$ Å). Most of the peaks were indexed in a hexagonal unit cell, with $a \approx 9.9$ and $c \approx 37$ Å. A sample exchanged with (NH_4^+) was used, as well as the as-synthesized sample, to obtain high-resolution powder patterns at the European Synchrotron Radiation Facility (ESRF) at Grenoble. The pattern of the (NH_4^+)-exchanged sample was indexed completely in space group $P4_2/n\,cm$, with $a = 9.8946$ and $c = 36.8715$ Å. The framework topology was determined by applying the FOCUS procedure, as described above. While a preliminary Rietveld refinement indicated that the topology was probably correct, this model was not pursued because the exchange was incomplete and the detection of (NH_4^+) in the presence of H_2O would be difficult.

The pattern of the as-synthesized material was indexed only in space group $P4_12_12$, with $a = 9.8837(1)$ and $c = 73.6505(6)$ Å, an approximate doubling along c compared to the (NH_4^+)-exchanged compound. Rietveld refinement was applied with restraints on the (Si, Zn)–O bond length, and the O–(Si, Zn)–O and (Si, Zn)–O–(Si, Zn) bond angles. A series of difference-Fourier maps based on iteratively improved models led to better agreement, but only as far as $R_F = 0.167$ and $R_{wp} = 0.453$.

By applying further chemical reasoning to the model, together with a new series of difference-Fourier maps and Rietveld refinement, gradual improvements in the model were obtained. The final refinement converged at $R_F = 0.069$, $R_{wp} = 0.147$, $R_e = 0.099$, and $\chi^2 = 2.2$. In all, 170 structural parameters were refined, which is the largest framework topology yet solved from powder data without manual intervention. Figure 9.16 shows profiles for the Rietveld refinement of the as-synthesized VIP-9.

The framework topology has seven T-sites, or nodes,[b,c] in the asymmetric unit. The framework can be described in terms of two types of layers linked by isolated tetrahedra, Figures 9.17 and 9.18. The simpler layer A is a 4.8^2 net, which is a two-dimensional string of undulating 4-membered rings (4-rings) and 8-rings, as shown by Figure 9.17. The building unit of layers B is a polyhedron consisting of a 3-ring with three bent 5-rings attached to it, which may be termed a $[5^33]$ structural unit. The polyhedra share 3-ring faces on one side, and 5-ring edges on the other side, thus forming infinite chains parallel to $\langle 110 \rangle$. Neighbouring,

[a] L. B. McCusker, R. W. Grosse-Kuntsleve, Ch. Baerlocher, M. Yoshikawa, and M. E. Davis, *Microporous Materials* **6**, 295 (1996).

[b] R. W. Grosse-Kuntsleve, Ph. D. Thesis (*Dissertation ETH No. 11422*), ETH, Zürich (1996).

[c] R. W. Grosse-Kuntsleve, L. B. McCusker, and Ch. Baerlocher, *Journal of Applied Crystallography* **32**, 536 (1999).

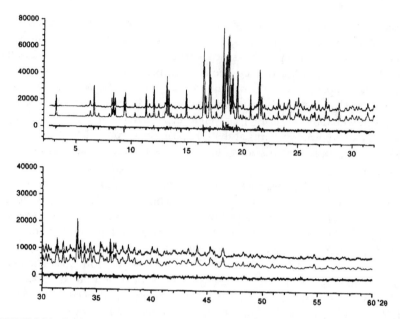

FIGURE 9.16.   Profile refinement for VIP-9, 0–30° and 30–60° in 2θ. In each diagram, the top, middle and bottom profiles refer to the observed, calculated, and difference patterns, respectively. [After McCusker et al., *loc. cit.* (1996).]

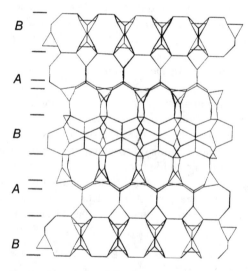

FIGURE 9.17.   The framework topology of VIP-9 viewed along [110]; oxygen atoms have been omitted for clarity. Layer B at $z = \frac{1}{2}$ is rotated by 90° with respect to those at $z = 0$ and 1. Similarly layer A at $z = \frac{3}{4}$ is rotated by 90° relative to that at $z = \frac{1}{4}$. [After McCusker, Grosse-Kuntsleve, Baerlocher, Yoshikawa, and Davis, *loc. cit.* (1996).]

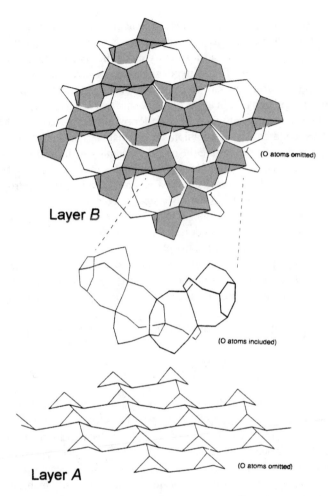

(O atoms omitted)

Layer B

(O atoms included)

(O atoms omitted)

Layer A

FIGURE 9.18. Layer-like building units in VIP-9: layer A, 4.8^2 net; layer B, chains of $[5^3 3]$ polyhedra. [After McCusker et al., *loc. cit.* (1996).]

parallel chains are shifted relative to one another by a half-chain length, and these structural units are new to zeolite data.

Monte Carlo Method

In this approach, a series of structural models in direct space is generated by random movements of a set of atoms in the unit cell, and each state of the system is evaluated on the basis of the agreement between the observed and calculated powder diffraction patterns. The models are postulated independently of the diffraction data, and once a satisfactory model has been obtained, it is refined by the Rietveld procedure. The atom positions may be chosen at random independently, as a group of atoms known to be part of the structure, particularly if the group forms a rigid body, or in terms of the connectivity of a molecule. In the latter case, the molecule is usually described in terms of internal coordinates, that is, bond lengths, bond angles, and torsion angles, that are convertible into Cartesian coordinates, and the variables are the orientation and positions of the molecule in the unit cell. In some cases, unknown torsion angles may be additional variables.

The Monte Carlo method itself is based on the well-known Metropolis algorithm,[a] and each state of the system is tested by calculating a residual R, rather than an energy term as in its original applications. An initial configuration of atoms x_i is displaced in a random manner but with constraints, relating, for example, to the amount of change permitted in the parameters of the set x_i. After each movement, the powder pattern is calculated, scaled to the observed pattern, and the whole-profile R-factor (9.18) calculated. Alternatively, the extracted intensities can be used as long as account is taken of the correlations between neighbouring reflections.[b]

Each trial structure is assessed on the basis of the difference Z, such that

$$Z = R(x_{\text{current}}) - R(x_{\text{previous}}) \tag{9.21}$$

If $Z \leq 0$, x_{current} is accepted; if $Z > 0$, x_{current} is accepted with a probability $\exp(-Z/S)$, where S is a scaling factor that operates like the energy parameter kT in the more conventional applications of the Metropolis algorithm.[c] It follows that the probability for rejection is $[1 - \exp(-Z/S)]$: if x_{current} is rejected, the previous trial structure now becomes "current." These stages are repeated (Figure 9.19), generating a Markov chain, that is, a sequence of events in which the outcome of each step is independent of the previous step; the probability of the change $x_i \to x_j$ depends only on the states i and j. Eventually the event showing the lowest R_{wp} value is subjected to Rietveld refinement. This technique has now been applied successfully to a number of structures.[d,e]

[a] N. Metropolis, A. W. Rosenbluth, M. N. Rosenbluth, A. H. Teller, and E. J. Teller, *Journal of Chemical Physics* **21**, 1087 (1957).

[b] W. I. F. David, K. Shankland, J. Cole, S, Maginn, W. D. S. Motherwell, and R. Taylor, *DASH User Manual*, Cambridge, UK, Cambridge Crystallographic Data Centre (2001).

[c] M. P. Allen and D. J. Tildesley, *Computer Simulation in Liquids*, Oxford University Press (1987).

[d] K. D. M. Harris and M. Tremayne, *loc. cit.* (1996).

[e] M. Tremayne, B. M. Kariuki and K. D. M. Harris, *Journal of Materials Chemistry* **6**, 1601 (1996).

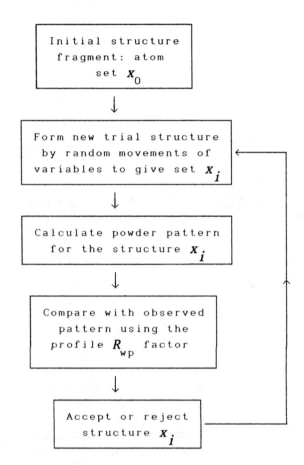

FIGURE 9.19. Monte Carlo cycling procedure. If a trial structure x_i is accepted it becomes x_{i+1} and is cycled back; otherwise x_i is returned for new random movements.

p-Bromophenylethanoic Acid

An interesting application of the Monte Carlo method elucidated the previously unknown structure of *p*-bromophenylethanoic acid, $BrC_6H_4CH_2CO_2H$.[a] By computer indexing of the first 20 lines of the powder pattern, the unit-cell dimension found were $a = 16.020, b = 4.607, c = 11.715$ Å, $\beta = 109.33°$. The systematic absences indicated space group $P2_1/c$, with $Z = 4$.

[a] K. D. M. Harris, M. Tremayne, P. Lightfoot, and P. G. Bruce, *Journal of American Chemical Society* **116**, 3543 (1994).

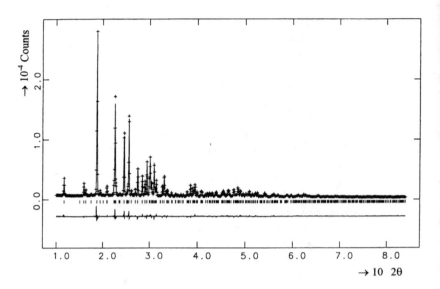

FIGURE 9.20.   Profile refinement for p-$BrC_6H_4CH_2CO_2H$; the experimental points are marked $+$ and the calculated profile is the full line. The lower line is the difference Fourier profile. [After Harris et al., *loc. cit.* (1994).]

The Monte Carlo technique was applied in two stages. In stage 1, the bromine atom alone was used. Of 1000 moves, one was found to be the most probable, at $R_{wp} = 45.2\%$; the value of R_{wp} for a totally random placement is *ca* 55% . In the second stage, the bromine atom was constrained in its best determined position, and the rigid C_7 fragment of the molecule rotated at random about an axis passing through the bromine atom. From the best fit model from this stage, the remaining atoms were obtained by difference-Fourier methods, and Rietveld refinement of the structure converged at $R_{wp} = 6.66\%$. Figure 9.20 illustrates the powder diffraction profiles from the Rietveld refinement. The difference between the stage 1 and final bromine atom positions was 0.2 Å, and other differences up to 1.3 Å were recorded for other atoms. This analysis shows well the power of this technique, which may now be considered one of the standard methods in the crystallographer's armoury.

Simulated Annealing

The simulated annealing procedure can be combined with the Monte Carlo method. In this case, the Metropolis algorithm is applied as before but with a systematic decreasing of the parameter S in the exponential function above, which is similar to a decrease in temperature of the energy quantity kT. The starting value of S is set such that all trial structures are accepted; then, as S is decreased, so the poorer fits (larger R_p values) are excluded. The final structure, the best fit, is then subjected to Rietveld refinement. In cases where the model does not contain the whole structure, the remainder can usually be located by difference-Fourier synthesis.

ESPOIR Program System

While Monte Carlo methods imply an element of random sampling, the fitting of a model to scattering data in this context is sometimes termed a reverse Monte Carlo procedure. ESPOIR* employs the reverse Monte Carlo technique coupled with simulated annealing for *ab initio* structure determination. It can use a completely random starting model, or else incorporate a structural entity of known geometry, in which case it functions similar to MR. The program fits the starting model to either $|F_o|$ data extracted from a powder pattern, or to single crystal data in the unlikely event that one of the single-crystal techniques fails. The author of the ESPOIR* program system has kindly consented to its inclusion with the suite of programs supplied with this book.

An innovative and important computer-time-saving feature of this program relates to the problem of overlapping peaks. Direct-space methods generally either fit the raw data to a model and derive measures of fit for each model, or fit some equation involving the extracted $|F_{obs}|$ data, taking into account overlapping of peaks. ESPOIR* follows a method intermediate between these two: instead of fitting the raw data, a pseudo powder pattern $P(2\theta)$ is reconstructed from the extracted $|F_{obs}|$ data. In this way no background, Lorentz, polarization, absorption, asymmetry, profile shape, or reflection multiplicity corrections have to be considered. A Gaussian shape function G is used for fast calculation and gross approximation to the overlapping, and the best fit between the pseudo pattern and the calculated pattern is judged from the parameter R_{PF}, where

$$R_{PF} = \frac{\sum |P(2\theta)_{obs} - K P(2\theta)_{calc}|}{\sum P(2\theta)_{obs}} \tag{9.22}$$

where $P(2\theta)_{obs} = G|F_{obs}|$ and K is a scale factor. Simulated annealing is introduced so as to reduce progressively the magnitude of the atoms' displacements. An additional variable parameter permits acceptance of Monte Carlo events that do not necessarily decrease R, so as to avoid false minima; about 40% of such events are retained typically. Nevertheless, at least 10 independent runs are recommended for a chance of success. With a more complex structural problem, more runs may be needed.

The ESPOIR* procedure is well illustrated with examples and problems in Chapter 11. It must be borne in mind that ESPOIR* is a program for obtaining a starting model for further refinement, either by the Rietveld technique or by traditional procedures. As with the standard procedures, ESPOIR* is not guaranteed to lead to success (Fr. espoir = hope).

α-Lanthanum Tungstate

α-Lanthanum tungstate, $La_2W_2O_9$, crystallizes in space group $P\bar{1}$, with $a =$ 7.2489(1), $b = 7.2878(1)$, $c = 7.0435(1)$ Å, $\alpha = 96.367(1)$, $\beta = 94.715(1)$, $\gamma =$ 70.286(1) deg and $Z = 2$. Diffraction patterns were obtained with both x-rays (Cu $K\alpha$) and neutrons ($\lambda = 1.5939$ Å). The pattern was indexed by TREOR, to give a single triclinic solution ($M_{20} = 24$). Using the x-ray data, intensities were extracted by the program FullProf and the structure solved by SHELX86 to locate two lanthanum and two tungsten sites. No full solution was found at this stage: the scattering from oxygen in the presence of lanthanum and tungsten is relatively weak.

The structure was solved with the neutron data using the program ESPOIR.[a] A WO_4 tetrahedron was first used as a rigid-body search fragment, rotating around fixed positions for tungsten and lanthanum. However, R_p would reduce no lower than 25%. Finally, the lanthanum and tungsten positions were fixed at the coordinates given by the x-ray study, and 9 oxygen atoms searched with ESPOIR by a random approach, using the neutron data. An R_p of 6.2% was achieved, and then the whole neutron pattern subjected to a final Rietveld refinement. Figure 9.21

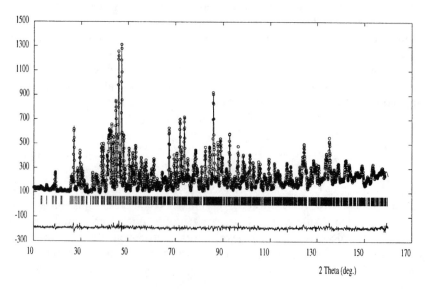

2 Theta (deg.)

FIGURE 9.21.   Final profile refinement for α-$La_2W_2O_9$: circles, observed pattern; full line, calculated pattern; vertical lines, reflection positions; bottom profile, difference pattern. [After Laligant, Le Bail, and Goutnenoire, *loc. cit.* (2001).]

[a] Y. Laligant, A. Le Bail, and F. Goutenoire, *Journal of Solid State Chemistry* **159** 223 (2001).

TABLE 9.5. Results from the X-ray and Neutron Data
Collection for α-La$_2$W$_2$O$_9$

	X-ray	Neutron
Number of reflections	1317	1366
Number of parameters refined	58	69
Peak-shape function	Pseudo-Voigt	Pseudo-Voigt
R_B	0.114	0.030
R_p	0.156	0.060
R_{wp}	0.186	0.069
R_e	0.064	0.019

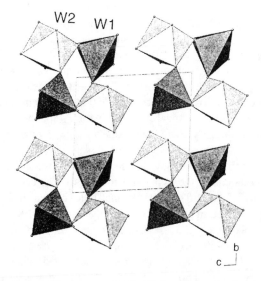

FIGURE 9.22. Projection of the structure of α-La$_2$W$_2$O$_9$ along a, showing the two environments of the tungsten atoms.

illustrates the final profile refinement, and Table 9.5 summarizes results from the x-ray and neutron studies.

The differences in the two tabulated sets probably relate to preferred orientation, which was considered to be the cause of the failure to find the oxygen atoms by difference-Fourier synthesis, thus leading to the poorer agreement indices. One aspect of the structure is shown in projection in Figure 9.22. The tungsten atoms are of two types: W1 is coordinated by five oxygen atoms forming a trigonal bipyramid, whereas W2 is coordinated octahedrally. This structure could explain the failure of the models based on WO$_4$ tetrahedral units. The sharing of corners builds up 4-rings to give [W$_4$O$_{18}$] structural units in a *cyclo*-tungstate structure. The two lanthanum atoms are coordinated, respectively, by nine and ten oxygen atoms.

9.8   Powder Diffraction with Proteins

X-ray diffraction analysis of the structure of a protein is often limited by the availability of suitable single crystals. Recently, however, it has been shown that the absence of single crystals may not present an insurmountable difficulty in this field any more than it does in materials science, as powder diffraction techniques have developed to the point where, as we have seen in the previous section, complex oxides, zeolites, and small organic molecular structures can often be solved from powder data alone.

T3R3 Zinc–insulin Complex

This fact has been demonstrated with the structure solution and refinement of a new variant of the T3R3 human zinc–insulin complex,[a] produced by mechanically grinding a polycrystalline sample.

High-resolution synchrotron x-ray powder-diffraction data were used to solve this crystal structure by molecular replacement adapted for Rietveld refinement. A complete Rietveld refinement of the 1630-atom protein structure, achieved by combining 7981 stereochemical restraints with a 4800-step ($d_{min} = 3.24$ Å) powder-diffraction pattern, yielded the residuals $R_{wp} = 3.73\%$, $R_p = 2.84\%$, and $R_F = 8.25\%$.

It was found that the grinding-induced phase change was accompanied by $9.5°$ and $17.2°$ rotations of the two T3R3 moieties in the crystal structure. The material reverts in 2–3 days to give the original T3R3 crystal structure. A Rietveld refinement of this 815-atom protein structure, by combining 3886 stereochemical restraints with a 6000-step ($d_{min} = 3.06$ Å) powder-diffraction pattern, yielded the residuals $R_{wp} = 3.46\%$, $R_p = 2.64\%$, and $R_F = 7.10\%$.

The ability, demonstrated by this work, to solve and refine a protein crystal structure from powder diffraction data suggests that this approach can be employed, for example, to examine structural changes in a series of protein derivatives in which the structure of one member is known from a single-crystal study.

9.9   Concluding Remarks

The examples of structure determination discussed in this chapter demonstrate the feasibility of solving crystal structures from powder diffraction data. It is to be expected that the method will become even more widely used in future, since it opens the way for the investigation of a wide spectrum of materials that have

---

[a] R. B. von Dreele, P. W. Stephens, G. D. Smith, and R. H. Blessing, *Acta Crystallographica* **D56**, 1549 (2000).

hitherto resisted detailed structural analysis. Not all of the current known procedures have been discussed here. For example, some structures have been solved by the *maximum entropy* method, in which the atoms of a molecule are placed in random positions in the asymmetric unit, and the structure solved by progressive removal of the randomness.[a]

Yet another procedure is based on *genetic algorithms*,[b,c] which are optimization techniques that are based on and invoke evolutionary principles. By natural selection, the 'fittest' structures in a population are allowed to survive and lead to improved members in subsequent populations, until the optimal individual, the correct structure, is attained.

The totality of techniques has been well reviewed very recently,[d] and many programs have been devised that address both the individual stages and the complete process of solving crystal structures from powder data. The serious powder analyst is strongly recommended to consult this literature source. In addition, there are several important references relating to work that has not been published elsewhere that can be accessed from web sites; they are listed in Table 9.6.

TABLE 9.6.   Crystallographic Data on Web Sites

Topic	Web site
Structure Determination from Powder Diffractometry Round Robin (SDPDRR)	http://www.cristal.org/SDPDRR/index.html http://www.ccp14.ac.uk/ccp/web-mirrors/armel/SDPDRR/index.html
Optimum Data Collection Strategy	http://www.ccp14.ac.uk/solution/ powder_data_collection.html http://www.ccp.14.ac.uk/solution/gsas/ convert_vct_data_to_gsas.html
Discussion on Variable Count Time (VCT) Data Collection	http://www.ccp14.ac.uk/solution/vct/index.html
General Crystallographic Studies	http://www.ccp14.ac.uk and http://www.ccp4.ac.uk

[a] M. Tremayne, P. Lightfoot, C. Gidwell, K. D. M. Harris, K. Shankland, C. J. Gilmore, G. Bricogne, and P. G. Bruce, *Journal of Materials Chemistry* **2**, 1301 (1992).

[b] B. M. Kariuki, K. Psallidas, K. D. M. Harris, R. L. Johnson, R. W. Lancaster, S. E. Staniforth, and S. M. Cooper, *Journal of the Chemical Society Chemical Communications* 1677 (1999).

[c] K. Shankland, W. I. F. David, T. Csoka, and L. McBride, *International Journal of Pharmacology* **165**, 117 (1998).

[d] See Bibliography (David, Shankland, McCusker, and Baerlocher).

Bibliography

DAVID, W. I. F., SHANKLAND, K., McCUSKER, L. B., and BAERLOCHER, Ch. (Editors), *Structure Determination from Powder Diffraction Data*, International Union of Crystallography/Oxford University Press, Oxford (2002).

PESCHAR, R., *Molecular Structure Solution Procedures*, **3**, 59 (1990).

PRINCE, E., *Structure and Statistics in Crystallography* (Edited, A. J. C. Wilson) Adenine Press (1985).

SCHENK, H., *Computational Crystallography* (Edited, D. Sayre) Oxford University Press (1982).

YOUNG, R. A., *The Rietveld Method*, International Union of Crystallography/Oxford University Press (1993).

Problems

9.1. A cylindrical powder camera has a radius of 57.30 mm. A given powder specimen is examined with Cu $K\alpha$ radiation. At what value of the Bragg angle θ would the $\alpha_1\alpha_2$ doublet begin to be resolved on the film, if the lines in that region of the film are of approximately 0.5 mm thickness? The α_1 and α_2 components have wavelengths 1.5405 Å and 1.5443 Å, respectively, and an intensity ratio $\alpha_1{:}\alpha_2 = 2$.

9.2. The following sequence of $\sin^2 \theta$ values was measured for lines on a powder photograph of a cubic substance taken with Cu $K\alpha$ radiation, $\lambda = 1.5148$ Å. Determine (a) the unit-cell type, (b) the indices of the lines, (c) a best value for the unit-cell dimension a. For (c), plot a, calculated for each line, against $f(\theta) = \frac{1}{2}(\cos^2 \theta / \sin \theta + \cos^2 \theta / \theta)$, with θ in radian, and extrapolate to $f(\theta) = 0$. This (Nelson–Riley) function tends to compensate for errors arising from absorption, specimen eccentricity, camera radius, and beam-divergence.

0.0465, 0.0635, 0.1717, 0.2486, 0.3712, 0.4170, 0.5394, 0.5544, 0.6609, 0.7368.

9.3. Confirm the values of the parameters for the reduced unit cell of magnesium tungstate discussed in Sections 9.3 and 9.3.1 with the program LEPAGE: use the default value of the collinearity parameter C, and also $C = 0.5°$.

9.4. The following unit-cell parameters were deduced from an indexed powder photograph: $a = 8.515$, $b = 8.515$, $c = 6.021$ Å, $\alpha = 135.0°$, $\beta = 69.3°$, $\gamma = 90.00°$. Determine the reduced and conventional unit cell, and list its parameters. What is the ratio of the volume of the conventional unit cell to that of the given unit cell?

9.5. The following values $(10^4 Q)$ were obtained for the first 40 lines on an accurate powder diffraction record of the microcrystalline single substance X. Deduce by induction a possible unit cell. Remember that the first three lines need not necessarily correspond with a^*, b^*, and c^*. Hence, index the lines on the basis of this unit cell. What can be said about the unit-cell type and space group for crystal X? Obtain the reduced and conventional unit cells, using the program LEPAGE. If the conventional unit cell is different from the unit cell first derived, transform the indices. What now are the conventional unit-cell type and the possible space group(s)?

Line number	$10^4 Q$	Line number	$10^4 Q$
1	83.1	21	892.4
2	89.1	22	916.0
3	172.2	23	962.3
4	249.8	24	981.7
5	332.6	25	999.1
6	356.1	26	1016.
7	416.0	27	1105.
8	421.5	28	1129.
9	439.3	29	1134.
10	516.9	30	1248.
11	559.8	31	1249.
12	642.9	32	1308.
13	648.6	33	1330.
14	683.3	34	1361.
15	688.8	35	1369.
16	732.1	36	1397.
17	748.4	37	1419.
18	801.5	38	1425.
19	837.2	39	1444.
20	884.5	40	1461.

9.6. In this problem and the next two, sets of data are provided to demonstrate the power and applicability of computer indexing, using the program ITO12*. The first data set has been included with the suite of computer programs, because it is vital to set up the data in the prescribed format. For those familiar with FORTRAN, the fields for numerical input are F10.5 (see also Chapter 11 for remarks on ITO12*). Crystal $XL1$: data are provided as values of $10^4 Q$; Q-values are sometimes *defined* as $10^4/d^2_{hkl}$. Use the program to determine the unit cell parameters and as much information as possible about the space group.

361.0	459.4	475.8	701.7	717.5	968.9	1310.	1312.
1059.	1088.	1363.	1411.	1428.	1444.	1653.	1724.
1785.	1838.	1902.	2039.	2081.	2145.	2300.	2332.
2380.	2451.	2485.	2639.	2656.	2675.	2807.	2871.
2996.	3185.	3212.	3250.	3265.	3402.	3507.	3428.

It matters that the first six lines, at least, are in numerical order; the remaining lines need not be so arranged.

9.7. Crystal $XL2$: data are provided as values of $10^4 Q$. Use the program to determine the unit-cell parameters and as much information as possible about the space group.

311.5	364.2	442.0	675.4	877.8	1020.3	1065.2	1111.0
1150.3	1246.3	1384.4	1456.3	1500.2	1534.1	1605.3	1610.0
1767.6	1773.4	1839.5	1954.9	2045.3	2157.1	2228.7	2242.5
2311.0	2318.4	2476.7	2702.3	2746.8	2773.3	2834.6	2865.6
2925.3	3003.0	3019.4	3062.3	3145.7	3198.0	3236.9	3276.7

9.8. Crystal $XL3$: data are provided as values of 2θ. Use the program to determine the unit-cell parameters and as much information as possible about the space group.

8.44	15.45	15.61	16.19	17.00	17.22	18.32	18.80
19.02	20.71	22.20	22.29	23.98	24.92	25.14	25.76
26.95	27.34	28.01	29.10	29.27	30.70	31.18	31.62
32.04	32.49	32.83	33.30	33.60	34.67	34.87	35.45
36.63	36.84	37.11	37.34	38.08	38.57	38.89	38.95

Problems on *structure-solving* from powder data are presented in Chapter 11, wherein ESPOIR is discussed further.

10

Proteins and Macromolecular X-ray Analysis

10.1 Introduction

In this chapter, we take a more detailed look at methods of x-ray analysis that are particularly applicable to large biological molecules. It will involve some useful reiteration of concepts and ideas discussed in previous chapters. We would also like to remind readers that although there are definite distinctions between large and small molecules in the crystallographic arena, there is no reason to exclude one from the other, and in fact, there are many advantages in being familiar with both. The major differences should become clearer as you progress through this chapter. It follows that while we mainly deal here with macromolecules, much of the information provided in this chapter is applicable to all areas of crystal structure analysis.

Traditionally the branch of x-ray crystallography which has developed around the study of macromolecular structures is thought of as being derived from the need in the 1940s and 1950s to extend the methods of analysis to include protein and enzyme structures, which range in relative molar mass from around 6,000 upwards. Structural studies of many other macromolecular types, including chromatin, DNA fragments, and other polynucleotides, present similar problems for the crystallographer, and consequently many of the methods currently in use for proteins are applicable. Specific topics to be covered include: methods for growing crystals, collection, measurement and use of the X-ray data, and methods for analyzing and interpreting the structures. Much of the emphasis here will be given to the practical aspects of this type of research; most of the background to the underlying theory having been covered in previous chapters. We have seen that x-ray crystallography is highly computational, and a great variety of software is available for performing the various stages of the calculations and graphics display. A guide to the availability of both equipment and software is included in Appendix A10, together with important Internet addresses. Many of

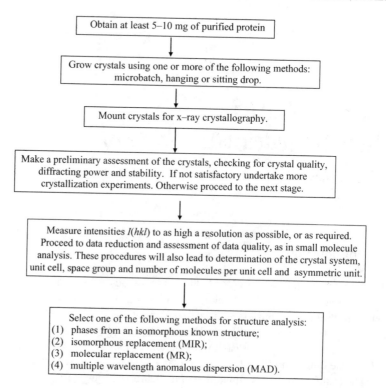

FIGURE 10.1.   Flow diagram for the preliminary stages in the x-ray analysis of a macromolecule.

the programs described are supported by the CCP4 Project, Daresbury Laboratory, UK,[a] and others mentioned in the text are readily available to prospective users. The flow chart in Figure 10.1 indicates the first stages in the x-ray analysis of a macromolecule.

10.1.1   What is a Protein

Although proteins are very large biological molecules, their chemical formulation is quite straightforward. Figure 10.2a shows some of the important features of the first three residues of a protein polypeptide chain: $C\alpha$ atoms are chiral carbons (see below); R_1, R_2, and R_3 are side-groups, each of which can be one of twenty[b] different chemical moieties that make up the "protein alphabet"; the

[a] ccp4@dl.ac.uk

[b] See for example, C. Branden and J. Tooze, *Introduction to Protein Structure*, 2nd ed., NY, Garland Publishing (1999).

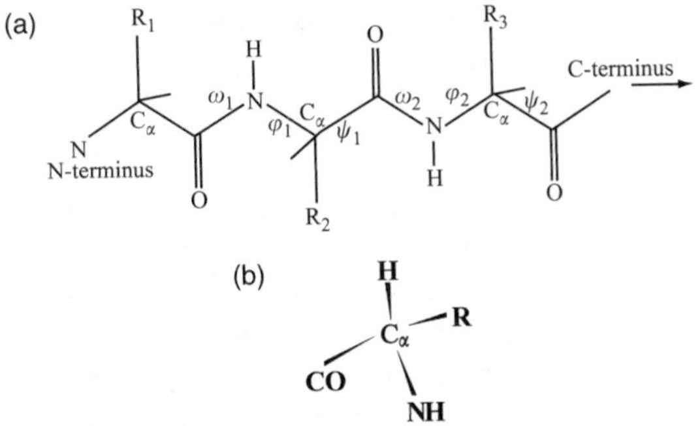

FIGURE 10.2. (a) Schematic diagram of a polypeptide chain found in proteins, showing the first two amino acid residues and the beginning of the third. The R-groups comprise one of twenty commonly occurring moieties. The peptide groups linked on either side to the $C\alpha$ atoms are predominantly planar, the torsion angle ω being usually in the range $\pm 20°$; two torsion angles ψ and ϕ define the polypeptide main-chain conformation (see Section 10.11.2). (b) The absolute configuration about a $C\alpha$ atom in the L-configuration, looking along H–C; the sequence C–O–R–N is encountered in a clockwise sense.

N-terminus (NH_2 or NH_3) is the beginning of the polypeptide chain and at some distance away will be the C-terminus group (CO_2H); the amino acid sequence is $R_1, R_2, R_3, \ldots, R_n$, where n is the total number of residues in the chain, and is known as the *primary structure*. Elements of *secondary structure* describe ways in which the polypeptide chain can fold; α-helices (right-handed) and β-sheets constitute two of the principle elements of secondary structure; *tertiary structure* describes how the elements of secondary structure are arranged to form the whole structure. Two proteins are said to be *homologous* (to a greater or lesser extent) if their amino acid sequences are similar; corresponding amino acids can be either exactly the same type or *conserved* by replacement of similar side-groups. The sequence N–$C\alpha$–C($= 0$)–N–$C\alpha$–C($= 0$)– $\cdots$ is known as the *main chain*. The bond labeled ω has limited torsional freedom with the torsion angle ω being $180° \pm 20°$ (maximum) for the usual *trans* peptide conformation, and $0° \pm 20°$ (maximum) for the less usual *cis* peptide; the bonds on either side of the $C\alpha$ atoms labeled with the torsion angles φ and ψ have greater but not unlimited rotational freedom. The folding of the protein chain depends on the values that φ and ψ take up along the chain to produce a minimum energy conformation, and this in turn will, of course, be dependent on the amino acid sequence. Some large proteins comprise more than one polypeptide chain, each of which acts as a subunit contributing to the *tertiary structure* of the protein; structurally equivalent subunits can be related either by exact crystallographic symmetry (the whole protein then occupies the corresponding symmetry element), by approximate or so-called *non-crystallographic symmetry* (NCS), or in no regular manner at all.

10.2 Crystallization of Proteins and Complexes for X-ray Analysis

10.2.1 Introduction

Bernal and Crowfoot[a] produced the first x-ray diffraction pattern of the protein crystal pepsin in 1934. Since then hundreds of biological macromolecules have been crystallized, at first using mainly unrefined methodology, and with little or no control of the outcome. Recently however it has been possible to rationalize the procedures, at least to some extent, through the use of hanging and sitting drops, which require minimal amounts of protein, and multiple sampling techniques, which enable a wide variety of conditions to be tried.

Once established, crystallization procedures are generally reproducible. However problems can arise as a result of one or more factors including: minor differences in protein composition or purity, for example, through extraction from different sources; or failure to reproduce exact crystallization conditions, perhaps due to poor reporting or recording protocols. Even the chance presence of a trace of dirt or grease can make all the difference, one way or the other. Special properties of protein crystals are listed below.

1. Protein crystals tend to be small compared to common crystals, being rarely greater than 1 mm on any edge.
2. In the vast majority of cases only one stereoisomer of any biological macromolecule exists in nature, so that their crystals contain only symmetry elements with *rotation* and/or *translation*; inversion and reflection are excluded as these symmetry operations involve a change of hand. As a consequence, the crystals themselves tend to exhibit fairly simple shapes compared to many minerals and other naturally occurring crystals, such as quartz or snowflakes.
3. Protein crystals are fragile and require ultra gentle handling. It arises from their high solvent content, which can be as much as 70%.
4. Protein crystals are extremely sensitive to pH, ionic strength, and temperature. Stability to low temperature can be improved through the use of *cryoprotectants*, forming a useful technique for improving the extent and quality of x-ray intensity data.
5. Protein crystals diffract x-rays both weakly, usually far short of atomic resolution; these effects are associated with the presence of large amounts of disordered or partially disordered solvent in the crystal.

[a] J. D. Bernal and D. Crowfoot, *Nature* **133**, 794 (1934).

10.2.2 Crystallization Conditions for Macromolecules

In keeping with aspects of crystal growth in general, the successful production of protein crystals is highly dependent on supersaturation. Macromolecular crystals usually form by nucleation at extremely high levels of supersaturation (100–1000%). In comparison, small molecule crystals usually nucleate at only a few percent supersaturation. Work with proteins usually means that the starting material is both expensive and sparse (20 mg is a large amount of protein), whereas most small molecule compounds are available in much larger quantities and consequently quite easy to crystallize.

Although high levels of supersaturation may be essential for promoting the nucleation of macromolecular crystals, there are general problems in terms of the formation of good quality crystals because supersaturated macromolecular solutions tend to produce amorphous precipitates. Consequently there is competition between crystals and precipitates at both nucleation and growth stages, and this competition is particularly acute because it is promoted at high levels of supersaturation. Because amorphous precipitates are kinetically favored, even though they are associated with higher energy states, they tend to dominate the solid phase and inhibit or even preclude crystal formation.

10.2.3 Properties of Protein Crystals

Protein crystals may contain one or more of the following, which may influence their size, quality, and x-ray diffraction characteristics: air pockets, disordered molecular deposits or clusters, inclusions, ordered and disordered solvent, precipitant ions, impurities, such as bound carbohydrate, inhibitors, non-covalently bound sugars, prosthetic groups or other ligands, covalent or non-covalently bound heavy atoms. The quality of sample homogeneity is another important factor.

10.2.4 Crystallization of Proteins

The crystallization of macromolecules depends on three factors:

1. Changing the relationship between the macromolecules and the solution components (water molecules and ions);
2. Altering the structure of the solvent so that the molecules are less well accommodated, thus promoting phase separation;
3. Enhancing the number and strength of favorable interactions between macromolecules. If the right conditions can be established the molecules will be continuously associating to form clusters and aggregates, to which new molecules are added more rapidly than old ones are lost. A crystal nucleus will then be born and growth will proceed.

10.2.5   Molecular Purity

In order to form crystals, the macromolecules have to be ordered in regular three-dimensional arrays. All forms of interference with regular packing will hinder crystallization. Lack of purity and homogeneity are major factors in causing unsuccessful and irreproducible crystallization experiments. It is advisable to carry out crystallization assays on fresh samples without mixing different batches of macromolecules. However, some micro-heterogeneities can be tolerated provided they do not occur in parts of the molecule involved in packing contacts in the crystal structure, a factor that is unpredictable prior to the structure being solved.

10.2.6   Practical Considerations

The high supersaturation of the molecules required for crystal nucleation can be achieved using a variety of *precipitants*. Widely used precipitants include ammonium sulphate, polyethylene glycol (PEG), methylpentanediol (MPD), and sodium chloride. When starting from non-saturated solutions, supersaturation can be reached by varying parameters such as temperature or pH. It should be noted that the conditions for optimal nucleation are not the same as those for optimal growth. Nucleation may be homogeneous or heterogeneous, occurring in the latter case on solid particles. This can lead to *epitaxial growth* (growth of crystals on other crystals). Interface or wall effects and the *shape* and *volume* of drops when using hanging or sitting drops (Section 10.2.9) can affect nucleation or growth. Therefore, the geometry of crystallization chambers or drops can be quite critical.

10.2.7   Batch Crystallization

This method is not normally used in current research, as large amounts of protein are required. As an exercise it can however be quite informative and serious students are encouraged to try it before moving to the smaller scale methods described below.

In this method protein solutions are prepared in milliliter quantities and crystallization is carried out in ordinary test tubes, typically of 5–10 ml capacity. The crystals grow on the walls of the test tubes and are harvested for storage when of suitable size. To grow crystals of the enzyme ribonuclease ($M_r \approx 13,000$), put 100 mg protein[a] into a test tube and dissolve in 2 ml water, cooled to 0°C; then gradually add 1 ml ice-cold absolute ethanol with stirring. Measure the pH and adjust to 5.0 by addition of a suitable buffer. Keep the tube at room temperature for 3 days. Re-cool and add a further 0.4 ml ethanol with stirring. The composition is now approximately 40% v/v. Small crystals should eventually appear on the walls

[a] Purchased from any large supplier of reagents.

of the tube, within 10 days. The crystals obtained should be monoclinic, space group $P2_1$.

10.2.8 Microbatch Screening

As discussed in Section 10.2.6 it is necessary to establish the correct crystallization conditions for a new protein. Consequently initial experiments are usually conducted at a microscale level with sample volumes in the range of a few microliters, with protein concentrations 5–10 mg ml^{-1}. Since it is difficult and expensive to prepare large quantities of highly purified protein, this strategy allows many different crystallization conditions to be screened. The experiment is conducted using microbatch plates. Each plate (Figure 10.3) contains a number of small wells in the form of a matrix, so that individual wells can be easily labeled for future reference. This is important as each well will correspond to a different set

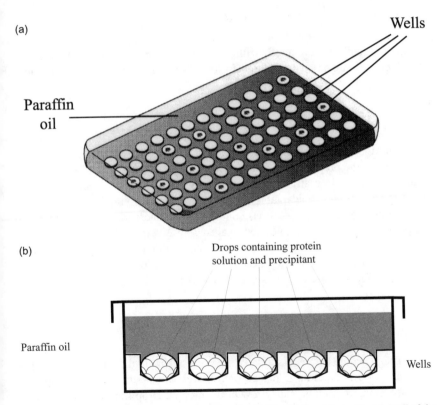

FIGURE 10.3. Microbatch plate. (a) The well matrix. (b) The profile of the wells and detail of the drops.

of conditions and crystals will only form in a very small number of wells, perhaps only one.

To set up the plate, drops are pipetted using a fine syringe under a layer of silicon oil (polydimethylsiloxane) or paraffin oil, or their 1 : 1 mixture. The drops contain protein and precipitant solution. All the reagents involved in the crystallization are present at a specific concentration and no significant concentration of the protein or of the reagents can occur in the drop. Reagent sampling kits are commercially available. Diffusion of water from the drop takes place through the oil thus changing the concentration in the drop, hopefully toward the required crystallization conditions. Once established, the localized well conditions can be refined in order to optimize the crystal size and quality for x-ray diffraction, using one of the techniques described below. Microbatch droplets usually dry up completely within a few weeks and consequently require to be carefully monitored.

10.2.9   Vapor Diffusion Techniques

The most popular and successful techniques for establishing crystallization conditions rapidly and efficiently with subsequent production of diffraction quality protein crystals are based on vapor diffusion. There are several practical variations in use.

Hanging Drop

This method is illustrated in Figure 10.4. Each drop is set up by being rapidly inverted over the prepared well where it hangs by surface tension. The wells are again in the form of a matrix, typically 4×6. A great variety of plates is available, many of which are also adaptable for the sitting drop technique. In the hanging drop method the droplet (5–20 μl) containing the macromolecule, a buffer and a precipitating agent is equilibrated against a reservoir (1–25 ml) containing a solution of the same precipitant at a higher concentration, say, by a factor of 2, than the droplet. Equilibration proceeds by evaporation of the volatile component until the vapor pressure of the droplet equals that of the reservoir. Crystals form in the droplet.

Sitting Drop

The principle here is essentially the same as that in the hanging drop technique except that the drop is placed on a bridge which sits over the precipitant reservoir. If successful, crystals grow in the droplet. Drop size and conditions of precipitants, buffers, and other factors similar to those used for hanging drops apply.

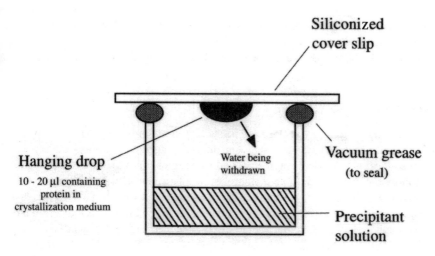

FIGURE 10.4.   Diagram of the hanging drop method of crystallization.

Vapor Diffusion Rate of Control

As in the microbatch technique, a layer of oil can be used with hanging or sitting drop techniques. This limits the rate of vapor diffusion; 200 μl of paraffin or silicon oil or a mixture of the two is applied over the reservoir solution. Varying the composition of the mixture provides additional control over the vapor diffusion rate.

Screening Crystallization Conditions

The above methods are all used with multiple screening protocols. A plate containing typically 4 × 6 wells provides a matrix of conditions. The method allows a broad range of salts, polymers and organic solvents over a wide range of pH to be sampled. When crystals are obtained, a second, finer screening around the relevant conditions can be used in order to optimize crystallization conditions to produce x-ray diffraction quality crystals. Ready-to-use reagents formulated from highly pure salts, buffers, and precipitants at various concentrations are commercially available.

Gel Crystallization Using Silica Hydrogel

Gels provide very efficient media for growing protein crystals. Silica gels, in particular are stable over a wide range of conditions, and compatible with additives and precipitants commonly used in macromolecular crystal growing. The gel forms a porous network in which the crystals can grow, minimizes convection, sedimentation, and nucleation, and therefore promotes the growth of large crystals without the strain usually imposed by the presence of a container. Silica hydrogel can be used for liquid–gel, liquid–gel–liquid, and vapor diffusion as well as dialysis techniques.

10.2.10   Co-Crystallization

Many proteins are required to be studied in the presence of ligands, which are usually quite small molecules and include sugars, inhibitors, coenzymes, nucleotides, nucleic acid fragments, and sometimes other macromolecules such as antibodies. When multiple isomorphous replacement (MIR) is being used as a phasing technique for structure analysis (see Sections 6.5.8 and 10.6.2), it is necessary to incorporate heavy-atom reagents into the crystals. One commonly used method of preparing these complex co-crystals is to add the material into the protein solution prior to crystallization. Crystals grown in this way are quite likely to suffer a change of crystal system, which is of course of no practical use in MIR, and can lead to unwanted complications in their applications. This is discussed further in Section 10.2.12.

10.2.11   How to Improve the Crystals

The following brief guide should help to improve the situation where poor (or no) crystals have appeared initially.

1. Vary the screening conditions: buffer, pH (possibly finer intervals), precipitant, protein concentration, drop size, method used, temperature (crystals will sometimes form in the fridge or cold room).
2. Use a finer matrix to vary the conditions around the wells with crystals. Try another method, for example, the hanging drop instead of the sitting drop. Try seeding—this works by transferring finely crushed crystal particles in the wells using a cat's whisker or hair.
3. One or more of the following additives may help the crystallization process: $Cu^{2+}, Zn^{2+}, Ca^{2+}, Co^{2+}$ ions, ethylenediamine tetraacetic acid, acetone, dioxane, phenol; for membrane proteins, n-octyl-β-D-glucopyranoside (up to critical micelle concentrations) and substrates, cofactors, inhibitors, or binding sugars.
4. With limited amounts of protein, concentrate on a few parameters known to be important: pH, initially in 0.5 intervals of pH unit, decreasing the interval as conditions are established; temperature, 4, 22, and 37 °C are most commonly useful; precipitants, ammonium sulfate, sodium chloride, polyethylene glycol, ethanol, or methylpentanediol are highly favored.

5. Crystals are obtained but their x-ray diffraction pattern looks like that of a small molecule (widely separated spots). Try the *click test* on one of the crystals: small molecule crystals such as ammonium sulphate (commonly mistaken for protein crystals) are usually physically hard and difficult to crush and will audibly "click" when poked with a needle.
6. Large crystals are obtained but the diffraction pattern disappears after 30 min exposure to the x-ray beam before a full set of data can be collected. Try freezing the crystals and book a session on a synchrotron facility.
7. Crystals are very soft and disintegrate when mounted in a glass capillary tube. Try being more careful when mounting; transfer through a larger pipette; use a larger diameter glass capillary.

10.2.12 Heavy-Atom Derivatives for MIR

A very wide range of heavy-atom reagents has been compiled for this purpose, mainly organic or inorganic compounds of mercury, gold, platinum, or uranium. The initial choice of possible compounds may be influenced by a knowledge of the amino acid sequence of the protein.[a] Lists of known useful compounds may also be consulted.[b,c] If good quality native crystals are already available, *soaking* them in heavy-atom solutions may be the best way to prepare derivatives.

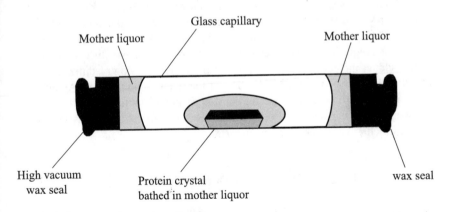

FIGURE 10.5. Capillary mounted protein crystal, as used for heavy-atom soaking or for x-ray data collection at room temperature.

[a] G. A. Petsko, *Methods in Enzymology* **14**, 147 (1985).
[b] T. L. Blundell and J. A. Jenkins, *Chemical Society Review* **6**, 139 (1977).
[c] D. E. McRee, *Practical Protein Crystallography.* Academic Press, San Diego & London (1993).

This can be carried out on single, mounted crystals, as in Figure 10.5 except that it will be necessary to open the capillary tube and wash the crystal at the end of the soaking period so as to stop uptake of heavy-atom material. The procedure should be carried out under controlled conditions of pH, temperature, time, and concentration. Heavy-atom concentrations in the range 0.5–20 M may be tried for anything between a few minutes to several days. The crystals may change color but should not undergo any other significant physical change. Detailed screening by x-ray techniques is necessary in order to establish isomorphism and to ensure that the heavy atoms have been incorporated into useful locations (see Chapter 6). Alternatively, co-crystallization can be tried. This involves setting up crystallization experiments using one of the methods described in the previous section, but with the addition of a heavy-atom reagent into the solution. It is essential to maintain the integrity of the solution prior to crystallization attempts. This may require careful adjustment of the conditions and a good deal of patience. If crystals form they will again require careful monitoring by x-rays so as to establish the desired incorporation of heavy atoms. Change of crystal unit cell and space group is more likely to occur in this method.

10.2.13   Protein Complex Crystals with Small Molecules

Many proteins associate with small molecules as part of their biological function, for example: enzyme–inhibitor complexes, protein–carbohydrate complexes (lectins), protein–nucleic acid fragments, protein–peptide complexes, such as cyclophillin/cyclosporin), and antibody–antigen complexes. Complex crystals may again be formed either by soaking or co-crystallization under controlled conditions. The methodology is similar to that used for the preparation of heavy-atom derivatives, requiring similar precautions and screening to establish incorporation of the adduct molecules. There is a slight preference for co-crystallization, as it is more likely to lead to full occupation of the ligand binding sites. Again, non-isomorphous derivatives may occur, but this problem should be easily overcome by using molecular replacement (MR) to determine the structure, assuming the native structure is known (see Chapter 7).

10.3   Crystal Mounting for X-ray Data Collection

The two methods currently in use for x-ray data collection depend principally upon the temperature, either room temperature or that of liquid nitrogen.

10.3.1   Mounting at Room Temperature

For x-ray data collection at room temperature, protein crystals are usually mounted in sealed thin-walled glass capillary tubes as shown in Figure 10.5.

The presence of a drop of mother liquor inside the tube allows a stable equilibrium to be established, which prevents the crystal from drying out and becoming denatured. Most proteins rapidly deteriorate once exposed to x-rays, through the formation of free radicals. Crystal lifetime can be anything from a few minutes to several days under normal laboratory conditions. Exposure to the highly intense beam from a synchrotron source usually decreases crystal lifetime dramatically. It is compensated, however, by an increased intensity, which permits shorter individual exposures and consequent increase in the quantity of data recorded.

10.3.2 Cryo-Crystallography

When the temperature of a crystal is lowered the thermal motion of the constituent atoms becomes less marked and the x-ray diffraction pattern can be improved in both intensity and resolution. These two factors are highly desirable in protein crystallography, as protein crystals are notoriously poor diffractors. Since the innovation of more efficient crystal cryo systems in the past 10 years [a popular and efficient model is the Oxford Cryosystems Cryostream Cooler[a]], there has been a marked increase in the number of protein structures determined at low temperature, and small molecule data is routinely collected on cooled crystals in some laboratories. However, for a new protein crystal, or a new derivative crystal, the method requires to be carefully set up empirically in order to establish the best conditions. The crystal is mounted and flash-cooled, usually to 100 K, free from ice, and set up in the x-ray beam. Damage from freezing is prevented through the use of a cryoprotectant liquid. The cryoprotectant may be incorporated in the mother liquor after crystallization or, less commonly, as a component of the crystallization reagents. The procedure involves supporting the crystal in a film of cryoprotected mother liquor in a small fibre loop (Figures 10.6a and 10.6b) which is then cooled in liquid nitrogen. The liquid surrounding the crystal must freeze as an amorphous glass to avoid crystal damage and diffraction from ordered ice crystals. Glycerol is most frequently used as a cryoprotectant in this type of work. The required glycerol concentration must be carefully established; unfortunately this may involve the loss of several crystals in the early stages of the experiment. For the low temperature data collection of mistletoe lectin I,[b] a complex mixture consisting of 0.1 M glycine buffer at pH 3.4, with 0.9 M ammonium sulfate, 0.05 M galactose and 30% glycerol, soaking for a few minutes prior to flash freezing. Once frozen, protein crystals are usually extremely stable, transportable, and can be stored and kept ready for subsequent x-ray diffraction experiments.

---

[a] http://www.oxford-diffraction.com
[b] H. Niwa, PhD Thesis, University of London (2001).

(a)

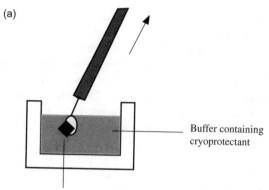

Buffer containing
cryoprotectant

Crystal is looped and drawn upwards.
It will stay in the loop by surface tension.

(b)

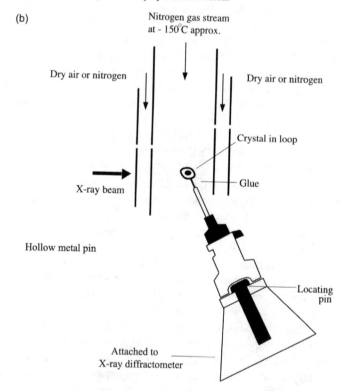

Nitrogen gas stream
at - 150°C approx.

Dry air or nitrogen

Dry air or nitrogen

Crystal in loop

X-ray beam

Glue

Hollow metal pin

Locating
pin

Attached to
X-ray diffractometer

FIGURE 10.6.   Cryo-crystallography. (a) The loop mounting of a crystal in a film of cryoprotected mother liquor, which is then cooled in a gaseous stream of evaporating liquid nitrogen. (b) Arrangement for aligning the crystal in the stream of cold gas. The goniometer head is usually a standard commercial device, with perpendicular translation slides, but normally no angular adjustments, because they are not needed for image-plate data collection.

10.4   Macromolecular Crystallography

10.4.1   Space Groups

Proteins are polymers of L-amino acids, the Cα atoms being specifically left-handed chiral centers (Figure 10.2(b)). Consequently, only *axes* of symmetry occur in protein crystals. This is because centres of symmetry, mirror planes, and glide planes would necessarily produce D-amino acids, which do not normally occur. Similar considerations in fact apply to the majority of biologically active molecules. As a consequence, protein crystals can occur in only 65 space groups (Table 10.1) out of the possible 230; space group determination is therefore a relatively simple procedure.

10.4.2   X-ray Diffraction from Macromolecular Crystals

Crystal Selection

We have seen, or implied in previous sections that, for single crystal analysis, each crystal to be used for measuring diffraction data should be well ordered and produce a single, clear diffraction pattern. Poor quality specimens that are not single crystals may suffer from a variety of faults including splitting, twinning, or the presence of slippage-disorder planes. Split or twinned crystals may diffract well but will produce multiple reflection spots that are difficult to interpret; such specimens should be discarded. Badly disordered crystals are characterized by the spreading out of each diffraction spot, often in an irregular manner, and by the poor overall resolution of the diffraction pattern, which rapidly fades away as a function of the radial distance from the center of the pattern. Faults of this type are particularly associated with protein crystals that are relatively soft (containing up to 70% solvent), and easily damaged through handling when being mounted for x-ray analysis. Even single crystals diffract poorly if heavily disordered. If a crystal proves to be unsuitable through any of the above faults it should be discarded and another one selected. Refinement of crystallization conditions is frequently necessary in order to produce better crystals.

Structure Factors and Temperature Factors

We have shown previously that the total x-ray scattering associated with a given *hkl* reflection is represented by the structure factor $F(hkl)$. This parameter and the atomic scattering factor have been discussed in detail in Chapter 3. The use of structure factor equations for macromolecules is no different from applications with other molecules, except of course that the number of atoms N per unit cell is greater than 1000 even for quite small proteins. The isotropic temperature factor B, or thermal displacement factor, has the following typical values for proteins: overall average, 20–30 $Å^2$; for main chain atoms, 10–20 $Å^2$; for

TABLE 10.1.   The 65 Enantiomorphic Space Groups Applicable to Protein Crystals (The Corresponding Laue Groups are Listed in Brackets)

Crystal system	Crystal class	Space-group symbol [Laue-group symbol]	Number Z of asymmetric units per unit cell
Triclinic	1	$P1$	1
		$[\bar{1}]$	
Monoclinic	2	$P2\ P2_1$	2
		$C2$	4
		$[2/m]$	
Orthorhombic	222	$P222\ P222_1\ P2_12_12$	4
		$P2_12_12_1$	4
		$C222\ C222_1$	8
		$I222\ I2_12_12_1$	8
		$F222$	16
		$[mmm]$	
Tetragonal	4	$P4\ P4_1\ P4_2\ P4_3$	4
		$I4\ I4_1$	8
	422	$[4/m]$	
		$P422\ P42_12\ P4_122$	8
		$P4_12_12\ P4_222\ P4_22_12$	8
		$P4_32_12\ P4_322$	8
		$I422\ I4_122$	16
		$[4/m\ mm]$	
Cubic	23	$P23\ P2_13$	12
		$I23\ I2_13$	24
		$F23\ [m3]$	48
	432	$P432\ P4_132\ P4_232$	24
		$P4_332$	24
		$I432\ I4_132$	48
		$F432\ F4_132$	96
		$[m3m]$	
	6	$P6\ P6_1\ P6_5\ P6_2\ P6_4$	6
Hexagonal	622	$P6_3$	6
		$[6/m]$	
			12
		$P622\ P6_122\ P6_522$	12
		$P6_222\ P6_422\ P6_322$	
		$[6/m\ mm]$	

TABLE 10.1.   (Continued)

Crystal system	Crystal class	Space-group symbol [Laue-group symbol]	Number Z of asymmetric units per unit cell
Trigonal[a]	3	$P3$ $P3_1$ $P3_2$	3
		$R3$	3 (9)
		$[\bar{3}]$	6
	$321, 312^b$	$P321$ $P3_121$ $P3_221$	6
		$R32$	6(18)
		$P312$ $P3_112$ $P3_212$	6
		$[\bar{3}m]$	

Notes:
[a] Trigonal crystals are referred to hexagonal axes; there is no separate trigonal unit cell. The same choice can be made for rhombohedral crystals, in which case there are three times the number of molecules per unit cell; but the cell shape is easier to handle.

[b] We mention, in passing, that while point groups 321 and 312 are identical, under a rotation of the symmetry elements by 30° with respect to the crystallographic axes, space groups $P321$ and $P312$ are different. In the infinite array of a space group, the two space groups relate to different arrangements of point in space, that is, differing sets of general equivalent positions. Similar arguments apply to the pairs of point groups $3m1/31m$ and $\bar{3}m1/\bar{3}1m$.

side-chain atoms 15–30 Å^2; for solvent and small ligand atoms, 25–45 Å^2. As a further rule of thumb it is common for an upper limit of around 80 Å^2 to be placed as a credibility indicator for any individual atom. Atoms with a temperature factor greater than this value after refinement (see Section 10.9) should be seriously re-examined. Values for the temperature factor of each atom evolve during the course of structure refinement.

Intensities and Phases

Experimental $|F_o(hkl)|$ values are derived from the measured intensities $I_o(hkl)$ by procedures that are discussed below. For a successful structure analysis, as many values as possible of $|F_o|(hkl)|$ must be measured to as high a resolution as can be achieved.

Values of the phase angles $\phi(hkl)$ required for calculation of electron density, are usually determined by application of one of the following techniques:

1. Direct calculation of phases, using a known isomorphous protein structure;
2. MIR (see Section 6.4.8);
3. MR (see Section 7.3, Patterson Search);
4. Multiple wavelength anomalous dispersion (MAD).

Calculated Structure Factors and R-Factors

Once a model of the structure has been proposed, values of the coordinates (x_j, y_j, z_j) for most of the atoms will be available. It is then possible to calculate values of the structure factor amplitudes and phases based on the model. The quality of the model can be tested by calculating a conventional R-factor:

$$R = [\Sigma ||F_o(hkl)| - |F_c(hkl)|]/ \Sigma |F_o(hkl)| \qquad (10.1)$$

where the summations are carried out over all reflections in the data set.

For a protein structure R is rarely less than 0.35 for the initial model, improving to 0.2 or better after refinement. The quality of any x-ray structure is restricted by the diffracting power of the crystal. The poor diffracting power of protein crystals restricts both the quality and the total number of data available compared to the large number of parameters involved in the calculation of the structure factors. It is consistent with the correspondingly poor R-factors attainable. By comparison the R-value for a good low molecular weight x-ray structure would be less than 7%.

Free R-Factor (R_{free})

When MR is used to determine a protein structure, a known homologous structure is fitted into the unit cell of the new (target) structure. It is common to use search structures which have as little as 40% identity with the sequence of the target structure and this is quite acceptable as long as the main chains of the protein fold in a similar way. The new structure is then refined by gradually transforming each amino acid in the sequence, in location and conformation, as necessary, in order to minimize the R-factor. Sometimes it is difficult to remove the initial bias, which is built into the new structure as a consequence of using the coordinates of another structure, a phenomenon known as feedback or "memory," which is built into the Fourier method.

In order to monitor this process 5–10% of the $|F_o|$ data are removed from the data set during the refinement process and are not allowed to contribute to the course of the analysis. This subset of data is used to calculate an R-factor called the Free R or R_{free}.[a,b] As the refinement proceeds, if $|F_c(hkl)|$ truly approaches $|F_o(hkl)|$, R_{free} will drop together with R. If the refined model fails to break away from the initial model, R will drop because its parameters are changing, but R_{free} will fail to improve because the model is not actually improving. With a correct model, R_{free} will usually decrease, but will remain a few percent greater than R.

10.4.3   Recording X-ray Diffraction from Macromolecular Crystals

The recording and measurement of x-ray diffraction patterns are topics which have been thoroughly covered in Chapter 4. Only points specifically of interest for macromolecular crystallography will be touched upon here.

[a] A. T. Brünger, *Nature* **355**, 472 (1992).
[b] I. J. Tickle, R. A. Laskowski, and D. S. Moss, *Acta Crystallographica* **D56**, 442–450 (2000).

Conventional X-ray Laboratory Sources

In conventional crystallography laboratories, university and industrial departments, rotating anode generators are reliable sources of intense x-radiation in most crystallographic applications. For macromolecular studies copper x-radiation in conjunction with a graphite monochromator and focusing mirrors give excellent results. It is worth noting, however, that rotating anode generators can be expensive to run, requiring constant pumping to high vacuum and frequent filament changes.

Synchrotron X-ray Sources

X-rays are generated at a synchrotron source when high-energy electrons are accelerated in a storage ring at relativistic speeds. The X-ray beam is narrow, extremely intense (100 times that of a conventional source), and the wavelength can be selected from a very wide range to match experimental requirements. In view of the high intensity and fine collimation, it is possible to record complete data sets even from small or weakly diffracting protein crystals in a matter of minutes. It is also possible to devise time-resolved experiments in order to monitor processes such as modified enzyme–substrate interactions. Synchrotron installations[a] are highly specialized research facilities. Experiments to be undertaken with synchrotron radiation (SR) require careful planning.

X-ray Cameras

The use of x-ray cameras can lead to rapid and reliable evaluation of crystal quality, symmetry and unit cell parameters, and is an excellent tool for establishing a sound understanding of the reciprocal lattice. While the availability of x-ray cameras in both university departments and other research laboratories is far less common nowadays, many still do retain them for both training and research. For protein work in particular, the precession camera is an invaluable tool for checking, in addition to the above uses, for changes in both the unit cell and the intensity pattern in applications involving MIR.

Diffractometers: Single Counter or Serial Diffractometers

Traditional diffractometers incorporate a mechanical goniometer to orientate the crystal into the correct position for each reflection and to rotate the counter, usually a scintillation counter to receive the scattered x-radiation from this single reflection. The energy is transformed electronically into a form suitable for conversion to intensity. Because each reflection is measured individually, with a count time typically of around 60 s, the process is very slow, particularly for

---

[a] See Appendix A10.

proteins, which routinely involve the measurement of tens of thousands of reflections. Whilst the accuracy attainable is better than for most of the other methods used for intensity measurement, the limited lifetime of protein crystals in the x-ray beam permits only a fraction of the available data to be recorded from one crystal. The use of several crystals for data collection introduces errors associated with the scale factors required to merge the various collections into a single data set. Further details of this method are to be found in Chapter 4.

Diffractometers: Area Detectors

The main disadvantages of single counter diffractometry, such as slow data collection rate and the requirement of several crystals for collection of a complete data set, with the attendant errors associated with scaling and crystal deterioration, have been largely overcome by the use of "electronic film" area detectors and image plates. They have enjoyed rapid development in recent years and have been discussed in Chapter 4. The principal advantages of this method with proteins are that data can be collected to a resolution of about 1.4 Å using Cu radiation and better with synchrotron source, and that a very wide range of intensities (approximately 10^5) can be recorded compared to an x-ray film (approximately 2×10^2).

10.4.4   Measurement of X-ray Diffraction from Macromolecular Crystals

Area Detectors in Practice

Data are usually collected by the oscillation method, with the oscillation or rotation axis perpendicular to the x-ray beam. Sophisticated programs such as STRATEGY and DENZO (see Appendix A10) are available for optimizing the experimental stages in practice, using a procedure known as autoindexing. This method allows time to be saved by using the most efficient data collection procedure for a given situation. These programs may be implemented on the basis of a single frame (or exposure), or better still, two frames of data collected at starting points 90° apart. This initial information leads to a derivation of approximate unit-cell parameters, Laue group, crystal orientation (the crystal is not necessarily mounted about a major crystallographic axis), total oscillation range, and oscillation range per frame and, hence, the total number of frames to be used for the experiment. For a given exposure the crystal is oscillated through a small angle $\Delta\phi$, set usually to a value between 0.1° and 1.5°, depending on the unit cell size. The next frame begins where the previous one ended, keeping $\Delta\phi$ at the same setting. The procedure is repeated until the required angular range has been covered. Once the x-ray intensities have been measured, structure analysis software usually assumes that the $I(hkl)$ data set covers a single asymmetric unit of the Laue group symmetry, together with the Friedel equivalents where anomalous dispersion is being utilized.

If time permits, it is a sound strategy to collect at least two asymmetric units of data and utilize the R_{merge} index to monitor data quality. For a crystal arbitrarily mounted, 180° of data frames will be more than sufficient for anything but a triclinic crystal. Such an arbitrary strategy can be wasteful of both diffractometer time and crystal lifetime and can be avoided using the programs DENZO and STRATEGY.

Image Plate Data Processing

While the procedure for acquiring raw data using an image plate is quite rapid, and this itself is a tremendous advantage bearing in mind the fact that protein crystals have notoriously short lifetimes in the x-ray beam, processing this data to produce the $|F_o|$ values can be very time consuming. Each frame of data contains information about the (X, Y, Z) coordinates of the diffraction maxima relative to the experimental set up, and digitized information from which a measure of the intensity can be derived. For a given crystal specimen the following procedures are carried out concomitantly:

1. Refinement of the orientation matrix to define the disposition of the crystal axes with respect to the x-ray beam direction;
2. Confirmation of the crystal system and possible space groups; measurement and refinement to the highest accuracy possible of the unit-cell parameters and their estimated standard deviations;
3. Determination of the indices hkl and intensities $I(hkl)$ for each recorded diffraction spot.

Partially Recorded Spots and Integration of Intensity Data

Because images are recorded from small oscillations of the crystal and the reciprocal lattice points for a protein crystal are close together, each image will contain a proportion of partially recorded reflections which are completed either on the previous or next record. To obtain the intensity reading, the two partials are added together. This effect will inevitably introduce errors in the intensity values. Recording of the diffraction image produces a series of digitized pixels, which can be viewed by computer graphics and processed to provide intensity estimates.

Conversion to an Intensity Reading

In *summation integration* a volume containing an individual spot is defined. Summation of the measured counts in all the pixels in the defined volume and subtraction of the background, determined by examining surrounding pixels, leads to an integrated intensity. This method is similar to the scanning method used in processing the data from serial diffractometer measurements (see Section 4.7.5).

In an alternative method of *profile fitting*, an empirically derived model reflection shape is scaled to the data and then integrated. This assumes that the reflection shape is independent of intensity. The observed profiles vary over the detector face, so that several model profiles are usually required, depending on the location on the detector face. Profile fitting is computationally expensive but produces more reliable results, less susceptible to random errors. The method was first proposed for one-dimensional profiles,[a] extended to two-dimensional profiles for precession films,[b] and later applied to oscillation photographs.[c]

Profile fitting depends on the assumption that strong and weak reflections share a common intensity profile, so that an observed reflection can be related to a "standard" profile for the area by a simple scale factor. In DENZO this is formulated as follows: The observed profile M_i is approximated by P_i

$$P_i = Cp_i + B_i \tag{10.2}$$

where C is a constant to be determined, B_i is the predicted value of the background and p_i is the predicted profile. The index i represents all pixels derived by scanning a spot in either a one-, two-, or three-dimensional profile. If the predicted profile is normalized, $\sum_i p_i = 1$, then the constant is the fitted intensity I, that is, $I = C$. Profile fitting minimizes the function

$$\frac{(M_i - P_i)^2}{v_i} \tag{10.3}$$

This solution means that each pixel provides an estimate of the spot intensity with variance v_i. A profile fitted intensity is then the weighted average of all observations, thus:

$$I = \frac{\sum_i (p_i^2/v_i)(M_i - B_i)/p_i}{\sum_i p_i^2/v_i} = \frac{\sum_i p_i(M_i - B_i)/v_i}{\sum_i p_i^2/v_i} \tag{10.4}$$

Post-Refinement

One of the aims in data collection is to measure the intensity of each indexed reflection as accurately as the experimental conditions allow. For partial reflections, accurate partiality and intensity fractions are necessary as described in the previous section. Parameters required for the measurement of intensities, such as cell dimensions and crystal orientation can be determined more accurately once an initial set of integrated intensities is available. This process of post-refinement,

[a] R. Diamond, *Acta Crystallographica* **A25**, 43 (1969).
[b] G. C. Ford, *Journal of Applied Crystallography* **7**, 555 (1974).
[c] M. G. Rossmann, *Journal of Applied Crystallography* **12**, 225 (1979).

allows a more reliable classification of reflections as full or partial, and gives sufficiently accurate estimates of the intensity fraction of partial reflections for them to be scaled and used. The parameters involved include:

1. Parameters determining the position of the reflection, such as cell dimensions, wavelength, and crystal orientation;
2. Reflection width, which is precisely formulated;[a]
3. A function relating the angular fraction to the intensity fraction for which a model function is employed.

Fully recorded reflections or sums of adjacent partials are used for the reference intensity of a partial reflection. Since these data and the flag that indicates whether a reflection is full or partial vary during refinement of the parameters mentioned above, as well as by scaling between frames, the post-refinement needs to be iterated with scaling several times. There is a wide selection of software available for data processing. A possible procedure using mainly CCP4 software is indicated in the flow diagram of Figure 10.7.

10.4.5 Problems with Data Collection and Suggested Cures

There are certain problems that may be encountered with data collection from protein crystals; the main points are enumerated below, together with possible solutions.

1. Diffraction is observed but the spots are wide apart and clear to very high resolution. This may indicate that the crystal is not protein but one of the small molecule salts used in crystallization, very often ammonium sulfate. To further establish this, the "click test" should be tried.
2. The crystals may appear to be well formed, optically clear, and with good morphology, and the diffraction pattern is typically protein with spots close together. Sometimes, however, the individual spots may be diffuse or partially diffuse, "tailed" or spread out in a tadpole shape, obviously split (twinned), or spread out in a particular direction, for example, a principal reciprocal lattice direction (statistical or systematic disorder). These symptoms are all indicative of poor crystal quality and in all such cases, it is not worth collecting data from these specimens. The crystallization conditions should be reviewed and possibly slowed down by cooling or adjusting the initial concentration. Crystallization at

[a] T. J. Greenhough and J. Helliwell, *Journal of Applied Crystallography* **15**, 493 (1982).

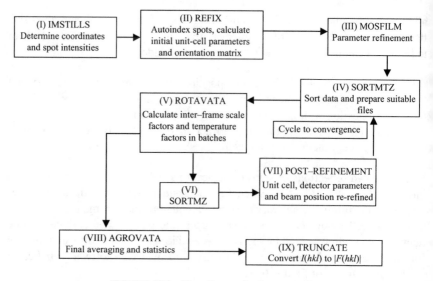

FIGURE 10.7.   Flow diagram for data processing.
Alternative software that may be used in the various stages of this scheme:
Stages (I), (II) and (III): DENZO;
Stage (V): SCALEPACK;
Stage (VIII): SCALEPACK2MTZ and CAD.

DENZO/XDISPLAY/SCALEPACK are part of the HKL suite: DENZO enables autoindexing, conversion to intensities and cell refinement; XDISPLAY displays the observed and calculated spot positions for comparison and validation; SCALEPACK carries out scaling, and merging on an iterative basis (post-refinement); MOSFILM is similar to DENZO but has the additional facility for spot size adjustment; REFIX allows MOSFILM to autoindex a single image; XDS or MARXDS include features of the programs mentioned previously that are required for MAR RESEARCH IPs; MARSCALE is an additional scaling program (see Appendix 10).

cold-room temperature or even in a refrigerator may be tried. Crystal-mounting routines should be considered, as handling protein crystals too vigorously will lead to damage. Physical contact with the crystals should be minimized. Other conditions should also be reviewed including storage time, because protein crystals can have a very short shelf life in crystallization trays, and also suffer from rough handling in transit.

3. Sometimes crystals may be very small and refuse to grow, have good morphology, give good diffraction spots, but the intensities are weak and θ_{max} is low. In such cases, the high-intensity SR should be tried, or cryo-cooling, or both.

4. If one or more unit-cell parameter is large (> 150 Å) diffraction spots may be weak and very close together on low intensity diffraction equipment. Again it would be better to use SR, which enables a larger crystal to

image plate distance to be used, resulting in better spot separation while
still maintaining satisfactory image strength.

5. If the crystals diffract well but their lifetime in the x-ray beam is too
   short to allow a full data set to be collected, again use of cryo-cooling
   to $-100\,°C$ or even $-150\,°C$ should be tried, together with synchrotron
   radiation and a CCD detector, so as to obtain very rapid collection. Cooling not only promotes better resolution in many cases but also improves
   crystal lifetime almost indefinitely. Crystals are usually loop mounted
   (Section 10.3.2) and can be stored and transported in a Dewar flask
   contained in an insulated box.

Cryo-cooling is the method of choice in many current applications and has
several advantages in addition to those listed above. They include elimination
of crystal slippage and drying out, both of which can occur with capillary tube
mounting.

10.4.6   Preliminary Structure Determination

Unit Cell, Laue Symmetry, and Systematically
Absent Reflections

Before serious consideration can be given to determination of the structure it
is necessary to characterize the crystal system, determine the unit-cell dimensions
and cell volume as accurately as possible, and identify possible space groups.
The unit-cell parameters will rarely be determined to better than 0.02 Å for most
protein crystals unless extreme care is taken and the diffraction pattern extends to
<1 Å resolution. Some image plates allow direct measurement of cell dimensions
or other lattice spacings. Subsequent confirmation and least-squares refinement
of the cell parameters using the data reduction software produces the final values
together with an estimate of the errors. The space group of the crystal is determined
from a consideration of the following.

Unit-Cell Parameter Values.   Table 2.3 lists the cell parameter restrictions associated with each of the seven crystal systems. It must be remembered that " $\not=$ "
means "not restricted by symmetry to equal," and consequently it is quite possible
for the unit cell parameters of a crystal to appear to have higher symmetry than is
actually the case. For example, a crystal which is actually monoclinic might have
values of α, β, and γ which are all 90° within experimental error; the crystal is
then *apparently* orthorhombic.

Laue Symmetry.   The diffraction pattern of a given crystal exhibits the symmetry of one of the 11 Laue Groups (Table 1.6). Provided that enough data have
been collected, Laue symmetry is usually reliably indicated as a further by-product
of data reduction. This serves to pinpoint the correct crystal system and possible

space groups (Table 10.1) that apply to the given crystal. Initially it is usually best to sample as large a portion of reciprocal space as possible in order to guarantee unambiguous determination of the Laue group.

Systematically Absent X-ray Reflections.   Table 10.1 shows that there are several possible space groups for a crystal belonging to a given crystal system. For example, a monoclinic protein crystal could have one of the three space groups $P2$, $P2_1$, or $C2$; similar considerations apply to the other crystal systems. Further consideration of the x-ray diffraction pattern may enable the exact space group to be indicated. This procedure depends on the recognition of systematic absences, and has been considered at length in Chapter 3; specific related problems can be found at the ends of Chapters 3 and 10. In general, screw axis absences (see Table 2.5) must be determined from a careful consideration of intensity values, and may be indicated from graphical plots of the reciprocal lattice using software such as HKLVIEW (see Appendix A10), which provides simulated precession photographs of the reciprocal lattice.

10.4.7   Ricin Agglutinin

Determination of the Space Group

Ricin Agglutinin[a] (RCA) belongs to the same family of proteins as the toxin Ricin,[b] both being derived from Castor beans. From biochemical evidence, it is known that Ricin comprises a toxic A-chain and a lectin protein B-chain; the two chains are linked by an S–S bond to form an A–B heterodimer; RCA has 2 A-chains and 2 B-chains that are highly homologous with the Ricin chains but are linked as B–A–A–B. The molecular weight (RMM) of RCA is thus about twice that of Ricin. There is a possibility that the B–A–A–B assembly could sit across a 2-fold axis in the crystal either an exact space group 2-fold axis or an approximate, non-crystallographic 2-fold axis. During data processing the unit-cell parameters for RCA found were $a = b = 100.05$, $c = 212.58$ Å, $\alpha = \beta = 90°$, $\gamma = 120°$, and the Laue symmetry is $\bar{3}$. Together these observations show that the crystals are trigonal, referred to hexagonal axes. In addition, limiting conditions in the diffraction pattern for $00l$, $l = 3n$, indicate the presence of a 3_1 or 3_2 screw axis parallel to c (Table 2.5). This information indicates that the space group is either $P3_1$ or $P3_2$ (Figure 10.8 and Table 10.1.). The x-ray analysis (see Section 10.6.3) resolved this ambiguity, the space group being shown conclusively to be $P3_2$ (Figure 10.8). Note that 3_1 and 3_2 screw operations are left-hand–right-hand opposites; it follows that only one of the enantiomorphic pair of space groups,

[a] E. C. Sweeney, A. G. Tonevitsky, D. E. Temiakov, I. I. Agapov, S. Saward, and R. A. Palmer, *Proteins, Structure, Function and Genetics* **28**, 586 (1997).

[b] W. Monfort, J. E. Villafranca, A. F. Monzingo, S. R. Ernst, B. Katkin, E. Rutenber, N. H. Xuong, R. Hamlin, and J. D. Robertus, *Journal of Biological Chemistry* **262**, 5398 (1987).

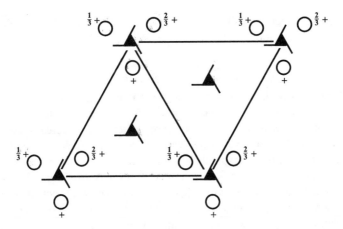

FIGURE 10.8.   Space group $P3_2$.

$P3_1$ or $P3_2$, can be correct for a given protein crystal. The question regarding the symmetry of the molecule was investigated using the self-rotation function and the result was confirmed using MR (see Section 10.6.3).

Resolution

The concept of resolution (see Section 7.9) is perhaps more relevant in macromolecular crystallography than in the case of well diffracting small molecule crystals, which usually diffract to atomic resolution. We may define d_{min} as the resolution of the x-ray diffraction pattern where, by the Bragg equation, $d_{min} = \lambda/2 \sin \theta_{max}$, θ_{max} being the maximum value of θ for reflections contained within the measured data set. Atomic resolution corresponds to a d_{min} of about 0.8 Å, for copper radiation, which is usually well out of reach for proteins. Most protein crystals fail to diffract to much better than 1.8 Å resolution. This is sufficient however to define the positions of nonhydrogen atoms and many of the solvent atoms in the structure. At 2.5 Å solvent atoms may be unreliable and at 3.5 Å it may be difficult to refine side-chain atoms beyond $C\beta$. Obviously the higher the resolution the better defined the structure will be, and it is usually worthwhile to expend the required effort in order to achieve it rather than to cut corners and end up with an inferior analysis. The resolution, as we shall see below, seriously affects the quality of the initial electron density map upon which the ensuing analysis is based.

Number of Reflections in and Completeness of the Data Set

For a given crystal it is possible to estimate the number of reciprocal lattice points within the range 0 to θ_{max}. For a crystal with a primitive lattice, ignoring the fact that some reflections will be related by symmetry, the number of reflections n is given by (see Section 6.4.1)

$$33.51 V_c \sin^3 \theta_{max}/\lambda^3 = 4.19 V_c/d_{min}^3. \tag{10.5}$$

Depending on how the x-ray data set has been measured, the number of reflections present may not be consistent with the number expected from the nominal value of θ_{max} or d_{min}. If the number of reflection data actually measured is significantly lower than the expected value it can be assumed that the resolution of the analysis will be correspondingly lower although it is not easy to estimate by how much. The cause may also be due to an anisotropic intensity distribution, associated with variation of diffracting power with direction. Some data processing programs put out a warning when such an effect is detected. The percentage completeness at nominal resolution will also be indicated. A value less than about 85% complete indicates the need to investigate the causes for the loss of data.

From the results so far for RCA, it follows that the unit-cell volume is 1.843×10^6 Å^3. A data set was recorded on a 180 mm diameter Mar Image Plate, with a crystal to detector distance 300 mm. The resolution d_{min} of the data was 3.6 Å for an x-ray wavelength of 1.542 Å. From the results above, the expected number of reflections is 165,512. This number includes all symmetry-related reflections. Since the unit cell is P and Laue symmetry $\bar{3}$, we divide this number of reflections by G (equal to 1, Table 3.1) and by the multiplicity of planes m, namely 6, to obtain an estimate of the number of unique data, that is, not related by symmetry, and which is therefore 27,585. Experimentally, 62,310 reflections were recorded. The data processing program reported that the multiplicity factor for the recorded data was 3.1, indicating that 62,310/3.1, or 20,100 reflections were unique. The data set is therefore said to be 20,100/27,585, or 73% complete at 3.6 Å resolution. We can assume that the effective resolution of the data is therefore less than the nominal resolution.

Internal Consistency, Space Group Ambiguities, and R_{merge}

As a test of data quality, data processing programs produce an internal consistency index, R_{merge} (also known as R_{int}), which is calculated by comparing $I(hkl)$ values that should be equal by virtue of symmetry. To be acceptable, R_{merge} for the complete measured data set for a protein should be 9% or better; for a well diffracting small molecule R_{merg} should be less than 7%, and is frequently 4% or better. Since the mean value of $I(hkl)$ decreases as the Bragg angle $\theta(hkl)$ increases, errors in the data will be more significant for the higher resolution data, that is, for smaller $d(hkl)$. For the outer shells of protein data, therefore, the local value of R_{merge} might be as high as 20%. Although this value may seem to be

inordinately high, such data may still contain a great deal of structural information and may serve a useful purpose in the analysis if retained, albeit with a low
weight.

In addition to providing an indication of data quality, the merging R-factor
can serve as a means of resolving space group ambiguities. By assigning a Laue
symmetry (Table 1.6) for the crystal that is, higher than true, such as *mmm* instead
of 222, falsely high values of R_{merge} will result. Dropping down to the correct
Laue symmetry should produce better consistency and lower R_{merge} values, but
only if a sufficient portion of reciprocal space has been covered during the data
collection. If this is not the case further data are needed.

Number of Molecules per Unit-Cell

Each of the possible protein space groups is associated with an expectation
number of molecules per unit cell given as Z in Table 10.1. If n is greater than the
expectation value, again usually by a simple factor, the protein structure is said
to be *oligomeric* and contains more than one molecule in the asymmetric unit,
not related to the other by crystal symmetry; in the case of proteins frequently
an approximate NCS relates such molecules. In this case the x-ray analysis will
involve the determination of the structures of this number of molecules rather than
just one. The molecules in this type of situation usually exhibit subtle differences
in molecular conformation, particularly in loop regions.

From Section 6.2, we express the crystal density as

$$D_c = nM_r u / V_c \tag{10.6}$$

where the parameters have the meanings as before. If V_c is expressed in Å^3,
D_c will have units of g cm^{-3}. In this relationship, both n and M_r, which involves
the unknown solvent component, are unknown. However, D_c can be assumed,
from experience, to lie in the range 1.2–1.4 g cm^{-3} for crystalline compounds
composed predominantly of the elements C, N, O, and H; the actual value will
depend heavily on the degree of solvation.

Let $M_r = M_P + sM_r$, where M_P is the relative molar mass of the protein, which is 133,000 and s is the solvent fraction of the cell mass. Thus
$M_r = M_P/(1-s)$, so that $D_c = nM_P u/[V_c(1-s)]$. If M_S is the *unknown* molecular mass incorporated in the solvent, we define the fraction s as $M_S/(M_P + M_S)$,
the solvent fraction of the unit-cell mass. Finally if the crystal space group is
known, as would be normal after data reduction, then since we know that n will
be a multiple or submultiple, say μ, of Z, the number of asymmetric units per
unit cell (Table 10.1), it follows that $n = \mu Z$. The density expression (10.6) then
becomes:

$$D_c = \mu Z M_P u / [V_c(1-s)] \tag{10.7}$$

It is known[a,b,c] that in crystalline proteins, s is usually in the range 0.27–0.65. Putting all of this information together it is usually possible to determine the appropriate value for μ, the number of protein molecules per asymmetric unit. These ideas are illustrated by the following example.

Analysis of the Solvent Content in Ricin Agglutinin

We now continue our study of RCA by investigating the possible values for the parameters μ and s. Since V_c is 1.843×10^6 Å^3, the value for u is simply 1.6605, Z (Table 10.1) is 3 for space group $P3_2$, and using the equation for D_c developed above and with $M_P = 133{,}000$, we have

$$D_c = 1.6605\mu Z M_P/[V_c(1-s)] = 0.36\mu/(1-s). \tag{10.8}$$

Assuming $\mu = 1$ RCA molecule per asymmetric unit ($3 \times$ B–A–A–B per unit cell) and $s = 0.65$ (a guess, at the top of the range because the crystals are wet and poorly diffracting), we find that D_c is 1.03 g cm^{-3}. Since this is a very low value for the density, we try $s = 0.7$, which gives $D_c = 1.20$ g cm^{-3}, a more realistic value. The physical wetness of the crystals and relatively poor diffracting power are consistent with the result that they are highly solvated. As we shall see in the next section, RCA has a non-crystallographic 2-fold axis relating the two halves of the B–A–A–B assembly, and this information is used in the x-ray analysis. In addition, the following alternative must be borne in mind as the x-ray analysis progresses: assuming μ is 2, D_c equals 1.20 g cm^{-3} for $s = 0.40$, which is approximately equal to the average solvation level found in protein crystals, and should therefore be considered reasonable until proved otherwise.

10.5 Types of Fourier Synthesis for Protein Analysis

10.5.1 Reconstruction of the Molecular Structure

We have seen (Chapter 5 onwards) that the calculation of electron density maps is a major objective of any single crystal x-ray analysis, and we recall the formula for calculating the electron density $\rho(xyz)$ at the general point with fractional coordinates (x, y, z):

$$\rho(xyz) = \frac{1}{V_c} \sum_{h_{min}}^{h_{max}} \sum_{k_{min}}^{k_{max}} \sum_{l_{min}}^{l_{max}} |F_o(hkl)| \cos 2\pi[(hx + ky + lz) - \phi(hkl)] \tag{10.9}$$

[a] B. W. Matthews in *The Proteins* 3rd ed., M. Neurath, and R. L. Hill, Academic Press, San Diego & London (1974).
[b] B. W. Matthews, *Journal of Molecular Biology* **82**, 513 (1974).
[c] M. L. Quillin, B. W. Matthews, *Acta Crystallographica* **D56**, 791 (2000).

For protein structures the phases can be determined using the experimentally intense method of MIR, or from a model derived by MR. The theory of MIR has been discussed in Chapter 6, and MR is treated in detail in Chapter 7; other important methods are to be found in this chapter. However, it is important to remember that the quality of an electron density map is strongly influenced by the accuracy of the phases. Poor phases will never produce a good map, whatever the resolution.

Most contemporary software uses the method of Fast Fourier Transform (FFT) for rapid calculation of electron density maps.[a,b,c] Related algorithms are used for calculation of structure factors. Programs for calculation and display of electron density and other Fourier series are listed in Appendix A10.

Properties of the Electron Density

We summarize here the important properties of the electron density function that we need to keep in mind when interpreting it in terms of a crystal structure:

1. When calculated at atomic resolution, $\rho(xyz)$ has local maximum values (peaks) at sites corresponding to atom centres (x_j, y_j, z_j).
2. Density values at these locations are approximately proportional to Z_j, the atomic number of the corresponding jth atom.
3. The locations of all atom positions together define the crystal structure. Atom positions not determined (for whatever reason) lower the quality of the structure model.
4. For proteins, where the extent of the x-ray data is usually far short of atomic resolution, the electron density tends to be blurred, or unfocused. Atom coordinates have to be inferred in this case, usually by model building, which makes extensive use of the known molecular geometry of bonds and groups involved in the structures.
5. Since clearly defined peaks are not a property of protein electron density, the magnitude of given density regions is assessed in terms of the overall root mean square (rms) value of $\rho(xyz)$. Significant density is taken to have $\rho(xyz) \geq 3\sigma_\rho$, although a value of $2\sigma_\rho$ (or even lower) can have useful features but requires very careful consideration.

10.5.2   Difference Electron Density

Difference electron density may be defined as

$$\Delta\rho(xyz) = \frac{1}{V_c} \sum_{h_{min}}^{h_{max}} \sum_{k_{min}}^{k_{max}} \sum_{l_{min}}^{l_{max}} \Delta|F(hkl)| \cos 2\pi[(hx+ky+lz)-\phi_c(hkl)] \quad (10.10)$$

[a] J. W. Cooley and J. W. Tukey, *Mathematical Computing* **19**, 297 (1965).

[b] A. N. Barrett and M. Zwick, *Acta Crystallographica* **A27**, 6 (1971).

[c] L. F. Ten Eyck, *Methods in Enzymology* **115**, 324 (1985).

where $\Delta|F(hkl)| = |F_o(hkl)| - |F_c(hkl)|$. This function exhibits positive density corresponding to atoms not included in the model, and negative density where atoms are missing from the model or badly placed in it.

10.5.3   The $2|F_o(hkl)| - |F_c(hkl)|$ Map

By adding $\rho(x,y,z) + \Delta\rho(x,y,z)$ we obtain a new type of map known simply as a $2|F_o(hkl)| - |F_c(hkl)|$ map represented by the expression

$$\frac{1}{V_c}\sum_{h_{min}}^{h_{max}}\sum_{k_{min}}^{k_{max}}\sum_{l_{min}}^{l_{max}}\{2|F_o(hkl)|-|F_c(hkl)|\}\cos 2\pi[(hx+ky+lz)-\phi_c(hkl)] \quad (10.11)$$

Maps of this type are very useful for locating parts of the structure which are not part of the current model; they show positive density corresponding to missing atoms and negative density corresponding to incorrect atoms *against the background of the original model*. It provides a very useful basis for making corrections and additions to the structure model.

Omit Maps

An electron density function where part of the model has been deliberately left out is called an *omit map*. Reasons for such omissions include disordered or apparently disordered regions, areas of poor stereochemistry, ambiguous side-chains (some pairs of amino acids are very similar), or simply parts of the model that are difficult to explain. The omitted parts of the structure should be returned in a subsequent electron density calculation as peaks of about half the height of comparable regions included in the phasing. An example of a contoured difference electron density map is given later in Figure 10.16.

10.5.4   Patterson Function

The Patterson function has been discussed in detail including its use in MIR for location of heavy atoms in Chapter 6, and for locating the position of the search model in an unknown structure in the MR method in Chapter 7. Further discussion on the use of the Patterson function for these two methods are given below.

10.6   Determination of the Phases for Protein Crystals

10.6.1   Introduction

There are two commonly used methods for phase determination in macro-molecular crystallography, namely, MIR and MR. In the early years of protein structure analysis (the 1950s–1970s) MIR was the primary choice for protein structure analysis, and during this period a core of good quality protein structures was established. During this time MR became increasingly popular, as the number of well-defined structures from which to select a search model gradually increased. There are currently over 19,000 protein structures deposited in the Protein Data Bank (see Appendix A10). MR is thus the current method of choice for any protein having good structural homology, as judged initially by the amino acid sequence, with a known structure. The more experimentally demanding MIR technique is still used for proteins where this condition does not apply. The advent of fast, large capacity computers, and improved software has also contributed significantly to the ease of application of MR.

10.6.2 Isomorphous Replacement (MIR)

We know that by incorporating one or more heavy atoms into a protein crystal, thus forming a heavy-atom derivative, changes can be induced in the x-ray intensities. For the derivative crystal to be isomorphous three conditions must hold:

1. The heavy atoms must not disturb the protein structure significantly.
2. The crystallographic space group must remain unaltered.
3. The unit-cell parameters must not differ significantly ($<0.5\%$ in any cell length).

Useful heavy atoms usually have atomic numbers 53 (iodine) or greater. The average fractional change in intensity $\Delta I(hkl)/I(hkl)$ at $\sin\theta/\lambda = 0$ can be estimated as: $\Sigma Z^2_H/\Sigma Z^2_P$ where the summations are over the number of heavy atoms N_H per unit cell and the number of non-heavy atoms N_P per unit cell, respectively. For a protein crystal containing a protein of relative molar mass 20,000 with 1 molecule per unit cell, N_P would be eqivalent to about 3200 carbon atoms. If N_H represents two mercury atoms per unit cell, then $\Delta I(hkl)/I(hkl) = (2 \times 80^2)/(3200 \times 6^2)$, or 11%. In practice some $I(hkl)$ values are increased and some are decreased; intensity differences of the order illustrated here would be useful for phasing.

Heavy Atoms and Compounds for Isomorphous Replacement

There is a very large repertoire of compounds[a,b,c] known to produce heavy-atom derivatives of proteins that are suitable for MIR. Heavy atoms are usually non-covalently linked to the native protein molecule, often in surface pockets or other

[a] T. L Blundell and J. A. Jenkins, *Chemical Society Review* **6**, 139 (1977).
[b] G. A. Petsko, *Methods in Enzymology* **14**, 147 (1985).
[c] S. P. Wood, *Protein Purification Applications: A Practical Approach*, Edited Harris and Angal, pp. 45–59, IRL Press, Oxford (1990).

easily accessible regions of the protein structure. The most popular heavy atoms are platinum, mercury, gold, silver, and uranium. The selection of appropriate compounds to screen for useful heavy-atom derivatives depends to a great extent on knowledge of the amino acid sequence.

10.6.3   Preparation and Screening of Heavy-Atom Derivatives

Two methods are used for the preparation of heavy-atom derivatives for isomorphous replacement, namely, soaking pre-grown native crystals in heavy-atom solutions, and co-crystallization of the protein and heavy atoms together from solution.

Soaking Method

Soaking can be carried out on crystals in hanging or sitting drops or on crystals mounted for x-ray diffraction in glass capillaries (see Section 10.2). This method is highly sensitive and dependent on correct pH, concentration, temperature, and time. Heavy-atom solution concentrations around 0.5–10 M are frequently used with soaking times ranging from a few minutes to several days or longer. This method can cause (often-visible) deterioration of crystal quality, but is not likely to induce changes in unit cell and symmetry.

Co-Crystallization in the Presence of Heavy Atoms

Generally speaking the introduction of heavy atoms by co-crystallization will initially follow the method established for crystallization of the native protein (see Section 10.2). Co-crystallization can be undertaken *in situ* in hanging or sitting drops, or in test tubes where batch crystallization is used. Co-crystallization may produce better quality crystals than soaking, but symmetry and unit-cell changes are more likely to occur as a consequence of the different crystallization conditions (presence of heavy atoms) and is therefore not usually the first method of choice.

Screening Possible Derivatives

All possible heavy-atom derivatives, produced as described in the previous section, have to be tested for isomorphism, for intensity changes that would indicate successful incorporation of heavy atoms, and for crystal quality. This screening of possible heavy-atom derivative crystals can be time consuming. The easiest method of detecting changes in $I(hkl)$ is from one or more x-ray precession photographs (Chapter 4). The same photographs provide a sensitive check on symmetry and unit-cell dimensions and can be rapidly compared. In the absence of a precession camera it may be necessary to collect a low resolution (5–6 Å) data set but this will be time consuming, particularly if no heavy atoms have been introduced. Whichever method is used the technique is, inevitably, experimentally demanding.

How Many Derivatives are Required

Details of procedures for locating heavy atoms and deriving phase information in the MIR technique can be found in Chapter 6. It is advisable to use at least three good heavy-atom derivatives for phasing purposes. Useful phase information can be obtained with two derivatives and to a much lesser degree, with a single derivative, but the quality of the calculated electron density will suffer.

The Initial MIR Model

As in direct methods phasing in small molecule analysis, MIR phases are not calculated from the structure factors of a model, but from probability functions. MIR phases are thus, unlike phases from MR, free from model bias. As little or no information may be available about the secondary structure of the protein, unless ultra-high resolution intensity data have been measured (to 1 Å or better), which is unlikely, it is imperative that a detailed knowledge of the amino acid sequence of the protein is available. The MIR phases are used to calculate the electron density distribution, which may first be enhanced by modification techniques (see Section 10.8.2) and interpreted using molecular graphics (see Section 10.8.3). If the phases are of sufficiently high quality the main-chain density should be strong and continuous, and on average greater than $3\sigma_\rho$, thus enabling this part of the structure to be modeled. Once the ends of the polypeptide chain have been identified, its course through the electron density can be retraced and side-groups fitted to the model and added to the coordinate list. This lengthy process will then be followed by further refinement (see Sections 10.8 and 10.9). Figure 10.9 is a suggested flow diagram for MIR.

10.6.4   Molecular Replacement

Introduction

In contrast to MIR, MR is not such an experimentally intense method of phase determination. The method requires

1. A set of $|F_o(hkl)|$ values for the protein crystal under investigation (the *target structure*);
2. The coordinates of a good quality structure (the *search structure*) expected to be similar to the target structure, usually having different space group and unit cell from the target structure;
3. A sound understanding of the principles and practices involved;
4. State-of-the art software and hardware.

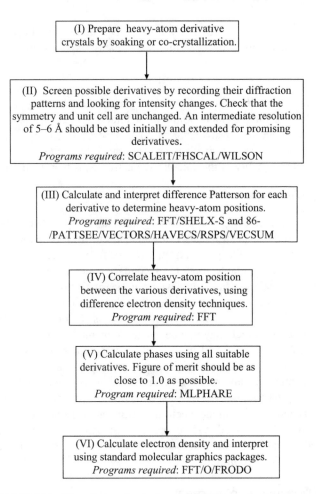

FIGURE 10.9.   Flow diagram for single or multiple isomorphous replacement.

The reader may wish to review Section 7.3 at the stage. In essence MR[a] involves a critical and quantitative comparison of the Patterson functions of the target and search models. In contemporary software for carrying out MR applications (AmoRe and X-PLOR) the method is strengthened through the availability of other crystallographic techniques, described below, which are now within the capabilities of modern computers.

For complex structures like proteins, interatomic vectors are densely packed in the unit cell and most will not be resolved in the Patterson map; lack of atomic

[a] M. G. Rossman and D. M. Blow, *Acta Crystallographica* **14**, 641(1961).

resolution in the x-ray data will also cause a further blurring of the vector distribution density. In the first (rotation) stage of MR, the shorter intramolecular interatomic vectors are used to determine possible orientations for the search molecule in the unit cell of the target molecule; the best subset of solutions is retained. The second (translation) stage tries to optimize the actual location of each of these selected possible molecular orientations in the target unit cell. Again the best solutions are retained for further development. Ideally there will be an *outstanding* solution, which will eventually yield a refineable structure for the target protein.

Self-Rotation Function and Non-Crystallographic Symmetry

A complication can arise in practice for target crystal structures containing more than one protein molecule in the asymmetric unit. A clue to the presence of NCS may already have been given by the preliminary analysis of unit-cell contents (see Section 10.4.6), but ambiguities can arise, mainly because of uncertainties in solvent content. In practice it is necessary to establish the presence of NCS as early as possible in the MR analysis, as this can save time by allowing the correct procedures to be followed. Calculation of the self-rotation function may help to resolve, or at least to provide clues as to the actual situation with a given protein crystal.

Molecules related by non-crystallographic symmetry frequently differ in both *position* and *spatial orientation* with respect to each other and are often, but by no means always, related by pseudo 2-fold rotational symmetry. The pseudorotation axis will not coincide with any of the space group symmetry elements and, of course, will not involve an exact 180° rotation. Other types of rotational NCS can also occur. In such cases, the Patterson function will include, around the origin, a differently oriented copy of the molecular interatomic vectors for as many protein molecules as are involved in the NCS. Consequently if two exact copies of the Patterson function are superimposed origin to origin, then by rotating one copy of the replica correctly with respect to the other, the peak distributions can be made to coincide, thus determining the relative orientations of the NCS related molecules. For this correct relative orientation, the rotation function will have a significantly large local maximum value.

Programs are available for plotting the results of this type of self-rotation and for calculating the relative molecular orientations in three dimensions. Initially the crystal system is orthogonalized as shown in the example in Figure 10.10, and *spherical polar angles* with respect to this axial system are used to define orientation, instead of the usual Eulerian angles. In this way, as shown in Figure 10.11, it is possible to specify a single rotation χ (also called κ) about an axis whose direction is defined by the spherical polar angles ψ (also called ω) and ϕ. The point $\psi = 0°$ on the pole projects on to the center of the circular plot, and points on the equator with $\psi = 90°$ project on to the circumference. Points with $\phi = 0°$

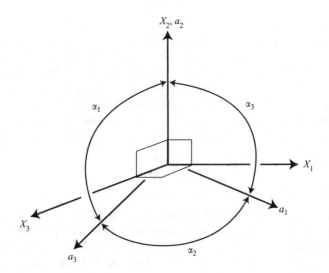

FIGURE 10.10.   Relationship of the orthogonal axes X_1, X_2, X_3, used in calculating the self-rotation function, to generalized crystallographic axes a_1, a_2, a_3: X_2 is along a_2, X_1 is normal to a_2 and a_3, and X_3 is normal to both X_1 and X_2.

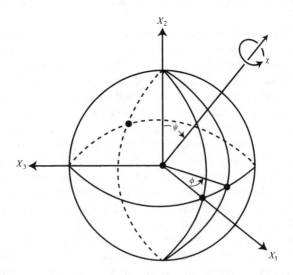

FIGURE 10.11.   Variables ψ and ϕ are spherical polar coordinates that specify a direction about which the axes may be rotated through an angle χ.

project on to the X_1-axis (meridian); ϕ increases in an anticlockwise direction. For orthogonal unit cells some programs place the crystallographic a-axis at $\phi = 0°$, the b-axis at $\phi = 90°$, and the c-axis at $\psi = 0°$. Other conventions are used, however and each program should be checked by prospective users. When working in a monoclinic space group it is convenient to place the unique y-axis along the polar direction. The use of spherical polar angles in this type of analysis enables sections of constant χ to be plotted.

The two most common cases of non-crystallographic symmetry that are found in practice are pseudo 2-fold axes, corresponding to $\chi = 180°$, and pseudo 3-fold axes, corresponding to $\chi = 120°$. Other sections of constant χ can be accommodated as required. In practice several self-rotation plots should be calculated by varying the radius of integration (see Section 7.3.7), and/or the resolution cutoff applied to the $|F_o(hkl)|$ input data. It is important to recognize features of the rotation function which persist under these different conditions as they are most likely to contain the information sought. Both of these parameters can be specified in the data input to the program. The use of plots with different radii of integration enables the effect of eliminating cross vectors, the longer interatomic vectors in the Patterson function, to be seen. It is of course not possible to eliminate all cross vectors, as the molecules in a crystal are packed close together. As a rule of thumb, for an approximately spherical molecule a search radius of 75–80% of the molecular diameter will include about 90% of the self-vectors. Nonspherical molecules present more of challenge.

10.6.5 Example of a Self-Rotation Function: Ricin Agglutinin

We have seen that RCA probably has a 2-fold axis across the B–A–A–B assembly of protein chains. Figure 10.12 shows the section of the self-rotation function at $\chi = 180°$, calculated with the program POLARRFN (part of the CCP4 suite of programs) and includes data to 4.0 Å resolution. For space group $P3_2$ the symmetry of this section must contain a 3-fold axis, corresponding to the symmetry of Laue group $\bar{3}$. The presence of non-crystallographic 2-fold axes (there are no crystallographic 2-fold axes in this space group) is exemplified by the very large peak (arrowed) on the stereogram circumference, with spherical polar coordinates of approximately $\phi = 20°$, $\psi = 90°$, and $\chi = 180°$. This peak corresponds to a pseudo 2-fold axis perpendicular to the c-axis, and at approximately 20° to b. This example is extended in the next section in which the rotation and translation functions are described.

Intermolecular Atomic Vectors and Translational Non-Crystallographic Symmetry

If two non-crystallographically related molecules differ only by a *non-lattice* translation, with no rotational component, the Patterson function will contain an outstandingly high peak at a position corresponding to the non-crystallographic

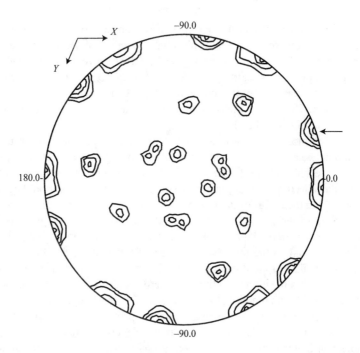

FIGURE 10.12.   Non-crystallographic symmetry: the section $\chi = 180°$ of the self-rotation function for the protein RCA, which crystallizes in space group $P3_2$, calculated with the program POLARRFN. The peak related by non-crystallographic 2-fold symmetry approximately perpendicular to the 3_2 axis, the *only* symmetry axis of this space group, is arrowed.

intermolecular vector, because all vectors between the two molecules will be lined up. Again it is important to determine these situations prior to embarking on the full MR analysis, by simply calculating the Patterson function and inspecting the regions of high density. Remember that the origin peak (0, 0, 0) always has the highest density in the Patterson. Most Patterson density will be a very small fraction of $P(000)$. We are therefore considering here Patterson density that is, a numerically significant fraction of $P(000)$, possibly as high as 20–50%.

10.6.6   Molecular Replacement in Practice

The Search Model

One of the following situations may provide a suitable opportunity for using MR:

1. Determination of a protein derivative structure containing a small ligand or other modification, crystallizing in a space group different from that

of the known native protein structure, for example, enzyme–inhibitor or lectin–carbohydrate complexes.

2. Determination of macromolecular complex combinations when one or both components have known structures. It may be possible to carry out successive MR searches in order to provide a starting model for the complete structural combination. If only one macromolecular component can be located by MR, lengthy Fourier development and refinement may follow. Examples include antibody–protein and protein–nucleic acid complexes.

3. Functionally similar proteins with high sequence homology, one member of the series having a known structure. Sequence homology as low as 20% may be sufficient especially if many of the sequence changes are conservative, that is, chemically similar. Sequence homology may also be used in less obvious cases of structural similarity, such as proteins not belonging to the same or related families, but this carries a greater risk of failure.

4. Proteins grown by a recombinant technique from a DNA template, or site-directed mutants (proteins grown from modified DNA) having known native structures. The modified proteins may sometimes crystallize isomorphously or otherwise in different space groups.

Data Base Searches

Several protein sequence data bases are readily available to accommodate sequence homology searches for suitable MR test structures; a list is given in Appendix A10. Depending on the type of protein involved it is often known in advance what likely target structures are available. For example, Ricin is the parent structure for the ribosome inactivating protein (RIP type II) family and would be the first choice to use in MR search for determination of another member of the family such as RCA. Sequence matching protocols and algorithms are under active development. Many of these routines are freely available on the Internet and can be accessed in combination with known protein sequences that are available through data bases such as BLAST. Where possible, well-defined structures should be chosen as search models. In simple terms, this means that the resolution should be 2 Å or better, and the R-factor should be 20% or less. It is more risky to use low resolution x-ray structures or nuclear magnetic resonance (NMR) derived models, both of which will suffer from a lack of refinement, but these may have to be used if no other models are available.

The Target Patterson Function

Calculation of the target Patterson function is straightforward in practice as the $|F_o(hkl)|$ are already available from diffraction measurements. It has been suggested that different resolution cutoff values should be tried, say, between 3 and 5 Å; in other words, it is best to employ medium to low resolution data in order to blur the differences between the search and target models. An inner low-resolution data cutoff should also be used, within 10–20 Å, as such data are influenced mainly by the solvent structure and therefore do not directly contribute to the protein structure at all. Facilities for easily setting up these cutoff values are available in most software. Again, different radii of integration should also be tried in order to pinpoint persistent solutions.

In the case of the target structure, it is more difficult to restrict the rotation vectors to the intramolecular type alone because the observed Patterson distribution around the origin contains both intra- and interatomic vectors. This problem can be overcome to some extent, by restricting the volume of the target Patterson used in the search. The radius of integration used for this purpose can be selected by the user in an attempt to optimize the signal to noise ratio. Several trials may be required in practice in order to achieve the desired result. It has been recommended[a] to start with a value for this radius of integration approximately equal to $0.5d_m$, where d_m is the approximate molecular diameter, increasing to about 0.75–$0.8d_m$ for further trials if necessary.

The Search Patterson Function

Several approaches are available for generating the Patterson function for the search model in macromolecular studies:

1. Construct the search Patterson as a vector array from the structure coordinates, as in procedures used for small molecules (Chapter 7). This type of approach generates an extremely large number of vectors and is computationally very costly and slow to apply but nevertheless is used effectively in the X-PLOR suite of programs.

2. In the program AmoRe the search model coordinates are transformed with respect to an artificial triclinic unit cell with angles of 90° and cell edges of pd_m, where d_m is again the molecular diameter and the factor p is selected by the user, frequently with a value of 2.0 or less. Values of $|F_c(hkl)|$ calculated from the triclinic coordinates are then used to calculate the search Patterson. This approach is computationally efficient and furthermore allows models derived from NMR studies or from model building to be used that do not correspond to any crystal structure. One or both of these procedures may be used in practice. In either case,

---

[a] I. J. Tickle and H. P. C. Driessen in *Methods in Molecular Biology, Vol 56: Crystallographic Methods and Protocols*, Edited C. Jones and M. Sanderson, pp. 173–203, Humana Press Inc., Totawa, NJ (1996).

the need to consider symmetry is eliminated, thereby ensuring that only intramolecular vectors are generated for the search model.

Recognition of the Correct Rotation Solution

Several powerful programs are currently available for MR analysis. Of these AmoRe, (a CCP4 supported program) and X-PLOR are commonly used. X-PLOR is a more general program suite, covering most aspects of macromolecular crystal structure analysis and refinement, whereas AmoRe is specific to MR. The use of these programs to locate the correct rotation function solution involves searching through a large number of trial rotation angle triplets (α, β, γ). Incremental values of 5° or 10° for each of these three angles are typical, resulting in a large number of trials. These programs optimize the rotation angles around local maximum regions of Patterson overlap. In AmoRe, each promising trial carries a correlation coefficient C_c indicating the level of agreement between the rotated Pattersons. The correct (or near correct) solution should have the highest C_c value. However at the rotation stage, C_c values do not always differ significantly between potential solutions, and the values tend to be much smaller, typically around 5–15%, than at the later translation and optimization stages. The AmoRe program also calculates the rms value σ_P for the Patterson density and gauges the significance of local density regions with respect to this value. A peak greater than $3\sigma_P$ is considered significant.

If different search models, resolutions, integration radii, or any of the other parameters available, have been varied over a series of searches, consistency in the occurrence of solutions for the angle triplets α, β, γ, with high C_c values, is a good indication that the given solution is worth carrying over to the translation stage. Because the C_c indicator is not absolutely reliable, all peaks greater than 50% of the maximum value, ranked in order, are normally retained for transfer to the translation stage. The density values in terms of σ_P are also printed out. In this way the probability of overlooking the correct rotation solution should be greatly decreased. The AmoRe system can accommodate the retained rotation solutions simultaneously into the translation routine. It also calculates an R-factor, R_f (not to be confused with R_{free}) based on observed and calculated $|F|$ values for each possible solution. Because the model is usually very incomplete at this stage, it is possible only to compare the relative values of R_f for the different solutions, but this practice has been found to be quite useful and effective.

The Rotation Function and Non-Crystallographic Symmetry

For structures that contain more than one molecule per asymmetric unit related by non-crystallographic symmetry, the rotation function would be expected to produce a corresponding number of solutions, equivalent, or nearly so, with respect to both the C_c index and *reproducibility* under different computational conditions such as resolution, radius of integration, search unit-cell size, and initial search model orientation. Only genuine solutions will, of course, produce acceptable translated models at the next stage of MR. There is no guarantee that such translated models will occupy their true relative positions in the unit cell, and this problem requires further attention, as explained below.

The Translation Function

It cannot be overemphasized that success with the MR technique depends heavily on the precision with which the correct rotation parameters have been established. Programs such as AmoRe and X-PLOR include optimization procedures at the rotation stage to improve the chances of success later on. For each of the potential rotation solutions carried over into this stage of MR a set of atomic coordinates is generated. It is necessary to place this oriented model into the correct location in the unit cell. Full attention to the symmetry elements and their relationship with each other and the *standard* origin must be maintained. In general only one of the retained rotation solutions will provide a successfully translated model. When NCS is present the user looks for the corresponding number of successfully translated models. Each test model, carried over from the rotation stage, is moved by incremental translations called (Δt_1, Δt_2, Δt_3) along three independent axial directions. The small steps used, typically 1 Å (or more initially) are further optimized, as with the rotation solutions, so as to obtain the most promising models. The translation function routines attempt to correlate, at least conceptually, the observed and calculated Patterson functions. As we have seen, optimal correspondence is measured in AmoRe by the correlation coefficient C_c, which approaches 1.0 for the best solution, but is usually much less than unity. In addition it is possible to compare $|F_o(hkl)|$ with $|F_c(hkl)|$ for each translated model and to calculate R_f, which is usually found to be lowest in value for the best solution. The AmoRe system employs both C_c and R_f in order to discriminate between potential solutions. As with the earlier rotation stage, consistency under different trial conditions is an important and extremely useful means of generating confidence in the validity of the solution of choice, prior to embarking on what can be very lengthy model building and refinement.

Rigid Body Refinement

Recent innovations in software are aimed at improving the likelihood of choosing the correct model resulting from an MR search. At the same time the accuracy of the three rotational and 3 translational parameters of the best solution are optimized and will therefore correspond to a model which will be easier to refine. To this end both AmoRe and X-PLOR include refinement routines which fine tune the rotation and translation parameters by adjusting the position of the search model without making changes to the geometry of the model. This technique s known as rigid body refinement. In this approach, the validity of the rotated and translated model is given a final check in terms of C_c, which should approach 1.0, and of R_f, which should minimize.

Subunits and Non-Crystallographic Symmetry

Individual molecules related by non-crystallographic symmetry may produce independent acceptable solutions as a result of the rotation and translation stages of MR. Because there is no guarantee that the coordinate sets for the individual molecules produced at this stage are properly correlated in the unit cell it is not usually possible to simply combine them to form the final trial structure. The main reason for this is concerned with the existence of more than one equivalent origin in many space groups. For example in space group $P2_1$ we know that there are four distinct 2_1 axes per unit cell, at $[0, y, 0]$, $[1/2, y, 0]$, $[0, y, 1/2]$, and $[1/2, y, 1/2]$. Both AmoRe and X-PLOR have the facility for allowing a rotated–translated model to be included as a rigid group, which is subtracted from the unknown structure in order to enhance the signal from the missing non-crystallographically related molecule or molecules. If successful, this process will result in the non-crystallographically related molecules being properly placed relative to one another in the unit cell. This is an extremely powerful facility and will produce a full trial structure to be developed further by refinement procedures.

Phases Derived from Molecular Replacement

We know that phase information derived from Patterson Search methods is initially derived through the calculation of structure factors, inspection of electron density maps, and least-squares refinement—the method of Fourier-least-squares. Similar methods apply to macromolecular analysis but with some important variations.

10.6.7 Application of the AmoRe Algorithms to Ricin Agglutinin

We continue with the study of RCA, which we introduced in Section 10.4.7, as an example of the use of the program AmoRe to solve a protein structure.[a] The amino acid sequences of the individual A- and B-chains in both Ricin itself and in RCA are very similar, the overall homology being about 85%. This suggests that,

[a] E. C. Sweeney, A. G. Tonevitsky, D. E. Temiakov, I. I. Agapov, S. Saward, and R. A. Palmer, *Proteins, Structure, Function and Genetics* **28**, 586 (1997).

although Ricin is only half the size of RCA, its structure[a,b,c,d] would be expected to make an ideal search molecule for RCA, with two final solutions corresponding to the two halves of the RCA molecule.

A number of trials was used to establish the rotation search for RCA:

1. The search-model Patterson maps were calculated using intensity data restricted to various resolution ranges, the most useful being 4.0–10 Å.
2. The Ricin search model was placed in an orthogonal cell of $P1$ symmetry, with unit-cell parameters 100, 80, 60 Å. Other search model unit cells were used with edges from 60 to 100 Å.
3. The radius of integration was set at values between 25 and 35 Å, and the rotation function was calculated at angular steps of 2.5°. An overall temperature factor applied to $|F_o|^2$ was set to -20 Å^2, which has the effect of sharpening the Patterson peaks (see Section 6.4.2). Carrying over the most promising peaks from the Rotation Stage and using space group $P3_2$, two different outstanding translation function solutions were produced as expected for the two halves of the RCA molecule. They had peak heights of $4.8\sigma_P$ and $4.5\sigma_P$ above the highest noise peak [a value greater than $3\sigma_P$ is normally considered to be outstanding] with R_f values of 49.2% and 54.0% and C_c 44.5% and 37.1%, respectively, corresponding to solutions 1 and 2, after application of the rigid body refinement protocol of AmoRe (Tables 10.2 and 10.3). Using both solutions simultaneously, all symmetry related molecules in the $P3_2$ unit cell for this solution were generated and examined graphically using the program MOLPACK. This program detected no inadmissibly short intermolecular contacts in the packing of this model. As anticipated however symmetry-related version of solution 2 (Ricin 2) was found to be situated in close proximity to solution 1 (Ricin 1). Closer examination strongly suggested that Ricin 1 and Ricin 2 were covalently linked, and after much further refinement the linkage was shown to be a disulfide S–S bond. The same rotation function results used with the enantiomorphic space group $P3_1$ produced no translation function peaks comparable with those in space group $P3_2$. It was concluded, therefore, that MR had produced a potentially refineable model for RCA, and subsequent development and

[a] W. Monfort, J. E. Villafranca, A. F. Monzingo, S. R. Ernst, B. Katzin, E. Rutenber, N. H. Xuong, R. Hamlin, and J. D. Robertus, *Journal of Biological Chemistry* **262**, 5398 (1987).

[b] E. Rutenber and J. Robertus, *Proteins* **10**, 260 (1991).

[c] B. J. Katzin, E. J., Collins, and J. D. Robertus, *Proteins, Structure, Function and Genetics* **10**, 251 (1991).

[d] The structure of Ricin can be viewed by accessing the Internet PDB [Protein Data Bank ID code 2AAI at http://www.rcsb.org/pdb/].

TABLE 10.2. AmoRe Rotation Function for
RCA in Space Group $P3_2$

	$\alpha°$	$\beta°$	$\gamma°$
1	**109.7**	**80.2**	**80.7**
2	**41.3**	**83.1**	**169.9**
	14.3	46.5	11.5
	70.0	141.7	103.0
	15.6	139.9	177.1

Data range 10 to 4 Å; radius of integration 35 Å; search cell 110, 110,
90 Å. The top five R-function peaks are shown. The correct solutions
1 and 2 are in bold type.
  Peaks 1 and 2 alone persist under other conditions, such as:
Data range 10 to 4 Å; radius of integration 32 Å; search cell 100, 90,
80 Å; Data range 10 to 4 Å; radius of integration 25 Å; search cell
100, 80, 60 Å; Data range 10 to 4 Å; radius of integration 37 Å; search
cell 100, 100, 80 Å. SWEENEY, E. C., *PhD Thesis*, University of
London (1996)

TABLE 10.3. Extension of the Rotations Solutions 1 and 2 in Table 10.2 to the
Translation Stage Using Space Groups $P3_2$ and $P3_1$

	$\alpha°$	$\beta°$	$\gamma°$	t_x	t_y	t_z	$C_c\%$	$R_f\%$	$\Delta\sigma$
$P3_2$ 1	109.7	80.2	80.7	0.6334	0.1427	0.5674	44.5	49.2	4.8
$P3_2$ 2	41.3	83.1	169.9	0.9679	0.0321	0.00	37.1	54.0	4.5
$P3_1$ 1	109.7	80.2	80.7	0.5128	0.3910	0.00	9.5	77.9	1.5
$P3_1$ 2	41.3	83.1	169.9	0.7372	0.3397	0.00	10.9	77.8	1.4

Compared to other solutions in Table 10.2, the solutions for space group $P3_2$ are outstanding in terms of C_c (highest)
and R_f (lowest), indicating the correctness of the solution. Rotation solutions 1 and 2 fail to provide any reasonable
translation solutions in space group $P3_1$, which is therefore eliminated. As a further test of the method, the starting
Ricin model was rotated randomly to a new starting position, and the rotation and translation functions recalculated.
Two outstanding peaks were again found that proved to be consistent with the above solutions 1 and 2. The packing
of the molecules from the solution in $P3_2$ shows no disallowed contacts, and reveals also how the two halves of the
Ricin molecule associate.

refinement of the $P3_2$ structure proved this to be the case. Figure 10.13
shows a view of the RCA double heterodimer molecule drawn with the
program RASMOL.

The Initial MR Model

Unlike phases from MIR, those produced with MR are calculated from the
structure factors of the re-positioned search model, and as such are subject to model
bias. As a consequence, the calculated electron density will tend to reproduce the

(a)

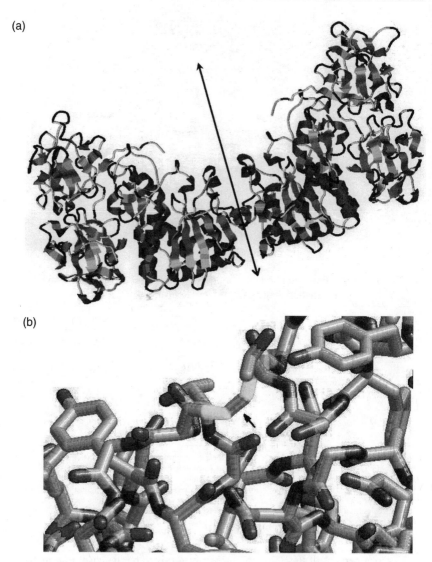

(b)

FIGURE 10.13.   (a) A view of the RCA double heterodimer molecule, drawn by the program RASMOL. The position of the pseudo-2-fold axis relating the two halves of the molecule that are joined by a disulfide bridge is shown by the arrow. This figure is a ribbon diagram showing the course of the main chain, and it is possible to pick out regions of helix structure, sheet structure, and random coils. (b) Stick-bond diagram of RCA; the central S—S bond is indicated by the arrow.

model even if it is incorrectly placed. However, there will usually be significant differences between the search model and the target model, particularly in terms of the amino acid sequence; such residues may have been omitted from the search model. The electron density calculated with model-based MR phases and the $|F_o(hkl)|$ data of the target protein should show electron density in these regions of the amino acid sequence, thus allowing the new side groups to be modeled and added to the atom list. Failure of the electron density to reveal these modifications is an indication of further problems. Another indication is a lack of correspondence between $|F_o(hkl)|$ and $|F_c(hkl)|$, resulting in a very high R-factor. However, as the model will be very incomplete at this stage, the R-factor alone should not be used to make decisions on the quality of the model. As in the case of MIR phasing, MR phases may first be enhanced by modification techniques (see Section 10.8.2). Interpretation of the electron density is again achieved with molecular graphics, after which further refinement of the model is then carried out (see Sections 10.8 and 10.9).

10.7   Siras and MAD Phasing

In Chapter 6, we discussed two methods that are applicable to solving the phase problem for proteins, both of which require the use of tunable SR for making anomalous scattering measurements. In the first of these methods, SIRAS (Single Isomorphous Replacement with Anomalous Scattering), two sets of intensity data are required: one from the native crystal, and a second from the heavy-atom derivative for which Friedel or Bijvoet pairs are measured, thus enabling the phase ambiguity of SIR to be resolved. In the second method, MAD, the anomalous dispersion measurements are made at two different wavelengths, so providing an extra data set. In both methods the wavelengths used are selected in order to optimize the anomalous dispersion effects from heavy atoms present in the crystals. The MAD technique is often used to solve the phase problem for proteins containing selenium-mutated methionine residues, produced by recombinant DNA technology, in which sulfur has been replaced with the heavier selenium atom ($Z_S = 16$, $Z_{Se} = 34$) in order to enhance the strength of the anomalous scattering signals. In such cases[a] the structure determined is often that of the modified protein.

10.8   Use of Phase Information and Density Modification

Successful application of one of the three methods MIR, MR, or MAD will provide at least approximate values of the phases. Any errors associated with

---

[a] See, for example, P. T. Erskine, N. Senior, S. Awan, R. Lambert, G. Lewis, I. J. Tickle, M. Sarwar, P. Spencer, P. Thomas, M. J. Warren, P. Shooling-Jordan, S. P. Wood, and J. Cooper, *Nature Structural Biology* **4**, 1025 (1997).

these derived phases will affect the quality of the calculated electron density and its interpretation in terms of the protein structure. To enable improvement in both the phases and electron density to be made at an early stage, density modification techniques have been developed to optimize the electron density map, particularly in cases where there is reason to believe the initial phasing is of poor quality, and also to enable a rational interpretation to be carried out. Some applications in which this technique might be necessary include SIR or MIR where the mean figure of merit is too low, say, less than 0.5, and in MR where C_c is too low, or R_f too high, or where no outstanding or persistent solution is observed.

10.8.1   Properties of $\rho(xyz)$ for Proteins

The inherently poor resolution of protein x-ray data means that, generally speaking, only heavy atoms, those with atomic number greater than 20, are expected to produce distinct peaks in the electron density map. The general features of protein electron density maps may be summarized as follows:

1. Main-chain density tends to be continuous and relatively strong. This is because atoms in the polypeptide chain are generally held firmly by interactions between neighboring atoms and are therefore less susceptible to thermal and statistical disorder (see Section 7.9). Recognition of strong, continuous density in the map can therefore enable the main polypeptide chain to be traced and the molecular envelope to be outlined.
2. Side-chain atoms tend to be less rigidly held in the structure than main-chain atoms, with correspondingly weak electron density.
3. Protein crystals can contain anything between 35% and 70% in solvent molecules. This part is the most disordered of the structure and will consequently be associated with very weak electron density, situated in the intermolecular interstices and solvent channels. Consideration of the regions of weak density should therefore enhance the recognition of the protein molecular boundary. Solvent molecules, such as water, closest to the protein are generally better ordered than bulk solvent atoms, and may be identified once some structure refinement has been effected.

10.8.2   Programs for Density Modification

We provide here a brief list of programs available for density modification (DM) in protein structure analysis.

DM[a] is a CCP4 supported program for carrying out density modification. The program applies constraints to the observed electron density, and derives new

[a] K. Cowtan, *Joint CCP4 and ESRF-EACBM Newsletter on Protein Crystallography* **31**, 34 (1994).

phases via Fourier transform techniques. Phases can be calculated for reflections not involved in the original phasing. However this procedure, known as phase extension, should be applied with caution, for example, it is unlikely that a 6 Å MIR map will rapidly expand to 2 Å, but the method might more realistically expand the phases from 2.5 Å to 2 Å.

SOLV[a] uses a method known as solvent flattening, which establishes uniformity of density in solvent regions, on the assumption that regions of disordered solvent are essentially without structure.

HIST[b] applies histogram matching based on the known characteristics of biological structures so as to predict the histogram of density values in the protein region. The current density map is then systematically modified according to the predicted histogram. This technique is complementary to solvent flattening. Between the two, the image of the protein structure should become much clearer and facilitate model building prior to further refinement (see below). The combined process of SOLV/HIST may require 10–20 iterations to converge.

SKEL uses skeletonization[c,d] to provide a sound basis for molecular graphics model building of the structure, by enhancing the connectivity of the electron density in main-chain regions.

SAYR[e] attempts to achieve phase improvement for data at 2 Å resolution or better by applying phase relationships adapted from a classical small molecule phasing method.

AVER[f,g] uses molecular averaging for structures where either NCS or exact crystallographic symmetry is present; it provides better phasing for initial model building. This constraint should, if possible, be released for the final structure.

Figure 10.14 is a possible flow diagram for use with the DM programs.

10.8.3   Preparing to Refine the Structure

At some stage in the determination of any structure using x-ray analysis it has to be decided whether the current model can be refined further. For small molecules, during the initial stages of Fourier refinement we are guided by a combination of good molecular geometry and the behavior of the R-factor, which should fall to 16–18% prior to least-squares refinement. Even a small protein of M_r approximately 15,000 contains about 1000 nonhydrogen atoms,

[a] B. C. Wang, *Methods in Enzymology* **115**, 90 (1985).

[b] K. Y. J. Zhang and P. Main, *Acta Crystallographica* **A46**, 377 (1990).

[c] D. Baker, C. Bystroff, R. Fletterick, and D. Agard, *Acta Crystallographica* **D49**, 429 (1994).

[d] S. Swanson, *Acta Crystallographica* **D50**, 695 (1994).

[e] D. Sayre, *Acta Crystallographica* **A30**, 180 (1974).

[f] G. Bricogne *Acta Crystallographica* **A30**, 395 (1974).

[g] D. Schuller, *Acta Crystallographica* **D52**, 425 (1996).

so that examination of the electron density is much more arduous and time consuming.

This work is carried out at a computer workstation, employing programs such as FRODO or its updated version O, TURBO-FRODO and SKEL, which allow a geometrically accurate protein model to be built into the density. Knowledge of the amino acid sequence is a necessary prerequisite. After each round of model building, structure factors should be calculated and the R-factor inspected. A very satisfactory R-factor for a protein structure at this stage would be around 25%; a value up to 30% could be tolerated but should be treated with caution. Further computational refinement may then be carried out, as described in the following discussion.

10.9   Macromolecular Structure Refinement and Solvent and Ligand Fitting

10.9.1   Refinement Techniques

The process of structure refinement involves optimization of the agreement between the observed and calculated diffraction patterns, represented by $|F_o(hkl)|$ and $|F_c(hkl)|$, and validation of the resulting molecular structure. If isomorphous replacement has been used to derive phase information, the initial structure model will be derived from an electron density map using molecular graphics. For an MR analysis, it is likely that rigid body refinement of the model will have been carried out at an earlier stage. Whichever method has been used for the initial structure analysis, refinement of the model should be undertaken by a combination of the following techniques:

1. Fourier refinement using successive Fourier synthesis;
2. Simulated annealing;
3. Least-squares analysis.

Further Details of Fourier Refinement

Lack of resolution in protein x-ray data results in poor definition in the electron density maps whether from MIR, MR, or any other phase determining method. It will be necessary to interpret this map by fitting the protein structure to the density. If MR has been used for the preliminary structure determination, electron density calculations will again play an important role initially, but the crystallographer will have prior knowledge of structural features of the search model that will help to establish the new structure. There are three aids to assist the initial process of interpreting the electron density:

1. *The amino acid sequence of the protein.*   If this is not known, interpretation of the crystal structure will necessarily be difficult and probably

not possible unless very good phasing at a high resolution has been achieved. If MR has been used, the target molecule will be subject to amino acid sequence changes; in addition, insertions and/or deletions may be required. These features have to be built into the density.

2. *Knowledge of the standard geometry*[a,b] *of proteins* (see Section 10.11) in terms of main-chain and side-chain bond lengths and angles, and of secondary structural features, particularly the α-helix and β-sheet. Software such as PROCHECK has been developed as an extremely useful aid to protein structure verification, including bond length, bond angle and conformational checks.

3. *Molecular graphics.*   Interpretation of electron density has been revolutionized over the past twenty years because of the development and availability of graphics software. These facilities enable structural features to be built into density maps, following standard geometry protocols, and provide the user with structure files for use in further analysis. Software available for carrying out these procedures include FRODO, O, and TURBO-FRODO.

The Fourier refinement process is represented in the flow diagram of Figure 10.14. It consists of successive cycles of electron density calculation, graphical interpretation, updating structure files, and calculation of structure factors, and R-factors. Improvement in the model and its fit to the electron density should cause significant decrease in the R-factors. This process terminates when no further enhancement is evident. The structure file at this stage consists of atom names, individual atom fractional coordinates (x, y, z), overall or average temperature factor B, for the whole crystal and a scale factor that converts $|F_o|$ values to an absolute scale. Individual B_j values are not refined at this stage of the analysis.

10.9.2   Simulated Annealing

After the initial model building of a protein structure it is to be expected that there will be some regions of the model that are some distance from their true positions. Such errors are difficult to correct with Fourier methods alone, as the detailed inspection of each map is a lengthy and difficult process. We have described the routine refinement of small molecule structures through the calculation of successive Fourier syntheses followed by least-squares analysis. The Fourier method has also been shown to be an essential tool for macromolecular structure analysis, and in the following sections of this chapter other techniques for refining large molecule structures will also be described. In order to apply least

---

[a] L. Pauling and R. B. Corey, *Proceedings of the National Acadamy of Sciences, USA* **37**, 729 (1951).
[b] R. A. Engh and R. Huber, *Acta Crystallographica* **A47**, 392 (1991).

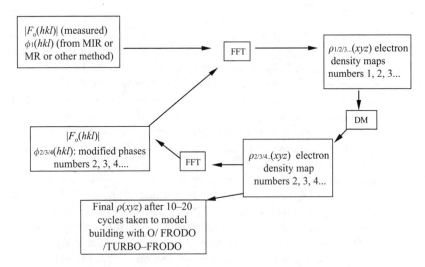

FIGURE 10.14.   Flow diagram for electron density modification (DM). The final electron density is used for model building, and for further refinement if of sufficient quality.

squares satisfactorily, the structure should not contain any gross errors; if it does the method will be unable to cope as it assumes that the structural parameters are close to their true values. Another difficulty is that the ratio of reflections in the data set to the number of parameters in the model (data/parameter ratio) is much smaller for a large molecule (see Problem 10.7). Least squares necessarily require the problem to be overdetermined to the extent of around 3–4 reflections per parameter or better, which is extremely difficult to achieve for proteins. In order to overcome these difficulties with the initial model the program X-PLOR or its updated version CNS can be used to carry out the process of *simulated annealing*, employing molecular dynamics.

The main objective of this method is to eliminate from the model regions of structure that are far from their true positions. Such regions may have been introduced through the initial search model in MR, or by faulty model building at a later stage. It is likely that such regions will be associated with bad interatomic contacts with inadmissibly high potential energy. In simulated annealing the structure is given a large perturbation ("heated to a high temperature"), yielding an ensemble of energetically allowed structures that is then allowed to recover ("cool") whilst preserving or re-establishing the correct minimum-energy molecular geometry. This process allows energy barriers that are associated with faults in the x-ray model to be overcome and corrected. The X-PLOR protocol involves minimization of an overall total potential energy term E_{tot}, which includes contributions from the empirical energy E_{emp} and the effective energy

E_{eff}, where $E_{eff} = E_{xray} + E_p + E_{nb}$. The terms $E_p + E_{nb}$ are included to take into account experimental information about phases and crystal packing, while E_{xray} is a "pseudo energy" that involves differences in the observed and calculated structure factors. Thus

$$E_{xray} = (W_a/N_a) \Sigma_{hkl} W_{hkl} (|F_o| - K|F_c|) \qquad (10.12)$$

where W_a puts the term on to the same basis as "energy" (established through a dummy dynamics run without E_{xray}), N_a is a normalization factor which renders W_a independent of resolution, W_{hkl} provides a weighting scheme, and K is a scale factor. X-PLOR is sometimes used without further refinement, and this can lead to a satisfactory structure. Individual isotropic thermal parameters B_j can be included in the refinement if the data collection resolution is sufficiently high, at least 2.5 Å. The conventional R-factor should drop to around 20% or lower unless there are big problems with the data, such as poor crystal quality or very weak diffraction, in which case it will not be possible to derive a highly significant structure with any refinement protocol. It may be desirable to undertake least-squares refinement of the structural parameters even after simulated annealing has been carried out. In any case routines such as PROCHECK or WHATCHECK (see Section 10.11) should be used to provide rigorous validation of the structure model.

10.9.3   Least-Squares Refinement: Constrained and Restrained Protocols

The method of least squares is used routinely in small molecule structure analysis, optimizing the fit between $|F_c|$ and $|F_o|$, and providing refined parameters and their standard deviations. Because most small molecule crystals diffract to around 0.8 Å the data set provides about 10 $|F_o|$ data values per parameter even when an anisotropic thermal displacement model is used. This degree of over-determination is not feasible with poorly diffracting protein crystals, for which there may be only one $|F_o|$ or less per parameter. Methods have therefore been developed, which attempt to economize on the number of parameters defining the structure model, and provide data in addition to the x-ray data that can contribute toward the refinement of the model. Software for carrying out such refinements is usually based on *constrained* or *restrained* protocols.

Constrained or Rigid Body Refinement

Rigid body refinement was mentioned in Section 10.6.3 as an option in some MR protocols. The method involves repositioning of the whole model in the unit cell without any further adjustment of individual atom coordinates. The method can be useful for refinement of molecules related by non-crystallographic symmetry by initially averaging the two or more molecules so as to improve the data/parameter ratio; this method and is available, for example, in the program X-PLOR. Constrained refinement can also be used for fixing rigidly predefined bond length and angle types and planarity of groups such as rings or peptide groups. While possibly effecting some degree of refinement in the structure model, constrained refinement will usually require further refinement by restrained least squares, where a degree of tolerance is built into the bond lengths and angles, if the data/parameter ratio allows. As an example of the savings in parameters that can be achieved by applying constraints, consider a flat six-membered ring. In terms of free atom parameters there are 18 positional parameters. By constraining the 6 bond lengths, 6 bond angles, and 3 cross-ring distances, only 15 parameters are required, a saving of 3 parameters. When applied repeatedly within a macromolecular structure, small savings like this can accumulate to a very worthwhile total, and at the same time ensure that the geometry of the refined structure will be acceptable. The input data set does, however, require a great deal of preparation in order to set up the required constraints.

Restrained Refinement

Many programs are available for carrying out restrained least-squares analysis of protein structures, two of which will be mentioned here, RESTRAIN and SHELXL-97. The method should not be used unless considerable effort has been applied to the Fourier refinement and possibly simulated annealing as well, to reduce the errors present in the structure and to include as many atoms as possible in the coordinate set. Restrained geometry refinement is more flexible than constrained refinement, each of the standard values of bond lengths, bond angles and other distances being tagged with a tolerance which specifies an acceptable range for the refined value of the parameter (see Section 10.11).

RESTRAIN. This program employs a least-squares algorithm, which uses terms involving differences $(\Delta F)^2$, or $(|F_o| - |F_c|)^2$, phase data if used, from MIR, standard geometry values (see Section 10.11), and the planarity of groups. Corrections to structure parameter values are derived by minimizing the function

$$M = \Sigma W_f(\Delta F)^2 + \Sigma W_\phi(\phi_o - \phi_c)^2 + \Sigma W_d(d_o - d_c)^2 + \Sigma W_a(a_o - a_c)^2$$
$$+ \Sigma W_b(b_o - b_c)^2 + \Sigma W_v V \qquad (10.13)$$

where ϕ_o is the MIR phase, d is a bond length, a is a cross-bond distance, such as $A-C$ in the sequence $A-B-C$ and therefore a measure of bond angle, b is a term for nonbonded distances, V is a planarity restraint, and the weights W apply to various types of terms and require very careful adjustment during the course of the refinement to ensure proper calculation of standard deviations for the refined parameters. Chirality is preserved by applying restraints to the edges of all chiral tetrahedra. Anisotropic thermal displacement parameters, nine for each freely refined atom, can be included if the data/parameter ratio allows. RESTRAIN is a CCP4 program: it employs non-FFT calculations for structure factors and partial derivatives and is consequently slow to perform each cycle, but requires fewer cycles to converge than FFT-based algorithms.

SHELXL-97.   SHELX programs are well known for their use in small molecule crystal structure analysis. The SHELXL-97 version has been adapted to accommodate protein refinement by incorporating the SHELXPRO interface. The geometry refinement options are similar to those described above and include additional facilities such as: anisotropic scaling; refinement progress display; and thermal displacement analysis. For conformational analysis, tabulation, display, and publication purposes the program carries out the following:

1. Checks for validity of main-chain conformations using a Ramachandran plot (see below);
2. Produces an electron density map file for the graphics program O;
3. Generates a Protein Data Bank (PDB) file for graphics display programs (see Appendix A10).

The PDB file is a standardized record of the structural parameters and is used as an input to programs for analyzing and displaying these results as well as for deposition at the PDB.

Practical Details.   Initially, small ligands and/or solvent can be excluded, but should be added through further use of electron density maps as the refinement progresses. Least-squares analysis predicts *corrections* for the structural parameters in the current model (coordinates, temperature factors and scale factor). It may be possible to refine individual B_j values if the data/parameter ratio is at least 2–3; anisotropic thermal displacement factors are unlikely to be refineable in all but very high-resolution protein structures. Selected groups of atoms can be given a single temperature factor, which provides an easy means of saving parameters.

Solvent Molecules and Small Ligands: Fitting to the Electron Density

The assignment of solvent atoms, usually water, into the crystal structure model is one of the last procedures to be carried out. These atoms make only small individual contributions to the x-ray intensities, but several hundred water molecules can influence considerably the values of $|F_c(hkl)|$. The level of solvation in protein crystals is usually around 50%, and most of the solvent is unstructured, forming very disordered, fluid regions in the intermolecular channels. However, within a layer closest to the protein molecule, it is to be expected that many solvent molecules will form strong interactions with the protein atoms, and consequently will acquire an ordered state approaching that of the protein itself. A measure of the ordering may be derived from the refinement if individual atomic thermal displacement parameters B_j are used, but this will be possible only for analyses employing a resolution better than about 1.8 Å. Generally, solvent atoms will have B_j values 10–20% larger than the protein atoms with which they are associated. Hydrogen-bonded interactions are responsible for ordering water molecules or other hydrophilic solvent molecules, such as ethanol. The interactions can involve any main-chain or side-chain atom which can act as a hydrogen-bond donor or acceptor. In order to locate possible water molecules it is necessary to inspect an $(|F_o| - |F_c|)$ or a $(2|F_o| - |F_c|)$ difference density map in detail. Possible water sites may be assigned to significant density regions, say, greater than $2\sigma_\rho$, and located within 2.5–3.3 Å (hydrogen-bonding distances) from one or more possible hydrogen-bond donors, usually −OH or >NH groups, or acceptor atoms, usually oxygen or nitrogen atoms; there should be no other close contacts present. Suitable electron density regions should ideally be spherical and small in volume. Since this procedure can be very time consuming, programs have been developed to expedite the process, including routines in O and SHELPRO. It should be emphasized that water sites assigned by such automatic procedures should be checked manually using molecular graphics and, if the analysis is at sufficiently high resolution, their B_j values should be refined. Sites with refined B_j greater than 80 Å^2 should be discarded. The inclusion of significant numbers of solvent atoms, which meet all these requirements should cause the R-factor to decrease by 1–2% or more. In many cases solvent atoms are functionally important and their recognition through the x-ray structure may be regarded as an added bonus.

The assignment of small ligands into the structural model follows a route similar to the above procedures for water molecules. Atom coordinates should be assigned and added to the data file for as many atoms as possible, paying full attention to the validity of the molecular geometry involving the new atoms. As much information as possible should be incorporated into the procedure, including known or preconceived stereochemistry of the group, known or expected binding region of the protein, proximity of protein atoms with ligand atoms, and validation of contact distances.

Figure 10.15 provides a possible flow diagram for the refinement process described above. An example of difference electron density showing a bound inhibitor molecule in the protein RNase T1[a] is shown in Figure 10.16.

---

[a] I. Zegers, PhD Thesis, Free University of Brussels (1994).

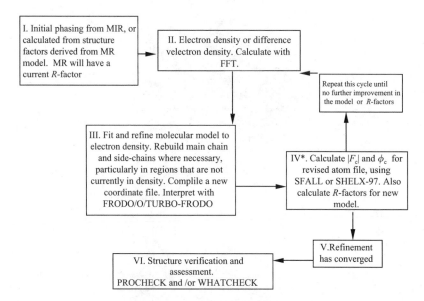

FIGURE 10.15.   Flow diagram for the refinement process. Initial refinement is carried out by successive Fourier synthesis. Once convergence has been reached, least squares (RESTRAIN or SHELX-97) or simulated annealing (X-PLOR) can be used in step IV. The cycle IV–II–III–IV is repeated until no further changes occur in the parameters of the structure and final convergence is reached.

10.10   Structure Validation: Final Checks

In this final section we mention some of the checks that can and should be made with respect to the protein structure when all of the above techniques have been exhausted. These concern both the technicalities of the x-ray analysis itself and the finished product.

10.10.1   *R*-Factors

One of the principal checks during the course of the analysis involves *R*-factors. While emphasizing that this should not be the only means of assessing the final structure, it does nevertheless provide the first indication that the analysis has progressed as planned. From the outset of the analysis, the overall quality of the intensity data will have been assessed from the value (or values) of the merging *R*-factor (see Section 10.4.6). This in turn provides an indication of how we can expect the *R*-factor of the structure to behave.

Over-refinement, that is, an unjustified fitting of too many parameters, can lead to unrealistically low *R*-factors, which should conform to the quality of

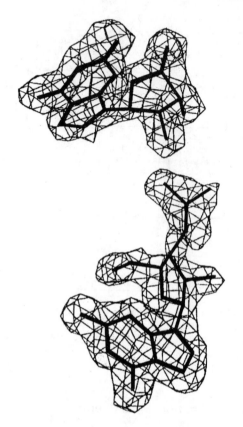

FIGURE 10.16. An example of an electron density plot for a protein structure. $(|F_o| - |F_c|)$ difference density in the active site region of RnaseT1.[a] Two moieties are clearly visble: a $3'$-guanosine monophosphate ($3'$-GMP, lower density region) and a guanosine subsite (upper density). The resolution is 1.7 Å and final R-value 14.5%

the x-ray data. Depending on the resolution attained, which is typically around 1.8–2.2 Å for a good protein structure, we can expect the R-factor to drop to 18–20% on completion of the refinement. For higher resolutions the R-factor should be less, because more of the structure, particularly water molecules, will have been fitted, and the structure will be less subjected to regions of disorder that are difficult to model. On the other hand, for lower resolutions the R-factor will tend to lie between 20% and 30%. In contemporary analyses it is expected that the R_{free} index will be determined as a check on model bias, particularly in structures based on MR. Because it is unbiased by the refinement process, as a general rule of thumb R_{free} should decrease as the R-factor falls but exhibit values 2–3% higher.

[a] I. Zegers, *PhD Thesis*, Free University of Brussels (1994).

10.10.2   Evaluation of Errors

Programs such as RESTRAIN and SHELX-97 provide estimated standard deviations for the refined structural parameters and these should be carefully reviewed to enable poorly refined parts of the structure to be identified. For a typical small molecule analysis a good estimate of the error in each of the refined parameters is calculated as a by-product of the least-squares procedure, the best estimates being from a full-matrix refinement. This is possible because of the high degree of overdetermination in small molecule analysis (data/parameter ratio> 10), whereas in protein analysis this ratio is more likely to be 2–3 and the estimated standard deviations will therefore not be so reliable. Several methods have been suggested for estimating the rms value of the atomic coordinate error of a protein structure $\sqrt{\overline{|\Delta r|^2}}$. The best method available to date[a] derives $\overline{|\Delta r|^2}$ from the slope of a plot of $\ln \sigma_A$ against $(\sin \theta / \lambda)^2$ in the formula

$$\ln \sigma_A = \frac{1}{2} \ln \left(\frac{\Sigma P}{\Sigma N} \right) - \frac{8\pi^3}{3} \overline{|\Delta r|^2} \left(\frac{\sin^2 \theta}{\lambda^2} \right) \tag{10.14}$$

where ΣN is Σf_j^2 summed over all atoms in the structure, and ΣP is Σf_j^2 summed over all atoms in the *partial* structure; σ_A is obtained from

$$(\sigma_A)^2 = \frac{\sum \left[(|E_{\text{obs}}|^2 - \overline{|E_{\text{obs}}|^2})(|E_{\text{calc}}|^2 - \overline{|E_{\text{calc}}|^2}) \right]}{\left\{ \sum (|E_{\text{obs}}|^2 - \overline{|E_{\text{obs}}|^2})^2 \sum (|E_{\text{calc}}|^2 - \overline{|E_{\text{calc}}|^2})^2 \right\}^{1/2}} \tag{10.15}$$

where $|E_{\text{obs}}|$ is obtained from $|F_o|$ (see Section 7.2.1), and $|E_{\text{calc}}|$ determined as given in Chapter 11; both $\overline{|E_{\text{obs}}|^2}$ and $\overline{|E_{\text{calc}}|^2}$ should equal unity (see Table 3.11). Values of σ_A estimated from this method are not absolute, and can only be considered to be on a relative basis for comparison purposes. According to current opinion, a value of σ_A around 0.3 Å or better is acceptable.

10.11   Geometry Validation: Final Checks

The geometrical features of proteins and related compounds have been subjected to intense study for over 50 years.[b,c] Information on the bond lengths, bond angles and torsion angles of smaller peptide structures are available in the Cambridge Crystallographic Data Base, and of proteins as such in the PDB (see Appendix A10). It is possible, therefore, to inspect a newly refined protein structure in detail and to make assessments of the geometry in terms of standard features. If either unusual features or possible errors are suspected, it will be necessary to

---

[a]R. J. Read, *Acta Crystallogr.* **A42**, 140 (1986).
[b]L. Pauling and R. B. Corey, *Proceedings of the National Academy of Sciences, USA* **37**, 729 (1951).
[c]R. A. Engh and R. Huber, *Acta Crystallographica* **A47**, 392 (1991).

look at them in more detail and to repeat some of the refinement protocols and make corrections where necessary.

10.11.1   Bond Lengths, Bond Angles, Planarity, and Chirality

Refinement programs, such as RESTRAIN, have built-in checks, which assess the derived geometry of the current model in terms of standard values. Average values of features such as peptide bonds are also calculated together with the spread of values found in the structure. Examination of these values provides a useful check on possible errors. Software such as PROCHECK and WHATCHECK have been developed for protein structure verification, including bond length, bond angle, and conformational checks, and should be used as a matter of routine. Known planar groups, such as phenyl rings, will probably have been constrained, and the effect of this should be checked to see if it has been successful. Chiral carbons are usually of known handedness and this can be restrained in some refinement programs. The chiral volume calculated as the scalar triple product of the vectors from the central atom to three attached atoms will have the correct sign only for the correct enantiomeric form. If a side group has been attached to a chiral centre incorrectly, the chiral volume will have the wrong sign and this can be corrected by rebuilding the model at this location.

10.11.2   Conformation

Main-Chain Conformation

The peptide torsion angle ω (see Section 10.1) is rarely outside the limits of $180° \pm 20°$ (*trans*) or $0° \pm 20°$ (*cis*). Deviations from these limits in the model should be examined carefully. The features of protein secondary structure that are governed by folding of the polypeptide in terms of the torsion angles ϕ and ψ are well documented and can be visualized and evaluated conveniently in terms of a Ramachandran plot.[a] This provides an overall picture of the (ϕ, ψ) values observed for the protein structure, all of which are plotted on to the same diagram. The shaded areas of Figure 10.17 show the regions of allowed conformation more or less as originally plotted. These regions were defined from studies of protein models in which the side-chains were all alanine (ala), which has a short methyl side group, and are based on those (ϕ, ψ) combinations that are free from steric clashes. Although protein structures are found to conform mainly to the allowed regions displayed here, some extension outside and between these areas is accepted on account of the large data base of known proteins now available. In Figure 10.17

[a] G. N. Ramachandran, C. Ramakrishnan, and V. J. Sasisekheran, *Journal of Molecular Biology* **7**, 95 (1963).

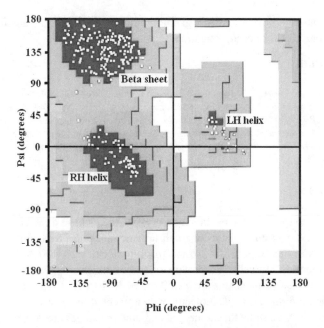

FIGURE 10.17. Ramachandran plot for the protein SNAII (*Sambucus nigra agglutinin II*), a Gal/GalNAc specific lectin extracted from the bark of the Elder, calculated with the program PROCHECK. The plot provides an overall picture of the (ϕ, ψ) coordinates (white squares) observed for this protein. (Gal = galactose; GalNAc = N-acetylgalactosamine) There are no unusual or disallowed conformations present.

the white squares denote the (ϕ, ψ) coordinates calculated for the refined structure of *Sambucus nigra* agglutinin II (SNAII),[a] a Gal/GalNAc specific lectin extracted from the bark of the elder, *Sambucus nigra*. All of the (φ, ϕ) values are seen to be within allowed regions, indicating that it is an acceptable structure. Any points derived from a new protein structure which are significantly outside the regions shown here should be carefully examined.

Side-Chain Conformations

In terms of stereochemistry there is a great variety amongst the 20 amino acids that occur in proteins, and their individual conformations are well characterized from known structures. As with all empirical data, tolerances in these conformations exist and should be borne in mind when evaluating a new structure. Programs such as PROCHECK and WHATCHECK provide a detailed analysis of

[a] H. Niwa, PhD Thesis, University of London (2001).

the side-chain conformations in a given protein structure and flag any unusual or dubious ones for further investigation.

Bibliography

BRANDEN, C., and TOOZE, J. *Introduction to Protein Structure*, 2nd ed., Garland, New York (1999).
CARTER, W., *Protein and Nuceic Acid Crystallization*, Academic Press, New York, London (1990).
DRENTH, J., *Principles of Protein X-Ray Crystallography*, 2nd ed., Springer, New York, Berlin (1999).
MCREE, D. E., *Practical Protein Crystallography*, Academic Press, San Diego, London (1993).
MCPHERSON, A., *Preparation and Analysis of Protein Crystals*, Krieger Publishing, Melbourne, FL (1989).

Problems

10.1. Typical specimens used in single crystal studies have dimensions between 0.1 and 0.5 mm. Crystals of small organic molecules have unit-cell dimensions in the range 5–100 Å and macromolecules 20–400 Å. Estimate the number of unit cells in a protein crystal of dimensions 0.2, 0.3, 0.4 mm if the unit-cell dimensions are $a = 30$, $b = 40$, $c = 50$ Å. Repeat the calculation for a small organic molecule with unit cell dimensions $a = 10$, $b = 12$, $c = 15$ Å. Assume that all crystal and unit cell angles are 90°. Comment on the relative diffracting power of the crystals.

10.2. Diffraction data are to be measured on an image plate using SR. The exact wavelength of the radiation can vary from time to time and needs to be calibrated. A smear of amorphous beeswax was placed where the crystal is normally mounted at a distance of 300 mm from the image plate. The wax has a dominant powder line of spacing 3.5 Å and produced a ring of diameter 140 mm on the image plate. Calculate the wavelength of the x-radiation.

10.3. Referring to Problem 10.2, it is required to collect data for a protein crystal with a known unit-cell length of about 300 Å. If a spot separation of at least 1 mm is required, comment on whether a crystal to detector distance of 450 mm would be appropriate.

10.4. During data processing for the crystal in Problems 10.2 and 10.3 the unit-cell parameters were found to be $a = b = 110.79$, $c = 308.53$ Å, $\alpha = \beta = 90°$, $\gamma = 120°$; the Laue symmetry was $\frac{6}{m}mm$. The only limiting conditions in the diffraction pattern were for *00l*, $l = 6n$. State the space group or possible space groups for this crystal.

10.5. For mistletoe lectin MLI[a], the space group is $P6_522$ with $Z = 12$, M_P is 63,000 and the unit-cell parameters are $a = b = 110.79$, $c = 308.53$ Å,

[a] E. C. Sweeney, A. G. Tonevitsky, R. A. Palmer, H. Niwa, U. Pfüller, J. Eck, H. Lentzen, I. I. Agapov, and M. P. Kirpichnikov, *FEBS Letters* **431**, 367 (1998).

$\alpha = \beta = 90°, \gamma = 120°$. Investigate possible values for μ (the number of molecules per asymmetric unit) and s, the fractional solvent content.

10.6. For mistletoe lectin MLI (Problem 10.5), there are 21,000 *recorded* reflections in a data set having $d_{min} = 2.9$ Å for wavelength $\lambda = 0.8$ Å (SR). Estimate the total number of reflections *expected* in the data set. If 21,000 unique reflections were recorded what is the percentage completeness of the data set?

10.7. A protein of relative molar mass 27,000 to be studied by x-ray crystallography, crystallizes in space group $P2_1$ with 2 molecules per unit cell. Estimate the number of nonhydrogen atoms to be located in the analysis given that the hydrogen content of the protein is 10% and assuming that of the 40% by weight of solvent (water) in the unit cell only 10% is ordered. If the unit-cell dimensions are $a = 58.2, b = 38.3, c = 54.2$ Å and $\beta = 106.5°$, estimate the data/parameter ratios corresponding to 6, 2.5, and 1.0 Å resolution data sets, and comment on these values.

10.8. Derive the general equivalent positions for space group $P3_2$. What conditions limit reflections in this space group? (Appendix A4 may be helpful.)

10.9. The choice of methods for the determination of protein crystal structures includes (a) MR, (b) MIR, and (c) MAD. Discuss the circumstances under which each of these methods would be the most appropriate to use for a given protein.

11

Computer-Aided Crystallography

11.1 Introduction

The title of this chapter emphasizes the need for the pre-knowledge gained from a study of the earlier chapters of this book. The programs supplied on the CD and described here are complementary to that work and designed to enable the reader to gain practical experience of the concepts and methods in x-ray structure analysis.

Computing is an essential feature in any modern crystallographic investigation. Here, we can provide only a flavor of what is available, but enough, we hope, to demonstrate the great importance of computational methods to this subject. We use the programs here in three main ways:

1. To study the derivation of point groups;
2. To carry out systematic point-group recognition using crystal or molecular models;
3. To simulate the procedures and calculations involved in the determination of crystal structures by x-ray diffraction data from both single-crystal and powder specimens.

On the basis of a familiarity with these programs, it should become possible to proceed, when needed, with the more comprehensive and detailed crystallographic software to which we have referred in the text, and which is referenced in Appendix A10.

The programs here are supplied as IBM/WINDOWS-compatible.EXE files operating in a DOS window. The complete suite of programs, together with the data files supplied, should be loaded from the CD supplied with this book into the PC. The same materials are also available at the web address www.wkap.nl/subjects/crystallography from where they can be downloaded according to the publisher's instructions. The programs at *that* source may be amended from time to time, as improvements and additions are applied to them. The first set will be dated 1 April 2003 (Version 1), so that any date after that implies a revision or addition to those programs, as will be notified there. The programs ITO12*, ESPOIR*, and LEPAGE* will not be part of such changes unless they are revised by *their* authors.

All programs can be executed by a double-click on the program .EXE icon; each program name appears after the section head that introduces it. The asterisk after a program name distinguishes it from all other programs mentioned, as part of the CD suite supplied. It is generally convenient to increase the size of a DOS window by clicking on the □ box at the top right-hand corner of the screen.

In most of these programs, the notation, $<X>$ is used to indicate an action to input the parameter X. Furthermore, KEY-IN $<X>$ implies that a key X is just depressed, whereas ENTER $<X>$ means that key X is depressed, followed by $<ENTER>$. Most but not all text data files carry the suffix .TXT but some need the suffix .DAT as will be indicated.

Except where otherwise noted, data may be prepared in free format, that is, numbers are terminated by one space or more. In certain programs, for example, XRAY* and ITO12*, the data format is specific and must be observed.

11.1.1 Structure of Program CD Suite

The CD is divided into four folders or directories: XRSYST contains the programs and data for the single-crystal techniques, PDSYST contains the programs for powder techniques, POWDER contains the programs and data for solving crystal structures from powder data, and GNSYST contains all other general programs referred to in the text and problem sections of the book. The structure of the CD suite is illustrated by Figure 11.1.

We now describe the different programs and their uses in detail, with the aid of examples.

11.2 Derivation of Point Groups (EULR*)

In Chapter 1, we discussed the fact that the crystallographic point groups consist of the symmetry operations R and $\bar{R}$ ($R = 1, 2, 3, 4, 6$), taken singly or in combinations. The 10 point groups that express a single symmetry operation are $1, 2, 3, 4, 6, \bar{1}, m (\equiv \bar{2}), \bar{3}, \bar{4}$, and $\bar{6}$, and the stereograms of these point groups are illustrated by Figure 1.36.

The first nontrivial combinations of symmetry operations follow from combining R with $\bar{1}$, and it is not difficult to show, with the aid of stereograms, that R and $\bar{1}$ together lead to R/m for $R = 2, 4$, and 6; for $R = 1$ and 3, point groups $\bar{R}$ already include $\bar{1}$ as an operator in the group. Thus, we have quickly derived 13 crystallographic point groups; $1, 2, 3, 4, 6, \bar{1}, m (\equiv \bar{2}), \bar{3}, \bar{4}, \bar{6}, 2/m, 4/m$ and $6/m$.

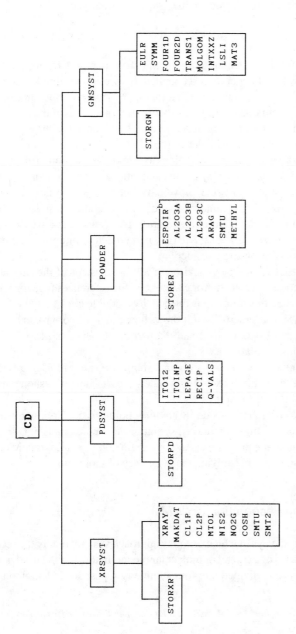

FIGURE 11.1.   Composition of the CD program suite. Each 'STOR' box contains a reserve copy of the programs on its immediate right. Each program name carries an asterisk in the body of the text, in order to distinguish these programs from others.

[a]There are also present six files, concerned with plotting, that must remain unaltered (five .DLL and one .CON).

[b]There is also a file RASMOL, concerned with drawing, together with two files for each substance, one with crystal data (.DAT) and one with the intensity data (.HKL); they, too, must remain unaltered.

The point groups that contain more than one symmetry operation display the essence of Euler's theorem on the combination of rotations. We have used this theorem implicitly in Chapter 1, and we may state it formally as

$$\mathbf{R}_2 \cdot \mathbf{R}_1 \equiv \mathbf{R}_3 \qquad (11.1)$$

which means that, from a given situation, symmetry operation $\mathbf{R}_1$ *followed by* operation $\mathbf{R}_2$ is equivalent to operation $\mathbf{R}_3$ applied to the original situation, and we know from the definition of point group that the three symmetry operations have at least one point in common. We saw this theorem in operation in the examples of point groups $mm2$ and $4mm$ (Figures 1.32 and 1.33), and it is the basis of a procedure for determining the remainder of the 32 crystallographic point groups. In general, the order of symmetry operations is important, although the result would not actually be affected with symmetry operations of degree 2 or less.

After the 13 point groups listed above have been noted, we next combine the operations R with another symmetry operation, say 2, and we need to know immediately the relative orientations of the rotation axes R and 2 that we use symbolically to represent these operations. Are they perpendicular or even coincident, and are there other possibilities to consider?

The program EULR* has been devised to follow the steps of the derivation of point groups described elsewhere, for example, in the references that appear on the monitor screen when this program is opened, by a double click on the EULR* icon. This program is not interactive, but it shows how Euler's theorem can be used with the combinations of operations 2 and the permitted values of R to develop six sets of orientations of rotation axes.

Then, independently of the program, one should consider replacing two of the rotation axes in each of the six sets by inversion axes. (Why not just one of the rotation axes, or all three of them?)

As a final step, we must consider if any *new* point groups are obtained by incorporating a center of symmetry into any point group where one is not already present. An extension of the program caters for certain non-crystallographic point groups that are encountered in studying the symmetry of molecules.

11.3   Point-Group Recognition (SYMM*)

There are several ways in which one may approach systematically the recognition of the point group of a crystal or molecular model. In the method used here,[a] molecules and crystals are divided into four symmetry types, dependent upon the

[a] M. F. C. Ladd, *International Journal of Mathematical Education in Science and Technology* **7**, 395 (1976).

TABLE 11.1. Crystallographic Point Groups Typed by
m and/or $\bar{1}$ or Neither

Neither m nor $\bar{1}$	Only m	Only $\bar{1}$	Both m and $\bar{1}$
$1, 2, 3, 4, \bar{4}, 6$	$m, mm2, 3m$	$\bar{1}, \bar{3}$	$2/m, mmm, \bar{3}m$
$222, 32, 422$	$4mm, \bar{4}2m$		$4/m, 4/m\ mm$
$622, 23, 432$	$\bar{6}m2, 6mm$		$6/m, 6/m\ mm$
	$\bar{6}m2, \bar{4}3m$		$m3, m3m$

presence of a center of symmetry and one mirror plane or more, or a center of symmetry alone, or one mirror plane or more but no center of symmetry, or neither of these symmetry elements; hence, the first step in the scheme is a search for these elements.

To demonstrate the presence of a center of symmetry, place the given model in any orientation on a flat surface; then, if the plane through the uppermost atoms (for a chemical species), or the uppermost face (for a crystal), is parallel to the plane surface supporting the model *and* the two planes in question are both equivalent and inverted across the center of the model, then a center of symmetry is present.

If a mirror plane is present, it divides the model into enantiomorphic (right-hand–left-hand) halves. A correct identification of these symmetry elements at this stage places the model into one of the four types listed in Table 11.1. The reader may care to examine a cube or a model of the SF_6 molecule, which shows both a center of symmetry and mirror planes, and a tetrahedron or a model of the CH_4 molecule, which shows mirror planes but no center of symmetry. Models of a cube and a tetrahedron may be constructed easily, as described below.

(a) *Cube.* On a thin card, draw a square of side a, say 40 mm. On each side of this square draw another identical square. Lightly score the edges of the first square, fold the other four squares in the same sense to form the faces of a cube, and fasten with "Selotape." There is an advantage in leaving the sixth face of the cube open, as we shall see, but we shall imagine its presence when needed.

(b) *Tetrahedron.* On similar card, draw an equilateral triangle of side $a\sqrt{2}$, where a is the length chosen for the side of the cube in (a). On each side of the triangle, draw another identical triangle. Lightly score the edges of the first triangle, fold the other three triangles in the same sense to meet at an apex, and fasten with "Selotape." On placing the tetrahedron inside the cube, it will be found that an edge of the tetrahedron is a face diagonal of the cube, thus aligning the symmetry elements common to both models.

If these models are to be used with the point-group recognition program, allocate model numbers 7 and 19 for the cube (or SF_6) and tetrahedron (or CH_4), respectively.

The identification of the point group of a model then proceeds along the lines indicated by the block diagram of Figure 11.2, on which the program SYMM* is based. In a preliminary and necessary study of the model, after assigning it to one of the four types, the principal rotation axis (the rotation axis of highest degree) is identified, the number of them if more than one, the presence and orientations of mirror planes, 2-fold rotation axes, and so on.

The program SYMM* may be executed as described above. It is interactive and the directions on the monitor screen should be followed. If an incorrect response is given during a path through the program, the user will be returned to that point in the program where the error occurred, for an alternative response to be made. Two such returns are allowed before the program rejects that particular examination for further (preliminary) study.

It is necessary that the crystal and molecular models to be used are allocated a model number appropriate to their symmetry, and Table 11.2 provides the necessary key, together with molecular examples or possible molecular examples of the point groups; a set of solid crystal models is equally satisfactory. It will be realized that some of the molecules listed are not rigid bodies, and will show the required symmetry only if their functional groups are orientated correctly. The program responds to the non-crystallographic point groups ∞m and $\infty/m(\bar{\infty})$ only. The ∞ symbol is replaced by the word *infinity* in the program itself, but the zero (0) is used to input ∞ to the program when asked for the point-group symbol.

11.4   Structure Determination Simulation (XRAY*)

The purpose of this program package is to facilitate an understanding of the practical applications of the techniques of structure determination by single-crystal x-ray diffraction that we have discussed in the earlier chapters, albeit here in two dimensions. Example structures have been selected that give, in projection, results that are readily interpretable in terms of chemical structures. It is all too easy, we believe, to use modern, sophisticated structure-solving packages without really understanding the nature of the calculations taking place within them. In particular, for reasons that we have discussed in Section 8.4.3, sometimes very subtle difficulties are encountered.

A subsidiary program, MAKDAT*, enables sets of primary data to be constructed in the precise format required by XRAY*. Alternatively, one may choose to prepare a data set independently, in which case the layout shown by the data sets provided should be followed exactly. It is stressed that the program does

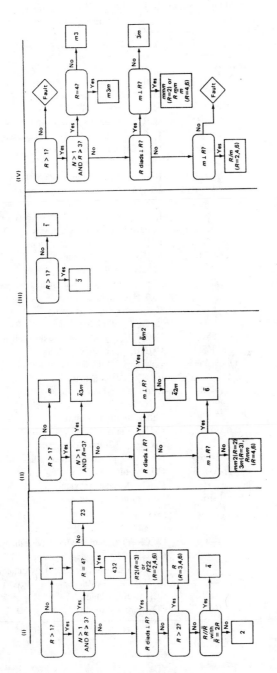

FIGURE 11.2. Flow diagram for the point-group recognition.

TABLE 11.2.   Point Groups and Model Numbers for the Program SYMM*,
with Molecular Examples or Possible Examples

Point group	Model number/s	Example or possible example
Crystallographic point groups		
1	91	CHBrClF, bromochlorofluoromethane
2	77	H_2O_2, hydrogen peroxide
3	84, 93	H_3PO_4, phosphoric acid
4	85, 94	$(CH_3)_4C_4$, tetramethylcyclobutadiene
6	88, 97	$C_6(CH_3)_6^{\bullet}$, hexamethylcyclohexadienyl
$\bar{1}$	78, 79	$C_6H_5CH_2CH_2C_6H_5$, dibenzyl
$\bar{3}$	48	$[Ni(NO_2)_6]^{4-}$, hexanitronickelate(II) ion
$\bar{4}$	86, 95	$[H_2PO_4]^-$, dihydrogen phosphate ion
$\bar{6}$	89, 98	$C_3H_3N_3(N_3)_3$, 1,3,5-triazidotriazine
$2/m$	68–70, 72–75, 80	CHCl=CHCl, *trans*-1,2-dichloroethene
$4/m$	56	$[Ni(CN)_4]^{2-}$, tetracyanonickelate(II) ion
$6/m$	37	$C_6(CH_3)_6$, hexamethylbenzene
$m\ (\bar{2})$	83, 92, 99	$C_6H_3Cl_3$, 1,2,4-trichlorobenzene
$mm2$	16, 64, 71, 76	C_6H_5Cl, chlorobenzene
$3m$	42	$CHCl_3$, trichloromethane
$4mm$	87, 96, 100	$[SbF_5]^{2-}$, pentafluoroantimonate(III) ion
$6mm$	81	$C_6(CH_2Cl)_6$, hexa(chloromethyl)benzene
222	67	C_8H_{12}, cycloocta-1,5-diene
32	43, 47	$[S_2O_6]^{2-}$, dithionate ion
422	55	$Co(H_2O)_4Cl_2$, tetraaquodichlorocobalt
622	36	$C_6(NH_2)_6$, hexaminobenzene
mmm	59–63, 65, 66	$C_6H_4Cl_2$, 1,4-dichlorobenzene
$\bar{6}m2$	44–46, 90	$[CO_3]^{2-}$, carbonate ion
$4/m\ mm$	49–54	$[AuBr_4]^-$, tetrabromoaurate(III) ion
$6/m\ mm$	29–35	C_6H_6, benzene
$\bar{4}2m$	57, 58	$ThBr_4$, thorium tetrabromide
$\bar{3}m$	38–41	C_6H_{12}, *chair*-cyclohexane
23	27	$C(CH_3)_4$, 2,2-dimethylpropane
$m3$	22–25	$[Co(NO_2)_6]^{3-}$, hexanitrocobaltate(III) ion
$\bar{4}3m$	17–21, 28	CH_4, methane
432	26	$C_8(CH_3)_8$, octamethylcubane
$m3m$	1–15	SF_6, sulfur hexafluoride
Non-crystallographic point groups		
$\bar{8}2m$	—	S_8, sulfur
5	—	$C_5(CH_3)_5^{\bullet}$, pentamethylcyclopentadienyl
$5m$	—	C_5H_5NiNO, nitrosylcyclopentadienylnickel
$\overline{10}m2$	—	$(C_5H_5)_2Ru$, *bis*-cyclopentadienylruthenium
$\bar{5}m$	—	$(C_5H_5)_2Fe$, *bis*-cyclopentadienyliron
∞m	101	HCl, hydrogen chloride
∞/m	102	CO_2, carbon dioxide

not teach the subject of structure determination. Rather, it provides the basic concepts with a medium for their exploration, and so relates closely to the techniques that are in use today. The program is interactive, and the messages on the monitor screen indicate the steps to be carried out by the user. Nevertheless, it will be useful to discuss here some of the features and capabilities of the system.

Each data set contains information about the unit cell, crystal symmetry in the appropriate projection, wavelength of radiation used, a set of $|F_o|$-values and data required in the calculation of Fourier maps, structure factors, and $|E|$-values. As the procedures are two dimensional, the symbols a, b, γ, h, k, x, y and so on are employed. Where the projection is other than that on (001), appropriate adjustments are made in setting up the primary data file; thus, true y and z can become x and y to the program.

The following basic procedures may be carried out within the XRAY* program package:

- Patterson function
- Superposition (minimum) function
- Structure factor calculation
- Least-squares refinement
- Electron density function
- Direct methods: calculation of $|E|$ values
- Direct methods: calculation of E-map
- Distances and angles calculation
- Scale and temperature factors by Wilson's method
- $|E|$ values calculated from the structure.

The program is entered by a double-click on the XRAY* icon. Several PAUSE situations occur throughout the routines in the system, so that material on the screen may be read. Continuation is effected by just depressing the Enter key. All primary data file names must be four-letter words plus the suffix .TXT, for example, NIS2.TXT, corresponding to a nickel o-phenanthroline complex, although only the first four letters of the name, NIS2, are used to call the data set; the program checks for the suffix .TXT. A name for the output file is chosen, say NISOUT, and entered at the keyboard, so that results of calculations may be written to it, and printed as required. In some routines, such as the calculation of $|E|$ values, other output files are organized, with appropriate messages to the screen. Certain other files pertaining to coordinates are created at an appropriate stage, and their significance will become clear later. The particular calculations available are then listed on the monitor screen, and we shall give a brief description of each routine in turn.

11.4.1   Patterson Function

The Patterson function $P(u, v)$ may be calculated using either $|F_o|^2$ or "sharpened" $|F_o|^2$ values as coefficients. In the sharpened mode, the coefficients used here are $\{|E|^2 - 1\}$ values, thus providing a sharpened, origin-removed Patterson map. It follows that, in the case of the sharpened Patterson, $|E|$ values must be first calculated, and the program so directs. After this has been done, the system returns to the Patterson routine, so that the appropriate messages can be followed.

The sharpening obtained from the use of $|E|$-values introduces a small number of spurious maxima, and it is always useful to compare sharpened and unsharpened maps.

The Patterson map, and all other Fourier maps, may be contoured on the screen. This facility enhances the interactive nature of the program system. It is possible, for example, to determine the coordinates of the heavy atom vectors directly from the map on the screen, then go to the structure factor routine and input the heavy atom x, y coordinates, so as to obtain partial phase information, and thence calculate a first electron density map. It should be possible then to recognize other atom peaks from this first Fourier map, so beginning the process of structure determination by successive Fourier synthesis.

At some stage it may become necessary to print and contour a Patterson or electron-density figure field, so as to get a clearer picture of the projection, particularly where contours below the relative level 10 are involved. All Fourier maps are scaled to a maximum of 100, and 10 is the lowest level that is, contoured on the screen.

It should be noted that, if the WINDOWS "screen saver" comes into operation while a plot is on the screen, owing to some delay, the white cross-wires might become fixed in position. On reactivating the plot, a second, mobile cross-wires will appear. To avoid these events, the "screen saver" should be deactivated.

11.4.2   Superposition Function

This routine calculates a minimum function $M(x, y)$ at each grid point of the projection (see Section 6.4.3):

$M(x, y) =$ Minimum of $\{P(x + \Delta u_1, y + \Delta v_1), P(x + \Delta u_2, y + \Delta v_2), \ldots, P(x + \Delta u_n, y + \Delta v_n)\}$, where $\Delta u_1, \Delta v_1, \Delta u_2, \Delta v_2, \ldots, \Delta u_n, \Delta v_n$ represent n displacement vectors, which may be either atom positions or a set of vectors including 0, 0, obtained from a partial interpretation of a Patterson map; symmetry-related positions should be entered. The grid points x and y are determined by the values set in the primary data. The minimum function, if successful, should indicate atomic positions that can then be used as discussed above.

11.4.3   Structure Factor Calculation

Each atom contributing to the structure factor calculation requires the following data:

Atom type identity number: a list is given at the start of the routine.
Fractional coordinates of the atoms: x, y.
Population parameter: 1, unless the atom is in a special position.

An overall (isotropic) temperature factor is given initially from the primary data; it may be altered by routine 10. In either case, the value will be allocated to each atom, in preparation for subsequent refinement.

In the calculation of structure factors, the coordinates may be entered either from a file or at the keyboard. If a file is chosen, it must be named XYS.TXT. The output from this routine is self-explanatory, and after an $|F_c|$ calculation and least-squares refinement, the current coordinates are retained in the file COORDS.TXT. This file may be invoked in a subsequent calculation, or edited as desired, before being used as input data. The file XYS.TXT remains unaltered by the program.

11.4.4   Least-Squares Refinement

The use of this routine implies that the positions of some atoms have been obtained. It may be just the heavy atoms, or a number of other atoms may be included as well. The routine uses the *diagonal* least-squares technique: it is an approximation to the ideal full-matrix procedure, but it is fast and satisfactory for emphasizing the principles involved. The x and y coordinates and the isotropic temperature factor B (initially the same for each atom, from the primary data) are refined. The changes $\delta x, \delta y$, and δB are determined and applied to each atom, and the R-factor listed together with other parameters. The cycle can be repeated until no further improvement is obtained, as judged by near-constancy in the (satisfactory) R-factor, or by very small changes in the δx and δy shifts. The scale factor is determined as $\sum |F_c| / \sum |F_o|$, and may be applied to the data at the end of any cycle of refinement. In this diagonal approximation, 60% of the calculated shifts are applied by the program. The value may be altered at any cycle, as desired. At each cycle, the new coordinates are stored in the file COORDS.TXT.

For the Ni-o-phenanthroline complex, the list of coordinates in Table 11.3, obtained by Patterson and electron density maps, will refine to about 9.8% by this least-squares routine, adjusting the x and y coordinates and the individual isotropic temperature factors.

11.4.5   Electron Density Maps

When some phase information becomes available, for example, from the partial solution of a Patterson function, an electron density summation with these

TABLE 11.3. Approximate Atomic
Coordinates for the Ni *o*-phenanthroline
Complex Derived from Successive
Fourier Syntheses

Atom ID	x	y	Pop
1	0.235	0.178	1.0
2	0.317	0.103	1.0
2	0.154	0.099	1.0
3	0.462	0.150	1.0
3	0.392	0.231	1.0
3	0.320	0.242	1.0
3	0.149	0.235	1.0
3	0.076	0.220	1.0
3	0.007	0.138	1.0
4	0.000	0.076	1.0
4	0.273	0.439	1.0
4	0.417	0.084	1.0
4	0.395	0.165	1.0
4	0.329	0.297	1.0
4	0.276	0.337	1.0
4	0.312	0.393	1.0
4	0.192	0.434	1.0
4	0.155	0.382	1.0
4	0.193	0.332	1.0
4	0.140	0.289	1.0
4	0.074	0.158	1.0

Note:
The total number of non-hydrogen atoms in the
molecule is 21. The ID numbers refer to atoms as
follow: $1 = Ni, 2 = S, 3 = N, 4 = C$.

phases allocated to the experimental $|F_o|$ data will reveal a portion of the structure, if the information from the partial structure is correct. Then, more atom positions can be interpreted and built into the next structure factor calculation, and so on. The same program routine will also calculate a difference electron density map, using $|F_o| - |F_c|$ as coefficients, provided that the R-factor is sufficiently low (<0.3). In the difference synthesis, atoms placed incorrectly appear as low or negative density regions, whereas unallocated atomic positions will show positive density, both relative to the general level of the figure field. It follows that when all atom positions have been determined correctly, the difference map figures should show a nearly level, ideally zero, figure field. Thus, a difference map may be used to make appropriate adjustments to atomic positions. Sometimes a fairly well-refined structure may show a significant positive region, which may indicate the presence of solvent of crystallization. We note also that incorrect temperature factors can lead to variations in the level of the figure field, although this effect would be expected to be small when individual temperature factors are applied, as from a least-squares refinement.

11.4.6   Direct Methods: Calculation of $|E|$ Values

We have shown earlier that the normalized structure factors $|E|$ may be calculated from the equation

$$|E|^2 = K^2 |F_0|^2 / \left\{ \varepsilon \sum_j f_j^2 \exp(-B\lambda^{-2} \sin^2 \theta) \right\} \qquad (11.2)$$

where the symbols have the meanings defined in earlier chapters, and the values of B and K may be obtained from a Wilson plot. However, a single isotropic B factor may not be representative of the structure and, as a consequence, a Wilson plot may deviate from linearity. In an alternative procedure,[a] we write

$$|E|^2 = K(s) |F_{\text{corr}}|^2 / \left\{ \varepsilon \sum_j f_j^2 \right\} \qquad (11.3)$$

where $K(s)$ is a factor that includes adjustments for the scaling of $|F_0|$ and temperature effects on f_j. In implementing this method, a number n of ranges is set up in equal increments of s^2, where $s = (\sin\theta)/\lambda$. For each range, $K(s)$ is computed as $\sum \varepsilon\sigma_2 / \sum |F_0|^2$, where each value of $|F_0|$ is given its appropriate multiplicity of planes. The $K(s)$ values, then as a function of s, are interpolated[b] so as to calculate $|F_{\text{corr}}|^2$ for each reflection. Then, $|E|^2$ is given by

$$|E|^2 = |F_{\text{corr}}|^2 / \varepsilon\sigma^2 \qquad (11.4)$$

It is desirable to enter the s and $K(s)$ values corresponding to the extreme ends of the data range, because computed extrapolation can be uncertain, particularly at the low θ end of the curve. Data are output to the monitor that enable these values to be estimated. A facility is provided for printing in the main output file any reflections excluded by the procedure. There should be *none*. If there are any, it means that the values entered for the extreme ends of the K-curve have not been chosen satisfactorily.

The $|E|$ values are written to a file EVALS, together with some statistics of the $|E|$ distribution; $|E|$ values greater than or equal to a chosen limit, ELIM, are written to the main output file, and a $\sum_2$ listing is set up in the file SIG2, in descending order of $|E|$ magnitude. The file EDATA contains the $|E|$ values greater than or equal to ELIM arranged in parity groups.

[a] J. Karle, H. Hauptman, and C. L. Christ, *Acta Crystallographica* **11**, 757 (1958).
[b] M. F. C. Ladd, *Zeitschift fur Kristallographic* **147**, 279 (1978).

The program comes to a halt at this stage, so that the reader can print the $\sum_2$ listing and develop signs for the $|E|$ values, along the lines already discussed in the text (see Section 7.2.2ff). Data sets are provided only for centrosymmetric structures, but similar procedures are used at the beginning of a determination of a non-centrosymmetric crystal structure. In practice, it may be necessary to re-run the E-values link with slightly lower value of the limit ELIM, if insufficient data are produced for a successful sign determination process. After a set of signs has been determined, the next program routine, 7, is used to prepare an E-map. Alternatively, where the plane group is $p2gg$, the program FOUR2D* (see Section 11.6.2) can be used.

11.4.7   Calculation of $|E|$ Maps

An $|E|$ map is an electron density map calculated, in the case of a centrosymmetric structure, with *signed* $|E|$ values. This routine in the program provides for a straightforward transfer of the $|E|$ values, with their signs as determined through the $\sum_2$ routine, to a Fourier calculation. As $|E|$ values are *sharpened* coefficients, a few spurious peaks may be anticipated. Thus, chemical knowledge has to be brought to bear on the extraction of a sensible chemical structure or fragment. Once this has been done, and shown to be satisfactory, electron density calculations may be carried out with normal $|F_0|$ coefficients.

This link of the program, however, permits modifications of signs, with further E-maps, without re-starting the system from scratch. A second call to link 7 lists the current set of data as h, k, and s. A value of 0 for s indicates an unsigned reflection which does not contribute to the E-map. To the question "Do you want to retain some of the current values of cos(phi)," the answer *no* implies a re-input of new signs. The answer *yes* implies that some changes are desired, and the opportunity to do that follows. The changes are then made, and a further E-map calculated.

From the E-map, atomic positions should be found that can be entered into the structure factor calculation, with or without a least-squares refinement, and an $|F_0|$ Fourier map then calculated (see above).

11.4.8   Bond Lengths and Bond Angles

This routine calculates bond lengths and bond angles, and distances between non-bonded atoms in the structures. The amount of information extracted depends on the distance limits input to the routine. There is normally little to be gained by calculating distances greater than that of the maximum van der Waals contact distances, typical values of which are listed in Table 11.4. In interpreting the results from this routine, it must be remembered that, in working in two dimensions, some variations from standard numerical values are to be expected, because of the distortion of the molecule in projection. However, in order to minimize this effect,

TABLE 11.4. Van der Waals Contact Distances
for Some Common Species

Atom	Radius/Å	Atom	Radius/Å
H	1.20	C	1.85
N	1.50	O	1.40
F	1.35	Si	2.10
P	1.90	S	1.83
Cl	1.80	As	2.00
Se	2.00	Br	1.95
Sb	2.20	Te	2.20
I	2.15	$-CH_3$	2.00
$>CH_2$	2.00	$-C_6H_5$	1.85^a

^{a} Half-thickness of phenyl ring.

structures have been selected in which the plane of the molecule lies nearly in the plane of the projection.

11.4.9 Scale and Temperature Factors by Wilson's Method

We have discussed a procedure based on Wilson's statistics in an earlier chapter, and this routine provides for this calculation. The data output contains a breakdown of the individual parts of the calculation; in particular, it lists the data for plotting the Wilson line, and thus checking on the linearity of the plot. It should be remembered that, for reliable statistics, the portion of reciprocal space under consideration should include all reflections other than systematic absences. Accidental absences, that is, reflections with intensities too low to be recorded, should be included at values 0.55 of the localized minimum $|F_o|$, that is, the minimum $|F_o|$ in a given range, for a centric distribution, and 0.66 of the localized minimum $|F_o|$ for an acentric distribution. Some of the data sets do not have the accidental absences included. The set NO2G.TXT is, however, complete in this respect: a check on this aspect of a data set is given by inspecting the number of reflections in each of the ranges of the Wilson plot routine; they should be approximately equal.

11.4.10 $|E|$ Values Calculated from the Structure

This link has been added for interest to show how $|E|$-values are calculated from a crystal structure. It follows that all atoms, preferably including hydrogen atoms, must be present in their correct locations. Then, $|E(hk)|$ is calculated from

$$|E(hk)| = \frac{1}{\sqrt{\varepsilon_{hk}\sigma_2}}(A_Z^2 + B_Z^2)^{1/2} \qquad (11.5)$$

where ε_{hk} is the epsilon factor for the hk reflection, σ_2 is given by

$$\sigma_2 = \sum_j Z_j^2 \tag{11.6}$$

and A_Z and B_Z are given by

$$A_Z = \sum_j Z_j \cos 2\pi(hx_j + ky_j) \quad B_Z = \sum_j Z_j \sin 2\pi(hx_j + ky_j) \tag{11.7}$$

and the sums are taken over all j atoms in a complete unit cell. It may be noted that no temperature factor is involved in the definition of $|E|$, because for a "point atom" $f_j = Z_j$ for all h, k. The phase associated with $|E|$ is given by $\tan^{-1}(B_Z/A_Z)$, having due regard to the sign of both A_Z and B_Z. It follows that the term $E(00)$ is given by $(\sum_j Z_j)/(\sum_j Z_j^2)^{1/2}$ which, for identical atoms, is $\sqrt{N}$, where N is the total number of atoms in the unit cell; the value of $E(00)$ is listed, with $|E|$ statistics in the results files EVALS.TXT and ECALC.TXT.

11.5   Crystal Structure Analysis Problems

As we indicated at the end of Chapter 8, problems additional to those listed there are given here, and extend the work of that chapter. These problems have been devised in conjunction with the XRAY* program system. The different data sets may not all operate equally well with all methods of structure solving provided by the system. Hence, although the operations available are indicated on the monitor screen during execution of the program, we suggest here those procedures by which good results may be obtained for each data set provided. Organic species mostly have been chosen because it is easier to find examples that show well resolved and interpretable projections than with inorganic species, except for simple structures like NaCl or CaF_2.

There are certain features associated with working in two dimensions that we should remember.

1. Because of the relatively small amount of data and a certain degree of inclination of the molecule to the plane of projection, some bond lengths and angles will not calculate to typical values.
2. Fourier maps will not necessarily be true to scale, and will not present the β-angle in oblique projections. However, the representations will be satisfactory insofar as they give good practice with the structure determining methods, and enable atoms to be located. It may be noted also that when the axis of projection is not perpendicular to the plane of projection, the true axes of the projection should be modified by an angular term. For example, for a monoclinic unit cell projected on to (100), the axes are b

and $(c \sin \beta)$. The $\sin \beta$ term may be important where the coordinates are measured from a map and when the β-angle is very different from $90°$. However, since coordinates are almost always refined by least squares, the correction may often be ignored.

3. It is not appropriate to be concerned with placing hydrogen atoms in structures determined from projections.

Except for the first structure, answers are not given. The correctness of the results should be judged according to the criteria of correctness already discussed.

11.5.1 Ni *o*-Phenanthroline Complex (NIS2)

In the directory containing the programs and data, double-click on the XRAY* icon. Enlarge the DOS window to a convenient size by clicking on the square box ☐ in the top right-hand corner of the screen. Press twice the Enter key after reading the information on the screen. Enter the data file name NIS2 and then a name for the general output file, say NISOUT.

The compound crystallizes in space group $P2_12_12_1$, and the plane group of the projection is $p2gg$. Open the menu, and select the Patterson link. Carry out Patterson and sharpened Patterson syntheses, and plot the maps on the screen. Use the print facility for at least one of the maps, so as to make for easy comparison with the other. In $p2gg$, the general equivalent positions are

$$\pm\left(x, y; \tfrac{1}{2} + x, \tfrac{1}{2} - y\right).$$

So that interatomic vectors will have the Patterson coordinates

$$\pm\left(2x, 2y; \tfrac{1}{2}, \tfrac{1}{2} - 2y; \tfrac{1}{2} - 2x, \tfrac{1}{2}\right).$$

On the screen, along the lines $x = \tfrac{1}{2}$ and $y = \tfrac{1}{2}$ there are two large peaks that may be taken as Ni–Ni vectors; they are double-weight (why?). From them we obtain the atomic coordinates as *ca* 0.24, 0.18; the peak corresponding to the $2x, 2y$ vector is not well developed in this projection. Other peaks indicate possible Ni–S vectors, but the results may not be completely satisfactory. It may be useful to keep copies of the Patterson maps for later reference.

Use the coordinates of the nickel atom in the asymmetric unit to calculate structure factors and then an electron density map. This map shows the Ni atom position, plus two other strong peaks that can be taken as the two S atom positions. Repeat the structure factor and electron density calculations with these three atoms, or first apply a least-squares refinement. The electron density map may not be easy to interpret in terms of all light atoms. If necessary, print the asymmetric unit of the Fourier map and contour it carefully; the lowest contour on these plots is 10 and the maximum is 100. It may help in this example to contour the figure field at the relative level 5. Search for peaks that would make up the picture of

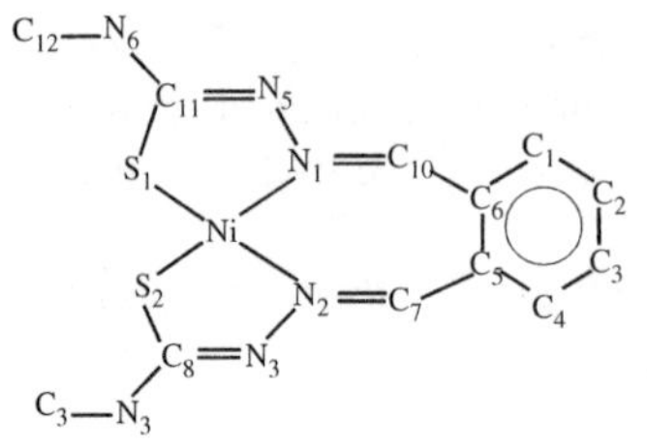

FIGURE 11.3. The molecular structure of $C_{12}H_{14}N_6S_2Ni$; the 14 hydrogen atoms are implicit.

the phenanthroline complex (see Figure 11.3 for the skeleton of the molecular structure). It may not be possible to find all the remaining atomic positions at this stage, but enough will be located to enable a better electron density map to be calculated.

When all 21 atoms, excluding hydrogen, have been found, several cycles of least-squares refinement will converge with an R-factor of about 9.8%. This is probably the best result that can be obtained with this data set. The difference electron density map at this stage will be almost featureless. A small, negative peak near the location of the nickel atom may indicate that the isotropic temperature factor is not a completely satisfactory approximation for this species, or that there are insufficient terms for true convergence of the Fourier series (see Sections 3.9.6 and 5.10.1).

Bond lengths and angles may be calculated. Because the c dimension is only 4.77 Å, the molecule is quite well resolved in this projection, and the lengths and angles should have fairly sensible chemical values. The results for the nickel and sulfur atoms are Ni–S(1) $= 1.953$ Å, Ni–S(2) $= 1.917$ Å, and S(1)–Ni–S(2) $= 95.08°$; small variations from these values may reflect the state of the refinement.

11.5.2   2-Amino-4,6-dichloropyrimidine (CL1P)

We shall consider the remaining problem examples in less detail. The title compound crystallizes in space group $P2_1/a$ with four molecules in the unit cell. The plane group of the projection on (010) is $p2$, doubled along the x-axis because of the translation of the a-glide plane. The data for this structure and for CL2P produce satisfactory Wilson plots.

There are two chlorine atoms in the asymmetric unit, not related by symmetry. Hence, these two chlorine atoms together with the two related by the twofold symmetry will give rise to eight non-origin Cl–Cl vectors:

Type 1, single weight for each : $\pm(2x_1, 2y_1)$ $\pm(2x_2, 2y_2)$

Type 2, double weight for each : $\pm(x_2 - x_1, y_2 - y_1)$ $\pm(x_2 + x_1, y_2 + y_1)$

where the single-weight vectors terminate at the corners of a parallelogram, and the double-weight vectors terminate at the mid-points of its sides.

Solve the Patterson projection for positions of the chlorine atoms, and then complete the structure determination for the non-hydrogen atoms. It will be helpful to print more than one copy of the Patterson map, and then to join them such that the origin is at the center of the composite. (Hint: the coordinates of one of the chlorine atoms are ca 0.16, 0.16.)

The data for this projection will refine to ca 11.1%. The bond lengths and angles from this projection indicate a tilt of the molecule out of the plane of projection.

11.5.3 2-Amino-4-methyl-6-dichloropyrimidine (CL2P)

Consider the unit cell data for this compound and that for the dichloro-pyrimidine just studied:

	2-Amino-4-methyl-6-chloropyrimidine	2-Amino-4,6-dichloropyrimidine
$a/\text{Å}$	16.426	16.447
$b/\text{Å}$	4.000	3.845
$c/\text{Å}$	10.313	10.283
$\beta/\deg$	109.13	107.97
Z	4	4
Space group	$P2_1/a$	$P2_1/a$

The two sets of unit cell data are sufficiently similar for the two pyrimidine derivatives to be treated as isomorphous. Hence, it should be possible to allocate trial atomic coordinates from the structure of the dichloropyrimidine; one of the chlorine atoms in the dichloro-compound has been replaced here by a methyl group.

Calculate Patterson maps for this compound and, by comparison with the previous structure, obtain atomic coordinates for a trial structure of this compound. Then refine the trial structure by successive Fourier syntheses and least squares.

This data set refines to an R-value of approximately 11.3%, but some of the bond lengths differ from the accepted values for this compound.

11.5.4 *m*-Tolidine Dihydrochloride (MTOL)

m-Tolidine dihydrochloride, crystallizes in space group $I2$, a non-standard setting of $C2$. Since there are two molecules per unit cell, the molecules occupy special positions on 2-fold axes. In the projection, on (010), the plane group is $c2$. Thus, the chlorine atoms are related by 2-fold symmetry to give a Patterson vector at $2x, 2z$. The projection can be solved by the heavy-atom method, and refined to ca 22% with the given data. The atoms are well resolved, albeit with some distortion arising from the disparity in the two cell dimensions for the projection. The bond lengths are at variance with standard values, because no account can be taken of the third dimension in their calculation from this projection.

11.5.5 Nitroguanidine (NO2G)

Nitroguanidine, $C(NH_2)_2NNO_2$, crystallizes in space group $Fdd2$, with the unit cell dimensions $a = 17.639, b = 24.873, c = 3.5903$ Å, and 16 molecules per unit cell. The small c dimension, approximately equal to the van der Waals non-bonded distance between light atoms, means that good resolution will arise in the projection on (001). The plane group of this projection is $p2gg$, with the a and b dimensions halved and four molecules in the transformed unit cell.

The data comprises the $|F(hk)|$ reflections and other data appropriate to the structure solving process in the projection. This structure is suitable for the direct methods procedure. In two dimensions, two reflections suffice to fix the origin provided they are chosen one from any two of the parity groups h even/k odd, h odd/k even, h odd/k odd; h even k even corresponds to a structure invariant and cannot be used in specifying the origin.

In $p2gg$, the sign relationships in reciprocal space may be summarized as $s(hk) = s(\bar{h}\bar{k}) = (-1)^{h+k} s(\bar{h}k)$, so that both positive and negative signs will be generated by the $\sum_2$ equation. One or more letters may be used as necessary, in order to aid the sign allocation process. In some cases, the signs attaching to such letters evolve during the procedure; otherwise, trial E-maps must be constructed with permuted values for the letter signs. This structure will refine to an R value of approximately 5%.

As an alternative procedure at the E-map stage, the data can be assembled as lines of $h, k, E, s(E)$, and used in conjunction with the program FOUR2D*, which has been written for plane group $p2gg$.

11.5.6 Bis(6-sulfanyloxy-1,3,5-triazin-2(1H)-one) (COSH)

This compound, $(C_3H_2N_3O_2S)_2$, crystallizes in space group $P2_1/m$, with two molecules per unit cell, so that the molecules occupy special positions on m-planes. In projection on (010), however, the plane group is $p2$ and the molecules then occupy general positions in the plane group.

The Patterson maps indicate more than one peak of similar height in the asymmetric unit, so that it may be necessary to investigate both of them in order to find a good trial structure. This structure responds well to the superposition technique, so that the solution is not as difficult as might have been expected. Refinement to ca 15.5% can be achieved.

11.5.7 2-S-methylthiouracil (SMTU)

2-S-methylthiouracil is triclinic, space group $P\bar{1}$, and $Z = 2$. The data supplied are appropriate to the (100) projection, for which the plane group is $p2$. This structure may present more difficulty than that in the previous example, because several peaks of similar height occur in the Patterson maps. The correct choice refines here to ca 13.0%; with three-dimensional data, $R = 7.3\%$ has been reported in the literature. The molecule is clearly indicated, but not all atoms, particularly the carbon attached to sulfur, are well resolved in this projection. Note that, for the (100) projection, the axes marked x and y on the plot are, strictly, $y \sin \gamma$ and $z \sin \beta$, respectively

This selection of problems provides good practice in the current, basic structure solving methods. Other problems can be built up as desired; data for suitable structures can be found in the early volumes of *Acta Crystallographica*.

11.6   General Crystal Structure and Other Programs

11.6.1   One-Dimensional Fourier Summation (FOUR1D*)

This program calculates and plots a one-dimensional Fourier summation, $\rho(x)$. The data comprise lines of index h and coefficients $A(h)$ and $B(h)$, and must be available in a file named ABDAT.TXT; the printer should be on-line for the plot of the function. If $\rho(x)$ is centrosymmetric, $B(h)$ should be entered as zero for each data line. The figure field for $\rho(x)$ appears on the screen, is established in the file RHOX.TXT and plotted, normalized to a maximum value of 50 in each case, while the file FUNCTN.TXT contains the true values of the data, *in a form suitable for the Fourier transform program* TRANS1*. The desired interval of subdivision N is entered at the keyboard; its maximum value in the program is 100.

11.6.2   Two-Dimensional Fourier Summation (FOUR2D*)

This program computes a two-dimensional Fourier summation for plane group $p2gg$. The data must be in a file named TWODAT.TXT as lines of h, k (integer numbers), $|F_o|$, s (real numbers), where s is the sign (± 1) that multiplies $|F_o|$. The interval of subdivision is 40 along both the x and y axes. The printer should be on-line. The output may be joined along the duplicated lines $y = 20$, and the plot contoured. The maximum of the figure field is normalized to 100.

The source code FOUR2D.F90 for this program, written in FORTRAN 90, is also supplied. Those conversant with FORTRAN 90 may wish to modify this program for other plane groups. The procedure is straightforward,[a] making use of the electron density equations given in the International Tables for X-ray Crystallography.[b] Consider, for example, plane groups $p2$: we can write $\rho(xy)$ as

$$\rho(xy) = K\left(\mathbf{C}_2^\mathrm{T}\mathbf{F}\mathbf{C}_1 - \mathbf{S}_2^\mathrm{T}\mathbf{G}\mathbf{S}_1\right)$$

---

[a] M. F. C. Ladd and M. Davies, *Zeitschift fur Kristallographic* **126**, 210 (1968).
[b] *Loc. cit.*

where K is a constant involving the normalization of the output results, $\mathbf{C}_2^T$ and $\mathbf{S}_2^T$ are the transposes of matrices of $\cos(2\pi h n_X)$ and $\sin(2\pi h n_Y)$, of order $h_{max} \times n_X$ ($n_X = n_Y = 0/40, 1/40, 2/40\ldots$, as set currently in the program), and $\mathbf{C}_1$ and $\mathbf{S}_1$ are matrices of $\cos(2\pi k n_Y/b)$ and $\sin(2\pi k n_Y/b)$, of order $k_{max} \times n_Y$. In plane group $p2$, $\mathbf{F}$ and $\mathbf{G}$ are matrices of order $h_{max} \times k_{max}$, with elements $[F_o(hk) + F_o(h\bar{k})]$ and $[F_o(hk) - F_o(h\bar{k})]$, respectively. A step to form these elements would need to be inserted into the program.

11.6.3   One-dimensional Fourier Transform (TRANS1*)

This program calculates the Fourier transform of a one-dimensional function $f(x)$. The function is divided into an *even* number of intervals, up to a maximum of 100, and contained, one datum to a line, in the file FUNCTN.TXT; only the values of the function are used as data. The number of data (the interval N of subdivision of the function) is entered at the keyboard, followed by the maximum frequency h_{max} for the output coefficients. Because of sampling conditions (see Section 5.7), if N is chosen as 30, h_{max} could be conveniently 10–15. The output in the file ABDAT.TXT can then be used with FOUR1D* to recreate the original function, $f(x)$. Note that, because FOUR1D* writes a file named FUNCTN.TXT, the original values of this function will be lost unless saved in another file.

11.6.4   Reciprocal Unit Cell (RECIP*)

This program determines the parameters of the reciprocal unit cell from those of the corresponding direct space unit cell, or vice versa, and the volumes of both cells. The input consists of the reciprocal constant K (unity or an x-ray wavelength, as required) and the parameters a, b, c, α, β, and γ. No particular length unit is assumed, but the angles are in degrees. The output is self-explanatory.

11.6.5   Molecular Geometry (MOLGOM*)

This program calculates bond lengths, bond angles, and torsion angles. It requires the unit cell dimensions input at the keyboard, and crystallographic coordinates from a file named MOLDAT.TXT. Each line of this file contains a sequential atom number and the x, y, and z coordinates of that atom. As prompted the atoms forming a torsion angle are entered at the keyboard. The convention relating to the sign of the torsion angle has been discussed (see Section 7.5.2 and Appendix A7). The results are listed in the file METRIC.TXT. Note that a program error at the torsion angle stage means that either the data is incorrect or that a torsion angle cannot be defined by the order of the atoms given.

11.6.6   Internal and Cartesian Coordinates (INTXYZ*)

This program converts the geometry of a molecule in terms of its internal coordinates, that is, bond lengths, bond angles and torsion angles, to a set of Cartesian coordinates for the molecule. The data must be supplied from a file named CART.TXT, and take the form of lines of atom code number, bond angle, torsion angle, bond length, as indicated on the monitor screen (with results) after the program is opened. The convention for torsion angles given in Appendix A7 applies. The first entry is always 0 0.0 0.0 0.0. In subsequent lines, the code number of the current atom is the atom number of a previous atom *to which the current atom is linked*. As an example of input data, consider the molecular fragment shown here, with its internal coordinates, or geometry as given.

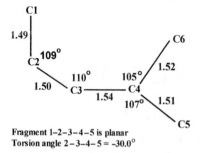

Fragment 1–2–3–4–5 is planar
Torsion angle 2–3–4–5 = –30.0°

Then the input data has the following format:

Atom number (not entered)	Input data for file CART.TXT			
1	0	0.	0.	0.
2	1	0.	0.	1.49
3	2	109.	0.	1.50
4	3	110.	0.	1.54
5	4	107.	180.	1.51
6	3	105.	–30.	1.52

Atom number 1 is at the origin, and the fragment 1–2–3–4–5 is planar. Does the 4–6 bond lie above or below this plane? The Cartesian coordinates given by the program are in the file GEOM.TXT as:

Atom	X	Y	Z
1	0.	0.	0.
2	–1.4900	0.	0.
3	–1.9784	1.4183	0.
4	–0.7815	2.3874	0.
5	–1.3472	3.7875	0.
6	–0.9042	2.2042	0.7341

11.6.7 Linear Least Squares (LSLI*)

This program determines the best-fit straight line to a series of data points, that must number at least 3 and, in this program, must not exceed 100. Data must be entered, in the order x_i, y_i, from the keyboard or from a data file; unit weights are assumed in this program. If the data are in a file, the file must be named LSSQ.TXT and its first two lines of the file are the *title* and the *number* of x, y data pairs that follow. For printer output, this device must be on-line.

It is assumed that errors in x are significantly smaller than those in y. The goodness-of-fit is reflected by the values of $\sigma(a), \sigma(b)$, and Pearsons's r coefficient. If the errors in the parameters a and b are to be propagated to another quantity z, then they follow the law

$$(\delta z)^2 = \sum_j (\delta z/\delta p_j)^2 (\delta p_j)^2 \tag{11.8}$$

where $z = f(p_j)(j = 1, 2, 3, \ldots n)$, and the sum extends over all n values of p_j; $(\delta z)^2$ may be regarded as the variance of the quantity z.

11.6.8 Matrix Operations (MAT3*)

This program accepts an input of two 3×3 matrices **A** and **B**, and forms $\mathbf{A} + \mathbf{B}, \mathbf{A} - \mathbf{B}, \mathbf{A} \times \mathbf{B}, \mathbf{A}^T, \mathbf{B}^T$, Trace(**A**), Trace(**B**), Det(**A**), Det(**B**), Cofactor(**A**), Cofactor(**B**), $\mathbf{A}^{-1}$, and $\mathbf{B}^{-1}$. If results are required on only one matrix, **A**, then **B** is entered as the unit matrix 1 0 0 /0 1 0 /0 0 1.

11.6.9 Q-Values (Q-VALS*)

This program is useful in conjunction with work on indexing powder diffraction patterns. Given the unit cell parameters a, b, c, α, β, and γ (in degrees), the program produces a set of values of $10^4 Q$; maximum values for h, k, and l are also entered at the keyboard. Provision is made for the indices k and l to take negative values. Thus, in any symmetry higher than triclinic, duplicate values of Q will be generated and may be discarded as required. The results are in the file INDEX.

11.6.10 Le Page—Unit Cell Reduction (LEPAGE*)

This program (see Appendix A10) was written by A. L. Spek of the University of Utrecht and kindly made available by him to the academic community. The program is entered by a double click on the LEPAGE* icon. For reduction, choose the D-option, and enter the unit cell parameters as indicated, one to a line. It may be desirable to vary the "2-axis criterion" by means of the C-option.

The program reports particularly the input unit cell, the reduced unit cell, and the conventional crystallographic unit cell, which may be the same as the reduced cell, and the transformation matrices for a, b, c and x, y, z. Other options are provided by the program, but they need not concern us here. Note that this program refers to the triclinic system by the symbol a: the alternative name for the triclinic system is *anorthic*.

A useful mnemonic for using a transformation matrix M and its inverse M^{-1} is indicated by the following diagram.

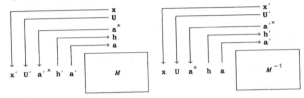

For example, if M is the matrix

	→a	
a'	⅔	⅓	⅓
	-⅓	⅓	⅓
	-⅓	-⅔	⅓

and **a** represents the triplet a, b, c, then writing $\mathbf{a}' = M\mathbf{a}$, we have

$$\mathbf{a}' = \tfrac{2}{3}\mathbf{a} + \tfrac{1}{3}\mathbf{b} + \tfrac{1}{3}\mathbf{c}$$
$$\mathbf{b}' = -\tfrac{1}{3}\mathbf{a} + \tfrac{1}{3}\mathbf{b} + \tfrac{1}{3}\mathbf{c}$$
$$\mathbf{c}' = -\tfrac{1}{3}\mathbf{a} - \tfrac{2}{3}\mathbf{b} + \tfrac{1}{3}\mathbf{c}$$

and if the inverse matrix M^{-1} is

	x'		
x	1	-1	0
	0	1	-1
	1	1	1

and $\mathbf{x}'$ represents the triplet x', y', z', then

$$x' = x + z$$
$$y' = -x + y + z$$
$$z' = -y + z$$

the arrow symbols $_{\mathbf{a}'}\upharpoonleft^{\mathbf{a}}$ should be interpreted as "$\mathbf{a}'$ in terms of $\mathbf{a}$," and so on. The second case is equivalent to writing $\mathbf{x}' = (M^{-1})^{\mathrm{T}}\mathbf{x}$, and then multiplying in the usual manner.

11.7   Automatic Powder Indexing (ITO12*)

The format of the input data is specific and must be followed. The data file must be named ITOINP.DAT, and set out as follows; the parentheses indicate FORTRAN formats.

Line 1 Title, up to 80 alphanumeric characters (A80).

Line 2 Leave blank; it is related to a number of parameters that take default values in the program, and which we need not discuss here.

Line 3 Four parameters: a *zeroshift*, 0.0 is recommended; a *print control* minimum value of M_{20} for a lattice to be printed, 4.0 is recommended; a *print control* minimum value of lines indexed for a lattice to be printed, 14.0 is recommended; the *number of data*, between 20 and 40. These four parameters are entered as real numbers, and terminate at character numbers 10, 20, 30, and 40 respectively, in the line (4F10.5).

Line 4 The data, $10^4 Q$ or $\sin^2 \theta$ or 2θ values (in ascending order), or d (in descending order), in n lines each containing eight values, the data ending at character numbers $10, 20, \ldots, 80$ in each line (8F10.5).

Line 5 Leave blank;

Line 6 The word END as its first three characters (A3).

An example input file ITOINP.DAT, which relates to Problem 9.6a, is provided with the program suite. The output file .DOC is generated by the program, and provides an echo of the input data file. The output file SUMRY contains the results of the indexing; it is mostly self-explanatory. However, the following column heads have meanings as follow:

NEWNR	New sequence number for the zone after evaluation (cp. *OLDNR*);
A	Provisional value of Q_A for the zone;
B	Provisional value of Q_B;
FMAAS	Provisional value of Q_F;
QUALITY	Measure of fit for the zone based on probability theory;
OLDNR	Old sequence number;
CNTR	Flag, equal to 0 for primitive zone or 1 for a centered zone;
NOBS & NCALC	Number of observed and calculated lines used in evaluating *QUALITY*;
ZERSHFT	Estimate of 2θ zero error for the zone.

The best results follow based on the first 20 indexed lines, and then those based on all lines in the data set, up to 40; finally the best, refined unit cells are printed. From a study of the observed reflections, deductions may be made about the space group of the crystal.

11.8 Automatic Powder Structure Solving (ESPOIR*)

We discussed briefly the method followed in this program in Chapter 9. Only the essentials needed to run a data set are included here, and the reader is referred to an original reference[a] for a fuller exposition of the features of this technique.

We follow through a sequence of instructions for using ESPOIR*. Because a large number of files can be generated in using this program, we have placed the program and data in a separate directory, named POWDER.

Below, we use C in the instructions to mean a single left-hand click on the mouse and DC to mean a double click, and we take first the example of aluminum oxide, α-Al_2O_3.

11.8.1 α-Alumina (Corundum)

In the folder POWDER there are three pairs of files for alumina, labeled Al_2O_3A, Al_2O_3B, and Al_2O_3C; for each of A, B, and C, one file contains crystal data and program settings, and the other contains the $h, k, l, |F_o|$ data. We consider first Al_2O_3A, in which the structure is treated from scratch, applying distance constraints for Al–Al of 3.0, Al–O of 1.6, and O–O of 1.6 Å, respectively. We know also the space group, $R\bar{3}c$, and that $Z = 6$. The results of two separate runs will be given, following the instruction below:

DC on the ESPOIR* program icon
C File; C Open File
DC on Al_2O_3A
C Run; C Espoir

A number of *tests* will now be executed, and one will (hopefully) have a low R value, indicating a good measure of fit. Assume test number 6 is the best fit. Close the "Progress View" window. Then:

C View: a number of options become available, but we shall be particularly concerned here with just two, Profile and Structure.

C Profile.

[a] M. Mileur and A. Le Bail, http://www.cristal.org/sdpd/espoir

This link shows the observed and calculated patterns (in red), and below them the difference pattern (in blue).

C View; C Structure.

The 10 test results are listed.

DC Al_2O_36

A line drawing of the structure according to the results from test 6 is shown.

C Display; C Ball & Stick
C Options; C Specular
C Options; C Labels

A ball and stick model with atoms labeled is displayed. Depress the right-hand mouse button to translate the image on the screen, and the left-hand button to rotate it. (C Display; C Spacefill also produces an instructive view.)

When a satisfactory view has been found:

C Export; C BMP

The selected view may then be placed in a chosen folder and printed at a later stage.

On returning to the original folder (POWDER), a number of new files relating to Al_2O_3A will be found. One of the files, "Al_2O_3Astru" (a .DAT file), contains structural data about the crystal and the view, together with a list of x, y, and z coordinates found from that run; the results for two such runs were:

		x	y	z
$A1$	Al	0.6667	0.3333	0.9812
	O	0.0000	0.6667	0.7500
$A2$	Al	0.3274	0.6725	0.8146
	O	0.6887	0.9995	0.2498

The two runs do not show apparently the same coordinates, because of the random nature of the process. We shall discuss these results shortly, but first we consider the other two data sets.

Next, carry out the procedure with Al_2O_3B. This data set is arranged to fit to the $|F_o|$ data rather than a regenerated pattern, and makes use of chosen values for the occupation numbers; the execution time is much shorter. One set of coordinates, the best fit, will be produced. Return to the data View Window, that is, C on View followed by C on Al_2O_3B.spf. Alternatively, return to the POWDER folder, and DC on $Al_2O_3 B$.SPF. The results of the two runs are listed below.

		x	y	z
B1	Al	0.3422	0.6702	0.8146
	O	0.6417	0.6716	0.9165
B2	Al	0.3414	0.6668	0.8146
	O	0.3424	0.9772	0.9159

Finally, repeat the second procedure now with Al_2O_3C. In this example, the constraints of the special positions of the type $0, 0, z$ for Al and $x, 0, \frac{1}{4}$ for O have been added to the data set. We obtain the coordinates and a plot of the best-fit structure. Two such runs are shown below:

		x	y	z
C1	Al	0.0000	0.0000	0.6470
	O	0.6940	0.0000	0.2500
C2	Al	0.0000	0.0000	0.6479
	O	0.3060	0.0000	0.2500

In order to interpret the totality of these results, we list the special positions for space group $R\bar{3}c$, with $Z = 6$, and the centers of symmetry in the unit cell:

$$\left(0,0,0; \tfrac{1}{3}, \tfrac{2}{3}, \tfrac{2}{3}; \tfrac{2}{3}, \tfrac{1}{3}, \tfrac{1}{3}\right)+$$

$$12 \text{ Al at } \pm\left(0,0,z; \ 0,0,\tfrac{1}{2}+z\right) \qquad 18 \text{ O at } \pm\left(x,0,\tfrac{1}{4}; \ 0,x,\tfrac{1}{4}; \ \bar{x},\bar{x},\tfrac{1}{4}\right)$$

$$\bar{1} \text{ at } \left(0,0,0; \ 0,0,\tfrac{1}{2}; \ 0,\tfrac{1}{2},0; \ \tfrac{1}{2},0,0; 0,\tfrac{1}{2},\tfrac{1}{2}; \ \tfrac{1}{2},0,\tfrac{1}{2}; \ \tfrac{1}{2},\tfrac{1}{2},0; \ \tfrac{1}{2},\tfrac{1}{2},\tfrac{1}{2}\right)$$

The program does not necessarily select all atoms from one and the same asymmetric unit, so that we have to consider the full implication of the space group symmetry and choice of origin. For example, *let us take results C2 as a norm.* Then, if O in C1 is moved across the center of symmetry at 1, 0, 0, that set then agrees with $C2$. In sets B, we add the translations $\tfrac{2}{3}, \tfrac{1}{3}, \tfrac{1}{3}$ in each case, which leads to $\approx 0, \approx 0, 0.1479 (\equiv 0.6479)$ and $\approx 0, \approx 0, 0.\bar{1}479 (\equiv 0.6479)$ for $B1$, and $\approx 0, \approx 0, 0.1479$ and $\approx 0, \approx 0, 0.2492$ for $B2$, both of which agree reasonably with sets C. In a similar way, sets A can be transformed to 0, 0, 0.6479 and 0, 0.3333, 0.2500, and $\approx 0, \approx 0, 0.1479 (\equiv 0.6479)$ and $0.3113, \approx 0, 0.2498$. Except for Al_2O_3C, where the constraints are strong, we would not always expect to get the same numerical values exactly in subsequent trials, because of the random nature of the movement of the atoms. Recently reported parameters for the α-alumina structure are:

	x	y	z
Al	0	0	0.6477
O	0.3064	0	$\frac{1}{4}$

11.8.2 Aragonite

As a second example, we consider the aragonite form of $CaCO_3$. The space group is *Pmcn*, a non-standard setting of *Pnma*. Since there are four formula entities per unit cell, the Ca and C atoms lie on special positions, but the oxygen atoms could occupy one set of general positions plus one set of special positions, or three sets of special positions. Both arrangements may need to be tried. We report here the successful choice, that is, with occupancies of $\frac{1}{2}$ for each of Ca, C, and O_1, and unit occupancy for O_2. A typical set of results is shown below, together with the appropriate symmetry operations of the space group, *Pmcn*:

	x	y	z		x	y	z
Ca	0.7552	0.9150	0.2602	$\bar{x}, \frac{1}{2}+y, \frac{1}{2}-z \longrightarrow$	0.2448	0.4150	0.2498
C	0.2498	0.7624	0.0859	$-x, y, z \longrightarrow$	0.2498	0.7624	0.0959
O_1	0.7553	0.0781	0.9074	$\bar{x}, \bar{y}, \bar{z} \longrightarrow$	0.2447	0.9219	0.0926
O_2	0.5289	0.3188	0.9123	$\bar{x}, \bar{y}, \bar{z} \longrightarrow$	0.4711	0.6812	0.0877

The known parameters for the Aragonite structure are

	x	y	z
Ca	$\frac{1}{4}$	0.4151	0.2405
C	$\frac{1}{4}$	0.7621	0.0852
O_1	$\frac{1}{4}$	0.9222	0.0956
O_2	0.4735	0.6807	0.0873

and since the structure model derived from the Monte Carlo program corresponds quite closely, it can be refined further by least squares.

11.8.3 2-*S*-Methylthiouracil

In the final practical example of this chapter, we look again at a structure given first above as an example with the XRAY* program. The following diagram shows the bond lengths and bond angles for this molecule, excluding those for the hydrogen atoms.

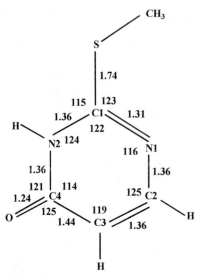

Bond lengths and bond angles for the 8-atom
planar skeleton of 2-S-methylthiouracil

Use the program INTXYZ* to obtain a set of Cartesian coordinates for the 8-atom planar fragment; the methyl group on S is not constrained to lie in the plane of the ring. Remember that the order of the atoms from the INTXYZ* results file will be the same as that of the input data. Insert these coordinates into the data set SMTU.DAT in the place of the row of asterisks there, and run ESPOIR*.

If the first set of runs does not contain a satisfactory trial structure, ESPOIR* should be run again. The output file .IMP contains a progress log, the final (best) set of coordinates for the run and the structure factor data. Take the final coordinates and extract the y, z values. Enter these values in the XRAY* system, using the SMTU.DAT in folder XRSYST data file, and call routine 3 so as to calculate $|F_c|$ data; note that the atom type numbers must match with the set-up of the SMTU.DAT file, that is, $C = 1, N = 2, O = 3$ and $S = 4$. Refine the data by least squares, routine 4 (it should converge at about 14.7%, as before) and obtain an electron density map. The electron density map may not refer to the same origin $(0, 0, 0)$ as that from the Patterson method.

A second set of data for this compound has been supplied in the directory XRSYST, for use with the XRAY* system, the file SMT2. These data apply to the x, z projection. Use the x and z coordinates from ESPOIR* to solve the x, z

projection; the refinement converges at $R \approx 9.4\%$. Hence, build up a three-dimensional picture of the SMTU crystal structure. It is convenient to use a scale drawing of the unit cell of the projection, say on the y, z plane, and plot the atom positions as different sized circles with the integral value of $100x$ adjacent to each atom.

We can see that the methyl carbon was not clearly resolved on the y, z projection because the S–CH$_3$ bond lies close to the direction of the x axis. Note also that the superposition of the ring atoms of the two molecules (because the center of the ring lies very close to a 2-fold rotation point) makes the exact location of these atoms difficult in this projection.

11.8.4   1-Methylfluorene

The Monte Carlo structure determination of 1-methylfluorene ($C_{14}H_{12}$) from powder data[a] gave inter alia the following results: $a = 14.2973, b = 5.7011, c = 12.3733$ Å, $b = 95.106°$, $P2_1/n$ (non-standard setting of $P2_1/c$), $Z = 4$, and the following bond lengths in Å:

C1–C2	1.398	C10–C9	1.400
C2–C3	1.439	C10–C11	1.381
C3–C4	1.411	C11–C12	1.513
C4–C5	1.386	C12–C13	1.492
C5–C6	1.486	C13–C1	1.392
C6–C7	1.387	C13–C5	1.409
C7–C8	1.408	C11–C6	1.397
C8–C9	1.401	C14–C1	1.507

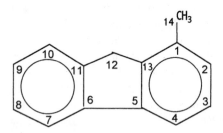

1-Methylfluorene, excluding the fluorenyl group hydrogen atoms

The data for ESPOIR* consists of the .HKL file of intensity data and the .DAT file of crystal data. The crystal data has been set up as a planar fluorenyl group (C_{13}) group and a methyl carbon. Although the fourteenth carbon atom could have

[a] M. Tremayne, B. M. Kariuki, and K. D. M. Harris, *Journal of Material Chemistry* **6**, 1601 (1996).

been fitted to the larger search fragment, it is functionally more efficient to leave it floating, because the fluorenyl group can fit the correct structure in two equivalent orientations.

Use the data supplied to solve the structure of 1-methylfluorene with ESPOIR*. If you consider that you have obtained a good fit, check it by calculating the bond lengths: set up the results in a MOLDAT.TXT file, as required by the program MOLGOM*. Compare the results from MOLGOM* with the bond lengths listed above for the correct structure.

Appendices

A1 Stereoviews and Crystal Models

A1.1 Stereoviews

Stereoviews of crystal structures began to be used to illustrate three-dimensional structures in 1926. Nowadays, this technique is quite commonplace, and computer programs exist that prepare the two views needed for producing a three-dimensional image of a crystal or molecular structure.

Two diagrams of a given object are necessary to form a three-dimensional visual image, and they must correspond to the views seen by the eyes in normal vision. Correct viewing of a stereoscopic diagram requires that each eye sees only the appropriate half of the complete illustration, and there are two ways in which it may be accomplished.

The simplest procedure is with a *stereoviewer*. A supplier of a stereoviewer that is relatively inexpensive is the Taylor-Merchant Corporation, 212 West 35th Street, New York, NY 10001, USA. (Tel: INT + 212 757 7700). The pair of drawings is viewed directly with the stereoviewer, whereupon the three-dimensional image appears centrally between the two given diagrams.

An alternative procedure involves training the unaided eyes to defocus, so that each eye sees only the appropriate diagram. The eyes must be relaxed and look straight ahead. This process may be aided by holding a white card edgeways between the two drawings. It may be helpful to close the eyes for a moment, then to open them wide and allow them to relax without consciously focusing on the diagram.

Finally, we give instructions whereby a simple stereoviewer can be constructed with ease. A pair of planoconvex or biconvex lenses each of the same focal length, approximately 100 mm, and diameter approximately 30 mm, is mounted between two opaque cards such that the centres of the lenses are approximately 63 mm apart. The card frame must be so shaped that the lenses may be brought close to the eyes. Figure A1.1 illustrates the construction of the stereoviewer.

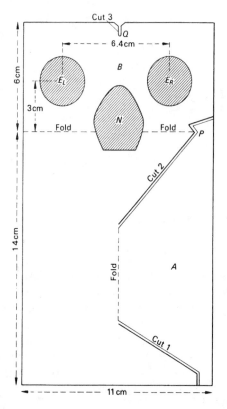

FIGURE A1.1. Simple stereoviewer. Cut out two pieces of card as shown, and discard the shaded portions. Make cuts along the double lines. Glue the two cards together with lenses E_L and E_R in position, fold the portions A and B backward, and fix the projection P into the cut at Q. Strengthen the fold with a strip of "Selotape." View from the side marked B. (This type of viewer is marketed by the Taylor-Merchant Corporation, New York.). It is helpful to obscure a segment on each lens of maximum depth *ca* 30% of the lens diameter, closest to the nose regions.

A1.2 Model of a Tetragonal Crystal

The crystal model illustrated in Figure 1.30 can be constructed easily. This particular model has been chosen because it exhibits a $\bar{4}$ axis, which is one of the more difficult symmetry elements to appreciate from drawings.

A good quality paper or thin card should be used for the model. The card should be marked out in accordance with Figure A1.2 and then cut out along the solid lines, discarding the shaded portions. Folds are made in the same sense along all dotted lines, the flaps *ADNP* and *CFLM* are glued internally, and the flap *EFHJ* is glued externally. What is the point group of the resulting model?

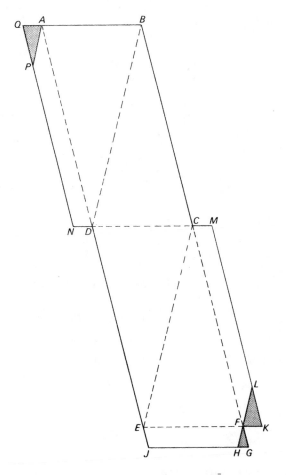

FIGURE A1.2.   Construction of a tetragonal crystal of point group $\bar{4}2m$:
$NQ = AD = BD = BC = DE = CE = CF = KM = 100$ mm;
$AB = CD = EF = GJ = 50$ mm;
$AP = PQ = FL = KL = 20$ mm;
$AQ = DN = CM = FK = FG = FH = EJ = 10$ mm.

A2   Analytical Geometry of Direction Cosines

A2.1   Direction Cosines of a Line

In Figure A2.1, let P_1 be any point x_1, y_1, z_1 referred to x, y, and z axes.
Draw lines from P_1 perpendicular to the x, y, and z axes to cut them at A, B, and
C, respectively. Thus, $OA = x_1, OB = y_1$, and $OC = z_1$. The direction cosines

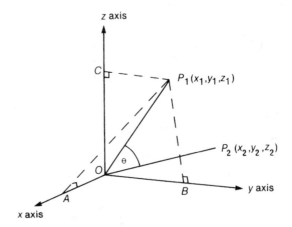

FIGURE A2.1.   Direction cosines of a line referred to the rectangular axes x, y and z.

of OP_1 are given by $\cos \chi_1 = x_1/OP_1$, $\cos \psi_1 = y_1/OP_1$, and $\cos \omega_1 = z_1/OP_1$. Hence,

$$\cos^2 \chi_1 + \cos^2 \psi_1 + \cos^2 \omega_1 = (x_1^2 + y_1^2 + z_1^2)/OP_1^2 \qquad (A2.1)$$

In the case that the x, y, and z axes are orthogonal then, since x_1, y_1, and z_1 are the projections of OP_1 on to the x, y, and z axes, it follows that

$$x_1^2 + y_1^2 + z_1^2 = OP_1^2 \qquad (A2.2)$$

Hence,

$$\cos^2 \chi_1 + \cos^2 \psi_1 + \cos^2 \omega_1 = 1 \qquad (A2.3)$$

A2.2   Angle between Two Lines

Given orthogonal axes and another line OP_2, let a point x_2, y_2, z_2 be marked off such that the lengths of OP_1 and OP_2 are equal, say r. We have for OP_1, from above,

$$x_1 = r \cos \chi_1, \quad y_1 = r \cos \psi_1, \quad \text{and} \quad z_1 = r \cos \omega_1$$

and for OP_2,

$$x_2 = r \cos \chi_2, \quad y_2 = r \cos \psi_2, \quad \text{and} \quad z_2 = r \cos \omega_2$$

where $\cos \chi_2$, $\cos \psi_2$, and $\cos \omega_2$ are the direction cosines of the line OP_2.

If the origin is shifted from O to P_1, then the coordinates of P_2 become

$$x_2' = x_2 - x_1, \quad y_2' = y_2 - y_1, \quad z_2' = z_2 - z_1 \qquad (A2.4)$$

so that the length $P_1 P_2$ is given by

$$(P_1 P_2)^2 = (x_2 - x_1)^2 + (y_2 - y_1)^2 + (z_2 + z_1)^2$$
$$= r^2 (\cos^2 \chi_1 + \cos^2 \chi_2 - 2 \cos \chi_1 \cos \chi_2$$
$$+ \cos^2 \psi_1 + \cos^2 \psi_2 - 2 \cos \psi_1 \cos \psi_2$$
$$+ \cos^2 \omega_1 + \cos^2 \omega_2 - 2 \cos \omega_1 \cos \omega_2) \qquad (A2.5)$$

Using (A2.3), we have

$$(P_1 P_2)^2 = 2r^2 [1 - (\cos \chi_1 \cos \chi_2 + \cos \psi_1 \cos \psi_2 + \cos \omega_1 \cos \omega_2)] \quad (A2.6)$$

In the isosceles triangle $OP_1 P_2$

$$P_1 P_2 / 2 = r \sin(\theta/2) \qquad (A2.7)$$

Therefore,

$$(P_1 P_2)^2 / 2r^2 = \sin^2(\theta/2) = 1 - \cos \theta \qquad (A2.8)$$

Comparing (A2.6) and (A2.8), we obtain

$$\cos \theta = \cos \chi_1 \cos \chi_2 + \cos \psi_1 \cos \psi_2 + \cos \omega_1 \cos \omega_2 \qquad (A2.9)$$

The same result can be achieved by following the general vector method in Section 7.5.1.

A3   Schönflies' Symmetry Notation

Theoretical chemists and spectroscopists use the Schönflies notation for describing point-group symmetry but, although both the crystallographic (Hermann–Mauguin) and Schönflies notations are adequate for point groups, only the Hermann–Mauguin system is satisfactory for space groups.

The Schönflies notation uses the rotation axis and mirror plane symmetry elements with which we are now familiar, but introduces the alternating axis of symmetry in place of the roto-inversion axis.

A3.1   Alternating Axis of Symmetry

A crystal is said to have an alternating axis of symmetry S_n of degree n, if it can be brought from one state to another indistinguishable state by the operation of rotation through $(360/n)$ degrees and reflection across a plane normal to that axis,

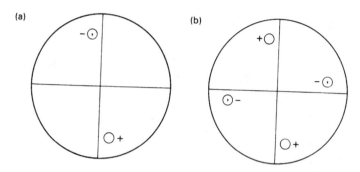

FIGURE A3.1. Stereograms of point groups: (a) S_2, (b) S_4.

overall a *single* symmetry operation. It must be stressed that this plane is *not* necessarily a mirror plane *in the point* group.[a] Operations S_n are nonperformable (see Section 1.4.1). Figure A3.1 shows stereograms of S_2 and S_4; we recognize them as $\bar{1}$ and $\bar{4}$,[b] respectively. The reader should consider what point groups are obtained if, additionally, the plane of the diagram *is* a mirror plane of the point group.

A3.2   Notation

Rotation axes are symbolized by C_n, where n takes the meaning of R in the Hermann–Mauguin system. Mirror planes are indicated by subscripts v, d, and h; v and d refer to mirror planes containing the principal axis, and h indicates a mirror plane normal to that axis. In addition, d refers to those vertical planes that are set diagonally, that is, between the crystallographic axes. The symbol D_n is introduced for point groups in which there are n 2-fold axes in a plane normal to the principal axis of degree n. The cubic point groups are represented through the special symbols T and O. Table A3.1 compares the Schönflies and Hermann–Mauguin symmetry notations.

A4   Rotation Matrices

We derive first a matrix for the rotation of a point X, Y, Z by an angle ϕ about an axis normal to the plane of x and y, with an angle γ between the x and y axes. Since the rotation axis is normal to the plane, the z coordinate of the point is unchanged.

---

[a] The usual Schönflies symbol for $\bar{6}$ is $C_{3h}(3/m)$. The reason that $3/m$ is not used in the Hermann–Mauguin system is that point groups containing the element $\bar{6}$ describe crystals that belong to the hexagonal system rather than to the trigonal system; $\bar{6}$ cannot operate on a rhombohedral lattice.

[b] Note that, among $\bar{R}$, $\bar{4}$ (S_4) is unique, in that it is not equivalent to any other symmetry element or combination of symmetry elements.

TABLE A3.1.   Schönflies and Hermann–Mauguin
Point-Group Symbols

Schönflies	Hermann–Mauguin[a]	Schönflies	Hermann–Mauguin[a]
C_1	1	D_4	422
C_2	2	D_6	622
C_3	3	D_{2h}	mmm
C_4	4	D_{3h}	$\bar{6}m2$
C_6	6		
C_i, S_2	$\bar{1}$	D_{4h}	$\dfrac{4}{m}mm$
C_s, S_1	$m(\bar{2})$		
S_6	$\bar{3}$		
S_4	$\bar{4}$	D_{6h}	$\dfrac{6}{m}mm$
C_{3h}, S_3	$\bar{6}$		
C_{2h}	$2/m$	D_{2d}	$\bar{4}2m$
C_{4h}	$4/m$	D_{3d}	$\bar{3}m$
C_{6h}	$6/m$	T	23
C_{2v}	$mm2$	T_h	$m3$
C_{3v}	$3m$	O	432
C_{4v}	$4mm$	T_d	$\bar{4}3m$
C_{6v}	$6mm$	O_h	$m3m$
D_2	222	$C_{\infty v}$	∞
D_3	32	$D_{\infty h}$	$\infty/m(\bar{\infty})$

[a] $2/m$ is an acceptable way of writing $\frac{2}{m}$, but $4/mmm$ is not as satisfactory as $\frac{4}{m}mm$; occasionally, we have written, for convenience, $4/m\,mm$.

In Figure A4.1, $\angle PQN = \pi - \gamma$, and $\angle QPN = \gamma - \pi/2$. Then, we have the ensuing analysis:

$$X = r\cos\theta - y\cos\gamma \quad Y = r\sin\theta/\sin\gamma \tag{A4.1}$$

It follows that

$$X' = r\cos(\theta + \phi)\sin\phi - Y'\cos\gamma \quad Y' = r\sin^{-1}\gamma\sin(\theta + \phi) \tag{A4.2}$$

Expanding (A4.2), substituting for $r\sin\theta$ and $r\cos\theta$ from (A4.1), and rearranging leads to

$$X' = X(\cos\phi - \cos\gamma\sin\phi/\sin\gamma) - Y(\sin\gamma\sin\phi + \cos^2\gamma\sin\phi/\sin\gamma) \tag{A4.3}$$

$$Y' = X(\sin\phi/\sin\gamma) + Y(\cos\phi) + \cos\gamma\sin\phi/\sin\gamma) \tag{A4.4}$$

Thus, we can write, in the usual notation,

$$\mathbf{X}' = \mathbf{S} \cdot \mathbf{X} \tag{A4.5}$$

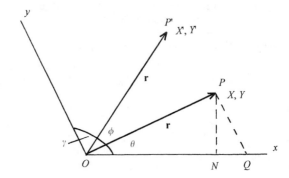

FIGURE A4.1.   Vector OP of length $|r|$ at a general angle θ to the x axis; OP' is the same vector after being rotated by an angle ϕ from OP. The general angle between the x and y axes is γ; for 3-fold and 6-fold symmetry, it has the value $120°$.

where the matrix **S** is given by

$$\mathbf{S} = \begin{bmatrix} (\cos\phi - \cos\gamma\,\sin\phi/\sin\gamma) & -(\sin\gamma\,\sin\phi + \cos^2\gamma\,\sin\phi/\sin\gamma) & 0 \\ (\sin\phi/\sin\gamma) & (\cos\phi + \cos\gamma\,\sin\phi/\sin\gamma) & 0 \\ 0 & 0 & 1 \end{bmatrix}$$
(A4.6)

For $\gamma = 120°$ and $\phi = 60°$, as in the hexagonal system, for example,

$$\mathbf{S} = \begin{bmatrix} 1 & \bar{1} & 0 \\ 1 & 0 & 0 \\ 0 & 0 & 1 \end{bmatrix}$$
(A4.7)

so that, from (A4.5), X, Y, Z rotates to $X - Y, X, Z$.

Matrix (A4.6) will suffice for all rotational operations that we meet in studying point groups except 3-fold rotation in the cubic system. However, we can see from the stereogram for point group 432 (Figure 1.36) that a 4-fold anticlockwise rotation about the x axis followed by a similar rotation about the z axis is equivalent to a 3-fold anticlockwise rotation about the direction [111].

Thus, from (A4.6) for a 4-fold rotation $\mathbf{R}_z$ about the z axis in the cubic (or tetragonal) system, $\phi = \gamma = 90°$ so that we obtain

$$\mathbf{R}_z = \begin{bmatrix} 0 & \bar{1} & 0 \\ 1 & 0 & 0 \\ 0 & 0 & 1 \end{bmatrix}$$

Similarly, for $\mathbf{R}_x$ we find

$$\mathbf{R}_x = \begin{bmatrix} 1 & 0 & 0 \\ 0 & 0 & \bar{1} \\ 0 & 1 & 0 \end{bmatrix}$$

Hence, $\mathbf{R}_{[111]} = \mathbf{R}_z \cdot \mathbf{R}_x$, that is,

$$\mathbf{R}_{[111]} = \begin{bmatrix} 0 & 0 & 1 \\ 1 & 0 & 0 \\ 0 & 1 & 0 \end{bmatrix} \qquad\qquad (A4.8)$$

A5 Spherical Trigonometry

A5.1 Spherical Triangle

Figure A5.1 shows a spherical triangle ABC formed on the surface of a sphere by the intersections of great circles of which AB, AC, and BC form parts. The arcs a, b, and c are the *sides* of the triangle, and A, B, and C are its *angles*; side a is measured by the angle $\angle BOC$, and the angle A by the angle between the tangents at A to the arc AB and AC, that is, the plane angle $\angle PAQ$. Similar definitions apply to the other four elements of the triangle. The following equations apply to

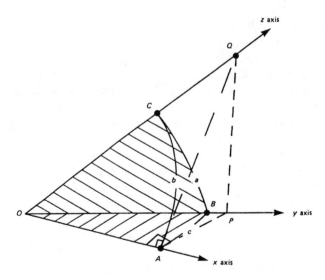

FIGURE A5.1.   Spherical triangle ABC formed by the intersections of three great circles on a sphere of centre O; AP and AQ are tangents to the sphere at A.

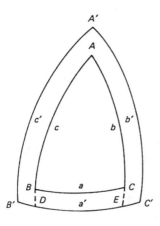

FIGURE A5.2. Spherical triangle ABC (see Figure A5.1) and its polar triangle $A'B'C'$: all points on the arc $B'C'$ are 90° away from A, the pole of the great circle represented by $B'C'$; similar relations apply to poles B and C.

spherical triangles, and are easily proved with the aid of Figure A5.1:

$$\cos a = \cos b \cos c + \sin b \sin c \cos A \qquad (A5.1)$$

$$\sin A / \sin a = \sin B / \sin b = \sin C / \sin c \qquad (A5.2)$$

Other examples of (A5.1) can be set down by cyclic permutation.

A5.2 Polar Triangle

The polar triangle and equations derived from it become important for the solution of triangles when applying Euler's theorem on the combination of rotations to the elucidation of point groups (see Section 1.4.2).

In Figure A5.2, ABC is again a spherical triangle. The arc $B'C'$ is drawn such that all points on it are 90° away from A; thus, A is the *pole* of the great circle of which $B'C'$ is an arc. Similarly, B and C are the poles of the arcs $A'C'$ and $A'B'$ respectively: $A'B'C'$ is defined as the *polar* triangle of triangle ABC. By reciprocity, ABC is the polar triangle of triangle $A'B'C'$.

The arcs AB and AC are produced to cut $B'C'$ in D and E, respectively. Since A is the pole of DE, DE is a measure of the angle A. But $B'E + C'D = B'C' + DE$ and, because B' and C' are the poles of CE and BD respectively, $B'E = C'D = 90°$. Thus, $B'C' + DE = a' + A = 180°$, so that

$$a = 180° - A' \qquad (A5.3)$$

and, since triangles ABC and $A'B'C'$ are polar to each other,

$$a' = 180° - A \qquad (A5.4)$$

with similar relationships for b, c, b', and c'. Using these relationships with (A5.1), we obtain

$$\cos a = (\cos A + \cos B \cos C)/(\sin B \sin C) \qquad (A5.5)$$

with similar results for the other four elements by cyclic permutation.

A6   Trigonometrical Formulae

The following formulae are often useful in manipulating the structure factor equation so as to obtain the geometrical structure factor equation.

$$\cos A + \cos B = 2\cos(A+B)/2 \ \cos(A-B)/2 \qquad (A6.1)$$

$$\cos A - \cos B = -2\sin(A+B)/2 \ \sin(A-B)/2 \qquad (A6.2)$$

$$\sin A + \sin B = 2\sin(A+B)/2 \ \cos(A-B)/2 \qquad (A6.3)$$

$$\sin A - \sin B = 2\cos(A+B)/2 \ \sin(A-B)/2 \qquad (A6.4)$$

$$\cos(A+B) = \cos A \cos B - \sin A \sin B \qquad (A6.5)$$

$$\cos(A-B) = \cos A \cos B + \sin A \sin B \qquad (A6.6)$$

$$\sin(A+B) = \sin A \cos B + \cos A \sin B \qquad (A6.7)$$

$$\sin(A-B) = \sin A \cos B - \cos A \sin B \qquad (A6.8)$$

$$\sin 2A = 2\sin A \ \cos A \qquad (A6.9)$$

$$\cos 2A = \cos^2 A - \sin^2 A = 2\cos^2 A - 1 \equiv 1 - 2\sin^2 A \qquad (A6.10)$$

$$\sin^2(2\pi n/4) = 0 \quad \text{for } n \text{ even or 1 for } n \text{ odd} \qquad (A6.11)$$

$$\cos^2(2\pi n/4) = 1 \quad \text{for } n \text{ even or 0 for } n \text{ odd} \qquad (A6.12)$$

$\cos 2\pi(\theta + n/2)$ is crystallographically equivalent to

$$\cos 2\pi(\theta - n/2) \qquad (A6.13)$$

where n is an integer.

A7   Cartesian Coordinates

In calculations that lead to results in absolute measure, such as bond distance and angle calculations, location of hydrogen-atom positions and so on, it is desirable to convert the crystallographic fractional coordinates x, y, z, which are dimensionless, to Cartesian (orthogonal) coordinates X, Y, and Z, in Å or nm.

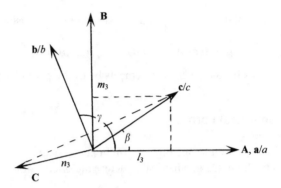

FIGURE A7.1. **A, B,** and **C** are unit vectors on Cartesian (orthogonal) axes X, Y, Z, and $\mathbf{a}/a, \mathbf{b}/b$, and $\mathbf{c}/c$ are unit vectors on conventional crystallographic axes x, y, z.

A7.1 Cartesian to Crystallographic Transformation and its Inverse

Instead of considering immediately the transformation $\mathfrak{A} = M\mathbf{a}$, it is simpler to consider first the inverse transformation $\mathbf{a} = M^{-1}\mathfrak{A}$, where M is the transformation matrix for the triplet $\mathfrak{A}(A, B, C)$ to the triplet $\mathbf{a}(a, b, c)$, because the components of $\mathbf{a}$ along the Cartesian axes are direction cosines (see Appendix A2).

Figure A7.1 illustrates the two sets of axes. Let A be a unit vector along $\mathbf{a}$, B a unit vector normal to $\mathbf{a}$ and in the a, b plane, and C a unit vector normal to both A and B. Then, we can write

$$\begin{bmatrix} \mathbf{a}/a \\ \mathbf{b}/b \\ \mathbf{c}/c \end{bmatrix} = \begin{bmatrix} l_1 & m_1 & n_1 \\ l_2 & m_2 & n_2 \\ l_3 & m_3 & n_3 \end{bmatrix} \begin{bmatrix} \mathbf{A} \\ \mathbf{B} \\ \mathbf{C} \end{bmatrix} \tag{A7.1}$$

From the figure, we can write down some of the elements of M^{-1}:

$$M^{-1} = \begin{bmatrix} 1 & 0 & 0 \\ \cos\gamma & \sin\gamma & 0 \\ \cos\beta & m_3 & n_3 \end{bmatrix} \tag{A7.2}$$

From the properties of direction cosines, we have

$$\cos\alpha = l_2 l_3 + m_2 m_3 + n_2 n_3$$
$$= \cos\beta \cos\gamma + m_3 \sin\gamma$$

so that

$$m_3 = (\cos\alpha - \cos\beta \cos\gamma)/\sin\gamma = -\cos\alpha^* \sin\beta \tag{A7.3}$$

Since the sums of the squares of the direction cosines is unity,

$$n_3^2 = 1 - \cos^2 \beta - \sin^2 \beta \cos^2 \alpha^* = \sin^2 \beta \sin^2 \alpha^*$$

so that

$$n_3 = \sin \beta \sin \alpha^* = v/\sin \gamma \qquad (A7.4)$$

since[a] $V = abc \sin \alpha^* \sin \beta \sin \gamma$, and v here refers to the unit parallelepiped $\mathbf{a}/a, \mathbf{b}/b, \mathbf{c}/c$, that is, $v = (1 - \cos^2 \alpha - \cos^2 \beta - \cos^2 \gamma + 2 \cos \alpha \cos \beta \cos \gamma)$. Hence, we can write the transformation in terms of the direct unit-cell parameters, multiplying the lines of the matrix by a, b, or c, as appropriate:

$$\begin{bmatrix} \mathbf{a} \\ \mathbf{b} \\ \mathbf{c} \end{bmatrix} = \begin{bmatrix} a & 0 & 0 \\ b\cos\gamma & b\sin\gamma & 0 \\ c\cos\beta & c(\cos\alpha - \cos\beta\cos\gamma)/\sin\gamma & cv/\sin\gamma \end{bmatrix} \begin{bmatrix} \mathbf{A} \\ \mathbf{B} \\ \mathbf{C} \end{bmatrix} \quad (A7.5)$$

which, in matrix notation, is $\mathbf{a} = M^{-1}\mathfrak{A}$. From the transformations discussed in Section 2.5ff, we have $\mathbf{X} = (M^{-1})^T \mathbf{x}$, or

$$\begin{bmatrix} X \\ Y \\ Z \end{bmatrix} = \begin{bmatrix} a & b\cos\gamma & c\cos\beta \\ 0 & b\sin\gamma & c(\cos\alpha - \cos\beta\cos\gamma)/\sin\gamma \\ 0 & 0 & cv/\sin\gamma \end{bmatrix} \begin{bmatrix} x \\ y \\ z \end{bmatrix} \quad (A7.6)$$

The deduction of M, the inverse of M^{-1}, is straightforward for a 3×3 matrix, albeit somewhat laborious, and can be found in most elementary treatments of vectors.[b] Thus, we have $\mathfrak{A} = M\mathbf{a}$ and $\mathbf{x} = M^T\mathbf{X}$, where

$$M = \begin{bmatrix} 1/a & 0 & 0 \\ -\cos\gamma/(a\sin\gamma) & 1/(b\sin\gamma) & 0 \\ (\cos\gamma\cos\alpha - \cos\beta)/(av\sin\gamma) & (\cos\gamma\cos\beta - \cos\alpha)/(bv\sin\gamma) & \sin\gamma/(cv) \end{bmatrix} \quad (A7.7)$$

The transfomation (A7.6) is employed in the program INTXYZ* (see Section 11.6.6) for the calculation of bond lengths, bond angles, and torsion angles from crystallographic parameters. The sign of a torsion angle is governed by a convention, as discussed in Section 7.5.2. For the sequence of atoms, P, Q, R, S (Figure A7.2), the torsion angle $+\chi_{PQRS}$ is the angle measured *clockwise* between QP and RS, as seen along the direction Q to R. A positive sign is equivalent to a right-handed screw in the sequence P–Q–R–S.

---

[a] M. J. Buerger, *X-ray Crystallography*, Wiley (1942).
[b] See, for example, *Symmetry and Group Theory in Chemistry*, Mark Ladd, Horwood Publishing Limited (1998).

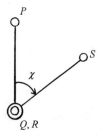

FIGURE A7.2. Convention for torsion angles: χ_{PQRS} is reckoned positive as shown, when the atom group $P-Q-R-S$ is viewed along QR.

A8 The integral $\int_0^\infty [(\sin x)/x]\,dx$

This integral was encountered in the study of Fourier transforms in Chapter 5. It is an important integral in diffraction theory, and we derive its value here.

We need to integrate the function $\exp(iz)/z$ over a semicircular path, such as $-R, -r, 0, r, R, S, -R$, where R is an arbitrary radius and z is a variable in the complex plane, Figure A8.1. However, we cannot integrate through the singularity at zero, so we excise a semicircular path $-r, 0, r, s, -r$, where r is another arbitrary radius.

Since $z = 0$ lies outside the now defined contour,

$$\oint \frac{\exp(iz)}{z}\,dz = 0 \tag{A8.1}$$

so that

$$\int_{-R}^{-r} \frac{\exp(ix)}{x}\,dx + \int_r^R \frac{\exp(ix)}{x}\,dx + \int_{-r,s,r} \frac{\exp(iz)}{z}\,dz + \int_{R,S,-R} \frac{\exp(iz)}{z}\,dz = 0 \tag{A8.2}$$

If we replace x in the first integral by $-x$, negate its limits, and combine it with the second integral, we can write

$$2i \int_r^R \frac{\sin x}{x}\,dx = -\int_{-r,s,r} \frac{\exp(iz)}{z}\,dz - \int_{R,S,-R} \frac{\exp(iz)}{z}\,dz \tag{A8.3}$$

We now let $r \to 0$ and $R \to \infty$. In the limit, the second integral on the right-hand side tends to zero (Cauchy's theorem), whereas the first integral, putting $z = r\exp(i\theta)$, becomes

$$-\lim_{r\to 0} \int_\pi^0 \frac{\exp[ir\exp(i\theta)]}{r\exp(i\theta)} ir\exp(i\theta)\,d\theta = -\lim_{r\to 0} \int_\pi^0 i\exp[ir\exp(i\theta)]\,d\theta$$

$$= -\int_\pi^0 i\,d\theta = i\pi \tag{A8.4}$$

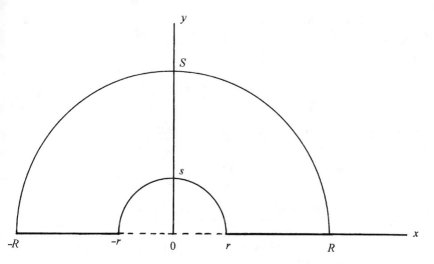

FIGURE A8.1.   Path of integration in the complex plane.

Hence,

$$\int_0^\infty \frac{\sin x}{x}\,dx = \frac{\pi}{2} \tag{A8.5}$$

A9   Gamma Function

The gamma function is useful in handling integrals of the type

$$\int_0^\infty x^n \exp(-ax^2)\,dx \tag{A9.1}$$

where a is a constant; they occur in several areas of intensity statistics. The gamma function $\Gamma(n)$ may be represented by the equation

$$\Gamma(n) = \int_0^\infty t^{n-1} \exp(-t)\,dt \tag{A9.2}$$

The following three results are useful:

1. For $n > 0$ and integral

$$\Gamma(n) - (n-1)! \tag{A9.3}$$

2. For $n > 0$

$$\Gamma(n+1) = n\Gamma(n) \tag{A9.4}$$

and if n is also integral

$$\Gamma(n+1) = n! \qquad\qquad\qquad\qquad\qquad (A9.5)$$

3. $$\Gamma\left(\tfrac{1}{2}\right) = \sqrt{\pi} \qquad\qquad\qquad\qquad\qquad (A9.6)$$

As an example, consider the solution of the integral

$$\mathcal{J} = \int_0^\infty x^4 \exp(-x^2/2)\,\mathrm{d}x$$

Let $x^2/2 = t$, so that $x = (2t)^{1/2}$ and $\mathrm{d}x = (2t)^{-1/2}\,\mathrm{d}t$. Then

$$\mathcal{J} = 2\sqrt{2}\int_0^\infty t^{3/2}\exp(-t)\,\mathrm{d}t = 2\sqrt{2}\,\Gamma(5/2) = 3(\pi/2)^{1/2}$$

A reduction formula (m integral) is also useful when working with these integrals:

$$\int x^m \exp(ax)\,\mathrm{d}x = \frac{x^m \exp(ax)}{a} - \frac{m}{a}\int x^{m-1}\exp(ax)\,\mathrm{d}x \qquad (A9.7)$$

A10 Crystallographic Software

The following list of software relates to those programs and program packages to which reference has been made in the forgoing text, and which are freely available to the academic community. The list is not exhaustive, and more packages are listed under the web reference http://www.ccp14.ac.uk. Most of the programs are mirrored by the Engineering and Physical Sciences Research Council (EPSRC) funded CCP14 Project at the above web reference, which has mirror sites in the USA and Canada. One contact name has been provided for each entry, together with Internet and/or Journal references as available.

A10.1 Single Crystal Suites

Most single crystal suites have a large variety of functionality; WinGX is an example of a suite linking to several other programs in a seamless manner via graphical user interfaces. In most cases, programs link to multiple versions of a solution program, such as SHELXS97 and SIR97.

Platon/System S for UNIX

A. L. Spek; SHELXS, DIRDIF, SIR and CRUNCH for solution; EXOR, DIRDIF, SIR and CRUNCH for autobuilding; SHELXL for refinement; E-mail >a.l.spek@chem.uu.nl< WWW >http://www.cryst.chem.uu.nl/platon/< Utrecht University, Utrecht, The Netherlands (1998).

SIR97 for MS-Windows and UNIX

C. Giacovazzo et al.; SIR97 for solution; CAOS for refinement (see also under A10.2); E-mail >sirmail@area.ba.cnr.it< WWW >http://www.irmec. ba.cnr.it/< *Journal of Applied Crystallography* **32**, 115–119 (1999).

WinGX for Windows

L. Farrugia; SHELXS, DIRDIF, SIR and PATSEE for solution; DIRDIF phases for autobuilding; SHELX for refinement; E-mail >louis@chem.gla.ac.uk < WWW >http://www.chem.gla.ac.uk/~louis/software/< *Journal of Applied Crystallography* **32**, 837–838 (1999).

A10.2   Single Crystal Structure Solution Programs

CAOS (Crystal Analysis Operating System) for MS-Windows and UNIX

R. Spagna et al.; Automated Patterson method; E-mail >spagna@isc. mlib.cnr.it< WWW >http://www.mlib.cnr.it/isc/caos/< *Journal of Applied Crystallography* **27**, 861–862 (1994).

CRYSTALS (11) for MS-Windows

David Watkin et al.; E-mail >david.watkui@chem.ox.ac.uk<WWW> http://www.xtl.ox.ac.uk/< Chemical Crystallography Laboratory, University of Oxford, Oxford (2000).

DIRDIF for PC/DOS and UNIX

(Windows version ported by L. Farrugia and available via the WinGX website) P. Beurskens et al.; Automated Patterson methods and fragment searching; E-mail: >ptb@sci.kun.nl< WWW >http://www-xtal.sci.kun.nl/xtal/documents/software/dirdif.html< Crystallography Laboratory, University of Nijmegen, The Netherlands (1999).

MULTAN source code

P. Main et al.; E-mail >pml@york.ac.uk< WWW (via Armel Le Bail's "Crystallography Source Code Museum") >http://sdpd.univ-lemans.fr/museum/< Université de York, Angleterre (1980).

PATSEE for PC/DOS and UNIX

(Windows version ported by L. Farrugia and available via the WinGX website) E. Egert et al.; Fragment searching methods; E-mail >bolte@indy1.org.chemie.uni-frankfurt.de< and >eg@indy3.org.chemie.uni-frankfurt.de< WWW >http://www.org.chemie.uni-frankfurt.de/egert/html/patsee. html< *Zeitschrift für Kristallographie* **216**, 565–572 (2001).

SHELXS86/SHELXS97/SHELXD

G. Sheldrick; Direct methods and Patterson; E-mail >gsheldr@shelx.uni-ac.gwdg.de< WWW>http://shelx.uni-ac.gwdg.de/SHELX/< Institüt für Anorganische Chemie der Universität, Tammanstrasse 4, D-3400 Göttingen, Germany (1998).

SIR88/92/SIR97/SIR2001

C. Giacovazzo et al.; Direct methods; E-mail >sirmail@area.ba.cnr.it< WWW >http://www.irmec.ba.cnr.it/< *Journal of Applied Crystallography* **32**, 115–119 (1999).

SnB (Shake"n"Bake)

C. M. Weeks et al.; Direct methods; E-mail >snb-requests@hwi.buffalo. edu< WWW >http://www.hwi.buffalo.edu/SnB/< *Journal of Applied Crystallography* **32**, 120–124 (1999).

A10.3   Single Crystal Twinning Software

TWIN 3.0 for Windows

Volker Kahlenberg et al., E-mail >vkahlen@uni-bremen.de< WWW >http://www.palmod.uni-bremen.de/FB5/kristall/vkalen/vkhomepage.html< *Journal of Applied Crystallography* **34**, 405 (2001).

TwinRotMac

A. L. Spek; E-mail >a.l.spek@chem.uu.nl< WWW >http://www.cryst. chem.uu.nl/platon/< (MS-Windows version ported by L. Farrugia, >http://www. chem.gla.ac.uk/~louis/software/platon/<) Utrecht University, Utrecht, The Netherlands (1988).

A10.4   Freestanding Structure Visualization Software

ORTEP-III

(Fortran source code); M. N. Burnett et al.; E-mail >ortep@ornl.gov <WWW>http://www.ornl.gov/ortep/ortep.html< (Oak Ridge National Laboratory Report ORNL-6 895 (1996).

A10.5   Powder Diffraction Data: Powder Indexing Suites (Dedicated and Other)

Checkcell for Windows

J. Laugier et al.; E-mail >jean.laugier2@wanadoo.fr< WWW >http://www. ccp14. ac.uk/tutorial/lmgp/< ENSP/Laboratoire des Matériaux et du Génie Physique, BP 46. 38,042 Saint Martin d'Hères, France. WWW >http://www.inpg. fr/LMGP; http://www.ccp14.ac.uk/tutorial/lmgp/<

CRYSFIRE for DOS

(Powder indexing suite, with programs *ITO, DICVOL, TREOR, TAUP, KOHL, LZON, LOSH*, and *FJZN*); R. Shirley; E-mail >r.shirley@surrey.ac.uk< WWW >http://www.ccp14.ac.uk/tutorial/crys/< (Manual: The Lattice Press, 41 Guildford Park Avenue, Guildford, Surrey GU2 7NL, England.)

DICVOL91

Daniel Louër; E-mail >Daniel.Louer@univ-rennes1.fr< WWW (via Armel Le Bail's website) >http://sdpd.univ-lemans.fr/also.html< *Journal of Applied Crystallography* **24**, 987–993 (1991).

FJZN

R. Shirley; E-mail >r.shirley@surrey.ac.uk< WWW >http://www.ccp14. ac.uk/tutorial/crys/<

ITO13 source code and MS-Windows binary

J. W. Visser, Henry Dunantlaan 81, 2614 GL Delft, The Netherlands; WWW (via Armel Le Bail's website) >http://sdpd.univ-lemans.fr/also.html< *Journal of Applied Crystallography* **2**, 89–95 (1969).

Kohl/TMO

F. Kohlbeck; E-mail >fkohlbec@pop.tuwien.ac.at< WWW >http://info. tuwien.ac.at/geophysik/Institute/Institute_e/Kohlbeck.htm< *Journal of Applied Crystallography* **11**, 60–61 (1978).

LOSH/LZON

R. Shirley et al.; E-mail >r.shirley@surrey.ac.uk< WWW >http://www. ccp14.ac.uk/tutorial/< *Acta Crystallographica* A**34**, S382 (1978).

TAUP/Powder

D. Taupin; E-mail >taupin@lps.u-psud.fr< WWW >ftp://hprib.lps.u-psud.fr/pub/powder/< and >http://www.ccp14.ac.uk/ccp/ccp14/ftp-mirror/taupin-indexing/pub/powder/< *Journal of Applied Crystallography* **6**, 380–385 (1973).

TREOR90 Source Code and MS-Windows Binary

P.-E. Werner; E-mail>pew@struc.su.se< WWW (via Armel Le Bail's website) >http://sdpd.univ-lemans.fr/also.html< *Journal of Applied Crystallography* **18**, 367–370 (1985).

A10.6   Powder Pattern Decomposition

ALLHKL

G. S. Pawley; *Journal of Applied Crystallography* **14**, 357–361 (1981).

WPPF

H. Toraya; *Journal of Applied Crystallography* **19**, 440 (1985).

A10.7   Structure Solution from Powder Diffraction Data

ESPOIR Source Code, DOS, Windows, and Linux Binaries

A. Le Bail et al.; E-mail >armel@fluo.univ-lemans.fr< WWW >http://sdpd.univ-lemans.fr/sdpd/espoir/< *Materials Science Forum* **378–381**, 65–70 (2001).

EXPO Source Code, Windows, and UNIX Binaries

C. Giacovazzo et al.; E-mail >sirmail@area.ba.cnr.it< WWW >http://www. irmec.ba.cnr.it/< *Journal of Applied Crystallography* **32**, 339–340 (1999); *Journal of Applied Crystallography* **34**, 704–709 (2000).

EXTRA (Included in EXPO)
FOCUS

R. W. Gross-Kuntsleve et al.; WWW >http://www.kristall.ethz.ch/LFK/software/< *Journal of Applied Crystallography* **30**, 985 (1997).

FullProf for DOS, Windows, and Linux

J. Rodriguez-Carvajal;   E-mail   >juan@bali.saclay.cea.fr<   WWW
>http:/www-llb.cea.fr/fullweb/powder.htm<   >ftp://charybde.saclay.cea.fr/pub/
divers/fullprof.2k/< *Abstract of 15th Conference of International Union of Crystallography* (Toulouse, France) p. 127, Satellite Meeting on Powder Diffraction (1990).

GSAS for Windows, Linux, and SGI IRIX

R. v Dreele et al.; E-mail >vondreele@lanl.gov< WWW >ftp://ftp.lanl.gov/
public/gsas/< Los Alamos National Laboratory Report LAUR 86–748 (1994).

Powder Solve (Commercial)

G. E. Engel et al.;   E-mail >webmaster@accelrys.com,   WWW
>http://www.accelrys.com/cerius2/powder.html< *Journal of Applied Crystallography* **32**, 1169–1179 (1999).

Profil for VMS, DOS and UNIX

J. Cockcroft; E-mail >cockcroft@gordon.cryst.bbk.ac.uk<WWW>http://
img.cryst.bbk.ac.uk/www/cockcroft/profil.htm< and >ftp://img.cryst.bbk.ac.uk/
pdpl/< *Zeitschriff für Kristallographie* **184**, 123–145 (1988).

Rietan GPL'D Fortran Source Code, Mac, UNIX, and Linux Binaries

Fujio Izumi et al.;   Email IZUMI.Fujio@nims.go.jp<   WWW >http:
//homepage.mac.com/fujioizumi/rietan/angle_dispersive/angle_dispersive.html <
Materials Science Forum, **321–324**, 198–203 (2000); *Materials Science Forum* **371–381**, 59–64 (2001).

SIRPOW

A. Altomare et al.; *Journal of Applied Crystallography* **28**, 842 (1995).

XRS-82/DLS source code

Ch. Baerlocher et al.; Email >ch.baerlocher@kristall.erdw.ethz.ch< WWW
>http://www.kristall.ethz.ch/LFK/software/< Institut für Kristallographie und
Petrographie, ETH Zürich.

A10.8   Software for Macromolecular Crystallography

This section contains general information software for structural studies on macromolecules; references are given as before. In order to be sure of keeping up with this fast moving research area the general websites >http://www.ccp4.ac.uk/main.html< and >http//sb.web.psi.ch/software.html< may be examined.

The Collaborative Computational project Number 4, or CCP4, deals with the software for macromolecular crystallography and also runs courses for those working in the field. Its website is listed above, and many of the programs below are available through and/or maintained by CCP4, from where further help may be sought.

Data Processing

General Reference Guide: L. Sawyer et al. (Editors), *Data Collection and Processing*; Proceedings of the CCP4 Study Weekend 29–30 January 1993 SERC Daresbury Laboratory DL/SC1/R34, Warrington, U.K.

HKL 4 (Includes DENZO, XDISPLAY, and SCALEPACK)
D. Gerwith, *The HKL Manual, 4th ed.* in Sawyer et al., *loc. cit.* (1993).

STRATEGY
R. B. G. Ravelli et al.; *Journal of Applied Crystallography* **30**, 551 (1997).

PREDICT
M. Noble; WWW>http://biop.ox.ac.uk/www/distrib/predict.html<

IMSTILLS, REFIX, MOSFILM, SORTMTZ, ROTAVATA, AGROVATA, POSTREFINEMENT, TRUNCATE, SCALEPACK, SCALEPACK2MTZ/CAD, XDISPLAYS, HKLVIEW, XDS, MARXDS, MARSCALE
CCP4 supported; WWW >http://www.ccp4.ac.uk/main.html<

Fourier and Structure Factor Calculations
SFALL (structure factors), FFT (fast Fourier calculation); WWW
>http://www.ccp4.ac.uk/main.html<

Molecular Replacement

AmoRe
J. Navaza; *Acta Crystallographica* A**50**, 157 (1994).

CNS 1.1
A. T. Brünger et al.; *Acta Crystallographica* D**54**, 905–921 (1998).

MOLREP
CCP4 supported; E-mail >alexei@ysbl.york.ac.uk<

REPLACE
WWW >http//sb.web.psi.ch/software.html<

Single and Multiple Isomorphous Replacement

SCALEIT, FHSCAL, WILSON, FFT, VECTORS HAVECS, RSPS, VECSUM, MLPHARE
CCP4 supported; WWW >http://www.ccp4.ac.uk/main.html<

SHELX86/-S
G. M. Sheldrick; Location of heavy-atom positions; University of Göttingen, Germany.

Software for Packing and Molecular geometry

MOLPACK
D. Wang et al.; *Journal of Molecular Graphics* **9**, 50 (1991).

PROCHECK
R. A. Laskowski et al.; *Journal of Applied Crystallography* **26**, 283 (1993).

WHATCHECK
R. W. W. Hooft et al.; *Nature* **381**, 272 (1996).

LIGPLOT
R. A. Laskowski; >http://www.biochem.ucl.ac.uk/bsm/ligplot/manual/index.html<

Software for Graphics and Model building
FRODO
T. A. Jones; *Acta Crystallographica* **A115**, 157–171 (1985).

O
T. A. Jones et al.; *Acta Crystallographica* **A47**, 110–119 (1991).

TURBO-FRODO
Bio-Graphics; >http://almb.curs.mrs.fr.TURBO-FRODO/turbo.html<

Software for Molecular Graphics and Display
RASMOL
R. Sayle; WWW >http://www.umass.edu/microbiol/rasmol/< Glaxo Research and Development, U.K. (1994).

SETOR
S. V. Evans; >http://scsg9.unige.ch.fln/eng/setorlic.html< *Journal of Molecular Graphics* **11**, 134 (1993).

MOLSCRIPT and BOBSCRIPT
>http://www.strubi.ox.ac.uk/bobscript/<

Software for Refinement
X-PLOR 3.1
A. T. Brünger; *Nature* **355**, 472 (1992).

CNS 1.1
A. T. Brünger et al.; *Acta Crystallographica* **D54**, 905–921 (1998); WWW >http//sb.web.psi.ch/software.html<

RESTRAIN
H. P. C. Driessen et al.; *Journal of Applied Crystallography* **22**, 510–516 (1989).

SHELXL-97
G. M. Sheldrick; WWW >http://linux.uni-ac.gwdg.de/SHELX< University of Göttingen, Germany (1997).

REFMAC 5
WWW >http//sb.web.psi.ch/software.html<

Software for Molecular Dynamics and Energy Minimization
SYBYL 6.4
WWW >http://www.tripos.com<

Data Bases

PDB (Protein Data Bank)
WWW >http://www.rscb.org/pdb< Status at 11 June 2002: 13,801 proteins; 684 nucleic acids; 14 carbohydrates; 7185 with structure factor files.

BLAST (Basic Local Alignment Search Tool)
Searches sequence data bases for protein and DNA structures and sequences; WWW >http://www.ncbi.nlm.nih.gov/BLAST<

CCDC (Cambridge Crystallographic Data Centre)
WWW >http://www.ccdc.cam.ac.uk< Status at 15 June 2002: > 250,000 small molecule entries.

ReLiBase
Finds all ligands for a particular protein family; WWW >http://www.pdb.bnl.gov:8081/home.http<

Synchrotrons Web Page
WWW >http://lmb.biop.ox.ac.uk/www/synchr.htt<

Tutorial Solutions

Solutions 1

Problems

1.1. Extend CA to cut the x' axis in H. All angles in the figure are easily calculated ($OA = OC = 1$). Evaluate OP, or a' (1.623), and OH (2.732). Express OH, the required intercept on the x' axis, as a fraction of a' (1.683). The intercept along b' (and b) remains unaltered, so that the fractional intercepts of the line CA are 1.683 and $\frac{1}{2}$ along x' and y respectively. Hence, CA is the line (0.5942, 2), or (1, 3.366), referred to the oblique axes.

1.2. (a) $h = a/(a/2) = 2, k = b/(-b/2) = \bar{2}, l = c/\infty = 0$; hence $(1\bar{2}0)$.
Similarly,
(b) (164) (c) $(00\bar{1})$ (d) $(3\bar{3}4)$ (e) $(0\bar{4}3)$ (f) $(\bar{4}2\bar{3})$

1.3. Set down the planes twice in each of the two rows, ignore the first and final indices in each row, and then cross-multiply, similarly to the evaluation of a determinant.

(a)   (1)   2   3   1   2   (3)

                  ×   ×   ×

      (0) $\bar{1}$ 1   0 $\bar{1}$ (1)

Hence, $U = 2 - (-3) = 5$ $V = 0 - 1 = -1$ $W = -1 - 0 = -1$ so that the zone symbol is $[5\bar{1}\bar{1}]$. If we had written the planes down in the reverse order, we would have obtained $[\bar{5}11]$. (What is the interpretation of this result?) Similarly:
(b) $[3\bar{5}2]$ (c) $[\bar{1}\bar{1}\bar{1}]$ (d) $[110]$

1.4. Following the procedure in 1.3, but with zone symbols leads to $(\bar{5}\bar{2}3)$. This plane and $(\bar{5}23)$ are parallel; $[UVW]$ and $[\bar{U}\,\bar{V}\,\bar{W}]$ are coincident.

1.5. Formally, one could write 422, $42\bar{2}$, $4\bar{2}2$, $4\bar{2}\bar{2}$, $\bar{4}22$, $\bar{4}2\bar{2}$, $\bar{4}\bar{2}2$, $\bar{4}\bar{2}\bar{2}$. However, the interaction of two inversion axes leads to an intersecting *pure* rotation axis, so that all symbols with one or three inversion axes are invalid. Now $\bar{4}2\bar{2}$

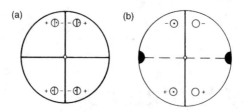

FIGURE S1.1.

and $\bar{4}\bar{2}2$ are equivalent under rotation in the x, y plane by 45°, so that there remain 422, $4\bar{2}\bar{2}$, and $\bar{4}2\bar{2}$ as unique point groups. Their standard symbols are 422, $4mm$ and $\bar{4}2m$, respectively. Note that if we do postulate a group with the symbol $42\bar{2}$, for example, it is straightforward to show, with the aid of a stereogram, that it is equivalent to, and a non-standard description of, $\frac{4}{m}mm$.

1.6. (a) mmm (b) $2/m$ (c) 1

1.7. See Figure S1.1. (a) $mmm; m \cdot m \cdot m \equiv \bar{1}$ (b) $2/m; 2 \cdot m \equiv \bar{1}$

1.8.

	{010}	{$\bar{1}$10}	{11$\bar{3}$}
$2/m$	2	4	4
$\bar{4}2m$	4	4	8
$m3$	6	12	24

1.9. (a) 1   (b) m (c) 2   (d) m (e) 1   (f) 2   (g) 6   (h) $6mm$ (i) 3   (j) $2mm$

(Did you remember to use the Laue group ion for each example?)

1.10. (a) From a thin card, cut out four irregular but identical quadrilaterals; when fitted together, they make a (plane) figure of symmetry 2.

1.11.
(a) $\bar{6}m2$ D_{3h} {$10\bar{1}0$} or {$01\bar{1}0$}
(b) $4/m\ mm$ D_{4h} {100} or {110}
(c) $m3m$ O_h {100}
(d) $\bar{4}3m$ T_d {111} or {1$\bar{1}$1}
(e) $3m$ C_{3v}
(f) 1 C_1

$\quad$ (g) $\quad 6/m\ mm \quad D_{6h}$
$\quad$ (h) $\quad mm2 \quad C_{2v}$
$\quad$ (i) $\quad mmm \quad D_{2h}$
$\quad$ (j) $\quad mm2 \quad C_{2v}$

1.12. Remember first to project the general form of the point group on to a plane of the given form, and then relate the projected symmetry to one of the two-dimensional point groups. In some cases, you will have more than one set of representative points in two dimensions.

(a) 2 $\quad$ (b) m $\quad$ (c) 1 $\quad$ (d) m $\quad$ (e) 1 $\quad$ (f) 1 $\quad$ (g) 3 $\quad$ (h) $3m$ $\quad$ (i) 3 $\quad$ (j) $2mm$

Solutions 2

2.1. The translations, equal to the lengths of the two sides of any parallelogram unit, repeat the molecule *ad infinitum* in the two dimensions shown. A 2-fold rotation point placed at any corner of a parallelogram is, itself, repeated by the same translations.

(i) The 2-fold rotation points lie at each corner, half-way along each edge and at the geometrical center of each parallelogram unit.

(ii) There are four unique 2-fold points per parallelogram unit: one at a corner, one at the center of each of two non-collinear edges and one at the geometrical center.

2.2. $\quad$ (i) $\quad\quad$ (ii)
$\quad$ (a) $4mm \quad\quad 6mm$
$\quad$ (b) Square $\quad$ Hexagonal
$\quad$ (c) If unit cell (i) is centered, then another square can be drawn to form a conventional unit cell of half the area of the centered unit cell. If unit cell (ii) is centered it is no longer hexagonal; each point is degraded to the $2mm$ symmetry of the rectangular system, and may be described by a conventional p unit cell. The transformation equations in each example are:

$$\mathbf{a}' = \mathbf{a}/2 + \mathbf{b}/2; \quad \mathbf{b}' = -\mathbf{a}/2 + \mathbf{b}/2$$

Note. A regular hexagon of "lattice" points with another point placed at its center is not a centered hexagonal unit cell: it represents three adjacent p hexagonal unit cells in different relative orientations. (Without the point at the center, the hexagon of points is not even a lattice.)

2.3. A C unit cell may be obtained by the transformations:

$$\mathbf{a}_C = \mathbf{a}_F; \quad \mathbf{b}_C = \mathbf{b}_F; \quad \mathbf{c}_C = -\mathbf{a}_F/2 + \mathbf{c}_F/2.$$

The new c dimension is obtained from evaluating the dot product $(-\mathbf{a}/2 + \mathbf{c}/2)\cdot(-\mathbf{a}/2 + \mathbf{c}/2)$, to give c' 5.7627 Å; a and b are unchanged. The angle β' in the transformed unit cell is obtained by evaluating $\cos\beta' = \mathbf{a}\cdot(-\mathbf{a}/2 + \mathbf{c}/2)/a'c' = (-a + c\cos\beta)/(2c')$, so that $\beta' = 139.29°$. $V_C(C \text{ cell})/V_F(F \text{ cell}) = \frac{1}{2}$. (Count the number of unique lattice points in each cell: each lattice point is associated with a unique portion of the volume.)

2.4. (a) The symmetry is no longer tetragonal, although the lattice is true (orthorhombic).

(b) The tetragonal symmetry is apparently restored, but the lattice is no longer true: the lattice points are not all in the same environment in the same orientation.

(c) A tetragonal F unit cell is formed and represents a true tetragonal lattice. However, tetragonal F is equivalent to tetragonal I (of smaller volume) under the transformation

$$\mathbf{a}_I = \mathbf{a}_F/2 + \mathbf{b}_F/2; \quad \mathbf{b}_I = -\mathbf{a}_F/2 + \mathbf{b}_F/2; \quad \mathbf{c}_I' = \mathbf{c}_F$$

2.5. F unit cell: $r^2_{[31\bar{2}]} = \mathbf{r}_{[31\bar{2}]}\cdot\mathbf{r}_{[31\bar{2}]} = 3^2a^2 + 1^2b^2 + 2^2c^2 - 2\cdot3\cdot(-2)\cdot6\cdot 8\cdot\cos110° = 826.00 \text{ Å}^2$, so that $r = 28.74$ Å. To obtain the value in the C unit cell, we could repeat this calculation with the dimensions of the C unit cell, leading to 28.64 Å. Alternatively, we could use the transformation matrix to obtain the F equivalent of $[31\bar{2}]_C$, and then use the original F cell dimensions on it. The matrix for F in terms of C is:

$$\mathbf{S} = \begin{bmatrix} 1 & 0 & 0 \\ 0 & 1 & 0 \\ 1 & 0 & 2 \end{bmatrix}$$

Hence,

$$(\mathbf{S}^{-1})^T = \begin{bmatrix} 1 & 0 & -\frac{1}{2} \\ 0 & 1 & 0 \\ 0 & 0 & \frac{1}{2} \end{bmatrix}$$

Then, $[UVW]_F = (\mathbf{S}^{-1})^T \cdot [UVW]_C = [41\bar{1}]_F$, so that $r_{[41\bar{1}]_F} = 28.64$ Å.

2.6. It is not an eighth system. The symmetry at each lattice point is $\bar{1}$, so that it is a special case of the triclinic system in which the γ angle is 90°.

			Origin on 2mm				
			$(0,0;\frac{1}{2},\frac{1}{2})+$				Limiting conditions
8	(f)	1	x,y;	$x,\bar{y}$;	$\bar{x},y$;	$\bar{x},\bar{y}$	$hk: h+k=2n$
4	(e)	m	$0,y$;	$0,\bar{y}$;			—
4	(d)	m	$x,0$;	$\bar{x},0$			—
4	(c)	2	$\frac{1}{4},\frac{1}{4}$;	$\frac{1}{4},\frac{3}{4}$			As above +
							$hk: h=2n,(k=2n)$
2	(b)	$2mm$	$0,\frac{1}{2}$				—
2	(a)	$2mm$	$0,0$				—

2.7. (a) Plane group $c2mm$ is shown in Figure S2.1.

(b) Plane group $p2mg$ is shown in Figure S2.2; this diagram also shows the minimum number of motifs P, V, and Z.

If the symmetry elements are arranged with 2 at the intersection of m and g, they do not form a group. Attempts to draw such an arrangement lead to continued halving of the "repeat" parallel to the g line.

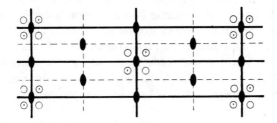

FIGURE S2.1.

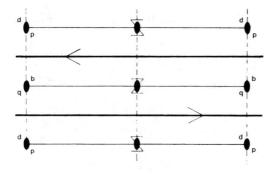

FIGURE S2.2.

2.8. (a)

			Origin on $\bar{1}$	Limiting conditions
4	(e)	1	$x, y, z;\ x, \frac{1}{2} - y, \frac{1}{2} + z$	hkl: None
			$\bar{x}, \bar{y}, \bar{z};\ \bar{x}, \frac{1}{2} + y, \frac{1}{2} - z$	$h0l: l = 2n$
				$0k0: k = 2n$
2	(d)	$\bar{1}$	$\frac{1}{2}, 0, \frac{1}{2};\ \frac{1}{2}, \frac{1}{2}, 0$	As above +
				$hkl: k + l = 2n$
2	(c)	$\bar{1}$	$0, 0, \frac{1}{2};\ 0, \frac{1}{2}, 0$	
2	(b)	$\bar{1}$	$\frac{1}{2}, 0, 0;\ \frac{1}{2}, \frac{1}{2}, \frac{1}{2}$	
2	(a)	$\bar{1}$	$0, 0, 0;\ 0, \frac{1}{2}, \frac{1}{2}$	

(100) $p2gg : b' = b, c' = c$ (010) $p2 : a' = a, c' = c/2$ (001) $p2gm : a' = a, b' = b$

Space group $P2_1/c$ is shown in Figure S2.3, on the (010) plane.

(b) Figure S2.4 shows the molecular formula of biphenyl, excluding the hydrogen atoms. The two molecules in the unit cell lie on any set of special positions (a)–(d), with the center of the C(1)–C(1)' bond on $\bar{1}$. Hence, the molecule is centrosymmetric and planar. The planarity imposes a conjugation on the molecule, including the C(1)–C(1)'. (This result is supported by the bond lengths C(1)–C(1)' $\approx$ 1.49 Å and C_{arom}–C_{arom} $\approx$ 1.40 Å. In the free-molecule state, the rings rotate about the C(1)–C(1)' bond to the energetically favourable conformation with the ring planes at approximately 45° to each other.)

2.9. Each pair of positions forms two vectors, between the origin and the points $\pm\{(x_2 - x_1), (y_2 - y_1), (z_2 - z_1)\}$. Thus, there is a single vector at each of the positions

$$2x, 2y, 2z;\ \ 2\bar{x}, 2\bar{y}, 2\bar{z};\ \ 2x, 2\bar{y}, 2z;\ \ 2\bar{x}, 2y, 2\bar{z}$$

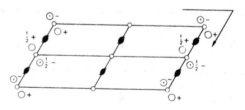

FIGURE S2.3.

FIGURE S2.4.

and two superimposed vectors at each of the positions

$$2x, \tfrac{1}{2}, \tfrac{1}{2} + 2z; \quad 0, \tfrac{1}{2} + 2y, \tfrac{1}{2}; \quad 2\bar{x}, \tfrac{1}{2}, \tfrac{1}{2} - 2z; \quad 0, \tfrac{1}{2} - 2y, \tfrac{1}{2}$$

Note. $- \left(2x, \tfrac{1}{2}, \tfrac{1}{2} + 2z \right) \equiv 2\bar{x}, \tfrac{1}{2}, \tfrac{1}{2} - 2z$

2.10.

$$
\begin{array}{ccc}
x, y, z & \xrightarrow{\quad b \quad} & 2p - x, \tfrac{1}{2} + y, z \\
\big\downarrow \bar{1} & & \big\downarrow a \\
\bar{x}, \bar{y}, \bar{z} & & \tfrac{1}{2} + 2p - x, 2q - \tfrac{1}{2} - y, z \\
2p - x, 2q - y, 2r - z & \xleftarrow{\quad n \quad} &
\end{array}
$$

Since $\bar{x}, \bar{y}, \bar{z}$ and $2p - x, 2q - y, 2r - z$ are one and the same point, $p = q = r = 0$, so that the three symmetry planes intersect in a center of symmetry at the origin.

Applying the half-translation rule, $T = a/2 + b/2 + a/2 + b/2 \equiv 0$. Hence, the center of symmetry lies at the intersection of the three symmetry planes.

2.11. Figure S2.5 shows space group *Pbam*.

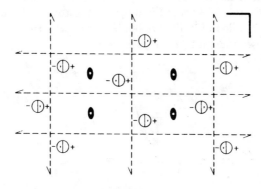

FIGURE S2.5.

Coordinates of general equivalent positions

$$x, y, z; \quad \tfrac{1}{2} - x, \tfrac{1}{2} - y, z; \quad \tfrac{1}{2} + x, \bar{y}, z; \quad \bar{x}, \tfrac{1}{2} + y, z;$$

$$x, y, \bar{z}; \quad \tfrac{1}{2} - x, \tfrac{1}{2} - y, \bar{z}; \quad \tfrac{1}{2} + x, \bar{y}, \bar{z}; \quad \bar{x}, \tfrac{1}{2} + y, \bar{z}$$

Coordinates of centers of symmetry

$$\tfrac{1}{4}, \tfrac{1}{4}, 0; \quad \tfrac{1}{4}, \tfrac{3}{4}, 0; \quad \tfrac{3}{4}, \tfrac{1}{4}, 0; \quad \tfrac{3}{4}, \tfrac{3}{4}, 0;$$

$$\tfrac{1}{4}, \tfrac{1}{4}, \tfrac{1}{2}; \quad \tfrac{1}{4}, \tfrac{3}{4}, \tfrac{1}{2}; \quad \tfrac{3}{4}, \tfrac{1}{4}, \tfrac{1}{2}; \quad \tfrac{3}{4}, \tfrac{3}{4}, \tfrac{1}{2}$$

Change of origin—to $\tfrac{1}{4}, \tfrac{1}{4}, 0$. (i) Subtract $\tfrac{1}{4}, \tfrac{1}{4}, 0$ from the above set of coordinates of general equivalent positions. (ii) Let $x_0 = x - \tfrac{1}{4}$, $y_0 = y$ and $z_0 = z$. (iii) After making all substitutions, drop the subscript, and rearrange to give:

$$\pm\{x, y, z; \quad \bar{x}, \bar{y}, \bar{z}; \quad \tfrac{1}{2} + x, \tfrac{1}{2} - y, \bar{z}; \quad \tfrac{1}{2} - x, \tfrac{1}{2} + y, \bar{z}\}$$

This result may be confirmed by redrawing the space group with the origin on $\bar{1}$.

2.12. Figure S2.6 shows two adjacent unit cells of space group Pn on the (010) plane. In the transformation to Pc, only the c spacing is changed:

$$\mathbf{c}_{P_c} = -\mathbf{a}_{P_n} + \mathbf{c}_{P_n}$$

Hence $Pn \equiv Pc$. By interchanging the labels of the x and z axes, which are not constrained by the 2-fold symmetry, we see that $Pc \equiv Pa$. Note that it is necessary to invert the sign on $\mathbf{b}$, so as to preserve a right-handed set of axes. In the case of Cm, the translations of $\tfrac{1}{2}$ along a and b mean that $Ca \equiv Cm$. Since there is no translation along c in Cm, Cm is not equivalent to Cc, although Cc is equivalent to Cn. If the x and z axes in Cc are interchanged, with due attention to $\mathbf{b}$, the symbol becomes Aa. (The *standard* symbols among these groups are Pc, Cm, and Cc.)

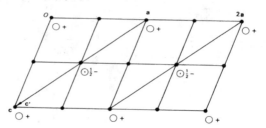

FIGURE S2.6.

2.13. $P2/c$ (a) $2/m$; monoclinic.

(b) Primitive unit cell; c-glide plane $\perp b$; 2-fold axis $\parallel b$.

(c) $h0l: l = 2n$.

$Pca2_1$ (a) $mm2$; orthorhombic.

(b) Primitive unit cell; c-glide plane $\perp a$; a-glide plane $\perp b$; 2_1 axis $\parallel c$;

(c) $0kl: l = 2n$; $h0l: h = 2n$.

$Cmcm$ (a) mmm; orthorhombic.

(b) C-face centered unit cell; m plane $\perp a$; c-glide plane $\perp b$; m plane $\perp c$.

(c) $hkl: h + k = 2n$; $h0l: l = 2n$.

$P\bar{4}2_1c$ (a) $\bar{4}2m$; tetragonal.

(b) Primitive unit cell; $\bar{4}$ axis $\parallel c$; 2_1 axes $\parallel a$ and b; c-glide planes $\perp [110]$ and $[1\bar{1}0]$.

(c) $hhl: l = 2n$; $h00: h = 2n$.

$P6_322$ (a) 622; hexagonal.

(b) Primitive unit cell; 6_3 axis $\parallel c$; 2-fold axes $\parallel a, b$, and u; 2-fold axes $30°$ to a, b and u, and in the (0001) plane.

$Pa3$ (a) $m3$; cubic.

(b) Primitive unit cell; a-glide plane $\perp b$ (equivalent statements are b-glide plane $\perp c$, c-glide plane $\perp a$); threefold axes $\parallel$ $[111], [1\bar{1}1], [\bar{1}11]$, and $[\bar{1}\bar{1}1]$.

(c) $0kl: k = 2n$; (equivalent statements are $h0l: l = 2n$; $hk0: h = 2n$.)

2.14. Plane group $p2$; the unit cell repeat along b is halved, and γ has the particular value of $90°$. Note that, because of the *contents* of the unit cell, it cannot belong to the rectangular two-dimensional system.

2.15. (a) Refer to Figure 2.24, number 10, for a cubic P unit cell (vectors **a**, **b**, and **c**).

(b) Tetragonal P $\mathbf{a}_P = \mathbf{b}/2 + \mathbf{c}/2$

$\mathbf{b}_P = -\mathbf{b}/2 + \mathbf{c}/2$

$\mathbf{c}_P = \mathbf{a}$

(c) Monoclinic C $\mathbf{a}_C = \mathbf{c}$

$\mathbf{b}_C = -\mathbf{b}$

$\mathbf{c}_C = \mathbf{a}$

(d) Triclinic, P $\mathbf{a}_T = \mathbf{a}$

$\mathbf{b}_T = \mathbf{b}/2 + \mathbf{c}/2$

$\mathbf{c}_T = -\mathbf{b}/2 + \mathbf{c}/2$

Other transformations may be acceptable for (d), provided that right-handed axes are maintained.

2.16.

$$\overset{\bar{4}\ \text{along } z}{\begin{bmatrix} 0 & 1 & 0 \\ \bar{1} & 0 & 0 \\ 0 & 0 & \bar{1} \end{bmatrix}} \qquad \overset{m \perp b}{\begin{bmatrix} 1 & 0 & 0 \\ 0 & \bar{1} & 0 \\ 0 & 0 & 1 \end{bmatrix}}$$
$$\quad \mathbf{R}_1 \qquad\qquad\qquad \mathbf{R}_2$$

$\mathbf{R}_2 \cdot \mathbf{R}_1 \cdot \mathbf{h} = \mathbf{h}'$. Forming first $\mathbf{R}_3 = \mathbf{R}_2 \cdot \mathbf{R}_1$, remembering the order of multiplication, we then evaluate

$$\underset{\mathbf{R}_3}{\begin{bmatrix} 0 & 1 & 0 \\ 1 & 0 & 0 \\ 0 & 0 & \bar{1} \end{bmatrix}} \cdot \underset{\mathbf{h}}{\begin{bmatrix} h \\ k \\ l \end{bmatrix}} = \underset{\mathbf{h}'}{\begin{bmatrix} k \\ h \\ \bar{l} \end{bmatrix}}$$

that is, $\mathbf{R}_3 \cdot \mathbf{h} = \mathbf{h}'$, so that $\mathbf{h}' = kh\bar{l}$; $\mathbf{R}_3$ represents a 2-fold rotation axis along [110].

2.17. The matrices are multiplied in the usual way, and the components of the translation vectors are added, resulting in

$$\begin{bmatrix} \bar{1} & 0 & 0 \\ 0 & \bar{1} & 0 \\ 0 & 0 & 1 \end{bmatrix} + \begin{bmatrix} \frac{1}{2} \\ \frac{1}{2} \\ \frac{1}{2} \end{bmatrix}$$

which corresponds to a 2_1 axis along $[\frac{1}{4}, \frac{1}{4}, z]$. The space group symbol is $Pna2_1$.

2.18. Since (collinear) $3 \cdot 2 \equiv 6$ (see, e.g., Figure 1.36, point group 6), we obtain

$$\begin{bmatrix} 1 & \bar{1} & 0 \\ 1 & 0 & 0 \\ 0 & 0 & 1 \end{bmatrix} + \begin{bmatrix} 0 \\ 0 \\ \frac{1}{2} \end{bmatrix}$$

for the 6_3 axis. Using the half-translation rule, we see that the center of symmetry must be displaced by $c/4$ from the point of intersection of 6_3 with m. Hence, for $\bar{1}$ at the origin, the matrix for m in this space group is

$$\begin{bmatrix} 1 & 0 & 0 \\ 0 & 1 & 0 \\ 0 & 0 & \bar{1} \end{bmatrix} + \begin{bmatrix} 0 \\ 0 \\ \frac{1}{2} \end{bmatrix}$$

that is, the m plane is located at $(x, y, \frac{1}{4})$. The coordinates of the general equivalent positions are $\pm\{x, y, z; \quad \bar{y}, x-y, z; \quad y-x, \bar{x}; z; \quad x, y, \frac{1}{2} - z;$

$\bar{y}; x - y, \frac{1}{2} - z; \quad y - x, \bar{x}, \frac{1}{2} - z\}$. (The 6-fold rotations can be checked with the general matrix given in Appendix A4.)

2.19. From Figure 2.11, it follows that

$$\mathbf{a_R} = 2\mathbf{a_H}/3 + \mathbf{b_H}/3 + \mathbf{c_H}/3$$
$$\mathbf{b_R} = -\mathbf{a_H}/3 + \mathbf{b_H}/3 + \mathbf{c_H}/3$$
$$\mathbf{c_R} = -\mathbf{a_H}/3 - 2\mathbf{b_H}/3 + \mathbf{c_H}/3$$

Following Section 2.2.3, we have $\mathbf{a_R} \cdot \mathbf{a_R} = (2\mathbf{a_H}/3 + \mathbf{b_H}/3 + \mathbf{c_H}/3) \cdot (2\mathbf{a_H}/3 + \mathbf{b_H}/3 + \mathbf{c_H}/3) = 3b^2/9 + c^2/9 = 12$ Å^2 so that $a = 3.464$ Å. Similarly, $\cos \alpha_R = (2\mathbf{a_H}/3 + \text{Å}^2, \mathbf{b_H}/3 + \mathbf{c_H}/3) \cdot (-\mathbf{a_H}/3 + \mathbf{b_H}/3 + \mathbf{c_H}/3)/a_R^2$, so that $\alpha_R = 51.32°$.

2.20. The transformation matrix $\mathbf{S}$ for $R_{hex} \rightarrow R_{obv}$ is given, from the solution to Problem 2.19, by

$$\mathbf{S} = \begin{bmatrix} \frac{2}{3} & \frac{1}{3} & \frac{1}{3} \\ -\frac{1}{3} & \frac{1}{3} & \frac{1}{3} \\ -\frac{1}{3} & -\frac{2}{3} & \frac{1}{3} \end{bmatrix}$$

and its inverse is

$$\mathbf{S}^{-1} = \begin{bmatrix} 1 & \bar{1} & 0 \\ 0 & 1 & \bar{1} \\ 1 & 1 & 1 \end{bmatrix}$$

so that the transpose becomes

$$(\mathbf{S}^{-1})^T = \begin{bmatrix} 1 & 0 & 1 \\ \bar{1} & 1 & 1 \\ 0 & \bar{1} & 1 \end{bmatrix}$$

Hence $(13^*\bar{1})_{hex}$ is transformed to $(32\bar{1})_{obv}$, and $[1\bar{2}^*3]_{hex}$ to $[405]_{obv}$.

2.21. Figure S2.7 illustrates the reflection of x, y, z across the plane (qqz), where $OC = OW = q$, so that $\angle OCW = 45°$; Q is the point $q - x, q - y, z$, and the remainder of the diagram is self-explanatory.

As an alternative procedure, we know that $4 \cdot \mathbf{m_y} = \mathbf{m_{diag}}$, since $4mm$ is a point group. Hence, x, y is transformed to $\bar{y}, \bar{x}$ by the operation $\mathbf{m_{diag}}$. If we now move the origin to the point $-q, -q$, it follows that $\bar{y}, \bar{x}$ then becomes $q - y, q - x$.

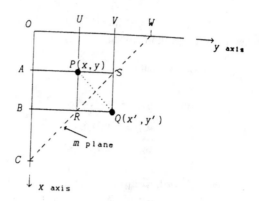

FIGURE S2.7.

2.22.

Diffraction symbol				Point group		
				2	m	$2/m$
$12/m1$	P	.	.	.	$P2$	Pm	$P2/m$
$12/m1$	P	.	c	.	Pc	$P2/c$
$12/m1$	P	.	2_1	.	$P2_1$		
$12/m1$	P	.	$2_1/c$	.			$P2_1/m$
							$P2_1/c$
$12/m1$	C	.	.	.	$C2$	Cm	$C2/m$
$12/m1$	C	.	c	.		Cc	$C2/c$

2.23. (a) From the matrix

$$\begin{bmatrix} \frac{1}{2} & 0 & 0 \\ 0 & 1 & 0 \\ 0 & 0 & 1 \end{bmatrix},$$

$(210) \rightarrow (110)$, and is confirmed by drawing.
(b) From the matrix

$$\begin{bmatrix} 1 & 0 & 0 \\ 0 & \frac{1}{2} & 0 \\ 0 & 0 & 1 \end{bmatrix},$$

$(210) \rightarrow (410)$, after clearing the fraction. By drawing, we see that the original (210) plane is now the second plane from the origin in the (410) family of planes; $d(410)_{new} = d(210)_{old}/2$ under the given transformation. In each case, the Miller index corresponding to the unit cell halving is also halved.

2.24. In $Cmm2$, the polar (2-fold) axis is normal to the centred plane, but parallel to it in $Amm2$. $Cmmm$ and $Ammm$ are equivalent by interchange of axes, so that they are not two distinct arrangements of points.

Solutions 3

3.1. $d\lambda/\text{Å} = 0.0243(1 - \cos 45°) = 0.00712$. Energy/J$= hc/(1 + 0.00712) = 1.97 \times 10^{-15}$.

3.2. Set an origin at the center of a line joining the two scattering centeres; then the coordinates are $\pm\lambda$. The amplitude of the separated points $A_\lambda = 2\cos(2\pi \times \lambda \times 2\lambda^{-1} \sin\theta \times \cos\theta) = 2\cos(2\pi \sin 2\theta)$, the angle between $\mathbf{r}$ and $\mathbf{S}$ being θ. For the two centers at one point ($r = 0$) the amplitue $A_p = 2$. Hence:

2θ	Ratio A_λ/A_p	Intensity
0	1	1
30, 150	-1	1
60, 120	0.666	0.444
90	1	1
180	0	0

3.3. Proved in the text: $f_{1s} = c_1^4/(c_1^2 + \pi^2 S^2)^2$, where $c_1 = (4 - 0.3)/0.529 = 6.994$ Å^{-1} and $S = 2(\sin\theta)/\lambda$. For the 2s contribution, the integral $\int_0^\infty x^3 \exp(-ax) \sin bx\, dx$ evaluates to $4(a^3 b - ab^3)/(a^2 + b^2)^4$, so that f_{2s}, becomes $[2\pi c_2^5/(96\pi S)] \times 3 \times 16\pi Sc_2[c_2^2 - (2\pi S)^2]/[c_2 + 4\pi^2 S^2]^4 = c_2^6[c_2^2 - 4\pi^2 S^2)/(c_2^2 + 4\pi^2 S^2)^4$, where $c_2 = (4 - 2.05)/0.529 = 3.686$ Å^{-1}. Hence:

θ	Scattering formula			Exponential formula
	f_{1s}	f_{2s}	f	f
0.0	1.000	1.000	4.000	4.002
0.2	0.938	0.116	2.109	2.060
0.5	0.692	-0.0082	1.368	1.360

3.4. Photon energy $= h\upsilon = hc/\lambda = hc/(hc/eV) = 1.6021 \times 10^{-19} \times 30000 = 4.806 \times 10^{-15}$ J.

3.5. $M_r(\text{CaSiO}_3) = 116.17$. $M_r(\text{Ca})/M_r(\text{CaSiO}_3) = 0.345$; $M_r(\text{Si})/M_r(\text{CaSiO}_3) = 0.242$; $M_r(\text{O})/M_r(\text{CaSiO}_3) = 0.408$. Hence, $\mu = 2720[0.345 \times 172) + (0.242 \times 60.3) + (0.408 \times 12.7)] = 21.5 \times 10^{-4}$ m^{-1}.

3.6. It is necessary to note carefully the changes in sign of both $A(hkl)$ and $B(hkl)$. Thus, the following diagram is helpful, together with the changes in sign of the argument of the trigonometric functions. For example, if both A and B change sign, ϕ is not unaltered by cancelling the signs, but becomes $\pi + \phi$.

$$\sin(-\theta) = -\sin(\theta)$$
$$\cos(-\theta) = \cos(\theta)$$
$$\tan(-\theta) = -\tan(\theta)$$

$P2_1$: Use equations (3.123)–(3.126) for k even and k odd.

$k = 2n$:
$$\phi(hkl) = -\phi(\bar{h}\bar{k}\bar{l}) = -\phi(h\bar{k}l) = \phi(\bar{h}k\bar{l}) \neq \phi(\bar{h}kl)$$
$$\phi(\bar{h}kl) = -\phi(h\bar{k}\bar{l}) = \phi(hk\bar{l}) = -\phi(\bar{h}\bar{k}l)$$

$k = 2n + 1$:
$$\phi(hkl) = -\phi(\bar{h}\bar{k}\bar{l}) = \pi - \phi(h\bar{k}l) = \pi + \phi(\bar{h}k\bar{l}) \neq \phi(\bar{h}kl)$$
$$\phi(\bar{h}kl) = -\phi(h\bar{k}\bar{l}) = \pi + \phi(hk\bar{l}) = \pi - \phi(\bar{h}\bar{k}l)$$

$Pma2$: Use equations (3.137)–(3.138) for h even and odd

h even: $\phi(hkl) = -\phi(\bar{h}\bar{k}\bar{l}) = \phi(\bar{h}kl) = \phi(h\bar{k}l) = -\phi(hk\bar{l}) = -\phi(h\bar{k}\bar{l})$

$$= -\phi(\bar{h}k\bar{l}) = \phi(\bar{h}\bar{k}l)$$

h odd: $\phi(hkl) = -\phi(\bar{h}\bar{k}\bar{l}) = \pi + \phi(\bar{h}kl) = \pi + \phi(h\bar{k}l) = -\phi(hk\bar{l})$

$$= \pi - \phi(h\bar{k}\bar{l}) = \pi - \phi(\bar{h}k\bar{l}) = \phi(\bar{h}\bar{k}l)$$

3.7. From the equations developed in Sections 3.4 and 3.4.1, but taking the reciprocal space constant K as the x-ray wavelength of 1.5418 Å, we find:

$a^* = 0.30314, b^* = 0.23115, c^* = 0.14096, \alpha^* = 60.182°, \beta^* = 55.878°,$
$\gamma^* = 47.591°.$

$V = 618.916 \text{ Å}^3; V^* = 5.9218 \times 10^{-3}.$

(The reciprocal cell lengths are dimensionless here, and V^* may be calculated as λ^3/V.)

3.8. If $\mathbf{r}_1$ and $\mathbf{r}_2$ are the distances of the two atoms from the origin, then we use $\mathbf{r}_1 = x_1\mathbf{a} + y_1\mathbf{b} + z_1\mathbf{c}$ and $\mathbf{r}_2 = x_2\mathbf{a} + y_2\mathbf{b} + z_2\mathbf{c}$. Then $r_1 = (\mathbf{r}_1 \cdot \mathbf{r}_1)^{1/2}$ (not forgetting the cross-products), and similarly for r_2. The angle θ at the origin is given by $\cos\theta = \mathbf{r}_1 \cdot \mathbf{r}_2/(r_1, r_2)$. Thus, the two distances are 2.660 and 2.983 Å, and 2.983 Å, and $\theta = 99.43°$.

3.9. The resultant $\mathbf{R}$ is obtained in terms of the amplitude $|R|$ and phase ϕ from $|R| = [(\Sigma_j A \cos\phi_j)^2 + (\Sigma_j B \sin\phi_j)^2]^{1/2} = [(-21.763)^2 + (-22.070)^2]^{1/2} = 30.995$, and $\phi = \tan^{-1}[(-22.070)/(-21.763)] = 45.40°$, but because both the numerator are denominator are negative the phase angle lies in the third quadrant, and $180°$ must be added to give $\phi = 225.40°$.

3.10. A-centering implies pairs of positions x, y, z and $x, \frac{1}{2} + y, \frac{1}{2} + z$. Hence, we write $F(hkl) = \sum_{j=1}^{n/2} f_j\{\exp[i2\pi(hx_j + ky_j + lz_j)] + \exp[i2\pi(hx_j + ky_j + lz_j + k/2 + l/2)]\}$. The terms within the braces $\{\}$ may be expressed as $\exp[i2\pi(hx_j + ky_j + lz_j)]\{1 + \exp[i2\pi(k/2 + l/2)]\}$ which is 2 for $(k + l)$ even, and zero for $(k + l)$ odd ($e^{in\pi} = 1/0$ for n even/odd). Hence, the limiting condition is $hkl: k + l = 2n$.

3.11. The coordinates show that the structure is centrosymmetric. Hence, $F(hk0) = A(hk0) = 2[g_P \cos 2\pi(hx_P + ky_P) + g_Q \cos 2\pi(hx_Q + ky_Q)]$

h k	A(hk)	h k	A(hk)	h k	A(hk)	h k	A(hk)
5 0	$2(-g_P + g_Q)$	0 5	$2(g_P - g_Q)$	5 5	$2(-g_P - g_Q)$	5 10	$2(-g_P + g_Q)$

For $g_P = 2g_Q, \phi(0\ 5) = 0, \phi(5\ 0) = \phi(5\ 5) = \phi(5\ 10) = \pi$.

3.12. $F(hk0) = 4g_U \cos 2\pi[ky_U + (h + k)/4] \cos 2\pi(h + k)/4$ which, because $(h + k)$ is even in the data, reduces to $F(hk0) = 4g_U \cos 2\pi ky_U$.

h k 0	$|F(hk0)_{y=0.10}|$	$|F(hk0)_{y=0.15}|$
0 2 0	86.5	86.5
1 1 0	258.9	188.1

Hence, 0.10 is the better value for y_U in terms of the two reflections given.

3.13. The shortest U–U distance d_{U-U} is from $0, y, \frac{1}{4}$ to $0, \bar{y}, \frac{3}{4}$, so that $d_{U-U} = [(0.20b)^2 + (0.5c)^2]^{1/2} = 2.76$ Å.

3.14. (a) $P2_1, P2_1/m$; (b) $Pa, P2/a$; (c) $Cc, C2/c$; (d) $P2, Pm, P2/m$.

3.15. (a) $P2_12_12$; (b) $Pbm2, Pbmm$; (c) $Ibm2, Ibmm$. Note that $Ibm2$, for example, might have been named $Icm2_1$: normally, where more than one symmetry element lies in a given orientation, the rules of precedence in naming is $m > a > b > c > n > d$ and $2 > 2_1$. In a few cases the rules may be ignored. For example, $I4cm$ could be named $I4bm$, but with the origin on 4, the c-glides pass through the origin, and the former symbol is preferred.

Writing example (c) with the redundancies indicated, we have

$hkl: h + k + l = 2n$
$0kl: k = 2n, (l = 2n)$ or $l = 2n, (k = 2n)$
$h0l: (h + l = 2n)$
$hk0: (h + k = 2n)$
$h00: (h = 2n)$

$0k0$: $(k = 2n)$
$00l$: $(l = 2n)$

3.16. (a)

  (i) $h0l$: $h = 2n$; $0k0 : k = 2n$.
  (ii) $h0l$: $l = 2n$
  (iii) hkl: $h + k = 2n$
  (iv) $h00$: $h = 2n$
  (v) $0kl$: $l = 2n$; $h0l : l = 2n$
  (vi) $h\,k\,l$: $h + k + l = 2n$; $h0l$: $h = 2n$

Other space groups with the same conditions: (i) None; (ii) $P2/c$; (iii) $C2, C2/m$; (iv) None. (v) $Pccm$; (vi) $Ima2(I2am)$

  (b) hkl: None
     $h0l$: $h + l = 2n$
     $0k0$: $k = 2n$
  (c) $C2/c$; $C222$

3.17. (a) In the given setting x' and **a** are normal to a c-glide, y' and $\bar{\mathbf{c}}$ are normal to an a-glide, and z' and **b** are normal to a b-glide. In the standard setting, x is along x' and the plane normal to has its glide in the new y direction, so that it is a b-glide; y is along z' and the plane normal to it is a glide now in the direction of z; z is along $-y'$ and the plane normal to it is now an a-glide. Thus, the symbol in the standard setting is $Pbca$.
(b) In $Pmna$ the symmetry leads to translations of $(c + a)/2$ and $a/2$, overall $c/2$, and in $Pnma$ the translations arising are $a/2, b/2$ and $c/2$. Hence, the full symbol for $Pmna$ is $P\frac{2}{m}\frac{2}{n}\frac{2_1}{a}$, whereas that for $Pnma$ is $P\frac{2_1}{n}\frac{2_1}{m}\frac{2_1}{a}$.

3.18. $\mu R = 2.00$, so that $A = 10.0$. Hence, $|F(hkl)|^2 = 56.3 \times 1.1547 \times 0.625 \times 10.0 = 406.3$.

3.19. For the given reflection, $(\sin\theta)/\lambda = 0.30$, for which $f_C = 2.494$. Hence, $\exp[-B(\sin^2\theta)/\lambda^2] = 0.5423$, so that $f_{C,27.55°} = 1.352$, which is 54.2% of what its value would be at rest. The root mean square displacement is $[6.8/(8\pi^2)]^{1/2} = 0.29$ Å. Since vibrational energy is proportional to $k_B T$, where k_B is the Boltzmann constant, a reduced temperature factor with concomitant enhanced scattering would be achieved by conducting the experiment at a low temperature.

3.20. (a) $\overline{|F|} = (2/\pi\Sigma)^{1/2} \int_0^\infty |F|\exp(-|F|^2/2\Sigma)d|F|$. Let $|F|^2/2\Sigma = t$, so that $d|F| = (\Sigma/2t)^{1/2}dt$.
    Then, $\overline{|F|} = (2\Sigma/\pi)^{1/2} \int_0^\infty t^0 \exp(-t)dt$. Since $t^0 = t^{(1-1)}$, the integral is $\Gamma(1)$, Hence, $\overline{|F|} = (2\Sigma/\pi)^{1/2}$.

$\overline{|F|^2} = (2/\pi\Sigma)^{1/2} \int_0^\infty |F|^2 \exp(-|F|^2/2\Sigma)\mathrm{d}|F|$. Making the above substitution, we have

$\overline{|F|^2} = (2\Sigma/\pi)^{1/2} \int_0^\infty t^{1/2} \exp(-t)\mathrm{d}t = (2\Sigma/\pi^{1/2})1/2\Gamma(1/2) = \Sigma$.
Thus, $M_c = (2/\pi\Sigma)/\Sigma = 2/\pi = 0.637$.

(b) $\overline{|E|^3} = (2/\pi)^{1/2} \int_0^\infty |E|^3 \exp(-|E|^2/2)\mathrm{d}|E|$. Let $|E|^2/2 = t$, so that $\mathrm{d}|E| = (2t)^{-1/2}\mathrm{d}t$. Then, $\overline{|E|^3} = (8/\pi)^{1/2} \int_0^\infty t\exp(-t)\mathrm{d}t = (8/\pi)\frac{1}{2}\Gamma(2) = 1.596$.

(c) $\overline{|E^2 - 1|} = 2\int_0^\infty |E^2 - 1|\,|E|\exp(-E^2)\mathrm{d}|E|$

By making the substitution $E^2 = t$, we have

$$\overline{|E^2 - 1|} = \int_0^1 (1 - t)\exp(-t)\mathrm{d}t + \int_1^\infty (t - 1)\exp(-t)\mathrm{d}t$$

$$= (-\mathrm{e}^{-t} \big|_0^1 + (t\mathrm{e}^{-t} \big|_0^1 + (\mathrm{e}^{-t} \big|_0^1 - (t\mathrm{e}^{-t} \big|_1^\infty - (\mathrm{e}^{-t} \big|_1^\infty + (\mathrm{e}^{-t} \big|_1^\infty$$

$$= 2/\mathrm{e} = 0.736$$

3.21. From Section 3.1.5, we see that the angular spread Δ of the radiation is given by $\Delta = m_e c^2/E$. Inserting the values for the constants and the given value of E, $\Delta = 2.55 \ 10^{-4}$ rad, of ca 53 seconds of arc. The maximum (characteristic) wavelength is given by $\lambda_c/\text{Å} = 18.64/[(B/\text{T})(E/\text{GeV})^2]$, so that $\lambda_c = 3.88$ Å.

3.22. For NaCl, $d_{111} = a/\sqrt{3} = 3.2487$ Å, so that $(\sin\theta_{111})/\lambda = 0.1539$ Å^{-1} and $(\sin\theta_{222})/\lambda = 0.3078$ Å^{-1}. Similarly, for KCl, $(\sin\theta_{111})/\lambda = 0.1379$ Å^{-1} and $(\sin\theta_{222})/\lambda = 0.1379$ Å^{-1}. Using the structure factor equation for the NaCl structure type, we have $F(111) = 4[f_{Na^+/K^+} + f_{Cl^-}\cos(3\pi)] = 4[f_{Na^+/K^+} - f_{Cl^-}]$, where as $F(222) = 4[f_{Na^+/K^+} + f_{Cl^-}]$. Thus, we obtain the following results:

	111		222	
	NaCl	KCl	NaCl	KCl
$(\sin\theta)/\lambda$	0.1539	0.1379	0.3078	0.2759
f_+	8.979	15.652	6.777	11.576
f_-	13.593	14.207	9.387	9.997
F	−18.46	1.445	64.66	86.29

Remembering that we measure $|F|^2$, it is clear that $|F(111)|$ for KCl is relatively vanishingly small.

Solutions 4

4.1. (a) The crystal system is tetragonal, and the Laue group is $\frac{4}{m}mm$; the optic axis lies along the needle axis (c) of the crystal. (b) The section is in extinction for any rotation in the x, y plane, normal to the needle axis; the section is optically isotropic. (c) For a general oscillation photograph with the x-ray beam normal to c, the symmetry is m. For a symmetrical oscillation photograph with the beam along a, b or any direction in the form $\langle 110 \rangle$ at the mid-point of the oscillation, the symmetry is $2mm$.

4.2. (a) The crystal system is orthorhombic. (b) Suitable axes may be taken parallel to three non-coplanar edges of the brick. (c) Symmetry m. (d) Symmetry $2mm$, with the m lines horizontal and vertical.

4.3. (a) Monoclinic, or possibly orthorhombic. (b) If monoclinic, p is parallel to the y axis. If orthorhombic, p is parallel to one of x, y, or z. (c) (i) Mount the crystal perpendicular to p, about either q or r, and take a Laue photograph with the x-ray beam parallel to p. If the crystal is monoclinic, symmetry 2 would be observed. If orthorhombic, the symmetry would be $2mm$, with the m lines in positions on the film that define the directions of the crystallographic axes normal to p. If the crystal is rotated such that the x-rays travel through the crystal perpendicular to p, a vertical m line would appear on the Laue photograph of either a monoclinic or an orthorhombic crystal. (ii) Use the same crystal mounting as in (i), but take a symmetrical oscillation photograph with the x-ray beam parallel or perpendicular to p at the mid-point of the oscillation. The rest of the answer is as in (i).

4.4. Following Section 4.4.2, we find $a = 9.00$, $b = 6.00$ and $c = 5.00$ Å. From Section 3.4, with $\kappa = 1.5$ Å, it follows that $a^* = 0.1667$, $b^* = 0.2500$, and $c^* = 0.3000$, all dimensionless.

The spacing $d(146) = 0.7261$ Å (Table 2.4); hence $\sin\theta_{146}$ [$= 1.5/(2 \times 0.7261)$] is greater that unity, so that reflection 146 cannot be recorded under the given conditions.

Each photograph would have a horizontal m line, conclusive of orthorhombic symmetry if the crystal is known to be biaxial. Otherwise, tests for higher symmetry would have to be applied.

4.5. (a) From Section 4.4.2, the repeat distance $t_{[110]}$ along [110] is given by $t_{[100]} = 2 \times 1.54/\{\sin[\tan-1(3.5/6.0)]\} = 6.113$ Å hence, $a(= b) = 6.113 \times /\sqrt{2} = 8.645$ Å. From similar arguments, $c = 7.506$ Å. (b) The maximum height of a layer line on the film is given as 4.0 cm. Thus, from Section 4.4.2, this corresponds to an angle of scatter ψ equal to $\tan^{-1}(4/3)$, or 53.13°. Thus, l_{max} is the integral part of $[c\sin(53.13)/\lambda]$, which is 3. (c) Symmetry 1 in (i), and symmetry m (horizontal) in (ii). (d) It would be identical, because the rotation axis is a direction of 4-fold symmetry.

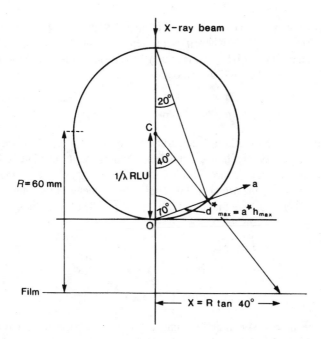

FIGURE S4.1.

4.6. Direct measurement between two distinct reciprocal lattice rows give the distance 42.75 mm, which corresponds to 85.5°. Since β is, by convention, obtuse, $\beta^* = 85.5°$ and $\beta = 94.5°$; the unit-cell volume is 1606.7 Å^3.

4.7. Refer to Figure S4.1. Let h_{max} represent the maximum value sought. Since we are concerned with a large d^* value, we take $\lambda \approx 0.2$ Å, the minimum value in the white radiation. Now $d^* = ha^* = (2/\lambda) \sin\theta$ and since, from the diagram, θ is the angle subtended at the circumference by d^*, $\theta = 20°$, so that h_{max} is the integral part of $(2/0.2\lambda)\sin(20°)$, which is 17. The X-coordinate on the film is $60\tan 40° = 50.35$ mm. The half-width of the film is 62.5 mm, so that (17,00) will be recorded on the film.

4.8. For the first film, we can write

$$I(hkl) + I(2h, 2k, 2l) = 300$$

and for the second film, after absorption, we have

$$0.35I(hkl) + 0.65I(2h, 2k, 2l) = 130$$

Solving these equations gives $I(hkl) = 216.7$ and $I(2h, 2k, 2l) = 83.3$.

4.9. From the geometry of the oscillation photograph, we derive $t = n\lambda/\{\sin[\tan^{-1}(h_n/R)]\}$, where t is the repeat distance along the axis, h_n is the height of the nth layer line above the zero-layer line and R is the camera radius. Since t is 8.642 Å, we obtain:

n	1	2	3
h_n/mm	2.48	5.00	7.64

4.10. Using a reciprocal lattice construct, we have for a repeat distance t in a lattice $t = n\lambda/\zeta_n$, where ζ_n, the dimensionless reciprocal lattice layer-line spacing, is obtained through $2h_n/D = \zeta_n/(1 - \zeta_n^2)^{1/2}$, $2h_n$ being the double distance of the nth layer line and D the diameter of the camera. Letting $(2h_n/D)^2 = p_n$, we find $\zeta_n = \sqrt{p_n/(1 + p_n)}$. Thus, we obtain:

Axis	[100]	[110]	[111]
p_n	0.5120	0.4306	0.8236
ζ_n	0.5819	0.5486	0.6720
n	3	2	1
t/Å	7.949	5.621	2.294

Hence, $a = 7.949$ Å, and the unit-cell type is I: because of halving, $t_{[111]}$ is $(a/2)/\sqrt{3}$ in I, but $a/\sqrt{3}$ in P and F; an oscillation photograph on [100] yields the correct value of a in each of these unit-cell types.

4.11. (a) For symmetry $2mm$ in Laue group $m3m$, the x-ray beam must be travelling along a $<110>$ direction (Table 1.6); we will choose [110], so that a and b lie in the horizontal plane (c is then the vertical direction). (b) We can use Figure 4.17, changing the sign of $-a^*$, and with $\phi = 45°$ because XO is [110] for the present problem. For an inner spot, it follows readily that $2\theta = \tan^{-1}(43.5/60.0)$, so that $2\theta = 35.94°$, and $\varepsilon = 27.03°$ (Figure 4.17). Hence, $\tan(27.03°) = 0.510 = h/k$, since $a = b$. In the given orientation, the reflections on the horizontal line are $hk0$ and, since the unit cell is F, h and k must be both even, with $k = 2h$, from above. Possible reflections are, therefore, $240, 480, 612, 0, \ldots$ It is straightforward to show that $\lambda = 2a\sin\theta/\sqrt{N}$, where $N = h^2 + k^2$. For 240, $\lambda = 0.746$ Å; for 480, $\lambda = 0.337$ Å, which is unreasonably small in crystallographic work. We note from the orientation of the a and b axes (a^* and b^*) that one of h and k must be negative; we can choose k. For an outer spot, we find in a similar manner that $\tan\varepsilon = 0.342$, so that $k = 3h$. Reasonable indices correspond to $h = 2$ and $k = 6$, again with one index negative; here, $\lambda = 0.753$ Å. To summarize:

The x-ray beam is along [110];
For the inner spots: $\theta = 17.97°$; $2\bar{4}0$ and $4\bar{2}0$; $\lambda = 0.746$ Å;
For the outer spots: $\theta = 26.13°$; $2\bar{6}0$ and $6\bar{2}0$; $\lambda = 0.753$ Å.

4.12. Since the crystal is uniaxial, it must be hexagonal, tetragonal, or trigonal. The Laue symmetry along axis 1 indicates that the crystal is trigonal, referred to hexagonal axes, and that axis 1 is therefore c. Following Section 4.2.2, we find for the repeat distances along the three axes:

Axis	1	2	3
Repeat/Å	15.65	8.264	4.772

The smallest repeat distance corresponds to the unit-cell dimension a, direction $[10\bar{1}0]$, Laue symmetry 2 (Figure 1.36 and Table 1.6). Axis 2, must be a direction in the x, y plane, and it is straightforward to show that it is the repeat distance along $[12\bar{3}0]$, Laue symmetry m. Thus, we have: $a = b = 4.772, c = 15.65$ Å; $\alpha = \beta = 90°, \gamma = 120°$; the Laue group is $\bar{3}m$.

4.13. Applying the Bragg equation, $\lambda^{-1} \sin\theta = 1/2d$ where $d = 6.696/2$ Å. Thus, (a) θ_{0002} (Cu) $= 13.312°$, and (b) θ_{0002} (Mo) $= 6.093°$.

4.14. (a) The data indicate a pseudo-monoclinic unit cell with γ unique. Following Section 2.4, we find $a = b = 6.418, c = 3.863$ Å. It would appear that the c dimension is true, and that the a, b plane is centered. It is straightforward to show that a and b are the half-diagonals of a rectangle with sides $\mathbf{a}' = \mathbf{a} - \mathbf{b}$ and $\mathbf{b}' = \mathbf{a} + \mathbf{b}$. Thus, the orthorhombic unit cell has the dimensions $a = 3.062, b = 12.465$ and $c = 3.863$ Å. The transformation can be written as $\mathbf{a}_{\text{true}} = \mathbf{M} \cdot \mathbf{a}_{\text{diff}}$, where

$$\mathbf{M} = \begin{bmatrix} 1 & -1 & 0 \\ 1 & 1 & 0 \\ 0 & 0 & 1 \end{bmatrix}$$

(b) The reciprocal cell is transformed according to $\mathbf{a}^*_{\text{true}} = (\mathbf{M}^{-1})^{\text{T}} \cdot \mathbf{a}^*_{\text{diff}}$. The transpose of the inverse matrix is

$$\begin{bmatrix} \frac{1}{2} & -\frac{1}{2} & 0 \\ \frac{1}{2} & \frac{1}{2} & 0 \\ 0 & 0 & 1 \end{bmatrix}$$

Hence, $a^* = .2321, b^* = 0.05701, c^* = 0.184$. These values may be confirmed by dividing the "true" values, for the orthorhombic cell, into the wavelength.

Solutions 5

5.1. $\int_{-c/2}^{c/2} \sin(2\pi mx/c) \cos(2\pi nx/c) dx = \int_{-c/2}^{c/2} \{\frac{1}{2} \sin[2\pi(m+n)x/c] + \frac{1}{2} \sin[2\pi(m-n)x/c]\} dx$, using identities from Appendix A6. Integration leads to $-[c/(2\pi(m + n)]\cos[2\pi(m + n)x/c]|_{-c/2}^{c/2} - [c/2\pi(m - n)]$ $\cos[2\pi(m - n)x/c]|_{-c/2}^{c/2}$. Since m and n are integers the integral is zero for $m \neq n$. For $m = n$, the original integral becomes $\int_{-c/2}^{c/2} \frac{1}{2} \sin(4\pi mx/c)dx$, which is also zero.

5.2. The plot of $\rho(x)$ as a function of x (in 40ths) shows peaks at 0, 20, and 40 for Mg (as expected), and at ca 8.25, 11.75, 28.25, and 31.75, for the 4 F atoms per repeat a; thus, x_F is $\pm(0.206; 0.706)$. Only the function to $a/4$ need be calculated, since there is m symmetry across the points $\frac{1}{4}$ (10/40), $\frac{1}{2}(20/40)$ and $\frac{3}{4}$ (30/40). (a) The first three terms alone are insufficient to resolve clearly the pairs of fluorine peaks that are closest in projection. (b) Changing the sign of the 600 reflection results in single peaks for fluorine at 10/40 and 30/40. The error in sign is evidently the more serious fault.

5.3. $G(S) = \int_{-p}^{p} a \exp(i2\pi Sx)dx = a\int_{-p}^{p} \cos(2\pi Sx) - ia \int_{-p}^{p} \sin(2\pi Sx)\,dx$. The second integral is zero, because the integrand is an odd function. Hence,

$$G(S) = a(2\pi S) \sin(2\pi Sx) |_{-p}^{p} = 2ap \sin(2\pi Sp)/(2\pi Sp)$$

we retain the parameters which would obviously cancel so as to preserve the characteristic $\sin(ax)/(ax)$ form. To obtain the original function, we evaluate

$$f(x) = (a/\pi) \int_{-\infty}^{\infty} (1/S) \sin(2\pi Sp) \exp(-i2\pi xS)dS$$

$$= (a/\pi) \int_{-\infty}^{\infty} (1/S) \sin(2\pi Sp) \cos(2\pi xS)dS,$$

where the sine term from the expanded integrand is zero as before. Using results from Appendix A6, the integral becomes

$$(a/2\pi)\{\int_{-\infty}^{\infty} (1/S) \sin[2\pi S(p + x)]dS$$
$$+ \int_{-\infty}^{\infty} (1/S) \sin[2\pi S(p - x)]\}dS \quad \text{or}$$
$$a(p + x) \int_{-\infty}^{\infty} \sin[2\pi S(p + x)]/[2\pi S(p + x)]dS$$
$$+a(p - x) \int_{-\infty}^{\infty} \sin[2\pi S(p - x)]/[2\pi (p - x)]dS.$$

From Appendix A8, $\int_{-\infty}^{\infty} \sin y/y\,dy = \pi$; hence, we derive

$$f(x) = (a/2)(p + x)/|p + x| + (a/2)(p - x)/|p - x|.$$

It is clear from this result that $f(x) = a$ for $|x| < p$, $f(x) = a/2$ for $x = \pm p$, and $f(x) = 0$ for $|x| = 0$, which correspond to the starting conditions.

5.4.

$$G(f) = A \int_{-\infty}^{\infty} \cos(2\pi f_0 t) \exp(-i2\pi f t) dt$$

$$= (A/2) \int_{-\infty}^{\infty} \{[\exp(i2\pi f_0 t) + \exp(-i2\pi f_0 t)] \exp(-i2\pi f t)\} dt$$

$$= (A/2) \int_{-\infty}^{\infty} \{\exp[-i2\pi (f - f_0)t] + \exp[-i2\pi (f + f_0)t]\} dt$$

$$= (A/2)\delta(f - f_0) + (A/2)\delta(f - f_0).$$

In the inversion, the δ-function repeats the function at $f = f_0$. Thus,

$$f(t) = (A/2) \int_{-\infty}^{\infty} [\delta(f + f_0) + \delta(f - f_0)] \exp(i2\pi f t) df$$

$$= (A/2)[\exp(i2\pi f_0 t) + \exp(i2\pi f_0 t)]$$

$$= A \cos(2\pi f_0 t).$$

5.5. The molecules have the displacements $\mathbf{p}$ and $-\mathbf{p}$ from the origin. Hence, the total transform $G_T(\mathbf{S})$ is given by

$$G_T(\mathbf{S}) = G_0(\mathbf{S}) \exp(i2\pi \mathbf{p} \cdot \mathbf{S}) + G_0^*(\mathbf{S}) \exp(-i2\pi \mathbf{p} \cdot \mathbf{S})$$

Using results from Section 3.5.3, we can write $G_0(\mathbf{S}) = |G_0| \exp(i\phi)$, and $G_0^*(\mathbf{S}) = |G_0| \exp(-\phi)$, where ϕ is a phase angle. Hence,

$$G_T(\mathbf{S}) = |G_0|\{\exp(i2\pi \mathbf{p} \cdot \mathbf{S} + \phi) + \exp(-i2\pi \mathbf{p} \cdot \mathbf{S} - \phi)$$

$$= 2|G_0| \cos(2\pi \mathbf{p} \cdot \mathbf{s} + \phi)$$

As discussed in Section 5.6.3, the maximum value of the transform is $2|G_0|$, at those points where $\cos(2\pi \mathbf{p} \cdot \mathbf{S} + \phi)$ is equal to unity. In this example, however, such points do not lie in planes and, consequently, the fringe systems are curved rather than planar.

5.6. The atoms related by the screw axis would have the fractional coordinates x, y, z and $\bar{x}, \frac{1}{2} + y, \frac{1}{2} - z$. From (5.43), we have

$$G(\mathbf{S}) = \sum_{j=1}^{n/2} f_j \{\exp[i2\pi (hx_j + ky_j + lz_j)]$$

$$+ \exp[i2\pi (-hx_j + ky_j - lz_j + k/2 + l/2)]\}$$

where the summation is over $n/2$ atoms in the unit cell not related by the 2_1 symmetry. Hence,

$$G(\mathbf{S}) = \sum_{j=1}^{n/2} f_j \{\exp[i2\pi k y_j \{\exp[i2\pi (h x_j + l z_j)$$
$$+ \exp[i2\pi(-h x_j - l z_j + k/2 + l/2)]\}$$

In a general transform, $h, k,$ and l could take any values. However, in a crystal they are integers, but in order to obtain a special condition, we must also consider the case that $h = l = 0$:

$$G(\mathbf{S})_{h=l=0} = 2 \sum_{j=1}^{n/2} f_j \exp(i2\pi k y_j [\exp(i\pi k)]$$

Then, we have $G(\mathbf{S})_{h=l=0} = 0$ for $k = 2n + 1$, that is, the $0k0$ reflections are systematically absent when k is odd.

5.7. Figure S5.1 indicates the nodal lines for the P–S fringe system. Since the transform is chosen to be positive at the origin, $\pm$ regions can be allocated

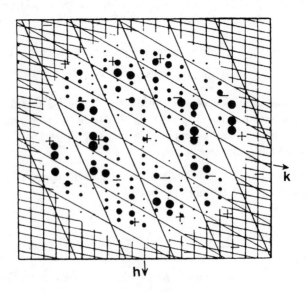

FIGURE S5.1.

to the transform, as shown. Hence, the intense reflections can be allocated signs, as follow:

240 −	250−	410 −	520 −
650 +	710+	720 +	820 +
$\bar{1}$30 +	$\bar{1}$40+	$\bar{2}$30 +	$\bar{2}$40 +
$\bar{3}$70 +	$\bar{4}$40−	$\bar{4}$70 +	$\bar{5}$30 −
$\bar{5}$40 −	$\bar{6}$70−	$\bar{7}$10 −	$\bar{7}$60 −
$\bar{9}$10 +	$\bar{9}$20+	$\bar{1}$0,00+	$\bar{1}$0,10+
$\bar{1}$0,20+			

5.8. In Figure S5.2, the three points are plotted in (a). A transparency is made of the structure in (a), inverted in the origin. The structure (a) is then drawn three times on the transparency, with each of the atoms of the inversion, in turn, over the origin of (a), and in the same orientation.

The completed diagram (b) is the required convolution: the six triangles outlined in (b) all produce the same set of nine vectors (three superimposed at the origin).

5.9. Figure S5.3 shows the contoured figure field of Figure P5.2. The same triangles are revealed, giving six sets of atom coordinates, as follow:

1	0.15, 0.10;	−0.15, −0.10;	−0.05, 0.30
2	0.05, 0.20;	−0.05, −0.20;	−0.20, −0.20
3	0.10, −0.10;	−0.10, 0.10;	−0.20, −0.30
4	0.05, −0.30;	0.15, 0.10;	−0.15, −0.10
5	0.25, 0.00;	0.05, 0.20;	−0.05, −0.20
6	0.10, −0.10;	−0.10, 0.10;	0.20, 0.30

5.10. The transform is positive in sign at the origin. Hence, by noting the succession of contours along the 00*l* row, we arrive at the following result:

001	002	003	004	005	006
+	−	+	+	−	−

5.11. The transform of

$$f(x) = f_T(x) = \frac{1}{\sqrt{2\pi}} \int_{-\infty}^{\infty} [\exp(-x^2/2) \exp(i2\pi\,Sx)]dx$$

$$= \frac{2}{\sqrt{2\pi}} \int_{0}^{\infty} [\exp(-x^2/2) \cos(2\pi\,Sx)]dx,$$

(a)

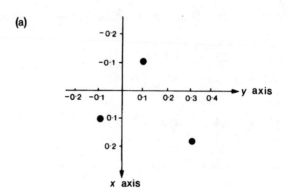

(b)

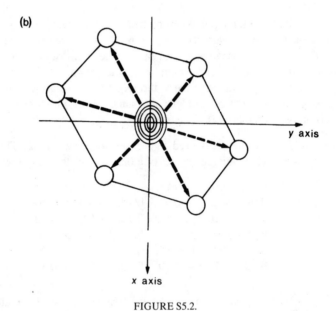

FIGURE S5.2.

because $f(x)$ is an even function. Hence, using standard tables of integrals,

$$f_T(x) = \frac{2}{\sqrt{2\pi}} \frac{\sqrt{\pi}}{2\sqrt{1/2}} \exp(-4\pi^2 S^2/2) = \exp(-2\pi^2 S^2).$$

The transform of $g(x) = g_T(x)$ is a δ-function with the origin at the point $x = 2$, so that $g_T(x) = \exp(i4\pi S)$, from Section 5.6.8. Hence, $c(x) = f_T(x)^* g_T(x) = \exp(i4\pi S - 2\pi^2 S^2)$.

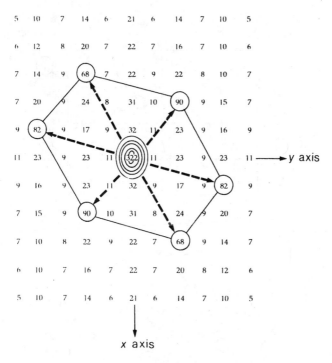

FIGURE S5.3.

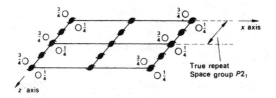

FIGURE S6.1.

Solutions 6

6.1. In $P2_1/c$, the general positions are $\pm(x, y, z; \ x, \frac{1}{2} - y, \frac{1}{2} + z)$, so that $A(hkl) = 2\{\cos 2\pi (hx + ky + lz) + \cos 2\pi (hx - ky + lz + k/2 + l/2)\} = 4 \cos 2\pi (hx + lz + k/4 + l/4) \cos 2\pi (ky - k/4 - l/4)$. Introducing the y-coordinate, $A(hkl) = 4 \cos 2\pi (hx + lz + k/4 + l/4) \cos 2\pi (l/4)$, so that the hkl reflections will be systematically absent for $l = 2n + 1$. The indication is that the c spacing should be halved, so that the true unit cell contains two species in space group $P2_1$ (see Figure S6.1). This problem illustrates the consequences of siting an atom on a glide plane: although

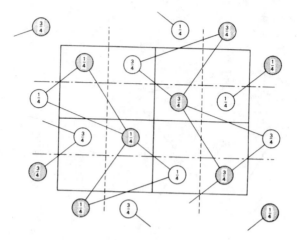

FIGURE S6.2.

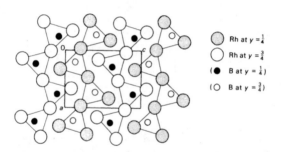

FIGURE S6.3.

we have considered here a hypothetical structure containing one atom in the asymmetric unit, in a multiatom structure, an atom may, by chance, be situated on a translational symmetry element.

6.2. Refer to Figures 2.37, S6.2, and S6.3.[a] There are eight rhodium atoms in the unit cell. If the atoms are in general positions, the minimum separation of atoms across any m plane is $\frac{1}{2} - 2y$. Even if $y = 0$, the distance would be too small to accommodate two rhodium atoms. Hence, they must occupy two sets of special positions. Positions on centeres of symmetry may be excluded on the same grounds as above. Thus, the atoms are located on two

---

[a] R. Mooney and A.J.E. Welch, *Acta Crystallographica* **7**, 49(1954).

sets of m planes as follow:

$$4\,Rh \pm \left(x_1, \tfrac{1}{4}, z_1; \quad \tfrac{1}{2} - x_1, \tfrac{3}{4}, \tfrac{1}{2} + z_1\right)$$

$$4\,Rh \pm \left(x_2, \tfrac{1}{4}, z_2; \quad \tfrac{1}{2} - x_2, \tfrac{3}{4}, \tfrac{1}{2} + z_2\right)$$

6.3. The space group is $P2_1/m$. The molecular symmetry cannot be $\bar{1}$, but it can be m. Hence, we can make the following assignments:
(a) Cl on m; (b) N on m; (c) two C on m, with four other C probably in general positions; (d) sixteen H in four sets of general positions, two H (in N–H groups) on m, and two H from CH_3 groups on m—those that have their C atoms on m. This arrangement is shown in Figure S6.4a. The species CH_3, H_1 and H_2 lie above and below the m plane. The alternative space group $P2_1$ was considered, but the full structure analysis[a] confirmed $P2_1/m$. Figure S6.4b illustrates $P2_1/m$, and is reproduced from the *International Tables for X-ray Crystallography*, Vol. I, by kind permission of the International Union of Crystallography.

6.4. $A(hhh) = 4\{g_{Pt} + g_K[\cos 2\pi(3h/4) + \cos 2\pi(9h/4)] + 6g_{Cl}[\cos 2\pi(hx) + \cos 2\pi(hx) + \cos 2\pi(hx)]\}$, where the factor 4 relates to an F unit cell (see Section 3.7.1). $B(hhh) = 0$, so that $F(hhh) = A(hhh)$, and $A(hhh)$ simplifies to $A(hhh) = 4\{g_{Pt} + 2g_K\cos(3\pi h/2) + 6g_{Cl}\cos(2\pi hx)\}$. We can now calculate $F(hhh)$ for the two values of x given:

		$x = 0.23$		$x = 0.24$	
hhh	$|F_o|$	$|F_c|$	$K_1|F_o|$	$|F_c|$	$K_2|F_o|$
111	491	340.6	314.7	317.4	329.5
222	223	152.2	142.9	159.5	149.6
333	281	145.2	180.1	190.8	188.6
$K_1 = 0.641$	$R_1 = 0.11$	$K_2 = 0.671$	$R_2 = 0.036$		

Clearly $x = 0.24$ is the preferred value. Pt—Cl = 2.34 Å. For a sketch and the point group, see Problem 1.11(a) and its solution.

6.5. $A_U(hkl) = 4\{\cos 2\pi(hx + ky + l/4) + \cos 2\pi(-hx + ky + l/4 + h/2 + k/2)\} = 4\{\cos 2\pi[ky + (h + k + l)/4] \cos 2\pi[hx - (h + k)/4]\}$. For 200, $A_U \propto |\cos 2\pi(2x - \tfrac{1}{2})|$ and, for this reflection to have zero intensity, $2\pi(2x - \tfrac{1}{2}) \approx (2n + 1)\pi/2$. For $n = 1$, $x \approx 3/8$ (by symmetry, the values 1/8, 5/8 and 7/8 are included). Generally, we choose the smallest

[a] J. Lindgren and I. Olovsson, *Acta Crystallographica* **B24**, 554 (1968).

(a)

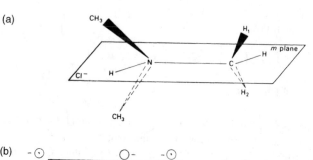

(b)

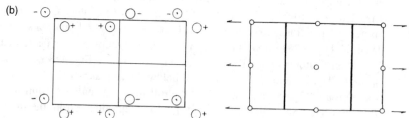

Origin at $\bar{1}$; unique axis b

4	f	1	$x, y, z;$ $\bar{x}, \bar{y}, \bar{z};$ $\bar{x}, \frac{1}{2} + y, z;$ $x, \frac{1}{2} - y, z.$

Limiting conditions
hkl: None
$h0l$: None
$0k0$: $k = 2n$

2	e	m	$x, \frac{1}{4}, z;$ $\bar{x}, \frac{3}{4}, \bar{z}$
2	d	$\bar{1}$	$\frac{1}{2}, 0, \frac{1}{2},$ $\frac{1}{2}, \frac{1}{2}, \frac{1}{2}$
2	c	$\bar{1}$	$0, 0, \frac{1}{2},$ $0, \frac{1}{2}, \frac{1}{2}.$
2	b	$\bar{1}$	$\frac{1}{2}, 0, 0;$ $\frac{1}{2}, \frac{1}{2}, 0.$
2	a	$\bar{1}$	$0, 0, 0;$ $0, \frac{1}{2}, 0.$

As above +
hkl: $k = 2n$

Symmetry of special projections

(001) pgm; $a' = a, b' = b$ (100) pmg; $b' = b, c' = c$ (010) $p2$; $c' = c, a' = a$

FIGURE S6.4.

of the symmetry-related values, that is, 1/8. For 111, and using the value for x, $A_U \propto \cos 2\pi(y + \frac{3}{4}) \cos 2\pi(1/4 - \frac{1}{2})$. For high intensity, $|\cos 2\pi (y + \frac{3}{4})| \approx 0$, $n\pi$. For $n = 0$, $y = -3/4 = \frac{1}{4}$ (by symmetry, the value $\frac{3}{4}$ is added. For $n = 1$, y is again $\frac{1}{4}$ (3/4). Proceeding in this manner with 231 leads to $y = 1/6$ (by symmetry, the values 1/3, 2/3 and 5/6 are included), and with 040 we find $y = 3/16$ (by symmetry, 5/16, 11/16 and 13/16 are included). The mean for the three value of y is $(1/4+1/6+3/16)/3$, or 0.20.

6.6. Since there are two molecules per unit cell in $P2_1/m$ in this structure, and the molecules cannot have $\bar{1}$ symmetry, the special positions sets $\pm(x, \frac{1}{4}, z)$ are selected. The B, C, and N atoms lie on m. Since the shortest distance

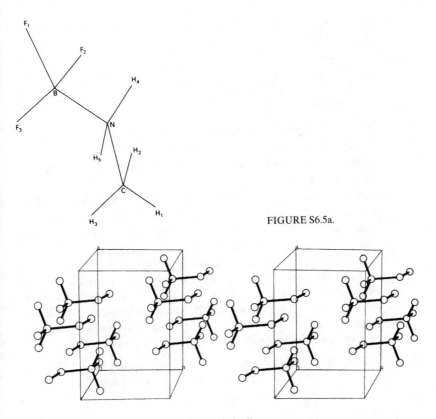

FIGURE S6.5a.

FIGURE S6.5b.

between m planes is 3.64 Å, the F_1, B, N, C, and H_1 atoms must lie on one and the same m plane (see Figure S6.5a). Hence, the remaining two F and four H atoms must be placed symmetrically across the same m plane. These conclusions were borne out by the structure analysis.[a] Figure S6.5b is a stereoview of the packing diagram for $CH_3NH_2BF_3$, showing the H_1, C, N, B, and three F atoms. The m plane is normal to the vertical direction in the diagram and the remaining two pairs of H atoms are disposed across the m plane as described above.

6.7. (a) (i) $|F(hkl)| = |F(\bar{h}\bar{k}\bar{l})|$ (ii) $|F(0kl)| = |F(0\bar{k}\bar{l})|$ (iii) $|F(h0l)| = |F(\bar{h}0\bar{l})|$

[a] S. Geller and J.L. Hoard, *Acta Crystallographica* **3**, 121 (1950).

(b) (i) $|F(hkl)| = |F(\bar{h}k\bar{l})| = F(h\bar{k}l)| $ (ii) $|F(0kl)| = |F(0\bar{k}\bar{l})| = |F(0\bar{k}l)|$ (iii) $|F(h0l)| = |F(\bar{h}0\bar{l})|$

(c) (i) $|F(hkl)| = |F(\bar{h}k\bar{l})| = |F(\bar{h}kl)| = |F(h\bar{k}l)| $ (ii) $|F(0kl)| = |F(0\bar{k}\bar{l})| = |F(0\bar{k}l)| $ (iii) $|F(h0l)| = |F(\bar{h}0\bar{l})| = |F(\bar{h}0l)|.$ Any combination of $\pm hkl$ not listed follows the pattern of (a)(i). In (b), for example, $|F(\bar{h}0l)| = |F(h0\bar{l})|.$

6.8. (a) In Pa, the symmetry element relates the sites x, y, z and $\frac{1}{2} + x, \bar{y}, z$, so that the Harker line is $[\frac{1}{2}, v, 0]$. In $P2/a$, the Harker section is $(u, 0, w)$ and the line $[\frac{1}{2}, v, 0]$. In $P222_1$, there are three Harker sections, $(0, v, w)$, $(u, 0, w)$ and $(u, v, \frac{1}{2})$.

(b) The Harker section $(u, 0, w)$ must arise through the symmetry-related sites x, y, z and $\bar{x}, y, \bar{z}$, which correspond to a 2-fold axis along y. Similarly, the line $[0, v, 0]$ arises from a mirror plane normal to y. Since the crystal is noncentrosymmetric, the space group must be $P2$ or Pm. If it is $P2$, there must be, by chance, closely similar y coordinates for many of the atoms in the structure. If it is Pm, chance coincidences occur between the x and z coordinates. [These conditions are somewhat unlikely, especially when many atoms are present, so that Harker sections and lines can be used to distinguish between space groups that are not determined by diffraction symmetry alone.]

6.9. (a) $P2_1/n$, a non-standard setting of $P2_1/c$ (see also Problem 2.11).

(b) The S–S vectors have the following Patterson coordinates:

(1) $\pm(\frac{1}{2}, \frac{1}{2} + 2y, \frac{1}{2})$ Double weight
(2) $\pm(\frac{1}{2} + 2x, \frac{1}{2}, \frac{1}{2} + 2z)$ Double weight
(3) $\pm(2x, 2y, 2z)$ Single weight
(4) $\pm(2x, 2\bar{y}, 2z)$ Single weight

Section $v = \frac{1}{2}$ Type 2 vector $x = 0.182, z = 0.235$
Section $v = 0.092$ Type 1 vector $y = 0.204$
Section $v = 0.408$ Type 3 or 4 vector $x = 0.183, y = 0.204, z = 0.234$

Thus we have four S–S vectors at: $\pm(0.183, 0.204, 0.235; 0.683, 0.296, 0.735)$

Any one of the other seven centers of symmetry, unique to the unit cell, may be chosen as the origin, whereupon the coordinates would be transformed accordingly. The sulfur atom positions are plotted in Figure S6.6. [Small differences in the third decimal places of the coordinates determined from the maps in Problems 6.9 and 6.10 are not significant.]

6.10. (a) By direct measurement, the sulfur atom coordinates are S $(0.266, 0.141)$ and S' $(-0.266, -0.141)$

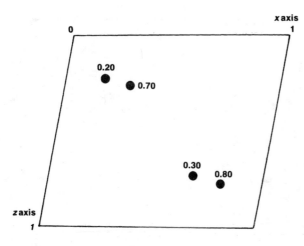

FIGURE S6.6.

(b) Draw an outline of the unit cell on tracing paper, and plot the position of −S on it. Place the tracing over the Patterson map (Figure P6.2), in the same orientation, with the position of −S over the origin of the Patterson map, and copy the Patterson map on to the tracing (Figure S6.7a). On another tracing, carry out the same procedure with respect to the position of −S′ (Figure S6.7b). Superimpose the two tracings (Figure S6.7c). Atomic positions correspond to positive regions of the two superimposed maps.

6.11. The summation to form $P(v)$ can be carried out[a] with program FOUR1D*. In using the program, each data line should contain k, $|F(0k0)|^2$ and 0.0, the zero representing the B coefficient of the Fourier series. $P(v)$ shows three non-origin peaks. If the highest of them is assumed to arise from the Hf–Hf vector, $y_{Hf} = 0.105$; the smaller peaks are Hf–Si vectors, from which we could obtain approximate y parameters for the silicon atoms. Their difference in height arises mainly from the fact that one of them is, in projection, close to the origin peak. However, the simplified structure factor equation for $F(0k0)$, based on the silicon atoms alone, is

$$F(0k0) \propto \cos(2\pi k y_{Hf})$$

so that the signs of the reflections, ignoring 012,0 and 016,0, are, in order, +−−++−. We can now calculate $\rho(y)$ with these signs attached to the $|F(0k0)|$ values. From the result, we obtain $y_{Hf} = 0.107$, $y_{Si_1} = 0.033$, and

---

[a] Beevers-Lipson strips, Tables P6.1 and P6.2, were an early aid to such summations.

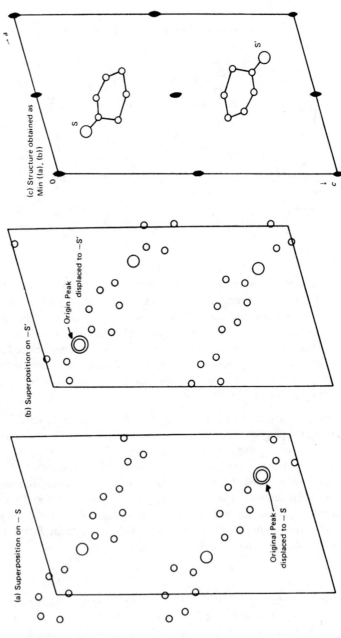

FIGURE S6.7.

$y_{Si_1} = 0.25$. These values for Y_{Si} lead to vectors which appear on $P(v)$. We conclude that the small peak on $\rho(y)$ at $y = 0.17$ is spurious, arising most probably from both the small number of data and experimental errors in them.

6.12. Since the sites of the replaceable atoms are the same in each derivative, and the space group is centrosymmetric, we can write $F(M_1) = F(M_2) + 4(f_{M_1} - fM_2)$. Hence, we can draw up the following table:

(a)

	\multicolumn{4}{c}{M}
h	NH_4	K	Rb	Tl
1	−	−	+	+
2	a	+	+	+
3	+	+	+	+
4	−	a	+	+
5	+	+	+	+
6	−	−	a	+
7	a	+	+	+
8	a	+	+	+

^{a}Indeterminate, because $|F|$ small or zero.

(b) The peak at 0 represents K and Al, superimposed in projection. The peak at 0.35 would then be presumed to be due to the S atom.

(c) The effect of the isomorphous replacement of S by Se can be seen at once in the increases in $|F(555)|$ and $|F(666)|$ and decrease in $|F(333)|$. These changes are not in accord with the findings in (b). Comparison of the two electron density plots shows that $d_{S/Se}$ must be 0.19 (the x coordinate is $d/\sqrt{3}$). The peak at 0.35 arises from a superposition of oxygen atoms in projection, and is not appreciably altered by the isomorphous replacement.

6.13. $A = 100 \cos 60° + (f_o + \Delta f') \cos 36° + 8 \cos 126° = 50 + 40.046 - 4.702 = 85.344$. $B = 100 \sin 60° + (f_o + \Delta f') \sin 36° + 8 \sin 126° = 86.603 + 29.095 + 6.472 = 122.17$ Hence, $|F(010)| = 149.0$, $\phi(010) = 55.06°$. For the $0\bar{1}0$ reflection, we have $A = 100 \cos 60° + (f_o + \Delta f') \cos 36° + 8 \cos 54° = 50 + 40.046 + 4.702 = 94.748$. $B = 100 \sin(-60°) + (f_o + \Delta f') \sin(-36°) + 8 \sin 54° = -86.603 - 29.095 + 6.472 = -109.226$ Hence, $|F(0\bar{1}0)| = 144.6$, $\phi(0\bar{1}0) = -49.06°$.

6.14. Draw a circle, at a suitable scale, to represent an amplitude $|F_P|$ of 858. From the center of this circle, set up a vector to represent $|F_{H1}| \exp(i\phi_1)$, where $|F_{H1}| = 141$ and $\phi_1 = (78 + 180)°$. At the termination of this vector,

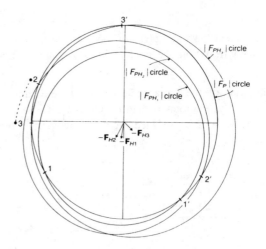

FIGURE S6.8.

draw a circle of radius 756 to represent $|F_{PH1}|$. Repeat this procedure for the other two derivatives (Figure S6.8). The six intersections 1–1′, 2–2′ and 3–3′ are strongest in the region indicated by ●- - -●. The required phase angle ϕ_M, calculated from (6.95), lies in this region. The centroid phase angle ϕ_B is biased slightly towards point 1 (see also Figure 6.41).

6.15. $\cos(hx - \phi)$ expands to $\cos hx \cos \phi + \sin hx \sin \phi$ which, for $\phi = \pi/2$, reduces to $\sin hx$. Hence, $\psi(x) = \pi/2 + 2\sum_{h=1}^{\infty}(1/h)\sin hx = \pi/2 + 2\sum_{h=1}^{\infty}(l/h)\cos(hx - \phi)$. This equation resembles closely a Fourier series (see Section 5.2).

6.16. (a) The total mass of protein per unit cell is $Z \times 1.8 \times 10^4 \times 1.6605 \times 10^{-24}$ g, where Z is the number of protein molecules per unit cell. Since there is an equal mass of solvent water in the unit cell, the density of the crystals is $(2Z \times 1.8 \times 1.6605 \times 10^{-20})/(40 \times 50 \times 60 \times 10^{-24} \sin 100°) = 0.506$ g cm^{-3}. Since the density of protein crystals is only slightly greater than 1 g cm^{-3}, $Z = 1.98$, or 2 to the nearest integer.

(b) In space group $C2$ there are four general equivalent positions (see Section 2.7.3). Since $Z = 2$, the protein molecule must occupy special positions on 2-fold axes, so that the molecule has symmetry 2.

6.17. In the notation of the text, we have for $F(hkl)$

$$F_H(+) = F'_H(+) + iF''_H(+)$$

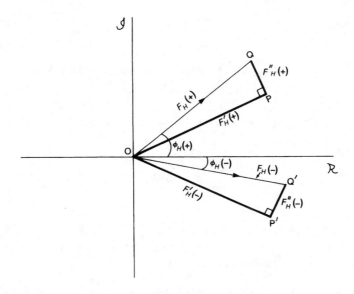

FIGURE S6.9.

and for $|F\bar{h}\bar{k}\bar{l})|$

$$F_H(-) = F'_H(-) + iF''_H(-)$$

where $F'_H(+)$ and $F'_H(-)$ are the structure factor components derived from the real part of (6.108), and $F''_H(+)$ and $F''_H(-)$ are its anomalous components. It is clear from Figure S6.9 that the moduli $|F_H(+)|$ and $|F_H(-)|$ are equal, but that $\phi_H(+) \neq \phi_H(-)$. In terms of the structure factor equations, we can write a single atom vector for $\mathbf{h}$ and $\bar{\mathbf{h}}$ as

$$F(\mathbf{h}) = (f' + i\Delta f'') \exp[i(2\pi\mathbf{h}\cdot\mathbf{r} + \pi/2)]$$
$$F(\bar{\mathbf{h}}) = (f' + i\Delta f'') \exp[-i(2\pi\mathbf{h}\cdot\mathbf{r} + \pi/2)]$$

from which it is clear that $|F(\mathbf{h})| = |F(\bar{\mathbf{h}})|$, but $\phi(\mathbf{h}) \neq \phi(\bar{\mathbf{h}})$ because $\pi/2$ acts in the same sense (positive) in each case.

6.18. In the notation of the text, and for a centrosymmetric structure, we have

$$F_{PH}(+) = A_P(+) + A'_H(+) + iA''_H(+)$$

where

$$A_P(+) = \sum_{j=1}^{N_P} f_j \cos 2\pi(hx_j + ky_j + lz_j)$$

$$A'_H(+) = \sum_{j=1}^{N_H} f'_j \cos 2\pi(hx_j + ky_j + lz_j)$$

$$A''_H(+) = \sum_{j=1}^{N_H} \Delta f''_j \cos 2\pi(hx_j + ky_j + lz_j)$$

Clearly, $|F(hkl)| = F(\bar{h}\bar{k}\bar{l})| = (A^2 + B^2)^{1/2}$, where $A = A_P(+) + A'_H(+)$ and $B = A''_H(+); \phi(hkl) = \phi(\bar{h}\bar{k}\bar{l}) = \tan^{-1} B/A$, and cannot equal 0 or π because of the finite value of $A''_H(+)$.

6.19. If, for a crystal of a given space group, Friedel's Law breaks down, then the diffraction symmetry reverts to that of the corresponding point group. Thus, we have

	$|F(khl)|$ equivalents	Bijvoet pairs
(a) $C2(2)$	$hkl/\bar{h}k\bar{l}$	$hkl/\bar{h}k\bar{l}$ with $h\bar{k}l/\bar{h}\bar{k}\bar{l}$
(b) $Pm(m)$	$hkl/h\bar{k}l$	$hkl/h\bar{k}l$ with $\bar{h}k\bar{l}/\bar{h}\bar{k}\bar{l}$
(c) $P2_12_12_1(222)$	$hkl/h\bar{k}\bar{l}/\bar{h}k\bar{l}/\bar{h}\bar{k}l$	$hkl/h\bar{k}\bar{l}/\bar{h}k\bar{l}/\bar{h}\bar{k}l$ with $\bar{h}\bar{k}\bar{l}/\bar{h}kl/h\bar{k}l/hk\bar{l}$
(d) $P4(4)$	$hkl/\bar{k}hl/\bar{h}\bar{k}l/k\bar{h}l$	$hkl/\bar{k}hl/\bar{h}\bar{k}l/k\bar{h}l$ with $k\bar{h}\bar{l}/hk\bar{l}/\bar{k}h\bar{l}/\bar{h}\bar{k}\bar{l}$

Strictly, pairs related as hkl and $\bar{h}\bar{k}\bar{l}$ should be discounted, as they are, of course, Friedel pairs.

6.20. From the Bragg equation, $1.25 = 2d(111) \sin \theta(111)$. Since $d(111) = a/\sqrt{3}, \theta(111) = 12.62°$. Differentiating the Bragg equation with respect to θ, we obtain $\delta\lambda = 2d(111) \cos \theta(111)\delta\theta$. Remembering that θ (and $\delta\theta$) must be measured in radian, $\delta\lambda = 0.0243$ Å.

6.21. For the NaCl structure type, we have $F(hkl) = 4[f_{Na^+} + (-1)^l f_{H^-/D^-}]$. Hence, the following results are obtained:

	(111)		(220)	
	NaH	NaD	NaH	NaD
X-rays	30.9	30.9	27.6	27.6
Neutrons	2.88	−1.28	−0.08	4.08

6.22. The number N of symmetry-independent reciprocal lattice points with a range $0 < \theta < \theta_{max}$ is $33.51\,V\sin^3\theta/(\lambda^3 Gm)$, from Chapter 4. The volume V of the unit cell is 6×10^4 Å^3, $G = 1$ for a P unit cell, and m, the number of symmetry-equivalent general reflections, is 8 for the Laue group mmm. Hence, $N = 74466.7\sin^3\theta_{max}$.

(a) $0° < \theta < 10°$: $\sin^3\theta_{max} = 5.236 \times 10^{-3}$, so that $N = 389$, or 779 if we consider the hkl and $\bar{h}\bar{k}\bar{l}$ reflections.

(b) $10° < \theta < 20°$: $\sin^3\theta_{max} = 4.001 \times 10^{-2}$, so that $N = 2979 - 389$, or 2590.

(c) $20° < \theta < 25°$: $\sin^3\theta_{max} = 7.548 \times 10^{-3}$, so that $N = 5620 - 2979$, or 2641.

The resolution, defined in terms of d_{min}, is $d_{min} = \lambda/(2\sin\theta_{max})$

(a) For $\theta_{max} = 10°$: $d_{min} = 4.3$ Å
(b) For $\theta_{max} = 20°$: $d_{min} = 2.2$ Å
(c) For $\theta_{max} = 25°$: $d_{min} = 1.8$ Å

Solutions 7

7.1. A possible set, with the larger $|E|$-values, is 705, $6\bar{1}\bar{7}$ and $8\bar{1}\bar{4}$. Reflection $42\bar{6}$ is a structure invariant, and 203 is linearly related to the pair $8\bar{1}\bar{4}$ and $6\bar{1}\bar{7}$. Reflection $4\bar{3}\bar{2}$ has a low $|E|$-value, so that triple relationships involving it would not have a high probability. Alternative sets are 705, 203, $8\bar{1}\bar{4}$ and 705, 203, $6\bar{1}\bar{7}$. A vector triplet exists between $8\bar{1}\bar{4}, 42\bar{6}$ and $4\bar{3}\bar{2}$.

7.2. The equations for A and B lead to the following relationships:

$$|F(hkl)| = |F(\bar{h}\bar{k}\bar{l})| = |F(h\bar{k}l)| = |F(\bar{h}k\bar{l})| \neq |F(\bar{h}kl)|;$$
$$|F(\bar{h}kl)| = |F(hk\bar{l})|$$

Because of the existence of the $k/4$ term, the phase relationships depend on the parity of k:

$$k = 2n : \phi(hkl) = -\phi(\bar{h}\bar{k}\bar{l}) = -\phi(h\bar{k}l)$$
$$= \phi(\bar{h}k\bar{l}) \neq \phi(\bar{h}kl); \; \phi(\bar{h}kl) = \phi(hk\bar{l})$$
$$k = 2n + 1 : \phi(hkl) = -\phi(\bar{h}\bar{k}\bar{l}) = \pi - \phi(h\bar{k}l)$$
$$= \pi + \phi(\bar{h}k\bar{l}) \neq \phi(\bar{h}kl); \; \phi(\bar{h}kl) = \pi + \phi(hk\bar{l})$$

7.3. Set (b) would be chosen: there is a redundancy in set (a) among 041, $\bar{1}62$ and $\bar{1}23$, because $F(041) = F(04\bar{1})$ in this space group. In space group

$C2/c, h + k$ is even, so that reflections $012, \overline{1}23, 162,$ and $\overline{1}62$ would not occur. The origin could be fixed by 223 and $13\overline{7}$, because there are only four parity groups for a C-centered unit cell.

7.4. Following the procedure given in Chapter 3, it will be found that $K = 4.0 \pm 0.4$, and $B = 6.6 \pm 0.3 \text{ Å}^2$. Since $B = 8\pi^2 \overline{U^2}$, the root mean square atomic displacement is $[6.6/(8\pi^2)]^{1/2}$, or 0.29 Å. (You were not expected to derive the standard errors in K and B; they are quoted in order to give an idea of the precision obtainable from a Wilson plot.)

7.5. A plot of the atoms in the unit cell and its environs shows that the shortest Cl...Cl contact distance is between atoms at $\frac{1}{4}, y, z$ and $\frac{3}{4}, \bar{y}, z$. Hence, $d^2(\text{Cl}...\text{Cl}) = a^2/4 + 4y^2 b^2$, so that $d(\text{Cl}...\text{Cl}) = 4.639$ Å. The superposition of errors (see Section 7.6) shows that the variance of $d(\text{Cl}...\text{Cl})$ is obtained from

$$[2d\sigma(d)]^2 = [2a\sigma(a)/4]^2 + [8y^2 b\sigma(b)]^2 + [8b^2 y\sigma(y)]^2$$

so that $\sigma(d) = 0026$ Å. (It may be noted that this answer calculates as 0.02637 Å to four significant figures; if we use only the third term, that in y, then the result is 0.02626 Å. Thus, the larger error in a distance between atoms is significantly the errors in the atomic coordinates.)

7.6. In the first instance we average the sum of $\phi_{\mathbf{k}}$ and $\phi_{\mathbf{h}-\mathbf{k}}$, namely, $(-37 - 3 - 24 + 38 + 13)/6$, or $-2.2°$. Applying the tangent expression, we have $\tan \phi_{\mathbf{h}} = (-2.648 - 0.267 - 1.830 + 1.662 + 0.832)/(3.514 + 5.093 + 4.111 + 3.3 + 2.128 + 3.605) = -0.103$, so that $\phi_{\mathbf{h}} = -5.9°$.

7.7. (a) When using Molecular Replacement in macromolecular crystallography the search and target molecules should be compatible in size as well as in their 3-D structures. If this is not the case problems may be encountered in obtaining a dominant solution to MR. The more possible solutions which have to be inspected, using Fourier methods, the more laborious the process becomes, maybe to the point where the analysis becomes untenable.

(b) For small molecule analysis it is more usual for the search "molecule" to be a fairly small fragment of the target molecule. In this case the search molecule must be as accurate as possible in bond lengths and angles because the data are at atomic resolution and the Patterson peaks similarly resolved. Programs such as PATSEE allow for complex search molecules to be used which have one degree of torsional freedom, thus increasing the size of the whole search fragment.

7.8. Vectors of the type labeled P1----P2 will not occur in the search Patterson as they involve atoms (in the region of P1) within the additional loop of the target molecule, absent in the search molecule. Only the search molecule will

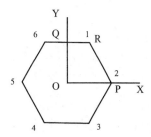

FIGURE S7.1.

be positioned by RT and the missing parts of the structure particularly in the loop, require to be located initially using Fourier and possibly least-squares in methods in small molecule analysis.

7.9. (a) It is not clear how the side chain comprising atoms 8–13 is oriented with respect to the rest of the molecule which is predominantly flat. The facility in PATSEE for vaying the linkage torsion angle could be used but was not necessary in practice because a suffuciently large independent fragment was available.

(b) By chance the molecular graphics program oriented the model which is perfectly flat, to be in the XY plane. Hence all z coordinates are zero in this plane.

(c) The CHEM-X program allows a chemical model of the molecule to be constructed ànd provides coordinates for the atoms. These coordinates are given not as fractional coordinates but as Å values with respect to the program's internal orthogonal axis sytem. To convert to fractional coordinates X,Y,Z values were each divided by 100 for all atoms. This set then belongs to an artificial unit cell with dimensions given in the question.

7.10. Considering the diagram, Figure S7.1 below: $OP = 1.400$ Å, $QR = 1.400 \cos 60° = 0.700$ Å, and $OQ = 1.400 \sin 60° = 1.212$ Å.
It follows that the Å coordinates are:

```
Atom 1: X= 0.700,    Y= 1.212,    Z=0.000
Atom 2: X= 1.40,     Y= 0.000,    Z=0.000
Atom 3: X= 0.700,    Y=-1.212,    Z=0.000
Atom 4: X=-0.700,    Y=-1.212,    Z=0.000
Atom 5: X=-1.400,    Y=0.000,     Z=0.000
Atom 6: X=-0.700,    Y=1.212,     Z=0.000
```

Fractional coordinates in a unit cell of a $=$ b $=$ c $=$ 100 Å and all angles $= 90°$:

```
Atom 1: X=  0.0070,     Y=  0.0121,     Z=0.000
Atom 2: X=  0.0140,     Y=  0.0000,     Z=0.000
Atom 3: X=  0.0070,     Y= -0.0121,     Z=0.000
Atom 4: X= -0.0070,     Y= -0.0121,     Z=0.000
Atom 5: X= -0.0140,     Y=  0.0000,     Z=0.000
Atom 6: X= -0.0070,     Y=  0.0121,     Z=0.000
```

7.11. From (3.172), dividing throughout by $\sum_{j=1}^{N} g_{j,\theta}^2$, taking ε as 1, gives $|E|^2 = 1 + (1/\sum_{j=1}^{N} g_{j,\theta}^2) \sum_j \sum_k g_{j,\theta} g_{k,\theta} \exp(i2\pi \mathbf{h} \cdot \mathbf{r}_{j,k})$. The second term on the right-hand side represents sharpened $|F|^2$ coefficients. The term in the Patterson function that creates the origin peak, $\sum_{j=1}^{N} g_{j,q}^2$, is now unity, so that a Patterson function with coefficients $(|E|^2 - 1)$ produces a sharpened Patterson function with the origin peak removed.

Solutions 8

8.1. (a) In space group $P2_1$, symmetry-related vectors have the coordinates $\pm(2x, \frac{1}{2}, 2z)$; the I–I vector in the half unit cell is easily discerned. By measurement on the map, $x_I = 0.422$ and $z_I = 0.144$, with respect to the origin O.

(b) The contribution of the iodine atoms, F_I, to the structure factors is given by $2f_I \cos 2\pi (0.422x_I + 0.144z_I)$. Hence, the following table:

hkl	$(\sin\theta)/\lambda$	$2f_I$	F_I	$|F_o|$
001	0.026	105	65	40
0014	0.364	67	67	37
106	0.175	88	−20	33
300	0.207	84	−8	35

The signs of 001,0014, and 106 are probably $+, +$ and $-$, respectively. The magnitude of $F_I(300)$ is a small fraction of $|F_o|$, and could easily be outweighed by the contribution from the rest of the structure. Thus, its sign remains uncertain from the data given. Small variations in the values determined for F_I are acceptable; they derive, most probably, from small variations in the graphical interpolation of the f_I values.

(c) The shortest I–I vector is that between the positions listed above. Hence, $d_{I-I} = \{[2 \times 0.422 \times 7.26]^2 + [0.5 \times 11.55]^2 + [2 \times 0.144 \times 19.22]^2 + [2 \times 0.422 \times 0.144 \times 7.26 \times 19.22 \cos(94.07)]\}^{1/2} = 10.02$ Å.

8.2. A Σ_2 listing is prepared as follows:

h	k	h − k	$\|E(\mathbf{h}) \,\|\, E(\mathbf{k}) \,\|\, E(\mathbf{h-k})\|$
0018	081	0817	9.5
011	024	035	5.0
	026	035	0.5
021	038	059	0.4
	0310	059	0.4
024	035	059	9.6
038	059	0817	7.2
	081	011,7	6.0
	081	011,9	10.2
0310	059	081	7.9
	081	011,9	9.2

Note the convention, that a two-figure Miller index takes a comma after it unless it is the third index.

In space group $P2_1/a, s(hkl) = s(\bar{h}\bar{k}\bar{l}) = (-1)^{h+k}s(h\bar{k}l)$, and $s(hk\bar{l}) = (-1)^{h+k}s(\bar{h}k l)$. With two-dimensional data, we need only two reflections to specify the origin, so we can take $s(081) = s(011,9) = +$. We proceed to the determination of signs:

h	k	h–k		Conclusion
011,9(+)	(Origin fixing)			
081(+)	(Origin fixing)			Origin at 0, 0
011,9(+)	081(+)	038		$s(038) = +$
011,9(+)	08$\bar{1}$(+)	0310		$s(0310) = +$
038(+)	08$\bar{1}$(+)	011,7		$s(011,7) = +$
0310(+)	081(+)	05$\bar{9}$		$s(059) = -$
059(−)	03$\bar{8}$(+)	0817		$s(0817) = -$
038(+)	059(−)	02$\bar{1}$		$s(021) = -$
0310(+)	059(−)	02$\bar{1}$		$s(021) = -$
0817(−)	08$\bar{1}$(+)	0018		$s(0018) = -$
			Let	$s(035) = a$
059(−)	035(a)	024		$s(024) = -a$
035(a)	024(−a)	011		$s(011) = -$
035(a)	01$\bar{1}$(+)	026		$s(026) = a$

The two indications for $s(021)$ and the single indication for $s(026)$ will have low probabilities, because of low $|E|$-values, and must be regarded as unreliable at this stage. Within the data set, no conclusion can be reached about $s(a)$; both + and − signs are equally likely. Reflection 0312 does not interact within the data set.

8.3. The space group is $P2_1/c$, from Table 8.4. Thus, $F(hkl) = F(\bar{h}\bar{k}\bar{l}) = (-1)^{k+l}F(h\bar{k}l)$; for the hk reflections it becomes $F(hk) = F(\bar{h}\bar{k}) =$

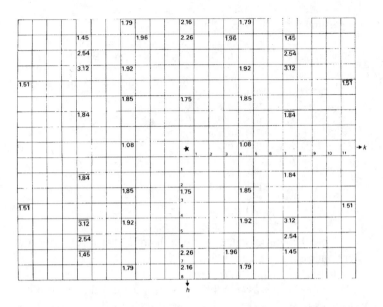

FIGURE S8.1.   The completed chart. Negative signs are shown as bars over the $|E|$ values.

$(-1)^k F(h\bar{k})$. Figure S8.1 shows the completed chart. A Σ_2 listing follows; an N indicates that no new relationships were derivable with the reflection so marked.

Number	h	k	h–k	$|E_{\mathbf{h}} \parallel E_{\mathbf{k}} \parallel E_{\mathbf{h-k}}|$
1	300	040	$3\bar{4}0$	3.5
2		840	$\bar{5}\bar{4}0$	6.0
3		570	$\bar{2}\bar{7}0$	10.0
4	700	570	$2\bar{7}0$	13.0
5	800	670	$2\bar{7}0$	10.1
6		340	$5\bar{4}0$	7.7
7		411,0	$4\bar{1}\bar{1}0$	4.9
8		040	$8\bar{4}0$	4.2
9	730	$0\bar{4}0$	770	3.1
10		$5\bar{4}0$	270	6.9
11	040	N		
12	340	$7\bar{7}0$	$\bar{4}11,0$	4.1
13	540	N		
14	840	N		
15	270	N		
16	570	N		
17	670	N		
18	770	N		
19	411,0	N		

An origin at 0, 0 may be chosen by specifying 270 (eoe, and occurring four times, and 540 (oee, and occurring three times) as both +. From the Σ_2 listing, we have:

Number	Conclusion	Comments
10	$s(730) = +$	
7	$s(800) = -$	$s(411, 0) = -s(4\overline{1}\overline{1}, 0)$
5	$s(670) = +$	
6	$s(340) = -$	Sign propagation has ended. Lets $s(040) = a$
1	$s(300) = -a$	
2	$s(840) = -a$	
3	$s(570) = -a$	
4	$s(700) = a$	
8	$s(840) = -a$	
9	$s(770) = a$	
11	$s(411, 0) = -a$	

The symbol a would be determined by calculating electron density maps with both $+$ and $-$ values, and assessing the results in terms of sensible chemical entities. In a more extended data set, the sign of a may evolve. No Σ_2 relationship is noticeably weak, and the above solution may be regarded as acceptable.

8.4. (a) The number of observed data with $I > 2\sigma(I)$ is far greater for form II, indicating that form I diffracts weakly. This effect may be associated with disorder, suggested by higher temperature factors, given that the two crystals were approximately equal in size. This, in spite of the fact that more data in total were measured for form I (18238) than for form II (12343).

(b) R_{int} for form II is much lower than for form I, indicating again that much of the intensity data for form I data is weak, and therefore inaccurate. However, the very low R_{int} value of 0.0341 for form II is in fact based on only a few hundred common, or symmetry-equivalent, reflections; 12343 were measured and 11521 were in the merged data set.

(c) The precision in both bond lengths and bond angles is better for form II than for form I. The differences are, however, somewhat marginal. In hindsight, we know that it arises because the disorder in form I was limited to a small number of locations on the CsH molecule.

(d) The R-factor for form II is considerably less than that for form I. While we have emphasized that the R-factor should used with care in assessing the quality of a structure, in combination with the other factors listed here there is no surprise in this result.

(e) The Flack parameter is not listed for form I. This is because it was unable to predict the hand of the structure in this case. These results emphasize the fact that the Flack parameter should be used with care for validation of the assignment of absolute configuration and should only be quoted with confidence for high quality structures. The esd for the Flack parameter for form I was in fact very high at 0.9, which is far too large to give a meaningful indication of absolute configuration. The result for form II is marginally acceptable but only because it backs up the biochemical results for assignment of configuration.

Solutions 9

9.1. Since $R = 57.30$ mm, 1 mm on the film is equal to $1°\theta$. Thus, 0.5 mm = $0.5° = 0.00873$ rad. The mean Cu $K\alpha$ wavelength is 1.5418 Å. Differentiating the Bragg equation with respect to θ, $\delta\lambda = 2d\cos\theta\,\delta\theta = \lambda\cot\theta\,\delta\theta$. Since $\delta\lambda = 0.0038$, we have $\cot\theta = 0.0038/(1.5418 \times 0.00873) = 0.2823$, so that $\theta \approx 74°$, the angle at which the $\alpha_1\alpha_2$ doublet would be resolved under the given conditions.

9.2. Perusal of the cubic unit-cell types leads us to expect that $(\sin^2\theta)/n$ for the first line (low θ region), where $n = 1, 2, 3, \ldots$, would result in a factor that would divide into all other experimental values of $\sin^2\theta$ to give integer or near-integer results. By trial, we find that, for the first line, $(\sin^2\theta)/3$ leads to a sequence of values that correspond closely to those for a cubic F unit cell. Thus, dividing all other values of $\sin^2\theta$ by 0.0155 we obtain:

Line no.	$\sin^2\theta/0.0155$	N	$a/\text{Å}$	hkl	θ/deg
1	3.00	3	6.192	111	12.45
2	4.10	4	6.118	200	14.60
3	11.08	11	6.170	311	24.48
4	16.04	16	6.185	400	29.91
5	23.95	24	6.199	422	37.54
6	26.90	27	6.203	333, 511	40.22
7	34.80	35	6.210	531	47.26
8	35.77	36	6.212	600, 442	48.12
9	42.64	43	6.218	533	54.39
10	47.54	48	6.222	444	59.13

(Some accidental absences appear in this sequence of lines.) Extrapolation by the method of least squares (program LSLI*) gives $a = 6.217$ Å. However, lines 1 and 2 produce significantly greater errors of fit than do the remaining eight lines. Since low-angle measurements tend to be less reliable, we can justifiably exclude lines 1 and 2. Least squares on lines 3 to 10 gives a best value, $a = 6.223$ Å.

9.3. The LEPAGE program gives the following results with the C-factor set at $1°$:

Reduced Cell	P	4.693, 4.929, 5.679 Å, 90.12, 90.01, 90.72°
Conventional cell	Orthorhombic P	4.693, 4.929, 5.679 Å, 90.12, 90.01, 90.72°

where all three angles are assumed to be $90°$ within experimental error. If we select the more stringent parameter $C = 0.5°$, we obtain

Reduced Cell	P	4.693, 4.929, 5.679 Å, 90.12, 90.01, 90.72°
Conventional cell	Monoclinic P	4.693, 5.679, 4.929 Å, 89.88, 90.72, 89.99°

where α and γ are taken now to be $90°$ within experimental error. The parameters are reordered so that β is the unique angle.

9.4. From the program LEPAGE*, we find:

Reduced Cell	P	6.021, 6.021, 8.515 Å, 110.70, 110.70, 90.01°
Conventional cell	Tetragonal I	6.021, 6.021, 14.75 Å, 90.00, 90.00, 90.00°

V(conventional unit cell)$/V$(given unit cell) $= 2$.

9.5. Let lines 1, 2, and 3 be 100, 010, and 001, respectively. Then, from multiples of their Q-values, we have $a^* = 0.09118$ (average of 100, 200, 300, and 400; lines 1, 5, 17, and 33, respectively), $b^* = 0.09437$ (average of 010, 020, 030, and 040; lines 2, 6, 18, and 38, respectively) and $c^* = 0.1312$ (average of 002 and 003; lines 3 and 15, respectively). Consider next the possible $hk0$ lines. $Q_{110} = Q_{100} + Q_{010} = 172.2$; this line has been allocated to 001, which is probably erroneous. Continue with the [001] zone:

$hk0$	Q_{hk0}	Line Number
110	172.2	3
120	439.5	9
130	885.0	20
210	421.5	8
220	688.8	15
230	1134.3	29
310	837.0	19
320	1104.3	27

This zone is well represented, and it follows that γ^* is $90°$. If line 4 is now taken as 001, then line 25 could be 002. We check this assignment by forming expected Q_{0kl} values: $Q_{011} = 338.9$, but there is no line at this Q-value, nor a pair of lines equidistant above and below this value, as there would be if the assignment is correct and $\alpha^* \neq 90°$. $Q_{021} = 606.2$, but this line fails the above test. It seems probable that line 4 is not 001. However, it must involve the l index and one of the indices h or k. If it is

TUTORIAL SOLUTIONS

101, then, for $\beta^* = 90°$, $Q_{001} = 249.8 - 83.1 = 166.7$, and if it is 011, then, for $\alpha^* = 90°$, $Q_{001} = 249.8 - 89.1 = 160.7$. For the second of these assignments, although a line at 160.7 is not present, there are the multiples 002 and 003 at lines 12 (642.8) and 39 (1446.3), respectively. With this assumption, line 4 is 011, line 12 is 002, and line 39 is 003. Confirmation arises from 012, 021, and 022 at lines 16 (731.9), line 10 (517.1), and line 25 (999.2). Thus, an average $c^* = 0.1267$ and $\alpha^* = 90°$. We now search for $h0l$ lines. For $\beta^* = 90°$, $Q_{101} = 243.8$; this line cannot be fitted into the pattern. $Q_{102} = 725.9$: there is no line at this value, but lines 11 and 21 are very nearly equidistant (166.1 and 166.5) from 725.9. Hence, the difference, 332.6 is $10^4(8c^*a^* \cos \beta^*)$, so that $\beta^* = 68.91°$. We have now a set of reciprocal unit cell diameters, from which, since two angles are 90°, the direct unit cell is calculated as $\beta = 111.09°$, $a = 1/(a^* \sin \beta) = 11.755$, $b = 1/b^* = 10.597$, $c = 1/(c^* \sin \beta) = 8.459$ Å. We make the conventional interchange of a and c, so that subject to β as the unique angle, and $c > b > a$, and now apply the further check of calculating the Q-values for this unit cell, using the program Q-VALS*, with the results listed in Table S9.1. In using this program, we remember that the unit cell appears to be monoclinic, so that we need to consider hkl and $hk\bar{l}$ reflections. From Table S9.1, it is evident that several reflections overlap, within the given experimental error. The unit cell type is P. The $h0l$ reflections are present only when h is an even integer. The reflections $30\bar{3}$ and 300, at the Q-values 1444.8 and 1444.9, respectively are probably not present and overlapped by the $23\bar{2}$ and 230. Hence, the space group is probably Pa (non-standard form of Pc) or $P2/a$ (non-standard form of $P2/c$). To consider if the symmetry is actually higher than monoclinic, the unit cell is reduced, using the program LEPAGE*. We find that the first unit cell is reduced, but the conventional unit cell is orthorhombic $B(\equiv C$ or $A)$, with a high degree of precision:

$$a = 8.459, b = 10.597, c = 21.935 \text{ Å}; \quad \alpha = \beta = \gamma = 90.00°$$

Since Miller indices transform as unit cell vectors, we find from the transformation matrix given by the program LEPAGE* that $h_B = -h$, $k_B = -k$ and $l_B = h + 2l$; the transformed indices are listed in Table S9.2. We note that the indices are listed as directly transformed. If we were dealing with structure factors, we could negate all the negative indices, because $|F(hkl)| = |F(\bar{h}\bar{k}\bar{l})| = |F(\bar{h}kl)| = |F(h\bar{k}l)| = |F(hk\bar{l})|$ in the orthorhombic system.

From an inspection of the hkl indices in Table S9.2 for the transformed unit cell, we find

hkl: $h + l = 2n$
$0kl$: None
$h0l$: $h = 2n$; ($l = 2n$)
$hk0$: $h = 2n$

TABLE S9.1. Observed and
Calculated Q-values for Substance X
and the hkl Indices of the Lines
Referred to the First Unit Cell

Q(obs)	$h\ k\ l$	Q(calc)
83.1	0 0 1	83.1
89.1	0 1 0	89.1
172.2	0 1 1	172.2
249.8	1 1 0	249.6
	1 1 $\bar{1}$	249.6
332.6	0 0 2	332.5
356.1	0 2 0	356.2
416.0	1 1 $\bar{2}$	415.8
	1 1 1	415.9
421.5	0 1 2	421.6
439.3	0 2 1	439.3
516.9	1 2 $\bar{1}$	516.7
	1 2 0	516.7
559.8	2 0 $\bar{1}$	559.0
642.9	2 0 $\bar{2}$	642.1
	2 0 0	642.2
648.6	2 1 $\bar{1}$	648.1
683.3	1 2 $\bar{2}$	683.0
	1 2 1	683.0
688.8	0 2 2	688.7
732.1	2 1 $\bar{2}$	731.2
	2 1 0	731.2
748.4	0 0 3	748.2
	1 1 $\bar{3}$	748.4
	1 1 2	748.4
801.5	0 3 0	801.5
837.2	0 1 3	837.3
884.5	0 3 1	884.6
892.4	2 0 $\bar{3}$	891.5
	2 0 1	891.6
916.0	2 2 $\bar{1}$	915.2
962.3	1 3 $\bar{1}$	962.0
	1 3 0	962.0
981.7	2 1 $\bar{3}$	980.6
	2 1 1	980.6
999.1	2 2 $\bar{2}$	998.3
	2 2 0	998.4
1016.	1 2 $\bar{3}$	1015.5
	1 2 2	1015.6
1015.	0 2 3	1104.4
1129.	1 3 $\bar{2}$	1128.2
	1 3 1	1128.3

TABLE S9.1. Continued

1134.	0 3 2	1134.0
1248.	1 1 $\bar{4}$	1247.2
	1 1 3	1247.2
1249.	2 2 $\bar{3}$	1247.7
	2 2 1	1247.8
1308.	2 0 $\bar{4}$	1307.2
	2 0 2	1307.3
1330.	0 0 4	1330.1
1361.	2 3 $\bar{1}$	1360.5
1369.	3 1 $\bar{2}$	1367.6
	3 1 $\bar{1}$	1367.6
1397.	2 1 $\bar{4}$	1396.2
	2 1 2	1396.3
1419.	0 1 4	1419.2
1425.	0 4 0	1424.8
1444.	2 3 $\bar{2}$	1443.6
	2 3 0	1443.6
1461.	1 3 $\bar{3}$	1460.8
	1 3 2	1460.8

giving the probable diffraction symbol as $B^{*}aa$, so that the space group is then either $B2_1aa$ or $Bmaa$. If the point group is $mm2$, then $B2_1aa$ (which is equivalent to $B2cm$) is the **cab** setting of the standard symbol $Abm2$. If the point group is mmm, then $Bmaa$ (which is equivalent to $Bmam$) is the **a$\bar{c}$b** setting of the standard symbol $Cmma$.

9.6. Crystal $XL1$: $a = 6.425$, $b = 9.171$, $c = 5.418$ Å, $\alpha = 90°$, $\beta = 90°$, $\gamma = 90°$. The unit cell is orthorhombic. The systematic absences indicate the diffraction symbol as $mmm\ Pn\,a\,.$, which corresponds to either $Pna2_1$ or $Pnam$. The latter is the **a$\bar{c}$b** setting of $Pnma$. (Reported: KNO_3; 9.1079, 6.4255, 5.4175 Å; $Pbnm$, which is the **cab** setting of $Pnma$.) The LEPAGE* reduction confirms the above cell as reduced and conventional, under reordering such that $a < b < c$. What is the space group now?

9.7. Crystal $XL2$: $a = 10.482$, $b = 11.332$, $c = 3.757$ Å, $\alpha = 90°$, $\beta = 90°$, $\gamma = 90°$. The unit cell is orthorhombic, with space group $Pbca$. The LEPAGE reduction confirms the above cell as reduced and conventional, under reordering such that $a < b < c$.

9.8. Crystal $XL3$: $a = 6.114$, $b = 10.722$, $c = 5.960$ Å, $\alpha = 97.59°$, $\beta = 107.25°$, $\gamma = 77.42°$. The unit cell is triclinic, space group $P1$ or $P\bar{1}$, The LEPAGE* reduction gives $a = 5.960$, $b = 6.114$, $c = 10.722$ Å, $\alpha = 77.42°$, $\beta = 82.41°$, $\gamma = 72.75°$. (Literature: $CuSO_45H_2O$: 6.1130, 10.7121, 5.9576 Å, 82.30°, 107.29°, 102.57°; $P\bar{1}$.)

TABLE S9.2. Transformation of the *hkl* Indices from the Monoclinic (first) Unit Cell to the Orthorhombic *B* Unit Cell

$h\ k\ l$	$h_B\ k_B\ l_B$	$h\ k\ l$	$h_B\ k_B\ l_B$
$0\ 0\ 1 \rightarrow$	$0\ 0\ 2$	$1\ 3\ \bar{1} \rightarrow$	$\bar{1}\ \bar{3}\ \bar{1}$
$0\ 1\ 0 \rightarrow$	$0\ \bar{1}\ 0$	$1\ 3\ 0 \rightarrow$	$\bar{1}\ \bar{3}\ 1$
$0\ 1\ 1 \rightarrow$	$0\ \bar{1}\ 2$	$2\ 1\ \bar{3} \rightarrow$	$\bar{2}\ \bar{1}\ \bar{4}$
$1\ 1\ 0 \rightarrow$	$\bar{1}\ \bar{1}\ 1$	$2\ 1\ 1 \rightarrow$	$\bar{2}\ \bar{1}\ 4$
$1\ 1\ \bar{1} \rightarrow$	$\bar{1}\ \bar{1}\ \bar{1}$	$2\ 2\ \bar{2} \rightarrow$	$\bar{2}\ \bar{2}\ \bar{2}$
$0\ 0\ 2 \rightarrow$	$0\ 0\ 4$	$2\ 2\ 0 \rightarrow$	$\bar{2}\ \bar{2}\ 2$
$0\ 2\ 0 \rightarrow$	$0\ \bar{2}\ 0$	$1\ 2\ \bar{3} \rightarrow$	$\bar{1}\ \bar{2}\ \bar{5}$
$1\ 1\ \bar{2} \rightarrow$	$\bar{1}\ \bar{1}\ \bar{3}$	$1\ 2\ 2 \rightarrow$	$\bar{1}\ \bar{2}\ 5$
$1\ 1\ 1 \rightarrow$	$\bar{1}\ \bar{1}\ 3$	$0\ 2\ 3 \rightarrow$	$0\ \bar{2}\ 6$
$0\ 1\ 2 \rightarrow$	$0\ \bar{1}\ 4$	$1\ 3\ \bar{2} \rightarrow$	$\bar{1}\ \bar{3}\ \bar{3}$
$0\ 2\ 1 \rightarrow$	$0\ \bar{2}\ 2$	$1\ 3\ 1 \rightarrow$	$\bar{1}\ \bar{3}\ 3$
$1\ 2\ \bar{1} \rightarrow$	$\bar{1}\ \bar{2}\ \bar{1}$	$0\ 3\ 2 \rightarrow$	$0\ \bar{3}\ 4$
$1\ 2\ 0 \rightarrow$	$\bar{1}\ \bar{2}\ 1$	$1\ 1\ \bar{4} \rightarrow$	$\bar{1}\ \bar{1}\ \bar{7}$
$2\ 0\ \bar{1} \rightarrow$	$\bar{2}\ 0\ 0$	$1\ 1\ 3 \rightarrow$	$\bar{1}\ \bar{1}\ 7$
$2\ 0\ \bar{2} \rightarrow$	$\bar{2}\ 0\ \bar{2}$	$2\ 2\ \bar{3} \rightarrow$	$\bar{2}\ \bar{2}\ \bar{4}$
$2\ 0\ 0 \rightarrow$	$\bar{2}\ 0\ 2$	$2\ 2\ 1 \rightarrow$	$\bar{2}\ \bar{2}\ 4$
$2\ 1\ \bar{1} \rightarrow$	$\bar{2}\ \bar{1}\ 0$	$2\ 0\ \bar{4} \rightarrow$	$\bar{2}\ 0\ \bar{6}$
$1\ 2\ \bar{2} \rightarrow$	$\bar{1}\ \bar{2}\ \bar{3}$	$2\ 0\ 2 \rightarrow$	$\bar{2}\ 0\ 6$
$1\ 2\ 1 \rightarrow$	$\bar{1}\ \bar{2}\ 3$	$0\ 0\ 4 \rightarrow$	$0\ 0\ 8$
$0\ 2\ 2 \rightarrow$	$0\ \bar{2}\ 4$	$2\ 3\ \bar{1} \rightarrow$	$\bar{2}\ \bar{3}\ 0$
$2\ 1\ \bar{2} \rightarrow$	$\bar{2}\ \bar{1}\ \bar{2}$	$3\ 1\ \bar{2} \rightarrow$	$\bar{3}\ \bar{1}\ \bar{1}$
$2\ 1\ 0 \rightarrow$	$\bar{2}\ \bar{1}\ 2$	$3\ 1\ \bar{1} \rightarrow$	$\bar{3}\ \bar{1}\ 1$
$0\ 0\ 3 \rightarrow$	$0\ 0\ 6$	$2\ 1\ \bar{4} \rightarrow$	$\bar{2}\ \bar{1}\ \bar{6}$
$1\ 1\ \bar{3} \rightarrow$	$\bar{1}\ \bar{1}\ \bar{5}$	$2\ 1\ 2 \rightarrow$	$\bar{2}\ \bar{1}\ 6$
$1\ 1\ 2 \rightarrow$	$\bar{1}\ \bar{1}\ 5$	$0\ 1\ 4 \rightarrow$	$0\ \bar{1}\ 8$
$0\ 3\ 0 \rightarrow$	$0\ \bar{3}\ 0$	$0\ 4\ 0 \rightarrow$	$0\ \bar{4}\ 0$
$0\ 1\ 3 \rightarrow$	$0\ \bar{1}\ 6$	$2\ 3\ \bar{2} \rightarrow$	$\bar{2}\ \bar{3}\ \bar{2}$
$0\ 3\ 1 \rightarrow$	$0\ \bar{3}\ 2$	$2\ 3\ 0 \rightarrow$	$\bar{2}\ \bar{3}\ 2$
$2\ 0\ \bar{3} \rightarrow$	$\bar{2}\ 0\ \bar{4}$	$1\ 3\ \bar{3} \rightarrow$	$\bar{1}\ \bar{3}\ \bar{5}$
$2\ 0\ 1 \rightarrow$	$\bar{2}\ 0\ 4$	$1\ 3\ 2 \rightarrow$	$\bar{1}\ \bar{3}\ 5$
$2\ 2\ \bar{1} \rightarrow$	$\bar{2}\ \bar{2}\ 0$		

Solutions 10

10.1. The number N of unit cells in a crystal is $V(\text{crystal})/V(\text{unit cell})$. Both crystals have the volume $V(\text{crystal}) = 2.4 \times 10^{-2}$ mm^3. The protein unit cell volume $V(\text{protein}) = 60,000$ Å^3, or $60,000 \times 10^{-21}$ mm^3. The total number of protein unit cells N_P is therefore $= 4 \times 10^{14}$. For the organic crystal unit cell, $V(\text{organic}) = 1800 \text{ Å}^3$, or $1800 \times 10^{-21} \text{ mm}^3$, so that the total number of organic unit cells N_0 is 1.333×10^{16}.

From (3.140) and (3.141), we write

$$E(hkl) = (I_0/\omega)(N^2\lambda^3)[e^4/m_e^2c^4)]LpA|F(hkl)|^2V(\text{crystal}) \quad \text{(S10.1)}$$

where N is the number of unit cells per unit volume of the crystal, L, p, and A are the Lorentz, polarization and absorption correction factors, and the other symbols have their usual meanings.

Historically, this equation was derived in 1914[a] and confirmed by careful measurements on a crystal of sodium chloride in 1921.[b] In (S10.1), $E(hkl)$ is the experimentally derived quantity and $|F(hkl)|$ is the term required in x-ray analysis, as discussed in Section 3.9. For our purposes, we write

$$E(hkl) \approx N^2 = [N_{\text{cells}}/V(\text{crystal})]^2 \quad \text{(S10.2)}$$

where N_{cells} is the total number of unit cells in the crystal volume $V(\text{crystal})$. Since diffraction power D is proportional to energy, we have for the two cases under discussion

$$D(\text{organic})/D(\text{protein}) = [N_{\text{cells}}(\text{organic})/N_{\text{cells}}(\text{protein})]^2$$
$$= [(1.333 \times 10^{16})/(4 \times 10^{14})]^2 = 1111$$

Based on these considerations alone, the organic crystal will diffract over 1000 times more powerfully than the protein crystal. However, most protein data sets are now collected with synchrotron radiation, the intensity of which more than makes up for the deficiency in diffracting power calculated above. Other factors affect the intensity: in particular, it follows from Section 3.10.1 that a local average value of $|F(hkl)|^2$ is proportional to $N_c f^2$ if, for simplicity, we assume an equal-atom structure, where N_c is here the number atoms per unit cell, which, to a first approximation, is proportional to the $V(\text{unit cell})$. Hence, the diffracting power of the crystal is directly proportional to $V(\text{unit cell})$, so that the above "squared effect" is somewhat diminished by the second factor.

10.2. The experimental arrangement and coordinate systems are shown in the diagram. For the powder ring, the two coordinates Y_d and Z_d will be the same that is, 70 mm, and the distance D is 300 mm. The angle subtended from O by the diffracted beam is 2θ so that $\tan 2\theta = 70/300$, or $\theta = 6.567°$. From the Bragg equation, $\lambda = 0.800 \text{ Å}$.

[a] C. G. Darwin *Philosophical Magazine* **27**, 315 (1914).

[b] W. L. Bragg, R. W. James and C. M. Bosanquet, *Philosophical Magazine* **42**, 1, (1921).

10.3. Let the separation of spots for the 300 Å spacing be ΔZ_d; then $\Delta Z_d/D =$ tan 2θ for a single diffraction order. Using the Braggs equation, we have $2 \times 300 \times \sin \theta = 0.8$ and $\theta = 0.0764°$. If $\Delta Z_d = 1$ mm then $D = 1/\tan 2\theta = 375$ mm. Using a value of D of 450 mm will be more than adequate. Note that the intensity falls off as the square of the distance, so that, in practice, moving the detector too far away will be costly in terms of lost data for a weakly diffracting protein crystal.

10.4. The information on limiting conditions indicates that there is either a 6_1 or a 6_5 screw axis in the crystal (Table 10.2). As the Laue symmetry is $\frac{6}{m}mm$, it follows from Table 10.1 that space group is either $P6_122$ or $P6_522$. Only the x-ray analysis can resolve this remaining ambiguity. Note that 6_1 and 6_5 screw operations are left-hand–right-hand opposites; only one can be correct for a given protein crystal.

10.5. The volume V_c is 3.28×10^6 Å^3. Substituting known values into the equation $D_c = \mu Z M_{\mathrm{P}} u/V_c(1-s)$ gives $0.38\mu/(1-s)$ for D_c, where μ is the number of molecules per asymmetric unit, and s is the fractional solvent content to be found by trial and error.

(a) Assuming that μ is 1 molecule per asymmetric unit and s is 0.68 (a guess), that is, the crystal contains 68% solvent by weight (the top of the known range), then it follows that $D_c = 1.20$ g cm^{-3}, a reasonable result. Note that we could make $s = 0.70$, slightly higher than normal, and this would give $D_c = 1.28$ g cm^{-3}, which is again quite acceptable. The important result for the structure analysis is that $\mu = 1$ so that $Z = 12$.

(b) As s from the above analysis is on the high side, we try $\mu = 2$ to see what this would imply. Then $D_c = 2 \times 0.38\mu/(1-s)$, or $0.76/(1-s)$ which, for $D_c = 1.4$ g cm^3, gives $s = 0.44$. This result is again reasonable, so that there is some ambiguity for this protein. All that can be done is to bear these results in mind during the x-ray analysis, and make use of any other facts which are known about the crystal. In the case of the protein MLI, it was known that the crystals diffracted x-rays only poorly, which is often a sign of high solvent content, and this fact is more consistent with $\mu = 1$.

10.6. The expected number of reflections $= 4.19V_c/d_{\mathrm{min}}^3$, or 563,442; this number includes all symmetry-related reflections. Since the Laue symmetry (Table 1.6) is $\frac{6}{m}mm$, the number of unique data 1/24 times the number in the complete sphere, namely, 23476. If only 21,000 reflections are recorded, the data set would be approximately 91% complete at the nominal resolution of 2.9 Å. This result corresponds more appropriately to 3.0 Å resolution (working backwards). Note that the above discussion is based on the number of reciprocal lattice points scanned in data collection and processing. Because protein crystals diffract poorly the number of reflections with significant intensities may well be as low as 50%. These weak data do actually contain structural information and will usually be retained in the working data set.

10.7. The asymmetric unit is one protein molecule. About 10% of the 27,000 Da is hydrogen leaving $27,000 - 2700 = 24,300$ Da, which is equivalent to 2025 carbon atoms. For the atoms in the water molecules to be located (oxygen) we add a further 30% of this number (oxygen is 16 Da compared to 12 Da for carbon). The total number of non-hydrogen atoms to be located is $2025 + 608$, or 2633. The number of parameters required for isotropic refinement (3 positional and 1 temperature factor per non-hydrogen atom) is $(2633 \times 4) + 1$ (scale factor), or 10,533. The unit cell volume is $58.2 \times 38.3 \times 54.2 \sin(106.5°)$, which equates to 1.158×10^5 Å^3. Using the equation for the number N of reciprocal lattice points in the whole sphere at a given resolution limit, $4.19V_c/d^3_{min}$, and dividing by a factor 4 for (Laue group $2/m$) we have the following results for the different resolutions:

	Reflections in 1 asymmetric unit	Data/parameter ratio
6 Å	$N = 2246/4 = 562$	0.05
2.5 Å	$N = 31066/4 = 7766$	0.74
1 Å	$N = 485400/4 = 121350$	11.5

Comments. The 6 Å stucture is completely unrefineable. The 2.5 Å structure is refineable, but only if heavily restrained. The 1 Å isotropic model structure should refine provided the data quality is adequate.

10.8. From the general expression (A4.6), with $\gamma = 120°$ and $\phi = 120°$, we derive the matrix

$$\begin{bmatrix} 0 & \bar{1} & 0 \\ 1 & \bar{1} & 0 \\ 0 & 0 & 1 \end{bmatrix} \quad \text{which, together with the translation vector} \quad \begin{bmatrix} 0 \\ 0 \\ \frac{2}{3} \end{bmatrix}$$

for the 3_2 screw axis leads to the general equivalent position set: $x, y, z; \bar{y}, x-y, \frac{2}{3}+z; y-x, \bar{x}, \frac{1}{3}+z$. The only condition limiting reflections is $000l : l = 3n$.

10.9. If the protein belongs to a family, or group, of proteins having similar functions or biological or other properties in common, and the structure of one member of the family is known, either from an *ab initio* or other structure determination, molecular replacement can be attempted. The method usually requires the two proteins involved in MR to have amino acid sequences which correspond either identically or which are of very similar types, that is, conserved, for at least 30% of their total lengths (30% homology). Note that if the two proteins crystallize in the same space group and have very similar unit cells, they are very likely to be isomorphous, and the new structure should be determinable initially by Fourier methods alone.

If the protein belongs to a new family for which no known structures exist, an *ab inito* method, MIR or MAD, has to be used for structure analysis. In the case of MAD, a tuneable source of synchrotron radiation is required.

INDEX

Ab initio methods, 548, 605, 795
Abnormal averages, 195
Absences in x-ray spectra: see also
  Accidental absences; Systematic
  absences
  accidental, 194
  systematic, 195
Absorption, 120, 228
  correction, 183
  edge, 121, 123, 342, 404
  measurement of, 183
Absorption coefficients, 121
Accidental absences, 162
  local average intensity for, 193
Acentric distribution: see Distribution,
  acentric
Airy disk, 306
Alternating axis of symmetry, 719
Alkali-metal halides, 1
Alums, crystal structure of, 415, 775
Amorphous substance, 4
Amplitude symmetry, 430; see also
  Phase symmetry
Analyser (of polarized light), 214
Angles: see also Bond lengths and bond
  angles; Interaxial angles; Interfacial
  angles; Phase angle
  between lines, 718
  between planes, 150
Angular frequency, 130
Anisotropic thermal vibration, 2, 189
Anisotropy, optical, 215, 218; see also
  Optically anisotropic crystals; Biaxial
  crystals; Uniaxial crystals
Anomalous dispersion: see Anomalous
  scattering
Anomalous scattering, 392ff
  and atomic scattering factor, 393

Anomalous scattering (cont.)
  and diffraction symmetry, 396
  and heavy atoms, 399
  and phasing reflections, 401
  and protein phasing, 403ff
  and structure factor, 399
  and symmetry, 396
Area detector: see Intensity
  measurement, by area detector
Argand diagram, 157, 435
Arndt-Wonnacott camera, 245
Assemblage, 23
Asymmetric unit, 26, 73
Asymmetry parameters, 498
Atom
  mass of, 336
  scattering by, 135
Atomic scattering factor, 136, 204, 303
  and anomalous scattering, 393
  corrections to, 393
  and electron density, 135
  exponential formula for, 138
  factors affecting, 187ff
  and spherical symmetry of atoms, 136,
    303
  temperature correction of, 190
  variation with $(\sin \theta)/\lambda$, 190, 228
Attenuation, 120
Averaging (Patterson) function, 346ff
Average intensity multiple: see Epsilon
  (ε) factor
Axes: see Cartesian axes; Coordinate
  axes; Crystallographic axes;
  Inversion axes; Oblique axes; Optic
  axis; Orthogonal axes; Rectangular
  axes; Reference axes; Rotation axes;
  Screw axis
Axial angles: see Interaxial angles

797